T0180038

Lecture Notes in Computer Science 12902

More information about this subseries at http://www.springer.com/series/7412

Marleen de Bruijne · Philippe C. Cattin ·
Stéphane Cotin · Nicolas Padoy ·
Stefanie Speidel · Yefeng Zheng ·
Caroline Essert (Eds.)

Medical Image Computing and Computer Assisted Intervention – MICCAI 2021

24th International Conference
Strasbourg, France, September 27 – October 1, 2021
Proceedings, Part II

Springer

Editors
Marleen de Bruijne
Erasmus MC - University Medical Center Rotterdam
Rotterdam, The Netherlands

University of Copenhagen
Copenhagen, Denmark

Stéphane Cotin
Inria Nancy Grand Est
Villers-lès-Nancy, France

Stefanie Speidel
National Center for Tumor Diseases (NCT/UCC)
Dresden, Germany

Caroline Essert
ICube, Université de Strasbourg, CNRS
Strasbourg, France

Philippe C. Cattin
University of Basel
Allschwil, Switzerland

Nicolas Padoy
ICube, Université de Strasbourg, CNRS
Strasbourg, France

Yefeng Zheng
Tencent Jarvis Lab
Shenzhen, China

ISSN 0302-9743 ISSN 1611-3349 (electronic)
Lecture Notes in Computer Science
ISBN 978-3-030-87195-6 ISBN 978-3-030-87196-3 (eBook)
https://doi.org/10.1007/978-3-030-87196-3

LNCS Sublibrary: SL6 – Image Processing, Computer Vision, Pattern Recognition, and Graphics

This Springer imprint is published by the registered company Springer Nature Switzerland AG
The registered company address is: Gewerbestrasse 11, 6330 Cham, Switzerland

Preface

The 24th edition of the International Conference on Medical Image Computing and Computer Assisted Intervention (MICCAI 2021) has for the second time been placed under the shadow of COVID-19. Complicated situations due to the pandemic and multiple lockdowns have affected our lives during the past year, sometimes perturbing the researchers work, but also motivating an extraordinary dedication from many of our colleagues, and significant scientific advances in the fight against the virus. After another difficult year, most of us were hoping to be able to travel and finally meet in person at MICCAI 2021, which was supposed to be held in Strasbourg, France. Unfortunately, due to the uncertainty of the global situation, MICCAI 2021 had to be moved again to a virtual event that was held over five days from September 27 to October 1, 2021. Taking advantage of the experience gained last year and of the fast-evolving platforms, the organizers of MICCAI 2021 redesigned the schedule and the format. To offer the attendees both a strong scientific content and an engaging experience, two virtual platforms were used: Pathable for the oral and plenary sessions and SpatialChat for lively poster sessions, industrial booths, and networking events in the form of interactive group video chats.

These proceedings of MICCAI 2021 showcase all 531 papers that were presented at the main conference, organized into eight volumes in the Lecture Notes in Computer Science (LNCS) series as follows:

- Part I, LNCS Volume 12901: Image Segmentation
- Part II, LNCS Volume 12902: Machine Learning 1
- Part III, LNCS Volume 12903: Machine Learning 2
- Part IV, LNCS Volume 12904: Image Registration and Computer Assisted Intervention
- Part V, LNCS Volume 12905: Computer Aided Diagnosis
- Part VI, LNCS Volume 12906: Image Reconstruction and Cardiovascular Imaging
- Part VII, LNCS Volume 12907: Clinical Applications
- Part VIII, LNCS Volume 12908: Microscopic, Ophthalmic, and Ultrasound Imaging

These papers were selected after a thorough double-blind peer review process. We followed the example set by past MICCAI meetings, using Microsoft's Conference Managing Toolkit (CMT) for paper submission and peer reviews, with support from the Toronto Paper Matching System (TPMS), to partially automate paper assignment to area chairs and reviewers, and from iThenticate to detect possible cases of plagiarism.

Following a broad call to the community we received 270 applications to become an area chair for MICCAI 2021. From this group, the program chairs selected a total of 96 area chairs, aiming for diversity — MIC versus CAI, gender, geographical region, and a mix of experienced and new area chairs. Reviewers were recruited also via an open call for volunteers from the community (288 applications, of which 149 were selected by the program chairs) as well as by re-inviting past reviewers, leading to a total of 1340 registered reviewers.

We received 1630 full paper submissions after an original 2667 intentions to submit. Four papers were rejected without review because of concerns of (self-)plagiarism and dual submission and one additional paper was rejected for not adhering to the MICCAI page restrictions; two further cases of dual submission were discovered and rejected during the review process. Five papers were withdrawn by the authors during review and after acceptance.

The review process kicked off with a reviewer tutorial and an area chair meeting to discuss the review process, criteria for MICCAI acceptance, how to write a good (meta-)review, and expectations for reviewers and area chairs. Each area chair was assigned 16–18 manuscripts for which they suggested potential reviewers using TPMS scores, self-declared research area(s), and the area chair's knowledge of the reviewers' expertise in relation to the paper, while conflicts of interest were automatically avoided by CMT. Reviewers were invited to bid for the papers for which they had been suggested by an area chair or which were close to their expertise according to TPMS. Final reviewer allocations via CMT took account of reviewer bidding, prioritization of area chairs, and TPMS scores, leading to on average four reviews performed per person by a total of 1217 reviewers.

Following the initial double-blind review phase, area chairs provided a meta-review summarizing key points of reviews and a recommendation for each paper. The program chairs then evaluated the reviews and their scores, along with the recommendation from the area chairs, to directly accept 208 papers (13%) and reject 793 papers (49%); the remainder of the papers were sent for rebuttal by the authors. During the rebuttal phase, two additional area chairs were assigned to each paper. The three area chairs then independently ranked their papers, wrote meta-reviews, and voted to accept or reject the paper, based on the reviews, rebuttal, and manuscript. The program chairs checked all meta-reviews, and in some cases where the difference between rankings was high or comments were conflicting, they also assessed the original reviews, rebuttal, and submission. In all other cases a majority voting scheme was used to make the final decision. This process resulted in the acceptance of a further 325 papers for an overall acceptance rate of 33%.

Acceptance rates were the same between medical image computing (MIC) and computer assisted interventions (CAI) papers, and slightly lower where authors classified their paper as both MIC and CAI. Distribution of the geographical region of the first author as indicated in the optional demographic survey was similar among submitted and accepted papers.

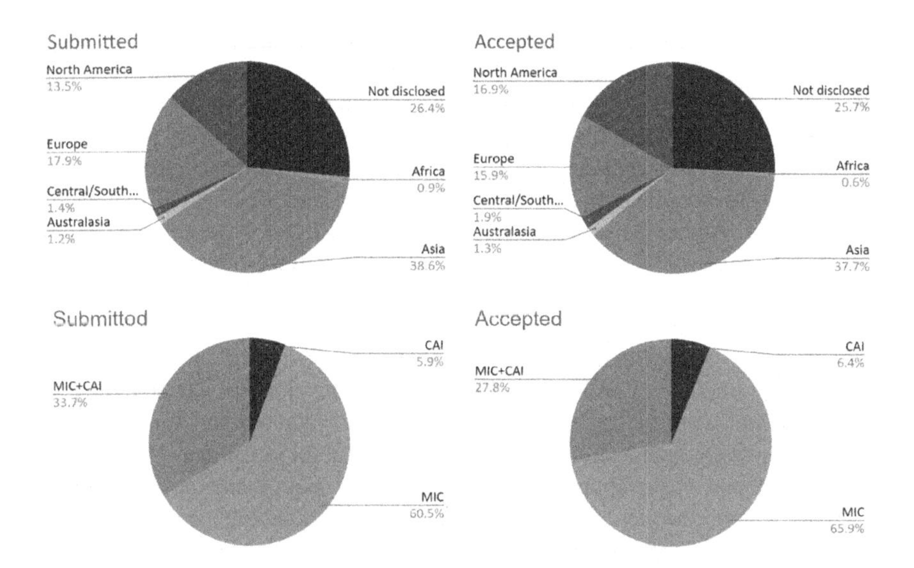

New this year, was the requirement to fill out a reproducibility checklist when submitting an intention to submit to MICCAI, in order to stimulate authors to think about what aspects of their method and experiments they should include to allow others to reproduce their results. Papers that included an anonymous code repository and/or indicated that the code would be made available were more likely to be accepted. From all accepted papers, 273 (51%) included a link to a code repository with the camera-ready submission.

Another novelty this year is that we decided to make the reviews, meta-reviews, and author responses for accepted papers available on the website. We hope the community will find this a useful resource.

The outstanding program of MICCAI 2021 was enriched by four exceptional keynote talks given by Alyson McGregor, Richard Satava, Fei-Fei Li, and Pierre Jannin, on hot topics such as gender bias in medical research, clinical translation to industry, intelligent medicine, and sustainable research. This year, as in previous years, high-quality satellite events completed the program of the main conference: 28 workshops, 23 challenges, and 14 tutorials; without forgetting the increasingly successful plenary events, such as the Women in MICCAI (WiM) meeting, the MICCAI Student Board (MSB) events, the 2nd Startup Village, the MICCAI-RSNA panel, and the first "Reinforcing Inclusiveness & diverSity and Empowering MICCAI" (or RISE-MICCAI) event.

MICCAI 2021 has also seen the first edition of CLINICCAI, the clinical day of MICCAI. Organized by Nicolas Padoy and Lee Swanstrom, this new event will hopefully help bring the scientific and clinical communities closer together, and foster collaborations and interaction. A common keynote connected the two events. We hope this effort will be pursued in the next editions.

We would like to thank everyone who has contributed to making MICCAI 2021 a success. First of all, we sincerely thank the authors, area chairs, reviewers, and session chairs for their dedication and for offering the participants and readers of these proceedings content of exceptional quality. Special thanks go to our fantastic submission platform manager Kitty Wong, who has been a tremendous help in the entire process from reviewer and area chair selection, paper submission, and the review process to the preparation of these proceedings. We also thank our very efficient team of satellite events chairs and coordinators, led by Cristian Linte and Matthieu Chabanas: the workshop chairs, Amber Simpson, Denis Fortun, Marta Kersten-Oertel, and Sandrine Voros; the challenges chairs, Annika Reinke, Spyridon Bakas, Nicolas Passat, and Ingerid Reinersten; and the tutorial chairs, Sonia Pujol and Vincent Noblet, as well as all the satellite event organizers for the valuable content added to MICCAI. Our special thanks also go to John Baxter and his team who worked hard on setting up and populating the virtual platforms, to Alejandro Granados for his valuable help and efficient communication on social media, and to Shelley Wallace and Anna Van Vliet for marketing and communication. We are also very grateful to Anirban Mukhopadhay for his management of the sponsorship, and of course many thanks to the numerous sponsors who supported the conference, often with continuous engagement over many years. This year again, our thanks go to Marius Linguraru and his team who supervised a range of actions to help, and promote, career development, among which were the mentorship program and the Startup Village. And last but not least, our wholehearted thanks go to Mehmet and the wonderful team at Dekon Congress and Tourism for their great professionalism and reactivity in the management of all logistical aspects of the event.

Finally, we thank the MICCAI society and the Board of Directors for their support throughout the years, starting with the first discussions about bringing MICCAI to Strasbourg in 2017.

We look forward to seeing you at MICCAI 2022.

September 2021

Marleen de Bruijne
Philippe Cattin
Stéphane Cotin
Nicolas Padoy
Stefanie Speidel
Yefeng Zheng
Caroline Essert

Organization

General Chair

Caroline Essert	Université de Strasbourg, CNRS, ICube, France

Program Chairs

Marleen de Bruijne	Erasmus MC Rotterdam, The Netherlands, and University of Copenhagen, Denmark
Philippe C. Cattin	University of Basel, Switzerland
Stéphane Cotin	Inria, France
Nicolas Padoy	Université de Strasbourg, CNRS, ICube, IHU, France
Stefanie Speidel	National Center for Tumor Diseases, Dresden, Germany
Yefeng Zheng	Tencent Jarvis Lab, China

Satellite Events Coordinators

Cristian Linte	Rochester Institute of Technology, USA
Matthieu Chabanas	Université Grenoble Alpes, France

Workshop Team

Amber Simpson	Queen's University, Canada
Denis Fortun	Université de Strasbourg, CNRS, ICube, France
Marta Kersten-Oertel	Concordia University, Canada
Sandrine Voros	TIMC-IMAG, INSERM, France

Challenges Team

Annika Reinke	German Cancer Research Center, Germany
Spyridon Bakas	University of Pennsylvania, USA
Nicolas Passat	Université de Reims Champagne-Ardenne, France
Ingerid Reinersten	SINTEF, NTNU, Norway

Tutorial Team

Vincent Noblet	Université de Strasbourg, CNRS, ICube, France
Sonia Pujol	Harvard Medical School, Brigham and Women's Hospital, USA

Clinical Day Chairs

Nicolas Padoy	Université de Strasbourg, CNRS, ICube, IHU, France
Lee Swanström	IHU Strasbourg, France

Sponsorship Chairs

Anirban Mukhopadhyay	Technische Universität Darmstadt, Germany
Yanwu Xu	Baidu Inc., China

Young Investigators and Early Career Development Program Chairs

Marius Linguraru	Children's National Institute, USA
Antonio Porras	Children's National Institute, USA
Daniel Racoceanu	Sorbonne Université/Brain Institute, France
Nicola Rieke	NVIDIA, Germany
Renee Yao	NVIDIA, USA

Social Media Chairs

Alejandro Granados Martinez	King's College London, UK
Shuwei Xing	Robarts Research Institute, Canada
Maxence Boels	King's College London, UK

Green Team

Pierre Jannin	INSERM, Université de Rennes 1, France
Étienne Baudrier	Université de Strasbourg, CNRS, ICube, France

Student Board Liaison

Éléonore Dufresne	Université de Strasbourg, CNRS, ICube, France
Étienne Le Quentrec	Université de Strasbourg, CNRS, ICube, France
Vinkle Srivastav	Université de Strasbourg, CNRS, ICube, France

Submission Platform Manager

Kitty Wong	The MICCAI Society, Canada

Virtual Platform Manager

John Baxter	INSERM, Université de Rennes 1, France

Program Committee

Name	Affiliation
Ehsan Adeli	Stanford University, USA
Iman Aganj	Massachusetts General Hospital, Harvard Medical School, USA
Pablo Arbelaez	Universidad de los Andes, Colombia
John Ashburner	University College London, UK
Meritxell Bach Cuadra	University of Lausanne, Switzerland
Sophia Bano	University College London, UK
Adrien Bartoli	Université Clermont Auvergne, France
Christian Baumgartner	ETH Zürich, Switzerland
Hrvoje Bogunovic	Medical University of Vienna, Austria
Weidong Cai	University of Sydney, Australia
Gustavo Carneiro	University of Adelaide, Australia
Chao Chen	Stony Brook University, USA
Elvis Chen	Robarts Research Institute, Canada
Hao Chen	Hong Kong University of Science and Technology, Hong Kong SAR
Albert Chung	Hong Kong University of Science and Technology, Hong Kong SAR
Adrian Dalca	Massachusetts Institute of Technology, USA
Adrien Depeursinge	HES-SO Valais-Wallis, Switzerland
Jose Dolz	ÉTS Montréal, Canada
Ruogu Fang	University of Florida, USA
Dagan Feng	University of Sydney, Australia
Huazhu Fu	Inception Institute of Artificial Intelligence, United Arab Emirates
Mingchen Gao	University at Buffalo, The State University of New York, USA
Guido Gerig	New York University, USA
Orcun Goksel	Uppsala University, Sweden
Alberto Gomez	King's College London, UK
Ilker Hacihaliloglu	Rutgers University, USA
Adam Harrison	PAII Inc., USA
Mattias Heinrich	University of Lübeck, Germany
Yi Hong	Shanghai Jiao Tong University, China
Yipeng Hu	University College London, UK
Junzhou Huang	University of Texas at Arlington, USA
Xiaolei Huang	The Pennsylvania State University, USA
Jana Hutter	King's College London, UK
Madhura Ingalhalikar	Symbiosis Center for Medical Image Analysis, India
Shantanu Joshi	University of California, Los Angeles, USA
Samuel Kadoury	Polytechnique Montréal, Canada
Fahmi Khalifa	Mansoura University, Egypt
Hosung Kim	University of Southern California, USA
Minjeong Kim	University of North Carolina at Greensboro, USA

Zhong Xue	Shanghai United Imaging Intelligence, China
Xin Yang	Huazhong University of Science and Technology, China
Jianhua Yao	National Institutes of Health, USA
Zhaozheng Yin	Stony Brook University, USA
Yixuan Yuan	City University of Hong Kong, Hong Kong SAR
Liang Zhan	University of Pittsburgh, USA
Tuo Zhang	Northwestern Polytechnical University, China
Yitian Zhao	Chinese Academy of Sciences, China
Luping Zhou	University of Sydney, Australia
S. Kevin Zhou	Chinese Academy of Sciences, China
Dajiang Zhu	University of Texas at Arlington, USA
Xiahai Zhuang	Fudan University, China
Maria A. Zuluaga	EURECOM, France

Reviewers

Alaa Eldin Abdelaal
Khalid Abdul Jabbar
Purang Abolmaesumi
Mazdak Abulnaga
Maryam Afzali
Priya Aggarwal
Ola Ahmad
Sahar Ahmad
Euijoon Ahn
Alireza Akhondi-Asl
Saad Ullah Akram
Dawood Al Chanti
Daniel Alexander
Sharib Ali
Lejla Alic
Omar Al-Kadi
Maximilian Allan
Pierre Ambrosini
Sameer Antani
Michela Antonelli
Jacob Antunes
Syed Anwar
Ignacio Arganda-Carreras
Mohammad Ali Armin
Md Ashikuzzaman
Mehdi Astaraki
Angélica Atehortúa
Gowtham Atluri
Chloé Audigier
Kamran Avanaki
Angelica Aviles-Rivero
Suyash Awate
Dogu Baran Aydogan
Qinle Ba
Morteza Babaie
Hyeon-Min Bae
Woong Bae
Junjie Bai
Wenjia Bai
Ujjwal Baid
Spyridon Bakas
Yaël Balbastre
Marcin Balicki
Fabian Balsiger
Abhirup Banerjee
Sreya Banerjee
Shunxing Bao
Adrian Barbu
Sumana Basu
Mathilde Bateson
Deepti Bathula
John Baxter
Bahareh Behboodi
Delaram Behnami
Mikhail Belyaev
Aicha BenTaieb

Camilo Bermudez
Gabriel Bernardino
Hadrien Bertrand
Alaa Bessadok
Michael Beyeler
Indrani Bhattacharya
Chetan Bhole
Lei Bi
Gui-Bin Bian
Ryoma Bise
Stefano B. Blumberg
Ester Bonmati
Bhushan Borotikar
Jiri Borovec
Ilaria Boscolo Galazzo
Alexandre Bousse
Nicolas Boutry
Behzad Bozorgtabar
Nathaniel Braman
Nadia Brancati
Katharina Breininger
Christopher Bridge
Esther Bron
Rupert Brooks
Qirong Bu
Duc Toan Bui
Ninon Burgos
Nikolay Burlutskiy
Hendrik Burwinkel
Russell Butler
Michał Byra
Ryan Cabeen
Mariano Cabezas
Hongmin Cai
Jinzheng Cai
Yunliang Cai
Sema Candemir
Bing Cao
Qing Cao
Shilei Cao
Tian Cao
Weiguo Cao
Aaron Carass
M. Jorge Cardoso
Adrià Casamitjana
Matthieu Chabanas
Ahmad Chaddad
Jayasree Chakraborty
Sylvie Chambon
Yi Hao Chan
Ming-Ching Chang
Peng Chang
Violeta Chang
Sudhanya Chatterjee
Christos Chatzichristos
Antong Chen
Chang Chen
Cheng Chen
Dongdong Chen
Geng Chen
Hanbo Chen
Jianan Chen
Jianxu Chen
Jie Chen
Junxiang Chen
Lei Chen
Li Chen
Liangjun Chen
Min Chen
Pingjun Chen
Qiang Chen
Shuai Chen
Tianhua Chen
Tingting Chen
Xi Chen
Xiaoran Chen
Xin Chen
Xuejin Chen
Yuhua Chen
Yukun Chen
Zhaolin Chen
Zhineng Chen
Zhixiang Chen
Erkang Cheng
Jun Cheng
Li Cheng
Yuan Cheng
Farida Cheriet
Minqi Chong
Jaegul Choo
Aritra Chowdhury
Gary Christensen

Daan Christiaens
Stergios Christodoulidis
Ai Wern Chung
Pietro Antonio Cicalese
Özgün Çiçek
Celia Cintas
Matthew Clarkson
Jaume Coll-Font
Toby Collins
Olivier Commowick
Pierre-Henri Conze
Timothy Cootes
Luca Corinzia
Teresa Correia
Hadrien Courtecuisse
Jeffrey Craley
Hui Cui
Jianan Cui
Zhiming Cui
Kathleen Curran
Claire Cury
Tobias Czempiel
Vedrana Dahl
Haixing Dai
Rafat Damseh
Bilel Daoud
Neda Davoudi
Laura Daza
Sandro De Zanet
Charles Delahunt
Yang Deng
Cem Deniz
Felix Denzinger
Hrishikesh Deshpande
Christian Desrosiers
Blake Dewey
Neel Dey
Raunak Dey
Jwala Dhamala
Yashin Dicente Cid
Li Ding
Xinghao Ding
Zhipeng Ding
Konstantin Dmitriev
Ines Domingues
Liang Dong
Mengjin Dong
Nanqing Dong
Reuben Dorent
Sven Dorkenwald
Qi Dou
Simon Drouin
Niharika D'Souza
Lei Du
Hongyi Duanmu
Nicolas Duchateau
James Duncan
Luc Duong
Nicha Dvornek
Dmitry V. Dylov
Oleh Dzyubachyk
Roy Eagleson
Mehran Ebrahimi
Jan Egger
Alma Eguizabal
Gudmundur Einarsson
Ahmed Elazab
Mohammed S. M. Elbaz
Shireen Elhabian
Mohammed Elmogy
Amr Elsawy
Ahmed Eltanboly
Sandy Engelhardt
Ertunc Erdil
Marius Erdt
Floris Ernst
Boris Escalante-Ramírez
Maria Escobar
Mohammad Eslami
Nazila Esmaeili
Marco Esposito
Oscar Esteban
Théo Estienne
Ivan Ezhov
Deng-Ping Fan
Jingfan Fan
Xin Fan
Yonghui Fan
Xi Fang
Zhenghan Fang
Aly Farag
Mohsen Farzi

Lina Felsner
Jun Feng
Ruibin Feng
Xinyang Feng
Yuan Feng
Aaron Fenster
Aasa Feragen
Henrique Fernandes
Enzo Ferrante
Jean Feydy
Lukas Fischer
Peter Fischer
Antonio Foncubierta-Rodríguez
Germain Forestier
Nils Daniel Forkert
Jean-Rassaire Fouefack
Moti Freiman
Wolfgang Freysinger
Xueyang Fu
Yunguan Fu
Wolfgang Fuhl
Isabel Funke
Philipp Fürnstahl
Pedro Furtado
Ryo Furukawa
Jin Kyu Gahm
Laurent Gajny
Adrian Galdran
Yu Gan
Melanie Ganz
Cong Gao
Dongxu Gao
Linlin Gao
Siyuan Gao
Yixin Gao
Yue Gao
Zhifan Gao
Alfonso Gastelum-Strozzi
Srishti Gautam
Bao Ge
Rongjun Ge
Zongyuan Ge
Sairam Geethanath
Shiv Gehlot
Nils Gessert
Olivier Gevaert
Sandesh Ghimire
Ali Gholipour
Sayan Ghosal
Andrea Giovannini
Gabriel Girard
Ben Glocker
Arnold Gomez
Mingming Gong
Cristina González
German Gonzalez
Sharath Gopal
Karthik Gopinath
Pietro Gori
Michael Götz
Shuiping Gou
Maged Goubran
Sobhan Goudarzi
Dushyant Goyal
Mark Graham
Bertrand Granado
Alejandro Granados
Vicente Grau
Lin Gu
Shi Gu
Xianfeng Gu
Yun Gu
Zaiwang Gu
Hao Guan
Ricardo Guerrero
Houssem-Eddine Gueziri
Dazhou Guo
Hengtao Guo
Jixiang Guo
Pengfei Guo
Xiaoqing Guo
Yi Guo
Yulan Guo
Yuyu Guo
Krati Gupta
Vikash Gupta
Praveen Gurunath Bharathi
Boris Gutman
Prashnna Gyawali
Stathis Hadjidemetriou
Mohammad Hamghalam
Hu Han

Liang Han
Xiaoguang Han
Xu Han
Zhi Han
Zhongyi Han
Jonny Hancox
Xiaoke Hao
Nandinee Haq
Ali Hatamizadeh
Charles Hatt
Andreas Hauptmann
Mohammad Havaei
Kelei He
Nanjun He
Tiancheng He
Xuming He
Yuting He
Nicholas Heller
Alessa Hering
Monica Hernandez
Carlos Hernandez-Matas
Kilian Hett
Jacob Hinkle
David Ho
Nico Hoffmann
Matthew Holden
Sungmin Hong
Yoonmi Hong
Antal Horváth
Md Belayat Hossain
Benjamin Hou
William Hsu
Tai-Chiu Hsung
Kai Hu
Shi Hu
Shunbo Hu
Wenxing Hu
Xiaoling Hu
Xiaowei Hu
Yan Hu
Zhenhong Hu
Heng Huang
Qiaoying Huang
Yi-Jie Huang
Yixing Huang
Yongxiang Huang
Yue Huang
Yufang Huang
Arnaud Huaulmé
Henkjan Huisman
Yuankai Huo
Andreas Husch
Mohammad Hussain
Raabid Hussain
Sarfaraz Hussein
Khoi Huynh
Seong Jae Hwang
Emmanuel Iarussi
Kay Igwe
Abdullah-Al-Zubaer Imran
Ismail Irmakci
Mobarakol Islam
Mohammad Shafkat Islam
Vamsi Ithapu
Koichi Ito
Hayato Itoh
Oleksandra Ivashchenko
Yuji Iwahori
Shruti Jadon
Mohammad Jafari
Mostafa Jahanifar
Amir Jamaludin
Mirek Janatka
Won-Dong Jang
Uditha Jarayathne
Ronnachai Jaroensri
Golara Javadi
Rohit Jena
Rachid Jennane
Todd Jensen
Won-Ki Jeong
Yuanfeng Ji
Zhanghexuan Ji
Haozhe Jia
Jue Jiang
Tingting Jiang
Xiang Jiang
Jianbo Jiao
Zhicheng Jiao
Amelia Jiménez-Sánchez
Dakai Jin
Yueming Jin

Bin Jing
Anand Joshi
Yohan Jun
Kyu-Hwan Jung
Alain Jungo
Manjunath K N
Ali Kafaei Zad Tehrani
Bernhard Kainz
John Kalafut
Michael C. Kampffmeyer
Qingbo Kang
Po-Yu Kao
Neerav Karani
Turkay Kart
Satyananda Kashyap
Amin Katouzian
Alexander Katzmann
Prabhjot Kaur
Erwan Kerrien
Hoel Kervadec
Ashkan Khakzar
Nadieh Khalili
Siavash Khallaghi
Farzad Khalvati
Bishesh Khanal
Pulkit Khandelwal
Maksim Kholiavchenko
Naji Khosravan
Seyed Mostafa Kia
Daeseung Kim
Hak Gu Kim
Hyo-Eun Kim
Jae-Hun Kim
Jaeil Kim
Jinman Kim
Mansu Kim
Namkug Kim
Seong Tae Kim
Won Hwa Kim
Andrew King
Atilla Kiraly
Yoshiro Kitamura
Tobias Klinder
Bin Kong
Jun Kong
Tomasz Konopczynski
Bongjin Koo
Ivica Kopriva
Kivanc Kose
Mateusz Kozinski
Anna Kreshuk
Anithapriya Krishnan
Pavitra Krishnaswamy
Egor Krivov
Frithjof Kruggel
Alexander Krull
Elizabeth Krupinski
Serife Kucur
David Kügler
Hugo Kuijf
Abhay Kumar
Ashnil Kumar
Kuldeep Kumar
Nitin Kumar
Holger Kunze
Tahsin Kurc
Anvar Kurmukov
Yoshihiro Kuroda
Jin Tae Kwak
Yongchan Kwon
Francesco La Rosa
Aymen Laadhari
Dmitrii Lachinov
Alain Lalande
Tryphon Lambrou
Carole Lartizien
Bianca Lassen-Schmidt
Ngan Le
Leo Lebrat
Christian Ledig
Eung-Joo Lee
Hyekyoung Lee
Jong-Hwan Lee
Matthew Lee
Sangmin Lee
Soochahn Lee
Étienne Léger
Stefan Leger
Andreas Leibetseder
Rogers Jeffrey Leo John
Juan Leon
Bo Li

Chongyi Li
Fuhai Li
Hongming Li
Hongwei Li
Jian Li
Jianning Li
Jiayun Li
Junhua Li
Kang Li
Mengzhang Li
Ming Li
Qing Li
Shaohua Li
Shuyu Li
Weijian Li
Weikai Li
Wenqi Li
Wenyuan Li
Xiang Li
Xiaomeng Li
Xiaoxiao Li
Xin Li
Xiuli Li
Yang Li
Yi Li
Yuexiang Li
Zeju Li
Zhang Li
Zhiyuan Li
Zhjin Li
Gongbo Liang
Jianming Liang
Libin Liang
Yuan Liang
Haofu Liao
Ruizhi Liao
Wei Liao
Xiangyun Liao
Roxane Licandro
Gilbert Lim
Baihan Lin
Hongxiang Lin
Jianyu Lin
Yi Lin
Claudia Lindner
Geert Litjens
Bin Liu
Chi Liu
Daochang Liu
Dong Liu
Dongnan Liu
Feng Liu
Hangfan Liu
Hong Liu
Huafeng Liu
Jianfei Liu
Jingya Liu
Kai Liu
Kefei Liu
Lihao Liu
Mengting Liu
Peng Liu
Qin Liu
Quande Liu
Shengfeng Liu
Shenghua Liu
Shuangjun Liu
Sidong Liu
Siqi Liu
Tianrui Liu
Xiao Liu
Xinyang Liu
Xinyu Liu
Yan Liu
Yikang Liu
Yong Liu
Yuan Liu
Yue Liu
Yuhang Liu
Andrea Loddo
Nicolas Loménie
Daniel Lopes
Bin Lou
Jian Lou
Nicolas Loy Rodas
Donghuan Lu
Huanxiang Lu
Weijia Lu
Xiankai Lu
Yongyi Lu
Yueh-Hsun Lu
Yuhang Lu

Imanol Luengo
Jie Luo
Jiebo Luo
Luyang Luo
Ma Luo
Bin Lv
Jinglei Lv
Junyan Lyu
Qing Lyu
Yuanyuan Lyu
Andy J. Ma
Chunwei Ma
Da Ma
Hua Ma
Kai Ma
Lei Ma
Anderson Maciel
Amirreza Mahbod
S. Sara Mahdavi
Mohammed Mahmoud
Saïd Mahmoudi
Klaus H. Maier-Hein
Bilal Malik
Ilja Manakov
Matteo Mancini
Tommaso Mansi
Yunxiang Mao
Brett Marinelli
Pablo Márquez Neila
Carsten Marr
Yassine Marrakchi
Fabio Martinez
Andre Mastmeyer
Tejas Sudharshan Mathai
Dimitrios Mavroeidis
Jamie McClelland
Pau Medrano-Gracia
Raghav Mehta
Sachin Mehta
Raphael Meier
Qier Meng
Qingjie Meng
Yanda Meng
Martin Menten
Odyssée Merveille
Islem Mhiri
Liang Mi
Stijn Michielse
Abhishek Midya
Fausto Milletari
Hyun-Seok Min
Zhe Min
Tadashi Miyamoto
Sara Moccia
Hassan Mohy-ud-Din
Tony C. W. Mok
Rafael Molina
Mehdi Moradi
Rodrigo Moreno
Kensaku Mori
Lia Morra
Linda Moy
Mohammad Hamed Mozaffari
Sovanlal Mukherjee
Anirban Mukhopadhyay
Henning Müller
Balamurali Murugesan
Cosmas Mwikirize
Andriy Myronenko
Saad Nadeem
Vishwesh Nath
Rodrigo Nava
Fernando Navarro
Amin Nejatbakhsh
Dong Ni
Hannes Nickisch
Dong Nie
Jingxin Nie
Aditya Nigam
Lipeng Ning
Xia Ning
Tianye Niu
Jack Noble
Vincent Noblet
Alexey Novikov
Jorge Novo
Mohammad Obeid
Masahiro Oda
Benjamin Odry
Steffen Oeltze-Jafra
Hugo Oliveira
Sara Oliveira

Arnau Oliver
Emanuele Olivetti
Jimena Olveres
John Onofrey
Felipe Orihuela-Espina
José Orlando
Marcos Ortega
Yoshito Otake
Sebastian Otálora
Cheng Ouyang
Jiahong Ouyang
Xi Ouyang
Michal Ozery-Flato
Danielle Pace
Krittin Pachtrachai
J. Blas Pagador
Akshay Pai
Viswanath Pamulakanty Sudarshan
Jin Pan
Yongsheng Pan
Pankaj Pandey
Prashant Pandey
Egor Panfilov
Shumao Pang
Joao Papa
Constantin Pape
Bartlomiej Papiez
Hyunjin Park
Jongchan Park
Sanghyun Park
Seung-Jong Park
Seyoun Park
Magdalini Paschali
Diego Patiño Cortés
Angshuman Paul
Christian Payer
Yuru Pei
Chengtao Peng
Yige Peng
Antonio Pepe
Oscar Perdomo
Sérgio Pereira
Jose-Antonio Pérez-Carrasco
Fernando Pérez-García
Jorge Perez-Gonzalez
Skand Peri
Matthias Perkonigg
Mehran Pesteie
Jorg Peters
Jens Petersen
Kersten Petersen
Renzo Phellan Aro
Ashish Phophalia
Tomasz Pieciak
Antonio Pinheiro
Pramod Pisharady
Kilian Pohl
Sebastian Pölsterl
Iulia A. Popescu
Alison Pouch
Prateek Prasanna
Raphael Prevost
Juan Prieto
Sergi Pujades
Elodie Puybareau
Esther Puyol-Antón
Haikun Qi
Huan Qi
Buyue Qian
Yan Qiang
Yuchuan Qiao
Chen Qin
Wenjian Qin
Yulei Qin
Wu Qiu
Hui Qu
Liangqiong Qu
Kha Gia Quach
Prashanth R.
Pradeep Reddy Raamana
Mehdi Rahim
Jagath Rajapakse
Kashif Rajpoot
Jhonata Ramos
Lingyan Ran
Hatem Rashwan
Daniele Ravì
Keerthi Sravan Ravi
Nishant Ravikumar
Harish RaviPrakash
Samuel Remedios
Yinhao Ren

Yudan Ren
Mauricio Reyes
Constantino Reyes-Aldasoro
Jonas Richiardi
David Richmond
Anne-Marie Rickmann
Leticia Rittner
Dominik Rivoir
Emma Robinson
Jessica Rodgers
Rafael Rodrigues
Robert Rohling
Michal Rosen-Zvi
Lukasz Roszkowiak
Karsten Roth
José Rouco
Daniel Rueckert
Jaime S. Cardoso
Mohammad Sabokrou
Ario Sadafi
Monjoy Saha
Pramit Saha
Dushyant Sahoo
Pranjal Sahu
Maria Sainz de Cea
Olivier Salvado
Robin Sandkuehler
Gianmarco Santini
Duygu Sarikaya
Imari Sato
Olivier Saut
Dustin Scheinost
Nico Scherf
Markus Schirmer
Alexander Schlaefer
Jerome Schmid
Julia Schnabel
Klaus Schoeffmann
Andreas Schuh
Ernst Schwartz
Christina Schwarz-Gsaxner
Michaël Sdika
Suman Sedai
Anjany Sekuboyina
Raghavendra Selvan
Sourya Sengupta
Youngho Seo
Lama Seoud
Ana Sequeira
Maxime Sermesant
Carmen Serrano
Muhammad Shaban
Ahmed Shaffie
Sobhan Shafiei
Mohammad Abuzar Shaikh
Reuben Shamir
Shayan Shams
Hongming Shan
Harshita Sharma
Gregory Sharp
Mohamed Shehata
Haocheng Shen
Li Shen
Liyue Shen
Mali Shen
Yiqing Shen
Yiqiu Shen
Zhengyang Shen
Kuangyu Shi
Luyao Shi
Xiaoshuang Shi
Xueying Shi
Yemin Shi
Yiyu Shi
Yonghong Shi
Jitae Shin
Boris Shirokikh
Suprosanna Shit
Suzanne Shontz
Yucheng Shu
Alberto Signoroni
Wilson Silva
Margarida Silveira
Matthew Sinclair
Rohit Singla
Sumedha Singla
Ayushi Sinha
Kevin Smith
Rajath Soans
Ahmed Soliman
Stefan Sommer
Yang Song

Youyi Song
Aristeidis Sotiras
Arcot Sowmya
Rachel Sparks
William Speier
Ziga Spiclin
Dominik Spinczyk
Jon Sporring
Chetan Srinidhi
Anuroop Sriram
Vinkle Srivastav
Lawrence Staib
Marius Staring
Johannes Stegmaier
Joshua Stough
Robin Strand
Martin Styner
Hai Su
Yun-Hsuan Su
Vaishnavi Subramanian
Gérard Subsol
Yao Sui
Avan Suinesiaputra
Jeremias Sulam
Shipra Suman
Li Sun
Wenqing Sun
Chiranjib Sur
Yannick Suter
Tanveer Syeda-Mahmood
Fatemeh Taheri Dezaki
Roger Tam
José Tamez-Peña
Chaowei Tan
Hao Tang
Thomas Tang
Yucheng Tang
Zihao Tang
Mickael Tardy
Giacomo Tarroni
Jonas Teuwen
Paul Thienphrapa
Stephen Thompson
Jiang Tian
Yu Tian
Yun Tian
Aleksei Tiulpin
Hamid Tizhoosh
Matthew Toews
Oguzhan Topsakal
Antonio Torteya
Sylvie Treuillet
Jocelyne Troccaz
Roger Trullo
Chialing Tsai
Sudhakar Tummala
Verena Uslar
Hristina Uzunova
Régis Vaillant
Maria Vakalopoulou
Jeya Maria Jose Valanarasu
Tom van Sonsbeek
Gijs van Tulder
Marta Varela
Thomas Varsavsky
Francisco Vasconcelos
Liset Vazquez Romaguera
S. Swaroop Vedula
Sanketh Vedula
Harini Veeraraghavan
Miguel Vega
Gonzalo Vegas Sanchez-Ferrero
Anant Vemuri
Gopalkrishna Veni
Mitko Veta
Thomas Vetter
Pedro Vieira
Juan Pedro Vigueras Guillén
Barbara Villarini
Satish Viswanath
Athanasios Vlontzos
Wolf-Dieter Vogl
Bo Wang
Cheng Wang
Chengjia Wang
Chunliang Wang
Clinton Wang
Congcong Wang
Dadong Wang
Dongang Wang
Haifeng Wang
Hongyu Wang

Hu Wang
Huan Wang
Kun Wang
Li Wang
Liansheng Wang
Linwei Wang
Manning Wang
Renzhen Wang
Ruixuan Wang
Sheng Wang
Shujun Wang
Shuo Wang
Tianchen Wang
Tongxin Wang
Wenzhe Wang
Xi Wang
Xiaosong Wang
Yan Wang
Yaping Wang
Yi Wang
Yirui Wang
Zeyi Wang
Zhangyang Wang
Zihao Wang
Zuhui Wang
Simon Warfield
Jonathan Weber
Jürgen Weese
Dong Wei
Donglai Wei
Dongming Wei
Martin Weigert
Wolfgang Wein
Michael Wels
Cédric Wemmert
Junhao Wen
Travis Williams
Matthias Wilms
Stefan Winzeck
James Wiskin
Adam Wittek
Marek Wodzinski
Jelmer Wolterink
Ken C. L. Wong
Chongruo Wu
Guoqing Wu
Ji Wu
Jian Wu
Jie Ying Wu
Pengxiang Wu
Xiyin Wu
Ye Wu
Yicheng Wu
Yifan Wu
Tobias Wuerfl
Pengcheng Xi
James Xia
Siyu Xia
Wenfeng Xia
Yingda Xia
Yong Xia
Lei Xiang
Deqiang Xiao
Li Xiao
Yiming Xiao
Hongtao Xie
Lingxi Xie
Long Xie
Weidi Xie
Yiting Xie
Yutong Xie
Xiaohan Xing
Chang Xu
Chenchu Xu
Hongming Xu
Kele Xu
Min Xu
Rui Xu
Xiaowei Xu
Xuanang Xu
Yongchao Xu
Zhenghua Xu
Zhoubing Xu
Kai Xuan
Cheng Xue
Jie Xue
Wufeng Xue
Yuan Xue
Faridah Yahya
Ke Yan
Yuguang Yan
Zhennan Yan

Changchun Yang
Chao-Han Huck Yang
Dong Yang
Erkun Yang
Fan Yang
Ge Yang
Guang Yang
Guanyu Yang
Heran Yang
Hongxu Yang
Huijuan Yang
Jiancheng Yang
Jie Yang
Junlin Yang
Lin Yang
Peng Yang
Xin Yang
Yan Yang
Yujiu Yang
Dongren Yao
Jiawen Yao
Li Yao
Qingsong Yao
Chuyang Ye
Dong Hye Ye
Menglong Ye
Xujiong Ye
Jingru Yi
Jirong Yi
Xin Yi
Youngjin Yoo
Chenyu You
Haichao Yu
Hanchao Yu
Lequan Yu
Qi Yu
Yang Yu
Pengyu Yuan
Fatemeh Zabihollahy
Ghada Zamzmi
Marco Zenati
Guodong Zeng
Rui Zeng
Oliver Zettinig
Zhiwei Zhai
Chaoyi Zhang
Daoqiang Zhang
Fan Zhang
Guangming Zhang
Hang Zhang
Huahong Zhang
Jianpeng Zhang
Jiong Zhang
Jun Zhang
Lei Zhang
Lichi Zhang
Lin Zhang
Ling Zhang
Lu Zhang
Miaomiao Zhang
Ning Zhang
Qiang Zhang
Rongzhao Zhang
Ru-Yuan Zhang
Shihao Zhang
Shu Zhang
Tong Zhang
Wei Zhang
Weiwei Zhang
Wen Zhang
Wenlu Zhang
Xin Zhang
Ya Zhang
Yanbo Zhang
Yanfu Zhang
Yi Zhang
Yishuo Zhang
Yong Zhang
Yongqin Zhang
You Zhang
Youshan Zhang
Yu Zhang
Yue Zhang
Yueyi Zhang
Yulun Zhang
Yunyan Zhang
Yuyao Zhang
Can Zhao
Changchen Zhao
Chongyue Zhao
Fenqiang Zhao
Gangming Zhao

He Zhao
Jun Zhao
Li Zhao
Qingyu Zhao
Rongchang Zhao
Shen Zhao
Shijie Zhao
Tengda Zhao
Tianyi Zhao
Wei Zhao
Xuandong Zhao
Yiyuan Zhao
Yuan-Xing Zhao
Yue Zhao
Zixu Zhao
Ziyuan Zhao
Xingjian Zhen
Guoyan Zheng
Hao Zheng
Jiannan Zheng
Kang Zheng
Shenhai Zheng
Yalin Zheng
Yinqiang Zheng
Yushan Zheng
Jia-Xing Zhong
Zichun Zhong
Bo Zhou
Haoyin Zhou
Hong-Yu Zhou
Kang Zhou
Sanping Zhou
Sihang Zhou
Tao Zhou
Xiao-Yun Zhou
Yanning Zhou
Yuyin Zhou
Zongwei Zhou
Dongxiao Zhu
Hancan Zhu
Lei Zhu
Qikui Zhu
Xinliang Zhu
Yuemin Zhu
Zhe Zhu
Zhuotun Zhu
Aneeq Zia
Veronika Zimmer
David Zimmerer
Lilla Zöllci
Yukai Zou
Lianrui Zuo
Gerald Zwettler
Reyer Zwiggelaar

Outstanding Reviewers

Neel Dey	New York University, USA
Monica Hernandez	University of Zaragoza, Spain
Ivica Kopriva	Rudjer Boskovich Institute, Croatia
Sebastian Otálora	University of Applied Sciences and Arts Western Switzerland, Switzerland
Danielle Pace	Massachusetts General Hospital, USA
Sérgio Pereira	Lunit Inc., South Korea
David Richmond	IBM Watson Health, USA
Rohit Singla	University of British Columbia, Canada
Yan Wang	Sichuan University, China

Honorable Mentions (Reviewers)

Mazdak Abulnaga	Massachusetts Institute of Technology, USA
Pierre Ambrosini	Erasmus University Medical Center, The Netherlands
Hyeon-Min Bae	Korea Advanced Institute of Science and Technology, South Korea
Mikhail Belyaev	Skolkovo Institute of Science and Technology, Russia
Bhushan Borotikar	Symbiosis International University, India
Katharina Breininger	Friedrich-Alexander Universität Erlangen Nürnberg, Germany
Ninon Burgos	CNRS, Paris Brain Institute, France
Mariano Cabezas	The University of Sydney, Australia
Aaron Carass	Johns Hopkins University, USA
Pierre-Henri Conze	IMT Atlantique, France
Christian Desrosiers	École de technologie supérieure, Canada
Reuben Dorent	King's College London, UK
Nicha Dvornek	Yale University, USA
Dmitry V. Dylov	Skolkovo Institute of Science and Technology, Russia
Marius Erdt	Fraunhofer Singapore, Singapore
Ruibin Feng	Stanford University, USA
Enzo Ferrante	CONICET/Universidad Nacional del Litoral, Argentina
Antonio Foncubierta-Rodríguez	IBM Research, Switzerland
Isabel Funke	National Center for Tumor Diseases Dresden, Germany
Adrian Galdran	University of Bournemouth, UK
Ben Glocker	Imperial College London, UK
Cristina González	Universidad de los Andes, Colombia
Maged Goubran	Sunnybrook Research Institute, Canada
Sobhan Goudarzi	Concordia University, Canada
Vicente Grau	University of Oxford, UK
Andreas Hauptmann	University of Oulu, Finland
Nico Hoffmann	Technische Universität Dresden, Germany
Sungmin Hong	Massachusetts General Hospital, Harvard Medical School, USA
Won-Dong Jang	Harvard University, USA
Zhanghexuan Ji	University at Buffalo, SUNY, USA
Neerav Karani	ETH Zurich, Switzerland
Alexander Katzmann	Siemens Healthineers, Germany
Erwan Kerrien	Inria, France
Anitha Priya Krishnan	Genentech, USA
Tahsin Kurc	Stony Brook University, USA
Francesco La Rosa	École polytechnique fédérale de Lausanne, Switzerland
Dmitrii Lachinov	Medical University of Vienna, Austria
Mengzhang Li	Peking University, China
Gilbert Lim	National University of Singapore, Singapore
Dongnan Liu	University of Sydney, Australia

Bin Lou	Siemens Healthineers, USA
Kai Ma	Tencent, China
Klaus H. Maier-Hein	German Cancer Research Center (DKFZ), Germany
Raphael Meier	University Hospital Bern, Switzerland
Tony C. W. Mok	Hong Kong University of Science and Technology, Hong Kong SAR
Lia Morra	Politecnico di Torino, Italy
Cosmas Mwikirize	Rutgers University, USA
Felipe Orihuela-Espina	Instituto Nacional de Astrofísica, Óptica y Electrónica, Mexico
Egor Panfilov	University of Oulu, Finland
Christian Payer	Graz University of Technology, Austria
Sebastian Pölsterl	Ludwig-Maximilians Universitat, Germany
José Rouco	University of A Coruña, Spain
Daniel Rueckert	Imperial College London, UK
Julia Schnabel	King's College London, UK
Christina Schwarz-Gsaxner	Graz University of Technology, Austria
Boris Shirokikh	Skolkovo Institute of Science and Technology, Russia
Yang Song	University of New South Wales, Australia
Gérard Subsol	Université de Montpellier, France
Tanveer Syeda-Mahmood	IBM Research, USA
Mickael Tardy	Hera-MI, France
Paul Thienphrapa	Atlas5D, USA
Gijs van Tulder	Radboud University, The Netherlands
Tongxin Wang	Indiana University, USA
Yirui Wang	PAII Inc., USA
Jelmer Wolterink	University of Twente, The Netherlands
Lei Xiang	Subtle Medical Inc., USA
Fatemeh Zabihollahy	Johns Hopkins University, USA
Wei Zhang	University of Georgia, USA
Ya Zhang	Shanghai Jiao Tong University, China
Qingyu Zhao	Stanford University, China
Yushan Zheng	Beihang University, China

Mentorship Program (Mentors)

Shadi Albarqouni	Helmholtz AI, Helmholtz Center Munich, Germany
Hao Chen	Hong Kong University of Science and Technology, Hong Kong SAR
Nadim Daher	NVIDIA, France
Marleen de Bruijne	Erasmus MC/University of Copenhagen, The Netherlands
Qi Dou	The Chinese University of Hong Kong, Hong Kong SAR
Gabor Fichtinger	Queen's University, Canada
Jonny Hancox	NVIDIA, UK

Nobuhiko Hata	Harvard Medical School, USA
Sharon Xiaolei Huang	Pennsylvania State University, USA
Jana Hutter	King's College London, UK
Dakai Jin	PAII Inc., China
Samuel Kadoury	Polytechnique Montréal, Canada
Minjeong Kim	University of North Carolina at Greensboro, USA
Hans Lamecker	1000shapes GmbH, Germany
Andrea Lara	Galileo University, Guatemala
Ngan Le	University of Arkansas, USA
Baiying Lei	Shenzhen University, China
Karim Lekadir	Universitat de Barcelona, Spain
Marius George Linguraru	Children's National Health System/George Washington University, USA
Herve Lombaert	ETS Montreal, Canada
Marco Lorenzi	Inria, France
Le Lu	PAII Inc., China
Xiongbiao Luo	Xiamen University, China
Dzung Pham	Henry M. Jackson Foundation/Uniformed Services University/National Institutes of Health/Johns Hopkins University, USA
Josien Pluim	Eindhoven University of Technology/University Medical Center Utrecht, The Netherlands
Antonio Porras	University of Colorado Anschutz Medical Campus/Children's Hospital Colorado, USA
Islem Rekik	Istanbul Technical University, Turkey
Nicola Rieke	NVIDIA, Germany
Julia Schnabel	TU Munich/Helmholtz Center Munich, Germany, and King's College London, UK
Debdoot Sheet	Indian Institute of Technology Kharagpur, India
Pallavi Tiwari	Case Western Reserve University, USA
Jocelyne Troccaz	CNRS, TIMC, Grenoble Alpes University, France
Sandrine Voros	TIMC-IMAG, INSERM, France
Linwei Wang	Rochester Institute of Technology, USA
Yalin Wang	Arizona State University, USA
Zhong Xue	United Imaging Intelligence Co. Ltd, USA
Renee Yao	NVIDIA, USA
Mohammad Yaqub	Mohamed Bin Zayed University of Artificial Intelligence, United Arab Emirates, and University of Oxford, UK
S. Kevin Zhou	University of Science and Technology of China, China
Lilla Zollei	Massachusetts General Hospital, Harvard Medical School, USA
Maria A. Zuluaga	EURECOM, France

Contents – Part II

Machine Learning - Semi-Supervised Learning

Machine Learning - Weakly Supervised Learning

Machine Learning - Self-Supervised Learning

SSLP: Spatial Guided Self-supervised Learning on Pathological Images

Jiajun Li[1,2], Tiancheng Lin[1,2], and Yi Xu[1,2](✉)

[1] Shanghai Jiao Tong University, Shanghai, China
xuyi@sjtu.edu.cn
[2] MoE Key Lab of Artificial Intelligence, AI Institute,
Shanghai Jiao Tong University, Shanghai, China

Abstract. Nowadays, there is an urgent requirement of self-supervised learning (SSL) on whole slide pathological images (WSIs) to relieve the demand of finely expert annotations. However, the performance of SSL algorithms on WSIs has long lagged behind their supervised counterparts. To close this gap, in this paper, we fully explore the intrinsic characteristics of WSIs and propose SSLP: Spatial Guided Self-supervised Learning on Pathological Images. We argue the patch-wise spatial proximity is a significant characteristic of WSIs, if properly employed, shall provide abundant supervision for free. Specifically, we explore three semantic invariance from 1) self-invariance: the same patch of different augmented views, 2) intra-invariance: the patches within spatial neighbors and 3) inter-invariance: their corresponding neighbors in the feature space. As a result, our SSLP model achieves 82.9% accuracy and 85.7% AUC on CAMELYON linear classification and 95.2% accuracy fine-tuning on cross-disease classification on NCTCRC, which outperforms previous state-of-the-art algorithm and matches the performance of a supervised counterpart.

Keywords: Self-supervised learning · Pathological images · Spatial proximity

1 Introduction

Nowadays, several pioneering studies have achieved diagnostic performance comparable to human experts via supervised deep learning models on pathological images [3,4,9,22,23]. However, the requirement of exhaustively annotated datasets significantly hampers the scalability and usability of these studies since the finely expert annotations are difficult to obtain. Specifically, interpreting WSIs is demanding [12,19] and curating annotations is time-consuming

J. Li and T. Lin—These authors have contributed equally.

Electronic supplementary material The online version of this chapter (https://doi.org/10.1007/978-3-030-87196-3_1) contains supplementary material, which is available to authorized users.

M. de Bruijne et al. (Eds.): MICCAI 2021, LNCS 12902, pp. 3–12, 2021.
https://doi.org/10.1007/978-3-030-87196-3_1

[24,25,27]. In order to relieve the demand of annotation, there is an urgent requirement of self-supervised learning (SSL) on WSIs without supervision.

Recently, SSL algorithms have made steady progress in natural images, some of which even outperform the supervised pre-training counterpart in several classic computer vision tasks [2,5,6,15,16]. According to [8], these SSL algorithms can be categorized into: 1) *Contrastive learning* [6,16,33] which attracts the positive sample pairs and repulses the negative sample pairs, 2) *Clustering* [5,17,29] which alternates between clustering the representations and learning to predict the cluster assignment and 3) *Consistency* [15,31] which directly predicts the output of one view from another view.

Inspired by the success of SSL algorithms in natural images, some studies [11,21] simply apply them, like CPC [26] and MoCo [7,16], to histopathological classification. These 'brute-force' extensions may be sub-optimal since WSIs are distinct from natural images in some characteristics. For example, rotation prediction [14] may bring limited benefit since WSIs have no dominant orientation. Recently, Xie et al. [30] design pretext tasks for accurate nuclei instance segmentation. Although their method has gained remarkable performance, the extremely requirement of task-specific knowledge to design these pretext tasks limits its universality. Abbet et al. [1] propose a complex framework of SSL on WSIs named DnR that consists of region of interest (RoI) detection, image reconstruction and contrastive learning with AND algorithm [17]. They expand the anchor neighborhoods defined in AND by the overlapping pathological image patches, forming the overlapping patches as positive pairs in contrastive learning. However, this definition leads to noisy supervision since the overlapping patches may contain the opposite semantic labels when an anchor is located on the boundary between tumor and normal tissues.

From this perspective, we hypothesize that it is desirable to design SSL algorithms on WSIs that are: 1) transferable to different downstream tasks and 2) properly utilizing some intrinsic characteristics of WSIs. Existing studies in this track have limitations in at least one of these two aspects. In this paper, we propose *Spatial Guided Self-supervised Leaning on Pathological Images* (SSLP) as a comprehensive way of deeply mining the semantic invariance in pathological images, where both spatial proximity and semantic proximity are explored to improve the performance of SSL in pathological image classification (Fig. 1). We argue the patch-wise spatial proximity is a significant characteristic of WSIs, which could be used to reveal the underlying patch-to-patch relationships. Originally, contrastive learning only investigates the relationship of different views from the same patch. With spatial proximity, we could further delve into the intra- and inter-invariance of the pathological image patches from the same and different WSIs. Moreover, contrastive learning critically relies on a pool of negatives to learn representations [13,33] and the vast majority of easy negatives are unnecessary for the training while those hardest negatives are harmful to performance. To make a trade-off, we propose a skewness semi-hard negative mining strategy by 'Beta Distribution'.

Extensive experiments are conducted on two public WSI datasets (*CAMELYON16* [3] and *NCTCRC* [18]). The results demonstrate the advantages of our SSLP algorithm over a wide variety of existing state-of-the-art SSL algorithms, showing a promising performance that matches the supervised counterpart.

2 Method

Given a set of unlabeled pathological image patches tiled from whole slide images (WSIs), our goal is to learn an encoder f_θ that maps these patches to $\mathcal{D}$-dimension embeddings v_i and implement v_i on several downstream tasks (Fig. 1(a)). v_i are stored in a Memory Bank [28] with a momentum coefficient t to relieve the limit of the mini-batch size. In order to learn discriminative v_i without supervision, traditionally, contrastive learning (i.e., Instance Discrimination [28]) aims to learn self-invariant representations (Sect. 2.1). For WSIs, through properly mining the patch-wise spatial proximity, we further explore the intra- and inter-invariance (Sects. 2.2 and 2.3) among the patches within spatial neighbors and their corresponding neighbors in the feature space to improve the performance of SSL on WSIs.

2.1 Self-invariance

For self-invariance learning, we focus on learning a feature representation that discriminates among all different images but keeps invariant for random augmented views of the same image (Fig. 1(c)). A natural solution is contrastive learning, which pulls random augmented views of the same image (positives) together and pushes different images (negatives) apart. Here, we consider an effective form of contrastive learning loss function called InfoNCE [26] as follows:

$$\mathcal{L}_{InfoNCE} = -\log \frac{\exp\left(v_+^T v_i/\tau\right)}{\exp\left(v_+^T v_i/\tau\right) + \sum_{v_- \in N_k} \exp\left(v_-^T v_i/\tau\right)} \tag{1}$$

where τ is a temperature hyper-parameter, v_+ is a positive sample for anchor v_i and N_k denotes a set of k negative samples v_- randomly selected from memory bank.

2.2 Intra-invariance

Based on spatial proximity of pathological image patches in a same WSI, the embeddings of spatial neighbors (adjacent patches) are more likely to share the same semantic label. This is called *intra-invariance*. A naive implement like $\mathcal{L}_{Divide}$ in DnR [1] is taking any two adjacent patches as positives:

$$\mathcal{L}_{Divide} = -\log \sum_{v_+ \in S_i} \frac{\exp\left(v_+^T v_i/\tau\right)}{\exp\left(v_+^T v_i/\tau\right) + \sum_{v_- \in N_k} \exp\left(v_-^T v_i/\tau\right)} \tag{2}$$

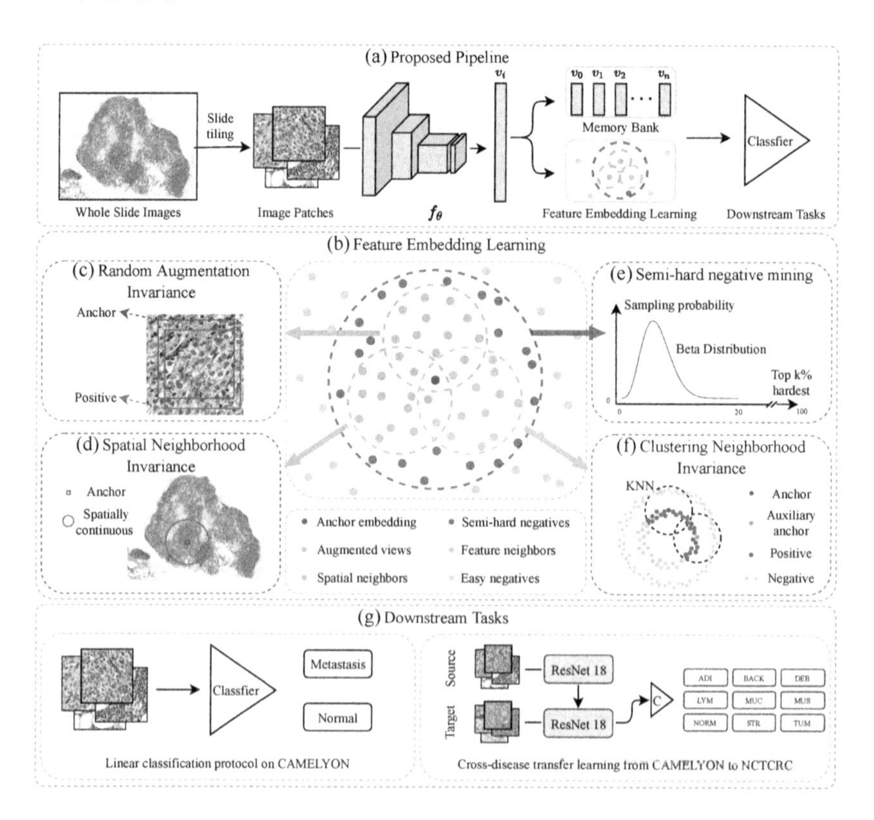

Fig. 1. Overview of the proposed SSLP method. **(a)** The pipeline of SSLP. **(b)** Feature embedding learning, containing 3 pretext tasks (**c,d,f**) and a hard-mining strategy (**e**). **(g)** Two downstream tasks utilized to evaluate the performance. (Color figure online)

where S_i is a set of spatial neighbors of v_i, defined as all patches whose spatial distance to the anchor patch (measured by Euclidean distance) is less than a certain threshold $\mathcal{T}$ (Fig. 1(d)). Equation 2 maximizes the 'similarity' of anchor v_i with each neighbor in S_i, equally. However, because of the tumor heterogeneity [10,32], the labels of adjacent patches are at the risk of inconsistency. As an improvement, we encourage the intra-invariance in an adaptive manner and formulate it as:

$$\mathcal{L}_{Spatial} = -\log \frac{\sum_{v_+ \in S_i} \exp\left(v_+^T v_i/\tau\right)}{\sum_{v_+ \in S_i} \exp\left(v_+^T v_i/\tau\right) + \sum_{v_- \in \mathcal{N}_{semi}} \exp\left(v_-^T v_i/\tau\right)} \tag{3}$$

where $\mathcal{N}_{semi}$ denotes a set of k semi-hard negatives. Overall, we propose two adjustments: 1) The numerator is defined as *a sum of 'similarity'* between the anchor v_i and the spatial neighbors, and 2) randomly selected negatives are replaced by semi-hard ones. The first adjustment changes the pretext task into maximizing the probability of v_i being recognized as a sample in S_i. Intuitively, this pretext task can be accomplished as long as partial patches with potential semantic consistency are found, without overfitting all neighbors in S_i. However, an excessively simple pretext task may cause the easy-mining problem [29], which hampers the efforts to learn discriminative representations. Therefore, the aim of the second adjustment is to set a proper base similarity scale. According to [13], the vast majority of easy negatives are unnecessary while those hardest negatives are harmful to performance. In order to set a trade-off, we design a novel semi-hard negative mining strategy (Fig. 1(e)), where the top 20% hardest negatives are sampled follows a skewness distribution, e.g. $P \sim \text{Be}(\alpha, \beta)$. Notably, this implement is just one specific way to formalize the mining strategy.

2.3 Inter-invariance

Besides intra-invariance, those patches from different WSIs also might share the same semantic label. This is named *inter-invariance.* Clustering is a simple way to determine semantic labels of patches under no supervision. We use k-means to cluster the embeddings into N groups $\mathbf{G} = \{G_1, G_2, \ldots, G_N\}$. However, clustering can be an arbitrary process. To explore more accurate feature neighbors (semantically consistent patches), we compute disparate clusterings multiple times, then intersect these clusters and only select those closest embeddings. Specifically, let $G^j_{g^j(v)}$ be the cluster of v in the jth clustering, where $g^j(v)$ is the cluster label. We can obtain the set of feature neighbors:

$$C_v = G^1_{g^1(v)} \bigcap G^2_{g^2(v)} \bigcap \cdots \bigcap G^H_{g^H(v)} \bigcap \mathcal{N}_K(v), \tag{4}$$

where H is the number of clusterings and $\mathcal{N}_K(v)$ denotes K closest embeddings of v (measured by Euclidean distance in feature space). Intuitively, a larger H and a smaller K improve the precision of feature neighbors at the cost of diversity. To reasonably expand the diversity, we can additionally encourage the label consistency among feature neighbors of different patches within the same spatial neighborhood. Given an anchor v_i, we further choose n 'similar' spatial neighbors as auxiliary anchors v_i^p, with $p \in \{1, 2, \ldots, n\}$. Thus, the set of feature neighbors for an anchor v_i can be reformulated as:

$$C_i = C_{v_i} \bigcup C_{v_i^1} \bigcup \cdots \bigcup C_{v_i^{n-1}} \bigcup C_{v_i^n}, \tag{5}$$

As illustrated in Fig. 1(f), by exploring the feature neighborhoods (dotted circles) for an anchor (red dot) and an auxiliary anchor (yellow dot) respectively, more feature neighbors with the same semantic labels are discovered. Finally, $\mathcal{L}_{Cluster}$ can be formulated to encourage label consistency among C_i:

$$\mathcal{L}_{Cluster} = -\log \sum_{v_+ \in C_i} \frac{\exp\left(v_+^T v_i / \tau\right)}{\exp\left(v_+^T v_i / \tau\right) + \sum_{v_- \in N_k} \exp\left(v_-^T v_i / \tau\right)} \tag{6}$$

2.4 Overall Loss Function

The overall loss function of SSLP is defined as:

$$\mathcal{L}_{SSLP} = \mathcal{L}_{InfoNCE} + w(t) \cdot (\mathcal{L}_{Spatial} + \mathcal{L}_{Cluster}) \tag{7}$$

where $w(t)$ is a time-dependent weighting function ramping up from 0 to 1 during the first 10 training epochs [20]. To further improve performance, randomly selected N_k is replaced by $\mathcal{N}_{semi}$ for $\mathcal{L}_{InfoNCE}$ and $\mathcal{L}_{Cluster}$ in the final implementation. Detailed experimental results can be seen in Table 1.

3 Experimental Results

Datasets. We study SSL algorithms performed in: 1) ***NCTCRC***: dataset contains pathological image patches with 224×224 pixels at 0.5 microns per pixel (MPP) from 9 tissue classes, and 2) ***CAMELYON***: dataset contains pathological image patches with 256×256 pixels at 0.5 MPP from 2 tissue classes. The details of the datasets are in Appendix A in the supplementary material. All these datasets will be publicly available sooner.

Experimental Setup. Following [16,33], we set hyper-parameters as $\mathcal{D} = 128, \tau = 0.07, t = 0.5, H = 3, K = 8192, n = 1, \alpha = 5, \beta = 15, k = \lfloor \frac{K}{1+n} \rfloor$. For all network structures, we use SGD with momentum of 0.9, weight decay of 1e-4 and batch size of 128 in 4 GPUs. Initial learning rate is set to 0.015, with a cosine decay schedule for 100 epochs. All SSL algorithms are trained on CAMELYON without any supervised labels. The data augmentation setting is similar to [16].

3.1 Linear Classification Protocol

Linear classification on frozen features is a common protocol of evaluating SSL algorithms. Specifically, we train a 2-way supervised linear classifier on top of the frozen 128-$\mathcal{D}$ feature vector extracted by ResNet-18 for 100 epochs with an initial learning rate of 0.03 and a weight decay of 0. The learning rate is multiplied by 0.1 at 60 and 80 epochs.

Table 1 demonstrates the linear classification results of SSLP on CAMELYON and gives the comparison with other state-of-the-art methods including a supervised baseline. '†' means the improvement of the original implementation of InstDisc by utilizing InfoNCE loss function. Our method SSLP outperforms all others, surpassing the state-of-the-art method MoCo v2 by a nontrivial margin and even exceeding the supervised counterpart: up to +3.3% Acc. and +1.5% AUC. Notably, MoCo v2 has employed several additional improvements on InstDisc†, such as stronger data augmentation, more negatives, a dynamic memory bank, a projection head and so on. These additional improvements are orthogonal to our method, and we believe that adding these additional improvements will further improve the performance of SSLP. Table 1 also demonstrates the comparision of two paradigms of intra-invariance learning ($\mathcal{L}_{Divide}$ vs. $\mathcal{L}_{Spatial}$), where we can observe that $\mathcal{L}_{Spatial}$ outperforms $\mathcal{L}_{Divide}$, corroborating the effectiveness of adaptive learning manner in exploring intra-invariance. Moreover, $\mathcal{L}_{Spatial}$, $\mathcal{L}_{Cluster}$ and $\mathcal{N}_{semi}$ show consistent performance improvement over InstDisc† (see the last three rows in Table 1), demonstrating the effectiveness of these critical components of SSLP.

Table 1. Pathological image patches classification. We report accuracy and AUC of linear classification results on CAMELYON, and accuracy of cross-disease classification finetuned on NCTCRC. For comparision, we report the performance of the supervised counterpart on these datasets (see appendix B for implement details). '†' denotes improved reproduction vs. original papers. In green are the performance improvement compared with InstDisc†.

Method	CAMELYON		NCTCRC
	Acc.(%)	AUC(%)	Acc.(%)
Supervised baseline	79.6	84.2	**95.9**
InstDisc† [28]	73.1	78.7	93.4
LocalAgg [33]	76.7	82.3	93.8
InstDisc† w/ $\mathcal{L}_{Divide}$	77.6	83.8	93.7
MoCo v1(k=4,096) [16]	75.7	81.9	93.8
MoCo v2(k=65,536) [7]	80.1	84.9	93.7
Our proposed method SSLP based on InstDisc†			
w/ $\mathcal{L}_{Spatial}$	80.0(+6.9)	84.4(+5.7)	94.0(+0.6)
w/ $\mathcal{L}_{Spatial}$, $\mathcal{L}_{Cluster}$	80.6(+7.5)	85.5(+6.8)	94.4(+1.0)
w/ $\mathcal{L}_{Spatial}$, $\mathcal{L}_{Cluster}$, $\mathcal{N}_{semi}$	82.9(+9.8)	**85.7(+7.0)**	95.2(+1.8)

3.2 Cross-Disease Transfer Learning

To investigate the transferability of SSL algorithms, we evaluate the fine-tuning performance of our method and other state-of-the-art methods on cross-disease classification task on NCTCRC. The results are also shown in Table 1. Specifically, SSLP still surpasses all others and achieves comparable performance with the supervised counterpart, indicating that SSLP can serve as a more effective initialization in cross-disease classification tasks on WSIs.

3.3 Ablation Study

Impact of Distance Threshold $\mathcal{T}$. Table 2(a) ablates the distance thresholds used in $\mathcal{L}_{Spatial}$ where $\mathcal{T} = 2$ surpasses $\mathcal{T} = 10$ by an extensive gap. We believe this is because: In $\mathcal{L}_{Spatial}$, under utilizing a fixed number of negatives, the difficulty to classify the anchor v_i as an embedding in *Spatial neighbors* S_i decreases with the increasing in number of positives. Therefore, $\mathcal{L}_{Spatial}$ may lose the capability of learning discriminative information when the number of positives is too large.

Impact of Semi-hard Negatives. Table 2(b) ablates the effects of hard-mining strategies. Obviously, the proposed semi-hard negative mining strategy based on 'Beta Distribution' gives an absolute performance improvement both on linear classification and transfer learning. Moreover, the first two lines in

Table 2. Ablation studies of SSLP. Rows with (*) indicates default values.

	CAMELYON		NCTCRC
	Acc.(%)	AUC(%)	Acc.(%)
(a) Distance threshold $\mathcal{T}$ in $\mathcal{L}_{Spatial}$			
$\mathcal{T} = 2$ (*)	**82.9**	**85.7**	**95.2**
$\mathcal{T} = 10$	77.7	84.2	94.2
(b) Hard-mining strategies in top 20% hardest negatives			
w/o hard-mining	80.6	84.4	94.4
Hardest negatives	76.1	81.4	93.7
Random sampling	81.0	85.5	94.0
Beta distribution (*)	**82.9**	**85.7**	**95.2**
(c) Number N of clusters in $\mathcal{L}_{Cluster}$			
$N = 10$	76.6	83.5	94.5
$N = 100$	81.1	**85.8**	**95.4**
$N = 1000$ (*)	**82.9**	85.7	95.2
(d) C_i in InstDisc† w/ $\mathcal{L}_{Spatial}$, $\mathcal{L}_{Cluster}$			
w/o Auxiliary anchor	80.5	85.2	94.2
w/ Auxiliary anchor (*)	**80.6**	**85.5**	**94.4**

Table 2(b), where utilizing the hardest negatives even causes performance degradation compared with not using hard-mining strategy, corroborates that those hardest negatives are harmful to performance.

Impact of the Number of Clusters. Table 2(c) ablates the number of clusters. Overall, similar as the trend observed in [5,33], the number of clusters has little influence as long as there are 'enough' clusters to decrease the skewness of each cluster.

Impact of Auxiliary Anchor. Table 2(c) ablates the effects of auxiliary anchor. In this paper, we adopt a conservative strategy where only the most similar auxiliary anchor in a relatively small spatial neighborhood ($\mathcal{T} = 2, n = 1$) is explored. This setting encourages the precision of C_i while constrains the diversity of C_i. Therefore, the limited improvement in performance indicates the diversity has not been fully utilized.

4 Conclusion

In this paper, we propose Spatial Guided Self-supervised Leaning on Pathological Images (SSLP) to improve the performance of SSL algorithms on whole slide pathological images (WSIs). Through properly mining the patch-wise spatial proximity, we further explore the intra- and inter-invariance which indicate the semantic invariance among patches within spatial neighbors and their corresponding neighbors in the feature space. Extensive experiments conducted on linear classification and cross-disease transfer learning demonstrate the superiority of SSLP over state-of-the-art methods.

Acknowledgement. This work was supported in part by Shanghai Municipal Science and Technology Major Project (2021SHZDZX0102), 111 project (BP0719010), Shanghai Science and Technology Committee (18DZ2270700) and Shanghai Jiao Tong University Science and Technology Innovation Special Fund (ZH2018ZDA17).

References

1. Abbet, C., Zlobec, I., Bozorgtabar, B., Thiran, J.P.: Divide-and-rule: self-supervised learning for survival analysis in colorectal cancer. In: Martel, A.L. et al. (eds.) Medical Image Computing and Computer Assisted Intervention–MICCAI 2020. MICCAI 2020. Lecture Notes in Computer Science, **12265**, 480–489. Springer, Cham (2020). https://doi.org/10.1007/978-3-030-59722-1_46
2. Bachman, P., Hjelm, R.D., Buchwalter, W.: Learning representations by maximizing mutual information across views (2019). arXiv preprint: arXiv:1906.00910
3. Bejnordi, B.E., et al.: Diagnostic assessment of deep learning algorithms for detection of lymph node metastases in women with breast cancer. JAMA **318**(22), 2199–2210 (2017)
4. Bulten, W., et al.: Automated deep-learning system for Gleason grading of prostate cancer using biopsies: a diagnostic study. Lancet Oncol. **21**(2), 233–241 (2020)
5. Caron, M., Misra, I., Mairal, J., Goyal, P., Bojanowski, P., Joulin, A.: Unsupervised learning of visual features by contrasting cluster assignments (2020). arXiv preprint: arXiv:2006.09882
6. Chen, T., Kornblith, S., Norouzi, M., Hinton, G.: A simple framework for contrastive learning of visual representations. In: International Conference on Machine Learning. pp. 1597–1607. PMLR (2020)
7. Chen, X., Fan, H., Girshick, R., He, K.: Improved baselines with momentum contrastive learning (2020). arXiv preprint: arXiv:2003.04297
8. Chen, X., He, K.: Exploring simple Siamese representation learning (2020). arXiv preprint: arXiv:2011.10566
9. Coudray, N., et al.: Classification and mutation prediction from non-small cell lung cancer histopathology images using deep learning. Nat. Med. **24**(10), 1559–1567 (2018)
10. Dagogo-Jack, I., Shaw, A.T.: Tumour heterogeneity and resistance to cancer therapies. Nat. Rev. Clin. Oncol. **15**(2), 81 (2018)
11. Dehaene, O., et al.: Self-supervision closes the gap between weak and strong supervision in histology (2020). arXiv preprint: arXiv:2012.03583
12. Egeblad, M., Nakasone, E.S., Werb, Z.: Tumors as organs: complex tissues that interface with the entire organism. Dev. Cell **18**(6), 884–901 (2010)
13. Frankle, J., et al.: Are all negatives created equal in contrastive instance discrimination? (2020). arXiv preprint: arXiv:2010.06682
14. Gidaris, S., Singh, P., Komodakis, N.: Unsupervised representation learning by predicting image rotations (2018). arXiv preprint: arXiv:1803.07728
15. Grill, J.B., et al.: Bootstrap your own latent: a new approach to self-supervised learning (2020). arXiv preprint: arXiv:2006.07733

16. He, K., Fan, H., Wu, Y., Xie, S., Girshick, R.: Momentum contrast for unsupervised visual representation learning. In: Proceedings of the IEEE/CVF Conference on Computer Vision and Pattern Recognition, pp. 9729–9738 (2020)
17. Huang, J., Dong, Q., Gong, S., Zhu, X.: Unsupervised deep learning by neighbourhood discovery (2019). arXiv preprint: arXiv:1904.11567
18. Kather, J.N., Halama, N., Marx, A.: 100,000 histological images of human colorectal cancer and healthy tissue (2018). https://doi.org/10.5281/zenodo.1214456
19. Kather, J.N., et al.: Multi-class texture analysis in colorectal cancer histology. Sci. Rep. **6**(1), 1–11 (2016)
20. Laine, S., Aila, T.: Temporal ensembling for semi-supervised learning (2016). arXiv preprint: arXiv:1610.02242
21. Lu, M.Y., Chen, R.J., Mahmood, F.: Semi-supervised breast cancer histology classification using deep multiple instance learning and contrast predictive coding (conference presentation). In: Medical Imaging 2020: Digital Pathology, vol. 11320, p. 113200J. International Society for Optics and Photonics (2020)
22. Mercan, C., et al.: Virtual staining for mitosis detection in breast histopathology. In: 2020 IEEE 17th International Symposium on Biomedical Imaging, pp. 1770–1774. IEEE (2020)
23. Mobadersany, P., et al.: Predicting cancer outcomes from histology and genomics using convolutional networks. Proc. Natl. Acad. Sci. **115**(13), E2970–E2979 (2018)
24. Ngiam, K.Y., Khor, W.: Big data and machine learning algorithms for health-care delivery. Lancet Oncol. **20**(5), e262–e273 (2019)
25. Niazi, M.K.K., Parwani, A.V., Gurcan, M.N.: Digital pathology and artificial intelligence. Lancet Oncol. **20**(5), e253–e261 (2019)
26. Oord, A.v.d., Li, Y., Vinyals, O.: Representation learning with contrastive predictive coding (2018). arXiv preprint: arXiv:1807.03748
27. Sahiner, B., et al.: Deep learning in medical imaging and radiation therapy. Med. Phys. **46**(1), e1–e36 (2019)
28. Wu, Z., Xiong, Y., Yu, S.X., Lin, D.: Unsupervised feature learning via non-parametric instance discrimination. In: Proceedings of the IEEE Conference on Computer Vision and Pattern Recognition, pp. 3733–3742 (2018)
29. Xie, J., Zhan, X., Liu, Z., Ong, Y.S., Loy, C.C.: Delving into inter-image invariance for unsupervised visual representations (2020). arXiv preprint: arXiv:2008.11702
30. Xie, X., et al.: Instance-aware self-supervised learning for nuclei segmentation. In: Martel, A.L. et al. (eds.) Medical Image Computing and Computer Assisted Intervention–MICCAI 2020. MICCAI 2020. Lecture Notes in Computer Science, **12265**, 341–350. Springer, Cham (2020). https://doi.org/10.1007/978-3-030-59722-1_33
31. Xie, Z., Lin, Y., Zhang, Z., Cao, Y., Lin, S., Hu, H.: Propagate yourself: exploring pixel-level consistency for unsupervised visual representation learning (2020). arXiv preprint: arXiv:2011.10043
32. Yuan, Y., et al.: Quantitative image analysis of cellular heterogeneity in breast tumors complements genomic profiling. Sci. Transl. Med. **4**(157), 157ra143 (2012)
33. Zhuang, C., Zhai, A.L., Yamins, D.: Local aggregation for unsupervised learning of visual embeddings. In: Proceedings of the IEEE International Conference on Computer Vision, pp. 6002–6012 (2019)

Segmentation of Left Atrial MR Images via Self-supervised Semi-supervised Meta-learning

Dani Kiyasseh[1(✉)], Albert Swiston[2], Ronghua Chen[2], and Antong Chen[2]

[1] University of Oxford, Oxford, UK
dani.kiyasseh@eng.ox.ac.uk
[2] Merck & Co., Inc., Kenilworth, NJ, USA

Abstract. Deep learning algorithms for cardiac MRI segmentation depend heavily upon abundant, labelled data located at a single medical centre. Clinical settings, however, contain abundant, *unlabelled* and scarce, labelled data located across distinct medical centres. To account for this, we propose a unified pre-training framework, entitled self-supervised semi-supervised meta-learning (S^{4}ML), that exploits distributed labelled and unlabelled data to quickly and reliably perform cardiac MRI segmentation given scarce, labelled data from a potentially different distribution. We show that S^{4}ML outperforms baseline methods when adapting to data from a novel medical centre, cardiac chamber, and MR sequence. We also show that this behaviour holds even in extremely low-data regimes.

Keywords: Meta-learning · Self-supervision · Semi-supervision · Cardiac MRI segmentation

1 Introduction

Deep learning algorithms have been successful in delineating the structures of the heart [3,9,35]. Often, these algorithms are trained on abundant, labelled data found in a centralized location (e.g., hospital). However, an increasingly common scenario within healthcare is characterized by the following. First, cardiac MRI data are located at distinct medical centres (e.g., health system with geographically disparate centres) [13,16]. Second, some medical centres contain specific MR sequence data (e.g., cine MRI) that are abundant and either a) labelled with segmentation maps of a cardiac chamber (e.g., LV Endo), or b) unlabelled

Dani Kiyasseh—Work done as intern at Merck & Co., Inc., Kenilworth, NJ, USA.
Ronghua Chen—Contributed while at Merck & Co., Inc., Kenilworth, NJ, USA.

Electronic supplementary material The online version of this chapter (https://doi.org/10.1007/978-3-030-87196-3_2) contains supplementary material, which is available to authorized users.

M. de Bruijne et al. (Eds.): MICCAI 2021, LNCS 12902, pp. 13–24, 2021.
https://doi.org/10.1007/978-3-030-87196-3_2

with these maps due to the high cost associated with providing annotations. Third, other centres contain scarce data (from another MR sequence, e.g., LGE) that are labelled with segmentation maps of a different cardiac chamber (e.g., LA Endo). Such scarcity could be due to, for example, a rare or novel medical condition, or limited medical imaging infrastructure. We therefore address the following question:

How do we exploit both labelled and unlabelled cardiac MRI data located across medical centres in order to quickly and reliably perform cardiac MRI segmentation of a previously-unseen cardiac chamber from a distinct MR sequence and medical centre?

Previous work has dealt with the exploitation of abundant, unlabelled and limited, labelled data [1,15]. However, such studies are limited in two main ways. First, they do not explore the notion of adapting a segmentation model to a *previously-unseen* chamber. Second, and more importantly, they assume that data are centralized and thus do not explore how to leverage data located at distinct medical centres.

In contrast, we design a unified pre-training framework, entitled self-supervised semi-supervised meta-learning (S^{4}ML, pronounced "Squad ML"), which formulates a multitude of self-supervised and supervised tasks based on data located *across* medical centres in order to perform cardiac MRI segmentation of a *previously-unseen* chamber of the heart from a distinct MR sequence and medical centre. Specifically, we exploit cine MRI data of the left ventricular endocardium (LV Endo) and adapt a network to LGE MRI data of the left atrial endocardium (LA Endo). We show that S^{4}ML allows for reliable adaptation even in low-data regimes. Such a system has broad applicability, allowing researchers and medical practitioners to exploit any and all datasets at their disposal (labelled, unlabelled, and regardless of size and MR sequence). It also adds value to settings with limited computational resources and data (particularly that which is currently inaccessible in silos and thus not exploited). This, in turn, can contribute to improving patient outcomes.

2 Related Work

Cardiac Segmentation with Reduced Supervision. Bai *et al.* [1] pre-train a model using rotation prediction before performing cardiac MRI segmentation, illustrating its superiority over training a network from a random initialization. Although recent work has achieved strong performance on LA Endo segmentation [12,17,33], none has explored this in the context of reduced supervision and distributed settings. Others have focused on style transfer between cardiac scans of different modalities in attempt to mitigate distribution shift [2,4,8]. For instance, Chen *et al.* [4] transfer from cardiac MRI scans to CT scans. Their work operates under the assumption that the target domain is known *a priori*. In our work, we do not make this assumption and instead propose a framework that allows for transfer across medical centres, cardiac chambers, and MR sequences.

Fast adaptation for Medical Image Segmentation. Several researchers have proposed to deploy unsupervised and semi-supervised learning for few-shot segmentation [10,19,22,23,30]. For instance, Feyjie *et al.* [10] design a semi-supervised skin-lesion segmentation model that incorporates unlabelled data through the use of a pretext task. Zhao *et al.* [34] propose to learn transformations to apply to brain MR images in the context of one-shot segmentation and Wang *et al.* [27] resort to atlas-based methods to perform one-shot segmentation of brain MRI. In contrast to our work, the aforementioned methods do not account for distributed data. Most similar to our work is that of Yu *et al.* [29] where they implement MAML for fast adaptation when performing cardiac motion estimation. Instead, we explore semi-supervision and distributed settings, and evaluate our methods in the context of cardiac MRI segmentation.

3 Methods

We design a framework capable of exploiting both labelled and unlabelled data at distinct medical centres to quickly perform cardiac MRI segmentation given limited, labelled data. We begin by formally outlining the task of segmentation. We then provide a brief primer on contrastive learning which allows us to extract rich and transferable representations from the unlabelled data. Lastly, we introduce meta-learning as a way to explicitly deal with data at distinct medical centres and to facilitate fast adaptation of the network.

Segmentation with Supervision. We have an encoder, $f_\theta : \boldsymbol{x} \in \mathbb{R}^{H\times W} \rightarrow \boldsymbol{v} \in \mathbb{R}^{E}$, that maps a cardiac MRI image, $\boldsymbol{x}$, with height and width, H and W, respectively, to an E-dimensional representation, $\boldsymbol{v}$. We also have a decoder, $g_\phi : \boldsymbol{v} \in \mathbb{R}^{E} \rightarrow \hat{\boldsymbol{y}} \in \mathbb{R}^{C\times H\times W}$, that maps the representations, $\boldsymbol{v}$, to a segmentation map, $\hat{\boldsymbol{y}}$, where each pixel comprises C classes. Given an instance, $\boldsymbol{x}_i$, the per-pixel output, $\hat{\boldsymbol{y}}_i^{h,w}$, and the corresponding ground-truth label, $c_i^{h,w}$, for $h \in [1, H]$, $w \in [1, W]$, we learn the composite function, $p = g_\phi \circ f_\theta$, by optimizing the categorical cross-entropy loss for a mini-batch of size, K.

$$\mathcal{L}_{CE} = -\sum_{i=1}^{K}\sum_{h,w} \log p(y_i^{h,w} = c_i^{h,w}|\boldsymbol{x}_i) \tag{1}$$

Contrastive Learning. To extract rich and transferable representations from the abundant, unlabelled data at our disposal, we exploit the contrastive learning framework, which comprises a sequence of attractions and repulsions. Formally, we augment an instance, $\boldsymbol{x}_i$, to generate $\boldsymbol{x}_i^A = T^A(\boldsymbol{x}_i)$ and $\boldsymbol{x}_i^B = T^B(\boldsymbol{x}_i)$, using two stochastic transformation operators, T^A and T^B. We encourage the respective representations, $\boldsymbol{v}_i^A = f_\theta(\boldsymbol{x}_i^A)$ and $\boldsymbol{v}_i^B = f_\theta(\boldsymbol{x}_i^B)$, of this pair to be similar to one another, using a similarity metric, s, such as cosine similarity, and dissimilar from representations of other instances. In the process, we learn representations that are invariant to class-preserving perturbations and which have been shown to benefit downstream classification tasks [6]. To capture this behaviour, we optimize the InfoNCE loss [21] with a temperature parameter, τ.

$$\mathcal{L}_{NCE}(\boldsymbol{v}^A, \boldsymbol{v}^B) = \frac{e^{s(\boldsymbol{v}_i^A, \boldsymbol{v}_i^B)}}{\sum_j e^{s(\boldsymbol{v}_i^A, \boldsymbol{v}_j^B)}} \qquad s(\boldsymbol{v}_i^A, \boldsymbol{v}_i^B) = \frac{\boldsymbol{v}_i^A \cdot \boldsymbol{v}_i^B}{\|\boldsymbol{v}_i^A\| \|\boldsymbol{v}_i^B\|} \frac{1}{\tau} \tag{2}$$

Meta-learning. To learn a network that can explicitly deal with data across distinct medical centres *and* which adapts quickly to a new task, we exploit the meta-learning framework, which comprises a meta-training and meta-testing stage. In the former, a network first solves a multitude of tasks (defined next). We formulate a distinct task for each medical centre such that for a centre with labelled data, we perform a segmentation task, and for one with unlabelled data, we perform the contrastive learning task. Therefore, for two centres, we would have $N = 2$ tasks. A network achieves these tasks *independently* of one another by starting with the same set of parameters, θ_0, and being updated locally (data from same centre) via gradient descent to obtain *task-specific* parameters, θ_t. As such, for these two tasks, we would obtain two task-specific parameters, θ_1 and θ_2.

$$\theta_1 \leftarrow \theta_0 - \eta \frac{\partial \mathcal{L}_{NCE}}{\partial \theta_0} \qquad \theta_2 \leftarrow \theta_0 - \eta \frac{\partial \mathcal{L}_{CE}}{\partial \theta_0} \tag{3}$$

To obtain a unified set of parameters from the task-specific set of parameters, a parameter aggregation step is required. This step differs from one meta-learning framework to the next. For example, with **Reptile** [20], the objective is to minimize the average distance between the parameters, θ_0, and each of the task-specific parameters, θ_n (see Eq. 4). With **LEAP** [11], the objective is to minimize the length, d, of the path traversed by the network on the loss surface (manifold, M) of each task (see Eq. 5). Note that α reflects the learning rate and ∇ denotes partial derivatives with respect to the parameters, θ_0. After a single aggregation step, the entire process is repeated.

$$\theta_0 \leftarrow \theta_0 - \alpha \sum_{n=1}^{N} (\theta_0 - \theta_n) \tag{4}$$

$$\theta_0 \leftarrow \theta_0 - \alpha \nabla \sum_{n=1}^{N} [d(\theta_n, M)] \tag{5}$$

Self-supervised Semi-supervised Meta-learning ($\mathbf{S^4ML}$). Having discussed segmentation, contrastive learning, and meta-learning, we now outline two variants of how to incorporate these concepts into a unified framework (see Fig. 1).

Vanilla $\mathbf{S^4ML}$. During meta-training, each medical centre is associated with a distinct task. The network performs segmentation for a medical centre with labelled data, $\mathcal{D}_L$, and independently performs contrastive learning for a medical centre with unlabelled data, $\mathcal{D}_U$. Given these tasks, we exploit meta-learning as outlined earlier. Hereafter, we refer to this approach as Vanilla S^4ML. We note that optimizing for a supervised and self-supervised task simultaneously (due to aggregation step) can hinder convergence on the downstream segmentation task of interest. This realization motivates our next variant.

Sequential S4ML. To learn more task-specific parameters, we propose a sequential meta-training procedure. First, the network exclusively performs contrastive learning (within meta-training) on unlabelled data, $\mathcal{D}_U$. It is then fine-tuned by exclusively performing segmentation (still within meta-training) on labelled data, $\mathcal{D}_L$. We choose this order based on findings by [7,32] that illustrated the benefit of learning task-specific parameters. Hereafter, we refer to this approach as Sequential S4ML.

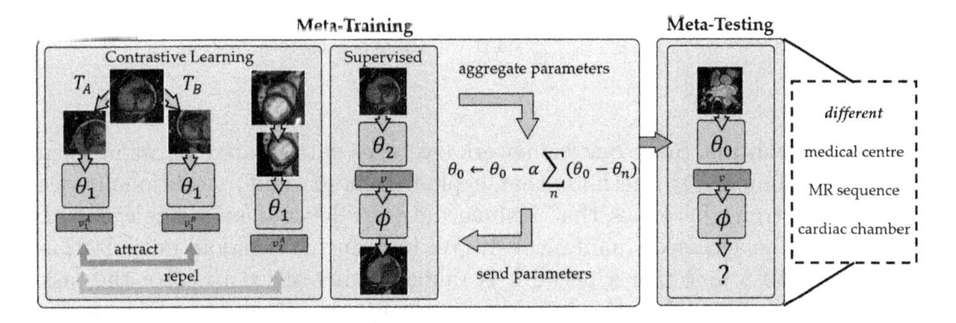

Fig. 1. Illustration of S4ML pipeline. During **meta-training**, the network solves $N = 2$ tasks (self-supervised contrastive learning and supervised segmentation) independently of one another to obtain task-specific parameters, θ_1 and θ_2. We aggregate parameters to obtain a unified set of parameters, θ_0, and send them to be iteratively updated. During **meta-testing**, the learned parameters θ_0 are transferred to solve the downstream task.

We note that our proposed approaches are flexible enough to 1) incorporate almost any self-supervised loss into the objective function, and 2) extend to the diverse scenarios in which each task entails both labelled and unlabelled data from different domains.

4 Experimental Design and Results

Datasets. In light of our emphasis on datasets in distinct medical centres, we exploit $\mathcal{D}_1$ = **2020 Cardiac M&M**[1] - a dataset of cardiac cine MRIs, across four medical centres and three vendors, pertaining to 175 patients with segmentation maps corresponding to three cardiac structures: LV Endo, left ventricular myocardium (LV Myo), and right ventricular endocardium (RV Endo). We also exploit $\mathcal{D}_2$ = **2018 Atrial Segmentation Challenge** [28] - a dataset of cardiac LGE MRIs from 99 patients with segmentation maps corresponding to LA Endo. Note that these datasets differ in terms of the medical centre to which they belong, the MR sequence, and the cardiac chamber. The number of instances used for training, validation, and testing are shown in Table 1.

[1] https://www.ub.edu/mnms/.

Table 1. Number of instances in each of the training, validation, and test splits for each dataset. We meta-train on $\mathcal{D}_1$ and meta-test on $\mathcal{D}_2$.

Dataset	Centre number	Number of 2D slices			Labelled
		Train	Validation	Test	
$\mathcal{D}_1$	1	783	247	248	✓
	2	464	159	155	✓
	3	250	81	81	✓
	4	340	114	120	✗
$\mathcal{D}_2$	-	3,551	1,095	1,215	✓

Baseline Methods. Since our framework leverages data located across medical centres, we compare to methods that exploit meta-learning. Additionally, we compare to strong baselines that assume data are located in a single centre, and can thus be accessed simultaneously. We introduce 1) random initialization (**Random Init.**) in which a network is randomly initialized to solve the task and 2) joint learning (**Joint**) where the upstream supervised tasks have access to all data at once and are trained together before transferring the parameters to the downstream task. Note that joint learning violates the assumption that data are located across disparate medical centres. We also introduce 1) supervised meta-learning (**SML**) where a network is exposed to only labelled data during meta-training, and 2) self-supervised meta-learning (**SSML**) where a network is exposed to only unlabelled data. These methods allow us to evaluate the potential added value of semi-supervision.

Implementation Details. We conduct all experiments using PyTorch and an implementation of the meta-learning algorithms[2]. **Meta-training** - we use data from three medical centres and thus $N = 3$. We use stochastic gradient descent for the local and aggregation steps with a mini-batch size of 8 and an image dimension of 192×192. For self-supervised tasks, we extract representations from the U-Net encoder [25] and apply spatial pooling to generate a 512-dimensional vector. We apply the same transformations as those in SimCLR [6] with $\tau = 0.1$. For supervised tasks, we perform a binary segmentation of the LV Endo, as this is one of the most commonly labelled structures. We also perform early stopping on the aggregate loss on the query sets with a patience of 50 epochs. **Meta-testing** - we choose to perform a binary segmentation of the LA Endo, and

[2] https://github.com/flennerhag/metalearn-leap.

optimize the binary cross-entropy loss using ADAM with a learning rate of 10^{-4} and a mini-batch size of 40. We perform early stopping on the validation loss with a patience of 5 epochs.

Results. We begin by evaluating our framework's ability to transfer across medical centres, chambers, and sequences. Specifically, after having meta-trained our model using a combination of self-supervised and supervised tasks with cardiac cine MRI, we transfer these parameters to solve the task of LA Endo segmentation using cardiac LGE MRI. Scans of this modality, which display the contrast between normal and abnormal cardiac tissues [26], are challenging to segment. This is due to the higher level of intensity heterogeneity and lower level of contrast between the tissues and blood that they exhibit relative to cine MRI scans [31]. In Fig. 2 (top and centre), we illustrate the Dice coefficient and Hausdorff distance, respectively, on the LA Endo segmentation task as we vary the fraction of available labelled data during meta-testing, $F \in [0.1, 0.25, 0.5, 1.0]$. We also indicate whether methods are statistically significantly different from one another (full list can be found in Appendix A).

We find that Joint training hinders downstream performance. This is seen, particularly at low fractions, by the lower Dice score relative to that of a randomly initialized network. For example, at $F = 0.1$, Dice ≈ 0.835 and ≈ 0.845 for the two methods, respectively. We also find that, with sufficient labelled data for meta-testing, semi-supervised approaches outperform their counterparts when transferring parameters across medical centres, cardiac chambers, and MR sequences. For example, at $F = 1.0$, Vanilla S^4ML Leap and SML achieve a Dice ≈ 0.87 and 0.86, respectively. A similar observation holds for the Hausdorff distance. We also find that meta-training with LEAP is more advantageous than doing so with Reptile. This is supported by the higher Dice and lower Hausdorff values exhibited by the former.

We qualitatively evaluate S^4ML by presenting an LGE MRI scan overlaid with generated segmentation maps (see Fig. 2 bottom). Please see Appendix B for further maps. We find that Sequential S^4ML produces more complete segmentation map. We also notice that Random Init. and SML produce patchy maps, and Joint misses the left-most component of the cardiac structure. Although Vanilla S^4ML also misses this component, it reliably captures the top boundary of the map.

Convergence of S^4ML. We also illustrate the validation loss incurred by various frameworks on the LA Endo segmentation task (see Fig. 3). We show that networks pre-trained with supervised and semi-supervised meta-learning adapt faster than those initialized randomly or via self-supervision. This can be seen by the lower initial loss (≈ 0.15 vs. > 0.20) and the fewer number of epochs (10 vs. 35) required for convergence of the two sets of methods, respectively. This represents a three-fold speed-up in training.

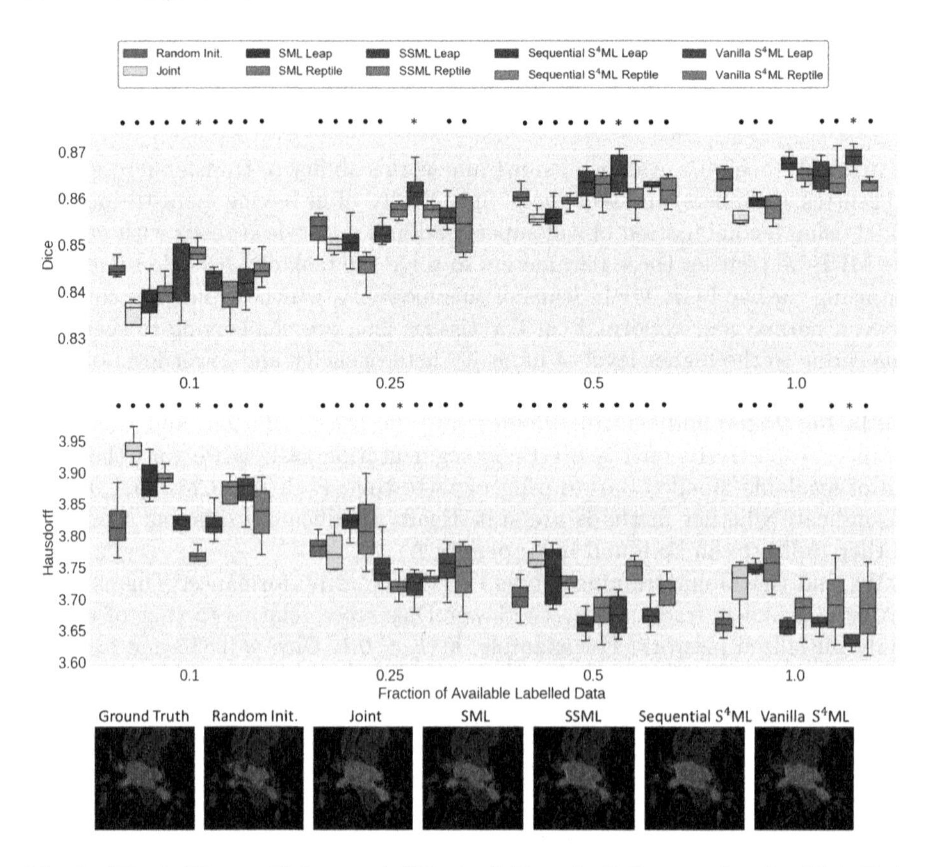

Fig. 2. **(Top)** Dice coefficient and **(Centre)** Hausdorff distance (mm) of LA Endo segmentation after model pre-training. Results are shown across five seeds for $F = (0.1, 0.25, 0.5, 1.0)$ during meta-testing. Asterisk (*) denotes top method and dot ($\cdot$) reflects significant difference ($p < 0.01$) to top method. **(Bottom)** Segmentation maps (yellow) overlaid on cardiac LGE MRI. We show that Vanilla and Sequential S^{4}ML are on par with, or better than, baselines in capturing the entirety of the structure.

Effect of Architecture. To quantify the effect of the neural architecture on our framework, we conduct experiments with a relatively compact network, MobileNetv3 [14], and one more heavily parameterized, DeepLabv3 [5] (see Table 2). We find that U-Net outperforms the two other architectures regardless of the pre-training method. For example, when pre-training with Vanilla S^{4}ML, the U-Net achieves Dice = 0.830 compared to 0.744 and 0.797 for MobileNetv3 and DeepLabv3, respectively. Such an outcome points to the relative effectiveness of the U-Net in the low-data regime. We also find that semi-supervision is either the best or second-best performing method regardless of the architecture used. This finding suggests that S^{4}ML is architecture-agnostic.

Effect of Self-supervised Task. We also evaluate the effect of the self-supervised task on the performance of our framework on LA Endo segmen-

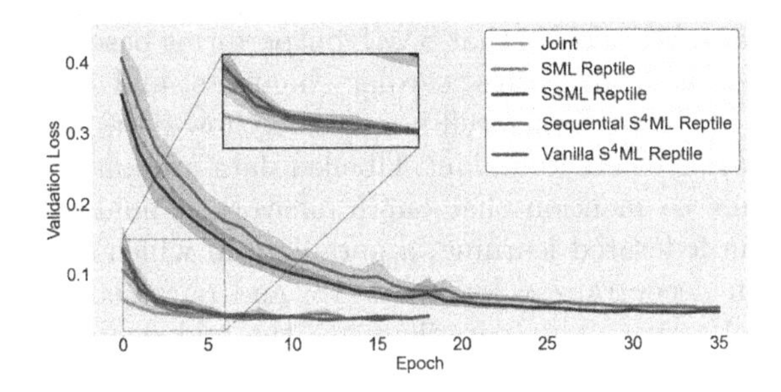

Fig. 3. Validation loss of networks pre-trained differently and fine-tuned for LA Endo segmentation ($F = 0.1$). Shaded area reflects one standard deviation from the mean across five seeds. We show that meta-learning methods learn faster (fewer epochs).

Table 2. Effect of architecture on Dice of LA Endo task at $F = 0.1$. Results are averaged across five seeds. Bold indicates the top-performing method and underline reflects the second-best methods.

	MobileNetv3	U-Net	DeepLabv3
# of params.	3.3M	17.3M	59.3M
without meta-training			
Random Init.	0.666	0.826	0.808
Joint	0.728	0.814	**0.822**
with meta-training			
SML	0.740	0.822	0.753
SSML	**0.763**	0.826	0.724
Sequential S^4ML	<u>0.754</u>	<u>0.828</u>	0.740
Vanilla S^4ML	0.744	**0.830**	<u>0.797</u>

Table 3. Effect of self-supervised task on Dice of LA Endo task at $F = 0.1$. Results are averaged across five seeds.

Task	Sequential S^4ML	Vanilla S^4ML
SimCLR	0.828	0.830
Rotation	0.836	**0.848**
Hflip	**0.844**	0.842

tation. Additional tasks include predicting whether an image has been rotated by some angle, $\alpha \in [0, 90, 180, 270]$, **(Rotation)** or horizontally flipped **(Hflip)** (see Table 3). We find that Hflip is more beneficial than SimCLR [6], achieving a Dice = 0.844 and 0.828, respectively, with Sequential S^4ML. We hypothesize that such an outcome is due to the deterministic nature of Hflip and its ability to better capture realistic invariances in the data. We leave an exhaustive comparison of such tasks for future work.

5 Discussion and Future Work

In this work, we proposed a unified pre-training framework, S^4ML, that exploits both labelled and unlabelled data across distinct medical centres in order to learn a model that quickly and reliably performs cardiac MRI segmentation given scarce, labelled data. We showed that S^{4}ML outperforms baseline methods when transferring across medical centres, cardiac chambers, and MR sequences. We also showed that S^{4}ML performs well in extremely low-data regimes, indicating its utility in scenarios where abundant, labelled data are unavailable.

It is worthwhile to mention that our framework is analogous to federated aggregation [24] in federated learning, a paradigm in which models are trained locally on data in decentralized locations [18] and parameters are aggregated at a single central server. This paradigm has the added benefit of preserving patient privacy by limiting a model's direct access to sensitive patient health information. Moving forward, designing domain-specific self-supervised tasks, investigating the effect of various combinations of such tasks, and extending our framework for transfer across organs are paths we will explore.

Acknowledgements. We would like to thank Warda Al-Jazairia for lending us her voice. For discussions on cardiac segmentation, we thank Asad Abu Bakar Ali and Smita Sampath, Genome and Biomarker Sciences, MSD, Singapore. For discussions on statistical analysis, we thank Nana Wang, Biometrics Research, Merck & Co., Inc., Kenilworth, NJ, USA.

References

1. Bai, W., et al.: Self-supervised learning for cardiac MR image segmentation by anatomical position prediction. In: Shen, D., et al. (eds.) MICCAI 2019. LNCS, vol. 11765, pp. 541–549. Springer, Cham (2019). https://doi.org/10.1007/978-3-030-32245-8_60
2. Chen, C., et al.: Unsupervised multi-modal style transfer for cardiac MR segmentation. In: Pop, M., et al. (eds.) STACOM 2019. LNCS, vol. 12009, pp. 209–219. Springer, Cham (2020). https://doi.org/10.1007/978-3-030-39074-7_22
3. Chen, C., et al.: Deep learning for cardiac image segmentation: a review. Front. Cardiovasc. Med. **7**, 25 (2020)
4. Chen, C., Dou, Q., Chen, H., Qin, J., Heng, P.A.: Synergistic image and feature adaptation: towards cross-modality domain adaptation for medical image segmentation. In: Proceedings of the AAAI Conference on Artificial Intelligence, vol. 33, pp. 865–872 (2019)
5. Chen, L.C., Papandreou, G., Schroff, F., Adam, H.: Rethinking atrous convolution for semantic image segmentation (2017). arXiv preprint: arXiv:1706.05587
6. Chen, T., Kornblith, S., Norouzi, M., Hinton, G.: A simple framework for contrastive learning of visual representations. In: International Conference on Machine Learning, pp. 1597–1607. PMLR (2020)
7. Chen, T., Kornblith, S., Swersky, K., Norouzi, M., Hinton, G.: Big self-supervised models are strong semi-supervised learners. In: Proceedings of Conference on Neural Information Processing Systems (NeurIPS) (2020)
8. Dou, Q., et al.: PNP-ADANET: plug-and-play adversarial domain adaptation network at unpaired cross-modality cardiac segmentation. IEEE Access **7**, 99065–99076 (2019)
9. Duan, J., et al.: Automatic 3d bi-ventricular segmentation of cardiac images by a shape-refined multi-task deep learning approach. IEEE Trans. Med. Imaging **38**(9), 2151–2164 (2019)

10. Feyjie, A.R., Azad, R., Pedersoli, M., Kauffman, C., Ayed, I.B., Dolz, J.: Semi-supervised few-shot learning for medical image segmentation (2020). arXiv preprint: arXiv:2003.08462
11. Flennerhag, S., Moreno, P.G., Lawrence, N.D., Damianou, A.: Transferring knowledge across learning processes. In: Proceedings of International Conference on Learning Representations (ICLR) (2018)
12. Ghosh, S., Ray, N., Boulanger, P., Punithakumar, K., Noga, M.: Automated left atrial segmentation from magnetic resonance image sequences using deep convolutional neural network with autoencoder. In: 2020 IEEE 17th International Symposium on Biomedical Imaging (ISBI), pp. 1756–1760. IEEE (2020)
13. Grimson, W., et al.: Federated healthcare record server–the synapses paradigm. Int. J. Med. Inf. **52**(1–3), 3–27 (1998)
14. Howard, A., et al.: Searching for mobilenetv3. In: Proceedings of the IEEE International Conference on Computer Vision, pp. 1314–1324 (2019)
15. Jamaludin, A., Kadir, T., Zisserman, A.: Self-supervised learning for spinal MRIs. In: Cardoso, M.J., et al. (eds.) DLMIA/ML-CDS -2017. LNCS, vol. 10553, pp. 294–302. Springer, Cham (2017). https://doi.org/10.1007/978-3-319-67558-9_34
16. Kairouz, P., et al.: Advances and open problems in federated learning (2019). arXiv preprint: arXiv:1912.04977
17. Li, L., Weng, X., Schnabel, J.A., Zhuang, X.: Joint left atrial segmentation and scar quantification based on a DNN with spatial encoding and shape attention. In: Martel, A.L., et al. (eds.) MICCAI 2020. LNCS, vol. 12264, pp. 118–127. Springer, Cham (2020). https://doi.org/10.1007/978-3-030-59719-1_12
18. McMahan, B., Moore, E., Ramage, D., Hampson, S., y Arcas, B.A.: Communication-efficient learning of deep networks from decentralized data. In: Artificial Intelligence and Statistics, pp. 1273–1282. PMLR (2017)
19. Mondal, A.K., Dolz, J., Desrosiers, C.: Few-shot 3d multi-modal medical image segmentation using generative adversarial learning (2018). arXiv preprint: arXiv:1810.12241
20. Nichol, A., Schulman, J.: Reptile: a scalable metalearning algorithm **2**(3), 4 (2018). arXiv preprint: arXiv:1803.02999
21. Oord, A.v.d., Li, Y., Vinyals, O.: Representation learning with contrastive predictive coding (2018). arXiv preprint: arXiv:1807.03748
22. Ouyang, C., Kamnitsas, K., Biffi, C., Duan, J., Rueckert, D.: Data efficient unsupervised domain adaptation for cross-modality image segmentation. In: Shen, D., et al. (eds.) MICCAI 2019. LNCS, vol. 11765, pp. 669–677. Springer, Cham (2019). https://doi.org/10.1007/978-3-030-32245-8_74
23. Panfilov, E., et al.: Improving robustness of deep learning based knee MRI segmentation: mixup and adversarial domain adaptation. In: Proceedings of the IEEE International Conference on Computer Vision Workshops (2019)
24. Pillutla, K., Kakade, S.M., Harchaoui, Z.: Robust aggregation for federated learning (2019). arXiv preprint: arXiv:1912.13445
25. Ronneberger, O., Fischer, P., Brox, T.: U-Net: convolutional networks for biomedical image segmentation. In: Navab, N., Hornegger, J., Wells, W.M., Frangi, A.F. (eds.) MICCAI 2015. LNCS, vol. 9351, pp. 234–241. Springer, Cham (2015). https://doi.org/10.1007/978-3-319-24574-4_28
26. Tseng, W.Y.I., Su, M.Y.M., Tseng, Y.H.E.: Introduction to cardiovascular magnetic resonance: technical principles and clinical applications. Acta Cardiol. Sin. **32**(2), 129–144 (2016)

27. Wang, S., Cao, S.e.a.: Lt-net: Label transfer by learning reversible voxel-wise correspondence for one-shot medical image segmentation. In: Proceedings of the IEEE/CVF Conference on Computer Vision and Pattern Recognition, pp. 9162–9171 (2020)
28. Xiong, Z., Fedorov, V.V., Fu, X., Cheng, E., Macleod, R., Zhao, J.: Fully automatic left atrium segmentation from late gadolinium enhanced magnetic resonance imaging using a dual fully convolutional neural network. IEEE Trans. Med. Imaging **38**(2), 515–524 (2018)
29. Yu, H., et al.: Foal: Fast online adaptive learning for cardiac motion estimation. In: Proceedings of the IEEE/CVF Conference on Computer Vision and Pattern Recognition, pp. 4313–4323 (2020)
30. Yu, L., Wang, S., Li, X., Fu, C.-W., Heng, P.-A.: Uncertainty-aware self-ensembling model for semi-supervised 3D left atrium segmentation. In: Shen, D., et al. (eds.) MICCAI 2019. LNCS, vol. 11765, pp. 605–613. Springer, Cham (2019). https://doi.org/10.1007/978-3-030-32245-8_67
31. Yue, Q., Luo, X., Ye, Q., Xu, L., Zhuang, X.: Cardiac segmentation from LGE MRI using deep neural network incorporating shape and spatial priors. In: Shen, D., et al. (eds.) MICCAI 2019. LNCS, vol. 11765, pp. 559–567. Springer, Cham (2019). https://doi.org/10.1007/978-3-030-32245-8_62
32. Zhai, X., Oliver, A., Kolesnikov, A., Beyer, L.: S4l: self-supervised semi-supervised learning. In: Proceedings of the IEEE International Conference on Computer Vision, pp. 1476–1485 (2019)
33. Zhang, X., Noga, M., Martin, D.G., Punithakumar, K.: Fully automated left atrium segmentation from anatomical cine long-axis MRI sequences using deep convolutional neural network with unscented Kalman filter. Med. Image Anal. **68**, 101916 (2021)
34. Zhao, A., Balakrishnan, G., Durand, F., Guttag, J.V., Dalca, A.V.: Data augmentation using learned transformations for one-shot medical image segmentation. In: Proceedings of the IEEE Conference on Computer Vision and Pattern Recognition, pp. 8543–8553 (2019)
35. Zotti, C., Luo, Z., Lalande, A., Jodoin, P.M.: Convolutional neural network with shape prior applied to cardiac MRI segmentation. IEEE J. Biomed. Health Inf. **23**(3), 1119–1128 (2018)

Deformed2Self: Self-supervised Denoising for Dynamic Medical Imaging

Junshen Xu[1(✉)] and Elfar Adalsteinsson[1,2]

[1] Department of Electrical Engineering and Computer Science, MIT, Cambridge, MA, USA
junshen@mit.edu

[2] Institute for Medical Engineering and Science, MIT, Cambridge, MA, USA

Abstract. Image denoising is of great importance for medical imaging system, since it can improve image quality for disease diagnosis and downstream image analyses. In a variety of applications, dynamic imaging techniques are utilized to capture the time-varying features of the subject, where multiple images are acquired for the same subject at different time points. Although signal-to-noise ratio of each time frame is usually limited by the short acquisition time, the correlation among different time frames can be exploited to improve denoising results with shared information across time frames. With the success of neural networks in computer vision, supervised deep learning methods show prominent performance in single-image denoising, which rely on large datasets with clean-vs-noisy image pairs. Recently, several self-supervised deep denoising models have been proposed, achieving promising results without needing the pairwise ground truth of clean images. In the field of multi-image denoising, however, very few works have been done on extracting correlated information from multiple slices for denoising using self-supervised deep learning methods. In this work, we propose Deformed2Self, an end-to-end self-supervised deep learning framework for dynamic imaging denoising. It combines single-image and multi-image denoising to improve image quality and use a spatial transformer network to model motion between different slices. Further, it only requires a single noisy image with a few auxiliary observations at different time frames for training and inference. Evaluations on phantom and *in vivo* data with different noise statistics show that our method has comparable performance to other state-of-the-art unsupervised or self-supervised denoising methods and outperforms under high noise levels.

Keywords: Image denoising · Dynamic imaging · Deep learning · Self-supervised learning

Electronic supplementary material The online version of this chapter (https://doi.org/10.1007/978-3-030-87196-3_3) contains supplementary material, which is available to authorized users.

M. de Bruijne et al. (Eds.): MICCAI 2021, LNCS 12902, pp. 25–35, 2021.
https://doi.org/10.1007/978-3-030-87196-3_3

1 Introduction

Noise is inevitable in medical images. A variety of sources lead to noisy images, such as acquisition with better spatial resolution in MRI [12] and reduction of radiation dose PET [30] and CT [27]. Further complicating the task is the complex noise statistics in different medical imaging modalities, which is not limited to additive white Gaussian noise. For example, the noise can be related to pixel intensity like Rician noise in magnitude images of MRI [31] or it can be affected by geometry parameters of the scanner like in CT [23]. Thus, a robust method for noise reduction plays an important role in medical image processing and also serves as a core module in many downstream analyses.

In many applications, more than one image is acquired during the scan to capture the dynamic of the subject, e.g., cine images for cardiac MRI [20], abdominal dynamic contrast-enhanced (DCE) MRI [11], and treatment planning 4D thoracic CT [4]. Image denoising is more necessary for dynamic imaging methods, since they often adopt fast imaging techniques to improve temporal resolution, which may reduce signal-to-noise ratio of each time frame. Dynamic imaging also provides more information for denoising as the images at different time frames have similar content and often follow the same noise model.

In recent years, a number of deep learning based methods have been proposed for image denoising [6,10,15,32]. Similar methods are also applied to medical image denoising, e.g., low-dose PET [29], CT [5] and MRI [28] denoising. These methods train a convolution neural network (CNN) that maps the noisy image to its clean counterpart. However the supervised training process requires a large-scale dataset with clean and noisy image pairs which can be difficult to acquire, especially in the field of medical imaging. Recently, there are several studies on learning a denoising network with only noisy images. The Noise2Noise [16] method trains the model on pairs of noisy images with the same content but different noises so that the network estimates the expectation of the underlaying clean image. However, in practice, it is difficult to collect a set of different noisy realizations of the same image. The Noise2Void [14] and Noise2Self [2] methods try to learn denoising network using a dataset of unpaired noisy images and use the blind-spot strategy to avoid learning a identity mapping. Yet, to achieve good performance, these two methods still require the test images to be similar to the training images in terms of image content and noise model. To address this problem, some denoising methods have been developed that only train networks with internal information from a single noisy image and do not rely on large training set. Ulyanov *et al.* introduced deep image prior (DIP) for single-image recovery [26]. Recently, Quan *et al.* proposed Self2Self [21] method, where it uses dropout [25] to implement the blind-spot strategy.

In terms of denosing methods for dynamic imaging, Benou *et al.* proposed a spatio-temporal denoising network for DCE MRI of brain, where the motion is negligible. Another category of methods first apply conventional registration method to register the images and then perform denoising on the registered images [18,23]. However, traditional optimization methods are time consuming [1] and registering noisy images directly may reduce the accuracy of registration.

In this work, we propose a deep learning framework for dynamic imaging denoising, named Deformed2Self, where we explore similarity of images in dynamic imaging by deforming images at different time frames to the target frame and utilize the fact that noises of different observations are independent and following similar noise model. Our method has the following features: 1) The whole pipeline can be trained end-to-end, which is efficient for optimization. 2) Our method is fully self-supervised, i.e., we only need noisy images without ground-truth clean images. 3) The model can be trained on a single image (with a few auxiliary observations) and has no prerequisite on large training dataset, making it suitable for applications with scarce data.

2 Methods

2.1 Problem Formulation

Let y_0 be the noisy image we want to denoise, which is generated with some noise models, e.g., for additive noise,

$$y_0 = x_0 + n_0, \tag{1}$$

where x_0 denotes the unknown clean image, and n_0 denotes the random measurement noise. The goal of single-image denoising is to find a mapping f that can recover x_0 from y_0, i.e., $\hat{x}_0 = f(y_0) \approx x_0$.

In dynamic imaging, multiple images are acquired for the same subject at different time frames. Suppose we have another N frames besides the target frame y_0. The noisy observations of these N frames and their unknown clean counterparts are denoted as $\{y_1, ..., y_N\}$ and $\{x_1, ..., x_N\}$ respectively, where $y_k = x_k + n_k$, $k = 1, ..., N$. For dynamic imaging denoising, information from different time frames are aggregated to estimate the clean image at target frame (frame 0), $\hat{x}_0 = f(y_0, y_1, ..., y_N)$. In many cases, the motion of subject occur during the scan is not negligible. Let ϕ_k be the deformation field between the target frame and frame k, $x_0 = x_k \circ \phi_k$. The noisy observation y_k can be rewritten as

$$y_k = x_k + n_k = x_0 \circ \phi_k^{-1} + n_k, \quad k = 0, 1, ..., N. \tag{2}$$

where $\phi_0 = \mathbf{I}$ is the identity mapping. Equation 2 indicates that the observed noisy images $\{y_k\}_{k=0}^{N}$ is generated from the target clean image x_0 following certain motion models and noise statistics. Therefore the auxiliary images from different time frames provide information for estimating the clean image at target frame.

2.2 The Deformed2Self Framework

Inspired by the data model above, we proposed a self-supervised denoising framework for dynamic imaging named Deformed2Self (Fig. 1), which consists of three

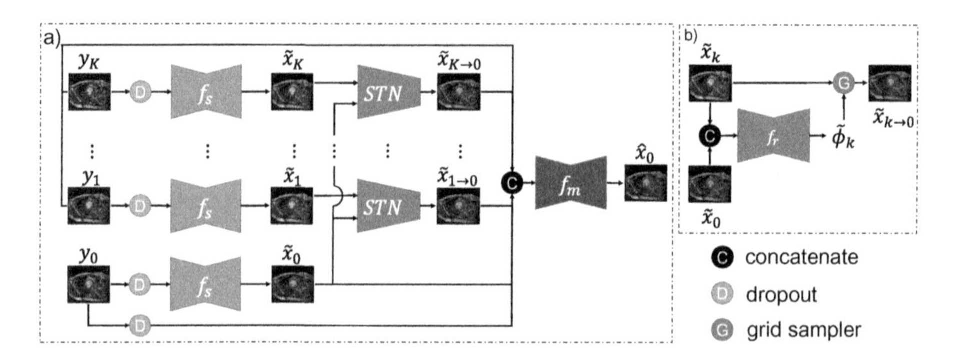

Fig. 1. Overview of the proposed self-supervised denoising framework. a) The architecture of Deformed2Self, where f_s and f_m are single- and multi-image denoising networks respectively. b) Details of the STN module, where f_r is a registration network.

modules: 1) a single-imaging denoising network to provide coarse estimation of the clean image, 2) a spatial transformer network (STN) [9] for matching spatial content of images from other time frames ($k = 1, ..., N$) to the target frame, and 3) a multi-image denoising network to generate a refined estimation of the clean image at target frame.

Single-Image Denoising: The first step of the proposed framework is single-image denoising, where we use a UNet-based [22] network f_s to denoise each frame separately. The benefit of this step is two-fold: 1) It provides a coarse estimation of the clean image utilizing only the information of each frame, which serves as a good initialization for the multi-image denoising to accelerate convergence. 2) It also improves the registration accuracy in the following STN module, since the prediction of deformation field suffers heavily from image noise.

To denoise images with internal information of each frame, we adopt the dropout-based blind-spot strategy [14,21]. Given a noisy observation y_k, $k \in \{0, ..., N\}$, we use a dropout layer to mask out some pixels in y_k before feeding it to the network. The output can be written as $\tilde{x}_k = f_s(b_k \odot y_k)$, where b_k is a random binary mask and $\odot$ is the Hadamard product. The network f_s is trained to recover the missing pixels based on information from the remaining pixels. The network's best guess would be the expected pixel value [14], and therefore the output image $\tilde{x}_k$ can be considered as the denoised version of y_k.

The network f_s is trained by minimizing the following loss function, which is the mean squared error (MSE) between the network output and the noisy image on the masked pixels.

$$\mathcal{L}_s = \frac{1}{N+1} \sum_{k=0}^{N} ||(1 - b_k) \odot (\tilde{x}_k - y_k)||_2^2 \tag{3}$$

Spatial Transformer Network: Let $\tilde{n}_k$ be the residual between the denoised image $\tilde{x}_k$ and the underlying clean image x_k, i.e., $\tilde{n}_k = \tilde{x}_k - x_k$, and thus

$\tilde{x}_k = x_0 \circ \phi_k^{-1} + \tilde{n}_k$. Suppose we can estimate ϕ_k, the deformation field between frame k and the target frame, and apply it to the denoised image x_k, then $\tilde{x}_k \circ \phi_k = x_0 + \tilde{n}_k \circ \phi_k = x_0 + \tilde{n}'_k$, i.e., $\tilde{x}_k \circ \phi_k$ is an image that is spatially matched with x_0 but is corrupted by noise $\tilde{n}'_k = \tilde{n}_k \circ \phi_k$. Note that $\{\tilde{x}_0, \tilde{x}_1 \circ \phi_1, ..., \tilde{x}_N \circ \phi_N\}$ can be considered as a set of images that share the same spatial content but have different noise and can be used for multi-image denoising later.

We estimate the motion between the target frame and other frames using a STN module (Fig. 1b). A network f_r is used to predict the deformation field given pairs of moving image $\tilde{x}_k$ and target image $\tilde{x}_0$, $\tilde{\phi}_k = f_r(\tilde{x}_k, \tilde{x}_0)$. Then $\tilde{x}_k$ is deformed with a grid sampler, $\tilde{x}_{k \to 0} = x_k \circ \tilde{\phi}_k$. We adopt the architecture in [1] for f_r and optimize it to minimize the following loss function,

$$\mathcal{L}_r = \frac{1}{N} \sum_{k=1}^{N} ||\tilde{x}_{k \to 0} - \tilde{x}_0||_2^2 + \lambda ||\nabla \tilde{\phi}_k||_2^2, \tag{4}$$

where λ is a weighting coefficient. The first term is the MSE between the warped image $\tilde{x}_k \circ \tilde{\phi}_k$ and the target image $\tilde{x}_0$, which is the similarity metric. The second term is the L2 norm of spatial gradient of $\tilde{\phi}_k$, serving as a regularization for the deformation field [1].

Multi-image Denoising: We now have two sets of images, the denoised and deformed images $\{\tilde{x}_0, \tilde{x}_{1 \to 0}, ..., \tilde{x}_{N \to 0}\}$ and the original noisy images $\{y_0, y_1, ..., y_N\}$. In the final stage, we aggregate all these images to generate a refined estimation of the target clean image. We adopt the blind-spot method similar to the single-image denoising stage but concatenate all images as input and produce an estimation for the target frame, $\hat{x}_0 = f_m(\tilde{x}_0, \tilde{x}_{1 \to 0}, ..., \tilde{x}_{N \to 0}, b \odot y_0, y_1, ..., y_N)$, where $\hat{x}_0$ is the final estimation and f_m is the multi-image denoising network. Again, we use dropout to remove some pixels in y_0 to avoid learning a identity mapping. f_m shares the same architecture with f_s except for the number of input channels. Similar to Eq. 3, a masked MSE loss is used to train network f_m,

$$\mathcal{L}_m = ||(1 - b) \odot (\hat{x}_0 - y_0)||_2^2. \tag{5}$$

2.3 Training and Inference

As mentioned above, all the three modules in the proposed method is trained in unsupervised or self-supervised manners. Besides, we can train this pipeline end-to-end with gradient descent based methods. Another advantage of our framework is that it can be trained on a single noisy image (with several auxiliary observations) without large-scale external dataset. In summary, the total loss function for our proposed model is $\mathcal{L} = \lambda_s \mathcal{L}_s + \lambda_r \mathcal{L}_r + \mathcal{L}_m$, where λ_s and λ_r are weighting coefficients. During training, the networks are updated for N_{train} iterations to learning a specific denoising model for the input images. At each iteration, input images are randomly rotated for data augmentation and the dropout layers sample different realizations of binary masks $\{b, b_0, ..., b_N\}$ so that the denoising networks can learn from different parts of the image.

For inference, we run forward pass of the model for N_{test} times with dropout layers enabled and average output $\hat{x}_0$'s to generate the final prediction.

3 Experiments and Results

3.1 Datasets and Experiments Setup

We use the following two datasets to evaluate our proposed denoising framework.

PINCAT: The PINCAT phantom [17,24] simulates both cardiac perfusion dynamics and respiration with variability in breathing motion. The spatial matrix size of the phantom is 128×128, which corresponds to a resolution of $1.5\,\text{mm} \times 1.5\,\text{mm}$. The intensity of the phantom is normalized to $[0, 1]$. The middle frame of the series is selected as the target frame.

ACDC: The ACDC dataset [3] consists of short-axis cardiac cine MRI of 100 subjects. The in-plane resolution and gap between slices range from $0.7\,\text{mm} \times 0.7\,\text{mm}$ to $1.92\,\text{mm} \times 1.92\,\text{mm}$ and 5 mm to 10 mm respectively. For pre-processing, we linearly normalize the image intensity to $[0, 1]$ and crop or pad them to a fixed size of 224×224. For each subject, we extract the sequence of the middle slice and use the end-systole (ES) and end-diastole (ED) phases as target frames.

In all the experiments, we use another four noisy images from the same sequences as auxiliary observations (two adjacent frames before and after the target frame), i.e., $N = 4$. We set $\lambda = 0.1$, and $\lambda_s = \lambda_r = 1$ in the loss function. For dropout layers, a dropout rate of 0.3 is used. We train the model using an Adam optimizer [13] with a learning rate of 1×10^{-4}. N_{train} is set to 2000 for PINCAT dataset and 4000 for ACDC dataset. N_{test} is set to 100 for both datasets. The neural networks are implemented with PyTorch 1.5, and trained and evaluated on a Titan V GPU. For PINCAT dataset, we simulate Gaussian noise and Poisson noise at different noise levels. The standard deviation σ of Gaussian noise is set to 15%, 20% and 25%. The noisy observation y under Poisson noise is generated by $y = z/P$, where $z \sim \text{Pois}(Px)$, x is the truth intensity and P is a parameter to control noise level. We set $P = 40, 20$, and 10 in the experiments. For ACDC dataset, we simulate Gaussian noise and Rician noise [8], with $\sigma = 5\%, 10\%$, and 15%. We compare our proposed method (D2S) with other state-of-the-art denoising methods, including deep image prior (DIP) [26], Self2Self (S2S) [21], BM3D [7] and VBM4D [19]. For DIP and S2S, we use the same learning rate mentioned above and tune the number of iterations on our datasets. We adopt the peak signal to noise ratio (PSNR) and the structural similarity index measure (SSIM) as evaluation metrics. The reference PyTorch implementation for Deform2Self is available on GitHub[1].

[1] https://github.com/daviddmc/Deform2Self.

3.2 Results

Comparison with Other Approaches: Table 1 shows the quantitative results on PINCAT and ACDC dataset. The proposed method achieves comparable or even better performance than other methods. Specifically, D2S has similar results to VBM4D in terms of Gaussian noise, and outperforms VBM4D for other noise models, especially for high noise levels. Besides, D2S outperforms S2S consistently, indicating that information from other time frames can largely boost performance of denoising models.

Figure 2 and 3 show example slices under Gaussian and Rician noise with $\sigma = 15\%$ and the denoised results using different methods. The D2S method has not only better statistical but also better perceptual results compared with other methods. Single-image methods such as BM3D and S2S, only use single noisy image, and therefore have not enough information to recover details that are corrupted by noise, resulting in blurred estimation. The DIP method suffers from significant structural artifacts in high noise levels. Though retrieving some details from adjacent frames, VBM4D also brings subtle artifacts to the denoised images. D2S is able to recover more detail structures with higher image quality.

Table 1. Quantitative results on PINCAT dataset for Gaussian and Poisson noise at different noise levels. The best results are indicated in red.

PINCAT												
Method	Gaussian						Poisson					
	$\sigma = 15\%$		$\sigma = 20\%$		$\sigma = 25\%$		$P = 40$		$P = 20$		$P = 10$	
	PSNR	SSIM	PSNR	SSIM	PSNR	SSIM	PSNR	SSIM	PSNR	SSIM	PSNR	SSIM
Noisy	16.55	0.300	14.05	0.208	12.11	0.151	22.42	0.603	19.40	0.472	16.19	0.346
BM3D	29.97	0.918	27.98	0.881	26.38	0.843	32.56	0.954	30.38	0.930	27.63	0.890
VBM4D	31.36	0.936	29.65	0.913	28.28	0.886	32.35	0.953	29.92	0.930	27.65	0.899
DIP	28.28	0.879	26.85	0.837	24.63	0.759	31.96	0.949	30.99	0.935	26.54	0.868
S2S	30.27	0.928	28.04	0.900	27.68	0.883	33.05	0.962	31.25	0.951	30.55	0.939
D2S	31.77	0.946	30.14	0.919	29.10	0.891	35.13	0.978	33.74	0.969	31.67	0.951
ACDC												
Method	Gaussian						Rician					
	$\sigma = 5\%$		$\sigma = 10\%$		$\sigma = 15\%$		$\sigma = 5\%$		$\sigma = 10\%$		$\sigma = 15\%$	
	PSNR	SSIM	PSNR	SSIM	PSNR	SSIM	PSNR	SSIM	PSNR	SSIM	PSNR	SSIM
Noisy	26.02	0.769	20.00	0.518	16.48	0.369	25.70	0.742	19.66	0.513	16.07	0.368
BM3D	32.32	0.953	28.54	0.905	26.45	0.860	29.58	0.874	23.69	0.777	19.78	0.689
VBM4D	32.54	0.957	28.96	0.911	26.88	0.863	29.79	0.879	23.94	0.791	19.93	0.707
DIP	26.95	0.875	25.55	0.815	23.48	0.718	26.10	0.811	22.76	0.736	19.10	0.629
S2S	30.41	0.942	28.45	0.912	26.90	0.880	28.28	0.861	23.51	0.784	19.73	0.709
D2S	32.16	0.960	30.26	0.936	28.22	0.887	29.37	0.879	24.25	0.812	20.20	0.743

Ablation Study: We perform ablation studies to evaluate different components in our framework. We evaluate models that ablate the single-image denoising module and STN module respectively.

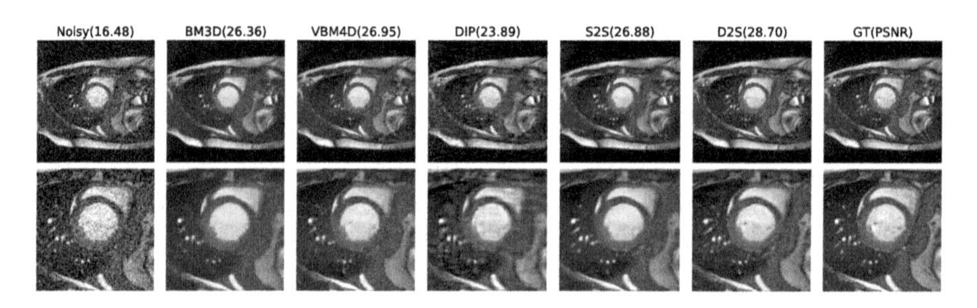

Fig. 2. Visualization results of different methods for ACDC dataset under Gaussian noise with $\sigma = 15\%$.

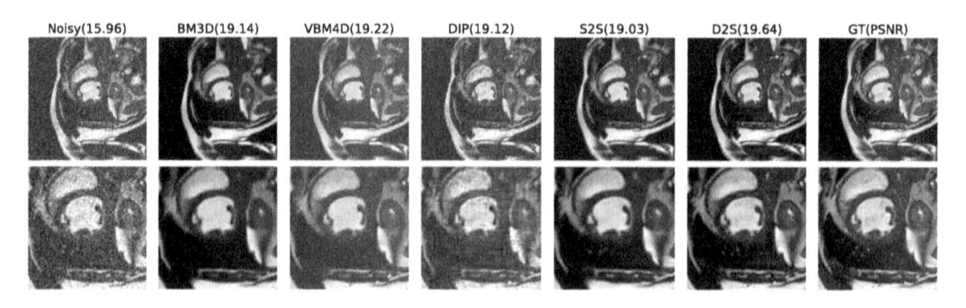

Fig. 3. Visualization results of different methods for ACDC dataset under Rician noise with $\sigma = 15\%$.

The ablated and full models are evaluated on the ACDC dataset with Gaussian and Rician noise ($\sigma = 10\%$). To investigate how single-image denoising improve registration accuracy and how image registration help match image content for the following multi-image denoising, we compute evaluation metrics PSNR and SSIM on the region of interest (ROI) that involves cardiac motion. The ROI masks include left ventricle, myocardium and right ventricle, which are annotated in the ACDC dataset [3].

Results of ablation studies (Table 2) indicate that the single-image denoising module improves the performance of the D2S model. It improves the accuracy of motion estimation in the registration network, and provides a good initialization for multi-image denoising. Besides, the registration module also makes contribution to the performance of our model. It registers images at different time frames to the same template so that the multi-channel input for the following denoising network is spatially matched, making the denoising task easier.

Table 2. Quantitative results of ablation studies.

Method	Gaussian ($\sigma = 10\%$)		Rician ($\sigma = 10\%$)	
	ROI-PSNR	ROI-SSIM	ROI-PSNR	ROI-SSIM
D2S	28.01	0.894	27.55	0.889
D2S w/o single-image denoising	27.81	0.888	27.34	0.884
D2S w/o image registration	27.68	0.887	27.24	0.883

4 Conclusions

In this work, we proposed Deformed2Self, a self-supervised deep learning method for dynamic imaging denoising, which explores the similarity of image content at different time frames by estimating the motion during imaging and improve image quality with sequential single- and multi-image denoising networks. In addition, the proposed method only relies on the target noisy image with a small number of observations at other time frames and has no prerequisite on a large training dataset, making it more practical for applications with scarce data. Experiments on a variety of noise settings show that our method has comparable or even better performance than other state-of-the-art unsupervised or self-supervised denoising methods.

References

1. Balakrishnan, G., Zhao, A., Sabuncu, M.R., Guttag, J., Dalca, A.V.: VoxelMorph: a learning framework for deformable medical image registration. IEEE Trans. Med. Imaging **38**(8), 1788–1800 (2019)
2. Batson, J., Royer, L.: Noise2Self: blind denoising by self-supervision. arXiv preprint arXiv:1901.11365 (2019)
3. Bernard, O., et al.: Deep learning techniques for automatic MRI cardiac multi-structures segmentation and diagnosis: is the problem solved? IEEE Trans. Med. Imaging **37**(11), 2514–2525 (2018)
4. Castillo, R., et al.: A framework for evaluation of deformable image registration spatial accuracy using large landmark point sets. Phys. Med. Biol. **54**(7), 1849 (2009)
5. Chen, H., et al.: Low-dose CT denoising with convolutional neural network. In: 2017 IEEE 14th International Symposium on Biomedical Imaging (ISBI 2017), pp. 143–146. IEEE (2017)
6. Chen, J., Chen, J., Chao, H., Yang, M.: Image blind denoising with generative adversarial network based noise modeling. In: Proceedings of the IEEE Conference on Computer Vision and Pattern Recognition, pp. 3155–3164 (2018)
7. Dabov, K., Foi, A., Katkovnik, V., Egiazarian, K.: Image denoising by sparse 3-D transform-domain collaborative filtering. IEEE Trans. Image Process. **16**(8), 2080–2095 (2007)
8. Gudbjartsson, H., Patz, S.: The Rician distribution of noisy MRI data. Magn. Reson. Med. **34**(6), 910–914 (1995)

9. Jaderberg, M., Simonyan, K., Zisserman, A., Kavukcuoglu, K.: Spatial transformer networks. arXiv preprint arXiv:1506.02025 (2015)
10. Jia, X., Liu, S., Feng, X., Zhang, L.: FOCNet: a fractional optimal control network for image denoising. In: Proceedings of the IEEE Conference on Computer Vision and Pattern Recognition, pp. 6054–6063 (2019)
11. Johansson, A., Balter, J.M., Cao, Y.: Abdominal DCE-MRI reconstruction with deformable motion correction for liver perfusion quantification. Med. Phys. **45**(10), 4529–4540 (2018)
12. Kale, S.C., Chen, X.J., Henkelman, R.M.: Trading off SNR and resolution in MR images. NMR Biomed. Int. J. Devoted Dev. Appl. Magn. Reson. Vivo **22**(5), 488–494 (2009)
13. Kingma, D.P., Ba, J.: Adam: a method for stochastic optimization (2017)
14. Krull, A., Buchholz, T.O., Jug, F.: Noise2Void-learning denoising from single noisy images. In: Proceedings of the IEEE Conference on Computer Vision and Pattern Recognition, pp. 2129–2137 (2019)
15. Lefkimmiatis, S.: Universal denoising networks: a novel CNN architecture for image denoising. In: Proceedings of the IEEE Conference on Computer Vision and Pattern Recognition, pp. 3204–3213 (2018)
16. Lehtinen, J., et al.: Noise2Noise: learning image restoration without clean data. arXiv preprint arXiv:1803.04189 (2018)
17. Lingala, S.G., Hu, Y., DiBella, E., Jacob, M.: Accelerated dynamic MRI exploiting sparsity and low-rank structure: kt SLR. IEEE Trans. Med. Imaging **30**(5), 1042–1054 (2011)
18. Lukas, S., Feger, S., Rief, M., Zimmermann, E., Dewey, M.: Noise reduction and motion elimination in low-dose 4D myocardial computed tomography perfusion (CTP): preliminary clinical evaluation of the ASTRA4D algorithm. Eur. Radiol. **29**(9), 4572–4582 (2019)
19. Maggioni, M., Boracchi, G., Foi, A., Egiazarian, K.: Video denoising, deblocking, and enhancement through separable 4-D nonlocal spatiotemporal transforms. IEEE Trans. Image Process. **21**(9), 3952–3966 (2012)
20. Malayeri, A.A., Johnson, W.C., Macedo, R., Bathon, J., Lima, J.A., Bluemke, D.A.: Cardiac cine MRI: quantification of the relationship between fast gradient echo and steady-state free precession for determination of myocardial mass and volumes. J. Magn. Reson. Imaging Off. J. Int. Soc. Magn. Reson. Med. **28**(1), 60–66 (2008)
21. Quan, Y., Chen, M., Pang, T., Ji, H.: Self2Self with dropout: learning self-supervised denoising from single image. In: Proceedings of the IEEE/CVF Conference on Computer Vision and Pattern Recognition, pp. 1890–1898 (2020)
22. Ronneberger, O., Fischer, P., Brox, T.: U-Net: convolutional networks for biomedical image segmentation. In: Navab, N., Hornegger, J., Wells, W.M., Frangi, A.F. (eds.) MICCAI 2015. LNCS, vol. 9351, pp. 234–241. Springer, Cham (2015). https://doi.org/10.1007/978-3-319-24574-4_28
23. Schirrmacher, F., et al.: Temporal and volumetric denoising via quantile sparse image prior. Med. Image Anal. **48**, 131–146 (2018)
24. Sharif, B., Bresler, Y.: Adaptive real-time cardiac MRI using paradise: validation by the physiologically improved NCAT phantom. In: 2007 4th IEEE International Symposium on Biomedical Imaging: From Nano to Macro, pp. 1020–1023. IEEE (2007)
25. Srivastava, N., Hinton, G., Krizhevsky, A., Sutskever, I., Salakhutdinov, R.: Dropout: a simple way to prevent neural networks from overfitting. J. Mach. Learn. Res. **15**(1), 1929–1958 (2014)

26. Ulyanov, D., Vedaldi, A., Lempitsky, V.: Deep image prior. In: Proceedings of the IEEE Conference on Computer Vision and Pattern Recognition, pp. 9446–9454 (2018)
27. Wolterink, J.M., Leiner, T., Viergever, M.A., Išgum, I.: Generative adversarial networks for noise reduction in low-dose CT. IEEE Trans. Med. Imaging **36**(12), 2536–2545 (2017)
28. Xie, D., et al.: Denoising arterial spin labeling perfusion MRI with deep machine learning. Magn. Reson. Imaging **68**, 95–105 (2020)
29. Xu, J., Gong, E., Ouyang, J., Pauly, J., Zaharchuk, G., Han, S.: Ultra-low-dose 18F-FDG brain PET/MR denoising using deep learning and multi-contrast information. In: Medical Imaging 2020: Image Processing, vol. 11313, p. 113131. International Society for Optics and Photonics (2020)
30. Xu, J., Gong, E., Pauly, J., Zaharchuk, G.: 200x low-dose pet reconstruction using deep learning. arXiv preprint arXiv:1712.04119 (2017)
31. You, X., Cao, N., Lu, H., Mao, M., Wanga, W.: Denoising of MR images with Rician noise using a wider neural network and noise range division. Magn. Reson. Imaging **64**, 154–159 (2019)
32. Zhang, K., Zuo, W., Chen, Y., Meng, D., Zhang, L.: Beyond a Gaussian denoiser: residual learning of deep CNN for image denoising. IEEE Trans. Image Process. **26**(7), 3142–3155 (2017)

Imbalance-Aware Self-supervised Learning for 3D Radiomic Representations

Hongwei Li[1,2], Fei-Fei Xue[4], Krishna Chaitanya[3], Shengda Luo[5], Ivan Ezhov[1], Benedikt Wiestler[6], Jianguo Zhang[4(✉)], and Bjoern Menze[1,2]

[1] Department of Computer Science, Technical University of Munich, Munich, Germany
hongwei.li@tum.de
[2] Department of Quantitative Biomedicine, University of Zurich, Zürich, Switzerland
[3] ETH Zurich, Zürich, Switzerland
[4] Department of Computer Science and Engineering, Southern University of Science and Technology, Shenzhen, China
zhangjg@sustech.edu.cn
[5] Faculty of Information Technology, Macau University of Science and Technology, Macao, China
[6] Klinikum rechts der Isar, Technical University of Munich, Munich, Germany

Abstract. Radiomics can quantify the properties of regions of interest in medical image data. Classically, they account for pre-defined statistics of shape, texture, and other low-level image features. Alternatively, deep learning-based representations are derived from supervised learning but require expensive annotations and often suffer from overfitting and data imbalance issues. In this work, we address the challenge of learning the representation of a 3D medical image for an effective quantification under data imbalance. We propose a *self-supervised* representation learning framework to learn high-level features of 3D volumes as a complement to existing radiomics features. Specifically, we demonstrate how to learn image representations in a self-supervised fashion using a 3D Siamese network. More importantly, we deal with data imbalance by exploiting two unsupervised strategies: a) sample re-weighting, and b) balancing the composition of training batches. When combining the learned self-supervised feature with traditional radiomics, we show significant improvement in brain tumor classification and lung cancer staging tasks covering MRI and CT imaging modalities. Codes are available in https://github.com/hongweilibran/imbalanced-SSL.

Electronic supplementary material The online version of this chapter (https://doi.org/10.1007/978-3-030-87196-3_4) contains supplementary material, which is available to authorized users.

M. de Bruijne et al. (Eds.): MICCAI 2021, LNCS 12902, pp. 36–46, 2021.
https://doi.org/10.1007/978-3-030-87196-3_4

1 Introduction

Great advances have been achieved in supervised deep learning, reaching expert-level performance on some considerably challenging applications [11]. However, supervised methods for image classification commonly require relatively large-scale datasets with ground-truth labels which is time- and resource-consuming in the medical field. Radiomics is a translational field aiming to extract objective and quantitative information from clinical imaging data. While traditional radiomics methods, that rely on statistics of shape, texture and others [1], are proven to be generalizable in various tasks and domains, their discriminativeness is often not guaranteed since they are low-level features which are not specifically optimized on target datasets.

Self-supervised learning for performing *pre-text tasks* have been explored in medical imaging [24,25], that serve as a proxy task to pre-train the deep neural networks. They learn representations commonly in a supervised manner on proxy tasks. Such methods depend on heuristics to design pre-text tasks which could limit the discriminativeness of the learnt representations. In this work, we investigate self-supervised *representation learning* which aims to **directly** learn the representation of the data without a proxy task.

Recent contrastive learning-based methods [6,15,20] learn informative representations *without* human supervision. However, they often rely on large batches to train and most of them work for 2D images. To this end, due to the high dimensionality and limited number of training samples in medical field, applying contrastive learning-based methods may not be practically feasible in 3D datasets. Specially, in this study, we identify two main differences required to adapt self-supervised representation learning for radiomics compared to natural image domain: i) Medical datasets are often multi-modal and three dimensional. Thus, learning representation methods in 3D medical imaging would be computationally expensive. ii) heterogeneous medical datasets are *inherently imbalanced*, e.g. distribution disparity of disease phenotypes. Existing methods are built upon approximately balanced datasets (e.g. CIFAR [18] and ImageNet [10]) and do not assume the existence of data imbalance. Thus, how to handle data imbalance problem is yet less explored in the context of self-supervised learning.

Related Work. Radiomic features have drawn considerable attention due to its predictive power for treatment outcomes and cancer genetics in personalized medicine [12,23]. Traditional radiomics include shape features, first-, second-, and higher- order statistics features.

Self-supervised representation learning [3,6,7,13,15,22] have shown steady improvements with impressive results on multiple natural image tasks, mostly based on contrastive learning [14]. Contrastive learning aims to attract positive (or *similar*) sample pairs and rebuff negative (or *disimilar*) sample pairs. Positive sample pairs can be obtained by generating two augmented views of one sample, and the remaining samples in the batch can be used to construct the negative samples/pairs for a given positive pair. In practice, contrastive learning methods benefit from a large number of negative samples. In medical imaging, there are some existing work related to contrastive learning [2,17,21]. Chaitanya *et al.*'s work [5] is the most relevant to our study, which proposed a representation learning framework for image segmentation by exploring local and global similarities and dissimilarities. Though these methods are effective in learning representations, they require a large batch size and/or negative pairs, which make them difficult to apply to *3D* medical data. Chen *et al.* [8] demonstrates that a Siamese network can avoid the above issues on a 2D network. The Siamese network, which contains two encoders with shared weights, compares two similar representations of two augmented samples from one sample. Importantly, it neither uses negative pairs nor a large batch size. Considering these benefits, we borrow the Siamese structure and extend it to 3D *imbalanced* medical datasets.

Contributions. Our contribution is threefold: (1) We develop a 3D Siamese network to learn self-supervised representation which is high-level and discriminative. (2) For the first time, we explore how to tackle the data imbalance problem in self-supervised learning without using labels and propose two effective unsupervised strategies. (3) We demonstrate that self-supervised representations can complement the existing radiomics and the combination of them outperforms supervised learning in two applications.

2 Methodology

The problem of interest is how to learn high-level, discriminative representations on 3D imbalanced medical image datasets in a self-supervised manner. The schematic view of the framework is illustrated in Fig. 1. First, a pre-trained 3D encoder network, denoted as E_a, takes a batch of original images X with batch size N as input and outputs N representation vectors. The details of the 3D encoder is shown in Table 4 of the Supplementary. The features are fed into the *RE/SE* module to estimate their individual weights or to resample the batch.

Then each image x in the batch X is randomly augmented into two images (or called an image pair). They are processed and compared by a 3D Siamese network, nicknamed *3DSiam*, which enjoys relatively low memory without relying on large training batch of 3D data. The proposed *3DSiam* extends original 2D Siamese network [8] from processing 2D images to 3D volumes while inherits its advantages. Since medical datasets are inherently imbalanced, by intuition sole *3DSiam* would probably suffer from imbalanced data distribution. In the following, we first introduce *RE/SE* module to mitigate this issue.

RE/SE Module to Handle Imbalance. Since there is no prior knowledge on the data distribution available, the way to handle imbalance must be *unsupervised*. The vectors mentioned above are fed into a *RE/SE* module before training the *3DSiam* network. The k-means algorithm is used first to cluster the representation vectors into k centers. We then proposed two simple yet effective strategies: a) sample re-weighting (RE), and b) sample selection (SE):

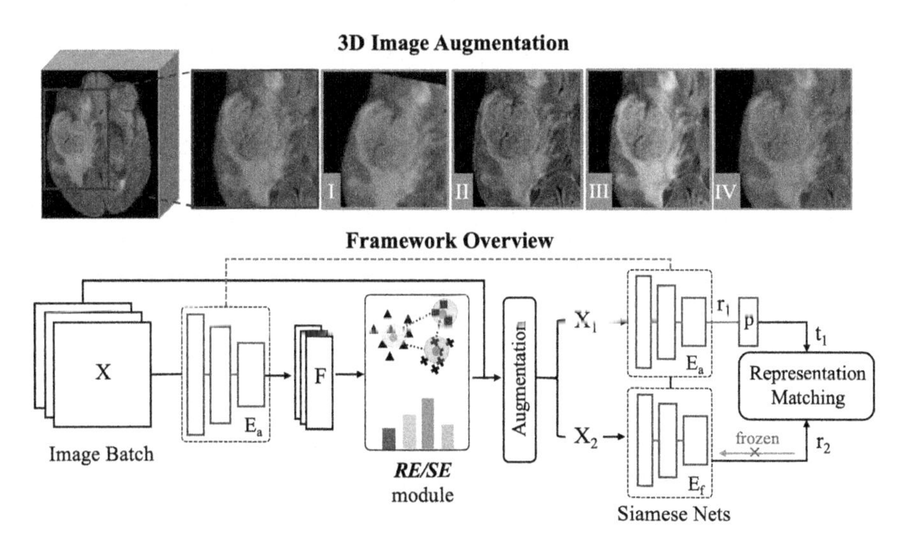

Fig. 1. Our proposed framework learns invariance from extensive 3D image augmentation within four categories: I) affine transform, II) appearance enhancement, III) contrast change, and IV) adding random noise. First, an image batch X is fed into an initialized 3D encoder to obtain its representation F. The RE/SE module first estimates its distribution by k-means based clustering and uses two strategies including sample re-weighting (RE) or sample selection (SE) to alleviate data imbalance issue. Each image is randomly augmented into two positive samples $\{X_1, X_2\}$ which are then used to train a 3D Siamese network by comparing their representations from two encoders $\{E_a, E_f\}$ with shared weights. p is a two-layer perceptron to transform the feature.

a) Sample re-weighting (RE). Denote a batch with N samples as $X = \{x_i | i = 1, 2, ..., N\}$. Given k clusters, denote the distribution of k clusters of features as $F = \{f_j | j = 1, 2, ..., k\}$ over N samples. f_j denotes the frequency of cluster j. Then we assign different weights to the samples in each cluster j. For each sample x_i, representation vector of which belongs to cluster j, we assign it a weight of N/f_j to penalize the imbalanced distribution during the batch training. In practice, we further normalize it by re-scaling to guarantee the minimum weight is 1 for each input batch.
b) Sample selection (SE). Denoting the clusters' centroids as $C = \{c_1, c_2, ...c_k\}$, we find the maximum Euclidean distance $max_{i,j \in [1,k], i \neq j} d(c_i, c_j)$ among all pairs of centroids. k is a hyper-parameter here. We hypothesize that the clusters with maximum centroid distance are representation vectors from different groups. To select m samples from the original N samples to form a new batch, denoted by $B_c = \{x_1, x_2, ..., x_m\}$, we sample $\frac{m}{2}$ nearest sample points centered on each of the selected maximum-distance centroids. m is set to be smaller than $\frac{N}{k}$ for low computation complexity and efficient sampling. The selected new batch is then used to train our *3DSiam* network. A motivation behind the selection strategy is outlined in Supplementary.

3D Siamese Network. The *3DSiam* takes as input two randomly augmented views x_1 and x_2 from a sample x. The two views are processed by two 3D encoder networks with *shared* weights. One of the encoder has frozen weights when training (denoted as E_f) and the other one is with active weights (denoted as E_a). Before training, E_f is always updated to the weights of E_a. E_a is followed by a two-layer perceptron called *predictor* to transform the features. The final objective is to optimize a matching score between the two *similar* representations $t_1 \triangleq p(E_a(x_1))$ and $r_2 \triangleq E_f(x_2)$. *3DSiam* minimizes their negative cosine similarity, which is formulated as:

$$S(t_1, r_2) = -\frac{t_1}{\|t_1\|_2} \cdot \frac{r_2}{\|r_2\|_2}, \tag{1}$$

where $\|\cdot\|_2$ is L_2-norm. Following [6], we define a symmetrized loss to train the two encoders, formulated as:

$$\mathcal{L} = \frac{1}{2} S(t_1, r_2) + \frac{1}{2} S(t_2, r_1), \tag{2}$$

where $t_2 \triangleq p(E_f(x_2))$, $r_1 \triangleq E_f(x_1)$. This loss is defined and computed for each sample with re-weighting in the batch X or the new batch B_c with equal weights. Notably the encoder E_f on x_2 receives no gradient from r_2 in the first term of Eq. (2), but it receives gradients from t_2 in the second term (and vice versa for x_1). This training strategy avoids collapsing solutions, i.e., t_1 and r_2 outputs a constant over the training process. When training is finished, r_2 is used as the final representation.

3 Experiments

Datasets and Preprocessing. The evaluation of our approach is performed on two public datasets: 1) a multi-center MRI dataset (*BraTS*) [4,19] including 326 patients with brain tumor. The MRI modalities include FLAIR, T1, T2 and T1-c with a uniform voxel size $1 \times 1 \times 1\,\text{mm}^3$. Only FLAIR is used in our experiment for simplicity of comparisons. 2) a lung CT dataset with 420 non-small cell lung cancer patients (*NSCLC-radiomics*) [1,9][1]. The effectiveness of the learnt representations is evaluated on two classification tasks (also called 'down-stream task'): a) discriminating high grade (H-grade) and low grade tumor (L-grade), and b) predicting lung cancer stages (i.e. I, II or III). The *BraTS* dataset is imbalanced in two aspects: a) the distribution of ground truth labels (H-grade *vs.* L-grade); b) the distribution of available scans among different medical centers. For *NSCLC-radiomics*, the distribution of ground truth labels are imbalanced as well, with ratio of 2:1:6 for stage I, II and III respectively. For *BraTS*, we make use of the segmentation mask to get the centroid and generate a 3D bounding box of $96 \times 96 \times 96$ to localize the tumour. If the bounding box exceeds the original volume, the out-of-box region was padded with background

[1] Two patients were excluded as the ground truth labels are not available.

intensity. For *NSCLC-radiomics*, we get the lung mask by a recent public lung segmentation tool [16], and then generate a $224 \times 224 \times 224$ bounding box to localize the lung. The lung volume was then resized to $112 \times 112 \times 112$ due to memory constraint. The intensity range of all image volumes was rescaled to [0, 255] to guarantee the success of intensity-based image transformations.

Configuration of the Training Schedule. We build a 3D convolutional neural network with two bottleneck blocks as the encoder for all experiments (details in Supplementary). In the beginning, we pre-train *3DSiam* for one epoch with a batch size of 6. Then we use it to extract features for the *RE/SE* module. After first epoch, the encoder from the *last iteration* is employed for *dynamic* feature extraction. For down-stream task evaluations, we use the last-layer feature of the encoder. For 3D data augmentation, we apply four categories shown in Fig. 1, including random rotations in $[-20, 20]$ degrees, random scale between $[0.7, 1.3]$, and random shift between $[-0.05, 0.05]$, Gamma contrast adjustment between [0.7, 1.5], image sharpening, and Gaussian blurring, considering the special trait of medical images. For optimization, we use Adam with 10^{-2} learning rate and 10^{-4} weight decay. Each experiment is conducted using one Nvidia RTX 6000 GPU with 24 GB memory. The number of cluster k is set to 3 in all experiments. Its effect is analyzed in the last section.

Computation Complexity. For *3DSiam* without *RE/SE* module, the training takes only around four hours for 50 epochs for the results reported for brain tumor classification task. We do not observe significant improvement when increasing the number of epochs after 50. We train *3DSiam* with *RE/SE* module for around 2000 iterations (not epochs) to guarantee that similar number of training images for the models of comparison are involved. In *RE/SE* module, the main computation cost is from k-means algorithm. We have observed that the overall computation time has increased by 20% (with i5-5800K CPU). It is worth noting that *RE/SE* module is not required during the inference stage, thus there is no increase of the computational cost in testing.

Feature Extraction and Aggregation. For each volume, we extract a set of 107 traditional radiomics features[2] including first- and second-order statistics, shape-based features and gray level co-occurrence matrix, denoted as f_{trad}. For the self-supervised learning one, we extract 256 features from the last fully connected layer of the encoder E_a, denoted as f_{SSL}. To directly evaluate the effectiveness of SSL-based features, we concatenate them to a new feature vector $f = [f_{trad}, f_{SSL}]$. Note that f_{trad} and f_{SSL} are always from the same subjects.

Evaluation Protocol, Classifier and Metrics. For evaluation, we follow the common protocol to evaluate the quality of the pre-trained representations by training a *supervised* linear support vector machine (SVM) classifier on the

[2] https://github.com/Radiomics/pyradiomics.

training set, and then evaluating it on the test set. For binary classification task (BraTS), we use the sensitivity and specificity as the evaluation metrics. For multi-class classification task (lung cancer staging), we report the overall accuracy and minor-class (i.e. stage II) accuracy considering all testing samples. We use *stratified* five-fold cross validation to reduce selection bias and validate each model. In each fold, we randomly sample 80% subjects from each class as the training set, and the remaining 20% for each class as the test set. Within each fold, we employ 20% of training data to optimize the hyper-parameters.

4 Results

Quantitative Comparison. We evaluate the effectiveness of the proposed self-supervised radiomics features on two classification tasks: a) discrimination of low grade and high grade of brain tumor and b) staging of lung cancer.

Table 1. Comparison of the performances of different kinds of features in two downstream tasks using stratified cross-validation. We further reduce 50% training data in each fold of validation to show the effectiveness against supervised learning. Our method outperforms supervised learning in both scenarios.

Methods	*BraTS*		*Lung cancer staging*	
	Sensitivity/Specificity		Overall/Minor-class accuracy	
	Full labels	*50% labels*	*Full labels*	*50% labels*
Trad. radiomics	0.888/0.697	**0.848**/0.697	0.490/0.375	0.481/0.325
Rubik's cube [25]	0.744/0.526	0.680/0.486	0.459/0.325	0.433/0.275
3DSiam	0.844/0.407	0.808/0.526	0.459/0.300	0.445/0.300
3DSiam+SE	0.848/0.513	0.824/0.566	0.471/0.350	0.443/0.325
3DSiam+RE	0.868/0.486	0.828/0.605	0.459/0.375	0.445/0.325
Trad.+3DSiam	0.904/0.645	0.804/0.566	0.495/0.350	0.486/0.350
Trad.+3DSiam+SE	0.916/**0.711**	**0.848/0.763**	**0.538**/0.375	**0.519**/0.350
Trad.+3DSiam+RE	**0.920/0.711**	0.804/0.763	0.524/**0.425**	0.502/**0.40**
Supervised learning	0.888/0.711	0.804/0.566	0.526/0.375	0.467/0.325

Effectiveness of RE/SE *Module.* From the first row of Table 1, one can observe that traditional radiomics itself brings powerful features to quantify tumor characteristics. On BraTS dataset, the comparison between traditional radiomics and vanilla self-supervised radiomics (*3DSiam*) confirms our hypothesis that features learned by vanilla self-supervised method behave poorly, especially on the minor class (poor specificity). However, self-supervised radiomics with *RE* or *SE* module surpasses *3DSiam* in specificity by a large margin. The aggregation of the vanilla self-supervised representation and traditional radiomics does not show significant improvement. More importantly, with *RE/SE* module added, the specificity increased by 6.6%, from 64.5% to 71.1%, which indicates a large boost in predicting the minor class (i.e. L-grade). Both comparisons of rows 4, 5, 6 and rows 7, 8, 9 demonstrate the success of our RE/SE module in tacking class imbalance, i.e., promoting the recognition of the minor class, while preserving the accuracy of the major class.

Comparison with State-of-the-Art. Our method (*Trad.+3DSiam+SE* in Table 2) outperforms the supervised one in two scenarios in two classification tasks, the result of which is achieved by using the same encoder backbone with a weighted cross-entropy loss. When it is trained with 50% less labels, the performance of supervised model decrease drastically. On lung cancer staging with three classes, although the overall accuracy of self-supervised radiomics is lower than the traditional one, with the *RE/SE* module, the combination of two kinds of radiomics achieves the topmost overall accuracy. This demonstrates the proposed self-supervised radiomics is complementary to existing radiomics. In the second row, we show the result of one self-supervised learning method trained by playing Rubik cubes [25] to learn contextual information with a same encoder. We observe that the representation learned in proxy task is less discriminative than the one directly from representation learning.

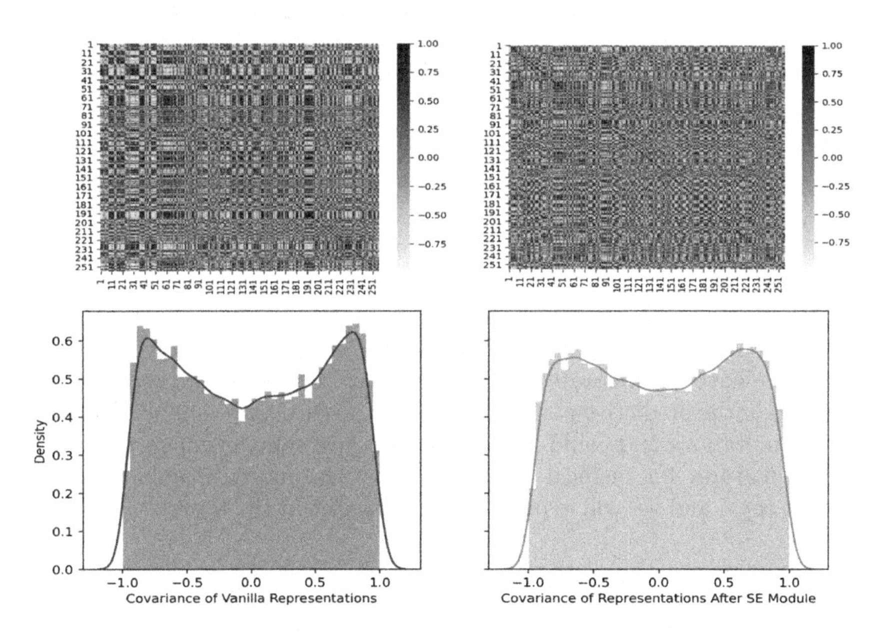

Fig. 2. Covariance analysis of the representations before and after the SE module. Across all 326 tumor patients, each feature was correlated with other ones, thereby generating the correlation coefficients. The density map show that the vanilla representation before SE module are more correlated (redundant) than the one after.

Analysis of Representations and Hyperparameters

Feature Covariance. For a better understanding of the role of the proposed module in relieving data imbalance problem, we further analyze the feature covariance to understand the role of the *SE module*. Consider two paired variables (x_i, x_j) in the representation R. Given n samples $\{(x_{i1}, x_{j1}), (x_{i2}, x_{j2}), ..., (x_{in}, x_{jn})\}$, Pearson's correlation coefficient $r_{x_i x_j}$ is defined as: $r_{x_i x_j} = \frac{cov(x_i, x_j)}{\sigma_{x_i} \sigma_{x_j}}$, where *cov* is the covariance and σ is the standard deviation. We found that the features after *SE* module become more compact as shown in Fig. 2 and more discriminative compared to the features without *SE* module.

Effect of the Number of Clusters k. The hyper-parameter k in the *SE* module is the number of clusters, which plays a vital role in constructing new batch. To evaluate its effect, we use different k to train *3DSiam* and evaluate it through classification task. To fairly compare different values of k, we keep the size m of the new batch B_c fixed to 6 which is also the batch size when $k = 0$ (without *SE* module). The initial batch size N is set to $\mathrm{k} \times \mathrm{q}$ where q is empirically set to 10 in the comparison. The AUC achieves the highest when $k = 3$. With $k = 5$, the AUC drops. This is probably because when k becomes large, the sampling may be biased when only considering a pair of clustering centers. For details, please refer to the curves of AUC over the number of clusters in Table 3 in Supplementary.

5 Conclusion

In this work, we proposed a 3D *self-supervised* representation framework for medical image analysis. It allows us to learn effective 3D representations in a self-supervised manner while considering the imbalanced nature of medical datasets. We have demonstrated that data-driven self-supervised representation could enhance the predictive power of radiomics learned from *large-scale* datasets without annotations and could serve as an effective compliment to the existing radiomics features for medical image analysis. Dealing with imbalance is an important topic and we will explore other strategies in the future.

Acknowledgement. This work was supported by Helmut Horten Foundation. I. E. was supported by the TRABIT network under the EU Marie Sklodowska-Curie program (Grant ID: 765148). S. L. was supported by the Faculty Research Grant (NO. FRG-18-020-FI) at Macau University of Science and Technology. B. W. and B. M. were supported through the DFG, SFB-824, subproject B12. K. C. was supported by Clinical Research Priority Program (CRPP) Grant on Artificial Intelligence in Oncological Imaging Network, University of Zurich.

References

1. Aerts, H.J., et al.: Decoding tumour phenotype by noninvasive imaging using a quantitative radiomics approach. Nat. Commun. **5**(1), 1–9 (2014)
2. Azizi, S., et al.: Big self-supervised models advance medical image classification. arXiv preprint arXiv:2101.05224 (2021)
3. Bachman, P., Hjelm, R.D., Buchwalter, W.: Learning representations by maximizing mutual information across views. In: Advances in Neural Information Processing Systems, pp. 15535–15545 (2019)
4. Bakas, S., et al.: Identifying the best machine learning algorithms for brain tumor segmentation, progression assessment, and overall survival prediction in the brats challenge. arXiv preprint arXiv:1811.02629 (2018)
5. Chaitanya, K., Erdil, E., Karani, N., Konukoglu, E.: Contrastive learning of global and local features for medical image segmentation with limited annotations. arXiv preprint arXiv:2006.10511 (2020)
6. Chen, T., Kornblith, S., Norouzi, M., Hinton, G.: A simple framework for contrastive learning of visual representations. arXiv preprint arXiv:2002.05709 (2020)
7. Chen, X., Fan, H., Girshick, R., He, K.: Improved baselines with momentum contrastive learning. arXiv preprint arXiv:2003.04297 (2020)
8. Chen, X., He, K.: Exploring simple siamese representation learning. arXiv preprint arXiv:2011.10566 (2020)
9. Clark, K., et al.: The cancer imaging archive (TCIA): maintaining and operating a public information repository. J. Digit. Imaging **26**(6), 1045–1057 (2013). https://doi.org/10.1007/s10278-013-9622-7
10. Deng, J., Dong, W., Socher, R., Li, L.J., Li, K., Fei-Fei, L.: ImageNet: a large-scale hierarchical image database. In: Computer Vision and Pattern Recognition, pp. 248–255 (2009)
11. Esteva, A., et al.: Dermatologist-level classification of skin cancer with deep neural networks. Nature **542**(7639), 115–118 (2017)
12. Gillies, R.J., Kinahan, P.E., Hricak, H.: Radiomics: images are more than pictures, they are data. Radiology **278**(2), 563–577 (2016)
13. Grill, J.B., et al.: Bootstrap your own latent-a new approach to self-supervised learning. In: Advances in Neural Information Processing Systems, vol. 33 (2020)
14. Hadsell, R., Chopra, S., LeCun, Y.: Dimensionality reduction by learning an invariant mapping. In: Computer Vision and Pattern Recognition, pp. 1735–1742 (2006)
15. He, K., Fan, H., Wu, Y., Xie, S., Girshick, R.: Momentum contrast for unsupervised visual representation learning. In: Computer Vision and Pattern Recognition, pp. 9729–9738 (2020)
16. Hofmanninger, J., Prayer, F., Pan, J., Röhrich, S., Prosch, H., Langs, G.: Automatic lung segmentation in routine imaging is primarily a data diversity problem, not a methodology problem. Eur. Radiol. Exp. **4**(1), 1–13 (2020). https://doi.org/10.1186/s41747-020-00173-2
17. Kiyasseh, D., Zhu, T., Clifton, D.A.: CLOCS: contrastive learning of cardiac signals. arXiv preprint arXiv:2005.13249 (2020)
18. Krizhevsky, A., Hinton, G., et al.: Learning multiple layers of features from tiny images (2009)
19. Menze, B.H., et al.: The multimodal brain tumor image segmentation benchmark (BRATS). IEEE Trans. Med. Imaging **34**(10), 1993–2024 (2014)
20. van den Oord, A., Li, Y., Vinyals, O.: Representation learning with contrastive predictive coding. arXiv preprint arXiv:1807.03748 (2018)

21. Vu, Y.N.T., Wang, R., Balachandar, N., Liu, C., Ng, A.Y., Rajpurkar, P.: MedAug: contrastive learning leveraging patient metadata improves representations for chest X-ray interpretation. arXiv preprint arXiv:2102.10663 (2021)
22. Ye, M., Zhang, X., Yuen, P.C., Chang, S.F.: Unsupervised embedding learning via invariant and spreading instance feature. In: Computer Vision and Pattern Recognition, pp. 6210–6219 (2019)
23. Yip, S.S., Aerts, H.J.: Applications and limitations of radiomics. Phys. Med. Biol. **61**(13), R150 (2016)
24. Zhou, Z., et al.: Models genesis: generic autodidactic models for 3D medical image analysis. In: Shen, D., et al. (eds.) MICCAI 2019. LNCS, vol. 11767, pp. 384–393. Springer, Cham (2019). https://doi.org/10.1007/978-3-030-32251-9_42
25. Zhuang, X., Li, Y., Hu, Y., Ma, K., Yang, Y., Zheng, Y.: Self-supervised feature learning for 3D medical images by playing a Rubik's cube. In: Shen, D., et al. (eds.) MICCAI 2019. LNCS, vol. 11767, pp. 420–428. Springer, Cham (2019). https://doi.org/10.1007/978-3-030-32251-9_46

Self-supervised Visual Representation Learning for Histopathological Images

Pengshuai Yang[1], Zhiwei Hong[2], Xiaoxu Yin[1], Chengzhan Zhu[3], and Rui Jiang[1](✉)

[1] Ministry of Education Key Laboratory of Bioinformatics, Bioinformatics Division, Beijing National Research Center for Information Science and Technology, Department of Automation, Tsinghua University, Beijing 100084, China
ruijiang@tsinghua.edu.cn

[2] Department of Computer Science and Technology, Tsinghua University, Beijing 100084, China

[3] Department of Hepatobiliary and Pancreatic Surgery, The Affiliated Hospital of Qingdao University, Qingdao 266000, Shandong, China

Abstract. Self-supervised learning provides a possible solution to extract effective visual representations from unlabeled histopathological images. However, existing methods either fail to make good use of domain-specific knowledge, or rely on side information like spatial proximity and magnification. In this paper, we propose CS-CO, a hybrid self-supervised visual representation learning method tailored for histopathological images, which integrates advantages of both generative and discriminative models. The proposed method consists of two self-supervised learning stages: cross-stain prediction (CS) and contrastive learning (CO), both of which are designed based on domain-specific knowledge and do not require side information. A novel data augmentation approach, stain vector perturbation, is specifically proposed to serve contrastive learning. Experimental results on the public dataset NCT-CRC-HE-100K demonstrate the superiority of the proposed method for histopathological image visual representation. Under the common linear evaluation protocol, our method achieves 0.915 eight-class classification accuracy with only 1,000 labeled data, which is about 1.3% higher than the fully-supervised ResNet18 classifier trained with the whole 89,434 labeled training data. Our code is available at https://github.com/easonyang1996/CS-CO.

Keywords: Self-supervised learning · Stain separation · Contrastive representation learning · Histopathological images

1 Introduction

Extracting effective visual representations from histopathological images is the cornerstone of many computational pathology tasks, such as image retrieval [26,29], disease prognosis [1,30], and molecular signature prediction [7,9,18].

M. de Bruijne et al. (Eds.): MICCAI 2021, LNCS 12902, pp. 47–57, 2021.
https://doi.org/10.1007/978-3-030-87196-3_5

Due to the powerful representation ability, deep learning-based methods gradually replace the traditional handcraft-feature extraction methods and become the mainstream. Deep learning-based methods usually rely on a large amount of labeled data to learn good visual representations, while preparing large-scale labeled datasets is expensive and time-consuming, especially for medical image data. Therefore, to avoid this tedious data collection and annotation procedure, some researchers take a compromise and utilize the ImageNet-pretrained convolutional neural network (CNN) to extract visual representations from medical images [9,30]. However, this compromise ignores not only the data distribution difference [21] between medical and natural images, but also the domain-specific information.

Considering the aforementioned dilemma, self-supervised learning is one of the feasible solutions, which attracts the attention of a growing number of researchers in recent years. Self-supervised learning aims to learn representations from large-scale unlabeled data by solving pretext tasks. In the past few years, research on self-supervised visual representation learning has made great progress. The existing self-supervised learning methods in computer vision field can be categorized into generative model-based approaches and discriminative model-based approaches in the light of the type of associated pretext tasks [16,22]. In earlier times, generative pretext tasks like image inpainting [24] and image colorization [31,32] are proposed to train an autoencoder for feature extraction; discriminative self-supervised pretext tasks such as rotation prediction [10], Jigsaw solving [23], and relative patch location prediction [8], are designed to learn high-level semantic features.

Recently, contrastive learning [13], which also belongs to discriminative approaches, achieves great success in self-supervised visual representation learning. The core idea of contrastive learning is to attract different augmented views of the same image (positive pairs) and repulse augmented views of different images (negative pairs). Based on this core idea, MoCo [14] and SimCLR [4] are proposed for self-supervised visual representation learning, which greatly shrink the gap between self-supervised learning and fully-supervised learning. The success of MoCo and SimCLR shows the superiority of contrastive learning. Furthermore, the following related work BYOL [12] and SimSiam [5] suggest that negative pairs are not necessary for contrastive learning, and they have become the new state-of-the-art self-supervised visual representation learning methods.

Above studies are about natural images. As for medical images, Chen et al. [3] developed a self-supervised learning model based on image context restoration and proved the effectiveness on several tasks. Specific to histopathological images, Gildenblat [11] and Abbet [1] utilized the unique spatial proximity information of whole slide images (WSIs) to establish self-supervised learning methods, relying on the plausible assumption that adjacent patches share similar content while distant patches are distinct. Xie [28] and Sahasrabudhe [25] also proposed self-supervised learning approaches based on magnification information, specially for histopathological image segmentation. However, using such

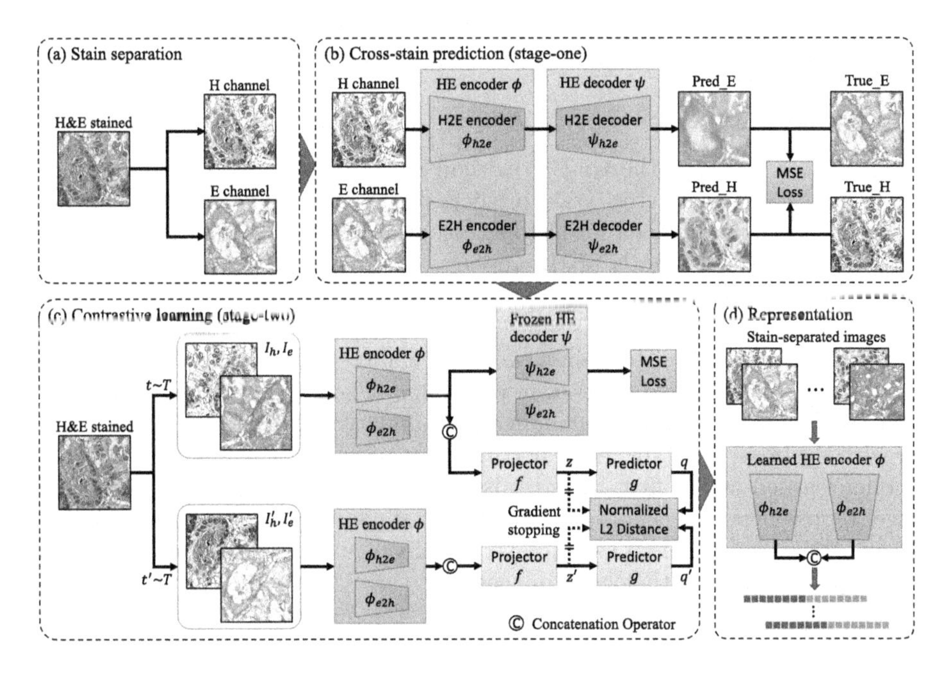

Fig. 1. The framework of the proposed CS-CO method.

side information limits the applicability of these methods. As far as we know, there is still a lack of universal and effective self-supervised learning methods for extracting visual representations from histopathological images.

In this paper, we present CS-CO, a novel hybrid self-supervised histopathological image visual representation learning method, which consists of **C**ross-**S**tain prediction (generative) and **CO**ntrastive learning (discriminative). The proposed method takes advantage of domain specific knowledge and does not require side information like image magnification and spatial proximity, resulting in better applicability. Our major contributions are summarized as follows.

- We design a new pretext task, i.e. cross-stain prediction, for self-supervised learning, aiming to make good use of the domain specific knowledge of histopathological images.
- We propose a new data augmentation approach, i.e. stain vector perturbation, to serve histopathological image contrastive learning.
- We integrate the advantages of generative and discriminative approaches and build a hybrid self-supervised visual representation learning framework for histopathological images.

2 Methodology

2.1 Overview of CS-CO

As illustrated in Fig. 1, the CS-CO method is composed of two self-supervised learning stages, namely cross-stain prediction and contrastive learning, both of which are specially designed for histopathological images. Before self-supervised learning, stain separation is firstly applied to original H&E-stained images to generate single-dye staining results, which are called H channel and E channel images respectively. These stain-separated images are used at the first self-supervised learning stage to train a two-branch autoencoder by solving the novel generative pretext task of cross-stain prediction. Then, at the second stage, the learned HE encoder is trained again in a discriminative contrastive learning manner with the proposed stain vector perturbation augmentation approach. The HE decoder learned at the first stage is also retained as a regulator at the second stage to prevent model collapse. After the two-stage self-supervised learning, the learned HE encoder can be used to extract effective visual representations from stain-separated histopathological images.

2.2 Stain Separation

In histopathology, different dyes are used to enhance different types of tissue components, which can be regarded as domain-specific knowledge implicit in histopathological images. For the commonly used H&E stain, cell nuclei will be stained blue-purple by hematoxylin, and extracellular matrix and cytoplasm will be stained pink by eosin [2]. The stain results of hematoxylin and eosin are denoted as H channel and E channel respectively. To restore single-dye staining results from H&E stain images and reduce the stain variance to some extent, we utilize the Vahadane method [27] for stain separation.

To be specific, for an H&E stained image, let $I \in \mathbb{R}^{m \times n}$ be the matrix of RGB intensity, $V \in \mathbb{R}^{m \times n}$ be the relative optical density, $W \in \mathbb{R}^{m \times r}$ be the stain color matrix, and $H \in \mathbb{R}^{r \times n}$ be the stain concentration matrix, where $m = 3$ for RGB images, r is the number of stains, and n is number of pixels. According to the Beer-Lambert law, the relation between V and H, W can be formulated as Eq. (1), where $I_0 = 255$ for 8-bit RGB images.

$$V = \log \frac{I_0}{I} = WH \tag{1}$$

Then, W and H can be estimated by solving the sparse non-negative matrix factorization problem as Eq. (2) proposed by [27].

$$\begin{aligned} \min_{W,H} \quad & \frac{1}{2}||V - WH||_F^2 + \lambda \sum_{j=1}^{r} ||H(j,:)||_1, \\ \text{s.t.} \quad & W, H \geq 0, \quad ||W(:,j)||_2^2 = 1 \end{aligned} \tag{2}$$

From the estimated stain concentration matrix H, the H channel and E channel images I_h and I_e can be restored as Eq. (3).

$$I_h = I_0 \exp(-H[0,:]), \quad I_e = I_0 \exp(-H[1,:]) \tag{3}$$

2.3 Cross-stain Prediction

At the first stage of the proposed self-supervised learning scheme, a deep neural network is trained to learn visual representations by solving the novel pretext task of cross-stain prediction. The deep model is composed of two independent autoencoders: one is for predicting E channel images from corresponding H channel images (H2E), and the other is the inverse (E2H). We denote the encoder and decoder of H2E branch as ϕ_{h2e} and ψ_{h2e}. The E2H branch is denotated similarly. For the sake of simplicity, we also denote the combination of ϕ_{h2e} and ϕ_{e2h} as HE encoder ϕ, and the combination of ψ_{h2e} and ψ_{e2h} as HE decoder ψ.

As shown in Fig. 1(b), restored H channel and E channel images are input into the two autoencoders separately, and the mean square error (MSE) losses are computed between the predicted and true images in both two branches.

$$I_{pred\_e} = \psi_{h2e}(\phi_{h2e}(I_h)), \quad I_{pred\_h} = \psi_{e2h}(\phi_{e2h}(I_e)) \tag{4}$$

$$\mathcal{L}_{cs} = MSELoss(I_{pred\_e}, I_e) + MSELoss(I_{pred\_h}, I_h) \tag{5}$$

By solving this proposed generative pretext task, the HE encoder can capture low-level general features from histopathological images. In addition, based on the characteristics of H&E stain mentioned in Sect. 2.2, the HE encoder is also expected to be sensitive to details which imply the correlation between nuclei and cytoplasm.

2.4 Contrastive Learning

Based on the two-branch autoencoder learned at the first stage, we adopt contrastive learning at the second stage to learn discriminative high-level features. Inspired by [5], we reorganize our model into the Siamese architecture, which is composed of the HE encoder ϕ, a projector f, and a predictor g. All parameters are shared across two branches. The HE decoder ψ is also kept in one branch as an untrainable regulator to prevent model collapse. The weights of ϕ and ψ learned at the first stage are loaded as initialization.

During contrastive learning, a pair of H channel and E channel images of the same H&E stained image is regarded as one data sample. As shown in Fig. 1(c), for each data sample, two transformations t and t' are sampled from the transformation family T for data augmentation. The transformation family includes stain vector perturbation, RandomFlip, RandomResizedCrop, and GaussianBlur. After transformation, the derived two randomly augmented views (I_h, I_e) and (I'_h, I'_e) are input into the Siamese network separately. For each augmented view, the contained H channel and E channel images are firstly encoded by ϕ_{h2e} and ϕ_{e2h} respectively. The outputs are pooled and concatenated together

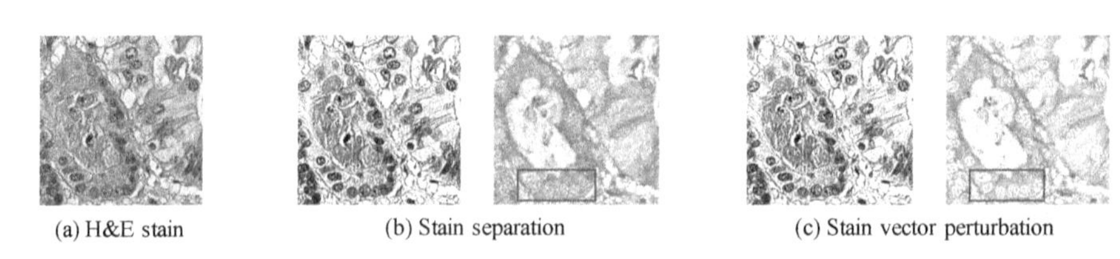

(a) H&E stain (b) Stain separation (c) Stain vector perturbation

Fig. 2. Stain vector perturbation. (a) The original H&E stain image. (b) Stain separation results using Vahadane [27] method. (c) Stain separation results with proposed stain vector perturbation ($\sigma = 0.05$). The red box outlines the visible differences. (Color figure online)

as one vector. Subsequently, the projector f and predictor g are applied to the vector sequentially.

For two augmented views, denoting the outputs of the projector f as $z \triangleq f(\phi(I_h, I_e))$ and $z' \triangleq f(\phi(I_h', I_e'))$ and the outputs of predictor g as $q \triangleq g(z)$ and $q' \triangleq g(z')$, we force q to be similar to z' and q' to be similar to z by minimizing the symmetrized loss:

$$\mathcal{L}_{co} = \frac{1}{2}||\tilde{q} - \tilde{z}'||_2^2 + \frac{1}{2}||\tilde{q}' - \tilde{z}||_2^2 \tag{6}$$

where $\tilde{x} \triangleq \frac{x}{||x||_2}$ and $||\cdot||_2$ is ℓ_2-norm. z and z' are detached from the computational graph before calculating the loss.

As for the frozen pretrained HE decoder ψ, it constrains the generalization of features extracted by the HE encoder ϕ by minimizing Eq. (5), so as to ensure no collapse occurs on the HE encoder ϕ. The total loss is formulated as Eq. (7), where α is the weight coefficient (in our implementation, $\alpha = 1.0$).

$$\mathcal{L}_{tot} = \mathcal{L}_{co} + \alpha \mathcal{L}_{cs} \tag{7}$$

Stain Vector Perturbation. Since the input images are gray, many transformations of colorful image cannot be used for contrastive learning. To guarantee the strength of transformation, we customize a new data augmentation approach called stain vector perturbation for histopathological images. Inspired by the error of stain vector estimation in stain separation, we disturb elements of the estimated W with $\epsilon \sim N(0, \sigma^2)$ to obtain the perturbed stain vector matrix W'. With W', another stain concentration matrix H' can be derived, and the corresponding H channel and E channel images can be restored from H'. The results of stain vector perturbation are shown in Fig. 2.

2.5 Representation Extraction

After two-stage self-supervised learning, the learned HE encoder ϕ can be used for visual representation extraction. As shown in Fig. 1(d), for an H&E stained image, the corresponding H and E channel images are firstly restored via stain separation and then input into the learned HE encoder ϕ. The outputs are pooled and concatenated together as the extracted visual representation.

3 Experimental Results

Dataset. We evaluate our proposed CS-CO method on the public dataset NCT-CRC-HE-100K [17]. The dataset contains nine classes of histopathological images of human colorectal cancer and healthy tissue. The predefined training set contains 100,000 images and the test set contains 7180 images. The overall nine-class classification accuracy on test set is 0.943 as reported in [19], which is achieved by fully-supervised learning with VGG19. It is worth noting that we exclude images belonging to the background (BACK) class for training and testing when we evaluate visual representation learning methods. The reason is that background is always non-informative and can be easily distinguished by simple threshold-based methods. The final sizes of training and test set are 89,434 and 6333 respectively, and the eight-class classification accuracy on the test set is reported as the evaluation metric.

Implementation Details. For the proposed CS-CO method, we use ResNet18 [15] as the backbone of the encoders ϕ_{h2e} and ϕ_{e2h}. The decoders ψ_{h2e} and ψ_{e2h} are composed of a set of upsampling layers and convolutional layers. The projector f and predictor g are both instantiated by the multi-layer perceptron (MLP), which consists of a linear layer with output size 4096 followed by batch normalization, rectified linear units (ReLU), and a final linear layer with output dimension 256. At the first training stage, we use SGD to train our model on the whole training set. The batch size is 64 and the learning rate is 0.01. At the second stage, for fast implementation, we train the model with Adam on 10,000 randomly selected training data. The batch size is 96, the learning rate is 0.001, and the weight decay is 1×10^{-6}. Early stopping is used at both stages for avoiding over-fitting.

According to the common evaluation protocol [20], a linear classifier is trained for evaluating the capacity of representation. The linear classifier is implemented by a single linear layer and trained with SGD. The batch size is 32 and the learning rate is 0.001. Early stopping is also used for avoiding over-fitting.

Method Comparison. We firstly train a ResNet18 model with the whole eight-class training data to establish the fully-supervised baseline. Then, we choose three types of methods to compared with our proposed CS-CO method. The first type contains two fixed ResNet18 models, one is random initialized, and the other is pretrained on ImageNet [6]. The second type contains two state-of-the-art contrastive learning methods: BYOL [12] and Simsiam [5]. The last type also contains two methods specifically proposed for medical images by Chen et al. [3] and Xie et al. [28] respectively. Except that the two ResNet18 models of

Table 1. Linear evaluation results (5-fold cross-validation) with different size (n) of training data.

Fully-supervised ResNet18	0.903 ± 0.015		
Methods	$\mathbf{n = 100}$	$\mathbf{n = 1000}$	$\mathbf{n = 10000}$
Random initialized ResNet18	0.134 ± 0.050	0.181 ± 0.070	0.427 ± 0.003
ImageNet pretrained ResNet18	0.628 ± 0.040	0.802 ± 0.012	0.844 ± 0.002
BYOL [12]	0.811 ± 0.011	0.898 ± 0.007	0.891 ± 0.005
SimSiam [5]	0.797 ± 0.029	0.897 ± 0.004	0.890 ± 0.005
Chen's method [3]	0.215 ± 0.067	0.661 ± 0.014	0.711 ± 0.003
Xie's method [28]	0.109 ± 0.042	0.507 ± 0.007	0.586 ± 0.009
CS-CO	**0.834 ± 0.018**	**0.915 ± 0.004**	**0.914 ± 0.002**

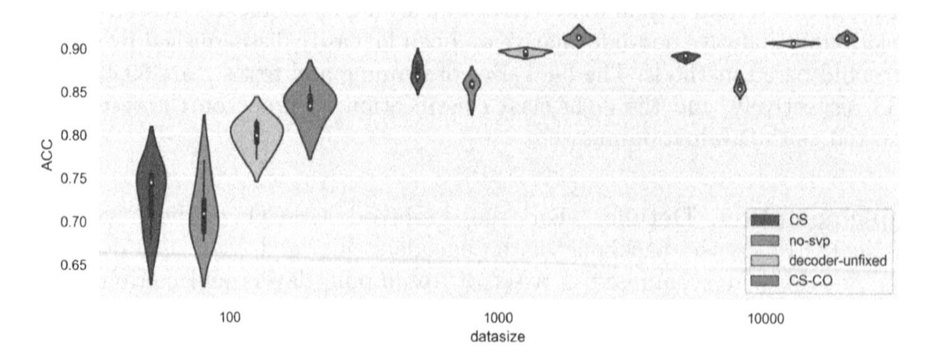

Fig. 3. Ablation study results (5-fold cross-validation).

the first type don't need to be trained, both our CS-CO method and the latter two types of self-supervised learning methods are firstly trained using the whole training data as unlabeled data.

Rather than using the whole training set, we randomly sample 100, 1000, and 10000 data from the training set and extract their visual representations with each method for the following linear classifier training. In this way, the impact of large data size can be stripped, and the classification accuracies on the test set can more purely reflect the representation capacity of each method. As shown in Table 1, our proposed CS-CO method demonstrates superior representation capacity compared to other methods. Furthermore, with only 1,000 labeled data and the linear classifier, our CS-CO method even outperforms the fully-supervised ResNet18 which is trained on the whole training set.

Ablation Study. We conduct ablation studies to explore the role of the following three key components of the proposed CS-CO method. 1) Contrastive learning: To verify whether the contrastive learning enhances the visual representation capacity, we do linear evaluation on the CS model, which is only trained by solving the cross-stain prediction task. In the cases of different amount of training data, the average test accuracies of CS model are 0.782, 0.873, and 0.892, which shows obvious gaps from the original CS-CO model. 2) Stain-vector perturbation: To demonstrate the effectiveness of stain-vector perturbation, we remove it from the transformation family of contrastive learning, and train another CS-CO model which is denoted as *no-SVP*. As shown in Fig. 3, the performance of *no-SVP* model is even worse than CS model, which suggests that stain-vector perturbation is crucial for contrastive learning. 3) Frozen-decoder: We also make the HE decoder trainable at the second training stage to train the CS-CO model, which is denoted as *decoder-unfixed*. As Fig. 3 shows, the *decoder-unfixed* model doesn't collapse but performs slightly worse than the original CS-CO model.

4 Conclusion

In this paper, we have proposed a novel hybrid self-supervised visual representation learning method specifically for histopathological images. Our method draws advantages from both generative and discriminative models by solving the proposed cross-stain prediction pretext task and doing contrastive learning with the proposed stain-vector perturbation augmentation approach. The proposed method makes good use of domain-specific knowledge and has good versatility. Linear evaluation results on dataset NCT-CRC-HE-100K suggest that our method outperforms current state-of-the-art self-supervised visual representation learning approaches. In future work, we intend to use the representations extracted by the proposed CS-CO method to solve downstream tasks such as disease prognosis and molecular signature prediction, so as to further prove the effectiveness of the proposed method in practice.

Acknowledgements. This work was partially supported by the National Key Research and Development Program of China (No. 2018YFC0910404), the National Natural Science Foundation of China (Nos. 61873141, 61721003), the Shanghai Municipal Science and Technology Major Project (No. 2017SHZDZX01), the Tsinghua-Fuzhou Institute for Data Technology, the Taishan Scholars Program of Shandong Province (No. 2019010668), and the Shandong Higher Education Young Science and Technology Support Program (No. 2020KJL005).

References

1. Abbet, C., Zlobec, I., Bozorgtabar, B., Thiran, J.-P.: Divide-and-rule: self-supervised learning for survival analysis in colorectal cancer. In: Martel, A.L., et al. (eds.) MICCAI 2020. LNCS, vol. 12265, pp. 480–489. Springer, Cham (2020). https://doi.org/10.1007/978-3-030-59722-1_46
2. Chan, J.K.: The wonderful colors of the hematoxylin-eosin stain in diagnostic surgical pathology. Int. J. Surg. Pathol. **22**(1), 12–32 (2014)
3. Chen, L., Bentley, P., Mori, K., Misawa, K., Fujiwara, M., Rueckert, D.: Self-supervised learning for medical image analysis using image context restoration. Med. Image Anal. **58**, 101539 (2019)

4. Chen, T., Kornblith, S., Norouzi, M., Hinton, G.: A simple framework for contrastive learning of visual representations. In: International Conference on Machine Learning, pp. 1597–1607. PMLR (2020)
5. Chen, X., He, K.: Exploring simple siamese representation learning. In: Proceedings of the IEEE/CVF Conference on Computer Vision and Pattern Recognition, pp. 15750–15758 (2021)
6. Deng, J., Dong, W., Socher, R., Li, L.J., Li, K., Fei-Fei, L.: ImageNet: a large-scale hierarchical image database. In: 2009 IEEE Conference on Computer Vision and Pattern Recognition, pp. 248–255. IEEE (2009)
7. Ding, K., Liu, Q., Lee, E., Zhou, M., Lu, A., Zhang, S.: Feature-enhanced graph networks for genetic mutational prediction using histopathological images in colon cancer. In: Martel, A.L., et al. (eds.) MICCAI 2020. LNCS, vol. 12262, pp. 294–304. Springer, Cham (2020). https://doi.org/10.1007/978-3-030-59713-9_29
8. Doersch, C., Gupta, A., Efros, A.A.: Unsupervised visual representation learning by context prediction. In: Proceedings of the IEEE International Conference on Computer Vision, pp. 1422–1430 (2015)
9. Fu, Y., et al.: Pan-cancer computational histopathology reveals mutations, tumor composition and prognosis. Nat. Cancer **1**(8), 800–810 (2020)
10. Gidaris, S., Singh, P., Komodakis, N.: Unsupervised representation learning by predicting image rotations. In: International Conference on Learning Representations (2018)
11. Gildenblat, J., Klaiman, E.: Self-supervised similarity learning for digital pathology. arXiv preprint arXiv:1905.08139 (2019)
12. Grill, J.B., et al.: Bootstrap your own latent: a new approach to self-supervised learning. arXiv preprint arXiv:2006.07733 (2020)
13. Hadsell, R., Chopra, S., LeCun, Y.: Dimensionality reduction by learning an invariant mapping. In: 2006 IEEE Computer Society Conference on Computer Vision and Pattern Recognition (CVPR 2006), vol. 2, pp. 1735–1742. IEEE (2006)
14. He, K., Fan, H., Wu, Y., Xie, S., Girshick, R.: Momentum contrast for unsupervised visual representation learning. In: Proceedings of the IEEE/CVF Conference on Computer Vision and Pattern Recognition, pp. 9729–9738 (2020)
15. He, K., Zhang, X., Ren, S., Sun, J.: Deep residual learning for image recognition. In: Proceedings of the IEEE Conference on Computer Vision and Pattern Recognition, pp. 770–778 (2016)
16. Jing, L., Tian, Y.: Self-supervised visual feature learning with deep neural networks: a survey. IEEE Trans. Pattern Anal. Mach. Intell. (2020)
17. Kather, J.N., Halama, N., Marx, A.: 100,000 histological images of human colorectal cancer and healthy tissue. https://doi.org/10.5281/zenodo.1214456
18. Kather, J.N., et al.: Pan-cancer image-based detection of clinically actionable genetic alterations. Nat. Cancer **1**(8), 789–799 (2020)
19. Kather, J.N., et al.: Predicting survival from colorectal cancer histology slides using deep learning: a retrospective multicenter study. PLoS Med. **16**(1), e1002730 (2019)
20. Kolesnikov, A., Zhai, X., Beyer, L.: Revisiting self-supervised visual representation learning. In: Proceedings of the IEEE/CVF Conference on Computer Vision and Pattern Recognition, pp. 1920–1929 (2019)
21. Liu, Q., Xu, J., Jiang, R., Wong, W.H.: Density estimation using deep generative neural networks. Proc. Natl. Acad. Sci. **118**(15), e2101344118 (2021)
22. Liu, X., et al.: Self-supervised learning: generative or contrastive **1**(2). arXiv preprint arXiv:2006.08218 (2020)

23. Noroozi, M., Favaro, P.: Unsupervised learning of visual representations by solving jigsaw puzzles. In: Leibe, B., Matas, J., Sebe, N., Welling, M. (eds.) ECCV 2016. LNCS, vol. 9910, pp. 69–84. Springer, Cham (2016). https://doi.org/10.1007/978-3-319-46466-4_5
24. Pathak, D., Krahenbuhl, P., Donahue, J., Darrell, T., Efros, A.A.: Context encoders: feature learning by inpainting. In: Proceedings of the IEEE Conference on Computer Vision and Pattern Recognition, pp. 2536–2544 (2016)
25. Sahasrabudhe, M., et al.: Self-supervised nuclei segmentation in histopathological images using attention. In: Martel, A.L., et al. (eds.) MICCAI 2020. LNCS, vol. 12265, pp. 393–402. Springer, Cham (2020). https://doi.org/10.1007/978-3-030-59722-1_38
26. Shi, X., Sapkota, M., Xing, F., Liu, F., Cui, L., Yang, L.: Pairwise based deep ranking hashing for histopathology image classification and retrieval. Pattern Recogn. **81**, 14–22 (2018)
27. Vahadane, A., et al.: Structure-preserving color normalization and sparse stain separation for histological images. IEEE Trans. Med. Imaging **35**(8), 1962–1971 (2016)
28. Xie, X., Chen, J., Li, Y., Shen, L., Ma, K., Zheng, Y.: Instance-aware self-supervised learning for nuclei segmentation. In: Martel, A.L., et al. (eds.) MICCAI 2020. LNCS, vol. 12265, pp. 341–350. Springer, Cham (2020). https://doi.org/10.1007/978-3-030-59722-1_33
29. Yang, P., et al.: A deep metric learning approach for histopathological image retrieval. Methods **179**, 14–25 (2020)
30. Yao, J., Zhu, X., Jonnagaddala, J., Hawkins, N., Huang, J.: Whole slide images based cancer survival prediction using attention guided deep multiple instance learning networks. Med. Image Anal. **65**, 101789 (2020)
31. Zhang, R., Isola, P., Efros, A.A.: Colorful image colorization. In: Leibe, B., Matas, J., Sebe, N., Welling, M. (eds.) ECCV 2016. LNCS, vol. 9907, pp. 649–666. Springer, Cham (2016). https://doi.org/10.1007/978-3-319-46487-9_40
32. Zhang, R., Isola, P., Efros, A.A.: Split-brain autoencoders: unsupervised learning by cross-channel prediction. In: Proceedings of the IEEE Conference on Computer Vision and Pattern Recognition, pp. 1058–1067 (2017)

Contrastive Learning with Continuous Proxy Meta-data for 3D MRI Classification

Benoit Dufumier[1,2(✉)], Pietro Gori[2], Julie Victor[1], Antoine Grigis[1], Michele Wessa[3], Paolo Brambilla[4], Pauline Favre[1], Mircea Polosan[5], Colm McDonald[6], Camille Marie Piguet[7], Mary Phillips[8], Lisa Eyler[9], Edouard Duchesnay[1], and the Alzheimer's Disease Neuroimaging Initiative

[1] NeuroSpin, CEA Saclay, Université Paris-Saclay, Gif-sur-Yvette, France
benoit.dufumier@cea.fr
[2] LTCI, Télécom Paris, IPParis, Paris, France
[3] Department of Neuropsychology, Johannes-Gutenberg University of Mainz, Mainz, Germany
[4] Department of Neurosciences, Fondazione IRCCS, University of Milan, Milan, Italy
[5] Université Grenoble Alpes, Inserm U1216, CHU Grenoble Alpe, Grenoble, France
[6] Centre for Neuroimaging and Cognitive Genomics (NICOG), Galway, Ireland
[7] Department of Neuroscience, University of Geneva, Geneva, Switzerland
[8] Department of Psychiatry, Western Psychiatric Institute, University of Pittsburgh, Pittsburgh, USA
[9] Department of Psychiatry, UC San Diego, San Diego, CA, USA

Abstract. Traditional supervised learning with deep neural networks requires a tremendous amount of labelled data to converge to a good solution. For 3D medical images, it is often impractical to build a large homogeneous annotated dataset for a specific pathology. Self-supervised methods offer a new way to learn a representation of the images in an unsupervised manner with a neural network. In particular, contrastive learning has shown great promises by (almost) matching the performance of fully-supervised CNN on vision tasks. Nonetheless, this method does not take advantage of available meta-data, such as participant's age, viewed as prior knowledge. Here, we propose to leverage continuous *proxy* metadata, in the contrastive learning framework, by introducing a new loss called y-Aware InfoNCE loss. Specifically, we improve the positive sampling during pre-training by adding more positive examples with similar *proxy* meta-data with the anchor, assuming they share similar discriminative semantic features. With our method, a 3D CNN model pre-trained on 10^4 multi-site healthy brain MRI scans can extract relevant features for three classification tasks: schizophrenia, bipolar diagnosis and Alzheimer's detection. When fine-tuned, it also outperforms 3D CNN trained from scratch on these tasks, as well as state-of-the-art self-supervised methods. Our code is made publicly available here.

Electronic supplementary material The online version of this chapter (https://doi.org/10.1007/978-3-030-87196-3_6) contains supplementary material, which is available to authorized users.

M. de Bruijne et al. (Eds.): MICCAI 2021, LNCS 12902, pp. 58–68, 2021.
https://doi.org/10.1007/978-3-030-87196-3_6

1 Introduction

Recently, self-supervised representation learning methods have shown great promises, surpassing traditional transfer learning from ImageNet to 3D medical images [29]. These models can be trained without costly annotations and they offer a great initialization point for a wide set of downstream tasks, avoiding the domain gap between natural and medical images. They mainly rely on a pretext task that is informative about the prior we have on the data. This proxy task essentially consists in corrupting the data with non-linear transformations that preserve the semantic information about the images and learn the reverse mapping with a Convolutional Neural Network (CNN). Numerous tasks have been proposed both in the computer vision field (inpainting [21], localization of a patch [9], prediction of the angle of rotation [10], jigsaw [18], etc.) and also specifically designed for 3D medical images (context restoration [4], solving the rubik's cube [30], sub-volumes deformation [29]). They have been successfully applied to 3D MR images for both segmentation and classification [24,26,29,30], outperforming the classical 2D approach with ImageNet pre-training. Concurrently, there has been a tremendous interest in contrastive learning [13] over the last year. Notably, this unsupervised approach almost matches the performance over fully-supervised vision tasks and it outperforms supervised pre-training [1,5,14]. A single encoder is trained to map semantically similar "positive" samples close together in the latent space while pushing away dissimilar "negative" examples. In practice, all samples in a batch are transformed twice through random transformations $t \sim \mathcal{T}$ from a set of parametric transformations $\mathcal{T}$. For a given reference point (anchor) x, the positive samples are the ones derived from x while the other samples are considered as negatives. Most of the recent works focus in finding the best transformations $\mathcal{T}$ that degrade the initial image x while preserving the semantic information [5,27] and very recent studies intend to improve the negative sampling [6,22]. However, two different samples are not necessarily semantically different, as emphasized in [6,28], and they may even belong to the same semantic class. Additionally, two samples are not always equally semantically different from a given anchor and so they should not be equally distant in the latent space from this anchor. In this work, we assume to have access to continuous *proxy* meta-data containing relevant information about the images at hand (*e.g.* the participant's age). We want to leverage these *meta-data* during the contrastive learning process in order to build a more universal representation of our data. To do so, we propose a new y-Aware InfoNCE loss inspired from the Noise Contrastive Estimation loss [12] that aims at improving the positive sampling according to the similarity between two *proxy* meta-data. Differently from [17], i) we perform contrastive learning with continuous meta-data (not only categorical) and ii) our first purpose is to train a generic encoder that can be easily transferred to various 3D MRI target datasets for classification or regression problems in the very small data regime ($N \leq 10^3$). It is also one of the first studies to apply contrastive learning to 3D anatomical brain images [2]. Our main contributions are:

- we propose a novel formulation for contrastive learning that leverages *continuous* meta-data and derive a new loss, namely the y-Aware InfoNCE loss
- we empirically show that our unsupervised model pre-trained on a large-scale multi-site 3D brain MRI dataset comprising $N = 10^4$ healthy scans reaches or outperforms the performance of CNN model fully-supervised on 3 classification tasks under the linear protocol evaluation
- we also demonstrate that our approach gives better results when fine-tuning on 3 target tasks than training from scratch
- we finally performed an ablation study showing that leveraging the meta-data improves the performance for all the downstream tasks and different set of transformations $\mathcal{T}$ compared to SimCLR [5].

2 Method

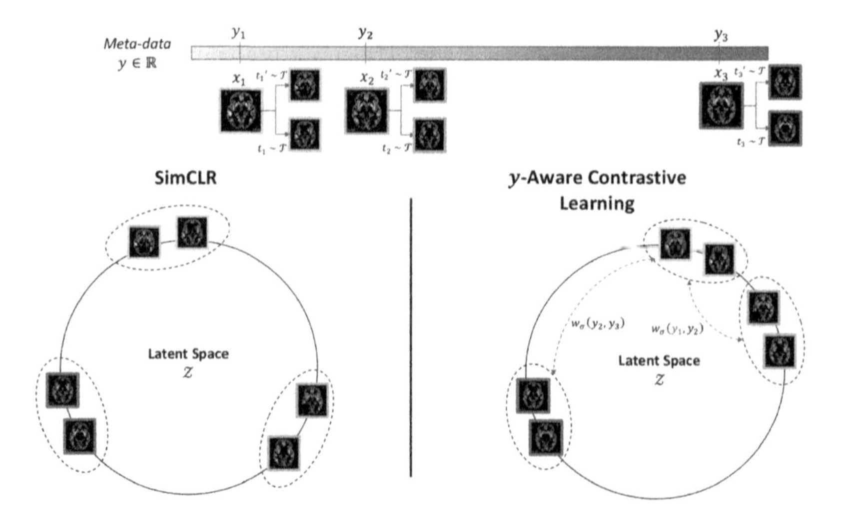

Fig. 1. Differently from SimCLR [5], our new loss can handle meta-data $y \in \mathbb{R}$ by redefining the notion of similarity between two images in the latent space $\mathcal{Z}$. For an image x_i, transformed twice through two augmentations $t_1, t_1' \sim \mathcal{T}$, the resulting views $(t_1(x_i), t_2(x_i))$ are expected to be close in the latent space through the learnt mapping f_θ, as in SimCLR. However, we also expect a different input $x_{k \neq i}$ to be close to x_i in $\mathcal{Z}$ if the two *proxy* meta-data y_i and y_k are similar. We define a similarity function $w_\sigma(y_i, y_k)$ that quantifies this notion of similarity.

Problem Formalization. In contrastive learning [5,17,27], one wants to learn a parametric function $f_\theta : \mathcal{X} \mapsto \mathbb{S}^{d-1} = \mathcal{Z}$ between the input image space $\mathcal{X}$ and the unit hypersphere, without meta-data. The goal of f_θ is to map samples to a representation space where semantically similar samples are "closer" than semantically different samples. To do that, each training sample $x_i \in \mathcal{X}$ is transformed twice through $t_1^i, t_2^i \sim \mathcal{T}$ to produce two augmented views of the same image $(v_1^i, v_2^i) := (t_1^i(x_i), t_2^i(x_i))$, where $\mathcal{T}$ is a set of predefined transformations. Then, for each sample i, the model f_θ is trained to discriminate between the "positive" pair (v_1^i, v_2^i), assumed to be drawn from the empirical joint distribution $p(v_1, v_2)$, and all the other "negative" pairs $(v_1^i, v_2^j)_{j \neq i}$, assumed to be drawn from the marginal distributions $p(v_1)p(v_2)$. With these assumptions, f_θ is an estimator of the mutual information I between v_1 and v_2 and it usually estimated by maximizing a lower bound of $I(v_1, v_2)$ called the InfoNCE loss [19] (Fig. 1):

$$\mathcal{L}_{NCE} = -\log \frac{e^{f_\theta(v_1^i, v_2^i)}}{\frac{1}{n}\sum_{j=1}^{n} e^{f_\theta(v_1^i, v_2^j)}} \tag{1}$$

where n is the batch size, $f_\theta(v_1, v_2) := \frac{1}{\tau} f_\theta(v_1)^T f_\theta(v_2)$ and $\tau > 0$ is a hyperparameter. f_θ is usually defined as the composition of an encoder network $e_{\theta_1}(x)$ and a projection head z_{θ_2} (e.g. multi-layer perceptron) which is discarded after training (here $\theta = \{\theta_1, \theta_2\}$). Outputs lie on the unit hypersphere so that inner products can be used to measure cosine similarities in the representation space. In Eq. 1, every sample $v_2^j|_{j \neq i}$ is considered *equally* different from the anchor v_1^i. However, this is hardly true with medical images since we know, for instance, that two young healthy subjects should be considered more similar than a young and an old healthy subject. If we suppose to have access to continuous proxy metadata $y_i \in \mathbb{R}$ (*e.g.* participant's age or clinical score), then two views v_1^i, v_2^j with similar metadata y_i, y_j should be also close in the representation space $\mathbb{S}^{d-1}$. Inspired by vicinal risk minimization (VRM) [3], we propose to re-define $p(v_1, v_2)$ by integrating the proxy metadata y, modeled as a random variable, such that a small change in y results in a negligible change in $p(v_1, v_2|y)$. Similarly to [8], we define the empirical joint distribution as:

$$p_{emp}^{vic}(v_1, v_2|y) = \frac{1}{n}\sum_{i=1}^{n}\sum_{j=1}^{n} \frac{w_\sigma(y_i, y_j)}{\sum_{k=1}^{n} w_\sigma(y_i, y_k)} \delta(v_1 - v_1^i)\delta(v_2 - v_2^j) \tag{2}$$

where $\sigma > 0$ is the hyperparameter of the Radius Basis Function (RBF) kernel w_σ. Based on Eq. 2, we can introduce our new y-Aware InfoNCE loss:

$$\mathcal{L}_{NCE}^{y} = -\sum_{k=1}^{n} \frac{w_\sigma(y_k, y_i)}{\sum_{j=1}^{n} w_\sigma(y_j, y_i)} \log \frac{e^{f_\theta(v_1^i, v_2^k)}}{\frac{1}{n}\sum_{j=1}^{n} e^{f_\theta(v_1^i, v_2^j)}} \tag{3}$$

In the limit case when $\sigma \to 0$, then we retrieve exactly the original InfoNCE loss, assuming that $y_i = y_j \Leftrightarrow x_i = x_j, \forall i, j \in [1..n]$. When $\sigma \to +\infty$, we assume that all samples $(x_i)_{i=1}^{n}$ belong to the same latent class.

Discrete Case. If the proxy meta-data $(y_i)_{i \in [1..N]}$ are discrete, then we can simplify the above expression by imposing $w_\sigma(y_i, y_k) = \delta(y_i - y_k)$ retrieving the Supervised Contrastive Loss [17]. We may see $\mathcal{L}_{NCE}^{y}$ as an extension of [17] in the continuous case. However, our purpose here is to build a robust encoder that can leverage meta-data to learn a more generalizable representation of the data.

Generalization. The proposed loss could be easily adapted to multiple metadata, both continuous and categorical, by defining one kernel per metadata. Other choices of kernel, instead of the RBF, could also be considered.

Choice of the Transformations $\mathcal{T}$. In our formulation, we did not specify particular transformations $\mathcal{T}$ to generate (v_1^i, v_2^i). While there have been recent works [4,27] proposing transformations on natural images (color distorsion, cropping, cutout [7], etc.), there is currently no consensus for medical images in the context of contrastive learning. Here, we design three sets of transformations that preserve the semantic information in MR images: cutout, random cropping and a combination of the two with also gaussian noise, gaussian blur and flip. Importantly, while color distortion is crucial on natural images [5] to avoid the model using a shortcut during training based on the color histogram, it is not necessarily the case for MR images (see Supp. 3).

3 Experiments

Datasets

- **Big Healthy Brains (BHB) dataset.** We aggregated 13 publicly available datasets[1] of 3D T1 MRI scans of healthy controls (HC) acquired on more than 70 different scanners and comprising $N = 10^4$ samples. We use this dataset only to pre-train our model with the **participant's age as the *proxy* metadata**. The learnt representation is then tested on the following four data-sets using as final task a binary classification between HC and patients.
- **SCHIZCONNECT-VIP**[2]**.** It comprises $N = 605$ multi-site MRI scans including 275 patients with strict schizophrenia (SCZ) and 330 HC.
- **BIOBD** [15,23]**.** This dataset includes $N = 662$ MRI scans acquired on 8 different sites with 356 HC and 306 patients with bipolar disorder (BD).
- **BSNIP** [25]**.** It includes $N = 511$ MRI scans with $N = 200$ HC, $N = 194$ SCZ and $N = 117$ BD. This independent dataset is used only at test time in Fig. 2b).
- **Alzheimer's Disease Neuroimaging Initiative (ADNI-GO)**[3]**.** We use $N = 387$ co-registered T1-weighted MRI images divided in $N = 199$ healthy controls and $N = 188$ Alzheimer's patients (AD). We only included one scan per patient at the first session (baseline).

  All data-sets have been pre-processed in the same way with a non-linear registration to the MNI template and a gray matter extraction step. The final spatial resolution is 1.5 mm isotropic and the images are of size $121 \times 145 \times 121$.

Implementation Details. We implement our new loss based on the original InfoNCE loss [5] with Pytorch [20] and we use the Adam optimizer during training. As opposed to SimCLR [5] and in line with [2], we only use a batch size

[1] Demographic information as well as the public repositories can be found in Supp. 1.
[2] http://schizconnect.org.
[3] http://adni.loni.usc.edu/about/adni-go.

of $b = 64$ as it did not significantly change our results (see Supp. 2). We also follow [5] by fixing $\tau = 0.1$ in Eq. 1 and Eq. 3 and we set the learning rate to $\alpha = 10^{-4}$, decreasing it by 0.9 every 10 epochs. The model e_{θ_1} is based on a 3D adaptation of DenseNet121[4] [16] and z_{θ_2} is a vanilla multilayer perceptron as in [5].

Evaluation of the Representation. In Fig. 2, we compare the representation learnt using our model f_θ with the ones estimated using i) the InfoNCE loss (SimCLR) [5], ii) Model Genesis [29], a SOTA model for self-supervised learning with medical images, iii) a standard pre-training on age using a supervised approach (i.e. l_1 loss for age prediction), iv) BYOL [11] and MoCo [14] (memory bank $K = 1024$), 2 recently proposed SOTA models for representation learning, v) a multi-task approach SimCLR with age regression in the latent space (SimCLR+Age) and a fully fine-tuned supervised DenseNet trained to predict the final task. This can be considered as an upper bound, if the training dataset is sufficiently big. For the pre-training of our algorithm f_θ, we only use the BHB dataset with the participant's age as *proxy* meta-data. For both contrastive learning methods and BYOL, we fix $\sigma = 5$ in Eq. 3 and Eq. 1 and only use random cutout for the transformations $\mathcal{T}$ with a black patch covering $p = 25\%$ of the input image. We use UNet for pre-training with Model Genesis and DenseNet121 for all other models.

In order to evaluate the quality of the learnt representations, we only added a linear layer on top of the frozen encoders pre-trained on BHB. We tune this linear layer on 3 different binary classification tasks (see Datasets section) with 5-fold cross-validation (CV). We tested two different situations: data for training/validation and test come either from the same sites (first row) or from different sites (second row). We also vary the size (i.e. number of subjects, N_{target}) of the training/validation set. For (a), we perform a stratified nested CV (two 5-fold CV, the inner one for choosing the best hyper-parameters and the outer one for estimating the test error). For (b), we use a 5-fold CV for estimating the best hyper-parameters and keep an independent constant test set for all N_{target} (see Supp. 4).

From Fig. 2, we notice that our method consistently outperforms the other pre-trainings even in the very small data regime ($N = 100$) and it matches the performance of the fully-supervised setting on 2 data-sets. Differently from age supervision, f_θ is less specialized on a particular proxy task and it can be directly transferred on the final task at hand without fine-tuning the whole network. Furthermore, compared to the multi-task approach SimCLR+Age, the features extracted by our method are less sensitive to the site where the MR images are coming from. This shows that our technique is the only one that efficiently uses the highly multi-centric dataset BHB by making the features learnt during pre-training less correlated to the acquisition sites.

Importance of σ and $\mathcal{T}$ in the Positive Sampling. In Fig. 3, we study the impact of σ in Eq. 3 on the final representation learnt for a given set of transfor-

[4] Detailed implementation in our repository.

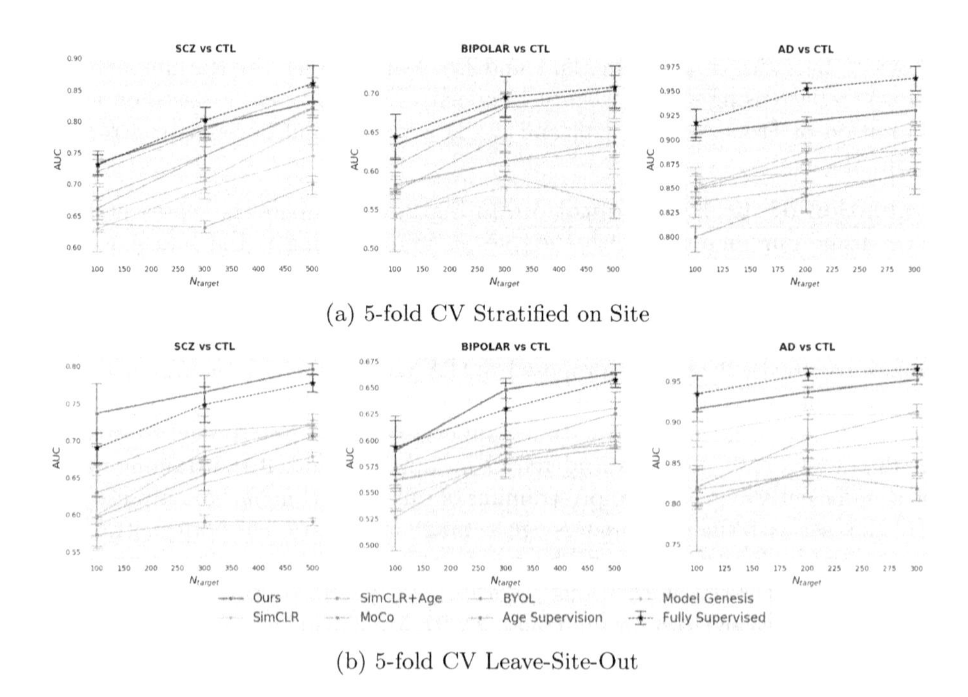

(a) 5-fold CV Stratified on Site

(b) 5-fold CV Leave-Site-Out

Fig. 2. Comparison of different representations in terms of classification accuracy (downstream task) on three different data-sets (one per column). Classification is performed using a linear layer on top of the pre-trained frozen encoders. (a) Data for training/validation and test come from the same acquisition sites (b) Data for training/validation and test come from different sites.

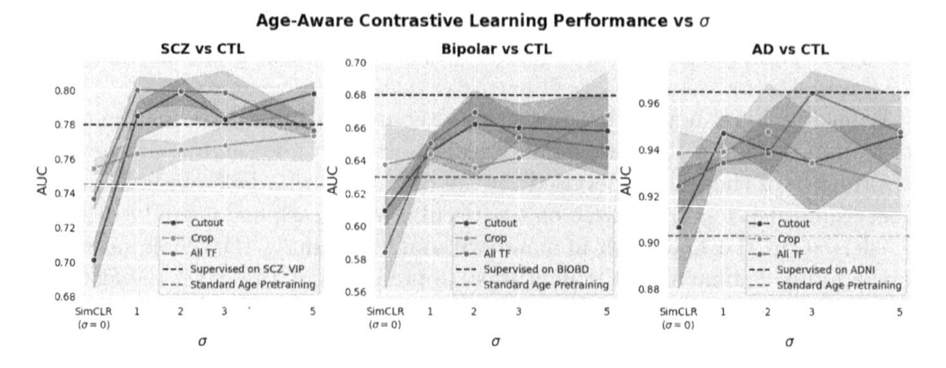

Fig. 3. Linear classification performance on three binary classification tasks with $N_{pretrained} = 10^4$. *All TF* includes crop, cutout, gaussian noise, gaussian blur and flip. The encoder is frozen and we only tune a linear layer on top of it. $\sigma = 0$ corresponds to SimCLR [5] with InfoNCE loss. As we increase σ, we add more positive examples for a given anchor x_i with close proxy meta-data.

mations $\mathcal{T}$. As highlighted in [5], hard transformations seem to be important for contrastive learning (at least on natural images), therefore we have evaluated three different sets of transformations $\mathcal{T}_1 = \{\text{Random Crop}\}$, $\mathcal{T}_2 = \{\text{Random Cutout}\}$ and $\mathcal{T}_3 = \{\text{Cutout, Crop, Gaussian Noise, Gaussian Blur, Flip}\}$. Importantly, we did not include color distorsion in $\mathcal{T}_3$ since i) it is not adapted to MRI images where a voxel's intensity encodes a gray matter density and ii) we did not observe significant difference between the color histograms of different scans as opposed to [5] (see Supp. 3). As before, we evaluated our representation under the linear evaluation protocol. We can observe that $\mathcal{T}_1$ and $\mathcal{T}_3$ give similar performances with $\sigma > 0$, always outperforming both SimCLR ($\sigma = 0$) and age supervision on BHB. It also even outperforms the fully-supervised baseline on SCZ vs HC. We also find that a strong cropping or cutout strategy is detrimental for the final performances (see Supp. 2). Since $\mathcal{T}_1$ is computationally less expensive than $\mathcal{T}_3$, we chose to use $\mathcal{T} = \mathcal{T}_1$ and $\sigma = 5$ in our experiments.

Fine-Tuning Results. Finally, we fine-tuned the whole encoder f_θ with different initializations on the 3 downstream tasks (see Table 1). To be comparable with Model Genesis [29], we also used the same UNet backbone for f_θ and we still fixed $\mathcal{T}_1 = \{\text{Random Cutout}\}$ and $\sigma = 5$. First, our approach outperforms the CNNs trained from scratch on all tasks as well as Model Genesis, even with the same backbone. Second, when using DenseNet, our pre-training remains better than using age supervision as pre-training for SCZ vs HC (even with the same transformations) and it is competitive on BD vs HC and AD vs HC.

Table 1. Fine-tuning results using 100 or 500 (300 for AD vs HC) training subjects. For each task, we report the AUC (%) of the fine-tuned models initialized with different approaches with 5-fold cross-validation. For age prediction, we employ the same transformations as in contrastive learning for the Data Augmentation (D.A) strategy. Best results are in **bold** and second bests are underlined.

Backbone	Pre-training	SCZ vs HC		BD vs HC		AD vs HC	
		N_{train} = 100	N_{train} = 500	N_{train} = 100	N_{train} = 500	N_{train} = 100	N_{train} = 300
UNet	None	$72.62_{\pm 0.9}$	$76.45_{\pm 2.2}$	$63.03_{\pm 2.7}$	$69.20_{\pm 3.7}$	$88.12_{\pm 3.2}$	$94.16_{\pm 3.9}$
	Model genesis [29]	$73.00_{\pm 3.4}$	$81.8_{\pm 4.7}$	$60.96_{\pm 1.8}$	$67.04_{\pm 4.4}$	$89.44_{\pm 2.6}$	$95.16_{\pm 3.3}$
	SimCLR [4]	$73.63_{\pm 2.4}$	$80.12_{\pm 4.9}$	$59.89_{\pm 2.6}$	$66.51_{\pm 4.3}$	$90.60_{\pm 2.5}$	$94.21_{\pm 2.7}$
	Age prediction w/D.A	$75.32_{\pm 2.2}$	$85.27_{\pm 2.3}$	$\mathbf{64.6}_{\pm \mathbf{1.6}}$	$\mathbf{70.78}_{\pm \mathbf{2.1}}$	$91.71_{\pm 1.1}$	$95.26_{\pm 1.5}$
	Age-aware contrastive learning (ours)	$\mathbf{75.95}_{\pm 2.7}$	$\mathbf{85.73}_{\pm 4.7}$	$63.79_{\pm 3.0}$	$70.35_{\pm 2.7}$	$\mathbf{92.19}_{\pm \mathbf{1.8}}$	$\mathbf{96.58}_{\pm \mathbf{1.6}}$
DenseNet	None	$73.09_{\pm 1.6}$	$85.92_{\pm 2.8}$	$64.39_{\pm 2.9}$	$70.77_{\pm 2.7}$	$92.23_{\pm 1.6}$	$93.68_{\pm 1.7}$
	None w/D.A	$74.71_{\pm 1.3}$	$86.94_{\pm 2.8}$	$64.79_{\pm 1.3}$	$72.25_{\pm 1.5}$	$92.10_{\pm 1.8}$	$94.16_{\pm 2.5}$
	SimCLR [5]	$70.80_{\pm 1.9}$	$86.35_{\pm 2.2}$	$60.57_{\pm 1.9}$	$67.99_{\pm 3.3}$	$91.54_{\pm 1.9}$	$94.26_{\pm 2.9}$
	Age prediction	$72.90_{\pm 4.6}$	$87.75_{\pm 2.0}$	$64.60_{\pm 3.6}$	$72.07_{\pm 3.0}$	$92.07_{\pm 2.7}$	$96.37_{\pm 0.9}$
	Age prediction w/D.A	$74.06_{\pm 3.4}$	$86.90_{\pm 1.6}$	$\mathbf{65.79}_{\pm 2.0}$	$73.02_{\pm 4.3}$	$\mathbf{94.01}_{\pm \mathbf{1.4}}$	$96.10_{\pm 3.0}$
	Age-aware contrastive learning (ours)	$\mathbf{76.33}_{\pm 2.3}$	$\mathbf{88.11}_{\pm 1.5}$	$65.36_{\pm 3.7}$	$\mathbf{73.33}_{\pm \mathbf{4.3}}$	$93.87_{\pm 1.3}$	$\mathbf{96.84}_{\pm \mathbf{2.3}}$

4 Conclusion

Our key contribution is the introduction of a new contrastive loss, which leverages continuous (and discrete) meta-data from medical images in a self-supervised setting. We showed that our model, pre-trained with a large heterogeneous brain MRI dataset ($N = 10^4$) of healthy subjects, outperforms the other SOTA methods on three binary classification tasks. In some cases, it even reaches the performance of a fully-supervised network without fine-tuning. This demonstrates that our model can learn a meaningful and relevant representation of healthy brains which can be used to discriminate patients in small data-sets. An ablation study showed that our method consistently improves upon SimCLR for three different sets of transformations. We also made a step towards a debiased algorithm by demonstrating that our model is less sensitive to the site effect than other SOTA fully supervised algorithms trained from scratch. We think this is still an important issue leading to strong biases in machine learning algorithms and it currently leads to costly harmonization protocols between hospitals during acquisitions. Finally, as a step towards reproducible research, we made our code public and we will release the BHB dataset to the scientific community soon.

Future work will consist in developing transformations more adapted to medical images in the contrastive learning framework and in integrating other available meta-data (*e.g.* participant's sex) and modalities (*e.g.* genetics). Finally, we envision to adapt the current framework for longitudinal studies (such as ADNI).

Acknowledgments. This work was granted access to the HPC resources of IDRIS under the allocation 2020-AD011011854 made by GENCI.

References

1. Caron, M., Misra, I., Mairal, J., Goyal, P., Bojanowski, P., Joulin, A.: Unsupervised learning of visual features by contrasting cluster assignments. In: NeurIPS (2020)
2. Chaitanya, K., Erdil, E., Karani, N., Konukoglu, E.: Contrastive learning of global and local features for medical image segmentation with limited annotations (2020)
3. Chapelle, O., Weston, J., Bottou, L., Vapnik, V.: Vicinal risk minimization. MIT (2001)
4. Chen, L., Bentley, P., Mori, K., Misawa, K., Fujiwara, M., Rueckert, D.: Self-supervised learning for medical image analysis using image context restoration. Med. Image Anal. **58**, 101539 (2019)
5. Chen, T., Kornblith, S., Norouzi, M., Hinton, G.: A simple framework for contrastive learning of visual representations. In: ICML (2020)
6. Chuang, C.Y., Robinson, J., Lin, Y.C., Torralba, A., Jegelka, S.: Debiased contrastive learning. In: Larochelle, H., Ranzato, M., Hadsell, R., Balcan, M.F., Lin, H. (eds.) NeurIPS (2020)
7. DeVries, T., Taylor, G.W.: Improved regularization of convolutional neural networks with cutout. arXiv preprint arXiv:1708.04552 (2017)

8. Ding, X., Wang, Y., Xu, Z., Welch, W.J., Wang, Z.J.: CcGAN: continuous conditional generative adversarial networks for image generation. In: ICLR (2021)
9. Doersch, C., Gupta, A., Efros, A.A.: Unsupervised visual representation learning by context prediction. In: ICCV (2015)
10. Gidaris, S., Singh, P., Komodakis, N.: Unsupervised representation learning by predicting image rotations. In: ICLR (2018)
11. Grill, J.B., et al.: Bootstrap your own latent - a new approach to self-supervised learning. In: NeurIPS (2020)
12. Gutmann, M., Hyvärinen, A.: Noise-contrastive estimation: a new estimation principle for unnormalized statistical models. In: AISTATS (2010)
13. Hadsell, R., Chopra, S., LeCun, Y.: Dimensionality reduction by learning an invariant mapping. In: CVPR (2006)
14. He, K., Fan, H., Wu, Y., Xie, S., Girshick, R.: Momentum contrast for unsupervised visual representation learning. In: CVPR (2020)
15. Hozer, F., et al.: Lithium prevents grey matter atrophy in patients with bipolar disorder: an international multicenter study. Psychol. Med. **51**(7), 1201–1210 (2021)
16. Huang, G., Liu, Z., Van Der Maaten, L., Weinberger, K.Q.: Densely connected convolutional networks. In: CVPR (2017)
17. Khosla, P., et al.: Supervised contrastive learning. In: NeurIPS (2020)
18. Noroozi, M., Favaro, P.: Unsupervised learning of visual representations by solving jigsaw puzzles. In: Leibe, B., Matas, J., Sebe, N., Welling, M. (eds.) ECCV 2016. LNCS, vol. 9910, pp. 69–84. Springer, Cham (2016). https://doi.org/10.1007/978-3-319-46466-4_5
19. van den Oord, A., Li, Y., Vinyals, O.: Representation learning with contrastive predictive coding. arXiv preprint arXiv:1807.03748 (2018)
20. Paszke, A., et al.: PyTorch: an imperative style, high-performance deep learning library. In: NeurIPS (2019)
21. Pathak, D., Krahenbuhl, P., Donahue, J., Darrell, T., Efros, A.A.: Context encoders: feature learning by inpainting. In: CVPR (2016)
22. Robinson, J., Chuang, C.Y., Sra, S., Jegelka, S.: Contrastive learning with hard negative samples. In: ICLR (2021)
23. Sarrazin, S., et al.: Neurodevelopmental subtypes of bipolar disorder are related to cortical folding patterns: an international multicenter study. Bipolar Disord. **20**(8), 721–732 (2018)
24. Taleb, A., et al.: 3D self-supervised methods for medical imaging. In: NeurIPS (2020)
25. Tamminga, C.A., Pearlson, G., Keshavan, M., Sweeney, J., Clementz, B., Thaker, G.: Bipolar and schizophrenia network for intermediate phenotypes: outcomes across the psychosis continuum. Schizophr. Bull. **40**, S131–S137 (2014)
26. Tao, X., Li, Y., Zhou, W., Ma, K., Zheng, Y.: Revisiting Rubik's cube: self-supervised learning with volume-wise transformation for 3D medical image segmentation. In: Martel, A.L., et al. (eds.) MICCAI 2020. LNCS, vol. 12264, pp. 238–248. Springer, Cham (2020). https://doi.org/10.1007/978-3-030-59719-1_24
27. Tian, Y., Sun, C., Poole, B., Krishnan, D., Schmid, C., Isola, P.: What makes for good views for contrastive learning? In: NeurIPS (2020)
28. Wei, C., Wang, H., Shen, W., Yuille, A.: CO2: consistent contrast for unsupervised visual representation learning. In: ICLR (2021)

29. Zhou, Z., Sodha, V., Pang, J., Gotway, M.B., Liang, J.: Models genesis. Med. Image Anal. **67**, 101840 (2021)
30. Zhuang, X., Li, Y., Hu, Y., Ma, K., Yang, Y., Zheng, Y.: Self-supervised feature learning for 3D medical images by playing a Rubik's cube. In: Shen, D., et al. (eds.) MICCAI 2019. LNCS, vol. 11767, pp. 420–428. Springer, Cham (2019). https://doi.org/10.1007/978-3-030-32251-9_46

Sli2Vol: Annotate a 3D Volume from a Single Slice with Self-supervised Learning

Pak-Hei Yeung[1(✉)], Ana I. L. Namburete[1], and Weidi Xie[1,2]

[1] Department of Engineering Science, Institute of Biomedical Engineering, University of Oxford, Oxford, UK
pak.yeung@pmb.ox.ac.uk, ana.namburete@eng.ox.ac.uk

[2] Visual Geometry Group, Department of Engineering Science, University of Oxford, Oxford, UK
weidi@robots.ox.ac.uk
https://pakheiyeung.github.io/Sli2Vol_wp/

Abstract. The objective of this work is to segment any *arbitrary* structures of interest (SOI) in 3D volumes by only annotating a *single* slice, (*i.e.* semi-automatic 3D segmentation). We show that high accuracy can be achieved by simply propagating the 2D slice segmentation with an affinity matrix between consecutive slices, which can be learnt in a self-supervised manner, namely slice reconstruction. Specifically, we compare our proposed framework, termed as **Sli2Vol**, with supervised approaches and two other unsupervised/self-supervised slice registration approaches, on 8 public datasets (both CT and MRI scans), spanning 9 different SOIs. Without any parameter-tuning, the same model achieves superior performance with Dice scores (0–100 scale) of over 80 for most of the benchmarks, including the ones that are unseen during training. Our results show *generalizability* of the proposed approach across data from different machines and with different SOIs: a major use case of semi-automatic segmentation methods where fully supervised approaches would normally struggle.

Keywords: Self-supervised learning · Semi-automatic segmentation

1 Introduction

Image segmentation is arguably one of the most important tasks in medical image analysis, as it identifies the structure of interest (SOI) with arbitrary shape (*i.e.* pixel level predictions), encompassing rich information, such as the position and size. In recent years, the development and application of different deep convolutional neural networks (ConvNet), for example U-Net [19], have significantly boosted the accuracy of computer-aided medical image segmentation.

Electronic supplementary material The online version of this chapter (https://doi.org/10.1007/978-3-030-87196-3_7) contains supplementary material, which is available to authorized users.

M. de Bruijne et al. (Eds.): MICCAI 2021, LNCS 12902, pp. 69–79, 2021.
https://doi.org/10.1007/978-3-030-87196-3_7

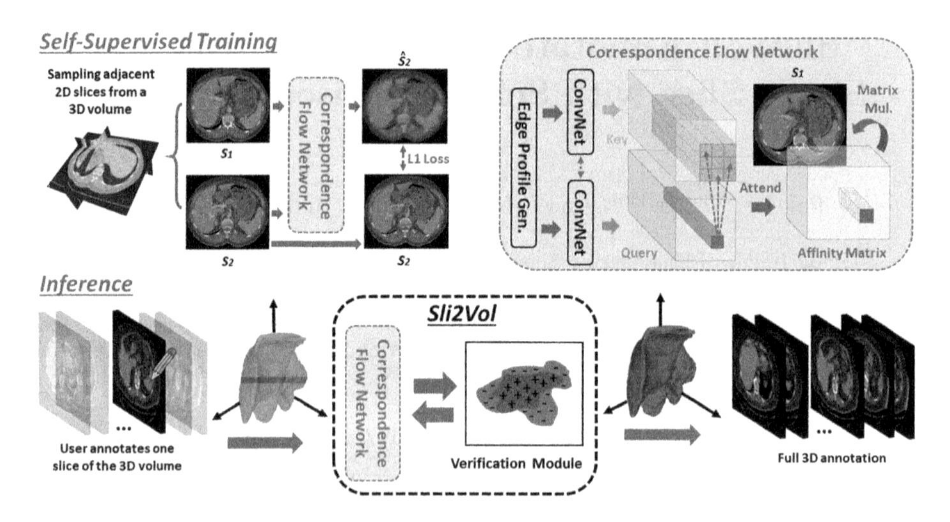

Fig. 1. Pipeline of our proposed framework. During *self-supervised training*, pair of adjacent slices sampled from 3D volumes are used to train a correspondence flow network. Provided with the 2D mask of a single slice of a volume, the trained network with the verification module can be used to propagate the initial annotation to the whole volume during *inference*.

Training fully automatic segmentation models comes with several limitations: *firstly*, annotations for the training volumes are usually a costly process to acquire; *secondly*, once domain shift appears, (*i.e.* from differences in scanner, acquisition protocol or the SOI varies during inference), the model may suffer a catastrophic drop in performance, requiring new annotations and additional fine-tuning. These factors have limited the use of the automatic segmentation approaches to applications with inter-vendor and inter-operator variance.

As an alternative, semi-automatic approaches are able to operate interactively with the end users: this is the scenario considered in this paper. Specifically, the goal is to segment any *arbitrary* SOIs in 3D volumes by only annotating a *single* slice within the volume, which may facilitate more flexible analysis of *arbitrary* SOIs with the desired generalizability (*e.g.* inter-scanner variability), and significantly reduce the annotating cost for fully supervised learning.

Similar tools have been developed with level set or random forest methods, which show excellent performance as reported in [5,7,17,26,27]. However, implementation of specific regularization and heavy parameter-tuning are usually required for different SOIs, limiting its use in practice. On the other hand, related work in medical image registration explores the use of pixelwise correspondence from optical flow [10,14] or unsupervised approaches [3,8,18], which in principle could be harnessed for the propagation of a 2D mask between slices within a volume. However, they are prone to error drift, *i.e.* error accumulation, introduced by inter-slice propagation of registration errors. In this work, we aim to overcome these limitations.

Here, we focus on the task of propagating the 2D slice segmentation through the entire 3D volume by matching correspondences between consecutive slices. Our work makes the following contributions: *firstly*, we explore mask propagation approaches based on unsupervised/self-supervised registration of slices, namely, naïve optical flow [6] and VoxelMorph [3], and our proposed self-supervised approach, called **Sli2Vol**, which is based on learning to match slices' correspondences [15,16] and using a newly proposed edge profile for information bottleneck. **Sli2Vol** is able to propagate the mask at a speed of 2.27 slices per second in inference. *Secondly*, to alleviate the error accumulation in mask propagation, we propose and exploit a simple verification module for refining the mask during inference time. *Thirdly*, we benchmark **Sli2Vol** on 8 public CT and MRI datasets [12,22,23,25], spanning 9 anatomical structures. Without any parameter-tuning, a *single* **Sli2Vol** model achieves Dice scores (0–100 scale) above 80 for most of the benchmarks, which outperforms other supervised and unsupervised approaches for all datasets in cross-domain evaluation. To the best of our knowledge, this is the first study to undertake cross-domain evaluation on such large-scale and diverse benchmarks for semi-automatic segmentation approaches, which shifts the focus to *generalizability* across different devices, clinical sites and anatomical SOIs.

2 Methods

In Sect. 2.1, we first formulate the problem setting in this paper, namely semi-automatic segmentation for 3D volume with *single* slice annotation. Next, we introduce the training stage of our proposed approach, **Sli2Vol**, in Sect. 2.2 and our proposed edge profile generator in Sect. 2.3. This is followed by the computations for inference (2.4), including our proposed verification module (2.5).

2.1 Problem Setup

In general, given a 3D volume, denoted by $\mathbf{V} \in \mathcal{R}^{H \times W \times D}$, where H, W and D are the height, width and depth of the volume, respectively, our goal is to segment the SOI in the volume based on a user-provided 2D segmentation mask for the *single* slice, *i.e.* $\mathbf{M}_i \in \mathcal{R}^{H \times W \times 1}$ with $1's$ indicating the SOI, and $0's$ as background. The outputs will be a set of masks for an individual slice, *i.e.* $\{\mathbf{M}_1, \mathbf{M}_2, ..., \mathbf{M}_D\}$.

Inspired by [15,16], we formulate this problem as learning feature representations that establish robust pixelwise correspondences between adjacent slices in a 3D volume, which results in a set of affinity matrices, $\mathbf{A}_{i \to i+1}$, for propagating the 2D mask between consecutive slices by *weighting and copying*. Model training follows a self-supervised learning scheme, where raw data is used, and only one slice annotation is required during inference time.

2.2 Self-supervised Training of Sli2Vol

In this section, we detail the self-supervised approach for learning the dense correspondences. Conceptually, the idea is to task a deep network for slice reconstruction by *weighting and copying* pixels from its neighboring slice. The affinity matrices used for weighting are acquired as a by-product, and can be directly used for mask propagation during inference.

During training, a pair of adjacent slices, $\{\mathbf{S}_1, \mathbf{S}_2\}, \mathbf{S}_i \in \mathcal{R}^{H \times W \times 1}$, are sampled from a training volume, and then fed to a ConvNet, parametrized by $\psi(\cdot, \theta)$ (as shown in the upper part of Fig. 1):

$$[\mathbf{k}_1,\ \mathbf{q}_2] = [\psi(g(\mathbf{S}_1);\ \theta),\ \psi(g(\mathbf{S}_2);\ \theta)] \tag{1}$$

where $g(\cdot)$ denotes an *edge profile generator* (details in Sect. 2.3) and $\mathbf{k}_1, \mathbf{q}_2 \in \mathcal{R}^{H \times W \times c}$ refer to the feature representation (c channels) computed from corresponding slices, termed as *key* and *query* respectively (Fig. 1). The difference in notation (*i.e.* $\mathbf{q}$ and $\mathbf{k}$) is just for emphasizing their functional difference.

Reshaping $\mathbf{k}_1$ and $\mathbf{q}_2$ to $\mathcal{R}^{HW \times c}$, an affinity matrix, $\mathbf{A}_{1 \to 2} \in \mathcal{R}^{HW \times \delta}$, is computed to represent the feature similarity between the two slices (Fig. 1):

$$\mathbf{A}_{1 \to 2}(u, v) = \frac{exp\langle \mathbf{q}_2(u, :), \mathbf{k}_1(v, :)\rangle}{\sum_{\lambda \in \Omega} exp\langle \mathbf{q}_2(u, :), \mathbf{k}_1(\lambda, :)\rangle} \tag{2}$$

where $\langle \cdot, \cdot \rangle$ is the dot product between two vectors and Ω is the window surrounding pixel v (*i.e.* in $\mathcal{R}^{H \times W}$ space) for computing local attention, with $n(\Omega) = \delta$.

Loss Function. During training, $\mathbf{A}_{1 \to 2}$ is used to *weight and copy* pixels from $\mathbf{S}_1$ (*i.e.* reshape to $\mathcal{R}^{HW \times 1}$) to reconstruct $\mathbf{S}_2$, denoted as $\hat{\mathbf{S}}_2$, by:

$$\hat{\mathbf{S}}_2(u, 1) = \sum_{v}^{\Omega} \mathbf{A}_{1 \to 2}(u, v) \mathbf{S}_1(v, 1). \tag{3}$$

We apply mean absolute error (MAE) between $\mathbf{S}_2$ and $\hat{\mathbf{S}}_2$ as the training loss.

2.3 Edge Profile Generator

Essentially, the basic assumption of the above-mentioned idea is that, to better reconstruct $\mathbf{S}_2$ via copying pixel from $\mathbf{S}_1$, the model must learn to establish reliable correspondences between the two slices. However, naïvely training the model may actually incur trivial solutions, for example, the model can perfectly solve the reconstruction task by simply matching the *pixel intensity* of $\mathbf{S}_1$ and $\mathbf{S}_2$.

In Lai *et al.* [15,16], the authors show that input color channel (*i.e. RGB* or *Lab*) dropout is an effective information bottleneck, which breaks the correlation between the color channels and forces the model to learn more robust correspondences. However, this is usually not feasible in medical images, as only single input channel is available in most of the modalities.

We propose to use a *profile of edges* as an *information bottleneck* to avoid trivial solution. Specifically, for each pixel, we convert its intensity value to a normalized edge histogram, by computing the derivatives along d different directions at s different scales, *i.e.* $g(\mathbf{S}_i) \in \mathcal{R}^{H\times W\times(d\times s)}$, followed by a *softmax* normalization through all the derivatives. Intuitively, $g(\cdot)$ explicitly represents the edge distributions centered each pixel of the slice $\mathbf{S}_i$, and force the model to pay more attentions to the edges during reconstruction. Experimental results in Sect. 4 verify the essence of this design in improving the model performance.

2.4 Inference

Given a volume, $\mathbf{V}$ and an initial mask at the i-th slice, $\mathbf{M}_i$, the affinity matrix, $\mathbf{A}_{i\to i+1}$, output from $\psi(\cdot,\theta)$ is used to propagate $\mathbf{M}_i$ iteratively to the whole $\mathbf{V}$.

In detail, two consecutive slices, $\{\mathbf{S}_i, \mathbf{S}_{i+1}\}$, are sampled from the volume $\mathbf{V}$ and fed into $\psi(g(\cdot),\theta)$ to get $\mathbf{A}_{i\to i+1}$, which is then used to propagate $\mathbf{M}_i$, using Eq. 3, ending up with $\hat{\mathbf{M}}_{i+1}$. This set of computations (Fig. 2 in the Supplementary Materials) is then repeated for the next two consecutive slices, $\{\mathbf{S}_{i+1}, \mathbf{S}_{i+2}\}$, in either direction, until the whole volume is covered.

2.5 Verification Module

In practice, we find that directly using $\hat{\mathbf{M}}_{i+1}$ for further propagation will potentially accumulate the prediction error after each iteration. To alleviate this drifting issue, and further boost the performance, we propose a simple verification module to correct the mask after each iteration of mask propagation.

Specifically, two regions, namely positive ($\mathbf{P} \in \mathcal{R}^{H\times W}$) and negative ($\mathbf{N} \in \mathcal{R}^{H\times W}$) regions, are constructed. $\mathbf{P}$ refers to the delineated SOI in $\mathbf{M}_i$, and $\mathbf{N}$ is identified by subtracting $\mathbf{P}$ from its own morphologically dilated version. Intuitively, the negative region denotes the thin and non-overlapping region surrounding $\mathbf{P}$ (Fig. 2 in the Supplementary Materials). We maintain the *mean intensity value* within each region:

$$p = \frac{1}{|P_i|}\langle P_i, S_i\rangle \qquad\qquad n = \frac{1}{|N_i|}\langle N_i, S_i\rangle$$

where $\langle\cdot,\cdot\rangle$ denotes Frobenius inner product, p and n refer to the positive and negative query values respectively.

During inference time, assuming $\hat{\mathbf{M}}_{i+1}$ is the predicted mask from the propagation, each of the proposed foreground pixels u in $\mathbf{S}_{i+1}$, is then compared to p and n and being re-classified according to its distance to the two values by:

$$\mathbf{M}^u_{i+1} = \begin{cases} 1, & \text{if } \hat{\mathbf{M}}^u_{i+1} = 1 \text{ and } \sqrt{(\mathbf{S}^u_{i+1} - p)^2} < \sqrt{(\mathbf{S}^u_{i+1} - n)^2} \\ 0, & \text{otherwise} \end{cases} \tag{4}$$

This set of computations is then repeated for the next round of propagation, where p and n are updated using the corrected mask, $\mathbf{M}_{i+1}$, and $\mathbf{S}_{i+1}$.

3 Experimental Setup

We benchmark our framework, **Sli2Vol**, on 8 different public datasets, spanning 9 different SOIs, and compare with a variety of fully supervised and semi-automatic approaches, using standard Dice coefficient (in a 0–100 scale) as the evaluation metrics. In Sect. 3.1, we introduce the datasets used in this paper. In Sect. 3.2, we summarize the experiments conducted for this study.

3.1 Dataset

Four training and eight testing datasets are involved. For **chest and abdominal CT**, a *single* model is trained on 3 unannotated dataset (*i.e.* C4KC-KiTS [9], CT-LN [21] and CT-Pancreas [20]) and tested on 7 other datasets (*i.e.* Sliver07 [25], CHAOS [12], 3Dircadb-01, 02 [23], and Decath-Spleen, Liver and Pancreas [22]).

For **cardiac MRI**, models are trained on the 2D video dataset from Kaggle [1], and tested on a 3D volume dataset (*i.e.* Decath-Heart [22]), which manifests large domain shift. Further details of the datasets are provided in Table 2 in the Supplementary Materials.

3.2 Baseline Comparison

Sli2Vol and a set of baseline approaches are tested, with their implementation details summarized in Table 3 in the Supplementary Materials.

First, we experiment with two approaches trained on fully annotated 3D data. **Fully Supervised (FS) - Same Domain** refers to the scenario where the training and testing data come from the *same* benchmark dataset. Results from both state-of-the-art methods [2,11,13,24] and 3D UNets trained by us are reported. On the other hand, **FS - Different Domain** aims to evaluate the generalizability of FS approaches when training and testing data come from *different* domains. Therefore, we train the 3D UNet (same architecture and settings as the **FS - Same Domain** for fair comparison) on a source dataset, and test it on another benchmark of the same task.

Second, we consider the case where only a single slice is annotated in each testing volume to train a 2D UNet (**FS - Single Slice**). For example, in *Sliver07*, the model trained on 20 slice annotations (single slice from each volume), is tested on the same set of 20 volumes. This approach utilizes the same amount of manual annotation as **Sli2Vol**, so as to investigate if a model trained on single slice annotations is sufficient to generalize to the whole volume.

Third, approaches based on registration of consecutive slices, namely **Optical Flow** [4,6], **VoxelMorph2D (VM) - UNet** and **VM - ResNet18Stride1**, are tested. The two VMs utilize a UNet backbone as proposed originally in [3] as well as the same backbone (*i.e.* ResNet18Stride1) as **Sli2Vol**, respectively.

For **Sli2Vol**, **FS - Single Slice**, **Optical Flow** and **VM**, we randomly pick one of the ± 3 slices around the slice with the largest groundtruth annotation as the initial mask. This simulates the process of a user sliding through the whole volume and roughly identifying the slice with the largest SOI to annotate, which is achievable in reality.

Table 1. Results (mean Dice scores ± standard deviation) of different approaches on different datasets and ROIs. Higher value represents better performance. In **row a**, results from both state-of-the-art methods [2, 11, 13] and 3D UNets trained by us (values in the bracket) are reported. Results in **row a** and **b** are only partially available in literature and they are reported just for demonstrating the approximated upper bound and limitation of fully supervised approaches, which are not meant to be directly compared to our proposed approach.

Modality	*MRI*	*Abdominal and Chest CT*													*Mean Results*
Training Dataset (for row e to j)	*Kaggle*	*C4KC-KiTS, CT-LN and CT-Pancreas*													
Testing Dataset	*Decath -Heart*	*Sliver07*	*CHAOS*	*Decath -Liver*	*Decath -Spleen*	*Decath-Pancreas*	*3D-IRCADb-01 and 3D-IRCADb-02*								
ROI	*Left Atrium*	*Liver*	*Liver*	*Liver*	*Spleen*	*Pancreas*	*Heart*	*Gall-bladder*	*Kidney*	*Surrenal-gland*	*Liver*	*Lung*	*Pancreas*	*Spleen*	
Number of Volumes	*20*	*20*	*20*	*131*	*41*	*281*	*3*	*8*	*17*	*11*	*22*	*12*	[illegible]	*7*	
Automatic (Trained with Fully Annotated Data)															
(a) Fully Supervised-same domain	92.7[11]	94.8[2] (93.9)	97.8[13] (92.8)	95.4[11] (91.0)	96.0[11]	79.3[11]	-	-	-	-	96.5[24]	-	-	-	-
(b) Fully Supervised-different domain	-	74.8 ±13.2	76.5 ±8.8	56.0 ±23.6	-	-	-	-	-	-	-	-	-	-	-
Semi-automatic															
(c) Fully Supervised-single slice	62.5 ±5.2	86.9 ±4.1	84.3 ±4.1	85.0 ±5.5	74.4 ±12.0	49.9 ±13.4	25.6 ±6.5	47.9 ±15.5	57.9 ±21.1	30.8 ±15.6	80.3 ±13.8	81.0 ±10.8	2[illegible].4 ±[illegible]7.9	58.6 ±4.7	60.4
(d) Optical Flow	51.1 ±7.4	65.2 ±8.8	72.0 ±9.9	47.0 ±15.9	72.9 ±14.5	25.1 ±8.2	32.2 ±11.6	24.6 ±12.4	73.6 ±14.6	22.1 ±12.9	68.4 ±9.4	33.6 ±18.0	21.9 ±12.6	70.8 ±17.5	48.6
(e) VoxelMorph2D-UNet	42.9 ±5.0	57.2 ±9.8	66.5 ±10.5	38.5 ±12.5	61.5 ±19.5	21.4 ±6.7	20.3 ±6.5	20.2 ±12.2	70.1 ±18.6	41.1 ±15.3	60.5 ±9.7	38.7 ±21.2	28.3 ±[illegible]1.0	54.1 ±12.4	44.4
(f) VoxelMorph2D-ResNet18NoStride	45.7 ±4.1	61.2 ±8.5	68.4 ±9.8	42.2 ±12.4	58.3 ±17.3	23.5 ±7.8	22.1 ±6.7	21.8 ±13.1	77.8 ±18.4	48.4 ±15.3	60.6 ±10.4	36.5 ±20.0	32.3 ±[illegible]3.3	60.0 ±12.1	47.5
Sli2Vol	*Ablation Studies*														
(g) Correspondence Flow Network	62.4 ±9.2	75.0 ±6.5	78.9 ±7.9	66.0 ±13.1	81.1 ±13.9	43.9 ±12.9	55.4 ±24.3	62.4 ±20.7	86.0 ±19.0	45.9 ±18.6	75.0 ±8.6	45.2 ±25.4	44.3 ±17.2	81.8 ±19.6	64.5
(h) Network + Edge Profile	56.8 ±8.4	74.8 ±7.4	77.8 ±8.4	64.4 ±14.1	83.6 ±13.2	48.9 ±11.2	49.4 ±12.3	68.5 ±13.8	86.8 ±15.7	**58.3 ±16.6**	73.9 ±8.5	48.8 ±26.4	53.[illegible] ±7.1	85.8 ±13.0	66.6
(i) Network + Verif. Module	**80.8 ±5.0**	81.1 ±5.0	83.4 ±6.3	72.0 ±8.9	79.1 ±17.3	37.3 ±13.6	50.9 ±11.6	70.7 ±12.7	83.3 ±21.4	47.5 ±20.8	78.8 ±6.9	79.8 ±29.3	45.[illegible] ±1[illegible].5	74.5 ±23.7	68.9
(j) Network + Verif. Module + Edge Profile	80.4 ±4.5	**91.3 ±3.2**	**91.0 ±2.9**	**86.8 ±7.2**	**88.4 ±10.9**	**54.2 ±10.0**	**75.9 ±10.9**	**68.9 ±9.9**	**91.4 ±4.8**	48.4 ±13.5	**88.2 ±3.0**	**81.4 ±28.5**	**58.[illegible] ±4.[illegible]**	**90.2 ±9.5**	**78.2**

4 Results and Discussion

The results of all the experiments are presented in Table 1, with qualitative examples shown in Fig. 3 in the Supplementary Materials. In Sect. 4.1, we explore the performance change of automatic approaches in the presence of domain shift, which leads to the analysis of the results of **Sli2Vol** in Sect. 4.2.

4.1 Automatic Approaches

As expected, although the state-of-the-art performance is achieved by the **FS-Same Domain** (**row a**), a significant performance drop (*i.e.* over 20 Dice) can be observed (by comparing **row b** and the values in the brackets in **row a**) for cross-domain (*i.e.* same SOI, different benchmarks) evaluation (**row b**).

Such variation of performance may be partially minimized by increasing the amount and diversity of training data, better design of training augmentation, and application of domain adaptation techniques. However, these may not always be practical in real-world scenarios, due to the high cost of data annotation and frequent domain shifts, for example variation of scanners and acquisition protocols in different clinical sites.

4.2 Semi-automatic Approaches

Sli2Vol, by contrast, does not need any annotated data for training, but only annotation of a single slice during inference to indicate the SOI to be segmented.

Single Slice Annotation. With the same amount of annotation, **Sli2Vol** (**row j**) clearly outperforms other baseline approaches (**row c - f**) on all benchmarks significantly ($p < 0.05$, t-test), with an average Dice score margin of over 18.

Propagation-Based Methods. Higher Dice score shown in **row g** over **row d - f** suggests that solely self-supervised correspondence matching may incur less severe error drift and, hence, be more suitable than **Optical Flow** and **VM** for mask propagation within a volume. Comparison of results in **row e, f** and **g** further verifies that the backbone architecture is not the determining factor for the superior performance achieved by **Sli2Vol**. Our proposed edge profile (**row h**) is shown to be a more effective bottleneck than using the original slice as input (**row g**) and it further boosts the marginal benefit of the verification module, which is manifested by comparing the performance gain from **row g** to **i** and that from **row h** to **j**.

Self-supervised Learning. Remarkably, **Sli2Vol** trained with self-supervised learning is agnostic to SOIs and domains. As for abdominal and chest CT, a *single* **Sli2Vol** model without any fine-tuning achieves a mean Dice score of 78.0 when testing on 7 datasets spanning 8 anatomical structures. As for the cardiac MRI experiments with large training-testing domain shift, **Sli2Vol** still performs reasonably well with a Dice score of 80.4 (**row j**). Under this scenario, **Sli2Vol** outperforms the fully supervised approaches significantly ($p < 0.05$, t-test), by more than 20 Dice scores (**row j** vs. **row b**), and the annotation efforts are much lower, *i.e.* only a single slice per volume.

5 Conclusion

In summary, we investigate on semi-automatic 3D segmentations, where any *arbitrary* SOIs in 3D volumes are segmented by only annotating a single slice. The proposed architecture, **Sli2Vol**, is trained with self-supervised learning to output affinity matrices between consecutive slices through correspondence matching, which are then used to propagate the segmentation through the volume. Benchmarking on 8 datasets with 9 different SOIs, **Sli2Vol** shows superior generalizability and accuracy as compared to other baseline approaches, agnostic to the SOI. We envision to provide end users with more flexibility to segment and analyze different SOIs with our proposed framework, which shows great potential to be further developed as a general interactive segmentation tool in our future works, to facilitate the community to study various anatomical structures, and minimize the cost of annotating large dataset.

Acknowledgments. PH. Yeung is grateful for support from the RC Lee Centenary Scholarship. A. Namburete is funded by the UK Royal Academy of Engineering under its Engineering for Development Research Fellowship scheme. W. Xie is supported by the UK Engineering and Physical Sciences Research Council (EPSRC) Programme Grant Seebibyte (EP/M013774/1) and Grant Visual AI (EP/T028572/1). We thank Madeleine Wyburd and Nicola Dinsdale for their valuable suggestions and comments about the work.

References

1. Data science bowl cardiac challenge data. https://www.kaggle.com/c/second-annual-data-science-bowl
2. Ahmad, M., et al..: Deep belief network modeling for automatic liver segmentation. IEEE Access **7**, 20585–20595 (2019)
3. Balakrishnan, G., Zhao, A., Sabuncu, M.R., Guttag, J., Dalca, A.V.: Voxelmorph: a learning framework for deformable medical image registration. IEEE Trans. Med. Imag. **38**(8), 1788–1800 (2019)
4. Bradski, G.: The OpenCV Library. Dr. Dobb's J. Softw. Tools **120**, 122–125 (2000)
5. Dawant, B.M., Li, R., Lennon, B., Li, S.: Semi-automatic segmentation of the liver and its evaluation on the MICCAI 2007 grand challenge data set. In: Proceedings of the 3D Segmentation in The Clinic: A Grand Challenge, pp. 215–221 (2007)
6. Farnebäck, G.: Two-frame motion estimation based on polynomial expansion. In: Bigun, J., Gustavsson, T. (eds.) SCIA 2003. LNCS, vol. 2749, pp. 363–370. Springer, Heidelberg (2003). https://doi.org/10.1007/3-540-45103-X_50

7. Foruzan, A.H., Chen, Y.-W.: Improved segmentation of low-contrast lesions using sigmoid edge model. Int. J. Comput. Assist. Radiol. Surg , **11**, 1–17 (2015). https://doi.org/10.1007/s11548-015-1323-x
8. Heinrich, M.P., Jenkinson, M., Brady, M., Schnabel, J.A.: MRI-based deformable registration and ventilation estimation of lung CT. IEEE Trans. Med. Imag. **32**(7), 1239–1248 (2013)
9. Heller, N., Sathianathen, N., Kalapara, A., et al.: C4kc kits challenge kidney tumor segmentation dataset (2019). https://doi.org/10.7937/TCIA.2019.IX49E8NX, https://wiki.cancerimagingarchive.net/x/UwakAw
10. Hermann, S., Werner, R.: High accuracy optical flow for 3d medical image registration using the census cost function. In: Klette, R., Rivera, M., Satoh, S. (eds.) PSIVT 2013. LNCS, vol. 8333, pp. 23–35. Springer, Heidelberg (2014). https://doi.org/10.1007/978-3-642-53842-1_3
11. Isensee, F., Petersen, J., Klein, A., et al.: nnU-net: Self-adapting framework for u-net-based medical image segmentation. arXiv preprint arXiv:1809.10486 (2018)
12. Kavur, A.E., et al.: Chaos challenge-combined (CT-MR) healthy abdominal organ segmentation. Med. Image Anal. **69**, 101950 (2021)
13. Kavur, A.E., Gezer, N.S., Barış, M., et al.: CHAOS challenge - combined (CT-MR) Healthy Abdominal Organ Segmentation, January 2020. https://arxiv.org/abs/2001.06535
14. Keeling, S.L., Ring, W.: Medical image registration and interpolation by optical flow with maximal rigidity. Journal of Mathematical Imaging and Vision **23**(1), 47–65 (2005)
15. Lai, Z., Lu, E., Xie, W.: Mast: a memory-augmented self-supervised tracker. In: Proceedings of the IEEE Conference on Computer Vision and Pattern Recognition, pp. 6479–6488 (2020)
16. Lai, Z., Xie, W.: Self-supervised learning for video correspondence flow. In: British Machine Vision Conference (2019)
17. Li, C., et al.: A likelihood and local constraint level set model for liver tumor segmentation from CT volumes. IEEE Trans. Biomed. Eng. **60**(10), 2967–2977 (2013)
18. Mocanu, S., Moody, A.R., Khademi, A.: Flowreg: fast deformable unsupervised medical image registration using optical flow. arXiv preprint arXiv:2101.09639 (2021)
19. Ronneberger, O., Fischer, P., Brox, T.: U-Net: convolutional networks for biomedical image segmentation. In: Navab, N., Hornegger, J., Wells, W.M., Frangi, A.F. (eds.) MICCAI 2015. LNCS, vol. 9351, pp. 234–241. Springer, Cham (2015). https://doi.org/10.1007/978-3-319-24574-4_28
20. Roth, H., Farag, A., Turkbey, E.B., Lu, L., Liu, J., Summers, R.M.: Data from pancreas-CT (2016). https://doi.org/10.7937/K9/TCIA.2016.TNB1KQBU, https://wiki.cancerimagingarchive.net/x/eIlX
21. Roth, H., et al.: A new 2.5D representation for lymph node detection in CT (2015). https://doi.org/10.7937/K9/TCIA.2015.AQIIDCNM, https://wiki.cancerimagingarchive.net/x/0gAtAQ
22. Simpson, A.L., Antonelli, M., Bakas, S., et al.: A large annotated medical image dataset for the development and evaluation of segmentation algorithms. arXiv preprint arXiv:1902.09063 (2019)
23. Soler, L., et al.: 3D image reconstruction for comparison of algorithm database: a patient specific anatomical and medical image database. Tech. Rep, IRCAD, Strasbourg, France (2010)

24. Tran, S.T., Cheng, C.H., Liu, D.G.: A multiple layer u-net, un-net, for liver and liver tumor segmentation in CT. IEEE Access **9**, 3752–3764 (2020)
25. Van Ginneken, B., Heimann, T., Styner, M.: 3D segmentation in the clinic: a grand challenge. In: MICCAI Workshop on 3D Segmentation in the Clinic: A Grand Challenge, vol. 1, pp. 7–15 (2007)
26. Wang, G., et al.: Slic-Seg: slice-by-slice segmentation propagation of the placenta in fetal MRI using one-plane scribbles and online learning. In: Navab, N., Hornegger, J., Wells, W.M., Frangi, A.F. (eds.) MICCAI 2015. LNCS, vol. 9351, pp. 29–37. Springer, Cham (2015). https://doi.org/10.1007/978-3-319-24574-4_4
27. Zheng, Z., Zhang, X., Xu, H., Liang, W., Zheng, S., Shi, Y.: A unified level set framework combining hybrid algorithms for liver and liver tumor segmentation in CT images. BioMed. Res. Int. **2018**, 3815346 (2018)

Self-supervised Longitudinal Neighbourhood Embedding

Jiahong Ouyang[1(✉)], Qingyu Zhao[1], Ehsan Adeli[1], Edith V. Sullivan[1], Adolf Pfefferbaum[1,2], Greg Zaharchuk[1], and Kilian M. Pohl[1,2]

[1] Stanford University, Stanford, CA 94305, USA
jiahongo@stanford.edu
[2] SRI International, Menlo Park, CA 94025, USA

Abstract. Longitudinal MRIs are often used to capture the gradual deterioration of brain structure and function caused by aging or neurological diseases. Analyzing this data via machine learning generally requires a large number of ground-truth labels, which are often missing or expensive to obtain. Reducing the need for labels, we propose a self-supervised strategy for representation learning named Longitudinal Neighborhood Embedding (LNE). Motivated by concepts in contrastive learning, LNE explicitly models the similarity between trajectory vectors across different subjects. We do so by building a graph in each training iteration defining neighborhoods in the latent space so that the progression direction of a subject follows the direction of its neighbors. This results in a smooth trajectory field that captures the global morphological change of the brain while maintaining the local continuity. We apply LNE to longitudinal T1w MRIs of two neuroimaging studies: a dataset composed of 274 healthy subjects, and Alzheimer's Disease Neuroimaging Initiative (ADNI, $N = 632$). The visualization of the smooth trajectory vector field and superior performance on downstream tasks demonstrate the strength of the proposed method over existing self-supervised methods in extracting information associated with normal aging and in revealing the impact of neurodegenerative disorders. The code is available at https://github.com/ouyangjiahong/longitudinal-neighbourhood-embedding.

1 Introduction

Although longitudinal MRIs enable noninvasive tracking of the gradual effect of neurological diseases and environmental influences on the brain over time [23], the analysis is complicated by the complex covariance structure characterizing a mixture of time-varying and static effects across visits [8]. Therefore, training deep learning models on longitudinal data typically requires a large amount of samples with accurate ground-truth labels, which are often expensive or infeasible to acquire for some neuroimaging applications [4].

Electronic supplementary material The online version of this chapter (https://doi.org/10.1007/978-3-030-87196-3_8) contains supplementary material, which is available to authorized users.

M. de Bruijne et al. (Eds.): MICCAI 2021, LNCS 12902, pp. 80–89, 2021.
https://doi.org/10.1007/978-3-030-87196-3_8

Recent studies suggest that the issue of inadequate labels can be alleviated by self-supervised learning, the aim of which is to automatically learn representations by training on pretext tasks (i.e., tasks that do not require labels) before solving the supervised downstream tasks [14]. State-of-the-art self-supervised models are largely based on contrastive learning [5,10,17,19,21], i.e., learning representations by teaching models the difference and similarity of samples. For example, prior studies have generated or identified similar or dissimilar sample pairs (also referred to as positive and negative pairs) based on data augmentation [6], multi-view analysis [21], and organizing samples in a lookup dictionary [11]. Enforcing such an across-sample relationship in the learning process can then lead to more robust high-level representations for downstream tasks [16].

Despite the promising results of contrastive learning on cross-sectional data [2,7,15], the successful application of these concepts to longitudinal neuroimaging data still remains unclear. In this work, we propose a self-supervised learning model for longitudinal data by exploring the similarity between 'trajectories'. Specifically, the longitudinal MRIs of a subject acquired at multiple visits characterize gradual aging and disease progression of the brain over time, which manifests a temporal progression trajectory when projected to the latent space. Subjects with similar brain appearances are likely to exhibit similar aging trajectories. As a result, the trajectories from a cohort should collectively form a smooth trajectory field that characterizes the morphological change of the brain development over time. We hypothesize that regularizing such smoothness in a self-supervised fashion can result in a more informative latent space representation, thereby facilitating further analysis of healthy brain aging and effects of neurodegenerative diseases.

To achieve the smooth trajectory field, we build a dynamic graph in each training iteration to define a neighborhood in the latent space for each subject. The graph then connects nearby subjects and enforces their progression directions to be maximally aligned (Fig. 1). As such, the resulting latent space captures the global complexity of the progression while maintaining the local continuity of the nearby trajectory vectors. We name the trajectory vectors learned from the neighborhood as *Longitudinal Neighbourhood Embedding (LNE)*.

We evaluate our method on two longitudinal structural MRI datasets: one consists of 274 healthy subjects with the age ranging from 20 to 90, and the second is composed of 632 subjects from ADNI to analyze the progression trajectory of Normal Control (NC), static Mild Cognitive Impairment (sMCI), progressive Mild Cognitive Impairment (pMCI), and Alzheimer's Diease (AD). On these datasets, the visualization of the latent space in a 2D space confirms that the smooth trajectory vector field learned by the proposed method encodes continuous variation with respect to brain aging. When evaluated on downstream tasks, we obtain higher squared-correlation (R2) in age regression and better balanced accuracy (BACC) in ADNI classifications using our pre-trained model compared to alternative self-supervised or unsupervised pre-trained models.

2 Method

We now describe LNE, a method that smooths trajectories in the latent space by local neighborhood embedding. While trajectory regularization has been explored in 2D or 3D spaces (e.g., pedestrian trajectory [1,9] and in non-rigid registration [3]), there are several challenges in the context of longitudinal MRI analysis: (1) each trajectory is measured on sparse and asynchronous (e.g., not aligned by age or visit time) time points; (2) the trajectories live in a high-dimensional space rather than a regular 2D or 3D grid space; (3) the latent representations are defined in a variant latent space that is iteratively updated. To resolve these challenges, we first propose a strategy to train based on pair-wise data and translate the trajectory-regularization problem to the estimation of a smooth vector field, which is then solved by longitudinal neighbourhood embedding on dynamic graphs.

Pairwise Training Strategy. As shown in Fig. 1, each subject is associated with a trajectory (blue vectors) across multiple visits (≥2) in the latent space. To overcome the problem of the small number of time points in each trajectory, we propose to discretize a trajectory into multiple vectors defined by pairs of images. Compared to using the whole trajectory of sequential images as a training sample as typically done by recurrent neural networks [18], this pairwise strategy substantially increases the number of training samples. To formalize this operation, let $\mathcal{X}$ be the collection of all MR images and $\mathcal{S}$ be the set of subject-specific image pairs; i.e., $\mathcal{S}$ contains all (x^t, x^s) that are from the same subject with x^t scanned before x^s. These image pairs are then the input to the Encoder-Decoder structure shown in Fig. 1. The latent representations generated by the encoder are denoted by $z^t = F(x^t)$, $z^s = F(x^s)$, where F is the encoder. Then, $\Delta z^{(t,s)} = (z^s - z^t)/\Delta t^{(t,s)}$ is formulated as the normalized trajectory vector, where $\Delta t^{(t,s)}$ is the time interval between the two scans. All $\Delta z^{(t,s)}$ in the cohort define the trajectory vector field. The latent representations are then used to reconstruct the input images by the decoder H, i.e., $\tilde{x}^t = H(z^t)$, $\tilde{x}^s = H(z^s)$.

Longitudinal Neighbourhood Embedding. Inspired by social pooling in pedestrian trajectory prediction [1,9], we model the similarity between each subject-specific trajectory vector with those from its neighbourhood to enforce the smoothness of the entire vector field. As the high-dimensional latent space cannot be defined by a fixed regular grid (e.g., a 2D image grid space), we propose to define the neighbourhood by building a directed graph $\mathcal{G}$ in each training iteration for the variant latent space that is iteratively updated. The position of each node is defined by the starting point z^t of the vector Δz and the value (representation) of that node is Δz itself. For each node i, Euclidean distances to other nodes $j \neq i$ are computed by $P_{i,j} = \| z_i^t - z_j^t \|_2$ while the N_{nb} closest nodes of node i form its 1-hop neighbourhood $\mathcal{N}_i$ with edges connected to i. The adjacency matrix A for $\mathcal{G}$ is then defined as:

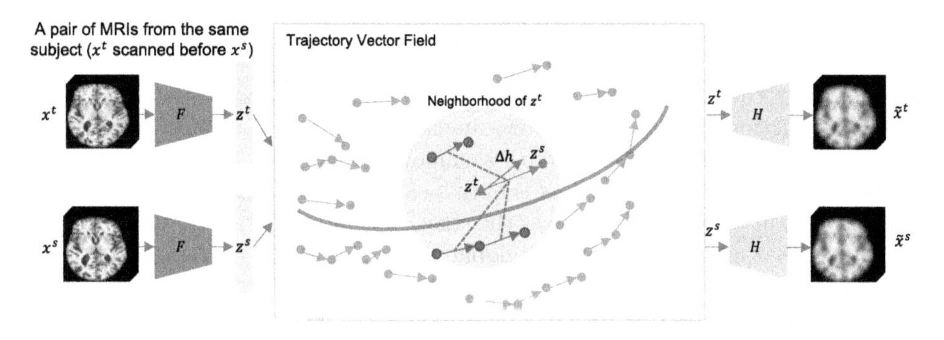

Fig. 1. Overview of the proposed method: an encoder projects a subject-specific image pair (x^t, x^s) into the latent space resulting in a trajectory vector (cyan). We encourage the direction of this vector to be consistent with Δh (purple), a vector pooled from the neighborhood of z^t (blue circle). As a result, the latent space encodes the global morphological change linked to aging (red curve). (Color figure online)

$$A_{i,j} := \begin{cases} exp(-\frac{P_{i,j}^2}{2\sigma_i^2}) & j \in \mathcal{N}_i \\ 0, & j \notin \mathcal{N}_i \end{cases}.$$
$$\text{with } \sigma_i := max(P_{i,j\in\mathcal{N}_i}) - min(P_{i,j\in\mathcal{N}_i})$$

Next, we aim to impose a smoothness regularization on this graph-valued vector field. Motivated by the graph diffusion process [13], we regularize each node's representation by a *longitudinal neighbourhood embedding* Δh 'pooled' from the neighbours' representations. For node i, the neighbourhood embedding can be computed by:

$$\Delta h_i := \sum_{j\in\mathcal{N}_i} A_{i,j} D_{i,j}^{-1} \Delta z_j,$$

where D is the 'out-degree matrix' of graph $\mathcal{G}$, a diagonal matrix that describes the sum of the weights for outgoing edges at each node. As shown in Fig. 1, the blue circle illustrates the above operation of learning the neighbourhood embedding that is shown by the purple arrow.

Objective Function. As shown in [24], the speed of brain aging is already highly heterogeneous within a healthy population, and subjects with neurodegenerative diseases may exhibit accelerated aging. Therefore, instead of replacing Δz with Δh, we define $\theta_{\langle \Delta z, \Delta h\rangle}$ as the angle between Δz and Δh, and only encourage $\cos(\theta_{\langle \Delta z, \Delta h\rangle}) = 1$, i.e., a zero-angle between the subject-specific trajectory vector and the pooled trajectory vector that represents the local progression direction. As such, it enables the latent representations to model the complexity of the global progression trajectory as well as the consistency of the local trajectory vector field. To impose the direction constraint in the autoencoder, we propose to add this cosine loss for each image pair to the standard mean squared error loss, i.e.,

$$L := \mathbf{E}_{(x^t,x^s)\sim\mathcal{S}}\left(\| x^t - \tilde{x}^t \|_2^2 + \| x^s - \tilde{x}^s \|_2^2 - \lambda \cdot \cos(\theta_{\langle\Delta z,\Delta h\rangle})\right),$$

with λ being the weighing parameter and $\mathbf{E}$ define the expected value. The objective function encourages the low-dimensional representation of the images to be informative while maintaining a smooth progression trajectory field in the latent space. As the cosine loss is only locally imposed, the global trajectory field can be non-linear, which relaxes the strong assumption in prior studies (e.g., LSSL [24]) that aging must define a globally linear direction in the latent space. Note, our method can be regarded as a contrastive self-supervised method. For each node, the samples in its neighbourhood serve as positive pairs with the cosine loss being the corresponding contrastive loss.

3 Experiments

Dataset. To show that LNE can successfully disentangle meaningful aging information in the latent space, we first evaluated the proposed method on predicting age from 582 MRIs of 274 healthy individuals with the age ranging from 20 to 90. Each subject had 1 to 13 scans with an average of 2.3 scans spanning an average time interval of 3.8 years. The second data set comprised 2389 longitudinal T1-weighted MRIs (at least two visits per subject) from ADNI, which consisted of 185 NC (age: 75.57 $\pm$ 5.06 years), 119 subjects with AD (age: 75.17 $\pm$ 7.57 years), 193 subjects diagnosed with sMCI (age: 75.63 $\pm$ 6.62 years), and 135 subjects diagnosed with pMCI (age: 75.91 $\pm$ 5.35 years). There was no significant age difference between the NC and AD cohorts (p = 0.55, two-sample t-test) as well as the sMCI and pMCI cohorts (p = 0.75). All longitudinal MRIs were preprocessed by a pipeline composed of denoising, bias field correction, skull striping, affine registration to a template, re-scaling to a $64 \times 64 \times 64$ volume, and transforming image intensities to z-scores.

Implementation Details. Let C_k denote a Convolution(kernel size of $3 \times 3 \times 3$)-BatchNorm-LeakyReLU(slope of 0.2)-MaxPool(kernel size of 2) block with k filters, and CD_k an Convolution-BatchNorm-LeakyReLU-Upsample block. The architecture was designed as C_{16}-C_{32}-C_{64}-C_{16}-CD_{64}-CD_{32}-CD_{16}-CD_{16} with a convolution layer at the top for reconstruction. The regularization weights were set to $\lambda = 1.0$. The networks were trained for 50 epochs by the Adam optimizer with learning rate of 5×10^{-4} and weight decay of 10^{-5}. To make the algorithm computationally efficient, we built the graph dynamically on the mini-batch of each iteration. A batch size $N_{bs} = 64$ and neighbour size $N_{nb} = 5$ were used.

Evaluation. Five-fold cross-validation (folds split based on subjects) was conducted with 10% training subjects used for validation. Random flipping of brain hemispheres, and random rotation and shift were used as augmentation during training. We first qualitatively illustrated the trajectory vector field (Δz) in 2D space by projecting the 1024-dimensional bottleneck representations (z^t and z^s) to their first two principal components. We then estimated the global trajectory of the vector field by a curve fitted by robust linear mixed effect model, which

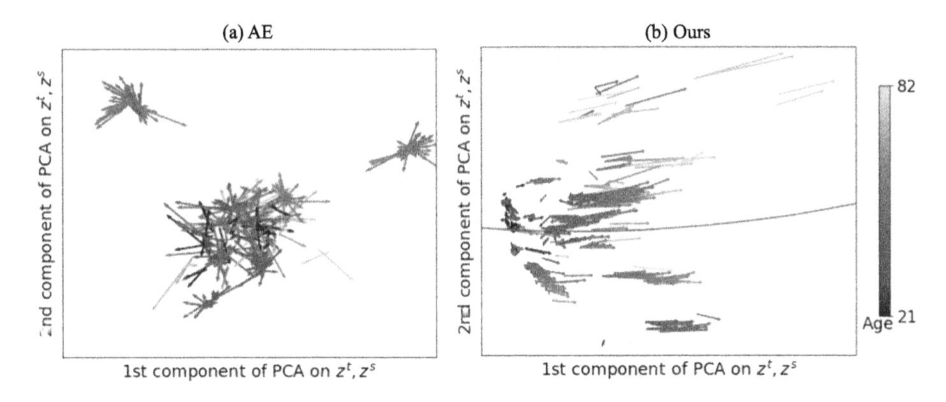

Fig. 2. Experiments on healthy aging: Latent space of AutoEncoder (AE) (a) and the proposed LNE (b) projected into 2D PCA space of z^t and z^s. Arrows represent Δz and are color-coded by the age of z^t. The global trajectory in (b) is fitted by robust linear mixed effect model (red curve). (Color figure online)

considered a quadratic fixed effect with random effect of intercepts. We further quantitatively evaluated the quality of the representations by using them for downstream tasks. Note, for theses experiments, we removed the decoder and only kept the encoder with its pre-trained weights for the downstream tasks. On the dataset of healthy subjects, we used the representation z to predict the chronological age of each MRI to show that our latent space was stratified by age. Note, learning a prediction model for normal aging is an emerging approach for understanding structural changes of the human brain and quantifying impact of neurological diseases (e.g. estimating brain age gap [20]). R2 and root-mean-square error (RMSE) were used as accuracy metrics. For ADNI, we predicted the diagnosis group associated with each image pair based on both z and trajectory vector Δz to highlight the aging speed between visits (an important marker for AD). In addition to classifying NC and AD, we also aimed to distinguish pMCI from sMCI, a significantly more challenging classification task.

The classifier was designed as a multi-layer perceptron containing two fully connected layers of dimension 1024 and 64 with LeakyReLU activation. In a separate experiment, we fine-tuned the LNE representation by incorporating the encoder into the classification models. We compared the BACC (accounting for different number of training samples in each cohort) to models using the same architecture with encoders pre-trained by other representation learning methods, including unsupervised methods (AE, VAE [12]), self-supervised method (SimCLR [6]. Images of two visits of the same subject with simple shift and rotation augmentation were used as a positive pair in SimCLR), and longitudinal self-supervised method (LSSL [24]).

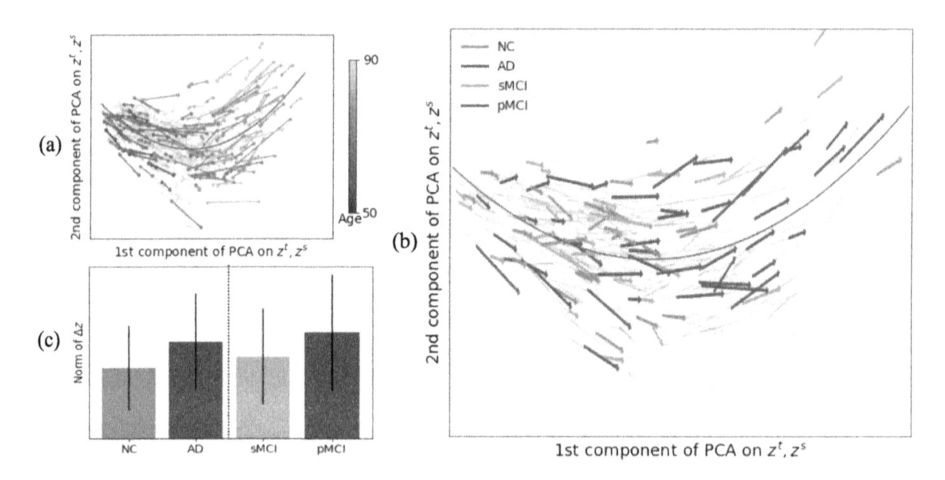

Fig. 3. Experiments on ADNI: (a) The age distribution of the latent space. Lines connecting z^t and z^s are color-coded by the age of z^t; Red curve is the global trajectory fitted by a robust linear mixed effect model. (b) Trajectory vector field color-coded by diagnosis groups; (c) The norm of Δz encoding the speed of aging for 4 different diagnosis groups. (Color figure online)

3.1 Healthy Aging

Figure 2 illustrates the trajectory vector field derived on one of the 5 folds by the proposed method (Fig. 2(b)). We observe LNE resulted in a smooth vector field that was in line with the fitted global trajectory shown by the red curve in Fig. 2(b). Moreover, chronological age associated with the vectors (indicated by the color) gradually increased along the global trajectory (red curve), indicating the successful disentanglement of the aging effect in the latent space. Note, such continuous variation in the whole age range from 20 to 90 was solely learned by self-supervised training on image pairs with an average age interval of 3.8 years (without using their age information). Interestingly, the length of the vectors tended to increase along the global trajectory, suggesting a faster aging speed for older subjects. On the contrary, without regularizing the longitudinal changes, AE did not lead to clear disentanglement of brain age in the space (Fig. 2(a)).

As shown in Table 1 (left), we utilized the latent representation z to predict the chronological age of the subject. In the scenario that froze the encoder, the proposed method achieved the best performance with an R2 of 0.62, which was significantly better ($p < 0.01$, t-test on absolute errors) than the second-best method LSSL with an R2 of 0.59. In addition to R2, the RMSE metrics are given in the supplement Table S1, which also suggests that LNE achieved the most accurate prediction. These results align with the expectation that a pre-trained self-supervised model with explicitly modeling of aging effect can lead to better downstream supervised age prediction. Lastly, when we fine-tuned the encoder during training, LNE remained as the most accurate method (both LNE and LSSL achieved an R2 of 0.74).

Table 1. Supervised downstream tasks in frozen or fine-tune scenarios. Left: Age regression on healthy subjects with R2 as an evaluation metric. Right: classification on ADNI dataset with BACC as the metric.

Methods	Health aging (**R2**)		ADNI (**BACC**)			
	Age		NC vs AD		sMCI vs pMCI	
	Frozen	Fine-tune	Frozen	Fine-tune	Frozen	Fine-tune
No pretrain	–	0.72	–	79.4	–	69.3
AE	0.53	0.69	72.2	80.7	62.6	69.5
VAE [12]	0.51	0.69	66.7	77.0	61.3	63.8
SimCLR [6]	0.56	0.73	72.9	82.4	63.3	69.5
LSSL [24]	0.59	**0.74**	74.2	82.1	69.4	71.2
Ours (LNE)	**0.62**	**0.74**	**81.9**	**83.6**	**70.6**	**73.4**

3.2 Progression of Alzheimer's Disease

We also evaluated the proposed method on the ADNI dataset. All 4 cohorts (NC, sMCI, pMCI, AD) were included in the training of LNE as the method was impartial to diagnosis groups (did not use labels for training). Similar to the results of the prior experiment, the age distribution in the latent space in Fig. 3(a) suggests a continuous variation with respect to brain development along the global trajectory shown by the red curve. We further illustrated the trajectory vector field by diagnosis groups in Fig. 3(b). While the starting points (z^t) of different diagnosis groups mixed uniformly in the field, vectors of AD (pink) and pMCI (brown) were longer than NC (cyan) and sMCI (orange). This suggests that LNE stratified the cohorts by their 'speed of aging' rather than age itself, highlighting the importance of using longitudinal data for analyzing AD and pMCI. This observation was also evident in Fig. 3(c), where AD and pMCI had statistically larger norm of Δz than the other two cohorts (both with $p < 0.01$). This finding aligned with previous AD studies [22] suggesting that AD group has accelerated aging effect compared to the NC group, and so does the pMCI group compared to the sMCI group.

The quantitative results on the downstream supervised classification tasks are shown in Table 1 (right). As the length of Δz was shown to be informative, we concatenated z^t with Δz as the feature for classification (classification accuracy based on z^t only is reported in the supplement Table S2). The representations learned by the proposed method yielded significantly more accurate predictions than all baselines ($p < 0.01$, DeLong's test). Note that the accuracy of our model with the frozen encoder even closely matched up to other methods after fine-tuning. This was to be expected because only our method and LSSL explicitly modeled the longitudinal effects which led to more informative Δz. In addition, our method that focused on local smoothness could capture the potentially non-linear effects underlying the morphological change along time, while the 'global linearity' assumption in LSSL may lead to information loss in the representations. It is worth mentioning that reliably distinguishing the subjects that will eventually develop AD (pMCI) from other MCI subjects (sMCI) is crucial for timely treatment. To this end, Supplement Table S3 suggests LNE improved over prior studies in classifying sMCI vs. pMCI, highlighting potential clinical values of our method. Ablation study on two important hyperparameters N_{nb} and λ is reported in the supplement Table S4.

4 Conclusion

In this work, we proposed a self-supervised representation learning framework, called LNE, that incorporates advantages from the repeated measures design in longitudinal neuroimaging studies. By building the dynamic graph and learning longitudinal neighbourhood embedding, LNE yielded a smooth trajectory vector field in the latent space, while maintaining a globally consistent progression trajectory that modeled the morphological change of the cohort. It successfully modeled the aging effect on healthy subjects, and enabled better chronological age prediction compared to other self-supervised methods. Although LNE was trained without the use of diagnosis labels, it demonstrated capability of differentiating diagnosis groups on the ADNI dataset based on the informative trajectory vector field. When evaluated for downstream task of classification, it showed superior quantitative classification performance as well.

Acknowledgement. This work was supported by NIH funding R01 MH113406, AA017347, AA010723, and AA005965.

References

1. Alahi, A., Goel, K., Ramanathan, V., Robicquet, A., Fei-Fei, L., Savarese, S.: Social LSTM: human trajectory prediction in crowded spaces. In: Proceedings of the IEEE Conference on Computer Vision and Pattern Recognition, pp. 961–971 (2016)
2. Balakrishnan, G., Zhao, A., Sabuncu, M.R., Guttag, J., Dalca, A.V.: An unsupervised learning model for deformable medical image registration. In: Proceedings of the IEEE Conference on Computer Vision and Pattern Recognition, pp. 9252–9260 (2018)
3. Balakrishnan, G., Zhao, A., Sabuncu, M.R., Guttag, J., Dalca, A.V.: Voxelmorph: a learning framework for deformable medical image registration. IEEE Trans. Med. Imaging **38**(8), 1788–1800 (2019)
4. Carass, A., et al.: Longitudinal multiple sclerosis lesion segmentation: resource and challenge. NeuroImage **148**, 77–102 (2017)
5. Caron, M., Misra, I., Mairal, J., Goyal, P., Bojanowski, P., Joulin, A.: Unsupervised learning of visual features by contrasting cluster assignments. arXiv preprint arXiv:2006.09882 (2020)
6. Chen, T., Kornblith, S., Norouzi, M., Hinton, G.: A simple framework for contrastive learning of visual representations. In: International Conference on Machine Learning, pp. 1597–1607. PMLR (2020)

7. Dalca, A.V., Yu, E., Golland, P., Fischl, B., Sabuncu, M.R., Eugenio Iglesias, J.: Unsupervised deep learning for Bayesian brain MRI segmentation. In: Shen, D., et al. (eds.) MICCAI 2019. LNCS, vol. 11766, pp. 356–365. Springer, Cham (2019). https://doi.org/10.1007/978-3-030-32248-9_40
8. Garcia, T.P., Marder, K.: Statistical approaches to longitudinal data analysis in neurodegenerative diseases: Huntington's disease as a model. Curr. Neurol. Neurosci. Rep. **17**(2), 14 (2017)
9. Gupta, A., Johnson, J., Fei-Fei, L., Savarese, S., Alahi, A.: Social GAN: Socially acceptable trajectories with generative adversarial networks. In: Proceedings of the IEEE Conference on Computer Vision and Pattern Recognition, pp. 2255–2264 (2018)
10. Hassani, K., Khasahmadi, A.H.: Contrastive multi-view representation learning on graphs. In: International Conference on Machine Learning, pp. 4116–4126. PMLR (2020)
11. He, K., Fan, H., Wu, Y., Xie, S., Girshick, R.: Momentum contrast for unsupervised visual representation learning. In: Proceedings of the IEEE/CVF Conference on Computer Vision and Pattern Recognition, pp. 9729–9738 (2020)
12. Kingma, D.P., Welling, M.: Auto-encoding variational Bayes. arXiv preprint arXiv:1312.6114 (2013)
13. Klicpera, J., Weißenberger, S., Günnemann, S.: Diffusion improves graph learning. arXiv preprint arXiv:1911.05485 (2019)
14. Kolesnikov, A., Zhai, X., Beyer, L.: Revisiting self-supervised visual representation learning. In: Proceedings of the IEEE/CVF Conference on Computer Vision and Pattern Recognition, pp. 1920–1929 (2019)
15. Li, H., Fan, Y.: Non-rigid image registration using self-supervised fully convolutional networks without training data. In: 2018 IEEE 15th International Symposium on Biomedical Imaging (ISBI 2018), pp. 1075–1078. IEEE (2018)
16. Liu, X., Zhang, F., Hou, Z., Wang, Z., Mian, L., Zhang, J., Tang, J.: Self-supervised learning: Generative or contrastive **1**(2). arXiv preprint arXiv:2006.08218 (2020)
17. Oord, A.v.d., Li, Y., Vinyals, O.: Representation learning with contrastive predictive coding. arXiv preprint arXiv:1807.03748 (2018)
18. Ouyang, J., et al.: Longitudinal pooling & consistency regularization to model disease progression from MRIs. IEEE J. Biomed. Health Inform. (2020)
19. Sabokrou, M., Khalooei, M., Adeli, E.: Self-supervised representation learning via neighborhood-relational encoding. In: Proceedings of the IEEE/CVF International Conference on Computer Vision, pp. 8010–8019 (2019)
20. Smith, S.M., et al.: Brain aging comprises many modes of structural and functional change with distinct genetic and biophysical associations. Elife **9**, e52677 (2020)
21. Tian, Y., Krishnan, D., Isola, P.: Contrastive multiview coding. arXiv preprint arXiv:1906.05849 (2019)
22. Toepper, M.: Dissociating normal aging from alzheimer's disease: a view from cognitive neuroscience. J. Alzheimer's Dis. **57**(2), 331–352 (2017)
23. Whitwell, J.L.: Longitudinal imaging: change and causality. Curr. Opin. Neurol. **21**(4), 410–416 (2008)
24. Zhao, Q., Liu, Z., Adeli, E., Pohl, K.M.: LSSL: Longitudinal self-supervised learning. arXiv preprint arXiv:2006.06930 (2020)

Self-supervised Multi-modal Alignment for Whole Body Medical Imaging

Rhydian Windsor[1(✉)], Amir Jamaludin[1], Timor Kadir[1,2], and Andrew Zisserman[1]

[1] Visual Geometry Group, Department of Engineering Science, University of Oxford, Oxford, UK
rhydian@robots.ox.ac.uk
[2] Plexalis Ltd., Thame, UK

Abstract. This paper explores the use of self-supervised deep learning in medical imaging in cases where two scan modalities are available for the same subject. Specifically, we use a large publicly-available dataset of over 20,000 subjects from the UK Biobank with both whole body Dixon technique magnetic resonance (MR) scans and also dual-energy x-ray absorptiometry (DXA) scans. We make three contributions: (i) We introduce a multi-modal image-matching contrastive framework, that is able to learn to match different-modality scans of the same subject with high accuracy. (ii) Without any adaption, we show that the correspondences learnt during this contrastive training step can be used to perform automatic cross-modal scan registration in a completely unsupervised manner. (iii) Finally, we use these registrations to transfer segmentation maps from the DXA scans to the MR scans where they are used to train a network to segment anatomical regions without requiring ground-truth MR examples. To aid further research, our code is publicly available (https://github.com/rwindsor1/biobank-self-supervised-alignment).

Keywords: Deep learning · Multi-modal imaging · Self-supervised learning

1 Introduction

A common difficulty in using deep learning for medical tasks is acquiring high-quality annotated datasets. There are several reasons for this: (1) using patient data requires ethical clearance, anonymisation and careful curation; (2) generating ground-truth labels may require expertise from clinicians whose time is limited and expensive; (3) clinical datasets are typically highly class-imbalanced with vastly more negative than positive examples. Thus acquiring sufficiently large datasets is often expensive, time-consuming, and frequently infeasible.

Electronic supplementary material The online version of this chapter (https://doi.org/10.1007/978-3-030-87196-3_9) contains supplementary material, which is available to authorized users.

M. de Bruijne et al. (Eds.): MICCAI 2021, LNCS 12902, pp. 90–101, 2021.
https://doi.org/10.1007/978-3-030-87196-3_9

As such, there is great interest in developing machine learning methods to use medical data and annotations efficiently. Examples of successful previous approaches include aggressive data augmentation [29] and generating synthetic images for training [10]. Alternatively, one can use *self-supervised pre-training* to learn useful representations of data, reducing annotation requirements for downstream learning tasks. This method has already shown much success in other areas of machine learning such as natural image classification [8,14,16] and natural language processing [6,9,25,28].

In this paper, we develop a self-supervised learning approach for cases where pairs of different modality images corresponding to the same subject are available. We introduce a novel pre-training task, where a model must to match together different-modality scans showing the same subject by comparing them in a joint, modality-invariant embedding space. If these modalities are substantially different in appearance, the network must learn semantic data representations to solve this problem.

In itself, this is an important task. Embeddings obtained from the trained networks allow us to check if two different scans show the same subject in large anonymised datasets (by verifying that their embeddings match). It also defines a notion of similarity between scans that has applications in population studies. However, the main reward of our method are the semantic *spatial* representations of the data learnt during training which can be leveraged for a range of downstream tasks. In this paper we demonstrate the embeddings can be used for unsupervised rigid multi-modal scan registration, and cross-modal segmentation with opposite-modality annotations.

The layout of this paper is as follows: Sect. 2 describes the cross-modal matching task in detail, including the network architecture, loss function, and implementation details, as well as experimental results from a large, publically-available whole body scan dataset. Section 3 introduces algorithms using the embeddings learnt in Sect. 2 for fast unsupervised multi-modal scan registration which are shown to succeed in cases where conventional registration approaches fail. In Sect. 3.1, we then use these registrations to transfer segmentation maps between modalites, showing that by using the proposed cross-modal registration technique, anatomical annotations in DXAs can be used to train a segmentation network in MR scans.

1.1 Related Work

Self-supervised representation-learning is an incredibly active area of research at the moment. The current dominant praxis is to train models to perform challenging self-supervised learning tasks on a large dataset, and then fine-tune learnt representations for specific 'downstream' tasks using smaller, annotated datasets. Major successes have been reported in image classification [4,7,8,11,16], video understanding [13,27] and NLP [17,25,28], with self-supervised approaches often matching or exceeding the performance of fully-supervised approaches.

Due to the existence of a few large, publically available datasets (such as [20]), yet lack of large annotated datasets suitable for most medical tasks, self-supervised learning shows great promise in the medical domain. For example, previous work has shown it can be used to improve automated diagnosis of intervertebral disc degeneration [19] and common segmentation tasks [33]. In [32], it also is shown that using multiple MR sequences in self-supervised learning improves performance in brain tumour segmentation.

Data with multiple modalities is a natural candidate for self-supervised approaches, as one can use information from one modality to predict information in the other. For example, previous work has shown self-supervised methods can benefit from fusion of the audio and visual streams available in natural video data [1–3,21,26]. In this paper we build on this multi-modal approach by extending it to explicit spatial registration across the modalities.

1.2 Dataset Information, Acquisition and Preparation

For the experiments in this paper we use data from the UK Biobank [31], a large corpus of open-access medical data taken from over 500,000 volunteer participants. A wide variety of data is available, including data related to imaging, genetics and health-related outcomes. In this study we focus on two whole body imaging modalities collected by the Biobank: (1) 1.5 T, 6-min dual-echo Dixon protocol magnetic resonance (MR) scans showing the regions from approximately the neck to the knees of the participant with variation due to the subject's height and position in the scanner; (2) Dual energy x-ray absorptiometry (DXA) scans showing the entire body. In total, at the time of data collection, the Biobank consisted of 41,211 DXA scans and 44,830 MR scans from unique participants.

Our collected dataset consists of pairs of same-subject multi-sequence MR and DXA scans, examples of which can be seen in Fig. 1. In total we find 20,360 such pairs. These are separated into training, validation and test sets with a 80/10/10% split (16,213, 2,027 and 2,028 scan pairs respectively). Scan pairs are constructed using (1) the fat-only and water-only sequences of the Dixon MR scans, and (2) the tissue and bone images from the DXA scans. For the purposes of this study, we synthesize 2D coronal images from the 3D MR scans by finding the mid-spinal coronal slice at each axial scan line using the method described in [36]. All scans are resampled to be isotropic and cropped to a consistent size for ease of batch processing (501×224 for MR scans and 800×300 for DXA scans). These dimensions maintain an equal pixel spacing of 2.2 mm in both modalities. The scans are geometrically related in that the MRI field of view (FoV) is a cropped, translated and slightly rotated transformation of the DXA scan's FoV. Both scans are acquired with the subjects in a supine position, and there can be some arm and leg movements between the scans.

2 Matching Scans Across Modalities

This section describes the framework used to match same-subject scans across the DXA and MRI modalities. As shown in Fig. 1, this is hard to perform manually with only a few scans. Differences in tissue types visible in DXA and MRI

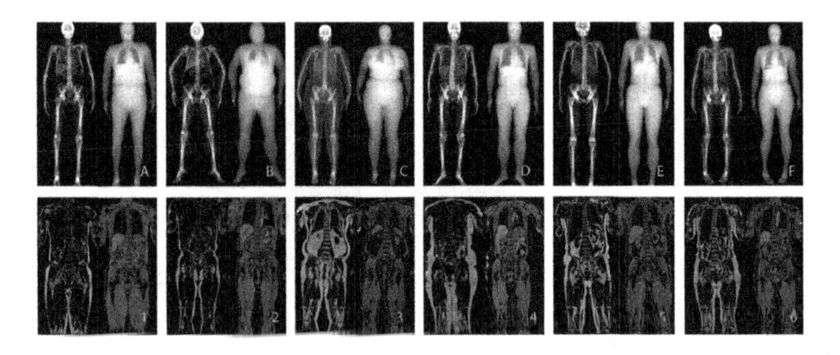

Fig. 1. Guess Who? Example scan pairs from our dataset. The top row shows bone (left) and tissue (right) DXA scans from the dataset. The bottom row shows synthesized mid-coronal fat-only (left) and water-only (right) Dixon MR slices. In this paper, semantic spatial representations of the scans are learnt by matching corresponding DXA and MR scan pairs. Can you match these pairs? ($A \rightarrow 5$, $B \rightarrow 3$, $C \rightarrow 4$, $D \rightarrow 2$, $E \rightarrow 1$, $F \rightarrow 6$)

mean many salient points in one modality are not visible at all in the other. Furthermore, the corresponding scans are not aligned, with variation in subject position, pose and rotation.

To tackle this problem, we use the dual encoder framework shown in Fig. 2, tasking it to determine the best alignment between the two scans such that similarity is higher for aligned same-subject scans than for aligned different-subject scans. Since both the DXA and MRI scans are coronal views and subject rotations relative to the scanner are very small, an approximate alignment requires determining a 2D translation between the scans. The similarity is then determined by a scalar product of the scans' spatial feature maps after alignment. In practice, this amounts to 2D convolution of the MRI's spatial feature map over the DXA's spatial feature map, and the maximum value of the resulting correlation map provides a similarity score.

The network is trained end-to-end by Noise Contrastive Estimation [12] over a batch of N randomly sampled matching pairs. If M_{ij} represents the similarity between the ith DXA and j^{th} MRI, where $i = j$ is a matching pair and $i \neq j$ is non-matching, and τ is some temperature parameter, the total loss for the k-th matching pair, ℓ_k, is given by

$$\ell_k = -\left(\log \frac{\exp(M_{kk}/\tau)}{\sum_{j=1}^{N} \exp(M_{kj}/\tau)} + \log \frac{\exp(M_{kk}/\tau)}{\sum_{j=1}^{N} \exp(M_{jk}/\tau)}\right) \tag{1}$$

2.1 Experiments

This section evaluates the performance of the proposed configuration on the cross-modal scan-matching task. To determine the relative importance of each MRI sequence and each DXA type, we train networks varying input channels

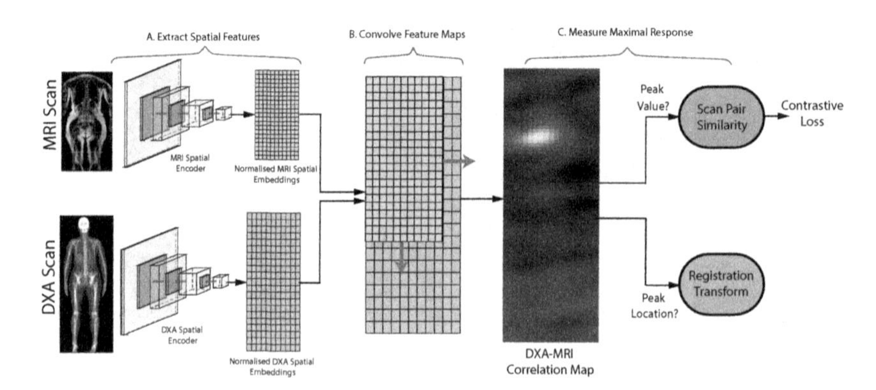

Fig. 2. The dual encoding configuration used for contrastive training. Two CNNs ingest scans of the respective modalities, outputting coarse spatial feature maps (A). The feature maps of each DXA-MRI pair are normalised and correlated to find the best registration (B). Using this registration, the maximum correlation is recorded as the similarity between the two scans (C). The architecture used for both spatial encoders is shown in the supplementary material.

to each modality's encoder (see Fig. 3b for the configurations used). To demonstrate the value of comparing spatial feature maps of scans instead of a single global embedding vector, we compare to a baseline network that simply pools the spatial feature maps into a scan-level descriptor, and is trained by the same contrastive method. Details of this baseline are given in the supplementary.

Implementation. Networks are trained with a batch size of 10 using an Adam optimizer with a learning rate of 10^{-5} and $\beta = (0.9, 0.999)$. A cross-entropy temperature of $T = 0.01$ is used (a study of the effect of varying this is given in the supplementary). Spatial embeddings are 128-dimensional. Training augmentation randomly translates the both scans by ± 5 pixels in both axis and alters brightness and contrast by $\pm 20\%$. Each model takes 3 days to train on a 24 GB NVIDIA Tesla P40 GPU. Networks are implemented in PyTorch v.1.7.0.

Evaluation Measures. We evaluate the quality of the learnt embeddings on the test set by assessing the ability of the system to: (1) retrieve the matching opposite modality scan for a given query scan based on similarity; (2) verify if a given scan pair is matching or not. In the latter case, positive matches to a scan are defined as those with similarities above a threshold, ϕ, and negative matches have similarity $\leq \phi$. Considering all possible DXA-MRI scan pairs (matching & non-matching), we can then generate an ROC curve by varying ϕ from -1 to 1. For the retrieval task, we report top-1 and top-10 recall based on similarity across all test set subjects, and the mean rank of matching pairs. For the verification task, we report the ROC area-under-curve (AUC) and the equal error rate (EER) (i.e. when $TPR = FPR$).

Results. Figure 3 shows the ROC curve and performance measures for varying input channels. All configurations vastly exceed the baseline's performance,

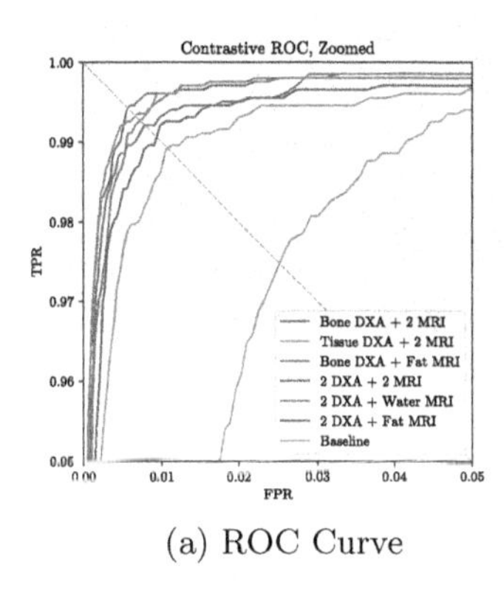

(a) ROC Curve

Input		Verification		Retrieval		
DXA	MRI	AUC	EER (%)	% Recall @1	% Recall @10	Mean Rank
Baseline		0.9952	2.57	26.3	78.7	9.246
B	F	0.9992	0.77	89.4	99.4	2.106
B	F,W	**0.9993**	0.84	87.7	99.4	2.079
T	F,W	0.9986	1.14	83.1	98.4	3.013
B,T	F	0.9989	0.98	85.8	98.7	2.569
B,T	W	**0.9993**	0.70	90.1	99.4	**1.920**
B,T	F,W	0.9992	**0.60**	**90.7**	**99.5**	2.526

(b) Retrieval and Verification Performance

Fig. 3. Verification and retrieval performance on the 2,028 scan test dataset with varying inputs of bone DXA (B), tissue DXA (T), fat-only MR (F), and water-only MR (W). (a) shows an ROC curve for the verification task. Table (b) reports performance statistics for the models, including equal error rate (EER), area under curve (AUC), recall at ranks 1 & 10 and the mean rank of matches.

indicating the benefit of spatial scan embeddings as opposed to scan-level descriptor vectors. The full model achieves a top-1 recall of over 90% from 2028 test cases. The tissue DXA-only model performs worst of all configurations suggesting bone DXAs are much more informative here. Extended results and recall at K curves are given in the supplementary material.

Discussion. The strong performance of the proposed method on the retrieval task by matching spatial (as opposed to global) features is significant; it suggests the encoders learn useful semantic information about specific regions of both scans. This has several possible applications. For example, one could select a query ROI in a scan, perhaps containing unusual pathology, calculate its spatial embeddings and find similar examples across a large dataset (see [30] for a more detailed discussion of this application). More conventionally, the learnt features could be also used for network initialization in downstream tasks on other smaller datasets of the same modality, potentially increasing performance and data efficiency. As a demonstration of the usefulness of the learnt features, the next section of this paper explores using them to register scans in a completely unsupervised manner.

3 Unsupervised Registration of Multi-modal Scans

A major advantage of this contrastive training method is that dense correspondences between multi-modal scans are learnt in a completely self-supervised manner. This is non-trivial; different tissues are shown in each modality, making intensity-based approaches for same- or similar-modality registration [23,35] ineffective. Here we explore this idea further, developing three methods for estimating rigid registrations between the modalities. Each method is assessed by

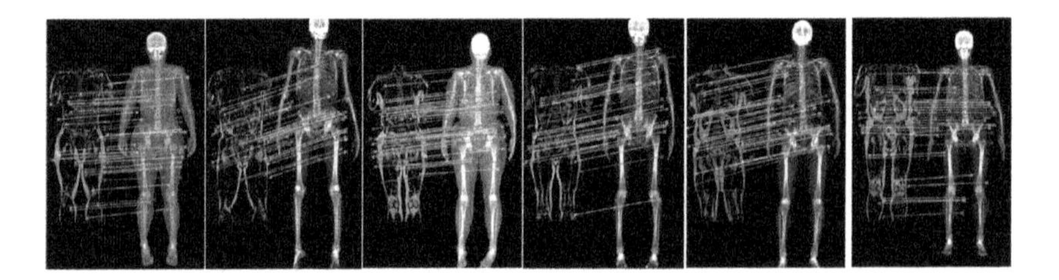

Fig. 4. Salient point correspondences between scan pairs found by Lowe's ratio test & RANSAC. The fat-only channel of MRI source image is shown on the left, with the target DXA bone scan shown on the right

measuring L2-distance error when transforming anatomical keypoints from the MRIs to the DXA scans. For each proposed registration method the aim is to estimate the three transformation parameters; a 2D translation and a rotation.

1. Dense Correspondences: During training, the contrastive framework attempts to align dense spatial feature maps before comparing them. We can use this to determine the registration translation by convolving the feature maps together and measuring the point of maximum response as the displacement between the images (as in Fig. 2, stages A, B). The rotation between the scans is found by rotating the MRI scan across a small range of angles, convolving the feature maps, and recording the angle which induces the greatest alignment.

2. Salient Point Matching: The dense correspondence method is slow, especially on a CPU, as it requires multiple convolution operations with large kernels. To speed up registration we need use only a few salient points between the feature maps. These can be found by matching pairs of points based on correlations and then employing Lowe's second nearest neighbour ratio test [22] to remove ambiguous correspondences, followed by RANSAC estimation of the transformation. Details of this algorithm are given in the supplementary material. Example correspondences found by this method are shown in Fig. 4.

3. Refinement Regressor: The previous approaches generate robust approximate registrations between the two images but are limited by the resolution of the feature maps they compare ($8\times$ downsample of a 2.2 mm pixel spacing original image). To rectify this issue we use a small regression network to refine predictions by taking the almost-aligned feature maps predicted by the aforementioned methods and then outputting a small refinement transformation. High-precision training data for this task can be generated 'for free' by taking aligned scan pairs from the salient point matching method, slightly misaligning them with a randomly sampled rotation and translation and then training a network to regress this random transformation. The regression network is trained on 50 aligned pairs predicted by the previous salient point matching method and manually checked for accuracy. For each pair, several copies are generated with slight randomly sampled translations and rotations at the original pixel resolution. For each transformed pair, the DXA and MRI spatial feature maps are then concate-

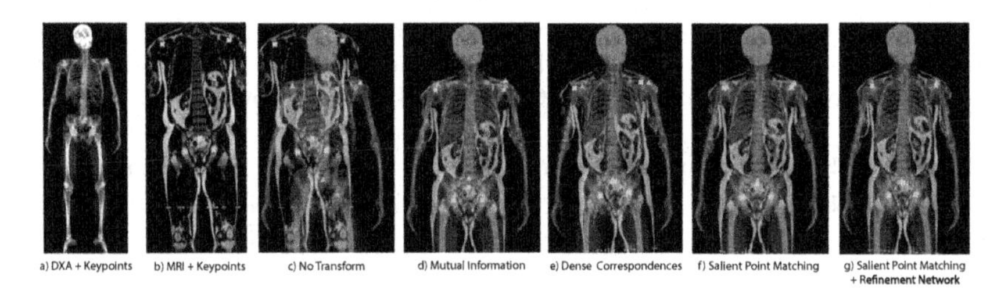

Fig. 5. Example results from each registration method. (a) & (b) show keypoints for the MRI and the DXA. The MRI & keypoints are registered to the DXA by: (c) no transform; (d) mutual information maximisation; (e) Dense correspondences as in pre-training; (f) Salient point matching via Lowe's ratio test; (g) Applying the refinement regressor to (f).

nated together, and used as input to a small CNN followed by a fully-connected network that estimates the three parameters of the transformation for each pair.

Experiments. To measure the quality of these registrations, 5 keypoints were marked in 100 test scan pairs: the femoral head in both legs (hip joints), humerus head in both arms (shoulder joints) and the S1 vertebra (base of the spine). MRI keypoints are annotated in 3D and then projected into 2D. These keypoints provide the ground truth for assessing the predicted transformations. Example annotations are shown in Fig. 5. We then measure the mean L2-localisation error when transferring the keypoints between modalities using rigid body transforms predicted by the proposed methods. We compare to baselines of (i) no transformation (i.e. the identity); and (ii) rigid mutual information maximisation[1]. To measure annotation consistency and error induced by change in subject pose, we also report the error of the 'best-possible' rigid transformation keypoints - that which minimises the mean L2 transfer error.

Results. Table 1 shows the localisation error achieved by each method. All methods yield accurate registrations between the images. The best method is found to salient point matching followed by the refinement network, which is also shown to be fast on both a GPU and CPU. We attempted to calculate SIFT and MIND features in both modalities and match them as proposed in [34] and [15] respectively however these approaches did not work in this case (see supplementary material).

Discussion. In this setting, our methods were found to outperform other approaches for multi-modal registration (mutual information, MIND and SIFT). We believe the reason for this is that DXA scans show mostly bony structures, whereas most visual content in MRI is due to soft tissues which can't be differentiated by DXA. As such, most pixels have no obvious intensity relation between scans. However, accurate registration between the scans is important as it allows collation of spatial information from both modalities. This can be exploited in

[1] Using MATLAB's `imregister` with `MattesMutualInformation` [24] as an objective.

Table 1. Keypoint transfer error for the proposed methods. We report the mean and median error for all keypoints combined and for the hip joints (HJ), shoulder joints (SJ) and S1 individually. Runtime on a GPU & CPU is also shown.

Method	Keypoint transfer error (cm)					Time (s)	
	HJ	S1	SJ	Median (all)	Mean (all)	GPU	CPU
No Trans.	22.1 ± 5.0	21.7 ± 5.3	22.3 ± 5.0	21.9	22.01 ± 5.1	0	0
Mut. Inf.	2.23 ± 1.3	2.67 ± 1.4	2.75 ± 2.2	2.21	2.52 ± 1.7	–	1.0
Dense Corr.	1.48 ± 0.8	1.52 ± 0.8	2.05 ± 1.2	1.52	1.72 ± 1.0	1.5	5.7
Sal. Pt. Mt.	1.34 ± 0.9	1.37 ± 1.0	2.04 ± 1.4	1.46	1.63 ± 1.3	0.4	1.1
Regressor	1.24 ± 0.8	1.30 ± 0.9	1.44 ± 0.9	1.12	1.32 ± 0.9	0.9	1.5
Best Poss.	0.84 ± 0.4	0.84 ± 0.5	0.87 ± 0.4	0.84	0.87 ± 0.4	–	–

at least two ways: (i) for joint features; registration allows shallow fusion of spatial features from both modalities. This could be useful in, for example, body composition analysis, conventionally done by DXA but which may benefit from the superior soft tissue contrast of MRI [5]. (ii) for cross-modal supervision; registration allows prediction of dense labels from one modality which can then be used a training target for the other. For example one could diagnose osteoporosis/fracture risk at a given vertebral level from MR using labels extracted from DXA by conventional methods.

3.1 Cross-Modal Annotation Transfer

A benefit of the demonstrated cross-modal registrations is that they allow the transfer of segmentations between significantly different modalities, meaning segmentation networks can be trained in both modalities from a single annotation set. This is useful in cases when a tissue is clearly visible in one modality but not the other. For example, here the pelvis is clearly visible in the DXA scan but not in the MRI slice. As an example of using cross-modal annotation transfer, the spine, pelvis and pelvic cavity are segmentated in DXA scans using the method from [18]. These segmentations are then transferred to the MRI scans by the refinement network from Sect. 3. where they act as pixel-wise annotations to train a 2D U-Net [29] segmentation network. Compared to manual segmentation of the spine performed in 50 3D MR scans and projected into 2D, this network achieves good performance, with a mean Dice score of 0.927 showing the quality of the transferred annotations. Examples are shown in the supplementary.

4 Conclusion

This paper explores a new self-supervised task of matching different-modality, same-subject whole-body scans. Our method to achieves this by jointly aligning and comparing scan spatial embeddings via noise contrastive estimation. On a test dataset of 2028 scan pairs our method is shown to perform exceptionally well with over 90% top-1 recall. We then show the learnt spatial embeddings can be used for unsupervised multi-modal registration in cases where standard approaches fail. These registrations can then be used to perform cross-modal annotation transfer, using DXA segmentations to train a MRI-specific model to segment anatomical structures. Future work will explore using the learnt spatial embeddings for other downstream tasks and extend this method to 3D scans.

Acknowledgements. Rhydian Windsor is supported by Cancer Research UK as part of the EPSRC CDT in Autonomous Intelligent Machines and Systems (EP/L015897/1). We are also grateful for support from a Royal Society Research Professorship and EPSRC Programme Grant Visual AI (EP/T028572/1).

References

1. Alwassel, H., Mahajan, D., Korbar, B., Torresani, L., Ghanem, B., Tran, D.: Self-supervised learning by cross-modal audio-video clustering. In: NeurIPS (2020)
2. Arandjelović, R., Zisserman, A.: Look, listen and learn. In: Proceedings of the ICCV (2017)
3. Arandjelović, R., Zisserman, A.: Objects that sound. In: Ferrari, V., Hebert, M., Sminchisescu, C., Weiss, Y. (eds.) ECCV 2018. LNCS, vol. 11205, pp. 451–466. Springer, Cham (2018). https://doi.org/10.1007/978-3-030-01246-5_27
4. Asano, Y.M., Rupprecht, C., Vedaldi, A.: Self-labelling via simultaneous clustering and representation learning. In: Proceedings of the ICLR (2020)
5. Borga, M.: MRI adipose tissue and muscle composition analysis–a review of automation techniques. Br. J. Radiol. **91**(1089), 20180252 (2018)
6. Brown, T., et al.: Language models are few-shot learners. In: NeurIPS (2020)
7. Caron, M., Misra, I., Mairal, J., Goyal, P., Bojanowski, P., Joulin, A.: Unsupervised learning of visual features by contrasting cluster assignments. In: NeurIPS (2020)
8. Chen, T., Kornblith, S., Norouzi, M., Hinton, G.: A simple framework for contrastive learning of visual representations. In: Proceedings of the ICLR (2020)
9. Devlin, J., Chang, M.W., Lee, K., Toutanova, K.: BERT: pre-training of deep bidirectional transformers for language understanding. In: Proceedings of the NAACL, pp. 4171–4186 (2019)
10. Ghorbani, A., Natarajan, V., Coz, D., Liu, Y.: DermGAN: synthetic generation of clinical skin images with pathology. In: Machine Learning for Health NeurIPS Workshop, pp. 155–170 (2019)
11. Grill, J.B., et al.: Bootstrap your own latent - a new approach to self-supervised learning. In: NeurIPS (2020)
12. Gutmann, M.U., Hyvärinen, A.: Noise-contrastive estimation of unnormalized statistical models, with applications to natural image statistics. J. Mach. Learn. Res. **13**(11), 307–361 (2012)
13. Han, T., Xie, W., Zisserman, A.: Self-supervised co-training for video representation learning. In: NeurIPS (2020)
14. He, K., Fan, H., Wu, Y., Xie, S., Girshick, R.: Momentum contrast for unsupervised visual representation learning. In: Proceedings of the CVPR (2020)
15. Heinrich, M.P., et al.: MIND: modality independent neighbourhood descriptor for multi-modal deformable registration. Med. Image Anal. **16**(7), 1423–1435 (2012)

16. Hénaff, O., et al.: Data-efficient image recognition with contrastive predictive coding. In: Proceedings of the ICLR (2020)
17. Howard, J., Ruder, S.: Universal language model fine-tuning for text classification. In: Proceedings of the ACL (2018)
18. Jamaludin, A., Kadir, T., Clark, E., Zisserman, A.: Predicting scoliosis in DXA scans using intermediate representations. In: Zheng, G., Belavy, D., Cai, Y., Li, S. (eds.) CSI 2018. LNCS, vol. 11397, pp. 15–28. Springer, Cham (2019). https://doi.org/10.1007/978-3-030-13736-6_2
19. Jamaludin, A., Kadir, T., Zisserman, A.: Self-supervised learning for spinal MRIs. In: Cardoso, M.J., et al. (eds.) DLMIA/ML-CDS-2017. LNCS, vol. 10553, pp. 294–302. Springer, Cham (2017). https://doi.org/10.1007/978-3-319-67558-9_34
20. Johnson, A.E.W., et al.: MIMIC-CXR, a de-identified publicly available database of chest radiographs with free-text reports. Sci. Data **6**(1), 317 (2019)
21. Korbar, B., Tran, D., Torresani, L.: Cooperative learning of audio and video models from self-supervised synchronization. In: NeurIPS, vol. 31 (2018)
22. Lowe, D.: Object recognition from local scale-invariant features. In: Proceedings of the ICCV, pp. 1150–1157, September 1999
23. Lowe, D.: Distinctive image features from scale-invariant keypoints. IJCV **60**(2), 91–110 (2004)
24. Mattes, D., Haynor, D.R., Vesselle, H., Lewellyn, T.K., Eubank, W.: Nonrigid multimodality image registration. In: Sonka, M., Hanson, K.M. (eds.) Medical Imaging 2001: Image Processing, vol. 4322, pp. 1609–1620. International Society for Optics and Photonics, SPIE (2001). https://doi.org/10.1117/12.431046
25. Mikolov, T., Sutskever, I., Chen, K., Corrado, G.S., Dean, J.: Distributed representations of words and phrases and their compositionality. In: NeurIPS (2013)
26. Owens, A., Efros, A.A.: Audio-visual scene analysis with self-supervised multisensory features. In: Ferrari, V., Hebert, M., Sminchisescu, C., Weiss, Y. (eds.) ECCV 2018. LNCS, vol. 11210, pp. 639–658. Springer, Cham (2018). https://doi.org/10.1007/978-3-030-01231-1_39
27. Qian, R., et al.: Spatiotemporal contrastive video representation learning. In: Proceedings of the CVPR (2021)
28. Radford, A., Wu, J., Child, R., Luan, D., Amodei, D., Sutskever, I.: Language models are unsupervised multitask learners. Technical report, OpenAI (2019)
29. Ronneberger, O., Fischer, P., Brox, T.: U-Net: convolutional networks for biomedical image segmentation. In: Navab, N., Hornegger, J., Wells, W.M., Frangi, A.F. (eds.) MICCAI 2015. LNCS, vol. 9351, pp. 234–241. Springer, Cham (2015). https://doi.org/10.1007/978-3-319-24574-4_28
30. Simonyan, K., Zisserman, A., Criminisi, A.: Immediate structured visual search for medical images. In: Fichtinger, G., Martel, A., Peters, T. (eds.) MICCAI 2011. LNCS, vol. 6893, pp. 288–296. Springer, Heidelberg (2011). https://doi.org/10.1007/978-3-642-23626-6_36
31. Sudlow, C., et al.: UK Biobank: an open access resource for identifying the causes of a wide range of complex diseases of middle and old age. PLoS Med. **12**(3), 1–10 (2015)
32. Taleb, A., Lippert, C., Klein, T., Nabi, M.: Multimodal self-supervised learning for medical image analysis. In: Feragen, A., Sommer, S., Schnabel, J., Nielsen, M. (eds.) IPMI 2021. LNCS, vol. 12729, pp. 661–673. Springer, Cham (2021). https://doi.org/10.1007/978-3-030-78191-0_51
33. Taleb, A., et al.: 3D self-supervised methods for medical imaging. In: NeurIPS (2020)

34. Toews, M., Zöllei, L., Wells, W.M.: Feature-based alignment of volumetric multi-modal images. In: Gee, J.C., Joshi, S., Pohl, K.M., Wells, W.M., Zöllei, L. (eds.) IPMI 2013. LNCS, vol. 7917, pp. 25–36. Springer, Heidelberg (2013). https://doi.org/10.1007/978-3-642-38868-2_3
35. Viola, P., Wells, W.: Alignment by maximization of mutual information. In: Press, I.C.S. (ed.) Proceedings of the ICCV, pp. 16–23, June 1995
36. Windsor, R., Jamaludin, A., Kadir, T., Zisserman, A.: A Convolutional approach to vertebrae detection and labelling in whole spine MRI. In: Martel, A.L., et al. (eds.) MICCAI 2020. LNCS, vol. 12266, pp. 712–722. Springer, Cham (2020). https://doi.org/10.1007/978-3-030-59725-2_69

SimTriplet: Simple Triplet Representation Learning with a Single GPU

Quan Liu[1], Peter C. Louis[2], Yuzhe Lu[1], Aadarsh Jha[1], Mengyang Zhao[1], Ruining Deng[1], Tianyuan Yao[1], Joseph T. Roland[2], Haichun Yang[2], Shilin Zhao[2], Lee E. Wheless[2], and Yuankai Huo[1(✉)]

[1] Vanderbilt University, Nashville, TN 37215, USA
yuankai.huo@vanderbilt.edu
[2] Vanderbilt University Medical Center, Nashville, TN 37215, USA

Abstract. Contrastive learning is a key technique of modern self-supervised learning. The broader accessibility of earlier approaches is hindered by the need of heavy computational resources (e.g., at least 8 GPUs or 32 TPU cores), which accommodate for large-scale negative samples or momentum. The more recent SimSiam approach addresses such key limitations via stop-gradient without momentum encoders. In medical image analysis, multiple instances can be achieved from the same patient or tissue. Inspired by these advances, we propose a simple triplet representation learning (SimTriplet) approach on pathological images. The contribution of the paper is three-fold: (1) The proposed SimTriplet method takes advantage of the multi-view nature of medical images beyond self-augmentation; (2) The method maximizes both intra-sample and inter-sample similarities via triplets from positive pairs, without using negative samples; and (3) The recent mix precision training is employed to advance the training by only using a single GPU with 16 GB memory. By learning from 79,000 unlabeled pathological patch images, SimTriplet achieved 10.58% better performance compared with supervised learning. It also achieved 2.13% better performance compared with SimSiam. Our proposed SimTriplet can achieve decent performance using only 1% labeled data. The code and data are available at https://github.com/hrlblab/SimTriplet.

Keywords: Contrastive learning · SimTriplet · Classification · Pathology

1 Introduction

To extract clinically relevant information from GigaPixel histopathology images is essential in computer-assisted digital pathology [15,22,25]. For instance, the Convolutional Neural Network (CNN) based method has been applied to depreciate sub-tissue types on whole slide images (WSI) so as to alleviate tedious manual efforts for pathologists [23]. However, pixel-wise annotations are resource extensive given the high resolution of the pathological images. Thus, the fully

M. de Bruijne et al. (Eds.): MICCAI 2021, LNCS 12902, pp. 102–112, 2021.
https://doi.org/10.1007/978-3-030-87196-3_10

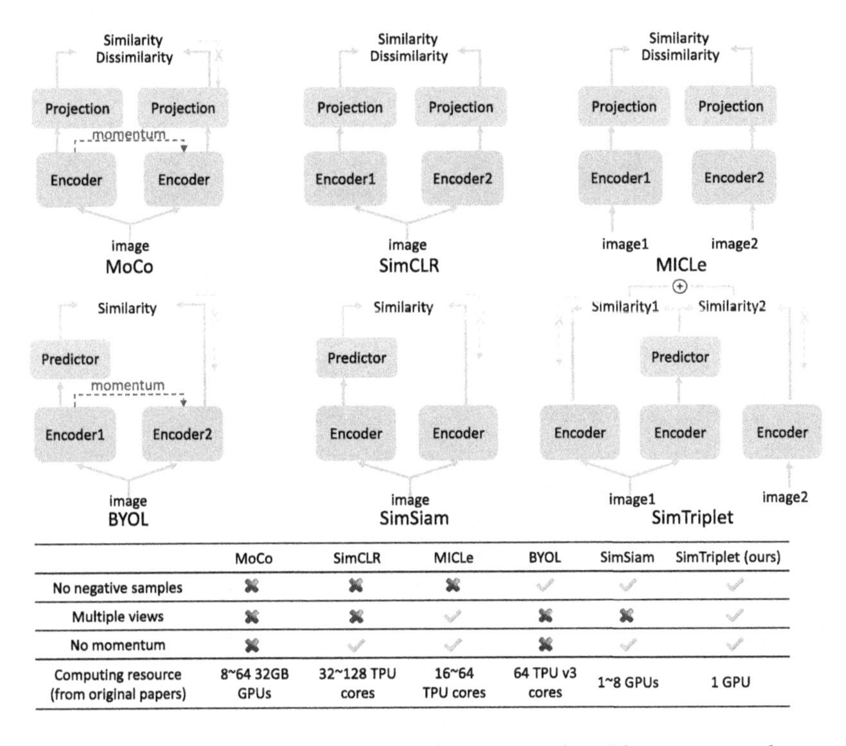

	MoCo	SimCLR	MICLe	BYOL	SimSiam	SimTriplet (ours)
No negative samples	✗	✗	✗	✓	✓	✓
Multiple views	✗	✗	✓	✗	✗	✓
No momentum	✗	✓	✓	✗	✓	✓
Computing resource (from original papers)	8~64 32GB GPUs	32~128 TPU cores	16~64 TPU cores	64 TPU v3 cores	1~8 GPUs	1 GPU

Fig. 1. Comparison of contrastive learning strategies. The upper panel compares the proposed SimTriplet with current representative contrastive learning strategies. The lower panel compares different approaches via a table.

supervised learning schemes might not be scalable for large-scale studies. To minimize the need of annotation, a well-accepted learning strategy is to first learn local image features through unsupervised feature learning, and then aggregate the features with multi-instance learning or supervised learning [14].

Recently, a new family of unsupervised representation learning, called contrastive learning (Fig. 1), shows its superior performance in various vision tasks [12,18,21,26]. Learning from large-scale unlabeled data, contrastive learning can learn discriminative features for downstream tasks. SimCLR [6] maximizes the similarity between images in the same category and repels representation of different category images. Wu et al. [21] uses an offline dictionary to store all data representation and randomly select training data to maximize negative pairs. MoCo [10] introduces a momentum design to maintain a negative sample pool instead of an offline dictionary. Such works demand large batch size to include sufficient negative samples (Fig. 1). To eliminate the needs of negative samples, BYOL [9] was proposed to train a model with a asynchronous momentum encoder. Recently, SimSiam [7] was proposed to further eliminate the momentum encoder in BYOL, allowing less GPU memory consumption.

To define different image patches as negative samples on pathological images is tricky since such a definition can depends on the patch size, rather than

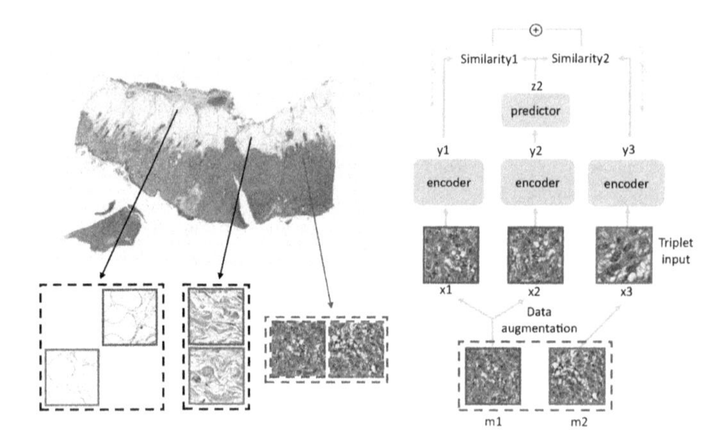

Fig. 2. Network structure of the proposed SimTriplet. Adjacent image pairs are sampled from unlabeled pathological images (left panel) for triplet representation learning (right panel). The GigaPixel pathological images provide large-scale "positive pairs" from nearby image patches for SimTriplet. Each triplet consists of two augmentation views from m_1 and one augmentation view from m_2. The final loss maximizes both inter-sample and intra-sample similarity as a representation learning.

semantic differences. Therefore, it would be more proper to use nearby image patches as multi-view samples (or called positive samples) of the same tissue type [20] rather than negative pairs. MICLe [2] applied multi-view contrastive learning to medical image analysis. Note that in [2,20], the negative pairs are still needed within the SimCLR framework.

In this paper, we propose a simple triplet based representation learning approach (SimTriplet), taking advantage of the multi-view nature of pathological images, with effective learning by using only a single GPU with 16 GB memory. We present a triplet similarity loss to maximize the similarity between two augmentation views of same image and between adjacent image patches. The contribution of this paper is three-fold:

- The proposed SimTriplet method takes advantage of the multi-view nature of medical images beyond self-augmentation.
- This method minimizes both intra-sample and inter-sample similarities from positive image pairs, without the needs of negative samples.
- The proposed method can be trained using a single GPU setting with 16 GB memory, with batch size = 128 for 224×224 images, via mixed precision training.

2 Methods

The principle network of SimTriplet is presented in Fig. 2. The original SimSiam network can be interpreted as an iterative process of two steps: (1) unsupervised

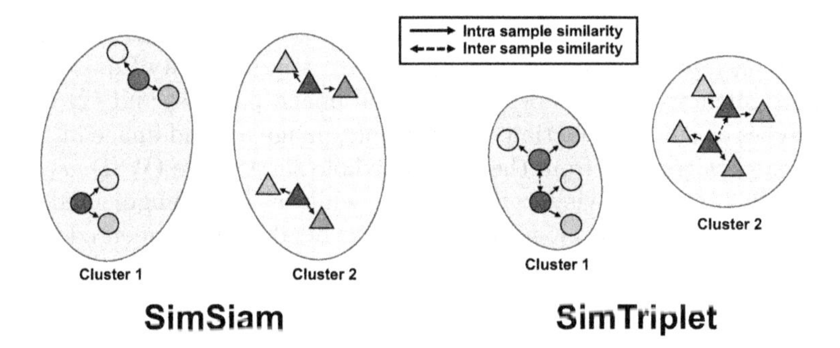

Fig. 3. Compare SimTriplet with SimSiam. SimSiam network maximizes intra-sample similarity by minimizing distance between two augmentation views from the same image. The proposed SimTriplet model further enforce the inter-sample similarity from positive sample pairs.

clustering and (2) feature updates based on clustering (similar to K-means or EM algorithms) [7]. By knowing the pairwise information of nearby samples, the SimTriplet aims to further minimize the distance between the "positive pairs" (images from the same classes) in the embedding space (Fig. 3). In the single GPU setting with batch size 128, SimTriplet provides more rich information for the clustering stage.

2.1 Multi-view Nature of Medical Images

In many medical image analysis tasks, multi-view (or called multi-instance) imaging samples from the same patient or the same tissue can provide complementary representation information. For pathological images, the nearby image patches are more likely belong to the same tissue type. Thus, the spatial neighbourhood on a WSI provide rich "positive pairs" (patches with same tissue types) for triplet representation learning. Different from [13], all samples in our triplets are positive samples, inspired by [7]. To train SimTriplet, we randomly sample image patches as well as their adjacent patches (from one of eight nearby locations randomly) as positive sample pairs from the same tissue type.

2.2 Triplet Representation Learning

Our SimTriplet network forms a triplet from three randomly augmented views by sampling positive image pairs (Fig. 2). The three augmented views are fed into the encoder network. The encoder network consists of a backbone network (ResNet-50 [11]) and a three-layer multi-layer perceptron (MLP) projection header. The three forward encoding streams share the same parameters. Next, an MLP predictor is used in the middle path. The predictor processes the encoder output from one image view to match with the encoder output of two other image views. We applies stop-gradient operations to two side paths. When computing loss between predictor output and image representation from encoder output, encoded representation is regarded as constant [7]. Two encoders on side paths will not be updated by back propagation. We used negative cosine similarity Eq. (1) between different augmentation views of (1) the same image patches, and (2) adjacent image patches as our loss function. For example, image m_1 and image m_2 are two adjacent patches cropped from the original whole slide image (WSI). x_1 and x_2 are randomly augmented views of image m_1, while x_3 is the augmented view of image m_2. Representation y_1, y_2 and y_3 are encoded from augmented views by encoder. z_1, z_2 and z_3 are the representation processed by the predictor.

$$\mathcal{C}(p,q) = -\frac{p}{\|p\|_2} \cdot \frac{q}{\|q\|_2} \tag{1}$$

$\mathcal{L}_{Intrasample}$ is the loss function to measure the similarities between two augmentation views x_1 and x_2 of image m_1 as seen in Eq. (2).

$$\mathcal{L}_{Intrasample} = \frac{1}{2}\mathcal{C}(y_1, z_2) + \frac{1}{2}\mathcal{C}(y_2, z_1) \tag{2}$$

$\mathcal{L}_{Intersample}$ is the loss function to measure the similarities between two augmentation views x_2 and x_3 of adjacent image pair m_1 and m_2 as in Eq. (3).

$$\mathcal{L}_{Intersample} = \frac{1}{2}\mathcal{C}(y_2, z_3) + \frac{1}{2}\mathcal{C}(y_3, z_2) \tag{3}$$

The triplet loss function as used in our SimTriplet network is defined as:

$$\mathcal{L}_{total} = \mathcal{L}_{Intrasample} + \mathcal{L}_{Intersample} \tag{4}$$

$\mathcal{L}_{Intrasample}$ minimizes the distance between different augmentations from the same image. $\mathcal{L}_{Intersample}$ minimizes the difference between nearby image patches.

2.3 Expand Batch Size via Mix Precision Training

Mix precision training [17] was invented to offer significant computational speedup and less GPU memory consumption by performing operations in half-precision format. The minimal information is stored in single-precision to retain the critical parts of the training. By implementing the mix precision to SimTriplet, we can extend the batch size from 64 to 128 to train images with 224×224 pixels, using a single GPU with 16 GB memory. The batch size 128 is regarded as a decent batch size in SimSiam [7].

3 Data and Experiments

3.1 Data

Annotated Data. We extracted image patches from seven melanoma skin cancer Whole Slide Images (WSIs) from the Cancer Genome Atlas (TCGA) Datasets (TCGA Research Network: https://www.cancer.gov/tcga). From the seven annotated WSIs, 4698 images from 5 WSIs were obtained for training and validation, while 1,921 images from 2 WSIs were used for testing. Eight tissue types were annotated as: blood vessel (353 train 154 test), epidermis (764 train 429 test), fat (403 train 137 test), immune cell (168 train 112 test), nerve (171 train 0 test), stroma (865 train 265 test), tumor (1,083 train 440 test) and ulceration (341 train 184 test).

Following [19,24] each image was a 512×512 patch extracted from $40\times$ magnification of a WSI with original pixel resolution 0.25 micron meter. The cropped image samples were annotated by a board certified dermatologist and confirmed by another pathologist. Then, the image patches were resized to 128×128 pixels. Note that the 224×224 image resolution provided 1.8% higher balance accuracy (based on our experiments) using the supervised learning. We chose 128×128 resolution for all experiments for a faster training speed.

Unlabeled Data. Beyond the 7 annotated WSIs, additional 79 WSIs without annotations were used for training contrastive learning models. The 79 WSIs were all available and usable melanoma cases from TCGA. The number and size of image patches used for different contrastive learning strategies are described in §**Experiment**.

3.2 Supervised Learning

We used ResNet-50 as the backbone in supervised training, where the optimizer is Stochastic Gradient Descent (SGD) [3] with the base learning rate $lr = 0.05$. The optimizer learning rate followed (linear scaling [8]) $lr \times$BatchSize/256. We used 5-fold cross validation by using images from four WSIs for training and image from the remaining WSI for validation. We trained 100 epochs and selected best model based on validation. When applying the trained model on testing images, the predicted probabilities from five models were averaged. Then, the class with the largest ensemble probability was used as the predicted label.

3.3 Training Contrastive Learning Benchmarks

We used the SimSiam network [6] as the baseline method of contrastive learning. Two random augmentations from the same image were used as training data. In all of our self-supervised pre-training, images for model training were resized to 128×128 pixels. We used momentum SGD as the optimizer. Weight decay was set to 0.0001. Base learning rate was $lr = 0.05$ and batch size equals 128. Learning rate was $lr \times$BatchSize/256, which followed a cosine decay schedule [16]. Experiments were achieved only on a single GPU with 16 GB memory. Models were pre-trained for $39,500/128 \times 400 \approx 127,438$ iterations. 79 unlabeled WSIs were used for self-supervised pre-training. We randomly cropped 500 images from each WSI and resized them to 128×128 pixels. 39,500 images in total serve as the original training data.

Following MICLe [2], we employed multi-view images as two inputs of the network. Since we did not use negative samples, multi-view images was trained

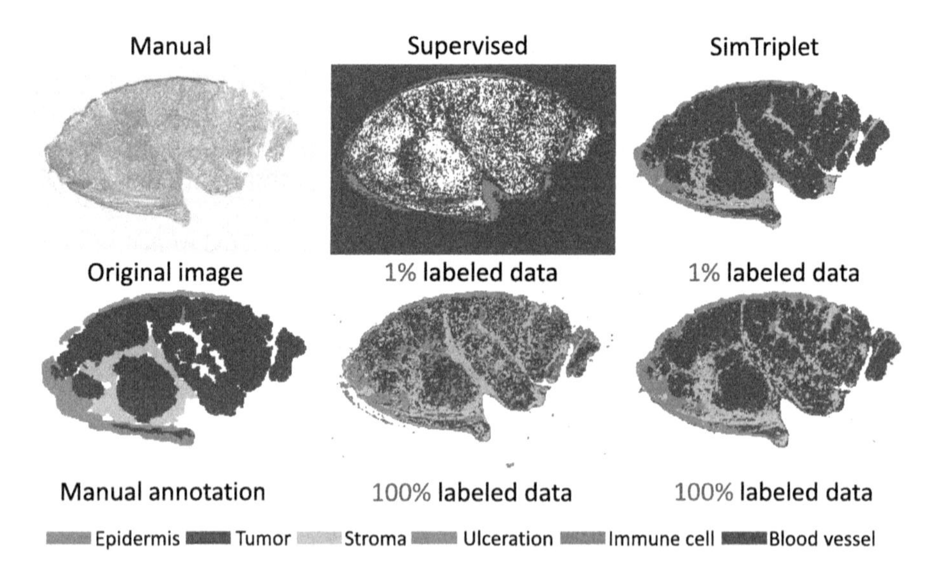

Fig. 4. Visualization of classification results. One tissue sample is manually segmented by our dermatologist (via QuPath software) to visually compare the classification results. The contrasting learning achieved superior performance compared with supervised learning, even using only 1% of all available labeled data.

by SimSiam network instead of SimCLR. For each image in the original training dataset, we cropped one patch which is randomly selected from its eight adjacent patches consisting of an adjacent images pairs. We had 79,000 images (39,500 adjacent pairs) as training data. Different from original SimSiam, network inputs were augmentation views of an adjacent pair. Referring to [7], we applied our data on SimSiam network. First, we used 39,500 images in original training dataset to pre-train on SimSiam. To see the impact of training dataset size, we randomly cropped another 39,500 images from 79 WSIs for training on a larger dataset of 79,000 images. We then used training data from the MICLe experiment to train the SimSiam network.

3.4 Training the Proposed SimTriplet

The same 79,000 images (39,500 adjacent pairs) were used to train the SimTriplet. Three augmentation views from each adjacent pair were used as network inputs. Two augmentation views were from one image, while the other augmentation view was augmented from adjacent images. Three augmentation views were generated randomly, where the augmentation settings were similar with the experiment on SimSiam [6]. Batch size was 128 and experiment run on a single 16 GB memory GPU.

3.5 Linear Evaluation (Fine Tuning)

To apply the self-supervised pre-training networks, as a common practice, we froze the pretrained ResNet-50 model by adding one extra linear layer which

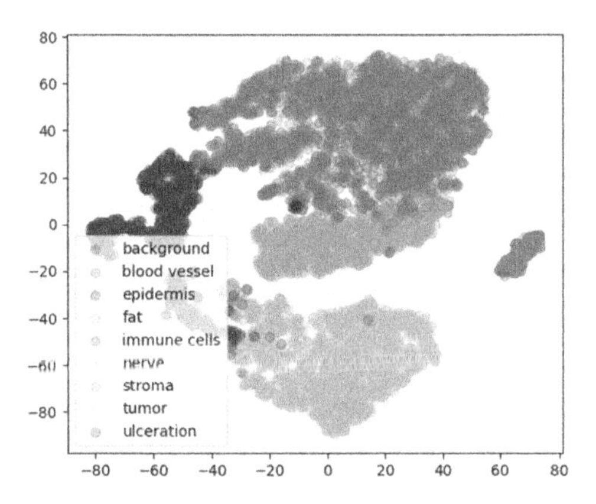

Fig. 5. t-SNE plot of abstracted feature by SimTriplet model. The abstracted feature is shown in t-SNE plot. Different color dots represent different tissue types.

followed the global average pooling layer. When finetuning with the annotated data, only the extra linear layer was trained. We used the SGD optimizer to train linear classifier with a based (initial) learning rate $lr = 30$, weight decay $= 0$, momentum $= 0.9$, and batch size $= 64$ (follows [7]). The same annotated dataset were used to finetune the contrastive learning models as well as to train supervised learning. Briefly, 4,968 images from 5 annotated WSIs were divided into 5 folders. We used 5-fold cross validation: using four of five folders as training data and the other folder as validation. We trained linear classifiers for 30 epochs and selected the best model based on the validation set. The pretrained models were applied to the testing dataset (1,921 images from two WSIs). As a multi-class setting, macro-level average F1 score was used [1]. Balanced accuracy was also broadly used to show the model performance on unbalanced data [4].

4 Results

Model Classification Performance. F1 score and balanced accuracy were used to evaluate different methods as described above. We trained a supervised learning models as the baseline. From Table 1, our proposed SimTriplet network achieved the best performance compared with the supervised model and SimSiam network [7] with same number of iterations. Compared with another benchmark SwAV [5], the F1 score and balanced accuracy of SwAV are 0.53 and 0.60, which are inferior compared with our SimTriplet (0.65 and 0.72) using the same batch size = 128 within 16 GB GPU memory. To show a qualitative result, a segmentation of a WSI from test dataset is shown in Fig. 4.

Model Performance on Partial Training Data. To evaluate the impact of training data number, we trained a supervised model and fine-tuned a classifier of the contrastive learning model on different percentages of annotated training

Table 1. Classification performance.

Methods	Unlabeled images	Paired inputs	F1 score	Balanced Acc
Supervised	0		0.5146	0.6113
MICLe [2]*	79k	✓	0.5856	0.6666
SimSiam [7]	39.5k		0.5421	0.5735
SimSiam [7]	79k	✓	0.6267	0.6988
SimSiam [7]	79k		0.6275	0.6958
SimTriplet (ours)	79k	✓	**0.6477**	**0.7171**

* We replace SimCLR with SimSiam.

Table 2. Balanced Acc of using different percentage of annotated data.

Methods	Percentage of used annotated training data			
	1%	10%	25%	100%
Supervised	0.0614	0.3561	0.4895	0.6113
SimSiam [7]	0.7085	0.6864	0.6986	0.6958
SimTriplet	**0.7090**	**0.7110**	**0.7280**	**0.7171**

data (Table 2). Note that for 1% to 25%, we ensure different classes contribute a similar numbers images to address the issue that the annotation is highly imbalanced.

5 Conclusion

In this paper, we proposed a simple contrastive representation learning approach, named SimTriplet, advanced by the multi-view nature of medical images. Our proposed contrastive learning methods maximize the similarity between both self augmentation views and pairwise image views from triplets. Moreover, our model can be efficiently trained on a single GPU with 16 GB memory. The performance of different learning schemes are evaluated on WSIs, with large-scale unlabeled samples. The proposed SimTriplet achieved superior performance compared with benchmarks, including supervised learning baseline and SimSiam method. The contrastive learning strategies showed strong generalizability by achieving decent performance by only using 1% labeled data.

Acknowledgement. Dr. Wheless is funded by grants from the Skin Cancer Foundation and the Dermatology Foundation.

References

1. Attia, M., Samih, Y., Elkahky, A., Kallmeyer, L.: Multilingual multi-class sentiment classification using convolutional neural networks. In: Proceedings of the Eleventh International Conference on Language Resources and Evaluation (LREC 2018) (2018)
2. Azizi, S., et al.: Big self-supervised models advance medical image classification (2021)
3. Bottou, L.: Large-scale machine learning with stochastic gradient descent. In: Lechevallier, Y., Saporta, G. (eds.) Proceedings of COMPSTAT'2010, pp. 177–186. Springer, Physica-Verlag (2010)
4. Brodersen, K.H., Ong, C.S., Stephan, K.E., Buhmann, J.M.: The balanced accuracy and its posterior distribution. In: 2010 20th International Conference on Pattern Recognition, pp. 3121–3124 (2010). https://doi.org/10.1109/ICPR.2010.764
5. Caron, M., Misra, I., Mairal, J., Goyal, P., Bojanowski, P., Joulin, A.: Unsupervised learning of visual features by contrasting cluster assignments. arXiv preprint arXiv:2006.09882 (2020)
6. Chen, T., Kornblith, S., Norouzi, M., Hinton, G.: A simple framework for contrastive learning of visual representations (2020)
7. Chen, X., He, K.: Exploring simple siamese representation learning. arXiv preprint arXiv:2011.10566 (2020)
8. Goyal, P., et al.: Accurate, large minibatch SOD: Training imagenet in 1 hour. arXiv preprint arXiv:1706.02677 (2017)
9. Grill, J.B., et al.: Bootstrap your own latent: a new approach to self-supervised learning. arXiv preprint arXiv:2006.07733 (2020)
10. He, K., Fan, H., Wu, Y., Xie, S., Girshick, R.: Momentum contrast for unsupervised visual representation learning. In: Proceedings of the IEEE/CVF Conference on Computer Vision and Pattern Recognition, pp. 9729–9738 (2020)
11. He, K., Zhang, X., Ren, S., Sun, J.: Deep residual learning for image recognition. In: Proceedings of the IEEE Conference on Computer Vision and Pattern Recognition, pp. 770–778 (2016)
12. Hjelm, R.D., et al.: Learning deep representations by mutual information estimation and maximization. arXiv preprint arXiv:1808.06670 (2018)
13. Hoffer, E., Ailon, N.: Deep metric learning using triplet network. In: Feragen, A., Pelillo, M., Loog, M. (eds.) SIMBAD 2015. LNCS, vol. 9370, pp. 84–92. Springer, Cham (2015). https://doi.org/10.1007/978-3-319-24261-3_7
14. Hou, L., Samaras, D., Kurc, T.M., Gao, Y., Davis, J.E., Saltz, J.H.: Patch-based convolutional neural network for whole slide tissue image classification. In: Proceedings of the IEEE Conference on Computer Vision and Pattern Recognition (CVPR), June 2016
15. Liskowski, P., Krawiec, K.: Segmenting retinal blood vessels with deep neural networks. IEEE Trans. Med. limag. **35**(11), 2369–2380 (2016)
16. Loshchilov, I., Hutter, F.: SGDR: Stochastic gradient descent with warm restarts (2017)
17. Micikevicius, P., et al.: Mixed precision training (2018)
18. Noroozi, M., Favaro, P.: Unsupervised learning of visual representations by solving jigsaw puzzles. In: Leibe, B., Matas, J., Sebe, N., Welling, M. (eds.) ECCV 2016. LNCS, vol. 9910, pp. 69–84. Springer, Cham (2016). https://doi.org/10.1007/978-3-319-46466-4_5

19. Raju, A., Yao, J., Haq, M.M.H., Jonnagaddala, J., Huang, J.: Graph attention multi-instance learning for accurate colorectal cancer staging. In: Martel, A.L., et al. (eds.) MICCAI 2020. LNCS, vol. 12265, pp. 529–539. Springer, Cham (2020). https://doi.org/10.1007/978-3-030-59722-1_51
20. Tian, Y., Krishnan, D., Isola, P.: Contrastive multiview coding. arXiv preprint arXiv:1906.05849 (2019)
21. Wu, Z., Xiong, Y., Yu, S., Lin, D.: Unsupervised feature learning via non-parametric instance-level discrimination (2018)
22. Xu, Y., et al.: Deep convolutional activation features for large scale brain tumor histopathology image classification and segmentation. In: 2015 IEEE International Conference on Acoustics, Speech and Signal Processing (ICASSP), pp. 947–951. IEEE (2015)
23. Xu, Y., et al.: Large scale tissue histopathology image classification, segmentation, and visualization via deep convolutional activation features. BMC Bioinform. **18**(1), 1–17 (2017)
24. Zhao, Y., et al.: Predicting lymph node metastasis using histopathological images based on multiple instance learning with deep graph convolution. In: Proceedings of the IEEE/CVF Conference on Computer Vision and Pattern Recognition, pp. 4837–4846 (2020)
25. Zhu, C., et al.: Retinal vessel segmentation in colour fundus images using extreme learning machine. Comput, Med. Imag. Graph. **55**, 68–77 (2017)
26. Zhuang, C., Zhai, A.L., Yamins, D.: Local aggregation for unsupervised learning of visual embeddings. In: Proceedings of the IEEE/CVF International Conference on Computer Vision, pp. 6002–6012 (2019)

Lesion-Based Contrastive Learning for Diabetic Retinopathy Grading from Fundus Images

Yijin Huang[1], Li Lin[1,2], Pujin Cheng[1], Junyan Lyu[1], and Xiaoying Tang[1(✉)]

[1] Department of Electrical and Electronic Engineering, Southern University of Science and Technology, Shenzhen, China
tangxy@sustech.edu.cn

[2] School of Electronics and Information Technology, Sun Yat-sen University, Guangzhou, China

Abstract. Manually annotating medical images is extremely expensive, especially for large-scale datasets. Self-supervised contrastive learning has been explored to learn feature representations from unlabeled images. However, unlike natural images, the application of contrastive learning to medical images is relatively limited. In this work, we propose a self-supervised framework, namely lesion-based contrastive learning for automated diabetic retinopathy (DR) grading. Instead of taking entire images as the input in the common contrastive learning scheme, lesion patches are employed to encourage the feature extractor to learn representations that are highly discriminative for DR grading. We also investigate different data augmentation operations in defining our contrastive prediction task. Extensive experiments are conducted on the publicly-accessible dataset EyePACS, demonstrating that our proposed framework performs outstandingly on DR grading in terms of both linear evaluation and transfer capacity evaluation.

Keywords: Contrastive learning · Self-supervised learning · Lesion · Diabetic retinopathy · Fundus image

1 Introduction

Diabetic retinopathy (DR) is one of the microvascular complications of diabetes, which may cause vision impairments and even blindness [12]. The major pathological signs of DR include hemorrhages, exudates, microaneurysms, and retinal neovascularization, as shown in Fig. 1. The color fundus image is the most widely-used photography for ophthalmologists to identify the severity of DR, which can clearly reveal the presence of different lesions. Early diagnoses and timely interventions are of vital importance in preventing DR patients from vision malfunction. As such, automated and efficient fundus image based DR diagnosis systems are urgently needed.

Electronic supplementary material The online version of this chapter (https://doi.org/10.1007/978-3-030-87196-3_11) contains supplementary material, which is available to authorized users.

M. de Bruijne et al. (Eds.): MICCAI 2021, LNCS 12902, pp. 113–123, 2021.
https://doi.org/10.1007/978-3-030-87196-3_11

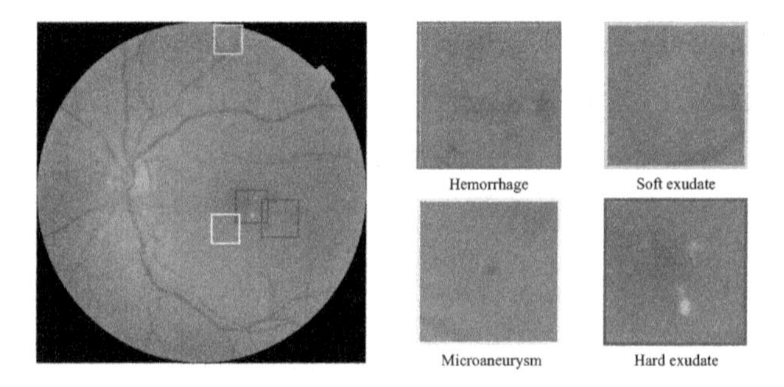

Fig. 1. A representative fundus image with four types of DR related lesions.

Recently, deep learning has achieved great success in the field of medical image analysis. Convolutional neural networks (CNNs) have been proposed to tackle the automated DR grading task [5,14,19]. The success of CNN is mainly attributed to its capability of extracting highly representative features. However, it usually requires a large-scale annotated dataset to train a network. The annotation process is time-intensive, tedious, and error-prone, and hence ophthalmologists bear a heavy burden in building a well-annotated dataset.

Self-supervised learning (SSL) methods [4,11,18,21] have been explored to learn feature representations from unlabeled images. As a representative SSL method, contrastive learning (CL) [1,2,8] has been very successful in the natural image field, which defines a contrastive prediction task that trying maximize the similarity between features from differently augmented views of the same image and simultaneously maximize the distance between features from different images, via a contrastive loss. However, the application of CL in the fundus image field [15,20] is relatively limited, mainly due to fundus images' high resolution and low proportion of diagnostic features. First, fundus images are of high resolution so as to clearly reveal structural and pathological details. And it is challenging for CL to train high-resolution images with large batch sizes which are nevertheless generally required by CL so as to provide more negative samples. Secondly, data augmentation is typically applied in CL to generate different views of the same image. However, some strong data augmentation operations applied to fundus images may destroy important domain-specific information, such as cropping and Cutout [3]. In a natural image, salient objects generally occupy a very large proportion and each part of them may contribute to the recognition of the object of interest, whereas the diagnostic features in a fundus image, such as lesions, may only occupy a small part of the whole image. This may result in that most cropped patches inputted to the CL framework have few or no diagnostic features. In this way, the network is prone to learn feature representations that are distinguishable for different views of the image but not discriminative for downstream tasks. CL in fundus image based DR grading is even rarer.

To address the aforementioned issues, we propose a lesion-based contrastive learning approach for fundus image based DR grading. Instead of using entire fundus images, lesion patches are taken as the input for our contrastive prediction task. By focusing on patches with lesions, the network is encouraged to learn more discriminative features for DR grading. The main steps of our framework are as follows. First, an object detection network is trained on a publicly-available dataset IDRiD [16] that consists of 81 fundus images with annotations of lesions. Then, the detection network is applied to the training set of EyePACS [9] to predict lesions with a relatively high confidence threshold. Next, random data augmentation operations are applied to the lesion patches to generate multiple views of them. The feature extraction network in our CL framework is expected to map inputted patches into an embedding feature space, wherein the similarity between features from different views of the same lesion patch and the distance between features from different patches are maximized, by minimizing a contrastive loss. The performance of our proposed method is evaluated based on linear evaluation and transfer capacity evaluation on EyePACS.

The main contributions of this paper are three-fold: (1) We present a self-supervised framework for DR grading, namely lesion-based contrastive learning. Our framework's contrastive prediction task takes lesion patches as the input, which addresses the problem of high memory requirements and lacking diagnostic features, as to common CL schemes. This design can be easily extended to other types of medical images with relatively weak physiological characteristics. (2) We study different data augmentation operations in defining our contrastive prediction task. Results show that a composition of cropping, color distortion, gray scaling and rotation is beneficial in defining pretext tasks for fundus images to learn discriminative feature representations. (3) We evaluate our framework on the large-scale EyePACS dataset for DR grading. Results from linear evaluation and transfer capacity evaluation identify our method's superiority. The source code is available at https://github.com/YijinHuang/Lesion-based-Contrastive-Learning.

2 Methods

2.1 Generation of Lesion Patches

Two datasets are used in this work. An object detection network is trained on one dataset with lesion annotations. Then, this detection network is used to generate lesion patches of fundus images from the other dataset for subsequent contrastive learning. Because of limited training samples in the first dataset, the detection network has a relatively poor generalization ability and cannot precisely predict lesions of fundus images from the other dataset. Therefore, a high confidence threshold is set to eliminate unconfident predictions. Then, we resize all fundus images to 512×512, and the bounding boxes of the patches are scaled correspondingly. After that, we expand every predicted bounding box to 128×128 with the lesion lying in the center and then randomly shift the box within a range such that the resulting box still covers the lesion. In this way, we increase the difficulty of the contrastive prediction task while ensure the training

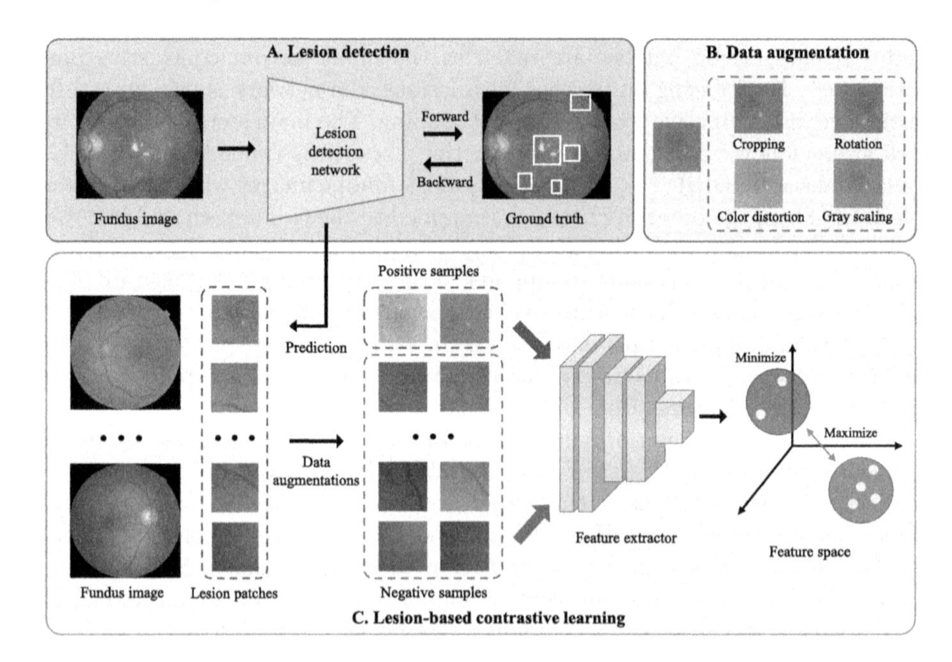

Fig. 2. The proposed framework. In part A, an object detection network is trained to predict lesion patches in fundus images for subsequent contrastive learning. Illustrations of data augmentation operations are provided in part B. In part C, lesion patches are processed by a composition of data augmentation operations to generate multiple views of the lesion patches. A feature extractor is applied to encode patches into an embedding feature space that minimizes a contrastive loss.

of our CL framework can be performed with a large batch size. Please note the lesion detection network is not involved in the testing phase.

2.2 Generation of Multiple Views of Lesion Patches

Rather than employing a carefully designed task to learn feature representations [4,11], CL tries to maximize the agreement between differently augmented views of the same image. Data augmentation, a widely used technique in deep learning, is applied to generate different views of a sample in CL. Some data augmentation operations that are commonly used in the natural image field may destroy important diagnostic features when transferred to the fundus image field. Therefore, as shown in part B of Fig. 2, four augmentation operations are considered in our work: cropping, color distortion, gray scaling, and rotation.

Let $D = \{x_i, i = 1, ..., N\}$ denote a randomly-sampled batch with a batch size of N. Two random compositions of data augmentation operators $(\mathcal{T}, \mathcal{T}')$ are applied to each data point x_i to generate two different views $(\mathcal{T}(x_i), \mathcal{T}'(x_i))$. Note that the parameters of these augmentation operators may differ from data point to data point. Now, we obtain a new batch $D' = \{\tilde{x}_i, i = 1, ..., 2N\}$. Given a patch $\tilde{x}_i$ that is generated from x_i, we consider $\tilde{x}_{j|j\neq i}$ that is also generated from x_i as a positive sample $\tilde{x}_i^+$ and every patch in the set $\Lambda_i^- = \{\tilde{x}_k\}_{k\neq i, k\neq j}$ as a negative sample $\tilde{x}_i^-$.

2.3 Lesion-Based Contrastive Learning

Given a data point $\tilde{x}_i$, we first use a feature extractor $f(\cdot)$ to extract its feature vector h_i. Specificly, $f(\cdot)$ is a CNN and h_i is the feature vector right before the fully connected layer of the network. Then, a projection head $g(\cdot)$ is applied to map the feature vector into an embedding space to obtain z_i. Given a batch D', we define $Z = \{z_i | z_i = g(f(\tilde{x}_i)), \tilde{x}_i \in D'\}$ and $\Omega^- = \{z_i^- | z_i^- = g(f(\tilde{x}_i^-)), \tilde{x}_i^- \in \Lambda_i^-\}$. For every z_i, our contrastive prediction task is to identify embedded feature $z_i^+ = g(f(\tilde{x}_i^+))$ from Z. To find z_i^+, we define the one having the highest cosine similarity with z_i as our prediction. To maximize the similarity of positive samples and to minimize that of negative samples, we define our contrastive loss as

$$\mathcal{L}(Z) = -\sum_{i=1}^{2N} \log \frac{\exp(\text{sim}(\boldsymbol{z}_i, \boldsymbol{z}_i^+)/\tau)}{\exp(\text{sim}(\boldsymbol{z}_i, \boldsymbol{z}_i^+)/\tau) + \sum_{z_i^- \in \Omega^-} \exp(\text{sim}(\boldsymbol{z}_i, \boldsymbol{z}_i^-)/\tau)}, \qquad (1)$$

where N denotes the batch size of Z, $\text{sim}(\boldsymbol{z}_i, \boldsymbol{z}_j) = \boldsymbol{z}_i^T \boldsymbol{z}_j / \|\boldsymbol{z}_i\| \|\boldsymbol{z}_j\|$, and τ is a temperature parameter. In the testing phase, we do not use the projection head $g(\cdot)$ but only the feature extractor $f(\cdot)$ for downstream tasks. Our framework of lesion-based CL is depicted in part C of Fig. 2.

2.4 Implementation Details

Data Augmentation Operations. For the cropping operation, we randomly crop lesion patches with a random factor in [0.8, 1.2]. For the gray scaling operation, each patch has a 0.2 probability of being gray scaled. The color distortion operation adjusts the brightness, contrast, and saturation of patches with a random factor in [-0.4, 0.4] and also changes the hue with a random factor in [-0.1, 0.1]. The rotation operation randomly rotates patches by an arbitrary angle.

Lesion Detection Network. Faster-RCNN [17] with ResNet50 [6] as the backbone is adopted as our lesion detection network. We apply transfer learning by initializing the network with parameters from a model pre-trained on the COCO [13] dataset. The detection network is trained with Adam optimizer for 100 epochs, with a 0.01 initial learning rate and getting decayed by 0.1 at the 50th epoch and the 80th epoch.

Contrastive Learning Network. We also use ResNet50 as our feature extractor. The projection head is a one-layer MLP with ReLU as the activation function, which reduces the dimension of the feature vector to 128. We adopt SGD optimizer with a 0.001 initial learning rate and cosine decay strategy to train the network. The batch size is set to be 768 and the temperature parameter τ is set to be 0.07. The augmented views of lesion patches are resized to 128×128 as the input to our contrastive learning task. All experiments are equally trained for 1000 epochs with a fixed random seed.

Table 1. The total number of lesion patches for the lesion-based CL under different confidence thresholds of the lesion detection network with a 31.17% mAP.

Confidence threshold	# Images	# Lesions
0.7	25226	88867
0.8	21578	63362
0.9	15889	35550

2.5 Evaluation Protocol

Linear Evaluation. Linear evaluation is a widely used method for evaluating the quality of the learned representations of a self-supervised model. The pre-trained feature extractor described in Sect. 2.3 is frozen and a linear classifier on top of it is trained in a fully-supervised manner. The performance of downstream tasks is then used as a proxy of the quality of the learned representations.

Transfer Capacity Evaluation. Pre-trained parameters of the feature extractor can be transferred to models used in downstream tasks. To evaluate the transfer learning capacity, we unfreeze and fine-tune the feature extractor followed by a linear classifier in supervised downstream tasks.

3 Experiment

3.1 Dataset and Evaluation Metric

IDRiD. IDRiD [16] consists of 81 fundus images, with pixel-wise lesion annotations of hemorrhages, microaneurysms, soft exudates, and hard exudates. These manual annotations are converted to bounding boxes to train an object detection network [7]. Microaneurysms are excluded in this work because detecting them is challenging and will lead to a large number of false positive predictions. In training the lesion detection network, 54 samples are used for training and 27 for validation.

EyePACS. 35k/11k/43k fundus images are provided in EyePACS [9] for training/validation/testing (the class distribution of EyePACS is shown in Fig. A1 of the appendix). According to the severity of DR, images are classified into five grades: 0 (normal), 1 (mild), 2 (moderate), 3 (severe), and 4 (proliferative). The training and validation sets without annotations are used for training our self-supervised model. The total number of lesion patches under different confidence thresholds of the detection network is shown in Table 1. Representative lesion detection results are provided in Fig. A2 of the appendix. Partial datasets are obtained by randomly sampling 1%/5%/10%/25%/100% (0.3k/1.7k/3.5k/8.7k/35k images) from the training set, together with the corresponding annotations. Images in the partial datasets and the test set are resized to 512×512 for training and testing subsequent DR grading models in both linear evaluation and transfer capability evaluation settings.

Evaluation Metric. We adopt the quadratic weighted kappa [9] for evaluation, which works well for unbalanced datasets.

3.2 Composition of Data Augmentation Operations

We evaluate the importance of an augmentation operation by removing it from the composition or applying it solely. As shown in the top panel of Table 2, it is insufficient for a single augmentation operation to learn discriminative representations. Even so, color distortion works much better than other operations, showing its importance in defining our contrastive prediction task for fundus images. This clearly indicates that DR grading benefits from color invariance. We conjecture it is because the EyePACS images are highly diverse, especially in terms of image intensity profiles. From the bottom panel of Table 2, we notice that an absence of any of the four augmentation operations leads to a decrease

Table 2. Impact of different compositions of data augmentation operations. The kappa is the result of linear evaluation on the 25% partial dataset, under a detection confidence threshold of 0.8.

Cropping	Rotation	Color distortion	Gray scaling	Kappa
✓				44.25
	✓			41.71
		✓		**53.83**
			✓	48.17
	✓	✓	✓	62.05
✓		✓	✓	57.94
✓	✓		✓	59.96
✓	✓	✓		61.55
✓	✓	✓	✓	**62.49**

Table 3. Linear evaluation and transfer capacity evaluation results on partial datasets.

Method	Confidence threshold	Partial dataset				
		1%	5%	10%	25%	100%
		Quadratic weighted kappa				
Linear evaluation						
SimCLR (128 × 128)	-	16.19	26.70	31.62	37.41	43.64
SimCLR (224 × 224)	-	12.15	26.56	29.94	37.86	55.32
Lesion-base CL (ours)	0.7	24.72	43.98	53.33	61.55	66.87
Lesion-base CL (ours)	0.8	26.22	48.18	56.30	62.49	66.80
Lesion-base CL (ours)	0.9	**31.92**	**56.57**	**60.70**	**63.45**	**66.88**
Transfer capacity evaluation						
Supervised	-	53.36	72.19	76.35	79.85	83.10
SimCLR (128 × 128)	-	63.16	72.30	76.33	79.59	82.72
SimCLR (224 × 224)	-	55.43	70.66	75.15	77.32	82.11
Lesion-base CL (ours)	0.7	**68.37**	**75.40**	77.34	80.34	82.80
Lesion-base CL (ours)	0.8	68.14	74.18	**77.49**	**80.74**	**83.22**
Lesion-base CL (ours)	0.9	66.43	73.85	76.93	80.27	83.04

in performance. Applying the composition of all four augmentation operations considerably increases the difficulty of our contrastive prediction task, but it also significantly improves the quality of the learned representations. Notably, this contrasts common CL methods; a heavy data augmentation would typically hurt the performance of disease diagnosis [15].

3.3 Evaluation on DR Grading

To evaluate the quality of the learned representations from fundus images, we perform linear evaluation and transfer capacity evaluation on the partial datasets. A state-of-the-art CL method SimCLR [2] is adopted as our baseline for comparison, which nevertheless takes an entire fundus image as the input for the contrastive prediction task. In Table 3, SimCLR (128 × 128) denotes the multiple views of fundus images are resized to 128 × 128 as the input for the contrastive prediction task in SimCLR, being consistent with the input size of our lesion-based CL framework. However, crop-and-resize transformation may critically change the pixel size of the input. SimCLR (224 × 224) experiments are conducted based on the consideration that aligning the pixel size of the input for CL and that for downstream tasks may achieve better performance. The ResNet50 model in the fully-supervised DR grading is initialized with parameters from a model trained on the ImageNet [10] dataset. Training curves are provided in Fig. A3 of the appendix.

As shown in Table 3, our lesion-based CL under a detection confidence threshold of 0.9 achieves 66.88% kappa on linear evaluation on the full training set. Significant improvements over SimCLR have been observed; 23.34% over SimCLR (128 × 128) and 11.56% over SimCLR (224 × , 224). The superiority of our method is more evident when a smaller training set is used for linear evaluation. Note that although using a higher confidence threshold results in fewer lesion patches for training, a better performance is achieved on linear evaluation. This implies that by improving the quality of lesions in the training set of the contrastive prediction task, the model can more effectively learn discriminative feature representations. For the transfer capacity evaluation, when fine-tuning on the full training set, there is not much difference between the fully-supervised method and CL methods. This is because feature representations can be sufficiently learned under full supervison, and thus there may be no need for CL based learning. Therefore, the advantage of our proposed method becomes more evident when the training samples for fine-tuning become fewer. With only 1% partial dataset for fine-tunning, the proposed method under a confidence threshold of 0.7 has a higher kappa by 15.01% than the fully-supervised method. Both linear evaluation and transfer capacity evaluation suggest that our proposed method can better learn feature representations, and thus is able to enhace DR grading, by exploiting unlabeled images (see also Fig. A4 in the appendix).

4 Conclusion

In this paper, we propose a novel self-supervised framework for DR grading. We use lesion patches as the input for our contrastive prediction task rather than entire fundus images, which encourages our feature extractor to learn representations of diagnostic features, improving the transfer capacity for downstream tasks. We also present the importance of different data augmentation operations in our CL task. By performing linear evaluation and transfer capacity evaluation on EyePACS, we show that our method has a superior DR grading performance, especially when the sample size of the training data with annotations is limited. This work, to the best of our knowledge, is the first one of its kind that has attempted contrastive learning on fundus image based DR grading.

References

1. Bachman, P., Hjelm, R.D., Buchwalter, W.: Learning representations by maximizing mutual information across views. In: NeurIPS (2019)
2. Chen, T., Kornblith, S., Norouzi, M., Hinton, G.: A simple framework for contrastive learning of visual representations. In: PMLR (2020)
3. DeVries, T., Taylor, G.W.: Improved regularization of convolutional neural networks with cutout. arXiv preprint (2017). arXiv:1708.04552
4. Gidaris, S., Singh, P., Komodakis, N.: Unsupervised representation learning by predicting image rotations. In: ICLR (2018)

5. He, A., Li, T., Li, N., Wang, K., Fu, H.: CABNet: category attention block for imbalanced diabetic retinopathy grading. IEEE Trans. Med. Imag. **40**(1), 143–153 (2020)
6. He, K., Zhang, X., Ren, S., Sun, J.: Deep residual learning for image recognition. In: CVPR (2016)
7. Huang, Y., et al.: Automated hemorrhage detection from coarsely annotated fundus images in diabetic retinopathy. In: IEEE 17th International Symposium on Biomedical Imaging (ISBI), pp. 1369–1372 (2020)
8. Jiao, J., Cai, Y., Alsharid, M., Drukker, L., Papageorghiou, A.T., Noble, J.A.: Self-supervised contrastive video-speech representation learning for ultrasound. In: Martel, A.L., Abolmaesumi, P., Stoyanov, D., Mateus, D., Zuluaga, M.A., Zhou, S.K., Racoceanu, D., Joskowicz, L. (eds.) MICCAI 2020. LNCS, vol. 12263, pp. 534–543. Springer, Cham (2020). https://doi.org/10.1007/978-3-030-59716-0_51
9. Kaggle diabetic retinopathy detection competition. https://www.kaggle.com/c/diabetic-retinopathy-detection
10. Krizhevsky, A., Sutskever, I., Hinton, G.E.: ImageNet classification with deep convolutional neural networks. In: NeruIPS (2012)
11. Lee, H.Y., Huang, J.B., Singh, M., Yang, M.H.: Unsupervised representation learning by sorting sequences. In: ICCV (2017)
12. Lin, L., Li, M., Huang, Y., Cheng, P., Xia, H., et al.: The SUSTech-SYSU dataset for automated exudate detection and diabetic retinopathy grading. Sci. Data **7**(1), 1–1 (2020)
13. Lin, T.Y., et al.: Microsoft coco: common objects in context. In: ECCV (2014)
14. Lin, Z., Guo, R., Wang, Y., Wu, B., Chen, T., Wang, W., Chen, D.Z., Wu, J.: A framework for identifying diabetic retinopathy based on anti-noise detection and attention-based fusion. In: Frangi, A.F., Schnabel, J.A., Davatzikos, C., Alberola-López, C., Fichtinger, G. (eds.) MICCAI 2018. LNCS, vol. 11071, pp. 74–82. Springer, Cham (2018). https://doi.org/10.1007/978-3-030-00934-2_9
15. Li, X., Jia, M., Islam, M.T., Yu, L., Xing, L.: Self-supervised feature learning via exploiting multi-modal data for retinal disease diagnosis. IEEE Trans. Med. Imag. **39**(12), 4023–4033 (2020)
16. Porwal, P., et al.: Indian diabetic retinopathy image dataset (IDRID): a database for diabetic retinopathy screening research. Data **3**(3), 25 (2018)
17. Ren, S., He, K., Girshick, R., Sun, J.: Faster r-CNN: towards real-time object detection with region proposal networks. IEEE Trans. Patt. Anal. Mach. Intell. **39**(6), 1137–1149 (2017)
18. Spitzer, H., Kiwitz, K., Amunts, K., Harmeling, S., Dickscheid, T.: Improving cytoarchitectonic segmentation of human brain areas with self-supervised siamese networks. In: Frangi, A.F., Schnabel, J.A., Davatzikos, C., Alberola-López, C., Fichtinger, G. (eds.) MICCAI 2018. LNCS, vol. 11072, pp. 663–671. Springer, Cham (2018). https://doi.org/10.1007/978-3-030-00931-1_76
19. Wang, Z., Yin, Y., Shi, J., Fang, W., Li, H., Wang, X.: Zoom-in-net: deep mining lesions for diabetic retinopathy detection. In: Descoteaux, M., Maier-Hein, L., Franz, A., Jannin, P., Collins, D.L., Duchesne, S. (eds.) MICCAI 2017. LNCS, vol. 10435, pp. 267–275. Springer, Cham (2017). https://doi.org/10.1007/978-3-319-66179-7_31

20. Zeng, X., Chen, H., Luo, Y., Ye, W.: Automated detection of diabetic retinopathy using a binocular siamese-like convolutional network. In: ISCAS (2019)
21. Zhuang, X., Li, Y., Hu, Y., Ma, K., Yang, Y., Zheng, Y.: Self-supervised feature learning for 3D medical images by playing a Rubik's cube. In: Shen, D., Liu, T., Peters, T.M., Staib, L.H., Essert, C., Zhou, S., Yap, P.-T., Khan, A. (eds.) MICCAI 2019. LNCS, vol. 11767, pp. 420–428. Springer, Cham (2019). https://doi.org/10.1007/978-3-030-32251-9_46

SAR: Scale-Aware Restoration Learning for 3D Tumor Segmentation

Xiaoman Zhang[1,2], Shixiang Feng[1], Yuhang Zhou[1], Ya Zhang[1,2](✉), and Yanfeng Wang[1,2]

[1] Cooperative Medianet Innovation Center, Shanghai Jiao Tong University, Shanghai, China
{xm99sjtu,fengshixiang,zhouyuhang,ya_zhang,wangyanfeng}@sjtu.edu.cn
[2] Shanghai AI Laboratory, Shanghai, China

Abstract. Automatic and accurate tumor segmentation on medical images is in high demand to assist physicians with diagnosis and treatment. However, it is difficult to obtain massive amounts of annotated training data required by the deep-learning models as the manual delineation process is often tedious and expertise required. Although self-supervised learning (SSL) scheme has been widely adopted to address this problem, most SSL methods focus only on global structure information, ignoring the key distinguishing features of tumor regions: local intensity variation and large size distribution. In this paper, we propose Scale-Aware Restoration (SAR), a SSL method for 3D tumor segmentation. Specifically, a novel proxy task, i.e. scale discrimination, is formulated to pre-train the 3D neural network combined with the self-restoration task. Thus, the pre-trained model learns multi-level local representations through multi-scale inputs. Moreover, an adversarial learning module is further introduced to learn modality invariant representations from multiple unlabeled source datasets. We demonstrate the effectiveness of our methods on two downstream tasks: i) Brain tumor segmentation, ii) Pancreas tumor segmentation. Compared with the state-of-the-art 3D SSL methods, our proposed approach can significantly improve the segmentation accuracy. Besides, we analyze its advantages from multiple perspectives such as data efficiency, performance, and convergence speed.

Keywords: Self-supervised learning · 3D model pre-training · 3D tumor segmentation

1 Introduction

Automatic and accurate tumor segmentation on medical images is an essential step in computer-assisted clinical interventions. Deep learning approaches have recently made remarkable progress in medical image analysis [10]. However, the success of these data-driven approaches generally demands massive amounts of

M. de Bruijne et al. (Eds.): MICCAI 2021, LNCS 12902, pp. 124–133, 2021.
https://doi.org/10.1007/978-3-030-87196-3_12

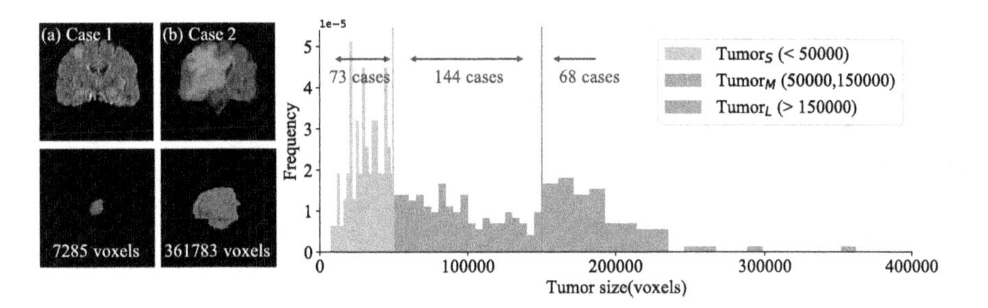

Fig. 1. Examples of brain MRI scans showing the large variations of shape and size of the tumor (left). Tumor size distribution for the BraTS2018 dataset (right).

annotated training data [12]. Unfortunately, obtaining a comprehensively annotated dataset in the medical field is extremely challenging due to the considerable labor and time costs as well as the expertise required for annotation.

To reduce the requirement on labeled training data, self-supervised learning (SSL) scheme has been widely adopted for visual representation learning [3,4,6]. A generic framework is to first employ predefined self-supervision tasks to learn feature representations capturing the data characteristics and further fine-tune the model with a handful of task-specific annotated data. For medical image analysis, many existing 3D SSL methods borrow ideas from proxy tasks proposed for natural images and extend them to 3D vision, such as rotating [3], jigsaw puzzles [16] and contrastive learning [14]. Another stream of 3D SSL methods [5,7] follows Model Genesis [17], one of the most famous SSL methods for medical images, which pre-trains the network with a transform-restoration process.

Despite their promises in medical image analysis, existing 3D SSL methods mainly have two drawbacks when applying to tumor segmentation. Firstly, different from ordinary semantic segmentation, tumor segmentation focuses more on the local intensity and texture variations rather than global structural or semantic information. Representations learned by SSL methods extended directly from computer vision tasks may not be effectively transferred to this target task.

Secondly, the tumor region in CT/MRI scans, as shown in Fig. 1, can vary in scale, appearance and geometric properties. In particular, the largest tumor in the dataset occupies more than hundreds of thousands of voxels, but the smallest one has even less than one thousand voxels. Such a large range of tumor scales increases the difficulty of training a segmentation model. Thus, learning to capture visual patterns of various scales is expected to beneficial for tumor segmentation, which is not taken into account by most existing methods.

In this paper, we propose Scale-Aware Restoration (SAR) learning, a self-supervised method based on the prior-knowledge for 3D tumor segmentation. Specifically, we propose a novel proxy task, i.e. scale discrimination, to pre-train 3D neural networks combined with the self-restoration task. Thus, the pre-trained model is encouraged to learn multi-scale visual representations and low-level detailed features. Moreover, since multiple unlabeled source datasets

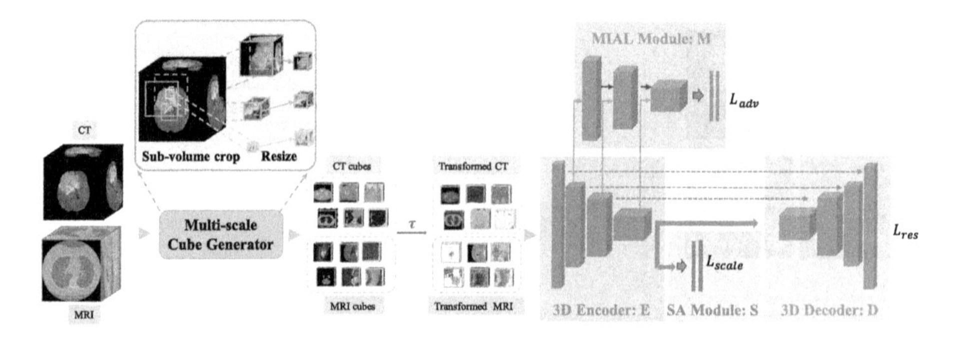

Fig. 2. Schematic diagram of the proposed method.

are used in the pre-training stage, we further introduce an adversarial learning module to learn modality-invariant representations.

2 Method

In this section, we introduce the proposed Scale-Aware Restoration (SAR) learning method in detail. As shown in Fig. 2, given an unlabeled image x, we extract multi-scale sub-volumes, denoted by x_i, which are then fed into the transformation module to get the transformed inputs $\hat{x}_i$. The goal of the proposed self-supervised proxy task, SAR, is to learn to predict the scale information (Scale-Aware) and recover the origin sub-volume (Restoration) of the transformed inputs $\hat{x}_i$. Moreover, we design an adversarial learning module to capture modality-invariant features as the pre-trained source datasets have various modalities and naively fusing multi-modal representations may lead to transfer performance degradation.

2.1 Self-restoration Framework

The goal of the self-restoration task is to learn to recover origin sub-volume x_i from transformed input $\hat{x}_i$, where $\hat{x}_i = \tau(x_i)$ and $\tau(\cdot)$ denotes the transformation function. In practice, we apply a sequence of image transformations to the input sub-volume. We follow [17] and adopt the same set of transformations, i.e. non-linear, local-shuffling, inner and outer-painting, corresponding to the appearance, texture and context, respectively. We take 3D U-Net, the most commonly used network structure for 3D medical image analysis, as the backbone. The Mean Square Error (MSE) loss is employed as the optimizer function for the restoration task, which is expressed as follows:

$$\min_{\theta_E,\theta_D} L_{res} = \sum_{x_i \in \mathcal{D}} \|D(E(\hat{x}_i)) - x_i\|_2, \tag{1}$$

where θ_E and θ_D are parameters of the encoder E and the decoder D, respectively.

2.2 Scale Aware Module

The presence of tumors of various sizes increases the difficulty of training a segmentation model, which motivates us to pre-train a model that can capture multi-scale visual patterns. Accordingly, we propose a novel self-supervised learning task, scale discrimination, in which the model learns to predict the scale information of multi-scale sub-volumes. The multi-scale cube generator in Fig. 2 shows the pre-processing process for input volumes. For each volume in the training dataset $\mathcal{D}$, we first randomly crop sub-volumes of three different scales: large, medium and small, corresponding to approximately $1/2, 1/4$ and $1/8$ of the side length of the original cube respectively. Since the smallest tumor occupies nearly $1/8$ of the whole volume while the largest has nearly $1/2$, we pick these scales to make sure the learnt representation is aware of the texture and intensity over the range of tumor size. The numbers of large scale, medium scale and small scale are in the ratio $1:1:2$ as the small scale cubes can be more diverse. The exact number of cubes are shown in Table 1. To fit the input size of the model, we then resize all the sub-volumes to the same size: $64 \times 64 \times 32$.

Let x_i denotes the resized input sub-volume and y_i is the corresponding scale label. We build a Scale-Aware (SA) module S to predict the scale of input sub-volumes, ensuring the learned representations containing multi-level discriminative features. Specifically, features generated from Encoder E are input to the SA module S for scale prediction. The optimizer function for the SA module is defined as follows:

$$\min_{\theta_E,\theta_S} L_{scale} = \sum_{(x_i,y_i)\in\mathcal{D}} L_{CE}(S(E(\hat{x}_i)), y_i), \tag{2}$$

where $L_{CE}(\cdot)$ represents cross entropy loss, $\hat{x}_i$ is the transformed input volume, y_i is the corresponding scale label, and θ_S are parameters of the SA module S.

2.3 Modality Invariant Adversarial Learning

Medical images in different modalities are quite different in characteristics. To avoid the performance degradation caused by cross-modality generalization, we develop a Modality Invariant Adversarial Learning (MIAL) module M for modality invariant representation learning. Different from the common domain discriminator, which uses $E(\hat{x})$ as input, we contact features of different layers in E with the corresponding layer in M. Such a densely connected operation improves the modality invariant feature learning by strengthening the connection between the encoder and discriminator. If the discriminator can successfully classify the modal of input sub-volumes, it means that the features extracted from the encoder still contain modality characteristics. Hence, we apply the min-max two-player game, where the MIAL module M is trained to distinguish the modality of input sub-volumes, and the encoder E is trained simultaneously to confuse the discriminator M.

Denote $\mathcal{D} = \{\mathcal{D}_{CT}, \mathcal{D}_{MRI}\}$ as the training dataset. The following adversarial loss is applied:

$$\min_{\theta_E} \max_{\theta_M} L_{adv} = \sum_{x_i^{CT} \in \mathcal{D}_{CT}} log(M(E(\hat{x}_i^{CT}))) + \sum_{x_i^{MRI} \in \mathcal{D}_{MRI}} log(1 - M(E(\hat{x}_i^{MRI}))), \tag{3}$$

where θ_E and θ_M are parameters of encoder E and MIAL module M, $\hat{x}_i^{CT}$ and $\hat{x}_i^{MRI}$ are the transformed input sub-volume of corresponding modality.

2.4 Objective Function

The final objective function for optimizing the pre-trained model can thus be formulated as follows:

$$\min_{\theta_E, \theta_D, \theta_S} \max_{\theta_M} L_{adv} + \alpha L_{scale} + \beta L_{res}, \tag{4}$$

where α and β are trade-off parameters weighting the loss functions, since the magnitudes of the three loss functions are inconsistent. In our experiments, we empirically set $\alpha = 1, \beta = 10$ to keep the order of magnitudes consistently.

3 Experiments

3.1 Pre-training Datasets

We collected 734 CT scans and 790 MRI scans from 4 public datasets, including LUNA2016 [11], LiTS2017 [2], BraTS2018 [1], and MSD (heart) [13], as source datasets to build the pre-trained model, SAR. These datasets are quite diverse, containing two modalities (i.e. MRI and CT) of data, from three different scan regions (i.e. brain, chest, and abdomen), and targeting four organs (i.e. lung, liver, brain, and heart). The detailed information about the source datasets is summarized in Table 1.

Table 1. Statistics of the pre-training datasets where cases represent volumes used in the pre-trained process and sub-volume number represents the number of cropped cubes at different scales.

Dataset	Cases	Modality	Organ	Median shape	Sub-volume number
LUNA2016	623	CT	Lung	$238 \times 512 \times 512$	(19936, 9968, 9968)
LiTS2017	111	CT	Liver	$432 \times 512 \times 512$	(7104, 3552, 3552)
BraTS2018	190×4	MRI	Brain	$138 \times 169 \times 138$	(24320, 12160, 12160)
MSD (heart)	30	MRI	Heart	$115 \times 320 \times 232$	(1920, 960, 960)

To alleviate the heterogeneity caused by multi-center (e.g. due to different scanners and imaging protocols), we process the data with spatial and intensity normalization. We resample all volumes to the same voxel spacing ($1 \times 1 \times 1\,\text{mm}^3$) by using third-order spline interpolation. To mitigate the side-effect of extreme pixel value, we clip the intensity values of CT volumes on the min (–1000) and max (+1000) interesting Hounsfield Unit range. And for the MRI volumes, all the intensity values are clipped on the min (0) and max (+4000) interesting range. Each volume is normalized independently to [0, 1].

3.2 Downstream Tasks

Brain Tumor Segmentation. We perform the brain tumor segmentation task using the training set of BraTS2018 [1]. The dataset contains 285 cases acquired with different MRI scanners. Following the experiment strategy in [17], we use FLAIR images and only segment the whole brain tumor. We randomly divide the training and test sets in the ratio of 2:1. Note that all the test images for target tasks are not exposed to model pre-training. For data preprocessing, each image is independently normalized by subtracting the mean of its non-zero area and dividing by the standard deviation. We report test performance through the mean and standard deviation of five trials.

Pancreas Organ and Tumor Segmentation. We perform the pancreas tumor segmentation task using the Pancreas dataset of MSD [13], which contains 282 3D CT training scans. Each scan in this dataset contains 3 different classes: pancreas (class 1), tumor (class 2), and background (class 0). For data preprocessing [8], we clip the HU values in each scan to the range $[-96, 215]$, and then subtract 77.99 and finally divide by 75.40. We report the mean and standard deviation of five-fold cross-validation segmentation results.

3.3 Implementation and Training Details

In the pre-training stage, the basic encoder and decoder follow the 3D U-Net [18] architecture with batch normalization. The MIAL module is constructed with three convolution layers with a global average pooling (GAP) layer and two fully connected (FC) layers. The SA module simply consists of a GAP layer and two FC layers. The network is trained using SGD optimizer, with an initial learning rate of 1e0, 1e–1, 1e–3 for 3D U-Net, SA module and MIAL module respectively. We use ReduceLROnPlateau to schedule the learning rate according to the validation loss.

In the fine-tuning stage, we transfer the 3D encoder-decoder for the target segmentation tasks with dice loss. We use the Adam optimizer [9] with an initial learning rate of 1e–3 for fine-tuning and use ReduceLROnPlateau to schedule the learning rate. Note that all of the layers in the model are trainable during fine-tuning. All experiments were implemented with a PyTorch framework on GTX 1080Ti GPU.

4 Results

Comparison with State-of-the-Art Methods. In this section, we compare SAR with training from scratch and the state-of-the-art 3D SSL methods including Model Genesis (Genesis) [15], Relative 3D patch location (3D-RPL), 3D Jigsaw puzzles (3D-Jig), 3D Rotation Prediction (3D-Rot) and 3D Contrastive Predictive Coding (3D-CPC) [14]. For a fair comparison, we re-trained all the SSL methods under the same experiment setup, including the network architecture and the pre-training datasets. For SSL tasks that use only the encoder during pre-training, we randomly initialize the decoder during fine-tuning.

Table 2. Comparison with state-of-the-art methods.

	Task 1.BraTS2018				Task 2.MSD (Pancreas)	
	Tumor_S	Tumor_M	Tumor_L	Tumor_{all}	Organ	Tumor
Scratch	65.41$_{\pm 7.82}$	78.30$_{\pm 3.33}$	74.90$_{\pm 4.72}$	74.35$_{\pm 3.95}$	73.42$_{\pm 1.21}$	25.19$_{\pm 4.27}$
Genesis	74.36$_{\pm 6.72}$	85.67$_{\pm 2.04}$	85.62$_{\pm 1.27}$	83.03$_{\pm 3.14}$	74.17$_{\pm 2.38}$	33.12$_{\pm 3.08}$
3D-RPL	72.77$_{\pm 1.59}$	82.22$_{\pm 0.29}$	85.62$_{\pm 1.27}$	80.15$_{\pm 0.47}$	71.84$_{\pm 2.75}$	28.30$_{\pm 3.47}$
3D-Jig	61.24$_{\pm 3.32}$	76.05$_{\pm 1.93}$	72.21$_{\pm 1.60}$	71.48$_{\pm 1.68}$	73.65$_{\pm 1.58}$	28.82$_{\pm 3.57}$
3D-Rot	63.06$_{\pm 3.08}$	77.18$_{\pm 1.06}$	72.65$_{\pm 1.36}$	72.66$_{\pm 1.33}$	70.99$_{\pm 1.31}$	24.91$_{\pm 0.63}$
3D-CPC	65.72$_{\pm 5.17}$	77.35$_{\pm 1.80}$	72.65$_{\pm 3.05}$	73.39$_{\pm 2.31}$	70.54$_{\pm 1.39}$	27.99$_{\pm 6.21}$
+ MIAL	76.31$_{\pm 1.30}$	85.99$_{\pm 0.68}$	85.20$_{\pm 0.83}$	83.46$_{\pm 0.92}$	75.18$_{\pm 1.78}$	32.12$_{\pm 2.30}$
+ SA	75.44$_{\pm 3.09}$	**87.09**$_{\pm 0.65}$	85.74$_{\pm 1.07}$	83.95$_{\pm 1.09}$	75.51$_{\pm 0.75}$	32.60$_{\pm 3.02}$
SAR	**78.10**$_{\pm 3.20}$	86.75$_{\pm 1.05}$	**87.84**$_{\pm 2.45}$	**84.92**$_{\pm 1.62}$	**75.68**$_{\pm 0.85}$	**33.92**$_{\pm 3.00}$

The overall results are summarized in Table 2. For BraTS2018, we divide the tumors into three scales according to their sizes so that we can better compare the results from a multi-scale perspective. Our proposed method outperforms several 3D SSL methods by a great margin and achieves a excellent performance of 84.92% in average Dice for BraTS2018. For pancreas and tumor segmentation of MSD (Pancreas), we achieve an average Dice of 75.68% for the pancreas and 33.92% for the tumor, superior to other methods. Besides, we perform independent two sample t-test between the SAR vs. others. All the comparison show statistically significant results ($p = 0.05$) except for MSD tumor (Genesis vs. SAR) Fig. 3 shows the training loss curves compared with other methods. SAR converges faster to a stable loss and achieves a lower dice loss. This suggests that the proposed representation learning scheme can speed up the convergence and boost performance.

Ablation Study. Firstly, we conduct an ablation experiment to evaluate the effectiveness of each component proposed in SAR. In Table 2, the performance is improved to 83.46%, 83.95% for BraTS2018 equipped with our proposed MIAL

Table 3. Ablation study results of pre-trained model using different scales of data.

	BraTS2018			
	Tumor_S	Tumor_M	Tumor_L	Tumor_{all}
1/2	$76.17_{\pm 2.55}$	$86.55_{\pm 0.41}$	$\mathbf{87.88}_{\pm 2.43}$	$84.36_{\pm 1.69}$
1/4	$75.31_{\pm 4.49}$	$85.97_{\pm 1.75}$	$86.16_{\pm 4.67}$	$83.44_{\pm 3.12}$
1/8	$76.88_{\pm 4.22}$	$86.29_{\pm 1.40}$	$87.06_{\pm 3.74}$	$84.20_{\pm 2.64}$
SAR	$\mathbf{78.10}_{\pm 3.20}$	$\mathbf{86.75}_{\pm 1.05}$	$87.84_{\pm 2.45}$	$\mathbf{84.92}_{\pm 1.62}$

Table 4. Comparison of experiment results for target tasks at different ratios of annotated data.

	BraTS2018 (Tumor_{all})					MSD (Average)				
Proportion	1/2	1/5	1/10	1/20	1/50	1/2	1/5	1/10	1/20	1/50
Scratch	68.37	57.48	53.57	41.10	25.36	41.09	33.24	30.29	25.37	8.91
SAR	79.48	69.18	67.13	44.86	31.95	47.76	42.26	34.87	29.85	16.82

module and SA module respectively. While combined with SA module, multi-scale information was integrated into our pre-trained model to learn comprehensive contextual and hierarchically geometric features at different levels. A significant performance gains up to 12.69% in Dice for small-size tumors in BraTS2018 confirms the effect of using multi-scale feature learning. For MSD (Pancreas), the further improvement to 75.68% and 33.92% of pancreas and tumor in SAR confirms the adoption of the two modules yields much better results than simply combined multiple datasets.

Secondly, we validate the benefits of multi-scale representation learning on BraTS2018 in Table 3 by conducting experiments using only single-scale sub-volumes for pre-training. When we use large sub-volumes (1/2) to pre-train, only an average Dice of 76.17% is achieved for small-scale tumors (Tumor_S), while the average Dice of large scale tumors (Tumor_L) reaches 87.88%. A similar situation occurs when we pre-training on the small-scale sub-volumes. Unsurprisingly, the highest overall results are obtained by applying multi-scale sub-volumes, proving the effectiveness of learning multi-scale informative features.

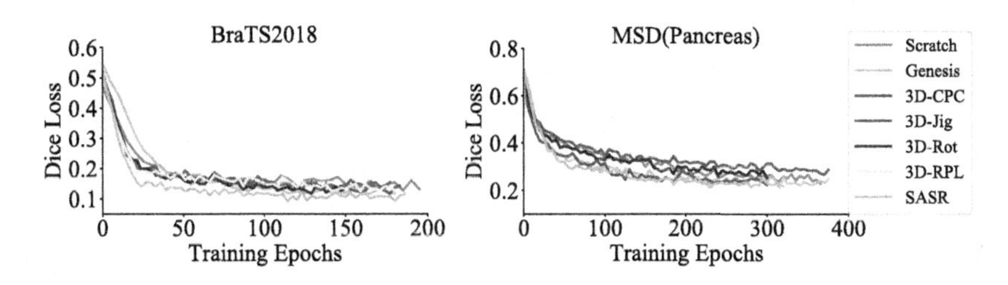

Fig. 3. Comparison of the training loss curves for target tasks.

Analysis of Annotation Cost To validate the gains in data-efficiency, we train the model with different proportions ($\frac{1}{2}$, $\frac{1}{5}$, $\frac{1}{10}$, $\frac{1}{20}$, $\frac{1}{50}$) of the training data and use the same test set to measure the performance. Table 4 displays the results compared with training from scratch. The model performance drops sharply when decreasing the training data due to the overfitting issue. Not surprisingly, fine-tuning on top of our pre-trained model can significantly improve performance on all data sizes. The pre-trained model leads to a reduction in the annotation cost by at least 50% on the target tasks compared with training from scratch. The results suggest that our SAR has the potential to improve the performance of 3D tumor segmentation with limited annotations by providing a transferable initialization.

5 Conclusion

In this paper, we propose Scale-Aware Restoration (SAR) Learning, a SSL method based on the prior-knowledge for 3D tumor segmentation. Scale discrimination combined with self-restoration is developed to predict the scale information of the input sub-volumes for multi-scale visual patterns learning. We also design an adversarial learning module to learn modality invariant representations. The proposed SAR successfully surpasses several 3D SSL methods on two tumor segmentation tasks, especially for small-scale tumor segmentation. Besides, we demonstrate its effectiveness in terms of data efficiency, performance, and convergence speed.

Acknowledgement. This work is supported partially by SHEITC (No. 2018-RGZN-02046), 111 plan (No. BP0719010), and STCSM (No. 18DZ2270700).

References

1. Bakas, S., et al.: Identifying the best machine learning algorithms for brain tumor segmentation, progression assessment, and overall survival prediction in the brats challenge. arXiv preprint arXiv:1811.02629 (2018)
2. Bilic, P., et al.: The liver tumor segmentation benchmark (lits). arXiv preprint arXiv:1901.04056 (2019)
3. Chen, L., Bentley, P., Mori, K., Misawa, K., Fujiwara, M., Rueckert, D.: Self-supervised learning for medical image analysis using image context restoration. Med. Image Anal. **58**, 101539 (2019)
4. Doersch, C., Gupta, A., Efros, A.A.: Unsupervised visual representation learning by context prediction. In: IEEE International Conference on Computer Vision, pp. 1422–1430 (2015)
5. Feng, R., et al.: Parts2whole: self-supervised contrastive learning via reconstruction. In: Domain Adaptation and Representation Transfer, and Distributed and Collaborative Learning, pp. 85–95 (2020)
6. Gidaris, S., Singh, P., Komodakis, N.: Unsupervised representation learning by predicting image rotations. arXiv preprint arXiv: 1803.07728 (2018)

7. Haghighi, F., Taher, M.R.H., Zhou, Z., Gotway, M.B., Liang, J.: Learning semantics-enriched representation via self-discovery, self-classification, and self-restoration. In: Medical Image Computing and Computer Assisted Intervention, pp. 137–147 (2020)
8. Isensee, F., Jäger, P.F., Kohl, S.A., Petersen, J., Maier-Hein, K.H.: Automated design of deep learning methods for biomedical image segmentation. arXiv preprint arXiv:1904.08128 (2019)
9. Kingma, D., Ba, J.: Adam: a method for stochastic optimization. In: International Conference on Learning Representations, vol. 42 (2014)
10. Litjens, G.J.S., et al.: A survey on deep learning in medical image analysis. Med. Image Anal. **42**, 60–88 (2017)
11. Setio, A.A.A., Jacobs, C., Gelderblom, J., van Ginneken, B.: Automatic detection of large pulmonary solid nodules in thoracic CT images. Med. Phys. **42**, 5642–5653 (2015)
12. Shin, H.C., et al.: Deep convolutional neural networks for computer-aided detection: CNN architectures, dataset characteristics and transfer learning. IEEE Trans. Med. Imag. **35**(5), 1285–1298 (2016)
13. Simpson, A.L., et al.: A large annotated medical image dataset for the development and evaluation of segmentation algorithms. arXiv preprint arXiv:1902.09063 (2019)
14. Taleb, A., et al.: 3D self-supervised methods for medical imaging. arXiv preprint arXiv:2006.03829 (2020)
15. Zhou, Z., Sodha, V., Pang, J., Gotway, M.B., Liang, J.: Models genesis. Med. Image Anal. **67**, 101840 (2021)
16. Zhuang, X., Li, Y., Hu, Y., Ma, K., Yang, Y., Zheng, Y.: Self-supervised feature learning for 3D medical images by playing a rubik's cube. In: Medical Image Computing and Computer Assisted Intervention, pp. 420–428 (2019)
17. Zongwei, Z., et al.: Models genesis: Generic autodidactic models for 3D medical image analysis. In: Medical Image Computing and Computer Assisted Intervention, pp. 384–393 (2019)
18. Çiçek, Ö., Abdulkadir, A., Lienkamp, S.S., Brox, T., Ronneberger, O.: 3D u-net: learning dense volumetric segmentation from sparse annotation. In: Medical Image Computing and Computer Assisted Intervention, pp. 424–432 (2016)

Self-supervised Correction Learning for Semi-supervised Biomedical Image Segmentation

Ruifei Zhang[1], Sishuo Liu[2], Yizhou Yu[2,3], and Guanbin Li[1,4](✉)

[1] Sun Yat-sen University, Guangzhou, China
liguanbin@mail.sysu.edu.cn
[2] The University of Hong Kong, Pokfulam, Hong Kong
[3] Deepwise AI Lab, Beijing, China
[4] Shenzhen Research Institute of Big Data, Shenzhen, China

Abstract. Biomedical image segmentation plays a significant role in computer-aided diagnosis. However, existing CNN based methods rely heavily on massive manual annotations, which are very expensive and require huge human resources. In this work, we adopt a coarse-to-fine strategy and propose a self-supervised correction learning paradigm for semi-supervised biomedical image segmentation. Specifically, we design a dual-task network, including a shared encoder and two independent decoders for segmentation and lesion region inpainting, respectively. In the first phase, only the segmentation branch is used to obtain a relatively rough segmentation result. In the second step, we mask the detected lesion regions on the original image based on the initial segmentation map, and send it together with the original image into the network again to simultaneously perform inpainting and segmentation separately. For labeled data, this process is supervised by the segmentation annotations, and for unlabeled data, it is guided by the inpainting loss of masked lesion regions. Since the two tasks rely on similar feature information, the unlabeled data effectively enhances the representation of the network to the lesion regions and further improves the segmentation performance. Moreover, a gated feature fusion (GFF) module is designed to incorporate the complementary features from the two tasks. Experiments on three medical image segmentation datasets for different tasks including polyp, skin lesion and fundus optic disc segmentation well demonstrate the outstanding performance of our method compared with other semi-supervised approaches. The code is available at https://github.com/ReaFly/SemiMedSeg.

Electronic supplementary material The online version of this chapter (https://doi.org/10.1007/978-3-030-87196-3_13) contains supplementary material, which is available to authorized users.

M. de Bruijne et al. (Eds.): MICCAI 2021, LNCS 12902, pp. 134–144, 2021.
https://doi.org/10.1007/978-3-030-87196-3_13

1 Introduction

Medical image segmentation is an essential step in computer-aided diagnosis. In practice, clinicians use various types of images to locate lesions and analyze diseases. An automated and accurate medical image segmentation technique is bound to greatly reduce the workload of clinicians.

With the vigorous development of deep learning, the FCN [15], UNet [19] and their variants [12,23] have achieved superior segmentation performance for both natural images and medical images. However, these methods rely heavily on labeled data, which is time-consuming to acquire especially for medical images. Therefore, many studies adopt semi-supervised learning to alleviate this issue, including GAN-based methods [9,24], consistency training [17,20], pseudo labeling [11] and so on. For instance, Mean Teacher (MT) [20] and its variants [13,22] employ the consistency training for labeled data and unlabeled data by updating teacher weights via an exponential moving average of consecutive student models. Recently, some works [1,14] integrate self-supervised learning such as jigsaw puzzles [16] or contrastive learning [4] to semi-supervised segmentation and achieve competitive results. However, few of them try to dig deeply into the context and structural information of unlabeled images to supplement the semantic segmentation.

In this work, we also consider introducing self-supervised learning to semi-supervised segmentation. In contrast to [1,14], we make full use of massive unlabeled data to exploit image internal structure and boundary characteristics by utilizing pixel-level inpainting as an auxiliary self-supervised task, which is combined with semantic segmentation to construct a dual-task network. As the inpainting of normal non-lesion image content will only introduce additional noise for lesion segmentation, we design a coarse-to-fine pipeline and then enhance the network's representations with the help of massive unlabeled data in the correction stage by only masking the lesion area for inpainting based on the initial segmentation result. Specifically, in the first phase, only the segmentation branch is used to acquire a coarse segmentation result, while in the second step, the masked and original images are sent into the network again to simultaneously perform lesion region inpainting and segmentation separately. Since the two tasks rely on similar feature information, we also design a gated feature fusion (GFF) module to incorporate complementary features for improving each other. Compared with the most related work [2] which introduces a reconstruction task for unlabeled data, their two tasks lack deep interaction and feature reuse, thus cannot collaborate and facilitate each other. Besides, our network not only makes full use of massive unlabeled data, but also explores more complete lesion regions for limited labeled data through the correction phase, which can be seen as "image-level erase [21]" or "reverse attention [3]".

Our contribution is summarized as follows. (1) We propose a novel self-supervised semi-supervised learning paradigm for general lesion region segmentation of medical imaging, and verify that the pretext self-supervised learning task of inpainting the lesion region at the pixel level can effectively enhance the feature learning and greatly reduce the algorithm's dependence on large-scale dense annotation. (2) We propose a dual-task framework for semi-supervised medical image segmentation. Through introducing the inpainting task, we create supervision signals for unlabeled data to enhance the network's representation learning of lesion regions and also exploit additional lesion features for labeled data, thus effectively correct the initial segmentation results. (3) We evaluate our method on three tasks, including polyp, skin lesion and fundus optic disc segmentation, under a semi-supervision setting. The experimental results demonstrate that our method achieves superior performance compared with other state-of-the-art semi-supervised methods.

2 Methodology

2.1 Overview

In this work, we adopt a coarse-to-fine strategy and propose a self-supervised correction learning paradigm for semi-supervised biomedical image segmentation. Specifically, we introduce inpainting as the pretext task of self-supervised learning to take advantage of massive unlabeled data and thus construct a dual-task network, as shown in Fig. 1. Our proposed framework is composed of a shared encoder, two decoders and five GFF modules placed on each layer of both decoders. We utilize ResNet34 [8] pretrained on the ImageNet [5] as our encoder, which consists of five blocks in total. Accordingly, the decoder branch also has five blocks. Each decoder block is composed of two Conv-BN-ReLU combinations. For the convenience of expression, we use E_{seg}, D_{seg} to represent the encoder and decoder of the segmentation branch, and E_{inp} and D_{inp} for those of the inpainting branch.

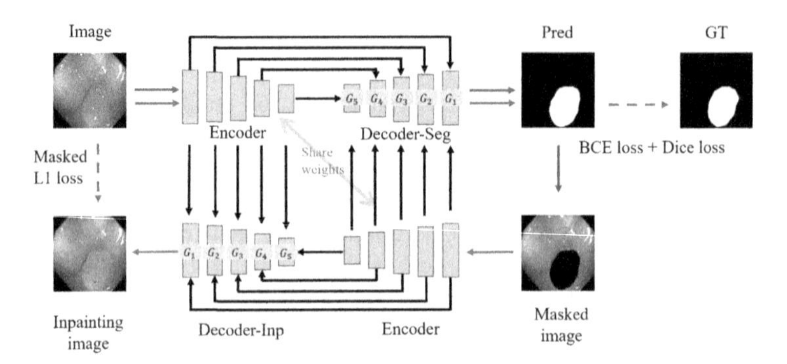

Fig. 1. The overview of our network. Both encoders share weights. G_1–G_5 represent five GFF modules. The red and blue arrows denote the input and output of our network in the first and second stage respectively. (Color figure online)

In the first step, given the image $x \in \mathbb{R}^{H \times W \times C}$, in which H, W, C are the height, width and channels of the image respectively, we use the segmentation branch E_{seg}, D_{seg} with skip-connections, the traditional U-shape encoder-decoder structure, to obtain a coarse segmentation map $\hat{y}_{coarse}$ and then mask the original input based on its binary result $\overline{y}_{coarse}$ by the following formulas:

$$\hat{y}_{coarse} = D_{seg}(E_{seg}(x)) \tag{1}$$

$$x_{mask} = x \times (1 - \overline{y}_{coarse}) \tag{2}$$

In the second phase, the original image x and the masked image x_{mask} are sent into E_{seg} and E_{inp} simultaneously to extract features e_{seg} and e_{inp}. Obviously, e_{seg} is essential for the inpainting task, and since the initial segmentation is usually inaccurate and incomplete, e_{inp} may also contain important residual lesion features for the correction of the initial segmentation. In order to adaptively select the useful features of e_{inp} and achieve complementary fusion of e_{seg} and e_{inp}, we design the GFF modules (G_1–G_5) and place them on each decoder layer of both branches. Specifically, for the ith layer, the features e^i_{seg} and e^i_{inp} are delivered into G_i through skip-connections to obtain the fusion $e^i = G_i(e^i_{seg}, e^i_{inp})$, and then sent to the corresponding decoder layer. Thus, both G_i of the two branches shown in Fig. 1 actually share parameters, taking the same input and generating the identical output. To enhance the learning of the GFF modules, we adopt a deep supervision strategy and each layer of the two decoder branches generate a segmentation result and an inpainting result respectively by the following formulas:

$$\hat{y}^i_{fine} = \begin{cases} D^i_{seg}([e^i, d^{i+1}_{seg}]), & i = 1, 2, 3, 4 \\ D^i_{seg}(e^i), & i = 5 \end{cases} \tag{3}$$

$$\hat{x}^i = \begin{cases} D^i_{inp}([e^i, d^{i+1}_{inp}]), & i = 1, 2, 3, 4 \\ D^i_{inp}(e^i), & i = 5 \end{cases} \tag{4}$$

Where $[\cdot, \cdot]$ denotes the concatenation process, and d^{i+1}_{seg}, d^{i+1}_{inp} represent the features from previous decoder layers. The deep supervision strategy can also avoid D_{inp} directly copying the features of the low-level e_{seg} to complete the inpainting task without in-depth lesion feature mining. The output of the last layer $\hat{y}^1_{fine}$ is the final segmentation result of our network.

2.2 Gated Feature Fusion (GFF)

To better incorporate complementary features and filter out the redundant information, we design the GFF modules placed on each decoder layer to integrate the features delivered from the corresponding encoder layer of two branches. The details are shown in Fig. 2. Our GFF module consists of a reset gate and a select gate. Specifically, for the G_i placed on the ith decoder layer, the value of two gates is calculated as follows:

$$r_i = \sigma(W_r \left[e^i_{seg}, e^i_{inp}\right]) \tag{5}$$

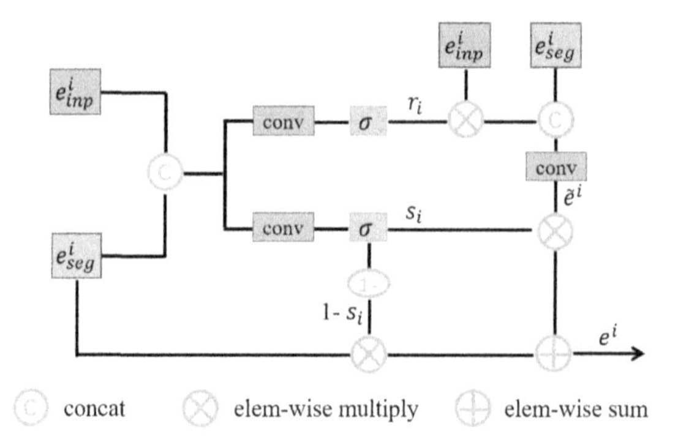

Fig. 2. Gated feature fusion module

$$s_i = \sigma(W_s \left[e^i_{seg}, e^i_{inp}\right]) \tag{6}$$

Where W_r, W_s denote the convolution process, taking the concatenation of e^i_{seg} and e^i_{inp} as input. σ represents the Sigmoid function. r_i and s_i represent the value of the reset gate and the select gate, respectively. Since the input of the inpainting branch is the masked image, the reset gate is necessary to suppress massive invalid background information. And then the select gate achieves adaptive and complementary feature fusion between the reintegrated features $\tilde{e}^i$ and the original segmentation feature e^i_{seg} by the following operations:

$$\tilde{e}^i = W\left[r_i \times e^i_{inp}, e^i_{seg}\right]) \tag{7}$$

$$e^i = s_i \times \tilde{e}^i + (1 - s_i) \times e^i_{seg} \tag{8}$$

where W also represents the convolution process to make the reintegrated features $\tilde{e}^i$ have the same dimension with e^i_{seg}.

2.3 Loss Function

We only calculate loss in the second stage. The labeled dataset and unlabeled dataset are denoted as D_l and D_u. For the labeled data $x_l \in D_l$, y_l is the Ground Truth. Since we adopt the deep supervision strategy, the overall loss is the sum of the combination of Binary CrossEntropy (BCE) Loss and Dice loss between each output and the Ground Truth:

$$\mathcal{L}_{seg}(x_l) = \sum_{i=1}^{5} L^i_{BCE}(\hat{y}^i_l, y^i_l) + L^i_{Dice}(\hat{y}^i_l, y^i_l) \tag{9}$$

where $\hat{y}^i_l$, y^i_l denote the segmentation map $\hat{y}^i_{fine}$ of the ith decoder layer and the corresponding down-sampling Ground Truth y_l.

For unlabeled data $x_u \in D_u$, the inpainting loss is the sum of $L1$ loss between each inpainting image and the original image in the masked region:

$$\mathcal{L}_{inp}(x_u) = \sum_{i=1}^{5} \overline{y}_u^i \times \left|\hat{x}_u^i - x_u^i\right| \tag{10}$$

where $\hat{x}_u^i$, x_u^i and $\overline{y}_u^i$ represent the inpainting image, down-sampling original image and binary segmentation result of the ith decoder layer, respectively. In the end, the total loss function is formulated as follows:

$$\mathcal{L} = \lambda_1 \sum_{x_l \in D_l} \mathcal{L}_{seg}(x_l) + \lambda_2 \sum_{x_u \in D_u} \mathcal{L}_{inp}(x_u) \tag{11}$$

where λ_1, λ_2 are weights balancing the segmentation loss and the inpainting loss. And we set $\lambda_1 = 2$ and $\lambda_2 = 1$ in our experiments.

3 Experimental Results

3.1 Dataset and Evaluation Metric

We conduct experiments on a variety of medical image segmentation tasks to verify the effectiveness and robustness of our approach, including polyp, skin lesion and fundus optic disc segmentation, respectively.

Polyp Segmentation. We use the publicly available kvasir-SEG [10] dataset containing 1000 images, and randomly select 600 images as the training set, 200 images as the validation set, and the remaining as the test set.

Skin Lesion Segmentation. We utilize the ISBI 2016 skin lesion dataset [7] to evaluate our method performance. This dataset consists of 1279 images, among which 900 are used for training and the others for testing.

Optic Disc Segmentation. The Rim-one r1 dataset [6] is utilized in our experiments, which has 169 images in total. We randomly split the dataset into a training set and a test set with the ratio of 8:2.

Evaluation Metric. Referring to common semi-supervised segmentation settings [13,22], for all datasets, we randomly use 20% of the training set as the labeled data, 80% as the unlabeled data and adopt five metrics to quantitively evaluate the performance of our approach and other methods, including "Dice Similarity Coefficient (Dice)", "Intersection over Union (IoU)", "Accuracy (Acc)", "Recall (Rec)" and"Specificity (Spe)".

3.2 Implementation Details

Data Pre-processing. In our experiments, since the image resolution of all datasets varies greatly, we uniformly resize all images into a fixed size of 320×320 for training and testing. And in the training stage, we use data augmentation, including random horizontal and vertical flips, rotation, zoom, and finally all the images are randomly cropped to 256×256 as input.

Training Details. Our method is implemented using PyTorch [18] framework. We set batch size of the training process to 4, and use SGD optimizer with a momentum of 0.9 and a weight decay of 0.00001 to optimize the model. A poly learning rate police is adopted to adjust the initial learning rate, which is $lr = init\_lr \times (1 - \frac{iter}{max\_iter})^{power}$, where $init\_lr = 0.001$, $power = 0.9$). The total number of epochs is set to 80.

Table 1. Comparison with other state-of-the-art methods and ablation study on the Kvasir-SEG dataset

Methods	Data	*Dice*	*IoU*	*Acc*	*Rec*	*Spe*
Supervised	600L (All)	89.48	83.69	97.34	91.06	98.58
Supervised	120L	84.40	76.18	96.09	85.35	98.55
DAN [24]	120L + 480U	85.77	78.12	96.37	86.86	98.53
MT [20]	120L + 480U	85.99	78.84	96.21	86.81	**98.79**
UA-MT [22]	120L + 480U	85.70	78.34	96.38	88.51	98.40
TCSM_V2 [13]	120L + 480U	86.17	79.15	96.38	87.14	98.76
MASSL [2]	120L + 480U	86.45	79.61	96.34	89.18	98.32
Ours	120L + 480U	**87.14**	**80.49**	**96.42**	**90.78**	97.89
Ours (add)	120L + 480U	85.59	78.56	96.12	87.98	98.26
Ours (concat)	120L + 480U	86.09	78.98	96.21	90.54	97.63

3.3 Comparisons with the State-of-the-Art

In our experiments, ResNet34 [8] based UNet [19] is utilized as our baseline, which is trained using all training set and our selected 20% labeled data separately in a fully-supervised manner. Besides, we compare our method with other state-of-the-art approaches, including DAN [24], MASSL [2], MT [20] and its variants (UA-MT [22], TCSM_V2 [13]). All comparison methods adopt ResNet34UNet as the backbone and use the same experimental settings for a fair comparison. On the **Kvasir-SEG dataset**, Table 1 shows that our method obtains the outstanding performance compared with other semi-supervised methods, with Dice of 87.14%, which is 2.74% improvement over the baseline only using the 120 labeled data, outperforming the second best method by 0.69%. On the **ISBI 2016 skin lesion dataset**, we obtain a 90.95% Dice score, which is superior to other semi-supervised methods and very close to the score of 91.38% achieved by the baseline using all training set images. On the **Rim-one r1 dataset**, we can conclude that our method achieves the best performance over five metrics, further demonstrating the effectiveness of our method. Note that detailed results on the latter two datasets are listed in the supplementary material due to the space limitation. Some visual segmentation results are shown in Fig. 3 (col. 1–8).

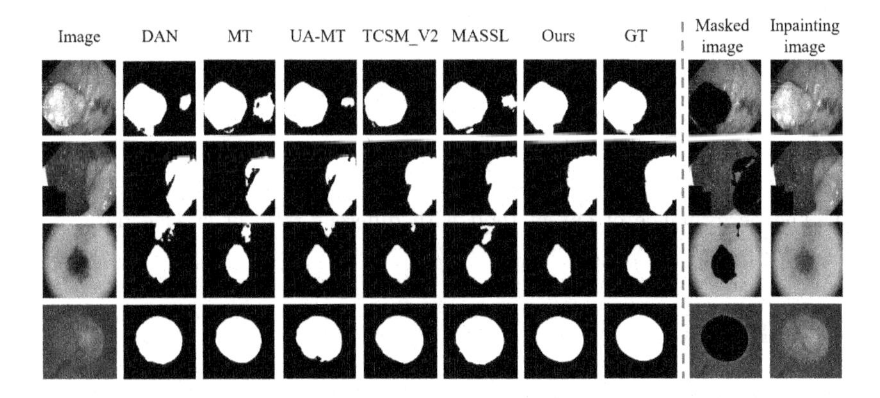

Fig. 3. Visual comparison of various lesion segmentation from state-of-the-art methods. Our proposed method consistently produces segmentation results closest to the ground truth. The inpainting result is shown in the rightmost column.

3.4 Ablation Study

Effectiveness of Our Approach with Different Ratio of Labeled Data. We draw the curves of Dice score under three settings in Fig. 4. To verify that our proposed framework can mine residual lesion features and enhance the lesion representation by GFF modules in the second stage, we conduct experiments and draw the blue line. The blue line denotes that our method uses the same labeled data with the baseline (the red line) to perform the two-stage process, without utilizing any unlabeled data. Note that we only calculate the segmentation loss for the labeled data. The performance gains compared with the baseline show that our network mines useful lesion information in the second stage. The green line means that our method introduces the remaining as unlabeled data for the inpainting task, further enhancing the feature representation learning of the lesion regions and improving the segmentation performance, especially when only a small amount of labeled data is used. When using 100% labeled data, the green line is equivalent to the blue line since no additional unlabeled data is utilized to do the inpainting task, thus maintaining the same results.

Effectiveness of the GFF Modules. To verify the effectiveness of the GFF modules, we also design two variants, which merge features by directly addition and concatenation, denoting as Ours (add) and Ours (concat) respectively. In Table 1, we can observe performance degradation by both approaches compared with our method, proving that the GFF module plays a significant role in filtering redundant information and improving the model performance.

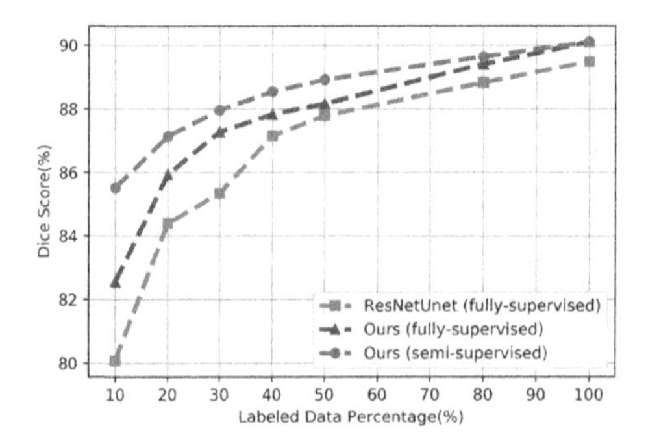

Fig. 4. The performance of our method with different ratio of labeled data on the Kvasir-SEG dataset.

4 Conclusions

In this paper, we believe that massive unlabeled data contains rich context and structural information, which is significant for lesion segmentation. Based on this, we introduce the self-supervised inpainting branch for unlabeled data, cooperating with the main segmentation task for labeled data, to further enhance the representation for lesion regions, thus refine the segmentation results. We also design the GFF module for better feature selection and aggregation from the two tasks. Experiments on various medical datasets have demonstrated the superior performance of our method.

Acknowledgement. This work is supported in part by the Key-Area Research and Development Program of Guangdong Province (No. 2020B0101350001), in part by the Guangdong Basic and Applied Basic Research Foundation (No. 2020B1515020048), in part by the National Natural Science Foundation of China (No. 61976250) and in part by the Guangzhou Science and technology project (No. 202102020633).

References

1. Chaitanya, K., Erdil, E., Karani, N., Konukoglu, E.: Contrastive learning of global and local features for medical image segmentation with limited annotations. arXiv preprint arXiv:2006.10511 (2020)
2. Chen, S., Bortsova, G., García-Uceda Juárez, A., van Tulder, G., de Bruijne, M.: Multi-task attention-based semi-supervised learning for medical image segmentation. In: Shen, D., et al. (eds.) MICCAI 2019. LNCS, vol. 11766, pp. 457–465. Springer, Cham (2019). https://doi.org/10.1007/978-3-030-32248-9_51
3. Chen, S., Tan, X., Wang, B., Hu, X.: Reverse attention for salient object detection. In: Ferrari, V., Hebert, M., Sminchisescu, C., Weiss, Y. (eds.) ECCV 2018. LNCS, vol. 11213, pp. 236–252. Springer, Cham (2018). https://doi.org/10.1007/978-3-030-01240-3_15
4. Chen, T., Kornblith, S., Norouzi, M., Hinton, G.: A simple framework for contrastive learning of visual representations. In: ICML, pp. 1597–1607. PMLR (2020)

5. Deng, J., Dong, W., Socher, R., Li, L.J., Li, K., Fei-Fei, L.: ImageNet: a large-scale hierarchical image database. In: CVPR, pp. 248–255. IEEE (2009)
6. Fumero, F., Alayón, S., Sanchez, J.L., Sigut, J., Gonzalez-Hernandez, M.: RIM-ONE: an open retinal image database for optic nerve evaluation. In: 24th International Symposium on Computer-Based Medical Systems (CBMS), pp. 1–6. IEEE (2011)
7. Gutman, D., et al.: Skin lesion analysis toward melanoma detection: a challenge at the international symposium on biomedical imaging (ISBI) 2016, hosted by the international skin imaging collaboration (ISIC). arXiv preprint arXiv:1605.01397 (2016)
8. He, K., Zhang, X., Ren, S., Sun, J.: Deep residual learning for image recognition. In: CVPR, pp. 770–778 (2016)
9. Hung, W.C., Tsai, Y.H., Liou, Y.T., Lin, Y.Y., Yang, M.H.: Adversarial learning for semi-supervised semantic segmentation. arXiv preprint arXiv:1802.07934 (2018)
10. Jha, D., et al.: Kvasir-SEG: a segmented polyp dataset. In: Ro, Y.M., De Neve, W., et al. (eds.) MMM 2020. LNCS, vol. 11962, pp. 451–462. Springer, Cham (2020). https://doi.org/10.1007/978-3-030-37734-2_37
11. Lee, D.H.: Pseudo-label: the simple and efficient semi-supervised learning method for deep neural networks. In: Workshop on Challenges in Representation Learning, ICML, vol. 3, p. 896 (2013)
12. Li, G., Yu, Y.: Deep contrast learning for salient object detection. In: CVPR, pp. 478–487 (2016)
13. Li, X., Yu, L., Chen, H., Fu, C.W., Xing, L., Heng, P.A.: Transformation-consistent self-ensembling model for semisupervised medical image segmentation. IEEE Trans. Neural Netw. Learn. Syst. **32**(2), 523–534 (2020)
14. Li, Y., Chen, J., Xie, X., Ma, K., Zheng, Y.: Self-loop uncertainty: a novel pseudo-label for semi-supervised medical image segmentation. In: Martel, A.L., et al. (eds.) MICCAI 2020. LNCS, vol. 12261, pp. 614–623. Springer, Cham (2020). https://doi.org/10.1007/978-3-030-59710-8_60
15. Long, J., Shelhamer, E., Darrell, T.: Fully convolutional networks for semantic segmentation. In: CVPR, pp. 3431–3440 (2015)
16. Noroozi, M., Favaro, P.: Unsupervised learning of visual representations by solving Jigsaw puzzles. In: Leibe, B., Matas, J., Sebe, N., Welling, M. (eds.) ECCV 2016. LNCS, vol. 9910, pp. 69–84. Springer, Cham (2016). https://doi.org/10.1007/978-3-319-46466-4_5
17. Ouali, Y., Hudelot, C., Tami, M.: Semi-supervised semantic segmentation with cross-consistency training. In: CVPR, pp. 12674–12684 (2020)
18. Paszke, A., et al.: Pytorch: an imperative style, high-performance deep learning library. In: NeurIPS, pp. 8026–8037 (2019)
19. Ronneberger, O., Fischer, P., Brox, T.: U-Net: convolutional networks for biomedical image segmentation. In: Navab, N., Hornegger, J., Wells, W.M., Frangi, A.F. (eds.) MICCAI 2015. LNCS, vol. 9351, pp. 234–241. Springer, Cham (2015). https://doi.org/10.1007/978-3-319-24574-4_28
20. Tarvainen, A., Valpola, H.: Mean teachers are better role models: weight-averaged consistency targets improve semi-supervised deep learning results. In: NeurIPS, pp. 1195–1204 (2017)
21. Wei, Y., Feng, J., Liang, X., Cheng, M.M., Zhao, Y., Yan, S.: Object region mining with adversarial erasing: a simple classification to semantic segmentation approach. In: CVPR, pp. 1568–1576 (2017)

22. Yu, L., Wang, S., Li, X., Fu, C.-W., Heng, P.-A.: Uncertainty-aware self-ensembling model for semi-supervised 3d left atrium segmentation. In: Shen, D., et al. (eds.) MICCAI 2019. LNCS, vol. 11765, pp. 605–613. Springer, Cham (2019). https://doi.org/10.1007/978-3-030-32245-8_67
23. Zhang, R., Li, G., Li, Z., Cui, S., Qian, D., Yu, Y.: Adaptive context selection for polyp segmentation. In: Martel, A.L., et al. (eds.) MICCAI 2020. LNCS, vol. 12266, pp. 253–262. Springer, Cham (2020). https://doi.org/10.1007/978-3-030-59725-2_25
24. Zhang, Y., Yang, L., Chen, J., Fredericksen, M., Hughes, D.P., Chen, D.Z.: Deep adversarial networks for biomedical image segmentation utilizing unannotated images. In: Descoteaux, M., Maier-Hein, L., Franz, A., Jannin, P., Collins, D.L., Duchesne, S. (eds.) MICCAI 2017. LNCS, vol. 10435, pp. 408–416. Springer, Cham (2017). https://doi.org/10.1007/978-3-319-66179-7_47

SpineGEM: A Hybrid-Supervised Model Generation Strategy Enabling Accurate Spine Disease Classification with a Small Training Dataset

Xihe Kuang[1], Jason Pui Yin Cheung[1], Xiaowei Ding[2], and Teng Zhang[1](✉)

[1] The University of Hong Kong, Pok Fu Lam, Hong Kong
u3006668@connect.hku.hk, {cheungjp,tgzhang}@hku.hk
[2] VoxelCloud Inc., Gayley Avenue 1085, Los Angeles, CA 90024, USA
xding@voxelcloud.io

Abstract. Most deep-learning based magnetic resonance image (MRI) analysis methods require numerous amounts of labelling work manually done by specialists, which is laborious and time-consuming. In this paper, we aim to develop a hybrid-supervised model generation strategy, called SpineGEM, which can economically generate a high-performing deep learning model for the classification of multiple pathologies of lumbar degeneration disease (LDD). A unique self-supervised learning process is adopted to generate a pre-trained model, with no pathology labels or human interventions required. The anatomical priori information is explicitly integrated into the self-supervised process, through auto-generated pixel-wise masks (using MRI-SegFlow: a system with unique voting processes for unsupervised deep learning-based segmentation) of vertebral bodies (VBs) and intervertebral discs (IVDs). With finetuning of a small dataset, the model can produce accurate pathology classifications. Our SpineGEM is validated on the Hong Kong Disc Degeneration Cohort (HKDDC) dataset with pathologies including Schneiderman Score, Disc Bulging, Pfirrmann Grading and Schmorl's Node. Results show that compared with training from scratch (n = 1280), the model generated through SpineGEM (n = 320) can achieve higher classification accuracy with much less supervision (~5% higher on mean-precision and ~4% higher on mean-recall).

Keywords: Self-supervised learning · Transfer learning · Pathology classification · MRI

1 Introduction

Magnetic resonance imaging (MRI) is widely used in Orthopaedics for the assessment of spine pathologies [1, 2]. Automated MRI analyses can have great clinical significance, by improving the efficiency and consistency of diagnosis [3]. For pathology classifications, the accuracy generated by a deep learning model can be comparable with human

M. de Bruijne et al. (Eds.): MICCAI 2021, LNCS 12902, pp. 145–154, 2021.
https://doi.org/10.1007/978-3-030-87196-3_14

specialists [4, 5]. However, numerous amounts of manual labels are required for the training process of the model, which is expensive, laborious, and time-consuming.

To reduce the amount of labelled data required in training, the idea of transfer learning [6, 7] has been adopted. The model is pretrained to learn the image features and patterns with a source task, and finetuned for the specific target task. Since the knowledge learnt from the pretraining is inherited, the model can achieve better performance on the target task with limited training data. Currently, many transfer learning based methods have been proposed, which can be divided into two categories, supervised pretraining and self-supervised pretraining. For supervised pretraining, the model will be trained from scratch on an existing large dataset with a similar task. The pretraining can utilize the dataset with natural images, such as ImageNet [8–13], or medical images [14]. However, the model pretrained with natural images cannot learn the 3D features contained by medical images, and pretraining with medical images in a supervised way still requires lots of manual annotation. To overcome this limitation, the self-supervised pretraining method is proposed, which allows the model to learn the image features on the unlabeled medical image directly [5, 15, 16]. Model Genesis [16] adopted a novel scheme to generate the generalizable source model for multiple tasks, which pretrained the model via restoring the image processed by specially designed transformations.

Inspired by Model Genesis, we propose a hybrid-supervised generation strategy, which consists of two steps, 1) self-supervised pretraining and 2) task specified finetuning. The deep learning model for the pathology classification consists of a CNN-based encoder and a classifier. The encoder is pretrained first with a self-supervised process, which explicitly integrates the anatomical information with the pixel-wise masks (generated through MRI-SegFlow [17], an unsupervised pipeline for the segmentation of spinal tissues) of VB and IVD. Then the model is finetuned for the specific classification task with pathology labels. We hypothesize that by explicitly introducing the anatomical priori knowledge in the pretraining process, the anatomy-related features, including anatomy shape, texture, and associated pixel intensity due to the different water amount contained in different tissues, can be learnt by the encoder, which enables the model to achieve higher classification accuracy with less training data.

In this paper, we aim to develop a hybrid-supervised strategy, called SpineGEM, to generate the deep learning model for classification of multiple LDD pathologies with a small training dataset, where G, E, and M represent Genesis (generate one model for solving multiple tasks), Economization (only a small dataset is required), and Model.

Our contributions include: 1) we developed a hybrid-supervised generation strategy called SpineGEM for the classification of multiple LDD pathologies, which explicitly utilized anatomical information in self-supervised pretraining; 2) we validated the performance of SpineGEM on an MRI dataset with clinical annotation of multiple LDD pathologies, and compared it with another state-of-the-art self-supervised method.

2 Methodology

The pathology classification process consisted of two steps: preprocessing and SpineGEM which included self-supervised pretraining and finetuning (Fig. 1). In preprocessing, an unsupervised pipeline MRI-SegFlow was adopted to generate the semantic segmentation of IVD and VB, and the regions of IVDs were identified based on the segmentation results. Then, the encoder of the deep learning model was pretrained in a self-supervised way to learn the anatomy-related features for the classification of different specific pathologies. A set of random transformations were applied on the MRI, which introduced the artefact with different image features including pixel intensity, texture, and shape. The encoder was pretrained with the task of recovering the original MRI from the processed image. The segmentation results were adopted in the transformation to provide anatomy information. No manual pathology annotation was required in the pretraining process. Then, the pretrained encoder was connected with a classifier and finetuned for the classification of a specific pathology. The finetuning was a normal supervised training process, however due to the pretrained encoder, much less manual annotated label was required.

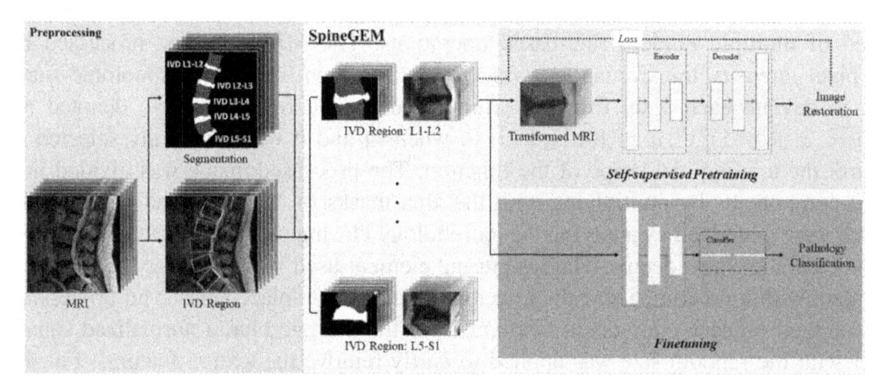

Fig. 1. The overview of SpineGEM. In preprocessing, the segmentation of IVD and VB were generated, and the IVD regions were extracted from the MRI. In pretraining, the MRI was processed with the random transformation, and an encoder-decoder network was trained to restore the processed image. Finally, the pretrained encoder was connected with a classifier and finetuned for the pathology classification.

2.1 Preprocessing

Unsupervised Semantic Segmentation. The segmentation of VBs and IVDs were generated through the published MRI-SegFlow [17], which was an unsupervised pipeline for the robust and accurate semantic segmentation of the lumbar MRI. The suboptimal region of interest (ROI) of each spinal tissue was identified first by a set of rule-based methods based on prior anatomy knowledge. The mistakes, including missing objectives, shape distortion, and false positives, were acceptable in the suboptimal ROIs. Then a unique voting process was adopted to combine all suboptimal ROIs and to automatically generate the high quality supervision. A CNN model was trained with the supervision to produce the accurate segmentation of VBs and IVDs.

Disc Region Extraction. Instead of using the MRI of the whole lumbar spine, for the analysis of each IVD, only a local region was served as the input of the model. The

IVD region was a 3D volume covering the whole IVD and its neighboring vertebral endplates. The circumscribed rectangle of the IVD segmentation in each MRI slice was generated first. The height (h_{IVD}), width (w_{IVD}), center point (x_{IVD}, y_{IVD}), and rotation angle (θ_{IVD}) of the IVD were defined as the maximum height, maximum width, average center point, and average rotation angle of all circumscribed rectangles. Then the IVD region was determined as a cuboid with the shape of $4h_{IVD} \times 1.5w_{IVD} \times 9$ centered at (x_{IVD}, y_{IVD}) rotated with θ_{IVD} in sagittal view, where 9 represented the mid-9-slices of each MRI. The size of IVD regions were standardized to $150 \times 200 \times 9$ with the bilinear interpolation before input to the CNN model.

2.2 Self-supervised Encoder Pretraining

Random Transformation. As illustrated in Fig. 2, the random transformation consisted of multiple random rule-based operations. The MRI was first processed by the pixel intensity transformation, which adopted a non-linear and monotonic function to deviate each pixel from its original value. The functions were denoted as: $f(x) = a + (-1)^a x^b, a \in \{0, 1\}, b > 0$, where a and b were randomly selected to control the tending and shape of the function. The processed image was divided into three components by multiplying with the area masks of VB, IVD and background, which were generated by applying the morphology closing on the segmentation results. It removed the shape details. The structuring element used in morphology closing was a square with a random size. Either average blur, mean replacement or no processing was applied on each component randomly. For the average blur, a normalized square filter with the random size was applied to partly remove the texture features. For the mean replacement, all pixel values were replaced with the mean value in the region, which eliminated all textural information. The mean replacement was not applied on the background, due to its wide pixel value range. Finally, the three components were merged with a distance-weighted sum, which was defined as:

$$I(x, y) = \frac{\sum d_i(x, y) C_i(x, y)}{\sum d_i(x, y)}, i \in \{1, 2, 3\} \quad (1)$$

Where C_i and I represented the image component of area i and merged image. $d_i(x, y)$ was the distance from (x, y) to the edge of area i, and if the (x, y) was not in the area i, the $d_i(x, y) = 0$. Due to the randomness, the specific transformation process could be different for different anatomical areas in one MRI case. However, for the different MRI slices of one anatomical structure, the process should be same to ensure the consistence between slices.

Encoder Pretraining. The encoder of the deep learning model was pretrained with the task of restoring the transformed image. To achieve this, a decoder with a symmetric structure to the encoder was developed, which was combined with the encoder, and formed a model with the encoder-decoder (E-D) architecture. The E-D model was trained to restoring the original MRI from the image processed by random transformation in an end-to-end way. No pathology label was required in the pretraining.

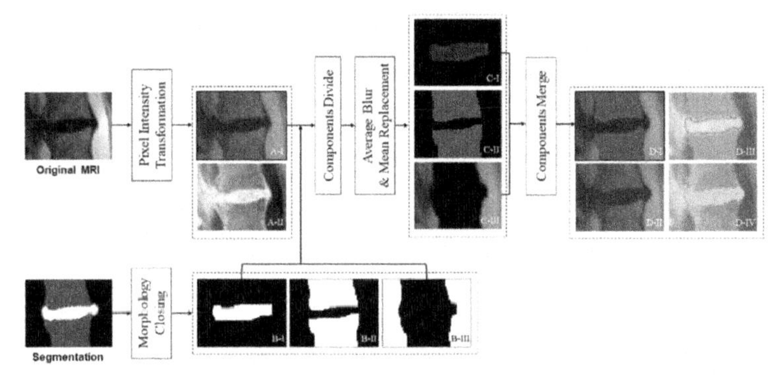

Fig. 2. The process of random transformation. A-I and A-II represented 2 versions of pixel value transformation. B-I–B-III were the area masks of IVD, VB, and background, and C-I–C-III were image components generated by A-I and B-I–B-III, which were processed by average blur (C-I, C-III) or mean replacement (C-II). D-I was the merging of C-I–C-III, and D-II–D-IV illustrated other 3 versions of random transformed MRI.

2.3 Task Specified Finetuning

The deep learning model for the classification of the specific pathology was developed by finetuning the pretrained encoder. To be specified, the decoder part of the trained E-D model was removed, and the encoder was connected with a classifier which consisted of several fully connected layers. Parameters in the classifier were initialized randomly. In the finetuning process, the deep learning model was trained with the pathology label manually annotated. Compared with training from scratch, the finetuning process based on the pretrained encoder required much less annotation. For the classification of different pathologies, the deep learning models could be finetuned from a same pretrained encoder, only the classifiers were reinitialized.

3 Implementation Details

3.1 Dataset

A dataset sourced from the Hong Kong Disc Degeneration Cohort (HKDDC) was adopted for the validation of our method. It consisted of 1600 sagittal lumbar T2-weighted MRI cases and each of them contained five IVDs from L1 to S1 (totally 8000 IVDs in the dataset). The MRIs were obtained with different machines and protocol therefore the image quality varied amongst different cases. The image resolutions were from 448×400 to 512×512, and the slice numbers were from 11 to 17, which demonstrated the diversity of the dataset. For each MRI, four pathologies of LDD including Schnciderman Score, Disc Bulging, Pfirrmann Grading and Schmorl's Node were annotated by a single expert radiologist. The distribution of labels for each pathology is presented in Table 1, where for Schneiderman Score, disc bulging and Pfirrmann Grading, a higher grading presents a more severe pathological condition.

Table 1. Label distributions of four LDD pathologies.

Schneiderman score					
Grading	0	1	2	3	
Percentage	7.4%	49.5%	40.7%	2.4%	
Disc bulging					
Grading	0	1	2	3	
Percentage	53.4%	43.9%	2.5%	0.2%	
Pfirrmann grading					
Grading	1	2	3	4	5
Percentage	1.2%	25.1%	35.4%	36.4%	1.9%
Schmorl's node					
Category	Absence (Normal)		Presence (Abnormal)		
Percentage	96.4%		3.6%		

3.2 Network Architecture

The encoder of the deep learning model adopted the basic architecture of the modified VGG-M [4, 18] and was improved to 3D version. It replaced all the original 2D operations, including convolution and max-pooling with the corresponding 3D operations to utilize the 3D features within the MR image. As illustrated in Fig. 3, the input MRI volume was processed by three Conv blocks, and each of them consisted of at least one 3D convolutional layer and one 3D max-pooling layer. The kernel sizes for the convolution in the three Conv blocks were $7 \times 7 \times 3$, $5 \times 5 \times 3$, and $3 \times 3 \times 3$ respectively. The classifier of the deep learning model contained two fully connected layers with 1024 hidden units and an output layer. The dimension of output vector varied according to the category number of different LDD pathologies. All convolutional layers in the encoder and fully connected layer in the classifier were activated with ReLU function except the output layer, which was activated with softmax function.

3.3 SpineGEM Configuration

Most parameters controlling the transformation of self-supervised pretraining process were randomly selected in a certain range. For the pixel intensity transformation, the range of a and b were $\{0, 1\}$ and $[1/2, 2]$. The size of structuring elements in morphology closing was chosen from $\{1, 3, 5, 7, 9\}$. For the image components of IVD and VB, the probability of being processed by average blur, mean replacement and no processing was 40%, 40% and 20% respectively, and for the components of background 50% of them was processed by average blur and the rest was not processed. The size of square filter in average blur was from 3 to 9.

The self-supervised pretraining process took 100 epochs, and for each epoch 1500 MRIs of IVD regions were randomly selected and transformed as input. Mean squared error (MSE) was adopted as the loss function for the restoration task, and stochastic gradient descent (SGD) was served as optimizer. Due to the randomness of transformation, no data augmentation was required.

The finetuning process took 150 epochs, and 320 MRI cases containing 1600 disc regions with pathology labels were used. The cross entropy was adopted as the loss function for the classification task, and SGD was the optimizer. The mini-batch strategy was adopted in both pretraining and finetuning with the batch size of 32. All programs are writing in python with the TensorFlow 2.0 framework. The Nvidia GeForce 2080 was used for the training and testing of the deep learning model.

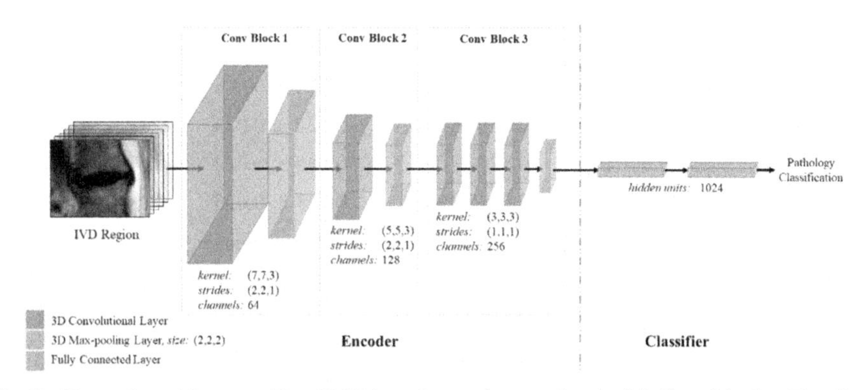

Fig. 3. Network architecture. The CNN-based encoder consisted of 3 Conv blocks. The Conv block 1 and 2 contained one 3D convolutional layer with the kernel size of $7 \times 7 \times 3$ and $5 \times 5 \times 3$ respectively. The Conv block 3 contained three $3 \times 3 \times 3$ convolutional layers.

4 Result and Discussion

Two validation metrics, including mean-precision (*mPre*), and mean-recall (*mRec*), were adopted in the paper to evaluate the performance of our method, which were defined as:

$$mPre = \frac{1}{N}\sum_{c=1}^{N}\frac{TP_c}{TP_c + FP_c} \tag{2}$$

$$mRec = \frac{1}{N}\sum_{c=1}^{N}\frac{TP_c}{TP_c + FN_c} \tag{3}$$

Where $c = 1, 2, \cdots, N$ represented different categories of the pathology. TP_c, FP_c and FN_c were the numbers of true positive, false positive and false negative samples of the pathology class c.

Table 2 showed the performance of model generated by SpineGEM, which compared with the unsupervised K-Means algorithm and models generated by supervised learning from scratch as well as the published self-supervised strategy proposed by A. Jamaludin [5]. All trained with 320 MRI scans. Given limited training data, SpineGEM outperformed the learning from scratch and published self-supervised method (Table 2). Additionally, unsupervised K-Means failed on the LDD pathology classification.

Table 2. Classification performance on 4 LDD Pathologies.

	Schneiderman score		Disc bulging	
	mPre	*mRec*	*mPre*	*mRec*
K-Means	0.291 ± 0.023	0.340 ± 0.014	0.274 ± 0.018	0.356 ± 0.026
Supervised	0.634 ± 0.028	0.519 ± 0.016	0.624 ± 0.024	0.575 ± 0.016
A. Jamaludin	0.699 ± 0.029	0.624 ± 0.012	0.645 ± 0.029	0.624 ± 0.022
SpineGEM	0.743 ± 0.022	0.631 ± 0.017	0.713 ± 0.025	0.682 ± 0.021
	Pfirrmann grading		Schmorl's node	
	mPre	*mRec*	*mPre*	*mRec*
K-Means	0.246 ± 0.011	0.302 ± 0.019	0.507 ± 0.025	0.553 ± 0.048
Supervised	0.571 ± 0.038	0.501 ± 0.018	0.651 ± 0.053	0.558 ± 0.036
A. Jamaludin	0.651 ± 0.047	0.614 ± 0.012	0.666 ± 0.028	0.652 ± 0.014
SpineGEM	0.659 ± 0.022	0.615 ± 0.018	0.733 ± 0.027	0.706 ± 0.027

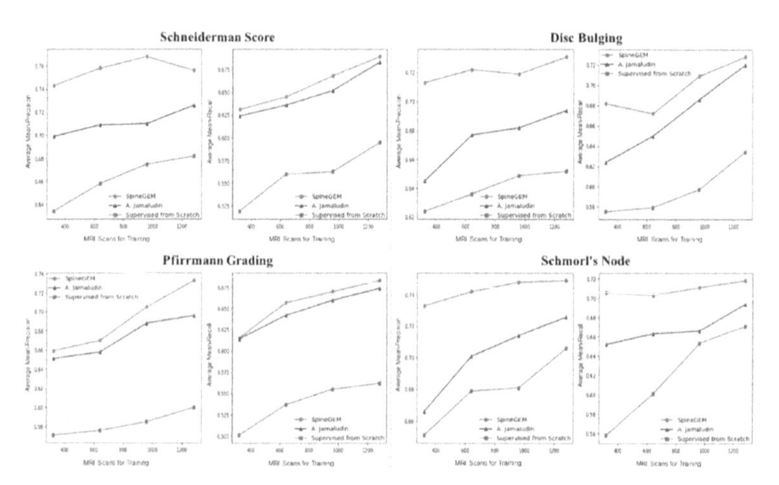

Fig. 4. The average mean-precision and mean-recall of LDD pathology classification achieved by models generated through 1) SpineGEM, 2) published self-supervised strategy proposed by A. Jamaludin [5], and 3) conventional supervised training from scratch.

Figure 4 showed the performance of models generated by different strategies, with 320, 640, 960 and 1280 MRI scans for training, respectively. It showed that with self-supervised pretraining, the model could achieve improved performance with reduced training data than training from scratch, and due to the explicit introduction of anatomical information in pretraining, the SpineGEM outperformed the previously published self-supervised method [5].

5 Conclusion

We have developed a hybrid-supervised strategy, called SpineGEM to generate the deep learning model for the classification of multiple LDD pathologies with a small training dataset. A self-supervised process is adopted first to pretrain the encoder of the deep learning model. The pretraining relies on the restoration task of the MRI processed by unique transformation, which explicitly integrates the anatomical information with pixel-wise masks of VB and IVD. The pretrained encoder will learn the anatomy-relative features, which is generalizable for the assessment of multiple pathologies. In finetuning, the deep learning model for the classification of the specific pathology is generated based on the pretrained encoder in the conventional supervised way. Our method is validated on the classification of four pathologies namely Schneiderman Score, Disc Bulging, Pfirrmann Grading and Schmorl's Node. The results demonstrate that compared with training from scratch, the model generated with our method can achieve higher accuracy with less labeled data.

References

1. Jensen, M.C., Brant-Zawadzki, M.N., Obuchowski, N., Modic, M.T., Malkasian, D., Ross, J.S.: Magnetic resonance imaging of the lumbar spine in people without back pain. N. Engl. J. Med. **331**(2), 69–73 (1994)
2. Pfirrmann, C.W., Metzdorf, A., Zanetti, M., Hodler, J., Boos, N.: Magnetic resonance classification of lumbar intervertebral disc degeneration. Spine **26**(17), 1873–1878 (2001)
3. Liu, J., et al.: Applications of deep learning to MRI images: a survey. Big Data Min. Anal. **1**(1), 1–18 (2018)
4. Jamaludin, A., Kadir, T., Zisserman, A.: SpineNet: automated classification and evidence visualization in spinal MRIs. Med. Image Anal. **41**, 63–73 (2017)
5. Jamaludin, A., Kadir, T., Zisserman, A.: Self-supervised learning for spinal MRIs. In: Cardoso, M. et al. (eds.) DLMIA 2017, ML-CDS 2017. LNCS, vol. 10553, pp. 294–302. Springer, Cham (2017). https://doi.org/10.1007/978-3-319-67558-9_34
6. Pan, S.J., Yang, Q.: A survey on transfer learning. IEEE Trans. Knowl. Data Eng. **22**(10), 1345–1359 (2009)
7. Weiss, K., Khoshgoftaar, T.M., Wang, D.: A survey of transfer learning. J. Big Data **3**(1), 1–40 (2016). https://doi.org/10.1186/s40537-016-0043-6
8. Shin, H.C., et al.: Deep convolutional neural networks for computer-aided detection: CNN architectures, dataset characteristics and transfer learning. IEEE Trans. Med. Imaging **35**(5), 1285–1298 (2016)
9. Guan, Q., Huang, Y.: Multi-label chest X-ray image classification via category-wise residual attention learning. Pattern Recogn. Lett. **130**, 259–266 (2020)

10. Gündel, S., Grbic, S., Georgescu, B., Liu, S., Maier, A., Comaniciu, D.: Learning to recognize abnormalities in chest X-rays with location-aware dense networks. In: Vera-Rodriguez, R., Fierrez, J., Morales, A. (eds.) CIARP 2018. LNCS, vol. 11401, pp. 757–765. Springer, Cham (2018). https://doi.org/10.1007/978-3-030-13469-3_88
11. Tang, Y., Wang, X., Harrison, A.P., Lu, L., Xiao, J., Summers, R.M.: Attention-guided curriculum learning for weakly supervised classification and localization of thoracic diseases on chest radiographs. In: Shi, Y., Suk, H.I., Liu, M. (eds.) MLMI 2018. LNCS, vol. 11046, pp. 249–258. Springer, Cham. https://doi.org/10.1007/978-3-030-00919-9_29
12. Ding, Y., et al.: A deep learning model to predict a diagnosis of Alzheimer disease by using 18F-FDG PET of the brain. Radiology **290**(2), 456–464 (2019)
13. Talo, M., Baloglu, U.B., Yıldırım, Ö., Acharya, U.R.: Application of deep transfer learning for automated brain abnormality classification using MR images. Cogn. Syst. Res. **54**, 176–188 (2019)
14. Chen, S., Ma, K., Zheng, Y.: Med3D: transfer learning for 3D medical image analysis. arXiv preprint arXiv:1904.00625 (2019)
15. Spitzer, H., Kiwitz, K., Amunts, K., Harmeling, S., Dickscheid, T.: Improving cytoarchitectonic segmentation of human brain areas with self-supervised siamese networks. In: Frangi, A., Schnabel, J., Davatzikos, C., Alberola-López, C., Fichtinger, G. (eds.) MICCAI 2018. LNCS, vol. 11072, pp. 663–671. Springer, Cham (2018). https://doi.org/10.1007/978-3-030-00931-1_76
16. Zhou, Z., Sodha, V., Pang, J., Gotway, M.B., Liang, J.: Models genesis. Med. Image Anal. **67**, 101840 (2021)
17. Kuang, X., Cheung, J.P., Wu, H., Dokos, S., Zhang, T.: MRI-SegFlow: a novel unsupervised deep learning pipeline enabling accurate vertebral segmentation of MRI images. In: 2020 42nd Annual International Conference of the IEEE Engineering in Medicine & Biology Society (EMBC), pp. 1633–1636. IEEE (2020)
18. Chatfield, K., Simonyan, K., Vedaldi, A., Zisserman, A.: Return of the devil in the details: delving deep into convolutional nets. arXiv preprint arXiv:1405.3531 (2014)

Contrastive Learning of Relative Position Regression for One-Shot Object Localization in 3D Medical Images

Wenhui Lei[1,2], Wei Xu[1], Ran Gu[1,2], Hao Fu[1,2], Shaoting Zhang[1,2], Shichuan Zhang[3], and Guotai Wang[1(✉)]

[1] School of Mechanical and Electrical Engineering, University of Electronic Science and Technology of China, Chengdu, China
guotai.wang@uestc.edu.cn
[2] SenseTime Research, Shanghai, China
[3] Sichuan Cancer Hospital and Institute, Chengdu, China

Abstract. Deep learning networks have shown promising performance for object localization in medical images, but require large amount of annotated data for supervised training. To address this problem, we propose: 1) A novel contrastive learning method which embeds the anatomical structure by predicting the Relative Position Regression (RPR) between any two patches from the same volume; 2) An one-shot framework for organ and landmark localization in volumetric medical images. Our main idea comes from that tissues and organs from different human bodies own similar relative position and context. Therefore, we could predict the relative positions of their non-local patches, thus locate the target organ. Our one-shot localization framework is composed of three parts: 1) A deep network trained to project the input patch into a 3D latent vector, representing its anatomical position; 2) A coarse-to-fine framework contains two projection networks, providing more accurate localization of the target; 3) Based on the coarse-to-fine model, we transfer the organ bounding-box (B-box) detection to locating six extreme points along x, y and z directions in the query volume. Experiments on multi-organ localization from head-and-neck (HaN) and abdominal CT volumes showed that our method acquired competitive performance in real time, which is more accurate and 10^5 times faster than template matching methods with the same setting for one-shot localization in 3D medical images. Code is available at https://github.com/HiLab-git/RPR-Loc.

Keywords: Contrastive learning · One-shot localization · Relative position

W. Lei and W. Xu—Contribute equally to this work.

Electronic supplementary material The online version of this chapter (https://doi.org/10.1007/978-3-030-87196-3_15) contains supplementary material, which is available to authorized users.

M. de Bruijne et al. (Eds.): MICCAI 2021, LNCS 12902, pp. 155–165, 2021.
https://doi.org/10.1007/978-3-030-87196-3_15

1 Introduction

Localization of organs and anatomical landmarks is crucial in medical image analysis, as it can assist in the treatment, diagnosis and follow-up of many diseases [23]. Nowadays, Deep Learning has achieved state-of-the-art performance in the localization of a broad of structures, including abdominal organs [22], brain tissues [19], etc. However, their success mainly relies on a large set of annotated images for training, which is expensive and time-consuming to acquire.

To reducing the demand on human annotations, several techniques such as weakly-supervised [15], semi-supervised [20], self-supervised [9,21,24] and one-shot learning (OSL) [2,5] attracted increasing attentions. OSL is especially appealing as it does not require annotations of the target during training, and only needs one annotated support image at the inference time. However, to the best of our knowledge, there have been very few works on one-shot object localization in volumetric medical images, such as Computed Tomography (CT) and Magnetic Resonance Images (MRI).

In this paper, we propose a one-shot localization method for organ/landmark localization in volumetric medical images, which does not require annotations of either the target organs or other types of objects during training and can directly locate any landmarks or organs that are specified by a support image in the inference stage. To achieve this goal, we present a novel contrastive learning [6,14,18] method called Relative Position Regression (RPR). Our main inspiration comes from the fact that the spatial distributions of organs (landmarks) have strong similarities among patients. Therefore, we could project every part of scans to a shared 3D latent coordinate system by training a projection network (Pnet) to predict the 3D physical offset between any two patches from the same image. We represent the bounding box of a target object by six extreme points, and locate these extreme points respectively by using Pnet to predict the offset from an agent's current position to the target position. Finally, we propose a coarse-to-fine framework based on Pnet that allows an agent to move mulitple steps for accurate localization, and introduce a multi-run ensemble strategy to further improve the performance. Experiments on multi-organ localization from head-and-neck (HaN) and abdominal CT volumes showed that our method achieved competitive performance in real time. Therefore, our method removes the need of human annotations for training deep learning models for object localization in 3D medical images.

2 Methodology

Relative Position Regression and Projection Network. Let $\boldsymbol{v}$ denote a volumetric image in the unannotated training set, and let $\boldsymbol{x}_q$ and $\boldsymbol{x}_s$ represent a moving (query) patch and a reference (support) patch with the same size $D \times H \times W$ in $\boldsymbol{v}$ respectively, we use a network to predict the 3D offset from $\boldsymbol{x}_q$ to $\boldsymbol{x}_s$. Assuming the pixel spacing of $\boldsymbol{v}$ is $\boldsymbol{e} \in R^3$ while $\boldsymbol{c}_q, \boldsymbol{c}_s \in R^3$ represent the centroid coordinates of $\boldsymbol{x}_q$ and $\boldsymbol{x}_s$, the ground truth offset $\boldsymbol{d}'_{qs}$ from $\boldsymbol{x}_q$ to $\boldsymbol{x}_s$ in the physical space is denoted as:

$$\boldsymbol{d}'_{qs} = (\boldsymbol{c}_s - \boldsymbol{c}_q) \circ \boldsymbol{e} \tag{1}$$

where $\circ$ represents the element-wise product.

As shown in Fig. 1, our Pnet is composed of two parts: convolution blocks to extract high-level features and fully connected layers mapping the features to a 3D latent vector. The details could be found in supplementary material. With Pnet, $\boldsymbol{x}_s$ and $\boldsymbol{x}_q$ are projected into $\boldsymbol{p}_s$ and $\boldsymbol{p}_q \in R^3$. Then the predicted offset $\boldsymbol{d}_{qs} \in R^3$ from the query patch $\boldsymbol{x}_q$ to the support patch $\boldsymbol{x}_s$ is obtained as:

$$\boldsymbol{d}_{qs} = r \cdot tanh(\boldsymbol{p}_s - \boldsymbol{p}_q) \tag{2}$$

where the hyperbolic tangent function $tanh$ and the hyper-parameter r together control the upper and lower bound of $\boldsymbol{d}_{qs}$, which is set to cover the largest possible offset. Finally, we apply the mean square error (MSE) loss function to measure the difference between $\boldsymbol{d}_{qs}$ and $\boldsymbol{d}'_{qs}$:

$$Loss = ||\boldsymbol{d}_{qs} - \boldsymbol{d}'_{qs}||^2 \tag{3}$$

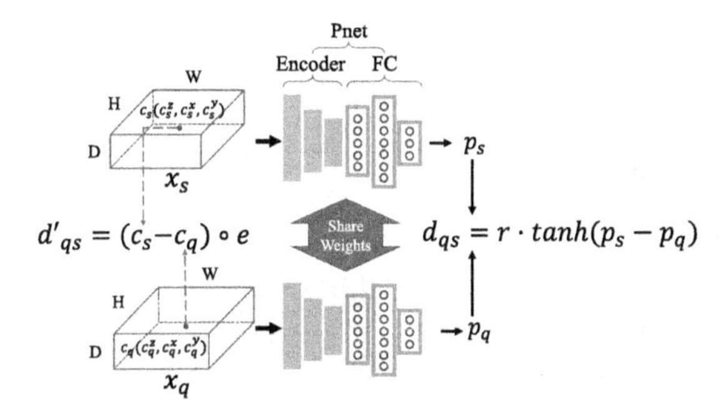

Fig. 1. Structure of our projection network (Pnet) for offset prediction. $\boldsymbol{x}_s$ and $\boldsymbol{x}_q$ are the support and query patches centered at $\boldsymbol{c}_s$ and $\boldsymbol{c}_q$. We use a shared Pnet to transform $\boldsymbol{x}_s$ and $\boldsymbol{x}_q$ to 3D latent vectors $\boldsymbol{p}_s$ and $\boldsymbol{p}_q$, respectively. The Pnet contains convolution blocks to extract features and fully connected layers for projection. We apply scale factor r and hyperbolic tangent function $tanh$ to get the predicted offset $\boldsymbol{d}_{qs}$, i.e., relative position from $\boldsymbol{x}_s$ to $\boldsymbol{x}_q$.

In the inference stage for landmark localization, we use an annotated support (reference) image to specify the desired landmark, and aim to find the position of the corresponding landmark in a test image. The problem is modeled as moving an agent from a random initial position to the target position. Therefore, we take a patch at a random position $\boldsymbol{c}_0 \in R^3$ in the test image as $\boldsymbol{x}_q$, which is the initial status of the agent, and take the patch centered at the specified landmark in the support image as the support patch, which is an approximation of the unknown target patch in the test image and serves as $\boldsymbol{x}_s$. By applying Pnet with $\boldsymbol{x}_s$ and $\boldsymbol{x}_q$, we obtain an offset $\boldsymbol{d}_{qs} \in R^3$, thus we move the agent with $\boldsymbol{d}_{qs}$, and obtain $\boldsymbol{c} = \boldsymbol{c}_0 + \boldsymbol{d}_{qs}$ as the detected landmark position in the test image.

Multi-step Localization. As moving the agent with only one step to the target position may not be accurate enough, our framework supports multi-step localization, i.e., Pnet can be applied several times given the agent's new position and the support patch. In this paper, we employ two-step inference strategy as that showed the best performance during experiment. The first step obtains a coarse localization, where the offset can be very large. The second step obtains a fine localization with a small offset around the coarse localization result.

To further improve the performance, we train two models for these two steps respectively. More specifically, for the coarse model $\boldsymbol{M}_c$, we crop $\boldsymbol{x}_q$ and $\boldsymbol{x}_s$ randomly across the entire image space and set r_c to cover the largest possible offset between $\boldsymbol{x}_q$ and $\boldsymbol{x}_s$. For the fine model $\boldsymbol{M}_f$, we first randomly crop the $\boldsymbol{x}_q$, then crop $\boldsymbol{x}_s$ around $\boldsymbol{x}_q$ within a small range of r_f mm, e.g., 30 mm. Therefore, $\boldsymbol{M}_c$ could handle large movements while $\boldsymbol{M}_f$ then focuses on the movement with small steps. During the inference stage, we use $\boldsymbol{M}_c$ to predict the coarse position, then use $\boldsymbol{M}_f$ to obtain the fine position. The RPR-Loc framework is shown in Fig. 2.

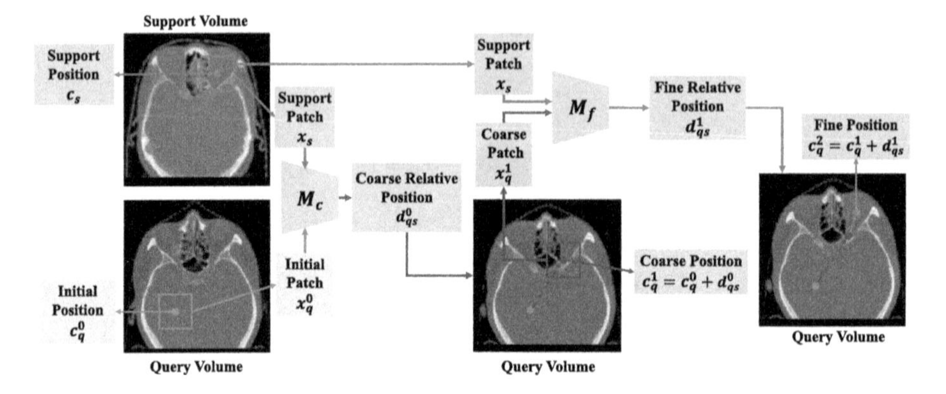

Fig. 2. Proposed one-shot localization method based on Relative Position Regression (RPR-Loc). The green box represents a support (reference) patch in the support image that specifies a landmark to locate in a test image, and the yellow box is an initial patch that represents the current position of an agent in the test image. $\boldsymbol{M}_c$ (coarse projection model) takes the support patch and initial patch as input to regress the offset from the initial patch to the desired landmark in the test image. The agent moves one step given by the offset, and its new position is represented by the red box, which is a coarse result of the localization. We apply the $\boldsymbol{M}_f$ (fine projection model) to obtain a fine localization result given the agent's new position and the support patch. (Color figure online)

Multi-run Ensemble. As the position of the agent during inference is initialized randomly, different initialization may obtain slightly different results. To obtain more robust results, we propose a multi-run ensemble strategy: we locate each target landmark K times with different random initializations, and average the localization results.

Organ Detection via Landmark Localization. We have described how to use Pnet to locate a landmark, and it can be easily reused for locating the bounding box of an organ without additional training. Specifically, for the organ B-box prediction, we transfer it to a set of landmark localization problems. As shown in Fig. 3, there are two possible methods to implement this. The first is to locate the maximum and minimum coordinates of B-box among three dimensions by predicting bottom-left corner and top-right corner $[\boldsymbol{D}_{min}, \boldsymbol{D}_{max}]$ [11], which is referred as "Diagonal Points" in Fig. 3(a). However, these two points may be located in meaningless areas (e.g., air background). As the case shown in Fig. 3(c), the minimum diagonal point D_{min} of the mandible is hard to detect due to lack of contextual information. Therefore, linking patches directly with the wanted organs would be more reasonable, and we propose to locate the six "Extreme Points" on the surface of an organ along x, y and z axes, which are denoted as $[\boldsymbol{Z}_{min}, \boldsymbol{Z}_{max}, \boldsymbol{X}_{min}, \boldsymbol{X}_{max}, \boldsymbol{Y}_{min}, \boldsymbol{Y}_{max}]$ in Fig. 3(b). As shown in Fig. 3(c), using the extreme points can ensure these landmarks be located in the body foreground, thus more contextual information can be leveraged to obtain more accurate position and scale of the associated B-box.

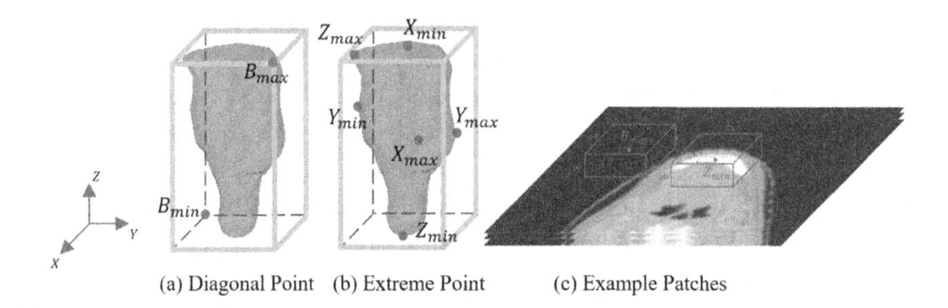

Fig. 3. Example of two methods transferring B-box estimation to landmark localization: (a) Directly locating the bottom-left corner and top-right corner of B-box; (b) Finding 6 extreme points along x, y, and z axes; (c) An example of the two different types of points and their surrounding patches of the mandible. The patch around D_{min} (minimum diagonal point) locates at meaningless area (air), while the one around Z_{min} (minimum extreme point along axial orientation) links directly to the bound of the mandible.

3 Experiments

Dataset and Evaluation Metrics. Our method was evaluated on two datasets: 1) A mixed head and neck (HaN) CT dataset containing 165 volumes from three sources: 50 patients from StructSeg 2019[1], 48 patients from MICCAI 2015 Head and Neck challenge [16] and 67 patients collected locally. We used four organs that were annotated in all of them: brain stem, mandible, left and right parotid gland.

The spacing was resampled to $3 \times 1 \times 1$ mm along axial, coronal, sagittal directions and cropped to remove the air background. The 165 volumes were splitted into 109 training, 19 validation and 38 testing samples; 2) A pancreas-CT dataset[17] with 82 volumes. The spacing was resampled to $3 \times 3 \times 3$ mm. We split the data into 50, 11 and 21 for training, validation and testing, respectively. The validation is used in training to select a model also by MSE loss.

The ground truth 3D B-boxes and extreme points in the validation and testing images were inferred from the 3D masks of each organ. While during the inference stage, we randomly select one image from the validation set as the support (reference) image, and use each image in the testing set as the query image, respectively. For the comparison of different methods, we calculated two metrics: (1) the Intersection over Union (IoU) between the predicted and ground-truth B-box; (2) the Absolute Wall Distance (AWD) following [7,22].

Implementation Details. Our model was trained with an NVIDIA GTX 1080 Ti GPU. The Adam optimizer [10] was used for training with batch size 6, initial learning rate 10^{-3} and 250 epochs. With regard to every query volume, we constrain each initial position in the non-air region. Fixed patch sizes of $16 \times 64 \times 64$ and $48 \times 48 \times 48$ pixels were used for the HaN and pancreas organ detection problem. The coarse and fine model $\boldsymbol{M_c}$, $\boldsymbol{M_f}$ are trained separately. We set $r_c = 300, r_f = 30$ for HaN and $r_c = 700, r_f = 50$ for pancreas.

Performance of Organ Localization and Multi-run Ensemble. We first compare the two different landmark representation strategies: "Diagonal Points" and "Extreme Points" and also investigate the effect of hyper-parameter K for multi-run ensemble on the localization performance. As shown in Table 1, with one initial patch, predicting extreme points rather than diagonal points can prominently increase IoU value and reduce AWD for all organs, especially for the mandible. Because in different volumes, the diagonal points on the bounding box may be in featureless regions such as the air background and difficult to locate, thus the performance of B-box prediction is limited in this strategy. Therefore, we apply the "extreme points" location strategy for all subsequent experiments.

Last five rows of Table 1 show that the average accuracy will also increase obviously as the number of runs for multi-run ensemble increases from 1 to 20. This is in line with our expectations that taking average of predictions from different start points would increase the precision and robustness.

[1] https://structseg2019.grand-challenge.org/Home/.

Table 1. The mean organ localization IoU (%, left) and AWD (mm, right) of four HaN OARs and pancreas under different strategies and different numbers of multi-run ensemble (MRE) with our coarse model. "Diagonal" and "Extreme" mean organ detection via localization of diagonal points and extreme points.

Strategy	MRE	Brain stem	Mandible	L parotid	R parotid	Mean	Pancreas
Diagonal	1	44.5, 7.06	32.3, 15.29	36.5, 9.94	35.7, 9.66	37.2, 10.49	43.3, 14.51
Extreme	1	45.9, 6.61	62.2, 8.04	38.6, 9.61	37.5, 9.09	46.1, 8.34	44.4, 13.85
Extreme	5	49.4, 5.63	62.7, 7.52	41.8, 8.66	36.8, 9.13	47.7, 7.73	47.2, 12.78
Extreme	10	50.0, 5.34	63.2, 7.40	42.4, 8.73	38.4, 8.88	48.7, 7.59	45.2, 13.53
Extreme	15	50.7, 5.34	**63.7, 7.28**	**43.1, 8.48**	**38.8, 8.79**	**49.1, 7.48**	46.0, 13.19
Extreme	20	**51.9, 5.16**	63.3, 7.45	42.6, 8.57	37.4, 9.03	48.8, 7.55	**48.1, 12.42**

Performance of Multi-step Localization. Based on the "Extreme Points" strategy for organ detection, we report the effectness of our multi-step model in Table 2 with $K = 15$ for HaN and $K = 20$ for pancreas. Moving the agent by two or three steps with coarse model $\boldsymbol{M_c}$ than just one step does not guarantee performance improvement. Because $\boldsymbol{M_c}$ is trained with a large r_c thus owning a sparse latent space, leading to its insensitivity about small step offset. In contrast, using $\boldsymbol{M_c}$ in the first step followed by using $\boldsymbol{M_f}$ in the second step achieved the best performance among HaN and pancreas tasks. It shows that the fine model $\boldsymbol{M_f}$ trained within small offset range r_f could deal with the movements of small steps more effectively.

Table 2. The mean localization IoU (%, left) and AWD (mm, right) of four HaN OARs and pancreas of different multi-step methods.

Method	Brain stem	Mandible	L parotid	R parotid	Mean	Pancreas
$\boldsymbol{M_c}$ (one step)	50.7, 5.34	63.7, 7.28	43.1, 8.48	38.8, 8.79	49.1, 7.48	48.1, 12.42
$\boldsymbol{M_c}$ (two steps)	51.4, 5.30	68.0, 6.03	45.0, 8.23	38.7, 8.89	50.7, 7.11	46.7, 13.46
$\boldsymbol{M_c}$ (three steps)	50.5, 5.52	68.7, 5.79	**45.3**, 8.30	38.9, 8.81	50.9, 7.10	46.0, 13.76
$\boldsymbol{M_c} + \boldsymbol{M_f}$	**61.5, 3.70**	**70.0, 5.39**	44.8, **7.74**	**44.2, 7.65**	**55.1, 6.12**	**49.5, 12.22**

Comparison with Other Methods. In this section, we compare our method with: 1) A state-of-the-art supervised localization method Retina U-Net[8] that was trained with the ground truth bounding boxes in the training set. 2) Template matching-based alternatives for one-shot organ localization under the same setting as our method. We crop patches around each extreme point from the support volume, then implement template matching method by sliding window operation with stride 2 along each dimension to find the patch in the test image that is the most similar to the support patch. We consider the following similarity-based methods for comparison: a) Gray Scale MSE (GS MSE), where mean square error of gray scale intensities is used for similarity measurement between the support and query patches; b) Gray Scale Cosine (GS Cosine), which means using cosine similarity as criteria for comparison; c) Gray Scale Normalized Cross Correlation (GS NCC). We use normalized cross correlation to evaluate the similarity; d) Feature Map MSE (FM MSE). We train an auto-encoder network [12] to transform a patch into a latent feature, and use MSE to measure the similarity between the latent features of $\boldsymbol{x}_q$ and $\boldsymbol{x}_s$; e) Feature Map Cosine (FM Cosine), where cosine similarity is applied to the latent features of $\boldsymbol{x}_q$ and $\boldsymbol{x}_s$.

With the same query volume, Table 3 shows comparison of these methods on accuracy and average time consumption for each organ. Our method outperforms GS MSE, GS Cosine, GS NCC and FM MSE by an average IoU score and AWD. Despite FM Cosine slightly outperformed our method in terms of IoU of the right parotids, its performance is much lower on the brain stem. In addition, our method is 1.7×10^5 times faster than FM cosine that uses sliding window scanning. Compared with the fully supervised Retina U-Net [8], our one-shot method still have a gap, but the difference is small for the mandible, and our method is much faster than Retina U-Net [8].

Table 3. Quantitative evaluation of different methods for 3D organ localization. For each organ, we measured the mean IoU (%, left) and AWD (mm, right). Note that Retina U-Net [8] requires full annotations, while the others does not need annotations for training.

Method	Brain stem	Mandible	L parotid	R parotid	Mean	Pancreas	Time(s)
Ours	**61.5, 3.70**	**70.0, 5.39**	**44.8, 7.74**	44.2, **7.65**	**55.1, 6.12**	**49.5, 12.22**	**0.15**
GS MSE [1]	39.3, 7.47	65.7, 7.20	39.8, 10.67	37.1, 10.67	45.5, 9.00	1.8, 90.23	1052
GS Cosine [13]	35.2, 7.83	67.4, 7.10	36.9, 11.07	36.7, 10.30	44.1, 9.08	2.7, 118.14	1421
GS NCC [4]	46.9, 7.35	58.9, 9.76	24.3, 18.78	23.9, 20.14	38.5, 14.01	3.5, 101.77	4547
FM MSE [25]	44.7, 6.52	67.7, 6.88	39.2, 10.43	40.6, 8.79	48.1, 8.16	7.9, 120.36	26586
FM Cosine [3]	50.2, 5.91	69.1, 5.53	44.7, 7.97	**44.6**, 8.22	52.2, 7.92	19.2, 56.27	25976
Retina U-Net [8]	68.9, 3.54	75.1, 6.17	60.5, 5.83	63.9, 4.75	67.1, 5.07	81.0, 3.23	4.7

As shown in Fig. 4, despite the large variation in shapes and low contrast to surrounding tissues, the proposed framework could accurately locate B-box in the query volume. This implies that our method is promising for removing the need of annotation for training deep learning models for object localization in 3D medical images.

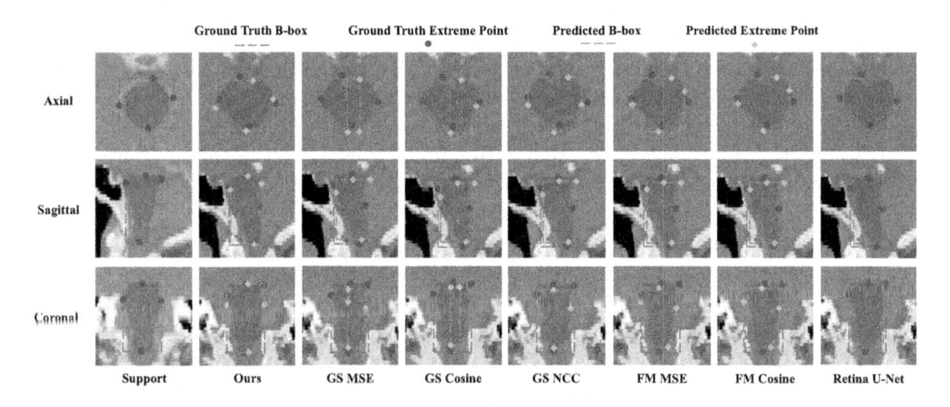

Fig. 4. Qualitative localization results of brain stem in a query volume on three dimensions. The proposed method achieves desirable results.

4 Discussion and Conclusions

In this work, we propose a relative position regression-based one-shot localization framework (RPR-Loc) for 3D medical images, which to our best knowledge is the first work of one-shot localization in volumetric medical scans. Note that our one-shot localization can be easily extended to few-shot localization given multiple support volumes. Opposed to the traditional time-consuming template matching methods, our framework is regression-based thus not sensitive to the volume size and more efficient. Our method does not need any annotation during the training stage and could be employed to locate any landmarks or organs contained in the training dataset during the inference stage. Results on multi-organ localization from HaN and pancreas CT volumes showed that our method achieved more accurate results and is thousand to million times faster than template matching methods under the same setting. This study demonstrates the effectiveness of our RPR-Loc in avoiding annotations of training images for deep learning-based object detection in 3D medical images, and it is of interest to apply it to other organs in the future.

Acknowledgements. This work was supported by the National Natural Science Foundations of China [61901084, 81771921] funding, key research and development project of Sichuan province, China [No. 20ZDYF2817].

References

1. Bankman, I.N., Johnson, K.O., Schneider, W.: Optimal detection, classification, and superposition resolution in neural waveform recordings. IEEE Trans. Biomed. Eng. **40**(8), 836–841 (1993)
2. Bart, E., Ullman, S.: Cross-generalization: learning novel classes from a single example by feature replacement. In: 2005 IEEE Computer Society Conference on Computer Vision and Pattern Recognition (CVPR 2005), vol. 1, pp. 672–679. IEEE (2005)

3. Bodla, N., Zheng, J., Xu, H., Chen, J.C., Castillo, C., Chellappa, R.: Deep heterogeneous feature fusion for template-based face recognition. In: 2017 IEEE Winter Conference on Applications of Computer Vision (WACV), pp. 586–595. IEEE (2017)
4. Briechle, K., Hanebeck, U.D.: Template matching using fast normalized cross correlation. In: Optical Pattern Recognition XII, vol. 4387, pp. 95–102. International Society for Optics and Photonics (2001)
5. Fei-Fei, L., Fergus, R., Perona, P.: One-shot learning of object categories. IEEE Trans. Pattern Anal. Mach. Intell. **28**(4), 594–611 (2006)
6. Hjelm, R.D., et al.: Learning deep representations by mutual information estimation and maximization. arXiv preprint arXiv:1808.06670 (2018)
7. Humpire Mamani, G., Setio, A., Ginneken, B., Jacobs, C.: Efficient organ localization using multi-label convolutional neural networks in thorax-abdomen CT scans. Phys. Med. Biol. **63** (2018)
8. Jaeger, P.F., et al.: Retina U-Net: Embarrassingly simple exploitation of segmentation supervision for medical object detection. In: Machine Learning for Health Workshop, pp. 171–183. PMLR (2020)
9. Jing, L., Tian, Y.: Self-supervised visual feature learning with deep neural networks: a survey. IEEE Trans. Pattern Anal. Mach. Intell. (2020)
10. Kingma, D.P., Ba, J.: Adam: a method for stochastic optimization. arXiv preprint arXiv:1412.6980 (2014)
11. Law, H., Deng, J.: CornerNet: detecting objects as paired keypoints. In: Ferrari, V., Hebert, M., Sminchisescu, C., Weiss, Y. (eds.) Computer Vision – ECCV 2018. LNCS, vol. 11218, pp. 765–781. Springer, Cham (2018). https://doi.org/10.1007/978-3-030-01264-9_45
12. Lei, W., Wang, H., Gu, R., Zhang, S., Zhang, S., Wang, G.: DeepIGeoS-V2: deep interactive segmentation of multiple organs from head and neck images with lightweight CNNs. In: Zhou, L., et al. (eds.) LABELS/HAL-MICCAI/CuRIOUS -2019. LNCS, vol. 11851, pp. 61–69. Springer, Cham (2019). https://doi.org/10.1007/978-3-030-33642-4_7
13. Liu, K., Wang, W., Wang, J.: Pedestrian detection with lidar point clouds based on single template matching. Electronics **8**(7), 780 (2019)
14. Oord, A.v.d., Li, Y., Vinyals, O.: Representation learning with contrastive predictive coding. arXiv preprint arXiv:1807.03748 (2018)
15. Oquab, M., Bottou, L., Laptev, I., Sivic, J.: Is object localization for free?-weakly-supervised learning with convolutional neural networks. In: Proceedings of the IEEE Conference on Computer Vision and Pattern Recognition, pp. 685–694 (2015)
16. Raudaschl, P.F., et al.: Evaluation of segmentation methods on head and neck CT: auto-segmentation challenge 2015. Med. Phys. **44**(5), 2020–2036 (2017)
17. Roth, H.R., et al.: DeepOrgan: multi-level deep convolutional networks for automated pancreas segmentation. In: Navab, N., Hornegger, J., Wells, W.M., Frangi, A.F. (eds.) MICCAI 2015. LNCS, vol. 9349, pp. 556–564. Springer, Cham (2015). https://doi.org/10.1007/978-3-319-24553-9_68
18. Tian, Y., Krishnan, D., Isola, P.: Contrastive multiview coding. arXiv preprint arXiv:1906.05849 (2019)
19. Vlontzos, A., Alansary, A., Kamnitsas, K., Rueckert, D., Kainz, B.: Multiple landmark detection using multi-agent reinforcement learning. In: Shen, D., et al. (eds.) MICCAI 2019. LNCS, vol. 11767, pp. 262–270. Springer, Cham (2019). https://doi.org/10.1007/978-3-030-32251-9_29

20. Wang, D., Zhang, Y., Zhang, K., Wang, L.: Focalmix: semi-supervised learning for 3d medical image detection. In: Proceedings of the IEEE/CVF Conference on Computer Vision and Pattern Recognition, pp. 3951–3960 (2020)
21. Wang, S., Cao, S., Wei, D., Wang, R., Ma, K., Wang, L., Meng, D., Zheng, Y.: LT-Net: label transfer by learning reversible voxel-wise correspondence for one-shot medical image segmentation. In: Proceedings of the IEEE/CVF Conference on Computer Vision and Pattern Recognition, pp. 9162–9171 (2020)
22. Xu, X., Zhou, F., Liu, B., Fu, D., Bai, X.: Efficient multiple organ localization in CT image using 3d region proposal network. IEEE Trans. Med. Imaging **38**(8), 1885–1898 (2019)
23. Zhang, J., Liu, M., Shen, D.: Detecting anatomical landmarks from limited medical imaging data using two-stage task-oriented deep neural networks. IEEE Trans. Image Process. **26**(10), 4753–4764 (2017)
24. Zhou, Z., et al.: Models genesis: generic autodidactic models for 3d medical image analysis. In: Shen, D., et al. (eds.) MICCAI 2019. LNCS, vol. 11767, pp. 384–393. Springer, Cham (2019). https://doi.org/10.1007/978-3-030-32251-9_42
25. Zou, W., Zhu, S., Yu, K., Ng, A.: Deep learning of invariant features via simulated fixations in video. Adv. Neural. Inf. Process. Syst. **25**, 3203–3211 (2012)

Topological Learning and Its Application to Multimodal Brain Network Integration

Tananun Songdechakraiwut[1(✉)], Li Shen[2], and Moo Chung[1]

[1] University of Wisconsin–Madison, Madison, USA
songdechakra@wisc.edu
[2] University of Pennsylvania, Philadelphia, USA

Abstract. A long-standing challenge in multimodal brain network analyses is to integrate topologically different brain networks obtained from diffusion and functional MRI in a coherent statistical framework. Existing multimodal frameworks will inevitably destroy the topological difference of the networks. In this paper, we propose a novel topological learning framework that integrates networks of different topology through persistent homology. Such challenging task is made possible through the introduction of a new topological loss that bypasses intrinsic computational bottlenecks and thus enables us to perform various topological computations and optimizations with ease. We validate the topological loss in extensive statistical simulations with ground truth to assess its effectiveness of discriminating networks. Among many possible applications, we demonstrate the versatility of topological loss in the twin imaging study where we determine the extent to which brain networks are genetically heritable.

Keywords: Topological data analysis · Persistent homology · Wasserstein distance · Multimodal brain networks · Twin brain imaging study

1 Introduction

In standard brain network modeling, the whole brain is usually parcellated into a few hundred disjoint regions [7,17,27]. For instance, well established, widely used Automated Anatomical Labeling (AAL) parcellates the brain into 116 regions [27]. These disjoint regions form nodes in a brain network. Subsequently, functional or structural information is overlaid on top of the parcellation to obtain brain connectivity between the regions. Structural connectivity is obtained from diffusion MRI (dMRI), which traces the white matter fibers in the brain. Strength of the structural connectivity is determined by the number of fibers connecting the parcellations. Resting-state functional connectivity

Electronic supplementary material The online version of this chapter (https://doi.org/10.1007/978-3-030-87196-3_16) contains supplementary material, which is available to authorized users.

M. de Bruijne et al. (Eds.): MICCAI 2021, LNCS 12902, pp. 166–176, 2021.
https://doi.org/10.1007/978-3-030-87196-3_16

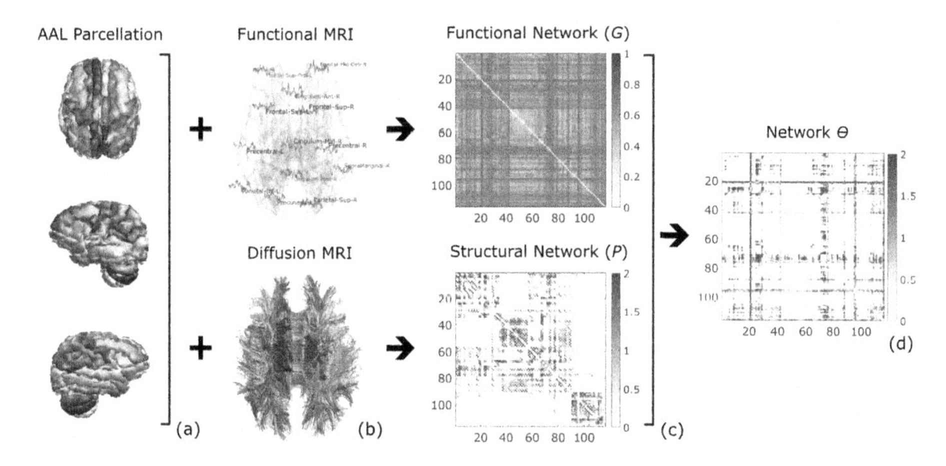

Fig. 1. Schematic of topological learning. (a) AAL partitions the human brain into 116 disjoint regions. (b, c) Functional network G is obtained from resting-state fMRI. The template structural network P is obtained from dMRI. The structural network P is sparse while the functional network G is densely connected with many cycles. (d) We learn network Θ that has the topological characteristics of both functional and structural networks.

obtained from functional MRI (fMRI) is often computed using the Pearson correlation coefficient between average fMRI time series in the parcellations [7]. While the structural connectivity provides information whether the brain regions are physically connected through the white matter fibers, the functional connectivity can exhibit relations between two regions without a direct neuroanatomical connection [14]. Thus, functional brain networks are often very dense with thousands of loops or cycles [7] while structural brain networks are expected to exhibit sparse topology without many cycles. Both the structural and functional brain networks provide topologically different information (Fig. 1). Nonetheless, not much research has been done thus far on integrating the brain networks at the localized connection level. Existing integration frameworks will inevitably destroy the topological difference in the process [16,30]. There is a need for a new multimodal brain network model that can integrate networks of different topology in a coherent statistical framework.

Persistent homology [7,9,15,17,25] provides a novel approach to the long-standing challenge in multimodal brain network analyses. In persistent homology, topological features such as connected components and cycles are measured across different spatial resolutions represented in the form of barcodes. It was recently proposed to penalize the barcodes as a loss function in image segmentation [15]. Though the method allows to incorporate topological information into the problem, it is limited to an image with a handful of topological features due to its expensive optimization process with $O(n^3)$. This is impractical in brain networks with a far larger number of topological features comprising hundreds of connected components and thousands of cycles. In this paper, we propose a more

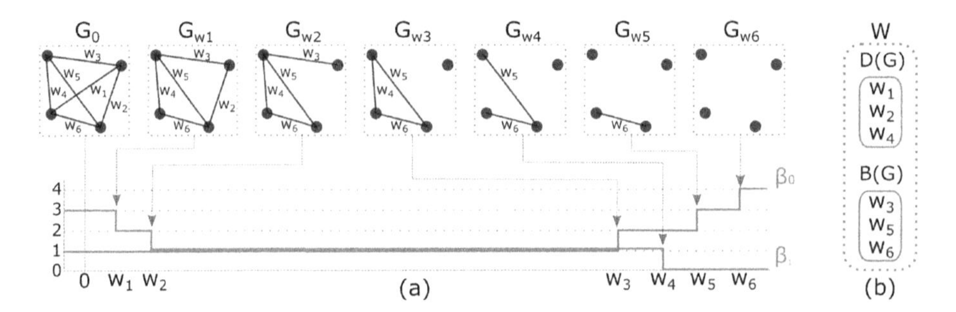

Fig. 2. (a) Graph filtration of G. β_0 is monotonically increasing while β_1 is monotonically decreasing over the graph filtration. Connected components are born at edge weights w_3, w_5, w_6 while cycles die at edge weights w_1, w_2, w_4. 0D barcode is represented by a set of birth values $B(G) = \{w_3, w_5, w_6\}$. 1D barcode is represented by a set of death values $D(G) = \{w_1, w_2, w_4\}$. (b) The weight set $W = \{w_1, ..., w_6\}$ is partitioned into 0D birth values and 1D death values: $W = B(G) \cup D(G)$.

principled and scalable *topological loss* with $O(n \log n)$. Our proposed method bypasses the intrinsic computational bottleneck and thus enables us to perform various topological computations and optimizations with ease.

Twin studies on brain imaging phenotypes provide a well established way to examine the extent to which brain networks are influenced by genetic factors. However, previous twin imaging studies have not been well adapted beyond determining heritability of a few brain regions of interest [2,4,12,21,24]. Measures of network topology are worth investigating as intermediate phenotypes that indicate the genetic risk for various neuropsychiatric disorders [3]. Determining heritability of the whole brain network is the first necessary prerequisite for identifying network based endophenotypes. With our topological loss, we propose a novel *topological learning* framework where we determine heritability of the functional brain networks while integrating the structural brain network information (Fig. 1). Our method increases statistical sensitivity to subtle topological differences, yielding more connections as genetic signals.

2 Method

Barcodes in Graph Filtration. Consider a network represented as a weighted graph $G = (V, w)$ comprising a node set V and edge weights $w = (w_{ij})$ with positive and unique weights. The number of nodes and edges are denoted by $|V|$ and $|E|$. Network G is a complete graph with $|E| = |V|(|V| - 1)/2$. The binary graph $G_\epsilon = (V, w_\epsilon)$ of G is defined as a graph consisting of the node set V and binary edge weight $w_{\epsilon,ij} = 1$ if $w_{ij} > \epsilon$ or 0 otherwise. The binary network G_ϵ is the 1-skeleton, a simplicial complex consisting of nodes and edges only [22]. In 1-skeleton, 0-dimensional (0D) holes are connected components and 1-dimensional (1D) holes are cycles [7]. The number of connected components and cycles in the binary network G_ϵ are referred to as the 0-th Betti number $\beta_0(G_\epsilon)$ and the 1-st Betti number $\beta_1(G_\epsilon)$. A *graph filtration* of G is defined as a collection of nested binary networks [7,17]: $G_{\epsilon_0} \supset G_{\epsilon_1} \supset \cdots \supset G_{\epsilon_k}$, where $\epsilon_0 < \epsilon_1 < \cdots < \epsilon_k$ are filtration values. By increasing the filtration value ϵ, we are thresholding at higher connectivity resulting in more edges being removed, and thus the 0-th and 1-st Betti numbers change.

Persistent homology keeps track of appearances (birth) and disappearances (death) of connected components and cycles over filtration values ϵ, and associates their *persistence* (lifetimes measured as the duration of birth to death) to them. Long lifetimes indicate global topological features while short lifetimes indicate small-scale topological features [11,20,29]. The persistence is represented by *0D* and *1D barcodes* comprising a set of intervals $[b_i, d_i]$, each of which tabulates a lifetime of a connected component or a cycle that appears at the filtration value b_i and vanishes at d_i (Fig. 2). Since connected components are born one at a time over increasing filtration values [7], these connected components will never die once they are born. Thus, we simply ignore their death values at ∞ and represent 0D barcode as a set of only birth values $B(G) = \cup_i \{b_i\}$. Cycles are considered born at $-\infty$ and will die one at a time over the filtration. Ignoring the $-\infty$, we represent 1D barcode as a set of only death values $D(G) = \cup_i \{d_i\}$.

Theorem 1. *The set of 0D birth values $B(G)$ and 1D death values $D(G)$ partition the weight set $W = \{w_{ij}\}$ such that $W = B(G) \cup D(G)$ with $B(G) \cap D(G) = \emptyset$. The cardinality of $B(G)$ and $D(G)$ are $|V| - 1$ and $1 + \frac{|V|(|V|-3)}{2}$ respectively.*

The proof is given in the supplementary material. Finding 0D birth values $B(G)$ is *equivalent* to finding edge weights of the *maximum spanning tree* (MST) of G using Prim's or Kruskal's algorithm [17]. Once B is computed, D is simply given as the remaining edge weights. Thus, the barcodes are computed efficiently in $O(|E| \log |V|)$.

Topological Loss. Since networks are topologically completely characterized by 0D and 1D barcodes, the topological dissimilarity between two networks can be measured through barcode differences. We adapt the Wasserstein distance to quantify the differences between the barcodes [9,15,23]. The Wasserstein distance measures the differences between underlying probability distributions on barcodes through the Dirac delta function [10]. Let $\Theta = (V, w^{\Theta})$ and $P = (V, w^{P})$ be two given networks. The *topological loss* $\mathcal{L}_{top}(\Theta, P)$ is defined as the optimal matching cost

$$\mathcal{L}_{top}(\Theta, P) = \min_{\tau_0} \sum_{b \in B(\Theta)} \big[b - \tau_0(b)\big]^2 + \min_{\tau_1} \sum_{d \in D(\Theta)} \big[d - \tau_1(d)\big]^2, \quad (1)$$

where τ_0 is a bijection from $B(\Theta)$ to $B(P)$ and τ_1 is a bijection from $D(\Theta)$ to $D(P)$. By Theorem 1, the bijections τ_0 and τ_1 always exist. The first term measures how close two networks are in terms of 0D holes (connected components) and is referred to as 0D topological loss $\mathcal{L}_{0D}$. The second term measures how close two networks are in terms of 1D holes (cycles) and is called 1D topological loss $\mathcal{L}_{1D}$. Connected components represent an integration of a brain network while cycles represent how strong the integration is [6]. The optimization can be done exactly as follows:

Theorem 2

$$\mathcal{L}_{0D} = \min_{\tau_0} \sum_{b \in B(\Theta)} \big[b - \tau_0(b)\big]^2 = \sum_{b \in B(\Theta)} \big[b - \tau_0^*(b)\big]^2, \tag{2}$$

where τ_0^* maps the i-th smallest birth value in $B(\Theta)$ to the i-th smallest birth value in $B(P)$ for all i.

$$\mathcal{L}_{1D} = \min_{\tau_1} \sum_{d \in D(\Theta)} \big[d - \tau_1(d)\big]^2 = \sum_{d \in D(\Theta)} \big[d - \tau_1^*(d)\big]^2, \tag{3}$$

where τ_1^* maps the i-th smallest death value in $D(\Theta)$ to the i-th smallest death value in $D(P)$ for all i.

The proof is given in the supplementary material. We can compute the optimal matchings τ_0^* and τ_1^* between Θ and P in $O\big(|B(\Theta)| \log |B(\Theta)|\big)$ and $O\big(|D(\Theta)| \log |D(\Theta)|\big)$ by sorting edge weights and matching them.

Topological Learning. Let $G_1 = (V, w^1), \cdots, G_n = (V, w^n)$ be observed networks used for training a model. Let $P = (V, w^P)$ be a network expressing a prior topological knowledge. In brain network analysis, G_k can be a functional brain network of k-th subject obtained from resting-state fMRI, and P can be a template structural brain network obtained from dMRI. The functional networks can then overlay the template network (Fig. 1).

We are interested in learning the model $\Theta = (V, w^\Theta)$ using both the functional and structural brain networks. At the subject level, we train Θ using individual network G_k by optimizing

$$\widehat{\Theta}_k = \arg\min_{\Theta} \mathcal{L}_F(\Theta, G_k) + \lambda \mathcal{L}_{top}(\Theta, P), \tag{4}$$

where the squared Frobenius loss $\mathcal{L}_F(\Theta, G_k) = ||w^\Theta - w^k||_F^2$ measures the goodness of fit between the model and the individual network. The parameter λ controls the amount of topological information of network P that is introduced to G_k. The larger the value of λ, the more we are learning toward the topology of P. If $\lambda = 0$, we no longer learn the topology of P but simply fit the model Θ to the individual network G_k.

In numerical implementation, $\Theta = (V, w^\Theta)$ can be estimated iteratively through gradient descent efficiently by Theorems 1 and 2. The *topological gradient* with respect to edge weights $w^\Theta = (w_{ij}^\Theta)$ is given as

$$\frac{\partial \mathcal{L}_{top}(\Theta, P)}{\partial w_{ij}^\Theta} = \begin{cases} 2\big(w_{ij}^\Theta - \tau_0^*(w_{ij}^\Theta)\big) & \text{if } w_{ij}^\Theta \in B(\Theta); \\ 2\big(w_{ij}^\Theta - \tau_1^*(w_{ij}^\Theta)\big) & \text{if } w_{ij}^\Theta \in D(\Theta). \end{cases} \tag{5}$$

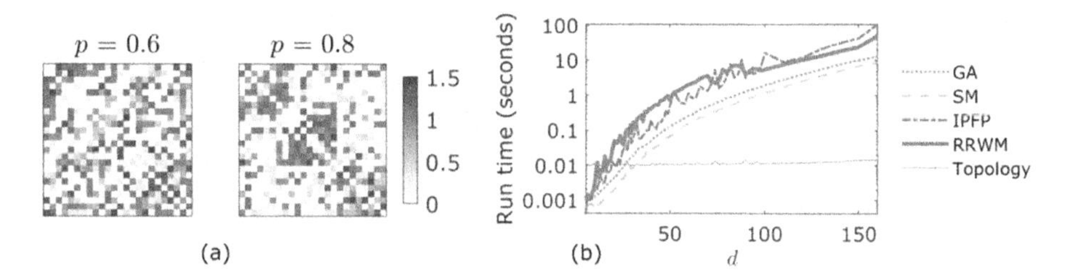

Fig. 3. (a) Two modular networks with $d = 24$ nodes and $c = 3$ modules were generated using $p = 0.6$ and 0.8. (b) The run time of graph matching cost between two modular networks of node size d plotted in the logarithmic scale. The run time of topological loss grows in a minuscule rate with the node size as opposed to the exponential run times of the graph matching algorithms.

By updating the edge weight w_{ij}^{Θ}, we adjust either a 0D birth value or a 1D death value, which changes topology of the model Θ. At each current iteration, we take a step in the direction of negative gradient with respect to an updated Θ from the previous iteration. As w_{ij}^{Θ} is moved through its optimal matching, the topology of Θ gets close to that of P while the Frobenius norm keeps Θ close to the observed network G_k. The time complexity of topological gradient is dominated by the computation of the MST with $O(|E| \log |V|)$.

3 Statistical Validation

We evaluated discriminative performance of the topological loss against four well-known graph matching algorithms: graduated assignment (GA) [13], spectral matching (SM) [18], integer projected fixed point method (IPFP) [19] and re-weighted random walk matching (RRWM) [5] using simulated networks.

We simulated random modular network $\mathcal{X}$ with d number of nodes and c number of modules where the nodes are evenly distributed among modules. We used $d = 12, 18, 24$ and $c = 2, 3, 6$. Each edge connecting two nodes within the *same* module was then assigned a random weight following a normal distribution $\mathcal{N}(\mu, \sigma^2)$ with probability p or Gaussian noise $\mathcal{N}(0, \sigma^2)$ with probability $1 - p$. Edge weights connecting nodes between *different* modules had probability $1 - p$ of being $\mathcal{N}(\mu, \sigma^2)$ and probability p of being $\mathcal{N}(0, \sigma^2)$. Any negative edge weights were set to zero. With larger value of within-module probability p, we have more pronounced modular structure (Fig. 3-a). The network $\mathcal{X}$ exhibits topological structures of connectedness. $\mu = 1$, $\sigma = 0.25$ and $p = 0.6$ were universally used as network variability.

We simulated two groups of random modular networks $\mathcal{X}_1, \cdots, \mathcal{X}_m$ and $\mathcal{Y}_1, \cdots, \mathcal{Y}_n$. If there is group difference in network topology, an average topological loss within group $\overline{\mathcal{L}}_W = \frac{\sum_{i<j} \mathcal{L}(\mathcal{X}_i, \mathcal{X}_j) + \sum_{i<j} \mathcal{L}(\mathcal{Y}_i, \mathcal{Y}_j)}{\binom{m}{2} + \binom{n}{2}}$ is expected to be smaller than an average topological loss between groups $\overline{\mathcal{L}}_B = \frac{\sum_{i=1}^{m} \sum_{j=1}^{n} \mathcal{L}(\mathcal{X}_i, \mathcal{Y}_j)}{mn}$. We measured the group disparity as the ratio statistic $\phi_{\mathcal{L}} = \overline{\mathcal{L}}_B \big/ \overline{\mathcal{L}}_W$. If $\phi_{\mathcal{L}}$ is large, the groups differ significantly in network topology. If $\phi_{\mathcal{L}}$ is small, it is likely

Table 1. Performance results are summarized as average p-values for various parameter settings of d (number of nodes) and c (number of modules).

d	c	GA	SM	RRWM	IPFP	$\mathcal{L}_{top}$
12 vs. 12	2 vs. 3	0.45 ± 0.27	0.48 ± 0.30	0.28 ± 0.31	0.34 ± 0.28	$\mathbf{0.08 \pm 0.16}$
	3 vs. 6	0.40 ± 0.29	0.35 ± 0.28	0.24 ± 0.26	0.35 ± 0.28	$\mathbf{0.06 \pm 0.13}$
18 vs. 18	2 vs. 3	0.25 ± 0.23	0.41 ± 0.26	0.26 ± 0.24	0.42 ± 0.28	$\mathbf{0.01 \pm 0.02}$
	3 vs. 6	0.28 ± 0.24	0.37 ± 0.31	0.21 ± 0.24	0.37 ± 0.30	$\mathbf{0.01 \pm 0.01}$
24 vs. 24	2 vs. 3	0.23 ± 0.25	0.30 ± 0.26	0.14 ± 0.20	0.31 ± 0.28	$\mathbf{0.00 \pm 0.01}$
	3 vs. 6	0.24 ± 0.26	0.29 ± 0.28	0.10 ± 0.13	0.37 ± 0.26	$\mathbf{0.00 \pm 0.00}$
12 vs. 12	2 vs. 2	0.49 ± 0.27	0.46 ± 0.30	0.51 ± 0.30	0.47 ± 0.28	0.53 ± 0.29
	3 vs. 3	0.45 ± 0.32	0.44 ± 0.26	0.47 ± 0.27	0.51 ± 0.30	0.46 ± 0.31
	6 vs. 6	0.57 ± 0.30	0.51 ± 0.28	0.56 ± 0.29	0.45 ± 0.26	0.58 ± 0.29

that there is no group difference. Similarly, we also defined the ratio statistic for the baseline algorithms. Since the distributions of the ratio statistics were unknown, the permutation test was used. In each simulation, we generated two groups with 10 modular networks each. We then computed 200000 permutations by shuffling group labels and obtained the p-values. The simulations were independently performed 50 times and the average p-value was reported.

The baseline graph matching algorithms are of polynomial time and not scalable compared to our method. For networks with $d = 100$ nodes, the run times of all the baselines are more than 100 times longer than that of topological loss (Fig. 3-b). When there is *network difference* (first three rows in Table 1), small p-value indicates that a method performs well at discriminating networks. In all the parameter settings, topological loss outperformed the other graph matching algorithms. Topological loss also consistently outperformed the baseline algorithms for other values of c, d and p. In the case of *no network difference* (last row in Table 1), small p-value indicates a method falsely detects the network difference when there is none. Since p-values of all the methods were not statistically significant, they all performed well. We also get similar results for other values of c, d and p. The graph matching algorithms are unable to detect topological differences while topological loss is able to easily detect such differences in subtle topological patterns with the minimal amount of run time. The MATLAB code for the simulation study is available at https://topolearn.github.io/topo-loss. The SM algorithm used in this simulation study and methods proposed in [1,26] rely on the same spectral graph theory and are expected to show analogous performance results.

4 Application to a Twin Imaging Study

Dataset and Preprocessing. dMRI and resting-state fMRI data were obtained from the Human Connectome Project [28]. fMRI went through further preprocessing including motion correction, scrubbing, bandpass filtering and

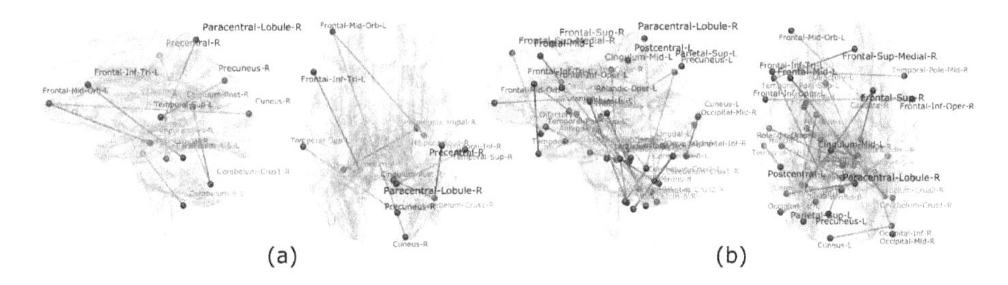

Fig. 4. Most heritable connections with 100% heritability using (a) Pearson correlation matrices and (b) topologically learned networks.

outlier removal among others. AAL was used to parcellate the brain into 116 regions [27]. fMRI were spatially averaged across voxels within each brain region resulting in 116 average fMRI time series per subject. There are 124 monozygotic (MZ) twin pairs and 70 same-sex dizygotic (DZ) twin pairs. For dMRI, about one million fiber tracts per subject were generated to compute biologically accurate brain connectivity [8]. AAL was used to parcellate the brain into 116 regions. The subject-level connectivity matrices were constructed by counting the number of tracts between the regions. The template structural network P was obtained by computing one sample t-statistic map over all the subjects and rescaling the t-statistic between 0 to 2 through the hyperbolic tangent function then adding 1 (Fig. 1). The t-statistic map from [8] is made publicly available at http://stat.wisc.edu/~mchung/softwares/dti.

Genetic Heritability. For the k-th subject, functional connectivity ρ_{ij}^k between regions i and j was computed using the Pearson correlation between time series. We converted the correlation into a metric through $w^k = (w_{ij}^k)$, where $w_{ij}^k = \sqrt{(1-\rho_{ij}^k)/2}$ and obtained a subject-level functional brain network $G_k = (V, w^k)$ [7]. The topological learning was applied to estimate the subject-level model Θ_k by minimizing the objective function (4) using the individual network G_k and the template structural network P. The model Θ_k was initialized to G_k. To determine an optimal subject-level λ, we searched over different λ's to find a value that minimized the total loss $\mathcal{L}_F + \mathcal{L}_{top}$ for each subject. The average of the optimal λ's across all the subjects was $\lambda = 1.0000 \pm 0.0002$, a highly stable result. Thus, we globally used $\lambda = 1$ for all the subjects. We then investigated if the learned networks $\widehat{\Theta}_k$ are genetically heritable. We used the ACE model where the heritability index (HI) is estimated using Falconer's formula [6].

Results and Discussion. We computed the HI using the initial Pearson correlation matrices as a baseline versus the topologically learned networks. Figure 4, which displays resulting HI thresholded at 100% heritability, shows far more connections for the learned networks as opposed to the Pearson correlation matrices. The learned networks are expected to inherit sparse topology without many cycles from the template network P (Fig. 1). This suggests that short-lived cycles were removed from the initial functional networks, improving the statistical sensitivity. For the learned networks, the connection with the highest HI is between left superior parietal lobule and left amygdala among many other connections with 100% heritability, suggesting that genes influence the development of these connections. Our findings can be used as a baseline for studying more complex relations between brain networks and other phenotypes.

Acknowledgments. We thank Shih-Gu Huang (National University of Singapore) and Gregory Kirk (University of Wisconsin–Madison) for assistance in preprocessing fMRI data. This study is funded by NIH R01 EB022856, EB02875, EB022574 and NSF MDS-2010778.

References

1. Becker, C.O., et al.: Spectral mapping of brain functional connectivity from diffusion imaging. Sci. Rep. **8**(1), 1–15 (2018)
2. Blokland, G., McMahon, K., Thompson, P., Martin, N., de Zubicaray, G., Wright, M.: Heritability of working memory brain activation. J. Neurosci. **31**, 10882–10890 (2011)
3. Bullmore, E., Sporns, O.: Complex brain networks: graph theoretical analysis of structural and functional systems. Nat. Rev. Neurosci. **10**(3), 186–198 (2009)
4. Chiang, M.C., et al.: Genetics of white matter development: a DTI study of 705 twins and their siblings aged 12 to 29. NeuroImage **54**, 2308–2317 (2011)
5. Cho, M., Lee, J., Lee, K.M.: Reweighted random walks for graph matching. In: Daniilidis, K., Maragos, P., Paragios, N. (eds.) ECCV 2010. LNCS, vol. 6315, pp. 492–505. Springer, Heidelberg (2010). https://doi.org/10.1007/978-3-642-15555-0_36
6. Chung, M.K., Huang, S.G., Gritsenko, A., Shen, L., Lee, H.: Statistical inference on the number of cycles in brain networks. In: 16th International Symposium on Biomedical Imaging, pp. 113–116. IEEE (2019)
7. Chung, M.K., Lee, H., DiChristofano, A., Ombao, H., Solo, V.: Exact topological inference of the resting-state brain networks in twins. Network Neurosci. **3**, 674 (2019)
8. Chung, M.K., Xie, L., Huang, S.-G., Wang, Y., Yan, J., Shen, L.: Rapid acceleration of the permutation test via transpositions. In: Schirmer, M.D., Venkataraman, A., Rekik, I., Kim, M., Chung, A.W. (eds.) CNI 2019. LNCS, vol. 11848, pp. 42–53. Springer, Cham (2019). https://doi.org/10.1007/978-3-030-32391-2_5
9. Clough, J.R., Oksuz, I., Byrne, N., Schnabel, J.A., King, A.P.: Explicit topological priors for deep-learning based image segmentation using persistent homology. In: Chung, A.C.S., Gee, J.C., Yushkevich, P.A., Bao, S. (eds.) IPMI 2019. LNCS, vol. 11492, pp. 16–28. Springer, Cham (2019). https://doi.org/10.1007/978-3-030-20351-1_2
10. Cohen-Steiner, D., Edelsbrunner, H., Harer, J., Mileyko, Y.: Lipschitz functions have Lp-stable persistence. Found. Comput. Math. **10**, 127 (2010). https://doi.org/10.1007/s10208-010-9060-6

11. Ghrist, R.: Barcodes: the persistent topology of data. Bull. Am. Math. Soc. **45**(1), 61–75 (2008)
12. Glahn, D., et al.: Genetic control over the resting brain. Proc. Nat. Acad. Sci. **107**, 1223–1228 (2010)
13. Gold, S., Rangarajan, A.: A graduated assignment algorithm for graph matching. IEEE Trans. Pattern Anal. Mach. Intell. **18**(4), 377–388 (1996)
14. Honey, C.J., Kötter, R., Breakspear, M., Sporns, O.: Network structure of cerebral cortex shapes functional connectivity on multiple time scales. Proc. Nat. Acad. Sci. **104**(24), 10240–10245 (2007)
15. Hu, X., Li, F., Samaras, D., Chen, C.: Topology-preserving deep image segmentation. In: Advances in Neural Information Processing Systems (2019)
16. Kang, H., Ombao, H., Fonnesbeck, C., Ding, Z., Morgan, V.L.: A Bayesian double fusion model for resting-state brain connectivity using joint functional and structural data. Brain Connectivity **7**(4), 219–227 (2017)
17. Lee, H., Kang, H., Chung, M.K., Kim, B.N., Lee, D.S.: Persistent brain network homology from the perspective of dendrogram. IEEE Trans. Med. Imag. **31**(12), 2267–2277 (2012)
18. Leordeanu, M., Hebert, M.: A spectral technique for correspondence problems using pairwise constraints. In: International Conference on Computer Vision, pp. 1482–1489. IEEE (2005)
19. Leordeanu, M., Hebert, M., Sukthankar, R.: An integer projected fixed point method for graph matching and map inference. In: Advances in Neural Information Processing Systems, pp. 1114–1122 (2009)
20. Marchese, A., Maroulas, V.: Signal classification with a point process distance on the space of persistence diagrams. Adv. Data Anal. Classification **12**(3), 657–682 (2017). https://doi.org/10.1007/s11634-017-0294-x
21. McKay, D., et al.: Influence of age, sex and genetic factors on the human brain. Brain Imag. Behav. **8**(2), 143–152 (2013). https://doi.org/10.1007/s11682-013-9277-5
22. Munkres, J.R.: Elements of Algebraic Topology. CRC Press, Boca Raton (2018)
23. Rabin, J., Peyré, G., Delon, J., Bernot, M.: Wasserstein barycenter and its application to texture mixing. In: Bruckstein, A.M., ter Haar Romeny, B.M., Bronstein, A.M., Bronstein, M.M. (eds.) SSVM 2011. LNCS, vol. 6667, pp. 435–446. Springer, Heidelberg (2012). https://doi.org/10.1007/978-3-642-24785-9_37
24. Smit, D., Stam, C., Posthuma, D., Boomsma, D., De Geus, E.: Heritability of small-world networks in the brain: a graph theoretical analysis of resting-state EEG functional connectivity. Hum. Brain Map. **29**, 1368–1378 (2008)
25. Songdechakraiwut, T., Chung, M.K.: Dynamic topological data analysis for functional brain signals. In: 17th International Symposium on Biomedical Imaging Workshops, pp. 1–4. IEEE (2020)
26. Surampudi, S.G., Naik, S., Surampudi, R.B., Jirsa, V.K., Sharma, A., Roy, D.: Multiple kernel learning model for relating structural and functional connectivity in the brain. Sci. Rep. **8**(1), 1–14 (2018)
27. Tzourio-Mazoyer, N., et al.: Automated anatomical labeling of activations in SPM using a macroscopic anatomical parcellation of the MNI MRI single-subject brain. NeuroImage **15**, 273–289 (2002)
28. Van Essen, D.C., et al.: The human connectome project: a data acquisition perspective. NeuroImage **62**(4), 2222–2231 (2012)

29. Xia, K., Wei, G.W.: Persistent homology analysis of protein structure, flexibility, and folding. Int. J. Numerical Methods Biomed. Eng. **30**(8), 814–844 (2014)
30. Xue, W., Bowman, F.D., Pileggi, A.V., Mayer, A.R.: A multimodal approach for determining brain networks by jointly modeling functional and structural connectivity. Front. Comput. Neurosci. **9**, 22 (2015)

One-Shot Medical Landmark Detection

Qingsong Yao[1], Quan Quan[1], Li Xiao[1], and S. Kevin Zhou[1,2(✉)]

[1] Key Lab of Intelligent Information Processing of Chinese Academy of Sciences (CAS), Institute of Computing Technology, CAS, Beijing 100190, China
yaoqingsong19@mails.ucas.edu.cn, {quanquan,xiaoli}@ict.ac.cn

[2] Medical Imaging, Robotics, and Analytic Computing Laboratory and Engineering (MIRACLE), School of Biomedical Engineering, Suzhou Institute for Advanced Research, University of Science and Technology of China, Suzhou 215123, China

Abstract. The success of deep learning methods relies on the availability of a large number of datasets with annotations; however, curating such datasets is burdensome, especially for medical images. To relieve such a burden for a landmark detection task, we explore the feasibility of using **only a single annotated image** and propose a novel framework named Cascade Comparing to Detect (CC2D) for one-shot landmark detection. CC2D consists of two stages: 1) Self-supervised learning (CC2D-SSL) and 2) Training with pseudo-labels (CC2D-TPL). CC2D-SSL captures the consistent anatomical information in a coarse-to-fine fashion by comparing the cascade feature representations and generates predictions on the training set. CC2D-TPL further improves the performance by training a new landmark detector with those predictions. The effectiveness of CC2D is evaluated on a widely-used public dataset of cephalometric landmark detection, which achieves a competitive detection accuracy of 86.25.01% within 4.0 mm, comparable to the state-of-the-art semi-supervised methods using a lot more than one training image. Our code is available at https://github.com/ICT-MIRACLE-lab/Oneshot_landmark_detection.

Keywords: Medical landmark detection · One-shot learning

1 Introduction

Accurate and reliable anatomical landmark detection is a fundamental first step in therapy planning and intervention, thus it has attracted great interest from academia and industry [28,30,35,36]. It has been proved crucial in many medical clinical scenarios such as knee joint surgery [29,31], bone age estimation [9], carotid artery bifurcation [32], orthognathic and maxillofacial surgeries [6], and pelvic trauma surgery [3]. Furthermore, it plays an important role in medical

Q. Yao and Q. Quan—Contribute equally in this paper.

Electronic supplementary material The online version of this chapter (https://doi.org/10.1007/978-3-030-87196-3_17) contains supplementary material, which is available to authorized users.

M. de Bruijne et al. (Eds.): MICCAI 2021, LNCS 12902, pp. 177–188, 2021.
https://doi.org/10.1007/978-3-030-87196-3_17

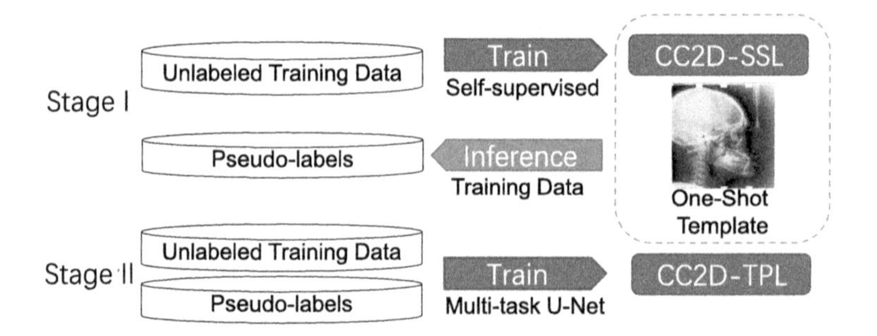

Fig. 1. Overview of the proposed Cascade Comparing to Detect (CC2D) framework. CC2D-SSL and CC2D-TPL represents self-supervised learning and training with pseudo-labels, respectively.

image analysis [13,14,17,19,20,22,35,36], e.g., initialization of registration or segmentation algorithms.

Recently, deep learning based methods have been developed to efficiently localize anatomical landmarks in radiological images such as computed tomography (CT) and X-rays [2,12,18,22]. Zhong et al. [33] use cascade U-Net to launch a two-stage heatmap regression. Chen et al. [6] regress the heatmap and coordinate offset maps at the same time. Liu et al. [18] improve the performance by utilizing relative position constraints. Li et al. [15] adapt cascade deep adaptive graphs to capture relationships among landmarks.

As is well known, expert-annotated training data are needed for all of those supervised methods. A common wisdom is that a model with a better generalization is learned from more data. However, it often consumes considerable cost and time for the radiologists to annotate sufficient training data, which might restrict further application in clinical scenarios. Several self-supervised learning attempts have been explored to break this limitation in classification and segmentation tasks, including patch ordering [39], image restoration [37,38], superpixel-wise [21] and patch-wise [4] contrastive learning. Nevertheless, training a landmark detection model with few annotated images is still challenging.

The landmarks are usually associated with anatomical features (e.g. the lips are halfway between the chin and nose in a cephalometric x-ray), and hence those features exhibit certain invariance among different images. Therefore, we pose an interesting question: Can we determine landmarks by learning from a few or even just one labeled image? In this paper, we challenge the hardest scenario: *Only one annotated medical image is available*, which defines the number of the landmarks and their corresponding locations of interests.

To tackle the challenge, we propose a novel framework named "Cascade Comparing to Detect **(CC2D)**" for one-shot landmark detection. CC2D consists of two stages: 1) Self-supervised learning **(CC2D-SSL)** and 2) Training with pseudo-labels **(CC2D-TPL)**. The CC2D-SSL idea is motivated by contrastive learning [7,34], which learns effective features for image classification. We instead propose to learn feature representations that embed consistent anatomical information for *image patch matching*. Our design is further inspired by our observation learned through interactions with clinicians that they firstly roughly locate the target regions through a coarse screening and then progressively refine the fine-grained location. Therefore, we propose to match the coarse-grained corresponding areas first, then gradually compare the finer-grained areas in the selected coarse area. Finally, the targeted landmark is localized precisely by comparing the cascade embeddings in a coarse-to-fine fashion.

In the second CC2D-TPL stage, we first use the CC2D-SSL model to generate predictions for the whole training set as *pseudo-labels*, followed by training a CNN based landmark detector using these pseudo-labels. This step brings two benefits. On one hand, the inference procedure becomes more concise. On the other hand, recent findings show that training an over-parameterized network from scratch tends to learn noiseless information firstly [1]. In our case, as we cannot predict every training point as accurate as ground truth in the SSL stage, a newly trained landmark detector can improve the performance by capturing the regular information hidden from the noisy labels produced by the CC2D-SSL stage.

We evaluate the performance of CC2D on the public dataset from the ISBI 2015 Grand-Challenge. CC2D achieves competitive detection accuracy of 86.25% within 4.0 mm, comparable to the state-of-the-art semi-supervised methods that use a lot more than one training image.

2 Method

In this section, we first introduce the mathematical details of the training and inference stage of the self-supervised learning part of CC2D in Sect. 2.1, respectively. Then, we illustrate how to train a new landmark detector from scratch with pseudo-labels in Sect. 2.2, which are the predictions of CC2D-SSL on the training set. The resulting detector is used to predict results for the test set.

2.1 Stage I: Self-supervised Learning (CC2D-SSL)

Training Stage: As shown in Fig. 2, for an input image X_r resized to 384×384, we arbitrarily select a target point $P_r = (x_r, y_r)$ and randomly crop a patch X_p with size 192×192 which contains P. Then we apply data augmentation to the content of X_p by rotation and color jittering. The chosen landmark is moved to $P_p = (x_p, y_p)$. Two feature extractors E_r and E_p project the input X_r and the patch X_p into a multi-scale feature space, resulting in a cascade of embeddings, denoted by F_r and F_p with a length of L, respectively. Here we mark the embedding of ith layer as F^i. Next, we extract the feature of the anchor point, denoted by F_a^i, from F_p^i for each layer, guided by its corresponding coordinates $P_a^i = (x_p/2^i, y_p/2^i)$ which are down-sampled for i times. To compare the features of input F_r with anchor features F_a, we compute cosine similarity maps for each layer:

$$s^i = \frac{\langle F_a^i \cdot F_r^i \rangle}{||F_a^i||_2 \cdot ||F_r^i||_2}, \tag{1}$$

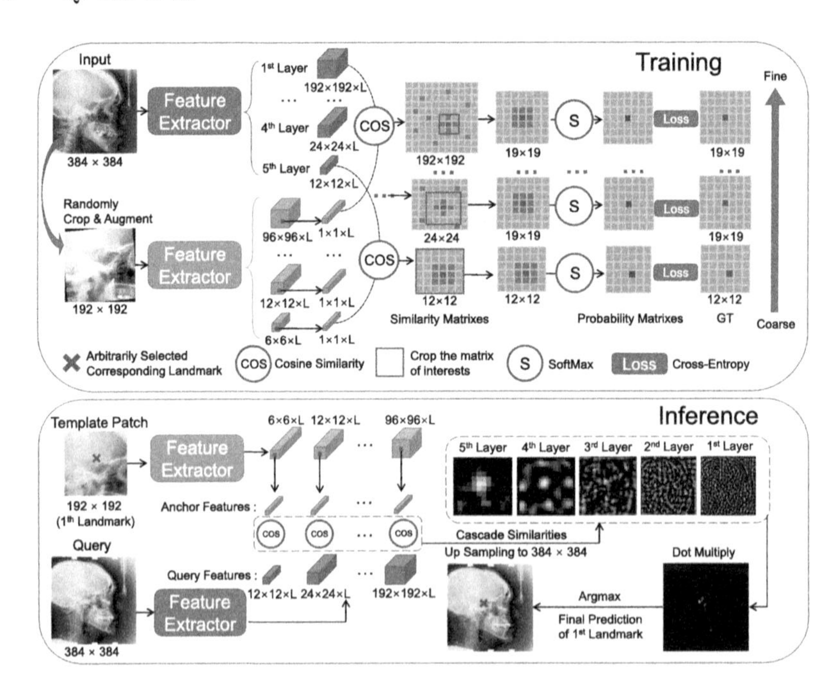

Fig. 2. Overview of the self-supervised learning part (CC2D-SSL). The two feature extractors are penalized to embed the corresponding landmarks to similar embeddings in the cascade feature space. The embeddings in the deepest layer select the coarse-grained area, while the embeddings in the inner layer improve the localization accuracy by comparing the finer areas in the selected coarse area. As consequence, the missing landmark is localized precisely in a coarse-to-fine fashion.

In CC2D-SSL, similarity maps at different scales have different roles. The deepest one (s^5 in this paper) differentiates the target coarse-grained area from every different area in the whole map, while the most shallow ones are only in charge of distinguishing the target pixel with adjacent pixels. The cascade similarity maps locate the target point in a **coarse-to-fine fashion**. Therefore, we set the matrix of interests $m^i = s^i$ if the ith layer is deepest. Otherwise, we crop the similarity map s^i to a square patch m^i centered on P_a^i of size α. As we aim at increasing the similarity of the correct pixels while decreasing others, we use softmax function to normalize m^i to probability matrix y^i with a temperature τ: $y^i = softmax(m^i * \tau)$. Note that the softmax operation is performed on all pixels. Correspondingly, we set the ground truth GT^i of each layer:

$$GT^i = \begin{cases} [\pi(x, y|x_p/2^i, y_p/2^i)]_{12\times 12} & \text{If layer } i \text{ is the deepest;} \\ [\pi(x, y|\alpha/2, \alpha/2)]_{\alpha\times\alpha} & \text{Otherwise,} \end{cases} \tag{2}$$

where $[\pi(x, y)]_{M\times N}$ is a matrix of size $M \times N$ and $\pi(x, y|x_0, y_0)$ is an indicator function that outputs 1 only if $x = x_0$ & $y = y_0$ and 0 otherwise. We use

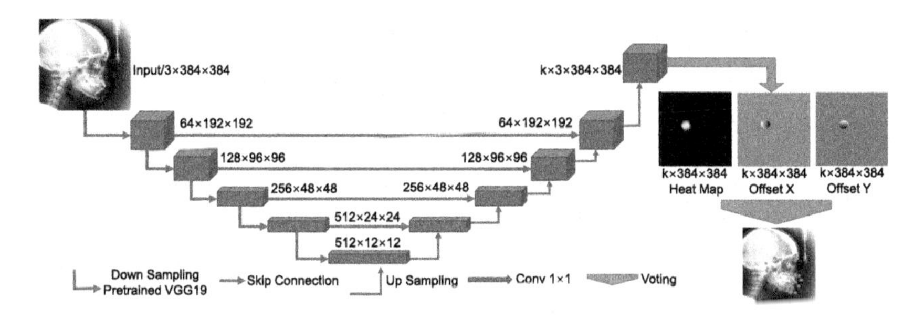

Fig. 3. The architecture of the multi-task U-Net used in CC2D-TPL. We fine-tune the encoder of U-Net initialized by the VGG19 network [25] pretrained on ImageNet [8].

cross-entropy loss L_{CE} to penalize the divergence between probability matrix y^i and ground-truth GT^i for each layer and summarize the losses as final loss L_{SSL}:

$$L_{SSL} = \sum_i L_{CE}(y^i, GT^i), \tag{3}$$

Inference Stage to Generate Pseudo-Labels: As shown in Fig. 2, we first extract template patches for landmarks defined in the annotated template image. Then we use E_p, E_r to embed the patches and query image, resulting in pixel-wise template features and query features. As well as the training stage, we extract the anchor features F_a according to the corresponding coordinate (the red point in Fig. 2). Next, we compute cascade cosine similarity maps s^i for all layers and clip to range $[0, 1]$. At last, we multiply the similarity maps s^i and generate the final prediction by *argmax* operator. In CC2D, we perform inference on the training set to generate pseudo-labels.

Feature Extractor: We set the encoder of U-Net [24] as a VGG19 [25]. Additionally, we use ASPP [5] and a convolution layer with a 1×1 kernel after every up-sampling layer to generate cascade feature embeddings. The detailed illustration can be found in the supplemental materials.

2.2 Stage II: Training with Pseudo-Labels (CC2D-TPL)

In this stage, we train a new CNN-based landmark detector from scratch with the pseudo-labels generated in Sect. 2.1, using the architecture in Fig. 3. We utilize the multi-task U-Net [30] g as our backbone, which predicts both heat map and offset maps simultaneously and has satisfactory performance [6]. Specifically, for the kth landmark located at (x_k, y_k) in an image X. We set the ground-truth heat map Y_k^h to be 1 if $\sqrt{(x - x_k)^2 + (y - y_k)^2} \leq \sigma$ and 0 otherwise. Then, the ground-truth x-offset map $Y_k^{o_x}$ predicts the relative offset vector $Y_k^{o_x} = (x - x_k)/\sigma$ from x to the corresponding landmark x_k. Similarly, its y-offset map $Y_k^{o_y}$ is defined [30]. The loss function L_k for the kth landmark consists of a binary cross-entropy loss, denoted by L^h, for punishing the divergence of predicted and ground-truth heatmaps, and the L_1 loss, denoted by L^o, for punishing the difference in coordinate offset maps.

$$L_k = L_k^h(Y_k^h, g_k^h(X)) + sign(Y_k^h) \sum_{o \in \{o_x, o_y\}} L_k^o(Y_k^o, g_k^o(X)), \tag{4}$$

where $g_k^h(X)$ and $g_k^o(X)$ are the networks that predict heatmaps and coordinate offset maps, respectively; $sign(\cdot)$ is a sign function which is used to make sure that only the area highlighted by heatmap is included for calculation. At last, we sum the losses L_k for all of the landmarks defined in the template image: $L_{TPL} = \sum_k L_k$. In the test phase, we conduct a majority-vote for candidate landmarks among all pixels with heatmap value $g_i^h(X, \theta) \geq 0.5$, according to their coordinate offset maps in $g_i^o(X)$. The winning position in the kth channel is the final predicted k^{th} landmark [6].

Table 1. Comparison of the state-of-the-art supervised approaches and our CC2D on the ISBI 2015 Challenge [27] testset. * represents the performances copied from their original papers. # represents the performances we re-implement with limited labeled images. We additionally evaluate the performance of CC2D-SSL on the test set. We train our CC2D and CC2D three times and report the standard deviation.

Model	Labeled images	MRE (↓) (mm)	SDR (↑) (%)			
			2 mm	2.5 mm	3 mm	4 mm
Ibragimov et al. [10]*	150	–	68.13	74.63	79.77	86.87
Lindner et al. [16]*	150	1.77	70.65	76.93	82.17	89.85
Urschler et al. [26]*	150	–	70.21	76.95	82.08	89.01
Payer et al. [23]*	150	–	73.33	78.76	83.24	89.75
Payer et al. [23]#	25	2.54	66.12	73.27	79.82	86.82
Payer et al. [23]#	10	6.52	49.49	57.91	65.87	75.07
Payer et al. [23]#	5	12.34	27.35	32.94	38.48	45.28
CC2D-SSL	1	2.58 ± 0.01	50.35 ± 0.65	59.92 ± 0.55	70.28 ± 0.43	82.24 ± 0.24
CC2D	1	2.36 ± 0.02	51.81 ± 0.73	63.13 ± 0.90	73.66 ± 0.49	86.25 ± 0.34

3 Experiments

3.1 Settings

Dataset: This study uses a widely-used public dataset for cephalometric landmark detection containing 400 radiographs, provided in IEEE ISBI 2015 Challenge [11,27]. Two expert doctors labeled 19 landmarks of clinical anatomical significance for each radiograph. We compute the average annotations by two doctors as the ground truth. The image size is 1935×2400 and the pixel spacing is 0.1 mm. According to the official website, the dataset is split to a training and a test subsets with 150 and 250 radiographs, respectively[1].

[1] We use the 126# labeled image in training set as the template image. Furthermore, we randomly select 10 template images from the training set, CC2D-SSL reaches 3.25 ± 0.62 mm MRE.

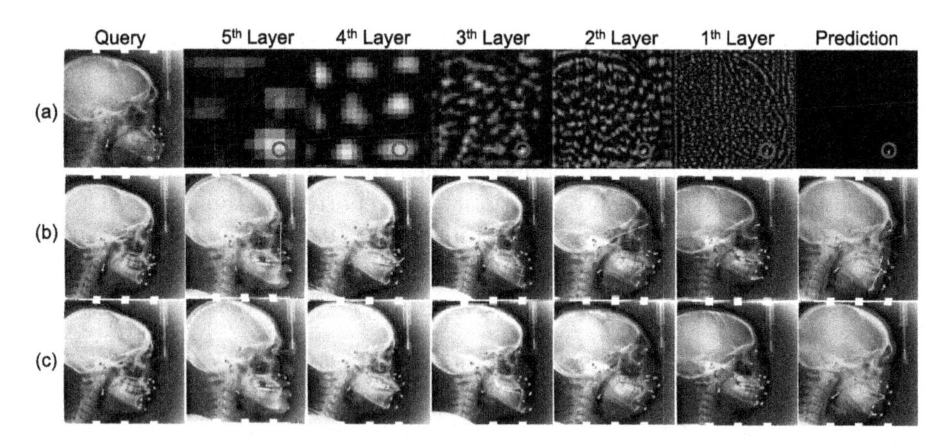

Fig. 4. (a) The inference procedure of CC2D-SSL for the 6th landmark in the query image. The similarity maps at different scales localize the target landmark in a coarse-to-fine fashion. We mark the correct positions by the red circles. (b) Visualizations of the testing images predicted by CC2D-SSL. The landmarks in green and red represent the predictions and ground truths, while the yellow lines mark their distances. (c) Visualizations of the testing images predicted by CC2D. The rightmost images in (b) and (c) are the failure cases. (Color figure online)

Metrics: Following the official challenge, we use mean radial error (MRE) to measure the Euclidean distance between prediction and ground truth, and successful detection rate (SDR) in four radii (2 mm, 2.5 mm, 3 mm, and 4 mm).

Implementation Details: All of our models are implemented in PyTorch, accelerated by an NVIDIA RTX3090 GPU. For self-supervised training, the two feature extractors are optimized by Adam optimizer for 3500 epochs with a learning rate of 0.001 decayed by half every 500 epochs, which takes about 6 h to converge with a batch size of 8. The embedding length L is set to 16 and the length α of m is 19. For training with pseudo-labels, the multi-task U-Net is optimized by Adam optimizer for 300 epochs with a learning rate of 0.0003, which takes about 50 min to converge with a batch size of 16.

3.2 The Performance of CC2D

During the inference stage of CC2D-SSL, the similarity map in the deepest layer detects the correct coarse area first, then the similarity maps in the inner layers gradually improve the localization accuracy. The procedure is illustrated in Fig. 4(a). Consequently, most of the landmarks in Fig. 2(b) are successfully detected. Moreover, after training by the predictions of CC2D-SSL on the training set, the new landmark detector in CC2D-TPL localizes the landmarks with better accuracy (as shown in Fig. 4(c) and Table 1).

We quantitatively compare our CC2D with the first [16] and second [10] place in the ISBI 2015 Challenge [27] in Table 1, as well as two recent state-of-the-art supervised method [23,26]. With one labeled image available, CC2D achieves the MRE of 2.36 mm and 4 mm SDR of 86.25%, which are competitive compared to the supervised methods. Furthermore, when the available annotated data is reduced, our CC2D shows more superiority. As reported in Table 1, our CC2D performs better than Payer et al. [22] retrained with 10 labeled images. However, CC2D localizes landmarks with more deviation if there is a drastic difference between the query (failure cases in Fig. 4) and template image (in Fig. 1).

3.3 Ablation Study and Hyper-Parameter Analysis

Hyper-Parameter Analysis: We study the influence of the embedding length L and the length α of the matrix of interests. According to Table 2, all of the experimental results fluctuate slightly when changing the two parameters, while setting $L = 16$ and $\alpha = 13$ leads to the best performance. For L, features with larger length may tend to overfit while smaller length may not represent the anatomical information effectively. For α, too small matrix has limited receptive field. On the contrary, too large matrix involves too many negative pixels, making the convergence of L_{SSL} (in Eq. 3) difficult.

Table 2. The performances of CC2D with different embedding lengths L and different α, which is the length of the matrix of interests m.

Para.	Value	MRE (↓) (mm)	SDR (↑) (%)			
			2 mm	2.5 mm	3 mm	4 mm
L	128	2.59	47.87	58.96	69.05	82.35
	64	2.82	42.27	51.60	63.70	79.03
	32	2.96	45.01	54.33	54.14	78.06
	16	**2.36**	**51.81**	**63.13**	**73.66**	**86.25**
	8	2.71	47.38	56.86	66.84	79.87
α	9	2.46	51.17	61.72	71.68	84.16
	11	2.40	53.53	63.32	72.71	84.58
	13	**2.36**	**51.81**	**63.13**	**73.66**	**86.25**
	15	2.45	49.41	60.02	71.01	85.09
	17	2.58	48.52	57.87	68.12	82.42
	19	2.56	45.85	56.92	68.42	84.48

Table 3. The performances of CC2D using cosine similarity maps in different layers.

Layer index					MRE (↓) (mm)	SDR (↑) (%)			
5	4	3	2	1		2 mm	2.5 mm	3 mm	4 mm
×	✓	✓	✓	✓	2.96	34.04	44.23	56.42	75.72
✓	×	✓	✓	✓	3.02	42.90	52.61	62.46	76.16
✓	✓	×	✓	✓	3.35	39.89	49.64	60.69	74.73
✓	✓	✓	×	✓	2.60	49.88	57.93	68.82	82.40
✓	✓	✓	✓	×	2.58	49.13	58.46	67.43	81.01
✓	✓	✓	✓	✓	**2.36**	**51.81**	**63.13**	**73.66**	**86.25**

Ablation Study: We compare the CC2D using similarity maps in different layers. As shown in (Table 3), the similarity map in the fifth (deepest) layer is most important which selects the coarsest and accurate area first, while other similarity maps also contribute to the final detection accuracy.

4 Conclusion and Future Work

In this paper, we propose Cascade Comparing to Detect (CC2D), a novel framework for building a robust landmark detection network with only one labeled image available. CC2D learns to map the consistent anatomical information into cascading feature spaces by solving a self-supervised patch matching problem. Using the self-supervised model, CC2D localizes the target landmark according to the one-shot template image in a coarse-to-fine fashion to generate pseudo-labels for training a final landmark detector. Extensive experiments evaluate the competitive performance of CC2D, comparable to the state-of-the-art semi-supervised methods. In future work, we plan to further improve the detection accuracy by considering the usage of the spatial relationships of different landmarks.

References

1. Arpit, D., et al.: A closer look at memorization in deep networks. In: ICML, pp. 233–242. PMLR (2017)
2. Bhalodia, R., Kavan, L., Whitaker, R.T.: Self-supervised discovery of anatomical shape landmarks. In: Martel, A.L., et al. (eds.) MICCAI 2020. LNCS, vol. 12264, pp. 627–638. Springer, Cham (2020). https://doi.org/10.1007/978-3-030-59719-1_61
3. Bier, B., et al.: X-ray-transform invariant anatomical landmark detection for pelvic trauma surgery. In: Frangi, A.F., Schnabel, J.A., Davatzikos, C., Alberola-López, C., Fichtinger, G. (eds.) MICCAI 2018. LNCS, vol. 11073, pp. 55–63. Springer, Cham (2018). https://doi.org/10.1007/978-3-030-00937-3_7

4. Chaitanya, K., Erdil, E., Karani, N., Konukoglu, E.: Contrastive learning of global and local features for medical image segmentation with limited annotations. arXiv preprint arXiv:2006.10511 (2020)
5. Chen, L.C., Papandreou, G., Schroff, F., Adam, H.: Rethinking atrous convolution for semantic image segmentation. arXiv preprint arXiv:1706.05587 (2017)
6. Chen, R., Ma, Y., Chen, N., Lee, D., Wang, W.: Cephalometric landmark detection by attentive feature pyramid fusion and regression-voting. In: Shen, D., et al. (eds.) MICCAI 2019. LNCS, vol. 11766, pp. 873–881. Springer, Cham (2019). https://doi.org/10.1007/978-3-030-32248-9_97
7. Chen, T., Kornblith, S., Norouzi, M., Hinton, G.: A simple framework for contrastive learning of visual representations. In: ICML, pp. 1597–1607. PMLR (2020)
8. Deng, J., Dong, W., Socher, R., Li, L.J., Li, K., Fei-Fei, L.: ImageNet: a large-scale hierarchical image database. In: CVPR, pp. 248–255 (2009)
9. Gertych, A., Zhang, A., Sayre, J., Pospiech-Kurkowska, S., Huang, H.: Bone age assessment of children using a digital hand atlas. Comput. Med. Imaging Graph. **31**(4–5), 322–331 (2007)
10. Ibragimov, B., Likar, B., Pernus, F., Vrtovec, T.: Computerized cephalometry by game theory with shape-and appearance-based landmark refinement (2015)
11. Kaggle: Cephalometric X-Ray Landmarks Detection Challenge (2015). https://www.kaggle.com/jiahongqian/cephalometric-landmarks/discussion/133268
12. Lang, Y., et al.: Automatic localization of landmarks in craniomaxillofacial CBCT images using a local attention-based graph convolution network. In: Martel, A.L., et al. (eds.) MICCAI 2020. LNCS, vol. 12264, pp. 817–826. Springer, Cham (2020). https://doi.org/10.1007/978-3-030-59719-1_79
13. Li, H., et al.: High-resolution chest X-ray bone suppression using unpaired CT structural priors. IEEE Trans. Med. Imaging **39**, 3053–3063(2020)
14. Li, H., Han, H., Zhou, S.K.: Bounding maps for universal lesion detection. In: Martel, A.L., et al. (eds.) MICCAI 2020. LNCS, vol. 12264, pp. 417–428. Springer, Cham (2020). https://doi.org/10.1007/978-3-030-59719-1_41
15. Li, W., et al.: Structured landmark detection via topology-adapting deep graph learning. In: Vedaldi, A., Bischof, H., Brox, T., Frahm, J.M. (eds.) ECCV 2020. LNCS, vol. 12354, pp. 266–283. Springer, Cham (2020). https://doi.org/10.1007/978-3-030-58545-7_16
16. Lindner, C., Cootes, T.F.: Fully automatic cephalometric evaluation using random forest regression-voting. Sci. Rep. **6**, 33581 (2016)
17. Liu, D., Zhou, S.K., Bernhardt, D., Comaniciu, D.: Search strategies for multiple landmark detection by submodular maximization. In: 2010 IEEE Conference on Computer Vision and Pattern Recognition (CVPR), pp. 2831–2838. IEEE (2010)
18. Liu, W., Wang, Yu., Jiang, T., Chi, Y., Zhang, L., Hua, X.-S.: Landmarks detection with anatomical constraints for total hip arthroplasty preoperative measurements. In: Martel, A.L., et al. (eds.) MICCAI 2020. LNCS, vol. 12264, pp. 670–679. Springer, Cham (2020). https://doi.org/10.1007/978-3-030-59719-1_65
19. Ji, W., Chen, W., Yu, S., Ma, K., Cheng, L., Shen, L., Zheng, Y.: Uncertainty quantification for medical image segmentation using dynamic label factor allocation among multiple raters. In: MICCAI on QUBIQ Workshop (2020)
20. Ji, W., et al.: Learning calibrated medical image segmentation via multi-rater agreement modeling. In: CVPR, pp. 12341–12351, June 2021

21. Ouyang, C., Biffi, C., Chen, C., Kart, T., Qiu, H., Rueckert, D.: Self-supervision with superpixels: training few-shot medical image segmentation without annotation. In: Vedaldi, A., Bischof, H., Brox, T., Frahm, J.-M. (eds.) ECCV 2020. LNCS, vol. 12374, pp. 762–780. Springer, Cham (2020). https://doi.org/10.1007/978-3-030-58526-6_45
22. Payer, C., Štern, D., Bischof, H., Urschler, M.: Regressing heatmaps for multiple landmark localization using CNNs. In: Ourselin, S., Joskowicz, L., Sabuncu, M.R., Unal, G., Wells, W. (eds.) MICCAI 2016. LNCS, vol. 9901, pp. 230–238. Springer, Cham (2016). https://doi.org/10.1007/978-3-319-46723-8_27
23. Payer, C., Štern, D., Bischof, H., Urschler, M.: Integrating spatial configuration into heatmap regression based CNNs for landmark localization. Med. Image Anal. **54**, 207–219 (2019)
24. Ronneberger, O., Fischer, P., Brox, T.: U-net: convolutional networks for biomedical image segmentation. In: Navab, N., Hornegger, J., Wells, W.M., Frangi, A.F. (eds.) MICCAI 2015. LNCS, vol. 9351, pp. 234–241. Springer, Cham (2015). https://doi.org/10.1007/978-3-319-24574-4_28
25. Simonyan, K., Zisserman, A.: Very deep convolutional networks for large-scale image recognition. In: ICLR (2015)
26. Urschler, M., Ebner, T., Štern, D.: Integrating geometric configuration and appearance information into a unified framework for anatomical landmark localization. Med. Image Anal. **43**, 23–36 (2018)
27. Wang, C.W., et al.: A benchmark for comparison of dental radiography analysis algorithms. Med. Image Anal. **31**, 63–76 (2016)
28. Yang, D., et al.: Automatic vertebra labeling in large-scale 3D CT using deep image-to-image network with message passing and sparsity regularization. In: IPMI, pp. 633–644 (2017)
29. Yang, D., Zhang, S., Yan, Z., Tan, C., Li, K., Metaxas, D.: Automated anatomical landmark detection ondistal femur surface using convolutional neural network. In: ISBI, pp. 17–21 (2015)
30. Yao, Q., He, Z., Han, H., Zhou, S.K.: Miss the point: targeted adversarial attack on multiple landmark detection. In: Martel, A.L., Abolmaesumi, P., Stoyanov, D., Mateus, D., Zuluaga, M.A., Zhou, S.K., Racoceanu, D., Joskowicz, L. (eds.) MICCAI 2020. LNCS, vol. 12264, pp. 692–702. Springer, Cham (2020). https://doi.org/10.1007/978-3-030-59719-1_67
31. Yao, Q., Xiao, L., Liu, P., Zhou, S.K.: Label-free segmentation of COVID-19 lesions in lung CT. IEEE Trans. Med. Imaging (2020)
32. Zheng, Y., Liu, D., Georgescu, B., Nguyen, H., Comaniciu, D.: 3D deep learning for efficient and robust landmark detection in volumetric data. In: Navab, N., Hornegger, J., Wells, W.M., Frangi, A.F. (eds.) MICCAI 2015. LNCS, vol. 9349, pp. 565–572. Springer, Cham (2015). https://doi.org/10.1007/978-3-319-24553-9_69
33. Zhong, Z., Li, J., Zhang, Z., Jiao, Z., Gao, X.: An attention-guided deep regression model for landmark detection in cephalograms. In: Shen, D., et al. (eds.) MICCAI 2019. LNCS, vol. 11769, pp. 540–548. Springer, Cham (2019). https://doi.org/10.1007/978-3-030-32226-7_60
34. Zhou, H.-Y., Yu, S., Bian, C., Hu, Y., Ma, K., Zheng, Y.: Comparing to learn: surpassing ImageNet pretraining on radiographs by comparing image representations. In: Martel, A.L., et al. (eds.) MICCAI 2020. LNCS, vol. 12261, pp. 398–407. Springer, Cham (2020). https://doi.org/10.1007/978-3-030-59710-8_39
35. Zhou, S.K., et al.: A review of deep learning in medical imaging: imaging traits, technology trends, case studies with progress highlights, and future promises. Proc. IEEE (2021)

36. Zhou, S.K., Rueckert, D., Fichtinger, G.: Handbook of Medical Image Computing and Computer Assisted Intervention. Academic Press, Cambridge (2019)
37. Zhou, Z., Sodha, V., Pang, J., Gotway, M.B., Liang, J.: Models genesis. Med. Image Anal. **67**, 101840 (2021)
38. Zhou, Z., et al.: Models genesis: generic autodidactic models for 3D medical image analysis. In: Shen, D., et al. (eds.) MICCAI 2019. LNCS, vol. 11767, pp. 384–393. Springer, Cham (2019). https://doi.org/10.1007/978-3-030-32251-9_42
39. Zhu, J., Li, Y., Hu, Y., Ma, K., Zhou, S.K., Zheng, Y.: Rubik's cube+: a self-supervised feature learning framework for 3D medical image analysis. Med. Image Anal. **64**, 101746 (2020)

Implicit Field Learning for Unsupervised Anomaly Detection in Medical Images

Sergio Naval Marimont[1(✉)] and Giacomo Tarroni[1,2]

[1] CitAI Research Centre, City, University of London, London, UK
{sergio.naval-marimont,giacomo.tarroni}@city.ac.uk
[2] BioMedIA, Imperial College, London, UK

Abstract. We propose a novel unsupervised out-of-distribution detection method for medical images based on implicit fields image representations. In our approach, an auto-decoder feed-forward neural network learns the distribution of healthy images in the form of a mapping between spatial coordinates and probabilities over a proxy for tissue types. At inference time, the learnt distribution is used to retrieve, from a given test image, a restoration, i.e. an image maximally consistent with the input one but belonging to the healthy distribution. Anomalies are localized using the voxel-wise probability predicted by our model for the restored image. We tested our approach in the task of unsupervised localization of gliomas on brain MR images and compared it to several other VAE-based anomaly detection methods. Results show that the proposed technique substantially outperforms them (average DICE 0.640 vs 0.518 for the best performing VAE-based alternative) while also requiring considerably less computing time.

Keywords: Anomaly detection · Unsupervised learning · Implicit fields · Occupancy networks

1 Introduction

Multiple deep learning methods have been proposed to automatically localize anomalies in medical images, with fully-supervised approaches being able to achieve high segmentation accuracies [4]. However, these methods 1) rely on the availability of large and diverse annotated datasets for training, and 2) they are specific to the anomalies annotated in the dataset and are therefore unable to generalize to previously unseen pathologies. On the other hand, the unsupervised learning paradigm is not affected by these limitations. Unsupervised approaches usually aim at learning the distribution of healthy/normal unannotated images and at classifying as anomalies the images that differ from the learnt distribution. Two categories of generative models, namely Variational Auto-Encoders (VAEs) and Generative Adversarial Networks (GANs), have been implemented in many techniques for unsupervised anomaly detection. However, comparative studies [1] show that their performance is still far from that of equivalent supervised methods.

M. de Bruijne et al. (Eds.): MICCAI 2021, LNCS 12902, pp. 189–198, 2021.
https://doi.org/10.1007/978-3-030-87196-3_18

Related Works: Generally, anomaly detection techniques make use of generative models to learn the distribution of healthy/normal images and leverage the learnt distribution to compute voxel-by-voxel anomaly scores (AS), which identify image areas that differ from the normal anatomy. The vanilla VAE-based approach assumes that a model trained on normal data will not be able to reconstruct anomalies, and consequently that the voxel-wise reconstruction loss can be used as AS. However, auto-encoders with enough capacity can also reconstruct abnormal samples, making VAE reconstruction loss a poor AS [1,14]. Several methods have tried to over-come this limitation. For instance, [14] proposed to leverage the KL divergence gradient w.r.t voxels as AS. [15] added context-encoding tasks to incentivise the VAE to properly generate restored (i.e. anomaly-free) images. In [2], the authors proposed to restore images by minimising a loss function composed by the VAE ELBO and a data consistency term. [8] proposed to generate restorations with a vector-quantized VAE by resampling low-probability latent variables. GAN-based approaches [10,11], rely on a similar ideas for restoration. However, GANs notably suffer from mode collapse, i.e. the tendency to learn to generate samples only from a subset of the normal image distribution. In addition, also GANs can generate anomalous samples [1]. Due to these issues, most of the approaches based on generative models have yielded limited anomaly detection performance, struggling to reach DICE scores of 0.5 in brain MRI datasets [1].

Techniques based on supervised learning using datasets with synthetically-generated anomalies [12] have been very recently proposed for anomaly detection. They have achieved high accuracy in the MICCAI 2020 Medical Out-of-Distribution challenge (MOOD) [16], which partially included synthetic anomalies in its test set. While promising, these approaches move the focus to the task of generating realistic anomalies, and their performance on datasets with real abnormalities remains largely unexplored.

Contributions: Recently, an approach referred to as implicit field learning (or occupancy net-works) has been introduced to reconstruct 3D shapes through learning their implicit surface representation [3,6]. Instead of using convolutional networks to learn a distribution over a dense set of voxels, in this approach a linear neural net-work learns to map continuous spatial coordinates to either object/background labels (binary classification) [3,6] or to the signed distance function with respect to the object surface [9]. Importantly, the authors of [9] also proposed to substitute the auto-encoder architecture with an auto-decoder architecture, removing the need for an encoder network to obtain the latent representation.

In this paper, we propose a novel approach to unsupervised anomaly detection that leverages the implicit field learning paradigm. Our main contributions are the following:

- We propose a modification of the implicit field learning technique that enables learning relevant anatomical features for unsupervised anomaly detection;

- We propose an anomaly detection neural network pipeline which overcomes the limitations of VAE models: by relying on an auto-decoder architecture, our network generates anomaly-free reconstructions. Additionally, the implicit field representation is detached from a specific input resolution and can be scaled seamlessly to deal with high resolution 3D medical images;
- We tested this approach in the task of unsupervised localization of gliomas on brain MR images and compared it to several other VAE-based approaches. The proposed technique substantially outperforms the competition both in terms of accuracy (average DICE 0.640 vs 0.518 for the best performing competitor) and computing speed (55.4 s vs 293.1 s, respectively).

2 Methods

Implicit Field Representation: While a 3D image is typically represented by the intensities of a set of discrete voxels, implicit field networks learn a continuous function f with spatial coordinates $\mathbf{p} = (x, y, z) \in \mathbb{R}^3$ as input. Instead of a binary label for object/background classification, we propose that this function maps to the probability distribution over C classes, each representing a range of voxel intensities (see next section). In addition to the spatial coordinates, the network receives as input a latent variable $\mathbf{z} \in \mathbb{R}^D$ which describes a specific 3D image:

$$f : \mathbb{R}^3 \times \mathbb{R}^D \rightarrow \{1, 2, ...C\} \tag{1}$$

The network f learns the posterior probability over intensity ranges for continuous spatial coordinates $\mathbf{p}$ and latent variables $\mathbf{z}$. We model the posterior probability using a *softmax* activation:

$$P(t = j \mid \mathbf{z}, \mathbf{p}) = \frac{\exp f^j(\mathbf{z}, \mathbf{p})}{\sum_i^C \exp f^i(\mathbf{z}, \mathbf{p})} \tag{2}$$

The latent variables $\mathbf{z}$ are obtained using the auto-decoder architecture proposed in [9]: as opposed to training an encoder network to produce the latent representation, in the auto-decoder approach each training 3D image is paired with a D-dimensional vector in an embedding space. During training, backpropagation optimizes not only the network parameters but also the latent vector representation of each 3D image. In the auto-decoder architecture (see Fig. 1), at inference time, the latent vector is initialized randomly and optimization is used to find the latent representation that better represents the test 3D image.

Specifically, during training the expression Eq. 3 is used to optimize latent codes and parameters θ of network f_θ by sampling K data points from N training 3D images:

$$\underset{\theta, \{\mathbf{z}_i\}_{i=1}^N}{\arg\min} \sum_{i=1}^{N} \left(\sum_{j=1}^{K} \mathcal{L}\left(f_\theta\left(\mathbf{z}_i, \mathbf{p}_{i,j}\right), t_{i,j}\right) + \frac{1}{\sigma^2} \|\mathbf{z}_i\|_2^2 \right) \tag{3}$$

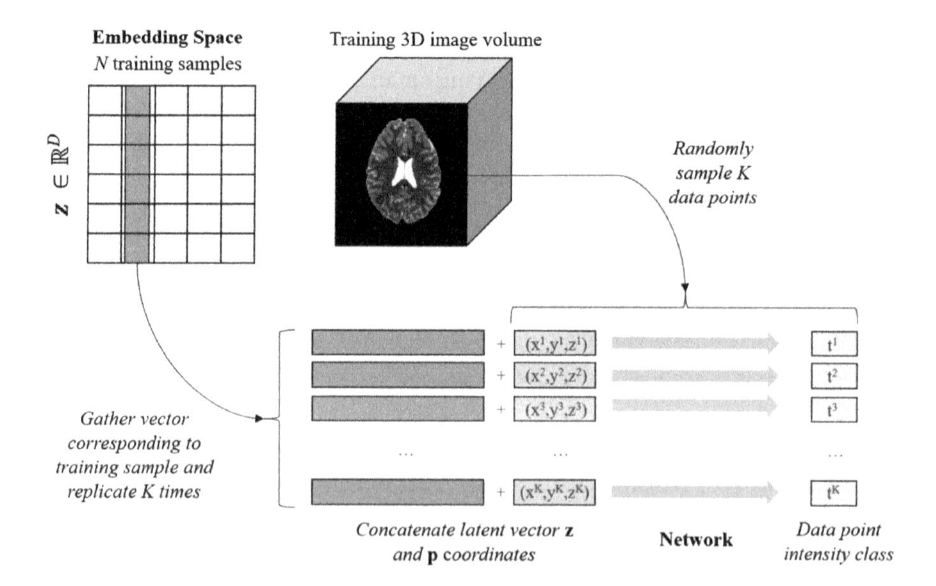

Fig. 1. Schema for auto-decoder training. A latent vector is obtained for a training sample and concatenated with coordinates sampled from the 3D image volume. The network learns the mapping (latent features, coordinates) $\rightarrow$ multiclass encoding of voxels intensities.

Note that $\mathcal{L}$ is the cross-entropy loss between network output and the true voxel class $t \in \{1, 2, \ldots C\}$. Similarly to [9], we assumed the prior for the latent codes distribution to be a spherical multivariate-Gaussian with covariance $\sigma^2 I$. The σ hyperparameter allows to modulate the amount of regularization in the latent distribution. During inference, the expression Eq. 3 is optimized only for $\mathbf{z}$ (fixing network parameters θ), obtaining the latent code that best describes a given test 3D image.

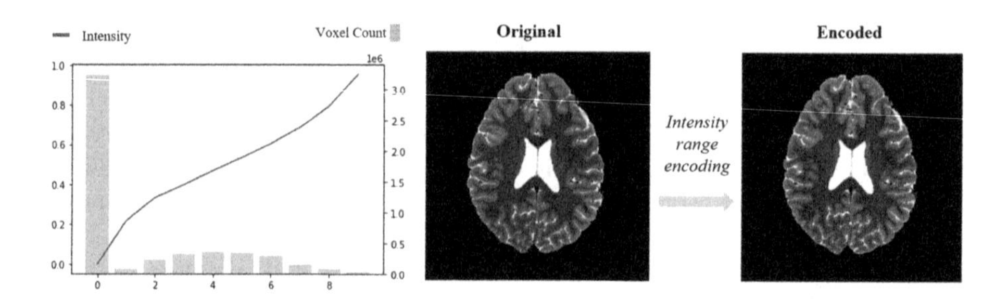

Fig. 2. Left: intensity and count of voxels per cluster (using kMeans, with k = 10, on 2 million voxels randomly sampled across multiple subjects). Right: effects of intensity range encoding on a sample image.

We also utilized the coordinate encoding function described in [7], by which each of the coordinates in $\mathbf{p}$ is first normalized to the range $[-1, 1]$ and later encoded using the expression 4 (shown for x. We used L = 10 in our experiments):

$$\gamma(x) = (\sin(2^0\pi x), \cos(2^0\pi x), ..., \sin(2^{L-1}\pi x), \cos(2^{L-1}\pi x)) \tag{4}$$

Intensity Range Encoding: Instead of modelling 3D image intensities as a continuous variable, we discretize the intensity values in C clusters, allowing the neural network to learn the probability distribution over C classes of intensities. We wish to define the intensity clusters so that the encoding preserves as much information as possible in the original volume. With this objective, we use k-Means, treating the number of clusters as an hyper-parameter (see Fig. 2).

The rationale behind intensity range encoding is two fold. First, it enables a rather straightforward extension of implicit field learning approaches to the task of image reconstruction. Second, it allows to untie the computation of the AS from distances in image intensities between original image and reconstructed version (which is the most common strategy for approaches based on generative models). Instead, in the proposed technique the AS is derived from the predicted probability over intensity clusters, which we assume to better represent different tissue types.

Mode-Pooling Smoothing: We found that smoothing and denoising the 3D images volumes slightly improved anomaly detection accuracy for our approach. In order to preserve the overall structures and only remove spurious intensity values, we propose a 3D *mode-pooling* layer which, for a 3-dimensional sliding window, returns the most common intensity cluster. We used a $2 \times 2 \times 2$ *mode-pooling* filter in our training set and $3 \times 3 \times 3$ in validation and test sets.

Voxel-Wise Anomaly Score (AS): At inference time, we aim at retrieving a healthy image from the model consistent with a test image. The retrieved image is called a *restoration*, as it preserves consistency with the test image but it belongs to the learnt distribution of healthy images. Anomalies are finally located by comparing the restoration with the test image. In order to generate a restoration, we move along latent space searching for the latent vector $\mathbf{z}$ that minimizes the following expression over K randomly sampled data points:

$$\arg\min_{\mathbf{z}}(\sum_{j=1}^{K} \mathcal{L}(f_\theta(\mathbf{z}, \mathbf{p}_j), t_j) + \frac{1}{\sigma^2}\|\mathbf{z}\|_2^2) \tag{5}$$

Minimization is performed with Adam optimizer for 700 steps with K = 16,200. Once a restoration is generated with the retrieved $\mathbf{z}$, we can compute a voxel-based anomaly score (AS). Specifically, we estimate the probability over intensity clusters for each voxel using the network and compute the voxel-wise cross-entropy loss between the test image and the restoration as anomaly score:

$$AS = -\log P(t = t_{GT} \mid \mathbf{z}, \mathbf{p}) \tag{6}$$

where t_{GT} is the true voxel intensity after intensity range encoding. The proposed voxel-wise AS is similar to the one in [11], replacing absolute residuals with cross-entropy to account for the intensity range encoding. We also perform post-processing to denoise the obtained AS obtained using a min-pooling layer (with filter size = 3) and average-pooling layer (with filter size = 15), both 3-dimensional.

3 Experiments and Results

Experimental Set-Up: We tested our approach by training the proposed technique on a dataset of brain MR images from healthy subjects and testing it on images with gliomas. As benchmarks, we trained and tested 3 VAE-based anomaly detection techniques. Since these methods have been originally presented with 2D architectures, we created a 2D version of our approach (which used MR slices as inputs) to enable a fairer comparison. In the 2D experiments we processed 1 every 4 axial slices, (i.e., 40 slices per volume). We then evaluated our approach in its native 3D implementation using 3D MR image volumes for training and testing.

Datasets and Data Pre-processing: We use two publicly available brain MRI datasets:

- The Human Connectome Project Young Adult (HCP) dataset [13] with images of 1,112 young and healthy subjects.
- The Multimodal Brain Tumor Image Segmentation Benchmark (BRATS) [5], 2018 edition dataset, consisting of images with annotated gliomas.

The training set consists of 1,055 images from HCP, the test set of 50 images randomly sampled from BRATS and the validation set of 11 (6 from BRATS and 5 from HCP), used for hyper-parameter tuning. In both HCP and BRATS we use the pre-processed, skull-stripped T2-weighted structural images. Additionally, in all experiments but one, both datasets were downsampled to $160 \times 160 \times 160$ resolution, intensities were clipped to the percentile 98 and later normalized to the range [0, 1]. In one experiment, we tested our approach using instead the original, high-resolution images, training at 300 voxel resolution in HCP and testing at 240 voxel resolution in BRATS. No augmentations were performed in training the proposed technique. Elastic transforms, scaling, rotations and random brightness and contrast were instead applied to all VAEs benchmark experiments. In VAE experiments, images are also normalized to have zero mean and unit standard deviation.

For our approach, k = 10 was chosen for intensity clustering after tuning.

Network Architecture and Implementation Details: We used the same network architecture and training details from [9] for all our experiments. The decoder is a feed-forward network composed of 8 fully-connected layers. Latent dimensionality is 256, all hidden layers have 512 units and use ReLU as activation and weight normalization. We apply dropout in all layers with probability 0.2. The embedding space is initialized with $N(0, 0.01^2)$ and the prior covariance hyper-parameter is set to $\sigma = 0.01$. Training lasted 2,000 epochs with Adam optimizer and we applied a learning rate decay. Training batches are composed of 97,200 randomly sampled points (16,200 points from 6 different volumes). Implementation, trained models and test sets identifiers are made publicly available in[1]. All experiments were run using a Nvidia GTX 1070 GPU.

In 2D experiments, we assign a latent code with 256 dimensions to each axial slice instead of the whole volume and prediction is also conditioned on the axial coordinate. At inference time, we obtain an AS for one axial slice every four. Each axial AS is replicated 4 times to return to the original axial resolution and the AS post-processing (min-pooling and average-pooling) is performed with 3D filters on the resulting volume. We also followed this methodology in all 2D VAE benchmarks.

Performance Evaluation: In order to assess the voxel-wise anomaly detection performance we followed the conventions used in [1]. We report the best possible DICE score [DICE] in our test set which is calculated as the maximum DICE taking into consideration all individual voxels in the test set. Additionally, we determine the optimal threshold for AS using the validation set and calculate the DICE score using this threshold for each subject in the test set. Mean and standard deviations of subject specific DICE scores are reported in Table 1. We also report Average Precision (AP), area under Receiver Operating Characteristics (AUROC), the False Positive Rate at 95% recall (FPR@95R) and inference time per image volume in seconds (IT (s)).

Table 1. Experimental results on BRATS 2018 dataset.

Method	[DICE]	DICE ($\mu \pm \sigma$)	AP	AUROC	FPR@95R	IT (s)
2 dimensional						
VAE[a]	0.472	0.447 ± 0.161	0.477	0.949	0.2229	0.1
VAE restoration[b]	0.417	0.390 ± 0.146	0.413	0.936	0.2448	79.1
VQ-VAE[c]	0.568	0.518 ± 0.188	0.593	0.972	0.1366	293.1
IF 2D (ours)	0.612	0.555 ± 0.178	0.665	0.991	0.0456	55.4
3 dimensional						
IF 3D (ours)	0.681	0.640 ± 0.177	0.733	0.992	0.0462	51.1
IF 3D* (ours)	**0.716**	**0.672 ± 0.155**	**0.771**	**0.994**	**0.0386**	64.1

[a]VAE with 10 latent dimensions, L1 reconstruction loss.
[b]VAE with 128 latent dimensions, 500 restoration steps as described in [2].
[c]VQ-VAE 20 × 20 latent, 8 restorations. Implementation and processing from [8].
*Train and test in original high-resolution (300 voxel HCP and 240 BRATS).

[1] https://github.com/snavalm/ifl_unsup_anom_det.

At 160 voxel resolution, the proposed 3D implicit fields (IF) method improved mean DICE score by 12 points (0.640 vs 0.518 of VQ-VAE). A moderate increase of 4 points was also observed when comparing to the 2D IF implementation. Importantly, our method can also seamlessly capitalize, without architectural modifications, on the 3D high resolution images, with the mean DICE score increasing to 0.672. While convolutional architectures require pre-specified resolutions, the implicit fields auto-decoder approach allowed us to train and test in different resolutions. Qualitative results are shown in Fig. 3.

These results show that our method improves on the state of the art in the task of localizing anomalies when training and testing using different datasets with different acquisition protocols. It is expected that image augmentation would help alleviate the differences produced by diverse acquisition pipelines, however augmentations were not applied in our method implementation. The auto-decoder architecture forces a latent code to be pre-assigned to each training sample, consequently the augmented images need to be consistent across training epochs.

Inference computing time is also moderate, with 51 s per volume for the 3D experiments. Note that for 2D methods, the time reported correspond to only one every four axial slices and consequently it is not directly comparable with IF 3D methods. The inference time is mostly associated with the optimization required to retrieve the latent test image representation. Adding an encoder network to the architecture would further reduce the inference time (either switching to an auto-encoder architecture or training an encoder after the auto-decoder, similarly to [10]).

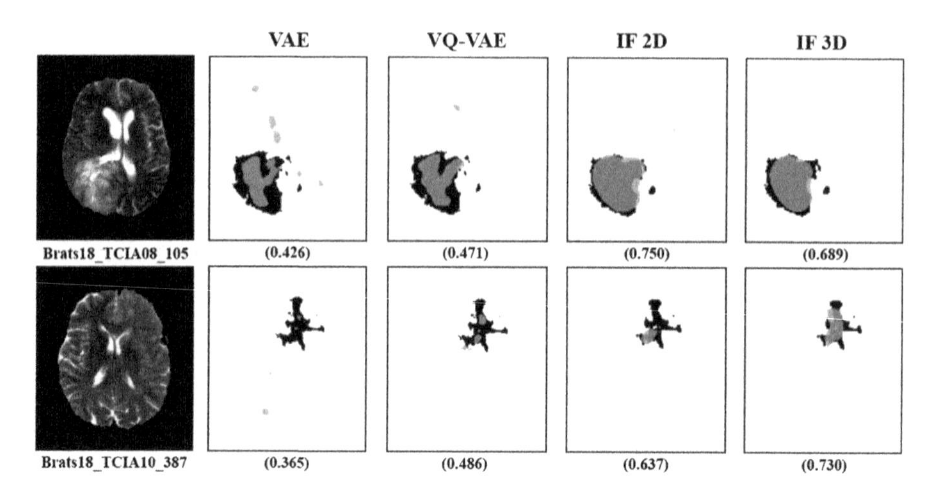

Fig. 3. Visual comparison of anomaly scores (pink) versus ground truth (black) for two different subjects (central axial slice). In brackets, DICE score for the whole subject. (Color figure online)

The optimization used to retrieve restorations translates in stochasticity at inference time because the minimization can converge to slightly different restorations. This could become a challenge if the algorithm converged to bad local minima, however it may present as well an opportunity to improve results by calculating the AS taking into consideration multiple restorations near the optimal **z**.

Finally, in our experiment VAE restoration underperformed VAE reconstruction loss, which is unexpected given the previous studies [1,2]. Differences in our implementation, namely deeper architectures with residual blocks, a smaller latent space, batch normalization and differences in image normalization, could have improved our VAE reconstruction or limited the effectivity of the restoration method.

4 Conclusion

We presented a novel unsupervised anomaly segmentation method based on implicit field learning that outperforms previous VAE-based approaches in glioma segmentation in brain MR images. In the future, we intend to perform further evaluations relative to other brain pathologies and medical image modalities.

References

1. Baur, C., Denner, S., Wiestler, B., Navab, N., Albarqouni, S.: Autoencoders for unsupervised anomaly segmentation in brain MR images: a comparative study. Med. Image Anal., 101952 (2021)
2. Chen, X., You, S., Tezcan, K.C., Konukoglu, E.: Unsupervised lesion detection via image restoration with a normative prior. In: Proceedings of The 2nd International Conference on Medical Imaging with Deep Learning, vol. 102, pp. 540–556. PMLR (2020)
3. Chen, Z., Zhang, H.: Learning implicit fields for generative shape modeling. In: Proceedings of the IEEE/CVF Conference on Computer Vision and Pattern Recognition, pp. 5939–5948 (2019)
4. Litjens, G., et al.: A survey on deep learning in medical image analysis. Med. Image Anal. **42**, 60–88 (2017)
5. Menze, B.H., et al.: The multimodal brain tumor image segmentation benchmark (BRATS). IEEE Trans. Med. Imaging **34**(10), 1993–2024 (2014)
6. Mescheder, L., Oechsle, M., Niemeyer, M., Nowozin, S., Geiger, A.: Occupancy networks: learning 3D reconstruction in function space. In: Proceedings of the IEEE/CVF Conference on Computer Vision and Pattern Recognition, pp. 4460–4470 (2019)
7. Mildenhall, B., Srinivasan, P.P., Tancik, M., Barron, J.T., Ramamoorthi, R., Ng, R.: NeRF: representing scenes as neural radiance fields for view synthesis. In: Vedaldi, A., Bischof, H., Brox, T., Frahm, J.-M. (eds.) ECCV 2020. LNCS, vol. 12346, pp. 405–421. Springer, Cham (2020). https://doi.org/10.1007/978-3-030-58452-8_24
8. Naval Marimont, S., Tarroni, G.: Anomaly detection through latent space restoration using vector quantized variational autoencoders. In: 2021 IEEE 18th International Symposium on Biomedical Imaging (ISBI), pp. 1764–1767. IEEE (2021)

9. Park, J.J., Florence, P., Straub, J., Newcombe, R., Lovegrove, S.: DeepSDF: learning continuous signed distance functions for shape representation. In: Proceedings of the IEEE/CVF Conference on Computer Vision and Pattern Recognition, pp. 165–174 (2019)
10. Schlegl, T., Seeböck, P., Waldstein, S.M., Langs, G., Schmidt-Erfurth, U.: f-AnoGAN: fast unsupervised anomaly detection with generative adversarial networks. Med. Image Anal. **54**, 30–44 (2019)
11. Schlegl, T., Seeböck, P., Waldstein, S.M., Schmidt-Erfurth, U., Langs, G.: Unsupervised anomaly detection with generative adversarial networks to guide marker discovery. In: Niethammer, M., et al. (eds.) IPMI 2017. LNCS, vol. 10265, pp. 146–157. Springer, Cham (2017). https://doi.org/10.1007/978-3-319-59050-9_12
12. Tan, J., Hou, B., Batten, J., Qiu, H., Kainz, B.: Detecting outliers with foreign patch interpolation. arXiv preprint arXiv:2011.04197 (2020)
13. Van Essen, D.C., et al.: The human connectome project: a data acquisition perspective. Neuroimage **62**(4), 2222–2231 (2012)
14. Zimmerer, D., Isensee, F., Petersen, J., Kohl, S., Maier-Hein, K.: Unsupervised anomaly localization using variational auto-encoders. In: Shen, D., Liu, T., Peters, T.M., Staib, L.H., Essert, C., Zhou, S., Yap, P.-T., Khan, A. (eds.) MICCAI 2019. LNCS, vol. 11767, pp. 289–297. Springer, Cham (2019). https://doi.org/10.1007/978-3-030-32251-9_32
15. Zimmerer, D., Kohl, S.A., Petersen, J., Isensee, F., Maier-Hein, K.H.: Context-encoding variational autoencoder for unsupervised anomaly detection. arXiv preprint arXiv:1812.05941 (2018)
16. Zimmerer, D., et al.: Medical out-of-distribution analysis challenge, March 2020 https://doi.org/10.5281/zenodo.3715870

Dual-Consistency Semi-supervised Learning with Uncertainty Quantification for COVID-19 Lesion Segmentation from CT Images

Yanwen Li[1], Luyang Luo[2], Huangjing Lin[1,2], Hao Chen[3](✉), and Pheng-Ann Heng[2,4]

[1] Imsight AI Research Lab, Shenzhen, China
liyanwen@imsightmed.com
[2] Department of Computer Science and Engineering, The Chinese University of Hong Kong, Shatin, Hong Kong, China
lyluo@cse.cuhk.edu.hk
[3] Department of Computer Science and Engineering, The Hong Kong University of Science and Technology, Kowloon, Hong Kong, China
jhc@cse.ust.hk
[4] Guangdong-Hong Kong-Macao Joint Laboratory of Human-Machine Intelligence-Synergy Systems, Shenzhen Institutes of Advanced Technology, Chinese Academy of Sciences, Shenzhen, China

Abstract. The novel coronavirus disease 2019 (COVID-19) characterized by atypical pneumonia has caused millions of deaths worldwide. Automatically segmenting lesions from chest Computed Tomography (CT) is a promising way to assist doctors in COVID-19 screening, treatment planning, and follow-up monitoring. However, voxel-wise annotations are extremely expert-demanding and scarce, especially when it comes to novel diseases, while an abundance of unlabeled data could be available. To tackle the challenge of limited annotations, in this paper, we propose an uncertainty-guided dual-consistency learning network (UDC-Net) for semi-supervised COVID-19 lesion segmentation from CT images. Specifically, we present a dual-consistency learning scheme that simultaneously imposes image transformation equivalence and feature perturbation invariance to effectively harness the knowledge from unlabeled data. We then quantify the segmentation uncertainty in two forms and employ them together to guide the consistency regularization for more reliable unsupervised learning. Extensive experiments showed that our proposed UDC-Net improves the fully supervised method by 6.3% in Dice and outperforms other competitive semi-supervised approaches by

Y. Li and L. Luo—The first two authors contributed equally.

Electronic supplementary material The online version of this chapter (https://doi.org/10.1007/978-3-030-87196-3_19) contains supplementary material, which is available to authorized users.

M. de Bruijne et al. (Eds.): MICCAI 2021, LNCS 12902, pp. 199–209, 2021.
https://doi.org/10.1007/978-3-030-87196-3_19

significant margins, demonstrating high potential in real-world clinical practice. (Code is available at https://github.com/poiuohke/UDC-Net).

Keywords: COVID-19 · Semi-supervised learning · Uncertainty · Segmentation

1 Introduction

By the end of 2020, the coronavirus disease 2019 (COVID-19) [36] characterized by atypical pneumonia has spread over 220 countries and areas, infected more than 81 million people, and caused near 1.8 million losses of lives[1]. For early screening of the COVID-19, chest computed tomography (CT) plays a vital role as a noninvasive and fast technique, which is reported to have high sensitivity for detecting COVID-19-related abnormal findings [1,6,7,13]. To improve the screening efficiency and alleviate radiologists' reading burden, various automatic COVID-19 chest CT analysis methods have been proposed from whole-volume classification and triaging [4,8,17,20,27], weakly-supervised lesion localization [16,31], to accurate segmentation of lesion regions [5,26]. Among previous studies, segmentation of COVID-19 often provides more accurate descriptions of the lesions, which has significant potential in assisting doctors with the diagnosis, treatment planning, and follow-up monitoring.

Currently, advanced segmentation methods are often fully supervised and heavily rely on pixel-wise or voxel-wise annotations. For novel diseases like COVID-19, acquiring such annotations is extremely expertise-demanded and time-consuming, while unlabeled data are often abundant due to increasing positive cases. Therefore, semi-supervised learning (SSL) that utilizes both labeled and unlabeled data is of great value to develop robust and accurate COVID-19 lesion segmentation algorithms. Thus far, many SSL approaches have been developed and successfully applied to various tasks [25]. Many works [2,9,14,19,23] adopts the smoothness assumption that two data samples that are close in the input space share the same label. This assumption is further expanded to the deep feature space, where similarities of feature maps are used for cluster assignment [21,28,29]. Despite the achievement, these approaches do not ensure robust learning from samples with low uncertainty. To reduce the influence of uncertain samples, uncertainty guidance has been introduced into the literature of SSL [15,30,33,34]. Nevertheless, semi-supervised segmentation of COVID-19 lesions remains a challenging task, of which the annotations are extremely scarce, and the lesions often have irregular and ambiguous contours.

To tackle the above challenges, we propose a novel deep neural network with a uncertainty-guided dual-consistency learning scheme for COVID-19 lesion segmentation from chest CT scan volumes. Specifically, we impose *image-level transformation equivalence* out of the observation that the prediction of a sample should obtain the same transformation of the input. Meanwhile, we adopt

[1] https://covid19.who.int.

feature-level perturbation invariance to a multi-decoder V-Net, where auxiliary decoder paths take perturbated features as inputs and form output consistency with a main decoder. Dual-consistency comprehensively enforces smoothness assumption into the SSL model from both input space and feature space, and hence the network could learn more invariant representations to diverse input or feature variants. Moreover, deep neural networks could memorize and easily overfit to noisy and uncertain contour points of COVID-19 lesions [35], which leads to poor generalization in real-world clinical practice. Hence, we further introduce a novel uncertainty guidance to the consistency learning process. Particularly, we quantify both the confidence uncertainty and the consensus uncertainty based on the multi-decoder structure. The estimated uncertainties are then used together in an indicator function to filter out uncertain samples during training. The proposed uncertainty-guided dual-consistency network (UDC-Net) is evaluated on a large-scale COVID-19 dataset with 852 whole-volume chest CT scans. Extensive experiments show that our approach outperforms other competitive SSL-based segmentation approaches, yielding state-of-the-art performance on semi-supervised COVID-19 lesion segmentation.

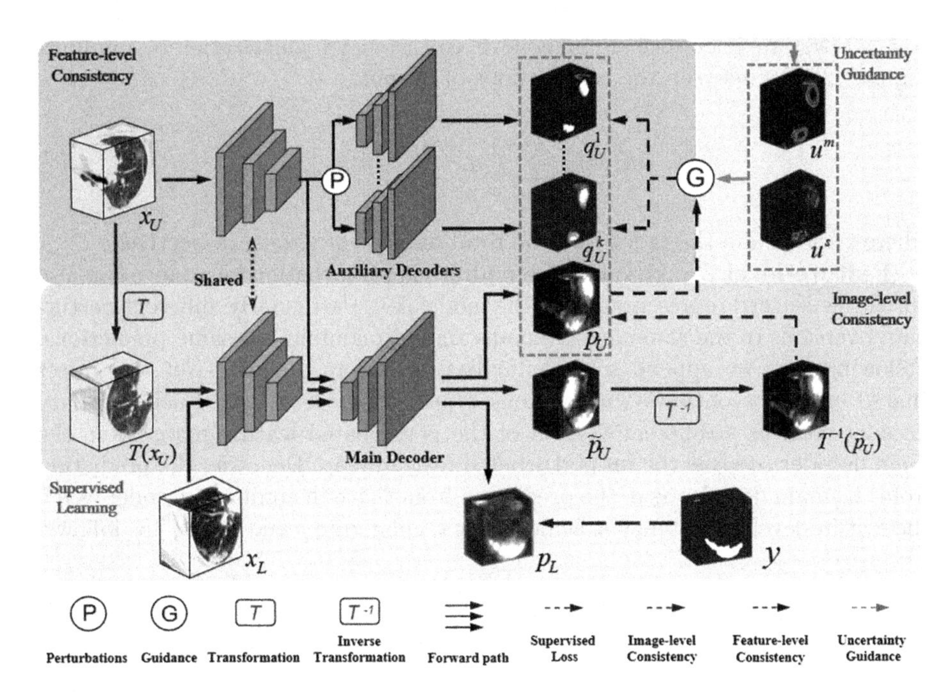

Fig. 1. Overview of UDC-Net. Feature-level consistency (in green) is formed by the main decoder's prediction $\mathrm{p_U}$ and auxiliary decoders' predictions $\{\mathrm{q_U^1}, \cdots, \mathrm{q_U^k}\}$. Image-level consistency (in blue) is formed by $\mathrm{p_U}$ and the prediction $\mathrm{\tilde{p}_U}$ of transformed image. The confidence uncertainty $\mathrm{u^m}$ and the consensus uncertainty $\mathrm{u^s}$ are quantified by mean and standard deviation of the multi-decoders' predictions, which are then used to guide the consistency learning (in red). A supervised loss is also used on the labeled data (in orange). (Color figure online)

2 Method

As shown in Fig. 1, our UDC-Net consists of a modified 3D multi-decoder V-Net [18] as its backbone. Apart from the supervised loss, our method makes full use of the unlabeled data by both feature-level and image-level consistency modules. Moreover, both the confidence uncertainty and the consensus uncertainty are estimated to guide more robust consistency learning.

2.1 Dual-Consistency Learning for Semi-supervised Segmentation

Image-level Consistency Learning via transformation equivalence of deep segmentation models f_{seg} indicates that while a transformation $T(\cdot)$ is applied to an input image x, there should be $f_{\text{seg}}(T(x)) = T(f_{\text{seg}}(x))$ [32]. We conduct random transformation on the images to get the perturbated version $T(x)$ as the input to our network. Subsequently, we have the corresponding prediction $f(T(x))$ given by the V-Net and the inverse transformation to the output $T^{-1}(f(T(x)))$, which should be consistent to the output of input data without transformation $f(x)$. Following the notations set before, let $p = f(x)$ and $\tilde{p} = f(T(x))$, we introduce an image-level consistency regularization by minimizing the L2 loss between the two versions of output:

$$\mathcal{L}_{\text{IC}} = \frac{1}{N} \sum_{i=1}^{N} \|p_i - [T^{-1}(\tilde{p})]_i\|_2^2 \tag{1}$$

where i and N are the index and the total number of voxels, respectively.

Feature-level Consistency Learning via perturbation invariance can also enrich the learned representation of the model [19]. Particularly, different perturbated versions of the same feature maps should maintain the same predictions. Following [21], we append several auxiliary decoders to the V-Net and inject shared encoder's outputs with various types of perturbations. Each auxiliary decoder receives a different version of the perturbated feature map, while the main decoder receives the un-perturbated feature map. Denoting the prediction from the main decoder as p, the prediction from the k-th auxiliary decoder as q^k, the feature-level consistency is achieved by regularizing p and each q^k as follows:

$$\mathcal{L}_{\text{FC}} = \frac{1}{N \cdot K} \sum_{i=1}^{N} \sum_{k=1}^{K} \|p_i - q_i^k\|_2^2 \tag{2}$$

where K is the total number of extra decoders. Following [21], seven types of feature perturbations, i.e., Feature noise, Feature dropout, Object masking, Context masking, Guided cutout, Intermediate VAT, and Random dropout, were introduced to seven auxiliary decoders, respectively. Detailed descriptions of each perturbation strategy can be found in the supplementary. All extra decoders were required to generate consistent prediction with the main decoder.

2.2 Dual Uncertainty Quantification for Robust Learning

The perturbation of the hidden representations during the consistency learning process could amplify the feature noises and uncertainty caused by the difficulty of accurately delineating the lesion contours of COVID-19. To this end, we propose to quantify both the confidence uncertainty and the consensus uncertainty of the multi-decoders, to guide more robust unsupervised learning.

Confidence Uncertainty indicates whether the model generates confident predictions. Previous works [15,34] used the entropy of the mean prediction of multiple perturbated inputs from self-ensembling models to estimate the prediction uncertainty. In our case, this form of uncertainty can be easily quantified using the main decoder and the K auxiliary decoders as below:

$$\mu_i = \frac{1}{K+1}\left[\left(\sum_{k=1}^{K} q_i^k\right) + p_i\right] \text{ and } \quad u_i^m = -\mu_i log \mu_i \tag{3}$$

where i indicates the voxel index, K is the total number of auxiliary decoders, μ is the mean prediction, and u^m is the estimated uncertainty. The higher u_i^m is, the less confidence the model is on its prediction.

Consensus Uncertainty indicates whether the model generates consistent predictions over multiple runs with perturbated data [9,11]. Supposing the average prediction of a suspicious infection area is high but the outputs from different branches vary severely, this means the area is sensitive to perturbation. By the smoothness assumption [3], the predictions for the target should be robust to perturbation, and the sensitive prediction hence highly suggests a noisy sample. Hence, we quantify the consensus uncertainty u^s as the standard deviation over the multi-decoders' predictions:

$$u_i^s = \frac{1}{K+1}\sqrt{\left[\sum_{k=1}^{K}(q_i^k - \mu_i)^2\right] + (p_i - \mu_i)^2} \tag{4}$$

Here, u^s essentially indicates the consensus among different decoders, which is complementary with u^m which measures the confidence of the model.

2.3 Uncertainty-Guided Dual-Consistency Learning for Segmentation

The quantified uncertainties are used to filter out uncertain voxels and consequently guide the model to learn from more reliable unlabeled data. Denoting i as the voxel index for the prediction volume, the reliable voxels are selected from a set $\Omega = \{i | u_i^s < \tau^s \; \& \; u_i^m < \tau^m\}$, where τ^s and τ^m are two thresholds. The cross consistency loss among decoders is then guided by:

$$\mathcal{L}_{\text{UFC}} = \sum_{k=1}^{K}\sum_{i\in\Omega} \|p_i - q_i^k\|_2^2 \tag{5}$$

Here, the uncertainty guidance is applied onto feature-level consistency learning as the uncertainties are generated with feature perturbations. Thus, the total loss for our uncertainty-guided dual-consistency learning UDC-Net for semi-supervised lesion segmentation is as follows:

$$\mathcal{L} = \mathcal{L}_{\mathrm{S}} + \alpha\mathcal{L}_{\mathrm{IC}} + \beta\mathcal{L}_{\mathrm{UFC}} \quad (6)$$

where $\mathcal{L}_{\mathrm{S}}$ is the supervised loss consists of a Dice loss and a cross entropy loss, α and β are two hyper-parameters weighing the contributions from different losses.

During training, we first trained a supervised V-Net and then added the extra decoders for finetuning with uncertainty-guided consistency learning. The training process was terminated if the Dice coefficient on the validation dataset stagnated. Adam [10] was used as the optimizer with an initial learning rate of 0.001 and a learning decay rate of 0.95 per epoch. As widely adopted by SSL works [21,24], α and β were set to be two sigmoid-shape monotonically functions of the training steps with maximum of 1. The threshold τ^m and τ^s were set to 0.34 and 0.12 after tuning on the validation set. For testing, we carried out sliding window inference and took only the main decoder's prediction. All implementation was done with Pytorch [22] on an NVIDIA TITAN X GPU.

3 Experiments

3.1 Datasets and Evaluation Metrics

Datasets. In total, 852 chest CT volumes acquired from December 2019 to April 2020 were collected and enrolled in this study, among which 144 were voxel-annotated by four experienced radiologists. The labeled data were divided into: (1) 65 cases as labeled training dataset; (2) 9 cases as the validation set; and (3) 70 cases as the testing set. The remained 708 chest CT scans were used as the unlabeled training data.

Evaluation Metrics. We adopted Dice Score (DSC), Jaccard similarity coefficient (Jaccard), and Average Symmetric Surface Distance (ASD) to evaluate the segmentation performance.

3.2 Ablation Study on Different Components

We conduct ablation studies to analyze the contributions of our proposed methods, and the quantitative results can be seen in Table 1. Regarding the testing set performance, image-level consistency (IC) shows increases of 2.4% in DSC, 2.5% in Jaccard, and 3.7 in ASD comparing to 3D V-Net. Meanwhile, feature-level consistency (FC) regularization leads to a large improvement of 4.5% in DSC, 5.5% in Jaccard, and 6.0 in ASD comparing to 3D V-Net. Unifying dual consistencies further improves DSC and Jaccard with about 1%, which demonstrates the effectiveness of learning from the unlabeled data. Further, introducing either the confidence uncertainty or the consensus uncertainty guidance consistently

Table 1. Ablation study of different components. All results are reported as validation/testing results. (FC: feature-level consistency; IC: image-level consistency; UM: confidence uncertainty computed by the mean of the multi-decoders' predictions; US: consensus uncertainty computed by the standard deviation of the multi-decoders' predictions)

Components				Evaluation metrics		
IC	FC	UM	US	DSC[%] ↑	Jaccard[%] ↑	ASD[mm] ↓
				70.0/71.1	56.5/56.8	12.1/12.1
✓				70.3/73.5	56.7/59.3	12.2/8.4
	✓			71.4/75.6	58.4/62.3	12.1/6.1
✓	✓			71.9/76.7	58.9/63.7	11.7/5.8
✓	✓	✓		72.2/77.0	59.0/64.0	11.3/**3.2**
✓	✓		✓	72.4/77.2	59.5/64.3	11.4/4.1
✓	✓	✓	✓	**72.7/77.4**	**59.9/64.5**	**10.9**/3.9

benefit the learning of the unlabeled data. Moreover, our method with dual uncertainty achieves better DSC and Jaccard with a comparable ASD to those of the single-uncertainty models, further demonstrating that dual uncertainties are complementary for guiding more robust learning.

3.3 Comparison with State-of-the-Art Methods

We compare our method against other state-of-the-art semi-supervised segmentation approaches. Several recent models were implemented, including Mean-Teacher (MT) [24], Uncertainty-aware mean teacher [34], Transformation-consistent Self-ensembling Model (TCSM) [12], and Cross Consistency Training (CCT) [21]. We run each methods four times with different random seeds.

Quantitative comparison results are reported in Table 2. For a fair comparison, we implemented all methods with the 3D V-Net as backbone. As observed, UDC-Net outperforms all other methods with at least 1.8% in Dice,

Table 2. Quantitative comparison with other semi-supervised methods.

Methods	Evaluation metrics		
	DSC[%] ↑	Jaccard[%] ↑	ASD[mm] ↓
V-Net [18]	71.1 ± 0.40	56.8 ± 0.45	12.1 ± 2.1
Mean teacher [24]	72.5 ± 0.25	58.2 ± 0.36	11.3 ± 1.8
UA-MT [34]	74.0 ± 0.11	60.1 ± 0.15	9.2 ± 0.9
TCSM [12]	72.9 ± 0.46	58.9 ± 0.58	9.1 ± 1.4
CCT [21]	75.6 ± 0.11	62.3 ± 0.19	6.1 ± 0.7
UDC-Net (ours)	**77.4 ± 0.14**	**64.5 ± 0.15**	**3.9 ± 0.5**

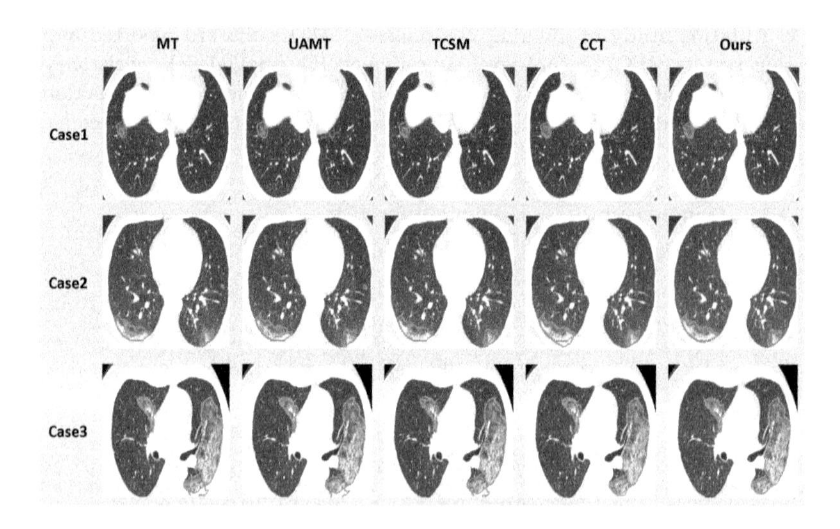

Fig. 2. Qualitative comparison. Green and orange curves delineate the model prediction and ground truth, respectively. Best viewed in color. (Color figure online)

2.2% in Jaccard, and 2.2 in ASD, showing outstanding unsupervised learning efficacy.

Qualitative comparison is illustrated by visualizing the segmentation results in Fig. 2. As demonstrated, Our UDC-Net delineates more accurate lesion contours than other methods regarding diverse shapes and sizes of lesion. Visualization of the two uncertainties can be found in the supplementary.

3.4 Analysis on Efficacy of Leveraging Unlabeled Data

We further evaluate our UDC-Net's effectiveness by varying the ratios of labeled and unlabeled training data. Table 3 shows that UDC-Net consistently improves the baseline V-Net with significant margins in both DSC, Jaccard, and ASD, whenever 32 or 65 labeled scans are provided. Moreover, the proposed approach consistently outperforms CCT [21] (the best model among those compared with ours) under all different scenarios. Notably, when less data are given, UDC-Net shows comparable or even better results than CCT. For instance, UDC-net achieves 75.0% DSC, 61.5% Jaccard, and 4.8 ASD with 32 labeled scans and 140 unlabeled scans (3rd row), which is comparable to the performance of CCT with double labeled scans (7th row). With 65 labeled scans and 140 unlabeled scans, UDC-Net (8th row) shows superior performance than CCT with 5 times unlabeled data (9th row). These findings demonstrate that our method enables more efficient unsupervised learning, suggesting

Table 3. Quantitative performance comparison under different numbers of training labeled/unlabeled data.

Method	# Scans used		Evaluation metrics		
	Labeled	Unlabeled	DSC[%] ↑	Jaccard[%] ↑	ASD[mm] ↓
V-Net [18]	32	0	70.4	56.0	4.3
CCT [21]	32	140	74.4	60.7	5.8
UDC-Net (ours)	32	140	**75.0**	**61.5**	**4.8**
CCT [21]	32	708	75.2	61.6	7.0
UDC-Net (ours)	32	708	**76.9**	**64.0**	**4.4**
V-Net	65	0	71.1	56.8	12.1
CCT [21]	65	140	75.1	61.8	5.8
UDC-Net (ours)	65	140	**76.6**	**63.5**	**5.4**
CCT [21]	65	708	75.6	62.3	6.0
UDC-Net (ours)	65	708	**77.4**	**64.5**	**3.9**

4 Conclusions

In this paper, we present an uncertainty-guided dual-consistency learning method for semi-supervised COVID-19 lesion segmentation from chest CT scans. Image-level transformation equivalence and feature-level perturbation invariance are both introduced to form dual consistency learning from unlabeled data. Meanwhile, the dual uncertainty mechanism further improves the learning process with more reliable and robust guidance. Extensive experiments on a large COVID-19 dataset demonstrate the efficiency of our method in real-world scenarios. Future work will include improving the method with more robust knowledge distillation and generalizing to other semi-supervised learning tasks.

Acknowledgement. This work was supported by Key-Area Research and Development Program of Guangdong Province, China (2020B010165004 and 2018B010109006), Hong Kong Innovation and Technology Fund (Project No. ITS/311/18FP and Project No. ITS/426/17FP.), and National Natural Science Foundation of China with Project No. U1813204.

References

1. Ai, T., et al.: Correlation of chest CT and RT-PCR testing in coronavirus disease 2019 (COVID-19) in China: a report of 1014 cases. Radiology, 200642 (2020)
2. Berthelot, D., Carlini, N., Goodfellow, I., Papernot, N., Oliver, A., Raffel, C.A.: MixMatch: a holistic approach to semi-supervised learning. In: NeurIPS, pp. 5049–5059 (2019)
3. Chapelle, O., Scholkopf, B., Zien, A.: Semi-supervised learning. IEEE TNNLS **20**(3), 542–542 (2009)

4. Di, D., Shi, F., Yan, F., Xia, L., Mo, Z., Ding, Z., et al.: Hypergraph learning for identification of COVID-19 with CT imaging. MedIA, 101910 (2020)
5. Fan, D.P., Zhou, T., Ji, G.P., Zhou, Y., Chen, G., Fu, H., et al.: Inf-Net: automatic COVID-19 lung infection segmentation from CT images. IEEE TMI **39**, 2626–2637 (2020)
6. Fang, Y., Zhang, H., Xie, J., Lin, M., Ying, L., Pang, P., et al.: Sensitivity of chest CT for COVID-19: comparison to RT-PCR. Radiology, 200432 (2020)
7. Huang, C., Wang, Y., Li, X., Ren, L., Zhao, J., Hu, Y., et al.: Clinical features of patients infected with 2019 novel coronavirus in Wuhan, China. Lancet **395**(10223), 497–506 (2020)
8. Jin, C., Chen, W., Cao, Y., Xu, Z., Tan, Z., Zhang, X., et al.: Development and evaluation of an artificial intelligence system for COVID-19 diagnosis. Nat. Commun. **11**(1), 1–14 (2020)
9. Ke, Z., Wang, D., Yan, Q., Ren, J., Lau, R.W.: Dual student: breaking the limits of the teacher in semi-supervised learning. In: ICCV, pp. 6728–6736 (2019)
10. Kingma, D.P., Ba, J.: Adam: a method for stochastic optimization. In: ICLR (2015)
11. Lee, J., Chung, S.Y.: Robust training with ensemble consensus. In: ICLR (2020). https://openreview.net/forum?id=ryxOUTVYDH
12. Li, X., Yu, L., Chen, H., Fu, C.W., Xing, L., Heng, P.A.: Transformation-consistent self-ensembling model for semisupervised medical image segmentation. IEEE TNNLS **32**, 523–534 (2020)
13. Liang, W., et al.: Early triage of critically ill COVID-19 patients using deep learning. Nat. Commun. **11**(1), 1–7 (2020)
14. Liu, Q., Yu, L., Luo, L., Dou, Q., Heng, P.A.: Semi-supervised medical image classification with relation-driven self-ensembling model. IEEE TMI **39**, 3429–3440 (2020)
15. Luo, L., Yu, L., Chen, H., Liu, Q., Wang, X., Xu, J., et al.: Deep mining external imperfect data for chest X-ray disease screening. IEEE TMI **39**(11), 3583–3594 (2020)
16. Ma, J., et al.: Active contour regularized semi-supervised learning for COVID-19 CT infection segmentation with limited annotations. Phys. Med. Biol. **65**, 225034 (2020)
17. Mei, X., Lee, H.C., Diao, K.y., Huang, M., Lin, B., Liu, C., et al.: Artificial intelligence-enabled rapid diagnosis of patients with COVID-19. Nat. Med., 1–5 (2020)
18. Milletari, F., Navab, N., Ahmadi, S.A.: V-Net: fully convolutional neural networks for volumetric medical image segmentation. In: 3DV, pp. 565–571. IEEE (2016)
19. Miyato, T., Maeda, S.i., Koyama, M., Ishii, S.: Virtual adversarial training: a regularization method for supervised and semi-supervised learning. IEEE TPAMI **41**(8), 1979–1993 (2018)
20. Oh, Y., Park, S., Ye, J.C.: Deep learning COVID-19 features on CXR using limited training data sets. IEEE TMI **39**, 2688–2700 (2020)
21. Ouali, Y., Hudelot, C., Tami, M.: Semi-supervised semantic segmentation with cross-consistency training. In: CVPR, pp. 12674–12684 (2020)
22. Paszke, A., et al.: Pytorch: an imperative style, high-performance deep learning library. In: NeurIPS, pp. 8026–8037 (2019)
23. Qiao, S., Shen, W., Zhang, Z., Wang, B., Yuille, A.: Deep co-training for semi-supervised image recognition. In: Ferrari, V., Hebert, M., Sminchisescu, C., Weiss, Y. (eds.) ECCV 2018. LNCS, vol. 11219, pp. 142–159. Springer, Cham (2018). https://doi.org/10.1007/978-3-030-01267-0_9

24. Tarvainen, A., Valpola, H.: Mean teachers are better role models: weight-averaged consistency targets improve semi-supervised deep learning results. NeurIPS **30**, 1195–1204 (2017)
25. van Engelen, J.E., Hoos, H.H.: A survey on semi-supervised learning. Mach. Learn. **109**(2), 373–440 (2019). https://doi.org/10.1007/s10994-019-05855-6
26. Wang, G., Liu, X., Li, C., Xu, Z., Ruan, J., Zhu, H., et al.: A noise-robust framework for automatic segmentation of COVID-19 pneumonia lesions from CT images. IEEE TMI **39**(8), 2653–2663 (2020)
27. Wang, L., Lin, Z.Q., Wong, A.: COVID-Net: a tailored deep convolutional neural network design for detection of COVID-19 cases from chest X-ray images. Sci. Rep. **10**(1), 1–12 (2020)
28. Wang, X., Chen, H., Ran, A.R., Luo, L., Chan, P.P., Tham, C.C., et al.: Towards multi-center glaucoma OCT image screening with semi-supervised joint structure and function multi-task learning. MedIA **63**, 101695 (2020)
29. Wang, X., Chen, H., Xiang, H., Lin, H., Lin, X., Heng, P.A.: Deep virtual adversarial self-training with consistency regularization for semi-supervised medical image classification. Med. Image Anal. **70**, 102010 (2021)
30. Wang, X., Tang, F., Chen, H., Luo, L., Tang, Z., Ran, A.R., et al.: UD-MIL: uncertainty-driven deep multiple instance learning for oct image classification. IEEE JBHI **24**, 3431–3442 (2020)
31. Wang, X., Deng, X., Fu, Q., Zhou, Q., Feng, J., Ma, H., et al.: A weakly-supervised framework for COVID-19 classification and lesion localization from chest CT. IEEE TMI **39**, 2615–2625(2020)
32. Worrall, D.E., Garbin, S.J., Turmukhambetov, D., Brostow, G.J.: Harmonic networks: deep translation and rotation equivariance. In: CVPR, pp. 5028–5037 (2017)
33. Xia, Y., Yang, D., Yu, Z., Liu, F., Cai, J., Yu, L., et al.: Uncertainty-aware multi-view co-training for semi-supervised medical image segmentation and domain adaptation. MedIA **65**, 101766 (2020)
34. Yu, L., Wang, S., Li, X., Fu, C.-W., Heng, P.-A.: Uncertainty-aware self-ensembling model for semi-supervised 3D left atrium segmentation. In: Shen, D., et al. (eds.) MICCAI 2019. LNCS, vol. 11765, pp. 605–613. Springer, Cham (2019). https://doi.org/10.1007/978-3-030-32245-8_67
35. Zhang, C., Bengio, S., Hardt, M., Recht, B., Vinyals, O.: Understanding deep learning requires rethinking generalization. In: ICLR (2017)
36. Zhu, N., Zhang, D., Wang, W., Li, X., Yang, B., Song, J., et al.: A novel coronavirus from patients with pneumonia in China, 2019. N. Engl. J. Med. **382**, 727–733 (2020)

Contrastive Pre-training and Representation Distillation for Medical Visual Question Answering Based on Radiology Images

Bo Liu, Li-Ming Zhan, and Xiao-Ming Wu(✉)

Department of Computing, The Hong Kong Polytechnic University, Kowloon, Hong Kong
{csbliu,cslmzhan,csxmwu}@comp.polyu.edu.hk

Abstract. One of the primary challenges facing medical visual question answering (Med-VQA) is the lack of large-scale well-annotated datasets for training. To overcome this challenge, this paper proposes a two-stage pre-training framework by learning transferable feature representations of radiology images and distilling a lightweight visual feature extractor for Med-VQA. Specifically, we leverage large amounts of unlabeled radiology images to train three teacher models for the body regions of brain, chest, and abdomen respectively via contrastive learning. Then, we distill the teacher models to a lightweight student model that can be used as a universal visual feature extractor for any Med-VQA system. The lightweight feature extractor can be readily fine-tuned on the training radiology images of any Med-VQA dataset, saving the annotation effort while preventing overfitting to small-scale training data. The effectiveness and advantages of the pre-trained model are demonstrated by extensive experiments with state-of-the-art Med-VQA methods on existing benchmarks. The source code and the pre-training dataset can be downloaded from https://github.com/awenbocc/cprd.

Keywords: Medical visual question answering · Contrastive learning · Representation distillation · Model compression

1 Introduction

Medical visual question answering (Med-VQA) has gained increasing attention over the past few years. Given a medical image and a clinical question about the image, it aims to find the correct answer by analyzing the visual information of the image. Med-VQA technology has great potential in medical and healthcare services. It can be used for computer-assisted diagnosis, intelligent medical guidance, clinical education and training, etc., which can help to significantly improve the quality of medical services and meet the increasing demand of the general public for medical resources.

M. de Bruijne et al. (Eds.): MICCAI 2021, LNCS 12902, pp. 210–220, 2021.
https://doi.org/10.1007/978-3-030-87196-3_20

While recent breakthroughs in image recognition and natural language processing have laid the foundation for the development of Med-VQA systems, the research progress of Med-VQA is impeded by the absence of large-scale well-annotated training datasets. The visual feature extraction module of existing Med-VQA models usually employs deep architectures and needs to be trained on a large collection of annotated radiology images, which however are often unavailable and costly to collect. To address this issue, a pioneering work [17] proposes mixture of enhanced visual features (MEVF) to pre-train the visual feature extraction module by constructing an auxiliary organ disease classification task on the radiology images of VQA-RAD [13] and observes positive effect. However, this approach cannot be transferred to other datasets, since the auxiliary pre-training task is designed based on the VQA-RAD dataset and requires extra effort for annotation.

In this paper, we tackle the data scarcity challenge by utilizing easily-available unannotated radiology image datasets for pre-training and representation distillation. First, we observe that the radiology images in current Med-VQA benchmarks mainly involve three human body regions – brain, chest, and abdomen, and there are large amounts of open-source unlabelled radiology images available for each region. Therefore, we propose to pre-train a visual feature extraction model (*teacher*) for each region respectively via contrastive learning. Second, to obtain a general and lightweight feature extractor, we distill the three teacher models into a small *student* model by contrastive representation distillation. The distilled model can be readily fine-tuned on any training dataset to facilitate the training of a Med-VQA system, without requiring further annotating process. Moreover, the small size of the distilled model can prevent overfitting to the training data, which typically only contains hundreds of radiology images.

To summarize, our contributions are two-fold. (1) We propose a new pre-training framework that leverages easily-acquired unannotated radiology images to pre-train and distill a general and lightweight visual feature extractor for Med-VQA, which can be easily adapted to small-scale training datasets. (2) We conduct extensive experiments with state-of-the-art Med-VQA methods on two benchmarks VQA-RAD [13] and SLAKE [14] to demonstrate the usefulness and benefits of the pre-trained model.

2 Related Work

Medical Visual Question Answering. Existing Med-VQA methods including [1,21,26] in ImageCLEF-Med challenge [2,11], often employ deep pre-trained architectures such as VGG [22] or ResNet [8] as the visual feature extraction module, which tend to cause overfitting due to limited training data in the Med-VQA domain. To overcome data limitation, MEVF [17] combines convolutional denoising auto-encoder (CDAE) [16] and meta-learning [24] to train a useful initialization for the visual feature extractor. Based on MEVF [17], conditional reasoning (CR) [29] further enhances the reasoning ability of the multimodal feature fusion module. Nevertheless, the pre-training process of MEVF requires additional data annotations on the training images, which requires medical expertise and is laborious and costly.

Contrastive Learning. Contrastive learning aims to learn high-quality feature representations by deriving self-supervision signals. CPC [18] pioneers in using the InfoNCE loss for contrastive learning on sequential tasks such as text or audio, which has been followed by many recent contrastive learning methods [5–7]. MoCo [7] utilizes a queue to efficiently store a large number of negative samples; SimCLR [5] explores the effectiveness of diverse image augmentation combinations; MoCo-v2 [6] takes advantages of both MoCo and SimCLR to enhance representation learning. These unsupervised methods have achieved promising results in learning image representations.

Model Compression. Knowledge distillation is introduced in [4,9] to compress a large model into a smaller one without losing too many generalization abilities, which is achieved by minimizing Kullback–Leibler divergence (KLD) between the probabilistic outputs of the large and the smaller models. A recent work [23] argues that the independence assumption in the KLD loss fails to retain important structural information of the large model and proposes to combine KLD with contrastive representation distillation to achieve better performance.

3 Contrastive Pre-training and Representation Distillation (CPRD)

In current Med-VQA benchmarks, the radiology images mainly involve three human body regions: brain, chest, and abdomen. For each region, unlabeled images can be easily obtained from many large-scale open-source datasets. Motivated by this observation, we propose to train three specialized teacher models to focus on different body region respectively and then teach a student model to learn both intra- and inter-region features for Med-VQA, as illustrated in Fig. 1.

3.1 Teachers: Intra-region Contrastive Pre-training

Let $\mathcal{D}_{brain}, \mathcal{D}_{chest}, \mathcal{D}_{abdomen}$ denote the set of radiology images for the three body regions respectively. Radiology images in each region have large diversity in terms of different organs and versatile imaging modalities, e.g., liver MRI, liver CT, and intestine CT in the abdomen region. Therefore, we employ Momentum Contrast [6], a self-supervised contrastive learning method, to train a *Teacher* model for each region with the corresponding dataset $\mathcal{D}_r$ ($r \in \{brain, chest, abdomen\}$) to implicitly model these differences. As shown in Fig. 1 (a), we sample an image x_i and a queue $q = \{x_j^-\}_{j=1}^M$ of M images different from x_i from $\mathcal{D}_r$. Then, data augmentation (such as resize, crop, color distort, and Gaussian blur), denoted as Aug, is applied on all the sampled images and produce:

$$\hat{\boldsymbol{x}}_i = Aug(x_i), \hat{\boldsymbol{x}}_i^+ = Aug(x_i),\ \hat{\boldsymbol{q}} = \{\hat{\boldsymbol{x}}_j^- = Aug(x_j^-)\}_{j=1}^M, \tag{1}$$

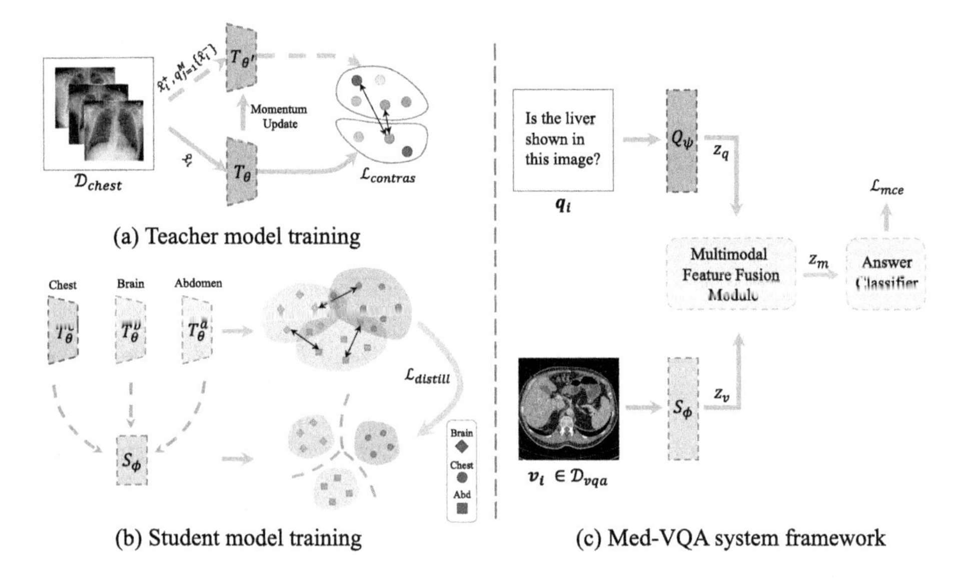

Fig. 1. Our proposed CPRD framework for Med-VQA. (a) Train a teacher model T_θ by self-supervised contrastive learning on the chest region. (b) Distill three teacher models into one student model S_ϕ. (c) Apply the student model S_ϕ for Med-VQA.

where $\hat{\boldsymbol{x}}_i$ and $\hat{\boldsymbol{x}}_i^+$ are two different views of x_i, generated by applying random augmentation on x_i twice. An encoder T_θ is used to learn the feature representation of $\hat{x}_i$, i.e., $\boldsymbol{z}_i = T_\theta(\hat{\boldsymbol{x}}_i)$. Another momentum encoder $T_{\theta'}$ is used to produce the representations of $\hat{\boldsymbol{x}}_i^+$ and $\hat{\boldsymbol{q}}$, i.e., $\{\boldsymbol{z}_i^+, \boldsymbol{z}_1^-, \boldsymbol{z}_2^-, ..., \boldsymbol{z}_M^-\}$. Since $\boldsymbol{z}_i$ and $\boldsymbol{z}_i^+$ are the representations of different views of x_i, $\boldsymbol{z}_i$ should be similar to $\boldsymbol{z}_i^+$ but dissimilar to the other M representations in $\hat{\boldsymbol{q}}$. The learning process can be guided by the InfoNCE contrastive loss [18]:

$$\mathcal{L}_{\boldsymbol{z}_i, \boldsymbol{z}_i^+, \{\boldsymbol{z}_j^-\}} = -\log \frac{exp(\boldsymbol{z}_i \cdot \boldsymbol{z}_i^+ / \tau)}{exp(\boldsymbol{z}_i \cdot \boldsymbol{z}_i^+ / \tau) + \sum_{j=1}^{M} exp(\boldsymbol{z}_i \cdot \boldsymbol{z}_j^- / \tau)}, \quad (2)$$

where τ is a temperature parameter [25] and $\cdot$ stands for dot product. In practice, the length of the queue q is usually much larger than the mini-batch size, making it costly to update $T_{\theta'}$ by gradient back-propagation. Following [6], we update it in an efficient way: $\theta' \leftarrow \beta\theta' + (1-\beta)\theta$, where β is the momentum coefficient. By optimizing the loss in Eq. (2), we obtain the teacher model T_θ for the region.

3.2 Student: Inter-region Representation Distillation

After obtaining the three teacher models: T_θ^a for $\mathcal{D}_{abdomen}$, T_θ^b for $\mathcal{D}_{brain}$, and T_θ^c for $\mathcal{D}_{chest}$, we design a lightweight *Student* model S_ϕ to distill representations of the teacher models, as shown in Fig. 1 (b). Let $\mathcal{D}_{all} = \{\mathcal{D}_{brain}, \mathcal{D}_{chest}, \mathcal{D}_{abdomen}\}$.

Inspired by the idea of contrastive representation distillation [23], for each region $\mathcal{D}_r \in \mathcal{D}_{all}$, for any image $x_i^r \in \mathcal{D}_r$, we randomly sample K images x_j^o ($j = \{1, \ldots, K\}$) from the other two datasets $\mathcal{D}_o = \mathcal{D}_{all} \backslash \mathcal{D}_r$. First, we make the student model inherit knowledge of each teacher by enforcing its representation of x_i^r, $S_\phi(x_i^r)$, to be similar to that of the corresponding teacher model, $T_\theta^r(x_i^r)$, by minimizing the loss function

$$\mathcal{L}_{sim} = -\frac{1}{N}\sum_{r=1}^{3}\sum_{i=1}^{L_r}\log\left(\frac{e^{T_\theta^r(x_i^r)\cdot S_\phi(x_i^r)/\tau}}{e^{T_\theta^r(x_i^r)\cdot S_\phi(x_i^r)/\tau} + \frac{K}{N}}\right), \tag{3}$$

where τ is the temperature parameter, L_r is the size of $\mathcal{D}_r$, and N is the size of $\mathcal{D}_{all}$ ($1 < K < N$). Meanwhile, we enable the student model to acquire the ability to distinguish the three regions by enforcing $S_\phi(x_i^r)$ to be dissimilar to $T_\theta^o(x_j^o)$, the representation of x_j^o (image of other regions) produced by the corresponding teacher model, by minimizing the loss function

$$\mathcal{L}_{dissim} = -\frac{1}{N \times K}\sum_{r=1}^{3}\sum_{i=1}^{L_r}\sum_{j=1}^{K}\log\left(1 - \left(\frac{e^{T_\theta^o(x_j^o)\cdot S_\phi(x_i^r)/\tau}}{e^{T_\theta^o(x_j^o)\cdot S_\phi(x_i^r)/\tau} + \frac{K}{N}}\right)\right). \tag{4}$$

Further, we train the student model to produce more discriminative representations by learning to identify the body region R of x_i^r. Note that the images are already grouped by regions in open-source databases so the region labels can be automatically generated. This is achieved by minimizing the classification loss

$$\mathcal{L}_{class} = -\frac{1}{N}\sum_{i=1}^{N}\log P(R = r | W S_\phi(x_i^r)), \tag{5}$$

where W is a linear classification layer, and P is the prediction probability of the target region. Finally, by combining Eqs. (3), (4) and (5), the student model is trained with the loss function

$$\mathcal{L}_{distill} = \alpha(\mathcal{L}_{dissim} + \mathcal{L}_{sim}) + (1 - \alpha)\mathcal{L}_{class}, \tag{6}$$

where α is a balancing parameter.

4 Applying CPRD for Med-VQA

The distilled student model can be used as a universal visual feature extractor for any Med-VQA system based on radiology images. Figure 1 (c) shows a typical Med-VQA pipeline. Given a radiology image v_i and a question q_i as inputs, the student model S_ϕ is applied on v_i to extract the visual features $\boldsymbol{z_v} = S_\phi(v_i)$, and a text encoder (e.g., LSTM [10] network) is used to extract the textual features q_i, i.e., $\boldsymbol{z_q} = Q_\psi(q_i)$. Then, $\boldsymbol{z_v}$ and $\boldsymbol{z_q}$ will be fused by some attention-based module (e.g., BAN [12]) to produce multimodal features $\boldsymbol{z_m}$.

Similar to general VQA, Med-VQA is also formulated as a classification problem [3]: predicting an answer from C fixed candidate answers in the training dataset. Note that there might be multiple correct answers for one question. As such, the multimodal features $\boldsymbol{z}_m$ will be fed to a classifier $\Phi(\cdot)$ (e.g., multilayer perceptron), to predict the probability of each candidate answer. All the model parameters, including those of the visual extractor S_ϕ, the text encoder Q_ψ, the feature fusion module and the classifier, are optimized in an end-to-end manner by minimizing the multi-label cross-entropy loss:

$$\mathcal{L}_{mce} = -\frac{1}{I}\sum_{i=1}^{I}\sum_{c=1}^{C}[l_i^c \log(\sigma^c(\Phi(\boldsymbol{z}_m))) + (1 - l_i^c)\log(1 - \sigma^c(\Phi(\boldsymbol{z}_m)))], \qquad (7)$$

where l_i is the multi-hot encoding of the answers for the current (v_i, q_i) pair, σ is the sigmoid function, and I is the size of the training dataset.

5 Experiments

In this section, we extensively evaluate the effectiveness of the visual feature extractor pre-trained by our proposed CPRD framework on the only two available manually-annotated Med-VQA datasets. We experiment with state-of-the-art Med-VQA methods and show that the pre-trained feature extractor can be used to significantly improve their performance.

5.1 Datasets

VQA-RAD [13] consists of 315 radiology images and 3,515 question-answer pairs. We follow the data splitting in [17]. **SLAKE** [14] is a recently proposed bi-lingual Med-VQA dataset. We use the English version, referred to as SLAKE-EN, which contains 642 radiology images and 7,033 question-answer pairs. We use the original data splitting. Besides, questions in VQA-RAD and SLAKE are both categorized into "closed-ended" questions whose answers are in limited choices, and "open-ended" questions whose answers are free-form text.

5.2 Experimental Setup

To train the teacher and student models, we randomly sample 22,995 unlabelled radiology images from open-resource databases[1], including 7,811 chest X-Rays, 7,592 abdomen CTs, and 7,592 brain CTs and MRIs. Our experiments are conducted on a Ubuntu server with 8 NVIDIA TITAN 12 GB Xp GPUs. All the hyper-parameters of the teacher and student models are chosen by cross validation via observing the loss in Eq. (2) and Eq. (6).

Teachers. For each region-focused teacher model, we use ResNet-50 to instantiate T_θ and $T_{\theta'}$ (Sect. 3.1) and train for 800 epochs with 4 GPUs for about 7 h. In each epoch, the mini-batch size is 128, and the queue length M is 1,024. The temperature parameter τ is set to be 0.2, 0.1, and 0.1 for brain, chest

[1] http://medicaldecathlon.com/.

and abdomen respectively. For model optimization, we use SGD optimizer with $1.5e^{-2}$ initial learning rate decayed by cosine schedule.

Student. We use ResNet-8 as the student model (Sect. 3.2) and train for 240 epochs with 1 GPU. We use SGD optimizer to minimize the loss $\mathcal{L}_{distill}$ with 0.05 initial learning rate decayed by cosine schedule. Besides, the queue length K is 8192, the temperature parameter τ is 0.07, and α in Eq. (6) is 0.9.

Med-VQA. After training the student model, we use the weights in the last epoch as initialization and fine-tune the model on a Med-VQA dataset for 100 epochs. We use Adamax optimizer with initial learning rate $2e^{-3}$ for model optimization. Following CR [29], we use accuracy as evaluation metric.

Table 1. Test accuracy of our method and baselines.

Models	VQA-RAD [13]			SLAKE-EN [14]		
	Overall	Open	Closed	Overall	Open	Closed
MFB fw. [28]	50.6	14.5	74.3	73.3	72.2	75.0
SAN fw. [27]	54.3	31.3	69.5	76.0	74.0	79.1
BAN fw. [12]	58.3	37.4	72.1	76.3	74.6	79.1
MEVF+SAN [17]	64.1	49.2	73.9	76.5	75.3	78.4
MEVF+BAN [17]	66.1	49.2	77.2	78.6	77.8	79.8
CPRD+BAN (ours)	**67.8**	**52.5**	**77.9**	**81.1**	**79.5**	**83.4**
MEVF+BAN+CR [29]	71.6	60.0	79.3	80.0	78.8	82.0
CPRD+BAN+CR (ours)	**72.7**	**61.1**	**80.4**	**82.1**	**81.2**	**83.4**

Table 2. Comparison of different visual modules in test accuracy and model size on VQA-RAD [13]. The number of parameters is calculated on the visual module only.

Visual modules	Overall (%)	Open (%)	Closed (%)	#Parameters (M)
VGG-16 [22] (ImageNet)	56.8	35.2	71.0	134.8
ResNet-50 [8] (ImageNet)	58.3	37.4	72.1	23.8
MEVF [17]	66.1	49.2	77.2	1.2
ResNet-8 (random init)	63.2	47.2	73.8	0.1
ResNet-8 (our CPRD)	**67.8**	**52.5**	**77.9**	**0.1**

5.3 Comparison with the State-of-the-Arts

We use our pre-trained model CPRD as the visual feature extractor, combined with the BAN attention mechanism [12] with or without the CR reasoning module [29] for Med-VQA. To demonstrate the necessity of domain-specific pre-training, we compare with general VQA frameworks including MFB [28],

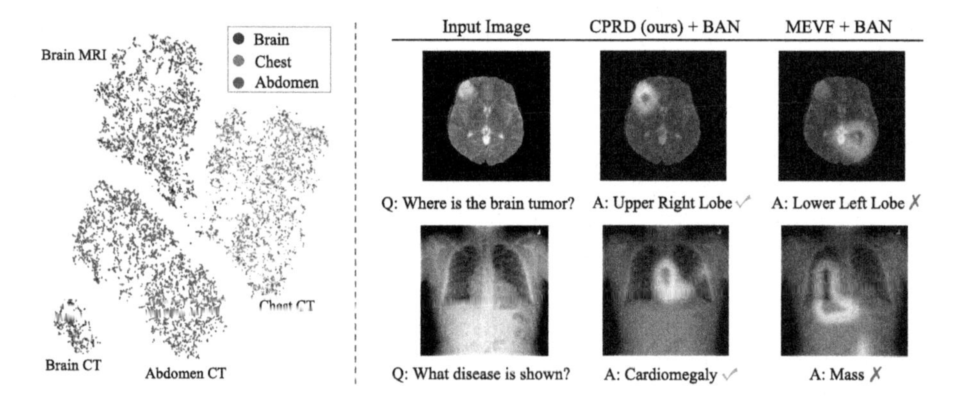

Fig. 2. (Left) t-SNE visualization of the representations learned by the student model; (Right) Grad-CAM maps from the visual modules of Med-VQA methods. ✓ and ✗ indicate the correctness of the answer given by each method.

SAN [27], and BAN [12].[2] Further, we compare with MEVF [17], which is the only baseline that uses a small model and pre-trains with medical images.

The results on VQA-RAD [13] and SLAKE [14] are reported in Table 1. For a fair comparison, all methods use a 1024-D LSTM network to extract textual features with word embeddings pre-trained by GloVe [19]. For MFB, SAN and BAN, we use ResNet-50 pre-trianed on ImageNet as the visual feature extractor. The following observations can be made. (1) Our method CPRD+BAN not only improves upon the performance of the strong baseline MEVF+BAN [17], but also achieves state-of-the-art results on the two benchmarks when further incorporating the CR [29] module. (2) Although MEVF+BAN [17] can significantly outperform the base framework BAN [12] on VQA-RAD, its performance gain on SLAKE is less significant (∼2%), far lower than the gain brought by our CPRD+BAN (∼5%). This demonstrates the generalization ability of our CPRD model on different datasets.

5.4 Ablation Analysis

We conduct an ablation study to analyze the impact of different pre-training strategies for the visual feature extraction module of Med-VQA. The results are summarized in Table 2. Specifically, we use BAN [12] as the multimodal feature fusion module and LSTM as the textual encoder for all methods in this subsection. Compared with the large models (i.e., VGG-16 and ResNet-50) pre-trained on ImageNet, it can be seen that lightweight models (i.e., MEVF and ResNet-8) perform better. Further, ResNet-8 pre-trained by our CPRD achieves better results than with random initialization, and outperforms the strongest baseline

[2] MFB, SAN, and BAN stand for the key reasoning module of the respective framework, where the visual and textual modules can be any applicable models.

MEVF with much fewer parameters. This again demonstrates the effectiveness and advantages of our CPRD model.

5.5 Visualization

The t-SNE [15] visualization of the representations learned by the ResNet-8 student model on the images of $\mathcal{D}_{all}$ (Sect. 3.2) is shown in Fig. 2 (left). It can be clearly seen that the student model learns discriminative representations for different regions. Further, the representations of brain CT and brain MRI are well separated, indicating that the student model also captures the differences among versatile imaging modalities for the same region. To demonstrate the visual evidence used in Med-VQA models for prediction, in Fig. 2 (right), we show the Grad-CAM [20] maps for visual modules based on the final predicted answers of our CPRD+BAN and a strong baseline MEVF+BAN. The first row is about a brain MRI image, and the second is about a chest X-Ray image, both from the test set of the SLAKE [14] dataset. It can be seen that our model can correctly answer the questions by locating the right visual evidence about the questions, which demonstrates the effectiveness of our visual module.

6 Conclusion

In this paper, we have proposed a two-stage pre-training framework to tackle the challenge of data scarcity in the Med-VQA domain. Our framework leverages large amounts of unannotated radiology images to pre-train and distill a lightweight visual feature extractor via contrastive learning and representation distillation. By applying this pre-trained model in current Med-VQA methods, we achieve new state-of-the-art performance on existing benchmarks.

Acknowledgment. This research was supported by the grant of P0030935 (ZVPY) funded by PolyU (UGC).

References

1. Abacha, A.B., Gayen, S., Lau, J.J., Rajaraman, S., Demner-Fushman, D.: NLM at imageclef 2018 visual question answering in the medical domain. In: Working Notes of CLEF 2018 - Conference and Labs of the Evaluation Forum. CEUR Workshop Proceedings, Avignon, France, vol. 2125. CEUR-WS.org (2018)
2. Abacha, A.B., Hasan, S.A., Datla, V.V., Liu, J., Demner-Fushman, D., Müller, H.: VQA-Med: overview of the medical visual question answering task at ImageCLEF 2019. In: Working Notes of CLEF 2019 - Conference and Labs of the Evaluation Forum. CEUR Workshop Proceedings, Lugano, Switzerland, vol. 2380. CEUR-WS.org (2019)
3. Antol, S., et al.: VQA: visual question answering. In: Proceedings of the IEEE International Conference on Computer Vision, pp. 2425–2433 (2015)

4. Buciluundefined, C., Caruana, R., Niculescu-Mizil, A.: Model compression. In: Proceedings of the 12th ACM SIGKDD International Conference on Knowledge Discovery and Data Mining, KDD 2006, New York, NY, USA. Association for Computing Machinery (2006)
5. Chen, T., Kornblith, S., Norouzi, M., Hinton, G.: A simple framework for contrastive learning of visual representations. In: International Conference on Machine Learning, pp. 1597–1607. PMLR (2020)
6. Chen, X., Fan, H., Girshick, R., He, K.: Improved baselines with momentum contrastive learning (2020)
7. He, K., Fan, H., Wu, Y., Xie, S., Girshick, R.: Momentum contrast for unsupervised visual representation learning (2020)
8. He, K., Zhang, X., Ren, S., Sun, J.: Deep residual learning for image recognition. In: Proceedings of the IEEE conference on Computer Vision and Pattern Recognition, pp. 770–778 (2016)
9. Hinton, G., Vinyals, O., Dean, J.: Distilling the knowledge in a neural network (2015)
10. Hochreiter, S., Schmidhuber, J.: Long short-term memory. Neural Comput., 1735–1780 (1997)
11. Ionescu, B., et al.: Overview of ImageCLEF 2018: challenges, datasets and evaluation. In: Bellot, P., et al. (eds.) CLEF 2018. LNCS, vol. 11018, pp. 309–334. Springer, Cham (2018). https://doi.org/10.1007/978-3-319-98932-7_28
12. Kim, J., Jun, J., Zhang, B.: Bilinear attention networks. In: Advances in Neural Information Processing Systems 31: Annual Conference on Neural Information Processing Systems, NeurIPS, Montréal, Canada, pp. 1571–1581. NeurIPS (2018)
13. Lau, J.J., Gayen, S., Abacha, A.B., Demner-Fushman, D.: A dataset of clinically generated visual questions and answers about radiology images. Sci. Data **5**, 1–10 (2018)
14. Liu, B., Zhan, L.M., Xu, L., Ma, L., Yang, Y., Wu, X.M.: SLAKE: a semantically-labeled knowledge-enhanced dataset for medical visual question answering (2021)
15. Van der Maaten, L., Hinton, G.: Visualizing data using t-SNE. J. Mach. Learn. Res. (2008)
16. Masci, J., Meier, U., Cireşan, D., Schmidhuber, J.: Stacked convolutional auto-encoders for hierarchical feature extraction. In: Honkela, T., Duch, W., Girolami, M., Kaski, S. (eds.) ICANN 2011, Part I. LNCS, vol. 6791, pp. 52–59. Springer, Heidelberg (2011). https://doi.org/10.1007/978-3-642-21735-7_7
17. Nguyen, B.D., Do, T.-T., Nguyen, B.X., Do, T., Tjiputra, E., Tran, Q.D.: Overcoming data limitation in medical visual question answering. In: Shen, D., et al. (eds.) MICCAI 2019, Part IV. LNCS, vol. 11767, pp. 522–530. Springer, Cham (2019). https://doi.org/10.1007/978-3-030-32251-9_57
18. Oord, A.v.d., Li, Y., Vinyals, O.: Representation learning with contrastive predictive coding. arXiv preprint arXiv:1807.03748 (2018)
19. Pennington, J., Socher, R., Manning, C.D.: Glove: global vectors for word representation. In: Proceedings of the 2014 Conference on Empirical Methods in Natural Language Processing, EMNLP, A meeting of SIGDAT, a Special Interest Group of the ACL, Doha, Qatar, pp. 1532–1543. ACL (2014)
20. Selvaraju, R.R., Cogswell, M., Das, A., Vedantam, R., Parikh, D., Batra, D.: Grad-CAM: visual explanations from deep networks via gradient-based localization. In: Proceedings of the IEEE International Conference on Computer Vision, pp. 618–626 (2017)

21. Shi, L., Liu, F., Rosen, M.P.: Deep multimodal learning for medical visual question answering. In: Working Notes of CLEF 2019 - Conference and Labs of the Evaluation Forum. CEUR Workshop Proceedings, vol. 2380, Lugano, Switzerland. CEUR-WS.org (2019)
22. Simonyan, K., Zisserman, A.: Very deep convolutional networks for large-scale image recognition. arXiv preprint arXiv:1409.1556 (2014)
23. Tian, Y., Krishnan, D., Isola, P.: Contrastive representation distillation. arXiv preprint arXiv:1910.10699 (2019)
24. Vuorio, R., Sun, S., Hu, H., Lim, J.J.: Multimodal model-agnostic meta-learning via task-aware modulation. In: Advances in Neural Information Processing Systems 32: Annual Conference on Neural Information Processing Systems, NeurIPS, Vancouver, BC, Canada, pp. 1–12 (2019)
25. Wu, Z., Xiong, Y., Yu, S., Lin, D.: Unsupervised feature learning via non-parametric instance-level discrimination. arXiv preprint arXiv:1805.01978 (2018)
26. Yan, X., Li, L., Xie, C., Xiao, J., Gu, L.: Zhejiang university at ImageCLEF 2019 visual question answering in the medical domain. In: Working Notes of CLEF 2019 - Conference and Labs of the Evaluation Forum. CEUR Workshop Proceedings, Lugano, Switzerland, vol. 2380. CEUR-WS.org (2019)
27. Yang, Z., He, X., Gao, J., Deng, L., Smola, A.J.: Stacked attention networks for image question answering. In: 2016 IEEE Conference on Computer Vision and Pattern Recognition, Las Vegas, NV, USA, pp. 21–29. IEEE Computer Society (2016)
28. Yu, Z., Yu, J., Fan, J., Tao, D.: Multi-modal factorized bilinear pooling with co-attention learning for visual question answering (2017)
29. Zhan, L.M., Liu, B., Fan, L., Chen, J., Wu, X.M.: Medical visual question answering via conditional reasoning. In: Proceedings of the 28th ACM International Conference on Multimedia, MM 2020, New York, NY, USA. Association for Computing Machinery (2020)

Positional Contrastive Learning for Volumetric Medical Image Segmentation

Dewen Zeng[1(✉)], Yawen Wu[2], Xinrong Hu[1], Xiaowei Xu[3], Haiyun Yuan[3], Meiping Huang[3], Jian Zhuang[3], Jingtong Hu[2], and Yiyu Shi[1(✉)]

[1] University of Notre Dame, Notre Dame, IN, USA
{dzeng2,yshi4}@nd.edu
[2] University of Pittsburgh, Pittsburgh, PA, USA
[3] Guangdong Provincial People's Hospital, Guangzhou, China

Abstract. The success of deep learning heavily depends on the availability of large labeled training sets. However, it is hard to get large labeled datasets in medical image domain because of the strict privacy concern and costly labeling efforts. Contrastive learning, an unsupervised learning technique, has been proved powerful in learning image-level representations from unlabeled data. The learned encoder can then be transferred or fine-tuned to improve the performance of downstream tasks with limited labels. A critical step in contrastive learning is the generation of contrastive data pairs, which is relatively simple for natural image classification but quite challenging for medical image segmentation due to the existence of the same tissue or organ across the dataset. As a result, when applied to medical image segmentation, most state-of-the-art contrastive learning frameworks inevitably introduce a lot of false negative pairs and result in degraded segmentation quality. To address this issue, we propose a novel positional contrastive learning (PCL) framework to generate contrastive data pairs by leveraging the position information in volumetric medical images. Experimental results on CT and MRI datasets demonstrate that the proposed PCL method can substantially improve the segmentation performance compared to existing methods in both semi-supervised setting and transfer learning setting. (Code available at github.com/dewenzeng/positional_cl).

1 Introduction

Deep neural networks (DNNs) play an important role in today's medical image segmentation [10,17,20,21,23]. To achieve state-of-the-art accuracy, most of the existing methods rely on supervised learning when large labeled datasets can be used for training. However, due to the extensive annotation effort and the

Electronic supplementary material The online version of this chapter (https://doi.org/10.1007/978-3-030-87196-3_21) contains supplementary material, which is available to authorized users.

M. de Bruijne et al. (Eds.): MICCAI 2021, LNCS 12902, pp. 221–230, 2021.
https://doi.org/10.1007/978-3-030-87196-3_21

requirement of expertise in the medical domain, acquiring such large labeled datasets is usually prohibitive. In the meantime, a large amount of unlabeled image data from modalities such as Computed Tomography (CT) and Magnetic Resonance Imaging (MRI) is generated every day all around the world. Therefore, it is desirable that the DNNs can leverage the numerous unlabeled data to achieve higher performance with limited annotations. Contrastive learning [3–5,8,14], as a self-supervised learning (SSL) method, has shown great success in learning image-level features from large-scale unlabeled data without using any human-annotated labels. The main idea of contrastive learning is to contrast the similarities of sample pairs in the representation space through contrastive loss, pulling the representations of similar pairs (a.k.a. positive pairs) together and pushing the representations of dissimilar pairs (a.k.a. negative pairs) apart. In SSL setting, an encoder is trained using contrastive loss with unlabeled data. After that, the trained encoder can be used as the initialization for training a supervised downstream task such as object detection and image segmentation. Many works have shown that the encoder learned by contrastive learning performs better than the encoder trained with supervised learning [3,8].

Most existing contrastive learning frameworks are for image classification where the instances in two different images have dissimilar features. When directly applying them to medical image segmentation where different images can have similar structures or organs, a large number of false negative pairs will be induced, leading to degraded performance. Recently, [2] attempted to address this issue through a global contrastive learning approach for 3D medical image segmentation. It divides each volume into several partitions and considers the slices of corresponding partitions in different volumes as positive pairs and those of different partitions as negative pairs. However, the last a few slices of a partition can be very similar to the first a few slices of the next partition as they are naturally adjacent, which may still result in many false negative pairs.

To alleviate the problem, we propose a novel positional contrastive learning (PCL) framework, which generates contrastive data pairs based on the position of a slice in volumetric medical images. Slices that are close are considered positive pairs while those that are far apart are considered negative. Such a strategy can better leverage the domain-specific cue of medical images as adjacent slices typically contain similar anatomical structures, thus reducing false negatives. We evaluate the proposed PCL framework on two CT datasets and two MRI datasets. The experimental results show that our method can achieve better performance compared with state-of-the-art baselines in both semi-supervised and transfer learning settings.

2 Related Work

Recent years have seen powerful self-supervised visual feature learning approaches with DNNs. By exploiting the information in large unlabeled datasets, a network can learn hierarchical features that can help the training of other downstream tasks, especially when the training labels of these tasks are limited. Early SSL methods are mostly based on the design of pretext tasks, in which pseudo labels are automatically generated for network training. As these methods rely on ad-hoc heuristics, the learned representation lack generality [3]. Contrastive learning has recently become a prevailing SSL method because of its superior performance. In contrastive learning, a contrastive loss [7] is used to enforce representations of positive pairs to be similar and those of negative pairs to be dissimilar [3,8,11,13,14,18]. MoCo [8] and SimCLR [3] are two different contrastive learning frameworks that yield state-of-the-art results. MoCo maintains a dictionary as a queue to store negative samples for training, while SimCLR explores the use of in-batch samples for negative sampling. Most of these works are based on image classification tasks, assuming that the instances in two different images have dissimilar features. This is not the case, however, for medical images, because the same target organ or structure usually exists in all the images across the dataset. For example, in ACDC MICCAI 2017 dataset [1], the target structures such as the left ventricle and the right ventricle appear in almost every slice of the volumetric image for all patients. As such, if we follow the method used in image classification tasks and treat the augmented images from different slices as negative, many of them will actually be false negatives.

The state-of-the-art contrastive learning method for medical image segmentation [2] attempted to address this issue through the partition of 3D medical images. However, it will still induce false negatives as discussed in Sect. 1. In contrast, the PCL method we propose uses the relative position of the slices in the volumes to decide whether they are positive pairs, thus the false negative issue can be alleviated. In addition, the method in [2] is only evaluated in semi-supervised setting where contrastive learning and downstream tasks are done on the same dataset. We extend the evaluation to transfer learning to test whether the features learned by PCL on one dataset are transferable to another, and show that PCL can do better than [2] in both settings.

3 Method

3.1 Framework Overview

In this work, for fair comparison we follow [2] and use 2D U-Net [17] to perform segmentation on 2D slices of 3D images, which has shown a remarkable success in many 3D image segmentation tasks [9,10,15,19,20]. The proposed method can also be readily generalized to patch-based 3D U-Net and 3D-2D hybrid U-Net approaches. Our PCL framework is shown in Fig. 1. In the pre-training stage, the input of the framework is a set of 2D slices in the xy plane sampled randomly from unlabeled volumetric medical images. These slices are then propagated to a U-Net encoder $f(\cdot)$ (also known as the feature extractor) followed by a shallow multilayer perceptron (MLP) projection head $g(\cdot)$. Let x_i denote an input 2D slice. Then $h_i = f(x_i)$ is the representation learned by the encoder $f(\cdot)$ and $z_i = g(f(x_i))$ is the embedding vector. A contrastive loss is employed on all the embeddings learned from the data in a mini-batch to perform contrastive learning. After contrastive learning finishes, $g(\cdot)$ is thrown away and $f(\cdot)$ is used as the initialization in the standard U-Net architecture to train the network on the limited labeled dataset by supervised learning in the fine-tuning stage.

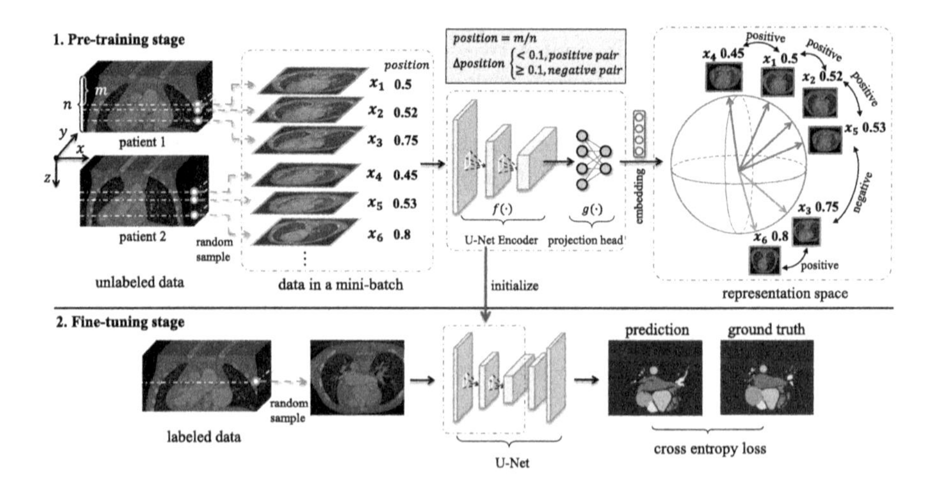

Fig. 1. Overview of the proposed PCL framework. In the pre-training stage, 2D slices (denoted as x_i) in the xy plane are extracted from volumetric medical images and fitted into a U-Net encoder for representation learning. The learned encoder is then used as initialization in the fine-tuning stage. We use *position* to denote the relative position of a slice along the z axis in a volume. Data pairs with small *position* difference (e.g., $\Delta position < 0.1$) are considered as positive pairs and those with large *position* difference are considered as negative pairs. Similar slices are marked/labeled with the same color.

3.2 Leveraging Structural Information in Medical Image

In medical images, similar anatomical structures often exist in all volumes of different patients across the dataset. In addition, we note the following two observations for volumetric medical images: 1) they have high spatial resolution along z axis so that adjacent 2D slices (e.g., x_1 and x_2 in Fig. 1) inside a volume usually have similar content; 2) if the volumes of different patients are perfectly aligned, the corresponding 2D slices in different volumes (e.g., x_2 and x_5 in Fig. 1) often contain similar anatomical information. In this paper, we utilize these two distinctive features in volumetric medical images to generate data pairs for contrastive learning.

To be specific, each 2D slice extracted from a volume is associated with a *position* variable. The *position*, which is between 0 and 1, represents the relative or normalized position of the slice along the z axis in the volume. Suppose m is the index of the 2D slice along the z axis and n is the total number of slices in the z axis (see Fig. 1). Then $position = m/n$. This allows the proper alignment between different volumes. Once each 2D slice in a mini-batch is assigned with its *position*, we can use the *position* difference to decide whether each data pair is similar or not. If the *position* difference of two slices is less than a threshold t (e.g., 0.1 in Fig. 1), they are likely to contain similar anatomical content and can be considered as positive pair. Otherwise, they are negative pair. The threshold t is a hyper-parameter that is different for different medical datasets. Note that this approach allows the positive and negative pairs to be formed on the fly instead of predefined such as in [2]. It is possible that (x_i, x_j) and (x_j, x_k) are positive pairs but (x_i, x_k) is a negative pair. We believe this can enforce the feature representation to be uniformly distributed on the representation hypersphere which may boost the contrastive learning performance [22].

As in [3,8], a pair of random transformations is applied for each sample in the mini-batch to help the encoder learn the spatial invariant feature of the target. The augmentations will not change the *position* value of the original sample, so our contrastive data pair generation strategy discussed above still works.

3.3 Contrastive Loss Function

Our contrastive learning loss function is based on [12]. For a set of N randomly sampled slices, $\{x_i\}_{i=1...N}$, the corresponding mini-batch consists of $2N$ samples after data augmentation, $\{\tilde{x}_i\}_{i=1...2N}$, in which $\tilde{x}_{2i}$ and $\tilde{x}_{2i-1}$ are two random augmentations of x_i. z_i represents the learned embedding of $\tilde{x}_i$. Then the loss function can be defined as:

$$\mathcal{L}^{PCL} = \sum_{i=1}^{2N} \mathcal{L}_i^{PCL}, \tag{1}$$

$$\mathcal{L}_i^{PCL} = -\frac{1}{|\Omega_i^+|} \sum_{j \in \Omega_i^+} log \frac{e^{sim(z_i, z_j)/\tau}}{\sum_{k=1}^{2N} \mathbb{1}_{i \neq k} \cdot e^{sim(z_i, z_k)/\tau}}. \tag{2}$$

where Ω_i^+ is the set of indices of positive samples to $\tilde{x}_i$. $sim(\cdot,\cdot)$ is the cosine similarity function that computes the similarity between two vectors in the representation space. τ is a temperature scaling parameter. Compared with the standard contrastive loss [3] that only has one positive pair on the numerator for any sample x_i, in Eq. 2 all positive pairs in a mini-batch (e.g., the augmented one and any of the remaining $2(N-2)$ samples whose *position* is close to x_i) contribute to the numerator, allowing better utilization of the proposed strategy.

4 Experiments and Results

Datasets: We evaluate the performance of the proposed PCL on four publicly available medical image datasets. **(1) The CHD dataset** is a CT dataset that consists of 68 3D cardiac images captured by a Simens biograph 64 machine [23]. The dataset covers 14 types of congenital heart disease and the segmentation labels include seven substructures: left ventricle (LV), right ventricle (RV), left atrium (LA), right atrium (RA), myocardium (Myo), aorta (Ao) and pulmonary artery (PA). **(2) The MMWHS dataset** was hosted in STACOM and MICCAI 2017 [24,25]. It consists of 20 cardiac CT and 20 MRI images and the annotations include the same seven substructures as the CHD dataset. **(3) The ACDC dataset** was hosted in MICCAI 2017 challenge [1]. The dataset has 100 patients with 3D cardiac MRI images. Each patient has around 15 volumes covering a full cardiac cycle, only volumes for the end-diastolic and end-systolic phase are labeled by an expert. The segmentation labels include three substructures: LV, RV, and Myo. **(4) The HVSMR dataset** was hosted in MICCAI 2016 challenge [16]. It has 10 3D cardiac MRI images captured in an axial view on a 1.5T scanner. Manual annotations of blood pool and Myo are provided.

Preprocessing: Following the work of [2], we first normalize the intensity of each 3D volume x to $[x_1, x_{99}]$, where x_p is the p-th intensity percentile in x. Then all 2D slices and the corresponding annotations are resampled to a fixed spatial resolution f_r and padded to a fixed image size f_s with 0. We do not apply cropping because it may remove important structure information in the original slice. The f_r and f_s for each dataset are defined as follows (1) CHD dataset: $f_r = 1.0 \times 1.0\,\text{mm}^2$ and $f_s = 512 \times 512$, (2) MMWHS dataset: $f_r = 1.0 \times 1.0\,\text{mm}^2$ and $f_s = 256 \times 256$, (3) ACDC dataset: $f_r = 1.25 \times 1.25\,\text{mm}^2$ and $f_s = 352 \times 352$, (4) HVSMR dataset: $f_r = 0.7 \times 0.7\,\text{mm}^2$ and $f_s = 352 \times 352$. No additional alignment technique is used for CHD and ACDC datasets because they are already roughly aligned as they are acquired.

4.1 Semi-supervised Learning

In this section, we test whether the proposed PCL can improve the performance in semi-supervised learning where contrastive learning and down-stream supervised learning (with limited annotation) are done on the same dataset.

Setup: We employ our PCL to pre-train a U-Net encoder on the whole CHD and ACDC, respectively, without using any human label. Note that for ACDC, each patient has more than 10 volumes covering a full cardiac cycle, only two of which have annotations. Since we do not need labels anyway, we use all the volumes from 100 patients for pre-training. Then the pre-trained model is used as the initialization to fine-tune a U-Net segmentation network with a small number of labeled samples on the same dataset. 5-fold cross-validation is used to evaluate the segmentation performance. Specifically, for each cross-validation fold on CHD, We randomly sample M patients from the 51 patients for fine-tuning, as if we only have the labels for these patients, and evaluate the results on the remaining 17 patients. We experiment with different values of M (e.g., 2, 6 and 10) to assess the influence of training set size in the fine-tuning stage on the contrastive learning performance. The same training strategy is also used for ACDC. We choose the threshold t to be 0.1 and 0.35 for CHD and ACDC because they have the best performance according to our experiment. The influence of thresholds on accuracy will be discussed in the supplementary. Data augmentations, including translation, rotation, and scale, are used in both the pre-training and fine-tuning stages. The pre-training is done on two NVIDIA GeForce GTX 1080 GPUs with 200 epochs. SGD is used as the optimizer and the learning rate is set to 0.1. We use cosine learning rate scheduler, batch size is set to 32. Temperature τ is set to 0.1 as in [3,8]. In the fine-tuning stage, we train the U-Net with cross-entropy loss for 100 epochs. The batch size is set to 5 and the learning rate is $5 \times e^{-5}$. Adam optimizer and cosine scheduler are used.

Table 1. Comparison of the proposed PCL method with baseline methods on CHD and ACDC. M is the number of patients used in the fine-tuning process. Results are reported in the form of mean (standard deviation) on 5-fold cross-validation. PCL provides better results than the baselines for all values of M.

			CHD (68 patients in total)				
Method	M=2	M=6	M=10	M=15	M=20	M=30	M=51
Random	0.184(.06)	0.508(.06)	0.584(.05)	0.627(.05)	0.658(.04)	0.693(.04)	0.754(.02)
Rotation [6]	0.171(.06)	0.488(.07)	0.575(.04)	0.625(.04)	0.651(.04)	0.691(.04)	0.749(.03)
PIRL [14]	0.196(.07)	0.504(.08)	0.617(.05)	0.658(.03)	0.674(.04)	0.714(.04)	0.761(.03)
SimCLR [3]	0.192(.06)	0.515(.06)	0.599(.06)	0.631(.05)	0.666(.05)	0.699(.05)	0.756(.03)
GCL [2]	0.255(.10)	0.564(.04)	0.646(.03)	0.669(.04)	0.697(.04)	0.725(.04)	0.766(.03)
PCL	**0.356(.08)**	**0.600(.06)**	**0.661(.05)**	**0.686(.05)**	**0.716(.04)**	**0.735(.05)**	**0.774(.03)**
			ACDC (100 patients in total)				
Method	M=2	M=6	M=10	M=15	M=20	M=30	M=80
Random	0.588(.07)	0.782(.03)	0.840(.03)	0.876(.01)	0.894(.01)	0.909(.01)	0.928(.00)
Rotation [6]	0.572(.08)	0.809(.03)	0.868(.02)	0.886(.01)	0.898(.01)	0.910(.01)	0.925(.00)
PIRL [14]	0.492(.03)	0.823(.04)	0.865(.01)	0.880(.02)	0.896(.02)	0.912(.01)	0.927(.00)
SimCLR [3]	0.352(.06)	0.725(.08)	0.824(.04)	0.869(.02)	0.894(.01)	0.913(.01)	0.927(.00)
GCL [2]	0.636(.05)	0.803(.04)	0.872(.01)	0.891(.01)	0.902(.01)	0.913(.01)	0.927(.01)
PCL	**0.671(.06)**	**0.850(.01)**	**0.885(.01)**	**0.904(.01)**	**0.909(.01)**	**0.919(.00)**	**0.929(.00)**

Baselines: We compare the performance of PCL with a random approach that does not use any pre-training as well as the following state-of-the-art baselines, all of which use the same unlabeled dataset in the pre-training and labeled dataset in the fine-tuning as PCL: (1) Rotation [6]: a pretext-based method that uses image rotation prediction to pre-train the encoder; (2) PIRL [14]: adopted from a contrastive learning scheme for natural image classification, which uses contrastive loss to learn pretext-invariant representations. (3) SimCLR [3]: adopted from another contrastive learning scheme for natural image classification, which constructs positive pairs for each sample only using two random augmentations; (4) GCL [2]: a contrastive learning scheme for 3D medical image segmentation which divides each volume into four partitions so that slices belonging to the same partition in different volumes are considered as positive pairs.

Results and Analysis: The results of the comparative study on both CHD and ACDC are shown in Table 1. We report the averaging Dice of 5-fold cross-validation results. From the table, we have the following observations. (1) Comparing PCL and GCL with other baselines, we can see that the performance improves

Table 2. Transfer learning comparison of the proposed PCL method with the baselines. Except for Random, all the methods are pre-trained on CHD and ACDC without labels and fine-tuned on MMWHS and HVSMR respectively.

CHD transferring to MMWHS (20 patients in total)						
Method	M=2	M=4	M=6	M=8	M=10	M=16
Random	0.232(.14)	0.661(.10)	0.732(.07)	0.769(.06)	0.808(.05)	0.834(.05)
Rotation [6]	0.247(.16)	0.659(.13)	0.751(.07)	0.768(.07)	0.803(.06)	0.850(.04)
PIRL [14]	0.251(.10)	0.670(.11)	0.755(.07)	0.774(.06)	0.821(.05)	0.851(.04)
SimCLR [3]	0.269(.17)	0.683(.10)	0.751(.07)	0.783(.06)	0.818(.05)	0.850(.04)
GCL [2]	0.262(.11)	0.703(.07)	0.768(.05)	0.805(.04)	0.820(.04)	0.851(.03)
PCL	**0.339(.15)**	**0.748(.08)**	**0.792(.05)**	**0.820(.05)**	**0.840(.04)**	**0.869(.03)**

ACDC transferring to HVSMR (10 patients in total)				
Method	M=2	M=4	M=6	M=8
Random	0.742(.06)	0.813(.05)	0.842(.03)	0.842(.04)
Rotation [6]	0.737(.07)	0.816(.06)	0.845(.03)	0.844(.03)
PIRL [14]	0.740(.05)	0.826(.04)	0.849(.03)	0.846(.03)
SimCLR [3]	0.700(.07)	0.779(.05)	0.808(.04)	0.815(.04)
GCL [2]	0.770(.05)	0.818(.05)	0.842(.03)	0.843(.03)
PCL	**0.781(.05)**	**0.832(.05)**	**0.857(.03)**	**0.857(.03)**

significantly ($\Delta Dice > 0.1$) in many settings for both CHD and ACDC, suggesting that by leveraging domain-specific structural information in volumetric medical images, the encoder can learn better task-related representation for segmentation. (2) The performance improvement of PCL and GCL are especially high when a very small number of training samples are used (e.g., 2 and 4). The gains become lesser when the number of training samples increases. This is because with more training samples, the information difference between the training set for fine-tuning and the training set for contrastive learning becomes small and the fine-tuning performance saturates. (3) SimCLR performs worse than Random on ACDC. This suggests that only using data augmentations to generate contrastive data pairs may lead to a large false negative rate for datasets like ACDC where the volumes have small z dimensions (around 10). (4) PCL performs better than GCL in all settings. The improvement in Dice can be up to 0.04. This shows that using the relative *position* difference instead of a hard partition strategy can better utilize the structural information in medical images and reduce false negatives to improve contrastive learning performance.

4.2 Transfer Learning

To assess whether the learned representations by PCL are transferrable, we use the encoder pre-trained on CHD and ACDC without labels as the initialization of a U-Net to fine-tune on MMWHS and HVSMR datasets respectively. The experiment setup and baselines are the same as in Sect. 4.1. Table 2 shows the comparison results. It can be seen that the proposed PCL framework outperforms all baselines on both datasets. The overall improvement on HVSMR is relatively smaller than MMWHS. This is because MMWHS is very similar to CHD which makes the features learned on CHD more helpful on MMWHS. On the other hand, ACDC and HVSMR are different in terms of acquisition view and image resolution, which limits the transfer learning performance. Visualization of the segmentation results on all datasets is shown in the supplementary.

5 Conclusion

In this paper, we propose a novel PCL framework for representation learning in volumetric medical images. The framework can effectively eliminate false negative pairs in existing contrastive learning methods for medical image segmentation. Experimental results on four 3D medical image datasets show that PCL significantly improves the segmentation performance in both semi-supervised setting and transfer learning setting.

Acknowledgements. This work is partially supported by NSF award IIS-2039538.

References

1. Bernard, O., et al.: Deep learning techniques for automatic MRI cardiac multi-structures segmentation and diagnosis: is the problem solved? IEEE Trans. Med. Imaging **37**(11), 2514–2525 (2018)
2. Chaitanya, K., Erdil, E., Karani, N., Konukoglu, E.: Contrastive learning of global and local features for medical image segmentation with limited annotations. In: Advances in Neural Information Processing Systems 33 (2020)
3. Chen, T., Kornblith, S., Norouzi, M., Hinton, G.: A simple framework for contrastive learning of visual representations. In: International Conference on Machine Learning, pp. 1597–1607. PMLR (2020)
4. Chen, T., Kornblith, S., Swersky, K., Norouzi, M., Hinton, G.: Big self-supervised models are strong semi-supervised learners. arXiv preprint arXiv:2006.10029 (2020)
5. Chen, X., Fan, H., Girshick, R., He, K.: Improved baselines with momentum contrastive learning. arXiv preprint arXiv:2003.04297 (2020)
6. Gidaris, S., Singh, P., Komodakis, N.: Unsupervised representation learning by predicting image rotations. In: International Conference on Learning Representations (2018)
7. Hadsell, R., Chopra, S., LeCun, Y.: Dimensionality reduction by learning an invariant mapping. In: 2006 IEEE Computer Society Conference on Computer Vision and Pattern Recognition (CVPR 2006), vol. 2, pp. 1735–1742. IEEE (2006)
8. He, K., Fan, H., Wu, Y., Xie, S., Girshick, R.: Momentum contrast for unsupervised visual representation learning. In: Proceedings of the IEEE/CVF Conference on Computer Vision and Pattern Recognition, pp. 9729–9738 (2020)
9. Isensee, F., Jaeger, P.F., Full, P.M., Wolf, I., Engelhardt, S., Maier-Hein, K.H.: Automatic cardiac disease assessment on cine-MRI via time-series segmentation and domain specific features. In: Pop, M., et al. (eds.) STACOM 2017. LNCS, vol. 10663, pp. 120–129. Springer, Cham (2018). https://doi.org/10.1007/978-3-319-75541-0_13

10. Isensee, F., et al.: nnU-Net: self-adapting framework for u-net-based medical image segmentation. arXiv preprint arXiv:1809.10486 (2018)
11. Jiao, J., Cai, Y., Alsharid, M., Drukker, L., Papageorghiou, A.T., Noble, J.A.: Self-supervised contrastive video-speech representation learning for ultrasound. In: Martel, A.L., et al. (eds.) MICCAI 2020. LNCS, vol. 12263, pp. 534–543. Springer, Cham (2020). https://doi.org/10.1007/978-3-030-59716-0_51
12. Khosla, P., et al.: Supervised contrastive learning. In: Advances in Neural Information Processing Systems 33 (2020)
13. Li, H., et al.: Contrastive rendering for ultrasound image segmentation. In: Martel, A.L., et al. (eds.) MICCAI 2020. LNCS, vol. 12263, pp. 563–572. Springer, Cham (2020). https://doi.org/10.1007/978-3-030-59716-0_54
14. Misra, I., Maaten, L.v.d.: Self-supervised learning of pretext-invariant representations. In: Proceedings of the IEEE/CVF Conference on Computer Vision and Pattern Recognition, pp. 6707–6717 (2020)
15. Nemoto, T., et al.: Efficacy evaluation of 2D, 3D U-Net semantic segmentation and atlas-based segmentation of normal lungs excluding the trachea and main bronchi. J. Radiat. Res. **61**(2), 257–264 (2020)
16. Pace, D.F., Dalca, A.V., Geva, T., Powell, A.J., Moghari, M.H., Golland, P.: Interactive whole-heart segmentation in congenital heart disease. In: Navab, N., Hornegger, J., Wells, W.M., Frangi, A.F. (eds.) MICCAI 2015. LNCS, vol. 9351, pp. 80–88. Springer, Cham (2015). https://doi.org/10.1007/978-3-319-24574-4_10
17. Ronneberger, O., Fischer, P., Brox, T.: U-Net: convolutional networks for biomedical image segmentation. In: Navab, N., Hornegger, J., Wells, W.M., Frangi, A.F. (eds.) MICCAI 2015. LNCS, vol. 9351, pp. 234–241. Springer, Cham (2015). https://doi.org/10.1007/978-3-319-24574-4_28
18. Tian, Y., Krishnan, D., Isola, P.: Contrastive multiview coding. arXiv preprint arXiv:1906.05849 (2019)
19. Ushinsky, A., et al.: A 3D–2D hybrid U-Net convolutional neural network approach to prostate organ segmentation of multiparametric MRI. Am. J. Roentgenol. **216**(1), 111–116 (2021)
20. Wang, T., et al.: MSU-Net: multiscale statistical U-Net for real-time 3D cardiac MRI video segmentation. In: Shen, D., et al. (eds.) MICCAI 2019. LNCS, vol. 11765, pp. 614–622. Springer, Cham (2019). https://doi.org/10.1007/978-3-030-32245-8_68
21. Wang, T., et al.: ICA-UNet: ICA inspired statistical UNet for real-time 3D cardiac cine MRI segmentation. In: Martel, A.L., et al. (eds.) MICCAI 2020. LNCS, vol. 12266, pp. 447–457. Springer, Cham (2020). https://doi.org/10.1007/978-3-030-59725-2_43
22. Wang, T., Isola, P.: Understanding contrastive representation learning through alignment and uniformity on the hypersphere. In: International Conference on Machine Learning, pp. 9929–9939. PMLR (2020)
23. Xu, X., et al.: Whole heart and great vessel segmentation in congenital heart disease using deep neural networks and graph matching. In: Shen, D., et al. (eds.) MICCAI 2019. LNCS, vol. 11765, pp. 477–485. Springer, Cham (2019). https://doi.org/10.1007/978-3-030-32245-8_53
24. Zhuang, X.: Challenges and methodologies of fully automatic whole heart segmentation: a review. J. Healthc. Eng. **4**(3), 371–407 (2013)
25. Zhuang, X., Shen, J.: Multi-scale patch and multi-modality atlases for whole heart segmentation of MRI. Med. Image Anal. **31**, 77–87 (2016)

Longitudinal Self-supervision to Disentangle Inter-patient Variability from Disease Progression

Raphaël Couronné(✉), Paul Vernhet, and Stanley Durrleman

Inria - Aramis Project-Team, Sorbonne Université, Institut du Cerveau - Paris Brain Institute - ICM, Inserm, CNRS, AP-HP, Hôpital de la Pitié Salpêtrière, Paris, France

Abstract. The problem of building disease progression models with longitudinal data has long been addressed with parametric mixed-effect models. They provide interpretable models at the cost of modeling assumptions on the progression profiles and their variability across subjects. Their deep learning counterparts, on the other hand, strive on flexible data-driven modeling, and additional interpretability - or, as far as generative models are involved, *disentanglement* of latent variables with respect to generative factors - comes from additional constraints. In this work, we propose a deep longitudinal model designed to disentangle inter-patient variability from an estimated disease progression timeline. We do not seek for an explicit mapping between age and disease stage, but to learn the latter solely from the ordering between visits using a differentiable ranking loss. Furthermore, we encourage inter-patient variability to be encoded in a separate latent space, where for each patient a single representation is learned from its set of visits, with a constraint of invariance under permutation of the visits. The modularity of the network architecture allows us to apply our model on various data types: a synthetic image dataset with known generative factors, cognitive assessments and neuroimaging data. We show that, combined with our patient encoder, the ranking loss for visits helps to exceed models with supervision, in particular in terms of disease staging.

Keywords: Longitudinal model · Disentanglement · Medical imaging

1 Introduction

Understanding the progression of diseases is essential for accurate early diagnosis, prognosis, and patient monitoring. Often, there is a strong interplay between

R. Couronné and P. Vernhet—Equal contribution.

Electronic supplementary material The online version of this chapter (https://doi.org/10.1007/978-3-030-87196-3_22) contains supplementary material, which is available to authorized users.

M. de Bruijne et al. (Eds.): MICCAI 2021, LNCS 12902, pp. 231–241, 2021.
https://doi.org/10.1007/978-3-030-87196-3_22

the pathological progression and the inter-subject variability, which makes it all the more necessary to characterize the contribution of each factor. Typically, in the context of neurodegenerative diseases, we may ask whether the atrophy of a particular brain region is predictive of a specific patient advancement in the disease, or rather can be dismissed as a specific characteristic of the individual.

Longitudinal data analysis has been usually addressed in the framework of parametric mixed-effect models. For instance, geometric approaches have been proposed either for the progression of biomarkers [21] or shape changes [4]. This family of models assumes that each subject follows a curve on a Riemannian manifold which translates from a common geodesic. They also assume that the direction of translation is orthogonal to the direction of the progression curve, which ensures that the changes due to the progression of the disease are disentangled from the effects of different physiological or anatomical characteristics of the patient. The family of progression profiles is constrained, e.g. sigmoid curves for biomarkers changes, and an affine function maps the age of the subject to a disease stage.

Generative models such as variational auto-encoders (VAE) [13] have been consistently used in deep learning as they offer a flexible learning framework, in which disentanglement may be enforced through soft constraints and optimization schemes, as in β-VAE and their extensions [10,12,18]. With time series, however, separating static and dynamic representations without inductive bias still remains a challenge. In most research works, authors disentangle time-varying from time-invariant information by leveraging time labels explicitly: in literature focused on style and content of videos [9,15], in face ageing progression [11] or medical data, where age is used for supervision [19,22]. These previous methods are not directly transferable to longitudinal data where duration between visits differs. In [1], authors seek an age direction in the latent *a posteriori*, while in [24] they estimate the latent age regression jointly with the reconstruction task in a supervised fashion. A Riemannian manifold learning point of view, in the spirit of parametric models is proposed in [17] as it estimates both a static representation and an affine time reparametrization per patient. All these methods assume that age at observation is a direct marker for the progression timeline, which is not the case for most neurodegenerative disorders. In the recent work closest to ours [25], the authors propose to learn the disease stage without relying on the patient age, in a self-supervised fashion. They use a cosine loss to enforce progression in a specific direction of the latent space, learned during optimization. They do not study the disentanglement of their model but rather focus on the correlation with a disease progression timeline.

In this paper, we propose a generic deep longitudinal model, designed to disentangle inter-patient variability from an estimated disease progression timeline. We learn a disease stage as a flexible function that does not rely on age, but solely on the individual order between visits using a differentiable ranking loss, leveraging a much weaker prior. The remaining latent space is further favored to produce representations independent of the progression thanks to a DeepSet network which acts as a permutation invariance function on visits. The main contributions of this paper are therefore (*i*) an architecture that is tailored to disease progression modeling and disentangles the changes due to progression from the changes due to phenotypic differences across subjects; (*ii*) a modular method with decoders adapted to data types; (*iii*) an application on synthetic and real datasets - including imaging and clinical data - showing that one direction of the latent space alone describes temporal progression.

2 Methodology

The proposed generic deep longitudinal model is summarized in Fig. 1.

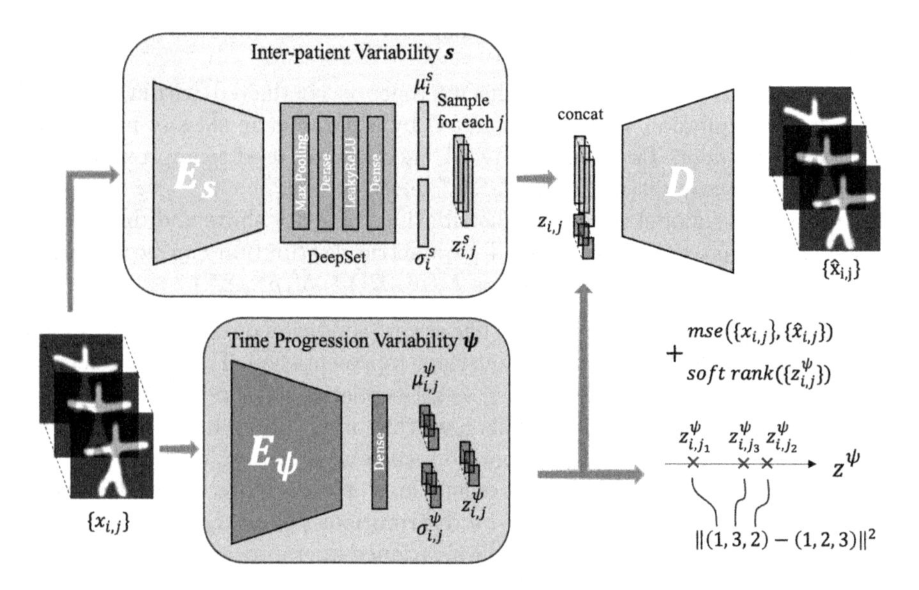

Fig. 1. Input data $\boldsymbol{x}_i = \{x_{i,j}, \forall j \in [1, n_i]\}$ is encoded simultaneously in a space encoder (Deepset) and a point-wise time encoder to get respectively (μ^s_i, σ^s_i) and $(\mu^\psi_{i,j}, \sigma^\psi_{i,j})$. (μ^s_i, σ^s_i) can be computed from *any* subset of visits, and in practice randomized fixed-size subsets of visits are drawn in the spirit of stochastic optimization. Note that we sample a $z^s_{i,j}$ per visit, but z^s_i could be sampled once for a patient. Decoder can be either agnostic, or specific (e.g., velocity fields for deformations).

2.1 Longitudinal Progression Model

In this section, we propose a temporal latent variable model that encodes the disease progression in a low-dimensional probabilistic space. It assumes that a sequence of observations is generated as the combination of an intrinsic code z^s (as in *space shift*) and a disease progression factor z^ψ (where we use ψ instead of t to clearly distinguish the stage from the temporality of visits).

Generative Disease Progression Model. Let $\{(t_{i,j}, x_{i,j})\}_{1\leq i\leq N}$ be a set of N subjects, each observed at the age of $t_{i,j}$ for $1 \leq j \leq n_i$ visits. We assume that the patient i's observations $\boldsymbol{x_i} = \{x_{i,j}, \forall j \in [1, n_i]\}$ are generated from a Bayesian generative model as follows:

$$x_{i,j} \overset{\text{iid}}{\sim} \mathcal{N}\Big\{\Phi\big(z_i^s,\ z_{i,j}^{\psi}\big);\ \epsilon^2\mathbb{I}\Big\} \text{ with } z_i^s \overset{\text{iid}}{\sim} \mathcal{N}(0, \lambda_s^2\mathbb{I}) \text{ and } z_{i,j}^{\psi} \overset{\text{iid}}{\sim} \mathcal{N}(0, \lambda_{\psi}^2) \quad (1)$$

Here Φ denotes an unknown non-linear transform from the strongly decoupled generative factor space, also called latent space, $\mathcal{Z} = \mathcal{Z}^s \times \mathcal{Z}^{\psi}$, towards our observation space of scores or images $\mathcal{X}$. The generative factor z^{ψ} is assumed to be independent from the individual variability z^s factor. Notice that we do not assume any relationship between the ages t_{ij} and the associated observations x_{ij}. Parameters ϵ^2, λ_{ψ}^2 and λ_s^2 are the diagonal Gaussian variance priors.

Variational Inference with VAE. The inference is conducted within the VAE paradigm. The function Φ is approximated by a parametric class of neural network Φ_{θ}, the decoder. Two neural network encoders are used to approximate the intractable posterior distribution $p(z_i^s, \boldsymbol{z_i^{\psi}}|\boldsymbol{x_i};\theta)$ with $\boldsymbol{z_i^{\psi}} = \{z_{i,j}^{\psi}, \forall j \in [1, n_i]\}$. They respectively model the latent distributions of space shifts and disease progression, such that the approximated parametric distribution can be factorized as a product of independent Gaussians $\mathcal{N}(\mu^s, \Sigma^s) \otimes \mathcal{N}(\mu^{\psi}, \Sigma^{\psi})$.

Set-Invariant Representation for $\mathcal{Z}^s$. The strong condition on $\mathcal{Z}^s$ is that it should extract from any time-series a time-invariant representation. To do so, we choose to learn the posterior $q_{\eta^s} \equiv \mathcal{N}(\mu^s, \Sigma^s)$ as a DeepSet encoder network [23] acting on any unordered subset of visits. This rewriting of our disentanglement hypothesis generalizes the use of simple operators such as averaging or maxing out of visits in an intermediate latent representation, or even more elaborate inverse Gaussian product of group-wise non-*iid* distributions [5]; all these can indeed be cast as specific choices of permutation-invariance operators, which were shown in [23] to be universally approximated by DeepSets.

Ranking Visits in $\mathcal{Z}^{\psi}$ as a Regularization Constraint. The remaining generative factor of our one-dimensional $\mathcal{Z}^{\psi}$ space must encode the dynamic of the progression preferentially. A disease progression constraint, $\mathcal{C}^{\text{ranking}}$, aims at favoring a natural ordering of visits along the temporal latent dimension in a self-supervised way. Unlike in [25], it builds upon the soft-ranking differentiable loss of [3] to enforce monotonic individual progression. The individual ranking errors of visits are penalized according to $\sum_j ||r(z_j^{\psi}) - j||_2^2$. Note that this loss depends on the number of visits of patient i, and may be rescaled accordingly (with e.g. the expectation assuming random ordering). We found it more practical to choose a fixed number of 3 visits per patient at each iteration. The stochastic gradient descent randomly selects these visits at each iteration. From a theoretical perspective, the minimization of this loss is similar to maximizing the Spearman correlation between different visits of a given subject i: it is therefore to be understood as a soft supervision which only relies on the ordering of visits and not on the times of observations, which are never seen by the model.

Final Objective. The final loss can be written as the sum of two terms: the evidence lower bound, written as the sum of the KL-divergence $\mathrm{KL}[q_\eta(\boldsymbol{z}_i^\psi, z_i^s|x_i)||p(\boldsymbol{z}_i)]$ and the data attachment term $-\mathbb{E}[\log p_\theta(\boldsymbol{x}_i|\boldsymbol{z}_i)]$ which is proportional to the ℓ^2 reconstruction loss, and the self-supervised ranking of visits $\mathcal{C}^{\text{ranking}}$ discussed previously with a weight γ to cross-validate (in practice we choose $\gamma = 0.1$):

$$L = \sum_{i=1}^{N} \mathrm{KL}[q_\eta(\boldsymbol{z}_i^\psi, z_i^s|\boldsymbol{x}_i)||p(\boldsymbol{z}_i)] - \textstyle\sum_{j=1}^{n_i} \mathbb{E}[\log p_\theta(x_{i,j}|z_{i,j})] + \gamma \mathcal{C}_i^{\text{ranking}} \quad (2)$$

2.2 Modularity

The model we propose in this paper can be seen as a "*meta*" architecture that can be instantiated according to the datatype: clinical data (1D), images (2D and 3D). We may even use decoders that are specifically designed for data, such as diffeomorphometry for brain grey matter. We detail the latter case, directly in the spirit of classical models with stationary velocity fields [7,14]. An additional parameter, a template $\mathcal{T}$, is learned at the centered reference disease stage $z^\psi = 0$. From this, any observation x_{ij} can be reconstructed from the latent code $(z_{i,j}^\psi, z_i^s)$ by a deformation field Φ_v, parametrized with a velocity decoder v, acting on $\mathcal{T}$.

3 Experimental Results

The network E_ψ is a classical CNN encoder with LeakyReLU non-linearities and a final dense layer toward $\mathcal{Z}^\psi$. The spatial encoder E_s is composed as a DeepSet, whose output does not depend upon the ordering of its inputs: it can be written as $\rho \circ \max_{j \sim \text{visits}}(f)$. We chose f to be a convolutional encoder network which outputs an intermediate representation (per observation), from which the max operator (over visits) retrieves a permutation invariant code. The latter is eventually mapped, *via* a MLP ρ, into the space shifts domain $\mathcal{Z}^s$. We choose a latent space dimension of $p = 5$ for Starmen and ADNI cognitive scores, and of $p = 8$ for ADNI MRIs. Inference was performed using Adam optimizer from PyTorch library version 1.7 with a learning rate of 0.01 and a batch size of 32. To promote fair comparison between the different models of the benchmark, the same architecture were used for encoders, as well as decoders.

3.1 Validation on Synthetic Data

To validate the disentangling ability of our model, we first generated a synthetic longitudinal dataset of starmen images, based on the longitudinal diffeomorphic model of [6]. From a given reference template y_0, the cross-sectional variability of our population is prescribed by a diffeomorphism localized at four control points: the head, right arm and legs. The common progression timeline, on the other hand, is generated through a displacement of the left arm only.

The dynamics of progression is given by an affine reparametrization of the age t_{ij} at visit j, characterized by individual onset τ_i and acceleration α_i factors, such that the true disease progression is given by $\psi^*_{ij} = t_0 + \alpha_i(t_{ij} - \tau_i - t_0)$. We sample variables in a similar fashion as in [6] to obtain a dataset of $N = 1000$ subjects, each with $n = 10$ visits (Fig. 2).

Table 1. Benchmark of proposed methods on Starmen dataset

Metric	β-VAE	ML-VAE	LR-AE	AR-VAE	LSSL	Ours	Ours (wD)	Ours (woR)
MSE (10^{-3})	7.90	22.7	10.9	8.26	7.32	8.83	**6.22**	14.2
	± 0.57	± 1.51	± 1.53	± 0.62	± 0.379	± 0.88	± 1.23	± 5.46
PLS z^ψ/z^s	-	0.660	0.137	0.125	0.098	**0.083**	0.083	0.149
	-	± 0.343	± 0.209	± 0.117	± 0.047	± 0.026	± 0.025	± 0.131
Staging ψ^*	0.263	0.030	0.971	0.984	0.994	**0.997**	0.996	0.524
	± 0.348	± 0.028	± 0.024	± 0.008	± 0.003	± 0.001	± 0.002	± 0.464

We benchmark contender approaches mentioned in the introduction on the Starmen dataset (see Table 1): β-VAE [10], ML-VAE [5], LSSL [25] and both supervised "Longitudinal Riemannian VAE" (LR-AE) [16] and "Age Regression VAE" (AR-VAE) [8]. We compare with our model in its generic iconic form on pixels and its diffeomorphic version (wD). The version without ranking loss (woR) is added for ablation purposes. Beyond the reconstruction quality (Mean Square Error), which reveals only LSSL and our model are below the baseline of β-VAE, we are interested in disentanglement capacity. It can be measured by correlations between the estimated staging z^ψ and the latent space code z^s with a partial least square regression analysis (PLS), so as to ensure the independence of $\mathcal{Z}^s \times \mathcal{Z}^\psi$ (2nd row). Among all methods, ours performs best: even though LSSL uses a similar loss, it does not constrain its orthogonal directions to be independent. Other methods, especially supervised, naturally learn correlated representations.

Fig. 2. Each row represents a synthetic subject across time.

Finally, the proper staging of ψ^* is evaluated by computing the Spearman ranking correlation between ψ^* and z^ψ: it evaluates the monotonicity of individual trajectories. Only methods with time supervision or ranking strategy (*ie* all but first two columns) manage to grasp a staging close to one (row 3).

In complement to the previous metrics, we visualized in Fig. 3 the effect of specific directions in the latent space *via* gradient maps. The β-VAE is a low baseline as it does not model the progression and only views the data as static representations: we plot the PCA in the whole space, and observe that no principal direction correlates with the left-arm progression. It is interesting to note that the benefit of the ranking loss (when no supervision is available) is made clear by the study of our model without it (woR), and ML-VAE. They both focus on group-structure only, and fail to grasp progression. LR-AE and AR-AE, because they use supervision of time, are displaying time-related correlations in a space shift direction (last row).

3.2 Application to Alzheimer's Disease

Cognitive Scores. We apply our model on four subtest scores of the ADAS-Cog scale obtained from the ADNI dataset, namely concentration, praxis, memory and language; normalized between 0 and 1, with higher values indicating lower performance. 248 patients with mild cognitive impairment (MCI) converting to Alzheimer's disease (AD) during the study are followed for an average of 6 visits over 3.5 years.

MSE (10^{-3}) on 5-fold cross validation yields 7.47 ± 0.778 for our model: slightly less than β-VAE (3.78 ± 0.562) and LSSL (3.64 ± 0.429), but on par with a parametric model of reference, Leaspy [21] (8.21 ± 0.155). Additionally, we can predict future visits from previous ones: in this scenario, our model reaches a lower MSE (10^{-3}) 29.1 ± 5.53 than LSSL 32.4 ± 5.93.

Figure 4 illustrates the estimated average time progression for each model, as well as the effects of orthogonal directions in the latent space. An agnostic β-VAE fails at extracting a consistent dimension for the time progression while still providing a good reconstruction of data. Our proposed model learns monotonicity in an unsupervised fashion, and is able to estimate a consistent time progression of the ADAS-Cog scores from longitudinal measurement in small number. Interestingly, despite the noisiness of cognitive scores, our model generates a progression very similar to that of Leaspy' sigmoid geodesics: in particular, the first PCA directions have the same effects on scores (cf degraded color effects).

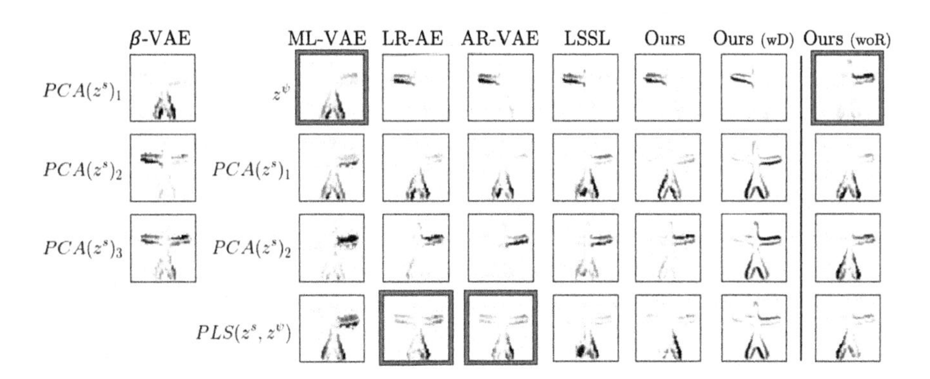

Fig. 3. Gradient directions in the latent space (extracted from a forward pass in the decoder). Row 1: gradient wrt to the latent space associated with disease progression. Rows 2 and 3: first two principal directions of the PCA in the orthogonal of the latent time ($\mathcal{Z}^s$ for us). 4th row: the direction in the orthogonal of the latent time that correlates most with it (PLS), as a way to challenge the model disentanglement.

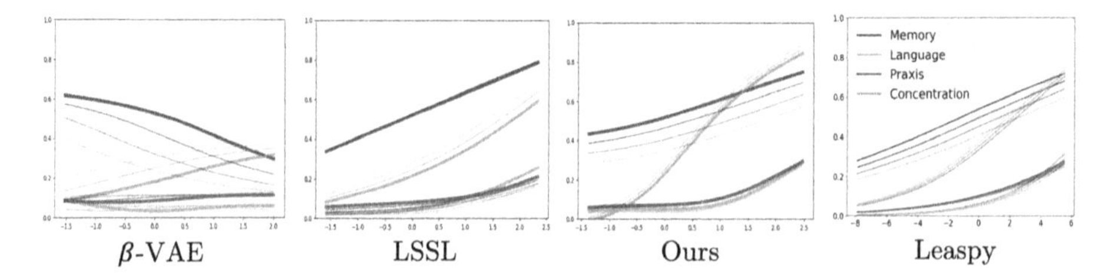

Fig. 4. Estimated average trajectory of scores. The effects of latent dimensions z^s (resp. z with β-VAE) are shown with degraded colors.

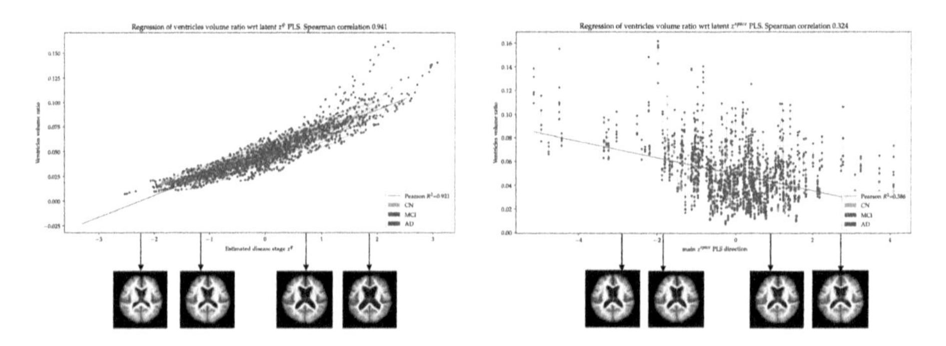

Fig. 5. PLS analysis with respect to ventricle volume ratio $\mathcal{V}$: z^ψ (left), z^s (right).

Neuroimaging Data. We selected 356 subjects MCI converters from ADNI, with a total of 1898 visits. The 1898 T1-weighted MRI were preprocessed using Clinica [20] (non-rigid alignment, skull stripping, intensity rescaling), and converted into mid axial slices of dimension 128×128.

We do not have access to the true disease progression, however the disease severity at the MCI stage can be monitored through cognitive scores (ADAS-Cog) or markers of the morphological evolution such as atrophy of the hippocampi and increase of ventricle volumes. We computed the ratio of ventricle volumes by brain volumes, as a covariate factor, noted $\mathcal{V}$, which we assimilate with a good proxy of the disease progression.

Figure 5 shows interactions between our two latent spaces and the disease stage proxy $\mathcal{V}$ *via* a correlation (PLS) analysis. First, z^ψ exhibits a quasi-linear regression fit associated with a high Spearman ranking (0.934 ± 0.025), suggesting that z^ψ has indeed captured the main disease progression trend $\mathcal{V}$ and the individual ordering of visits. On the other hand, the best correlated direction (measured as the PLS main direction) between z^s and $\mathcal{V}$ is not localized on the ventricle. This result further implies that $\mathcal{Z}^s$ has indeed captured the variability necessary to perform a good fit of the data without correlating significantly with the disease stage marker.

Furthermore, z_ψ behaves as a clinical score informative of the onset: its distribution is significantly earlier for women ($p < 3.83e^{-2} \pm 7.05e^{-2}$ for Mann-Whitney U test), as observed from clinical data. Its derivative $\frac{\partial z_\psi}{\partial t}$ correlates with the pace of the disease: it is significantly ($p < 4.82e^{-2} \pm 1.06e^{-2}$) faster for APOE4 carriers (1 or 2 alleles), a result in accordance with well-documented disease progression patterns in AD [2].

4 Conclusion

In this paper, we proposed a generative variational autoencoder architecture that leverages the repetition of measurements per individual to disentangle between the global disease timeline and inter-patient variability. The one-dimensional time variability is captured by a differentiable ranking loss, while a permutation invariant function reduces the remaining information in a representative space. As we further demonstrate, inductive biases on the data itself (such as using diffeomorphometry for structural medical imaging) are completely synergetic and improve the quality of the representations learned. A very interesting avenue would be to extend our method to account for non-monotonic progression priors.

Acknowledgements. This work has been partly funded by the European Research Council (ERC) under Grant Agreement No. 678304, European Union's Horizon 2020 research and innovation program under Grant Agreement No. 826421 (TVB-Cloud), and the program "Investissements d'avenir" ANR-10-IAIHU-06 (IHU-A-ICM) and ANR-19-P3IA-0001 (PRAIRIE 3IA Institute).

References

1. Berchuck, S.I., Mukherjee, S., Medeiros, F.A.: Estimating rates of progression and predicting future visual fields in glaucoma using a deep variational autoencoder. Sci. Rep. **9**(1), 18113 (2019)
2. Bigio, E., Hynan, L., Sontag, E., Satumtira, S., White, C.: Synapse loss is greater in presenile than senile onset Alzheimer disease: implications for the cognitive reserve hypothesis. Neuropathol. Appl. Neurobiol. **28**(3), 218–227 (2002)
3. Blondel, M., Teboul, O., Berthet, Q., Djolonga, J.: Fast differentiable sorting and ranking. In: International Conference on Machine Learning, pp. 950–959. PMLR (2020). ISSN 2640–3498
4. Bône, A., Louis, M., Martin, B., Durrleman, S.: Deformetrica 4: an open-source software for statistical shape analysis. In: ShapeMI @ MICCAI 2018, Granada, Spain, November 2018 (2018). https://hal.inria.fr/hal-01874752
5. Bouchacourt, D., Tomioka, R., Nowozin, S.: Multi-level variational autoencoder: learning disentangled representations from grouped observations. CoRR abs/1705.08841 (2017)
6. Bône, A., Colliot, O., Durrleman, S.: Learning distributions of shape trajectories from longitudinal datasets: a hierarchical model on a manifold of diffeomorphisms, pp. 9271–9280 (2018)

7. Dalca, A.V., Rakic, M., Guttag, J., Sabuncu, M.R.: Learning conditional deformable templates with convolutional networks. arXiv:1908.02738 [cs, eess] (2019). arXiv: 1908.02738
8. Gao, L., Pan, H., Liu, F., Xie, X., Zhang, Z., Han, J.: Brain disease diagnosis using deep learning features from longitudinal MR images. In: Cai, Y., Ishikawa, Y., Xu, J. (eds.) Asia-Pacific Web (APWeb) and Web-Age Information Management (WAIM) Joint International Conference on Web and Big Data. LNCS, vol. 10987, pp. 327–339. Springer, Cham (2018). https://doi.org/10.1007/978-3-319-96890-2_27
9. Grathwohl, W., Wilson, A.: Disentangling space and time in video with hierarchical variational auto-encoders. arXiv preprint arXiv:1612.04440 (2016)
10. Higgins, I., et al.: beta-VAE: learning basic visual concepts with a constrained variational framework (2016)
11. Hsu, W.N., Zhang, Y., Glass, J.: Unsupervised learning of disentangled and interpretable representations from sequential data. Adv. Neural Inf. Process. Syst. **30**, 1878–1889 (2017)
12. Kim, H., Mnih, A.: Disentangling by factorising. arXiv:1802.05983 [cs, stat] (2019). arXiv: 1802.05983
13. Kingma, D.P., Welling, M.: Auto-encoding variational bayes. arXiv preprint arXiv:1312.6114 (2013)
14. Krebs, J., Delingette, H., Ayache, N., Mansi, T.: Learning a generative motion model from image sequences based on a latent motion matrix. arXiv:2011.01741 [cs] (2020). arXiv: 2011.01741
15. Li, Y., Mandt, S.: Disentangled sequential autoencoder. arXiv preprint arXiv:1803.02991 (2018)
16. Louis, M., Charlier, B., Durrleman, S.: Geodesic discriminant analysis for manifold-valued data. In: Proceedings of the IEEE Conference on Computer Vision and Pattern Recognition Workshops, pp. 332–340 (2018)
17. Louis, M., Couronné, R., Koval, I., Charlier, B., Durrleman, S.: Riemannian geometry learning for disease progression modelling. In: Chung, A., Gee, J., Yushkevich, P., Bao, S. (eds.) Information Processing in Medical Imaging. IPMI 2019. LNCS, vol. 11492, pp. 542–553. Springer, Cham (2019). https://doi.org/10.1007/978-3-030-20351-1_42
18. Mathieu, E., Rainforth, T., Siddharth, N., Teh, Y.W.: Disentangling disentanglement in variational autoencoders. In: International Conference on Machine Learning, pp. 4402–4412. PMLR (2019)
19. Ravi, D., Alexander, D.C., Oxtoby, N.P.: Degenerative adversarial neuroimage nets: Generating images that mimic disease progression. In: Shen, D., et al. (eds.) Medical Image Computing and Computer Assisted Intervention – MICCAI 2019. MICCAI 2019. LNCS, vol. 11766, pp. 164–172. Springer, Cham (2019). https://doi.org/10.1007/978-3-030-32248-9_19
20. Routier, A., et al.: Clinica: an open source software platform for reproducible clinical neuroscience studies (2019). https://hal.inria.fr/hal-02308126
21. Schiratti, J.B., Allassonniere, S., Colliot, O., Durrleman, S.: Learning spatiotemporal trajectories from manifold-valued longitudinal data. In: Advances in Neural Information Processing Systems, pp. 2404–2412 (2015)
22. Xia, T., Chartsias, A., Tsaftaris, S.A.: Consistent brain ageing synthesis. In: Shen, D., et al. (eds.) Medical Image Computing and Computer Assisted Intervention – MICCAI 2019. LNCS, vol. 11767, pp. 750–758. Springer, Cham (2019). https://doi.org/10.1007/978-3-030-32251-9_82

23. Zaheer, M., Kottur, S., Ravanbakhsh, S., Poczos, B., Salakhutdinov, R.R., Smola, A.J.: Deep sets. In: Guyon, I. (eds.) Advances in Neural Information Processing Systems, vol. 30, pp. 3391–3401. Curran Associates, Inc. (2017)
24. Zhang, Z., Song, Y., Qi, H.: Age progression/regression by conditional adversarial autoencoder. In: 2017 IEEE Conference on Computer Vision and Pattern Recognition (CVPR), pp. 4352–4360. IEEE, Honolulu, HI (2017). https://doi.org/10.1109/CVPR.2017.463
25. Zhao, Q., Liu, Z., Adeli, E., Pohl, K.M.: LSSL: Longitudinal Self-Supervised Learning. arXiv:2006.06930 [cs, stat] (2020). http://arxiv.org/abs/2006.06930, arXiv: 2006.06930

Self-supervised Vessel Enhancement Using Flow-Based Consistencies

Rohit Jena[1(✉)], Sumedha Singla[2], and Kayhan Batmanghelich[2]

[1] Carnegie Mellon University, Pittsburgh, PA, USA
rjena@cs.cmu.edu, rjena@seas.upenn.edu
[2] University of Pittsburgh, Pittsburgh, PA, USA

Abstract. Vessel segmentation is an essential task in many clinical applications. Although supervised methods have achieved state-of-art performance, acquiring expert annotation is laborious and mostly limited for two-dimensional datasets with a small sample size. On the contrary, unsupervised methods rely on handcrafted features to detect tube-like structures such as vessels. However, those methods require complex pipelines involving several hyper-parameters and design choices rendering the procedure sensitive, dataset-specific, and not generalizable. We propose a self-supervised method with a limited number of hyper-parameters that is generalizable across modalities. Our method uses tube-like structure properties, such as connectivity, profile consistency, and bifurcation, to introduce inductive bias into a learning algorithm. To model those properties, we generate a vector field that we refer to as a *flow*. Our experiments on various public datasets in 2D and 3D show that our method performs better than unsupervised methods while learning useful transferable features from unlabeled data. Unlike generic self-supervised methods, the learned features learn vessel-relevant features that are transferable for supervised approaches, which is essential when the number of annotated data is limited.

1 Introduction

Tube-like structures, such as vessels and airways, are ubiquitous in studying human anatomy. Segmenting such structures is essential for characterizing the progression of many diseases [5,12]. Supervised deep learning methods have made significant progress for accurate segmentation [9,21,23], but annotated datasets are largely limited to 2D data and often have a small sample size. We develop a self-supervised task that incorporates the structure's key properties and learns optimal representation for the structure. Our method is applicable for both 2D and 3D, and it can be employed to bootstrap supervised methods when the number of *annotated* data is limited.

Electronic supplementary material The online version of this chapter (https://doi.org/10.1007/978-3-030-87196-3_23) contains supplementary material, which is available to authorized users.

M. de Bruijne et al. (Eds.): MICCAI 2021, LNCS 12902, pp. 242–251, 2021.
https://doi.org/10.1007/978-3-030-87196-3_23

Automatic vessel segmentation is a challenging problem, given that the vascular networks are complex multi-level tree structures with high variability in local geometry, curvature, and radius, which further varies across modalities and subjects. Recently, various deep learning (DL) based techniques have been proposed for various segmentation tasks of tube-like structures [9,21,23]. Training supervised DL algorithms requires many annotated images, which is particularly laborious for complex tube-like structures. Due to the lack of a large-scale annotated dataset to train a supervised method, *unsupervised* vessel segmentation methods are still popular, and there is a growing interest in deploying DL-based *unsupervised* and *self-supervised* methods [22,27].

An unsupervised pipeline specially designed for vessel segmentation varies across modalities and image dimensionality (e.g., 2D retinography [8] and 3D thoracic CT [18]). Most state-of-the-art pipelines rely on hand-crafted features based on different variants of classical Hessian-based scale-space filters [7,14,20]. Achieving state-of-the-art results requires post-processing using different techniques such as particle sampling [5] and region growing [25] with several design choices for each step with their corresponding hyper-parameters. Such design renders the procedure problem-specific and not transferable across domains. In contrast, we propose an end-to-end unsupervised vessel segmentation model that generalizes across modalities and dimensions.

Our work is inspired by the matched filter response (MFR) method [3] and scale-space approaches [6]. These methods model the vessel as piece-wise linear segments and use multiple Gaussian kernels to identify vessel-like structures. Our approach is a modern adaptation of MFR and scale-space using a fully convolutional network (FCN). This paper's central idea is to model the vessel with a *flow* which defines a continuous path; the profile of the tube-like structure matches with expected *template* as we *walk* along with the flow. We use the notion of walking along with the flow as an inductive bias for a self-supervised method that learns the set of optimal features. The FCN architecture, which is used to infer the flow, naturally processes the image in a multi-scale fashion.

The paper makes the following contributions. (1) It proposes a self-supervision inspired by the problem, in this case, segmenting tube-like structures. (2) The method is generic and can be deployed for 2D and 3D images for enhancing vessel in various modalities. (3) Unlike other unsupervised methods using fixed hand-crafted features, our method adapts the features as the dataset changes. When annotated data is limited, the trained features can also be transferred to boost a supervised method's performance. To the best of our knowledge, our work is the first unsupervised deep learning method that takes a raw image as input and outputs per-pixel vessel statistics as output, along with an associated per-pixel vesselness score. We evaluate the performance of our method on real 2D and 3D datasets. We also show the efficacy of incorporating context into self-supervised learning by comparing our method with state-of-the-art self-supervised pretraining tasks.

2 Method

We use some of the key properties of *tube-like* structures (*e.g.*, vessels) to define a self-supervised algorithm. We use vessel structure as a running example, however, the proposed method is general and can be applied to other tube-like structures such as airways. We consider the following three properties:

(1) Path Continuity: The path continuity assumes that the vessel trees' structure and arrangement can be viewed as a continuous path in space (represented by a *vector field*). Each vector in the *vector field* indicates the radius and direction of the vessel. We refer to this vector field as *vessel flow* field. **(2) Profile Consistency:** A Tube-like structure has similar profiles at different points when the planes are perpendicular to the structure's medial axis (i.e., the vector field). In other words, the orthogonal profile of the tube-like structure can be approximated by a predefined template, T. For a vessel, T is simply a unit disk. **(3) Bifurcation:** The possible bifurcation refers to the fact that the vessel may split into two sub-vessels. We use two vector fields to model the bifurcation. Wherever there is a bifurcation, the two vector fields point toward each bifurcation branch; otherwise, they are aligned with the vessel flow.

General Framework: Let $I : \Omega \rightarrow \mathbb{R}$ represent input image defined over the d-dimensional domain $\Omega \subseteq \mathbb{R}^d$. We consider $d \in \{2, 3\}$. We assume a network f_θ generates three outputs: $\boldsymbol{u}$, r, and $(\boldsymbol{b}_1, \boldsymbol{b}_2)$. Let $\boldsymbol{u} : \Omega \rightarrow \mathbb{S}^{d-1}$ denote the vessel flow, defined as a vector field where for any point $\boldsymbol{p} \in \Omega$, where $\mathbb{S}^{d-1}$ is a d-dimensional unit hypersphere. We define $\boldsymbol{b}_1 : \Omega \rightarrow \mathbb{S}^{d-1}$ and $\boldsymbol{b}_2 : \Omega \rightarrow \mathbb{S}^{d-1}$

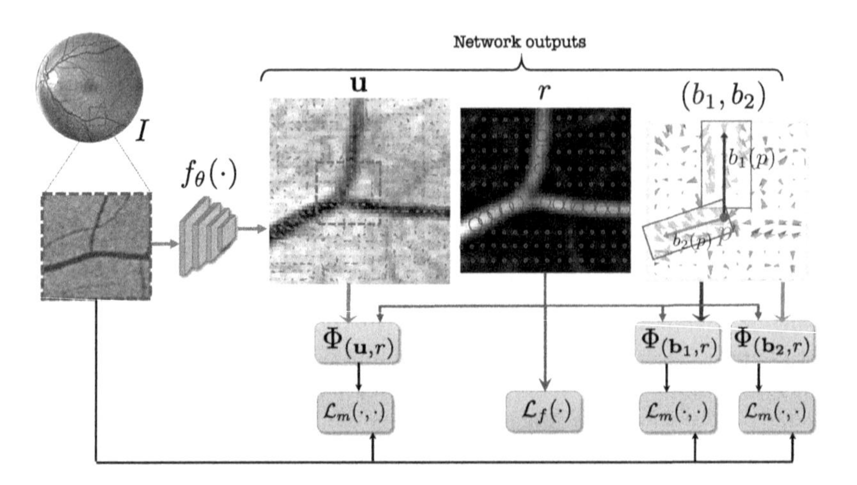

Fig. 1. The overall framework of the proposed method. The network f_θ generates a scalar field r specifying the local width of the vessel, vector field $\boldsymbol{u}$ denoting the direction of the vessel, and a tuple of two vector fields $\boldsymbol{b}_1, \boldsymbol{b}_2$ pointing towards branches in cases of bifurcation. $\Phi_{(\boldsymbol{u},r)}$ is a transformation parameterized by a flow field $\boldsymbol{u}$ and a radius field r that is composed with the image I. There are two losses, $\mathcal{L}_m(\cdot,\cdot)$, and $\mathcal{L}_f(\cdot)$ encouraging the properties of the tube-like structures.

to represent the two directions of the bifurcation. Furthermore, $r : \Omega \rightarrow \mathbb{R}^+$ is defined as a scalar field to represent the radius of the vessel. The idea has been shown in Fig. 1. We define the losses to enforce those properties. We first explain the profile symmetry followed by path continuity and bifurcation.

2.1 Profile Consistency

The vessel flow, $\boldsymbol{u}$, specifies the directionality of the vessel. At each point $\boldsymbol{p}$, the orthogonal plane to the vessel flow specifies the vessel's profile. Given a template T, one expects the resized vessel profile to match the template. We use $I \circ \Phi_{(\boldsymbol{u},r)}(\boldsymbol{p})$ to denote the profile of image I at point $\boldsymbol{p}$ along the direction $\boldsymbol{u}(\boldsymbol{p})$. The output $r(\boldsymbol{p})$ specifies the radial field of view of the profile. The image is resized by resampling according to $r(\boldsymbol{p})$. The $\Phi_{(\boldsymbol{u},r)}$ denotes this transformation and $\circ$ denotes the composition of the transformation with image I. Similar to image registration, we maximize a similarity metric ($\mathcal{S}(\cdot,\cdot)$) between the template (T) and the transformed profile ($I \circ \Phi_{(\boldsymbol{u},r)}(\boldsymbol{p})$) over the entire domain (Ω):

$$\mathcal{L}_m(\boldsymbol{u}, r; I, T) = -\int_{\Omega} \overbrace{\mathcal{S}(I \circ \Phi_{(\boldsymbol{u},r)}(\boldsymbol{p}), T)}^{V(\boldsymbol{p})} \, d\boldsymbol{p}. \quad (1)$$

Various choices are possible for $\mathcal{S}(\cdot,\cdot)$. We use Normalized Cross-Correlation (NCC) because of its robustness to changes in the illumination and contrast. The $\mathcal{L}_m(\boldsymbol{u}, r; I, T)$ indicates that the value of the loss is a function of vessel flow, $\boldsymbol{u}$ and vessel radius, r given the image and template. For other tube-like structures such as airways, one can change the template. We define $V(\boldsymbol{p})$ in the domain as a **vesselness** value for the pixel $\boldsymbol{p}$.

2.2 Path Continuity

We impose vessel flow continuity for two reasons. First, we assume the entire vessel structure is a connected component. Second, for a point $\boldsymbol{p}$, the vessel flow $\boldsymbol{u}(\boldsymbol{p})$ and $-\boldsymbol{u}(\boldsymbol{p})$ result in the same vessel profile. Such sign ambiguity may result in discontinuities in vector field $\boldsymbol{u}$. To prevent this, we propose an alignment loss that encourages $\boldsymbol{u}$ to have a consistent direction along the vessel. We do that by *walking* along the direction of the vessel flow. We formulate the walk by a stationary path integral. For a point $\boldsymbol{p}$, let's assume $\boldsymbol{q}_{\boldsymbol{p}}(t)$ is the new coordinate after walking for t from $\boldsymbol{p}$. The flow vector at the final point is $\boldsymbol{u}(\boldsymbol{q}_{\boldsymbol{p}}(t))$. If the vector field is consistent and continuous, the inner product between $\boldsymbol{u}(\boldsymbol{p})$ and $\boldsymbol{u}(\boldsymbol{q}_{\boldsymbol{p}}(t))$ should be large; i.e., they should be almost aligned. We define a loss to maximize that loss over the entire domain:

$$\dot{\boldsymbol{q}}_{\boldsymbol{p}} = \boldsymbol{u}(\boldsymbol{p}), \qquad \boldsymbol{q}_{\boldsymbol{p}}(t=0) = \boldsymbol{p},$$
$$\mathcal{L}_f(\boldsymbol{u}, r) = -\int_{\Omega}\int_0^{2r(\boldsymbol{p})} \langle \boldsymbol{u}(\boldsymbol{p}), \boldsymbol{u}(\boldsymbol{q}(t))\rangle \, dt d\boldsymbol{p}, \quad (2)$$

where $\boldsymbol{q}$ is the position of the walk, $\dot{\boldsymbol{q}}$ is the time derivative and the given point, $\boldsymbol{p}$ is the initial point in the path, and $\langle \cdot, \cdot \rangle$ represents inner product.

We set $2r(\boldsymbol{p})$ as an upper limit in the integral so that the length of the traversal path is proportional to the size of the vessel.

Table 1. Evaluation of unsupervised vessel segmentation on 2D datasets. Acc is the global accuracy over the entire image, LAcc is the local accuracy around the dilated vessel regions defined in [15]. There are the significant improvements in Dice score, which is a crucial metric in sparse segmentation.

Method	DRIVE				STARE			
	AUC	Acc	LAcc	Dice	AUC	Acc	LAcc	Dice
Hessian	0.55	52.90	60.31	0.25	0.53	46.84	59.24	0.21
Frangi	0.93	94.27	72.02	0.67	0.93	91.16	67.12	0.59
Sato	0.94	93.67	71.90	0.66	0.94	91.03	66.52	0.58
Meijering	0.94	93.83	72.70	0.67	0.94	90.29	68.11	0.58
Ours	**0.96**	**95.69**	**75.83**	**0.74**	**0.96**	**95.29**	**69.44**	**0.68**
	HRF				**RITE**			
Hessian	0.37	33.70	55.35	0.17	0.51	49.77	60.21	0.23
Frangi	0.93	93.90	70.59	0.63	0.94	94.26	73.10	0.68
Sato	0.94	93.73	70.25	0.62	0.95	94.01	72.86	0.67
Meijering	0.92	91.11	72.73	0.58	0.95	94.77	**77.32**	0.71
Ours	**0.95**	**94.96**	**72.78**	**0.68**	**0.97**	**96.07**	77.05	**0.75**

2.3 Bifurcation

A bifurcation point (BP) is where the main vessel splits into two branches. Similar to vessel flow $\boldsymbol{u}$, we predict bifurcation flow fields $\boldsymbol{b}_1$ and $\boldsymbol{b}_2$ as vector fields representing the two directions of the bifurcation. Although the BP does not match the canonical template T, the incoming vessel and the two branches should match T. We define the birfurcation loss as an extension of $\mathcal{L}_m$,

$$\mathcal{L}_b(\boldsymbol{b}_1, \boldsymbol{b}_2, r; I, T) = \mathcal{L}_m(\boldsymbol{b}_1, r; I, T) + \mathcal{L}_m(\boldsymbol{b}_2, r; I, T). \tag{3}$$

Note that, in the absence of bifurcation, $\boldsymbol{b}_1(\boldsymbol{p}) = \boldsymbol{b}_2(\boldsymbol{p}) = -\boldsymbol{u}(\boldsymbol{p})$ minimizes the same loss as $\mathcal{L}_m$ in the opposite direction of the vessel flow. Hence in practice, we can add the loss function in Eq. 3 to Eq. 1.

2.4 Implementation Details

The overall cost function is as follows:

$$\begin{aligned} \min_{\theta} \quad & \mathcal{L}_m(\boldsymbol{u}, r; I, T) + \lambda_1 \mathcal{L}_f(\boldsymbol{u}, r) + \lambda_2 \mathcal{L}_b(\boldsymbol{b}_1, \boldsymbol{b}_2, r; I, T) \\ \text{s.t:} \quad & (r, \boldsymbol{u}, \boldsymbol{b}_1, \boldsymbol{b}_2) = f_\theta(I), \end{aligned} \tag{4}$$

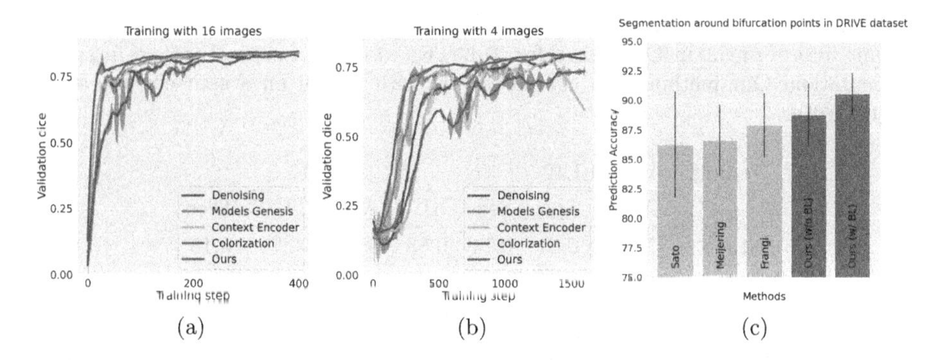

Fig. 2. Training curves for supervised vessel-segmentation on STARE dataset with **(a)** limited data (4 images), **(b)** more data (16 images), after self-supervised pretraining on DRIVE and **(c)** Bifurcation segmentation performance on DRIVE dataset.

where λ_i's are weighting hyper-parameters and $f_\theta(\cdot)$ is the network. To be consistent with the literature, we adopt the U-Net architecture for f_θ [4]. The models are trained with the Adam optimizer, with a learning rate of 0.001 for 2D and 0.0003 for 3D. For 2D images, we use the entire image as input, with a batch size of 4, and for 3D, we use a $64 \times 64 \times 64$ patch with a batch size of 1. During training, we augment the images by flipping and rotating in increments of 90 degrees. All our models are trained on NVIDIA Tesla V100-SXM2 GPUs.

3 Experiments

We perform three experiments to evaluate our method. (1) We compare the performance of our approach with commonly used unsupervised methods for vessel segmentation: Frangi [7], Sato [20], Hybrid Hessian [14] and Meijering [13] filters on four 2D datasets and two 3D datasets, all of which are publicly available. (2) We study the efficiency of learned representation for the downstream vessel segmentation task. To do that, we compare our method with existing self-supervised methods. (3) We examine the efficacy of the bifurcation loss in segmenting the regions around bifurcation points in 2D compared to methods which do not consider bifurcations.

Datasets: For the 2D experiments, we use four publicly available retinal image datasets: DRIVE [16], STARE [10], HRF [1] and RITE [11]. The DRIVE and RITE datasets consist of 40 images, divided into a training and testing set of 20 images. The STARE database consists of 20 images, each image with two sets of segmented images. The HRF dataset consists of 45 retinopathy images which we divide into a training set of 21 images and testing set of 24 images. For 3D vessel segmentation, we use the VESSEL12 dataset [19] consisting of 20 CT lung images from a variety of sources. The dataset also contains 3 images with sparsely annotated vessel and non-vessel locations along 3 axial slices, which we use as a test set. We also use the TubeTK dataset [2] which consists of 3D MRA images of 100

Table 2. Results on the VESSEL12 and TubeTK test images. Our method has significant improvement in Dice score for TubeTK, which is a critical metric in sparse segmentation. Our method also compares well with Frangi on a sparsely annotated ground truth.

Method	VESSEL12				TubeTK		
	Acc	Spec	Sens	AUC	Acc	AUC	Dice
Sato	79.1	0.81	0.74	0.88	91.17	0.74	0.15
Meijering	90.16	0.89	0.92	0.96	97.25	0.83	0.34
Frangi	**96.88**	**0.97**	0.96	0.97	98.79	0.90	0.42
Ours	95.49	0.92	**0.99**	**0.99**	**99.05**	**0.95**	**0.59**

healthy patients of size $448 \times 448 \times 128$. We use 42 images with ground truths as the test set, and the remaining images are used for training the network.

3.1 Comparison with Unsupervised Methods

We compare our model against popular vessel enhancement methods, including Frangi, Sato, Hessian, and Meijering filters. All the methods take raw images as input and produce a vessel-enhanced image as output. The enhanced image is segmented into a binary map using a hard threshold, which is selected to achieve a maximum dice score over the training dataset. The performance is then reported over the test set using the hard threshold. Table 1 presents a quantitative comparison of different vessel-segmentation methods as a binary classification problem. We reported results on five measures, namely, the area under the curve (AUC) of ROC curves, accuracy (acc), local-accuracy (LAcc) [15], and dice score.

For the VESSEL12 dataset, we drop the Dice score since we do not have access to a dense ground truth. Therefore, we treat the problem as a classification problem and compare sensitivity (sens) and specificity (spec) as well. For TubeTK, we compare the Dice score of the methods with the dense ground truth. The results are summarized in Table 2. Similar to 2D, our method performs consistently across datasets, and has a significantly higher Dice score. Without using any annotations for training, our method outperforms other commonly used unsupervised vessel enhancement methods. Since vessel segmentation is a sparse segmentation problem, the critical metrics are Dice score and Local Accuracy, on which our method has a significant improvement over baselines.

3.2 Efficacy of the Representation

Since our method is learning based, it can learn feature representations that are essential for vessel detection. This section compares the efficacy of the representation from different self-supervised tasks onto a downstream supervised vessel segmentation task. We compare our model with four self-supervision baselines, namely, context-encoder [17], image-denoising [24], image-colorization [26] and

Models Genesis [27]. First, we train multiple networks using different pretext tasks on the DRIVE dataset. These networks are then *finetuned* on a supervised vessel segmentation task on the STARE dataset. We consider a *limited-data* and a *high-data* scenario, where finetuning is done with only 4 and 16 images respectively. Figure 2(a,b) show the training dynamics in both cases. Our method takes fewer iterations to converge compared to the other methods and achieves the best validation dice score (Fig. 3).

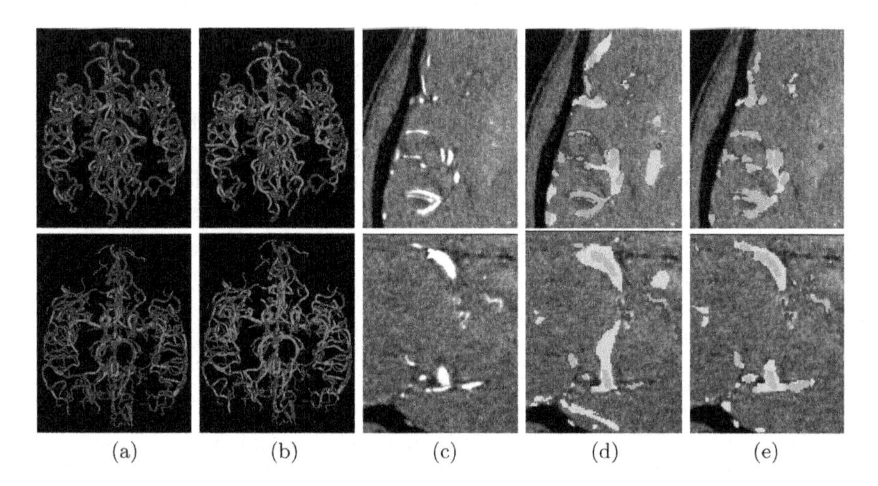

Fig. 3. Results on the TubeTK dataset. **(a)** Result of Frangi segmentation **(b)** Result of our segmentation (green denotes true positives and red denote false negatives). **(c)** Axial slice of the image containing vascular structure **(d)** Result of Frangi segmentation **(e)** Result of our segmentation. (Green denotes true positives, yellow denotes false positives and red denotes false negatives.) Our method has fewer yellow and red regions. (Color figure online)

3.3 Segmentation Around Bifurcation Points (BPs)

In this experiment, we demonstrate the importance of our predicted bifurcation flow fields ($\boldsymbol{b}_1, \boldsymbol{b}_2$) in vessel-segmentation performance around BPs. To quantitatively measure the segmentation, we manually annotated bounding boxes (BBs) at multiple bifurcation regions in the DRIVE dataset. We performed an ablation study, where we didn't consider bifurcation loss (BL) in our final formulation. Figure 2(c) reports the accuracy of identifying vessel pixels within the extracted BBs. Our proposed bifurcation loss significantly improves the segmentation at regions around BPs.

4 Conclusion

Our proposed self-supervised model demonstrates the ability to perform efficient vessel-segmentation on real 2D digital retinal images and 3D CT and MRA scans.

It does so by using critical properties of tube-like structures such as connectivity, profile consistency, and bifurcations to introduce inductive bias into deep learning and learn in a self-supervised setting. Our adaption of self-supervised task demonstrates robustness and generalizability in features in a downstream segmentation task. To summarize, our work is a step towards incorporating geometrical constraints of tube-like structures into a deep learning framework and providing a robust self-supervised model for vessel segmentation. Further work should explore improving the segmentation performance on thin, low contrast vessels. A prospective study may explore employing the vessel-segmentation to understand a disease manifestation and establishing clinical usage.

Acknowledgment. This work was partially supported by NIH Award Number 1R01HL141813-01, NSF 1839332 Tripod+X, SAP SE, and Pennsylvania Department of Health. We are grateful for the computational resources provided by Pittsburgh SuperComputing grant number TG-ASC170024.

References

1. Budai, A., Bock, R., Maier, A., Hornegger, J., Michelson, G.: Robust vessel segmentation in fundus images. Int. J. Biomed. Imag. (2013)
2. Bullitt, E., et al.: Vessel tortuosity and brain tumor malignancy: a blinded study1. Acad. Radiol. **12**(10), 1232–1240 (2005)
3. Chaudhuri, S., Chatterjee, S., Katz, N., Nelson, M., Goldbaum, M.: Detection of blood vessels in retinal images using two-dimensional matched filters. IEEE Trans. Med. Imaging **8**(3), 263–269 (1989)
4. Cortinovis, D.: Retina blood vessel segmentation with a convolution neural network (U-Net)
5. Estépar, R.S.J., et al.: Computed tomographic measures of pulmonary vascular morphology in smokers and their clinical implications. Am. J. Respir. Crit. Care Med. **188**(2), 231–239 (2013)
6. Estépar, R.S.J., Ross, J.C., Russian, K., Schultz, T., Washko, G.R., Kindlmann, G.L.: Computational vascular morphometry for the assessment of pulmonary vascular disease based on scale-space particles. In: 2012 9th IEEE International Symposium on Biomedical Imaging (ISBI), pp. 1479–1482. IEEE (2012)
7. Frangi, A.F., Niessen, W.J., Vincken, K.L., Viergever, M.A.: Multiscale vessel enhancement filtering. In: Wells, W.M., Colchester, A., Delp, S. (eds.) MICCAI 1998. LNCS, vol. 1496, pp. 130–137. Springer, Heidelberg (1998). https://doi.org/10.1007/BFb0056195
8. Fraz, M.M., et al.: Blood vessel segmentation methodologies in retinal images-a survey. Comput. Methods Programs Biomed. **108**(1), 407–433 (2012)
9. Guo, C., Szemenyei, M., Yi, Y., Wang, W., Chen, B., Fan, C.: SA-Unet: spatial attention U-Net for retinal vessel segmentation. arXiv preprint arXiv:2004.03696 (2020)
10. Hoover, A.D., Kouznetsova, V., Goldbaum, M.: Locating blood vessels in retinal images by piecewise threshold probing of a matched filter response. IEEE Trans. Med. Imaging **19**(3), 203–210 (2000)

11. Hu, Q., Abràmoff, M.D., Garvin, M.K.: Automated separation of binary overlapping trees in low-contrast color retinal images. In: Mori, K., Sakuma, I., Sato, Y., Barillot, C., Navab, N. (eds.) MICCAI 2013. LNCS, vol. 8150, pp. 436–443. Springer, Heidelberg (2013). https://doi.org/10.1007/978-3-642-40763-5_54
12. Junior, S.B., Welfer, D.: Automatic detection of microaneurysms and hemorrhages in color eye fundus images. Int. J. Comput. Sci. Inf. Technol. **5**(5), 21 (2013)
13. Meijering, E., Jacob, M., Sarria, J.C., Steiner, P., Hirling, H., Unser, M.: Design and validation of a tool for neurite tracing and analysis in fluorescence microscopy images. Cytometry A **58A**(2), 167–176 (2004)
14. Ng, C.-C., Yap, M.H., Costen, N., Li, B.: Automatic wrinkle detection using hybrid Hessian filter. In: Cremers, D., Reid, I., Saito, H., Yang, M.-H. (eds.) ACCV 2014. LNCS, vol. 9005, pp. 609–622. Springer, Cham (2015). https://doi.org/10.1007/978-3-319-16811-1_40
15. Nguyen, U.T.V., Bhuiyan, A., Park, L.A.F., Ramamohanarao, K.: An effective retinal blood vessel segmentation method using multi-scale line detection. Pattern Recogn. **46**(3), 703–715 (2013)
16. Niemeijer, M., Staal, J., van Ginneken, B., Loog, M., Abramoff, M.D.: Comparative study of retinal vessel segmentation methods on a new publicly available database. In: Medical Imaging 2004: Image Processing, vol. 5370, pp. 648–656. International Society for Optics and Photonics (2004)
17. Pathak, D., Krähenbühl, P., Donahue, J., Darrell, T., Efros, A.: Context encoders: feature learning by inpainting (2016)
18. Rudyanto, R.D., et al.: Comparing algorithms for automated vessel segmentation in computed tomography scans of the lung: the vessel12 study. Med. Image Anal. **18**(7), 1217–1232 (2014)
19. Rudyanto, R.D., Kerkstra, S., van Rikxoort, E.M., Fetita, C., Brillet, P.Y., et al.: Comparing algorithms for automated vessel segmentation in computed tomography scans of the lung: the vessel12 study. Med. Image Anal. **18**(7), 1217–1232 (2014)
20. Sato, Y., et al.: Three-dimensional multi-scale line filter for segmentation and visualization of curvilinear structures in medical images. Med. Image Anal. **2**(2), 143–168 (1998)
21. Soomro, T.A., et al.: Deep learning models for retinal blood vessels segmentation: a review. IEEE Access **7**, 71696–71717 (2019)
22. Taleb, A., et al.: 3d self-supervised methods for medical imaging. arXiv preprint arXiv:2006.03829 (2020)
23. Tetteh, G., et al.: DeepVesselNet: vessel segmentation, centerline prediction, and bifurcation detection in 3-d angiographic volumes. arXiv preprint arXiv:1803.09340 (2018)
24. Vincent, P., Larochelle, H., Bengio, Y., Manzagol, P.A.: Extracting and composing robust features with denoising autoencoders. In: Proceedings of the 25th International Conference on Machine Learning, pp. 1096–1103 (2008)
25. Yu, G., Li, P., Miao, Y., Bian, Z.: Multiscale active contour model for vessel segmentation. J. Med. Eng. Technol. **32**(1), 1–9 (2008)
26. Zhang, R., Isola, P., Efros, A.A.: Colorful image colorization. In: Leibe, B., Matas, J., Sebe, N., Welling, M. (eds.) ECCV 2016. LNCS, vol. 9907, pp. 649–666. Springer, Cham (2016). https://doi.org/10.1007/978-3-319-46487-9_40
27. Zhou, Z., et al.: Models genesis: generic autodidactic models for 3d medical image analysis. In: Shen, D., et al. (eds.) MICCAI 2019. LNCS, vol. 11767, pp. 384–393. Springer, Cham (2019). https://doi.org/10.1007/978-3-030-32251-9_42

Unsupervised Contrastive Learning of Radiomics and Deep Features for Label-Efficient Tumor Classification

Ziteng Zhao[1] and Guanyu Yang[1,2](✉)

[1] LIST, Key Laboratory of Computer Network and Information Integration (Southeast University), Ministry of Education, Nanjing, China
yang.list@seu.edu.cn

[2] Centre de Recherche en Information Biomédicale Sino-Français (CRIBs), Rennes, France

Abstract. Tumor classification is important for decision support of precision medicine. Computer-aided diagnosis by convolutional neural networks relies on a large amount of annotated dataset, which is costly sometimes. To solve the poor predictive ability caused by tumor heterogeneity and inadequate labeled image data, a self-supervised learning method combined with radiomics is proposed to learn rich visual representation about tumors without human supervision. A self-supervised pretext task, namely "Radiomics-Deep Feature Correspondence", is formulated to maximize agreement between radiomics view and deep learning view of the same sample in the latent space. The presented self-supervised model is evaluated on two public medical image datasets of thyroid nodule and kidney tumor and achieves high score on linear evaluations. Furthermore, fine-tuning the pre-trained network leads to a better score than the train-from-scratch models on the tumor classification task and shows label-efficient performance using small training datasets. This shows injecting radiomics prior knowledge about tumors into the representation space can build a more powerful self-supervised method.

Keywords: Self-supervised learning · Unsupervised contrastive learning · Radiomics · Tumor classification

1 Introduction

Deep convolutional neural networks (CNNs) have made major breakthroughs in the past few years, largely driven by increased computing power and massive labeled datasets. Benefitting from the huge advances of deep learning in image classification, computer-aided medical diagnostics has achieved great success [3]. Precise prediction of tumor type can help doctors recognize and interpret the subtle difference between different kinds of medical images. Moreover, it is critical to decision support of personalized cancer treatment for patients.

M. de Bruijne et al. (Eds.): MICCAI 2021, LNCS 12902, pp. 252–261, 2021.
https://doi.org/10.1007/978-3-030-87196-3_24

However, tumor classification using CNNs is still challenging due to (1) imaging data and tumor heterogeneity, and (2) the poor generalization ability of CNNs caused by inadequate labeled image data. Individual variability and the differences about acquisition protocols, contrast-agents, levels of contrast enhancements and scanner resolutions of medical image data lead to unpredictable size, shape and intensity diversity of tumors. Furthermore, up to date, most deep learning methods used in medical diagnosis are strongly supervised networks which require sufficiently large medical image datasets with expert annotations. Preparing large and labeled datasets is usually difficult or even impossible, with the result that CNNs cannot learn rich feature information and overfit severely.

To deal with complex data distribution and deficient annotated data, the self-supervised learning, a prominent pattern of unsupervised learning, is proposed to learn useful feature information from wide-ranging unlabeled data and then to solve the target task better with a small set of training data [1,12,13,15,18,21,23]. The key to self-supervised learning is to select suitable pretext tasks that can guide CNNs to extract high-quality visual features for the target tasks. Among many pretext tasks, unsupervised contrastive learning [1,8,23] is very popular in the field of natural images. It is a promising class of methods that build representations by learning to encode what makes two things similar or different. At a very high level, the intent of contrastive learning is to reduce feature dimensionality by maximizing agreement between different views of the same sample in the latent space. For example, CMC [19] learns invariant representations from various channels of one image; SimCLR [1] learns from differently augmented views of one image. For the applications of computer aided diagnosis, Jamaludin et al. [9] pre-trained a Siamese CNN distinguishing if a pair of images from different collection time is from the same patient. Jiao et al. [10] used cross-model contrastive learning to model the correspondence between video and audio of ultrasound. Therefore, it is evident that contrastive learning is a domain and task agnostic paradigm for self-supervised learning. It allows us to inject our prior knowledge about the structure in the data into the representation space and build more powerful self-supervised methods.

In tumor diagnosis, radiomics handcrafted quantitative features extracted from volumes of interest play an important role [4]. These features can describe a large number of phenotypic features, such as shape and texture [11]. From contrastive learning perspective, radiomics handcrafted features are a more effective view of medical images compared with the image augment transformation view because their feature dimensionality is low and they contain domain knowledge about diagnosis. Therefore, when contriving self-supervised pretext tasks, using contrastive learning of radiomics view and image view can help networks reduce feature dimensions and learn more discriminative features related to tumor diagnosis.

Therefore, we propose an unsupervised contrastive learning approach using radiomics and deep features for label-efficient tumor classification. We design a self-supervised pretext task, namely "Radiomics-Deep Feature Correspondence",

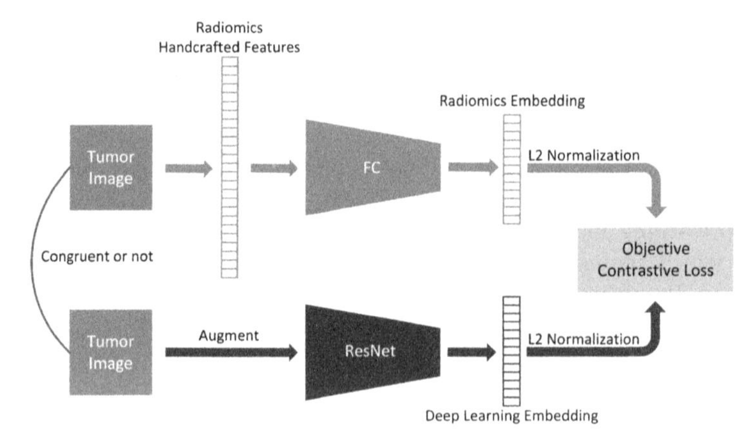

Fig. 1. Unsupervised contrastive learning by the pretext task of Radiomics-Deep Feature Correspondence. The pre-trained network is then fine-tuned to the target task.

to deeply exploit rich feature information from tumor areas and improve networks' ability to diagnose tumors in situations where labels are insufficient. We evaluate the presented method on two distinct public medical image datasets: thyroid nodule classification in ultrasound images and multiclass kidney tumor classification in CT. Experimental results show the proposed approach can learn better visual representation and its performance is superior to that of the network trained from scratch and is still good with less labeled training data. To the best of our knowledge, this is the first work achieving self-supervised learning by Radiomics-Deep Feature Correspondence pretext task.

2 Method

In this section, we begin with an overview of the self-supervised approach using Radiomics-Deep Feature Correspondence pretext task. We then introduce the details of the method, including contrastive loss of the self-supervised learning network and the full pipeline of the approach. The self-supervised learning network is illustrated in Fig. 1.

2.1 Radiomics-Deep Feature Correspondence

Our aim is to learn representations that hold information shared between radiomics view and deep learning view without human supervision. Radiomics handcrafted features extracted from tumor areas are the radiomics view's input and are processed by fully connected layers. Augmented Tumor images are the deep learning view's input and are processed by a CNN. After the above processing, we get radiomics features $r^{(n)}$ and deep learning features $d^{(n)}$ in the latent space and perform radiomics-deep feature correspondence pretext task by contrasting congruent and incongruent pairs. The self-supervised approach's label shows whether the two kinds of features, i.e., $r^{(n)}$ and $d^{(n)}$, are "corresponding". In other words, positive, $r^{(i)}$ and $d^{(i)}$ are from the same tumor area and should have similar distribution; negative, $r^{(i)}$ and $d^{(j)}$ are from different tumor areas. The pretext task can help networks learn a representation that maximize mutual information between the two views of the same tumor but is otherwise compact.

2.2 Contrastive Loss and Memory Bank

To achieve the pretext task's aim, we apply contrastive learning [2,23], where feature embeddings such that two views of the same tumor map to nearby points while views of different tumor to far apart points. The distance of points is measured with cosine similarity in representation space. In practice, we train InfoNCE [5,23] loss to correctly select d corresponding to r out of a set $S = \{r, r^{(1)}, r^{(2)}, \dots, r^{(k)}\}$ that contains k incongruent radiomics features, i.e., select a only positive sample from a set that contains k negative samples. This loss function $\mathcal{L}^{d,r}_{contrast}$ is shown below.

$$\mathcal{L}^{d,r}_{contrast} = -log \frac{exp(d^\top r)}{\sum_{i \in k} exp(d^\top r^i)} \tag{1}$$

Loss $\mathcal{L}^{d,r}_{contrast}$ in Eq. 1 treats deep learning view as anchor and enumerates over radiomics view. Symmetrically, we can get $\mathcal{L}^{r,d}_{contrast}$ by anchoring at radiomics view. We add the two loss up as the final loss:

$$\mathcal{L}(r,d) = \mathcal{L}^{d,r}_{contrast} + \mathcal{L}^{r,d}_{contrast} \tag{2}$$

Better representations using $\mathcal{L}(r,d)$ in Eq. 2 can be learnt by using lots of negative samples [17]. In order to reduce the computational cost of calculating a large number of negative samples, we maintain two memory banks to store two views' feature embeddings for each training sample following [23]. The memory banks are dynamically updated with latent features computed on the fly.

2.3 The Full Framework of the Approach

We first pre-train the network in Fig. 1 on radiomics-deep feature correspondence pretext task. Upon convergence, we add a fully-connected layer at the end of the pre-trained network and use it for tumor classification.

Radiomics Features Extraction. Around 1,000 radiomics handcrafted features are extracted from each tumor area by PyRadiomics [20]. PyRadiomics is a comprehensive open-source python package, which is able to extract reproducible handcrafted features through a large panel of hard-coded feature algorithms. These features which contain intensity-based features, shape-based features, texture-based features and higher-order features are based on domain knowledge and can characterize tumor heterogeneity to some extent. And then they are standardized and processed by three fully connected layers to 128-dimensional feature vectors.

Deep Learning Features Extraction. 2D ResNet-50 [6,7] is used to extract deep features from each tumor area. For the input of CNN, different sized tumor areas are resized to the average size and then normalized to $[0, 1]$. In addition, we apply an augmentation which is random cropping followed by resize back to the original size. In terms of network architecture, the backbone network can be any well-structured 2D CNN. Here we choose ResNet empirically, which is one of the best performing and generic networks. For the network's output, deep learning features are generated by global average pooling of the last convolutional layers of ResNet. And they are converted to 128-dimensional feature vectors, which are consistent with the dimension of radiomics features. Additionally, we constrain two views's embeddings by L_2 normalization before calculating the contrastive loss function, as suggested in [23].

3 Experiment

This section first introduces the two public medical image datasets and implementation details about experiments. We then evaluate our method using linear classification and transfer learning, and demonstrate that it's an effective self-supervised mechanism to classify tumor types.

3.1 Dataset

Thyroid Nodule Classification in Ultrasound Images. The challenge of Thyroid Nodule Segmentation and Classification in Ultrasound Images (TN-SCUI2020) [14] provide a public 2D dataset of thyroid nodule with over 3,644 patient cases from different ages, genders, and were collected in different sites using various ultrasound machines (e.g. Mindray DC-8, Philips-cx50, TOSHIBA Aplio300). Each ultrasound image is provided with its annotated class (benign or malignant) and a detailed delineation of the nodule. For pre-processing, we crop nodule areas from images and resize the areas to 196×160 in dimension. The dataset is randomly split to a train set (2,916 cases) and a test set (728 cases) by the ratio of $80 : 20$, while preserving the percentage of samples for each class. In self-supervised learning, the train set is used for the pretext task.

Multiclass Kidney Tumor Classification in CT. The challenge of 2019 Kidney and Kidney Tumor Segmentation (KiTS19) [22] released 210 3D abdominal CT images with kidney tumor subtypes and segmentations of kidney and kidney tumor. These CT images are from more than 50 institutions and scaned with different CT scanners and acquisition protocols. There are many subtypes of tumor in the dataset: clear cell renal cell carcinoma (RCC) (143 cases), papillary RCC (21 cases), chromophobe RCC (19 cases), oncocytoma (10 cases) and other smaller classes. We classify kidney tumors into four larger categories. In order to balance the quantity in each category, we randomly select 20 cases from the clear cell RCC category and combine them with the other three types of data to form a classification data set (70 cases). The remaining data (140 cases) in the KiTS19 dataset is used for self-supervised learning. For pre-processing, we uniformly select 10 slice images for each tumor area because of a wide range of slice thicknesses. If there are less than 10 slices, select slices repeatedly from the middle to the ends. And we resize these tumor slices to $10 \times 64 \times 64$, and then send them to CNN as a whole. For classification, we perform patient wise five-fold cross-validation.

3.2 Implementation Details

Our approach is implemented using PyTorch 1.7.1 and trained with a single NVIDIA GeForce GTX 2080Ti. And all the networks are optimized with stochastic gradient descent (SGD). In the self-supervised training phase, the model is trained up to 700 epochs with a learning rate of 3e–3. The capacity of the memory bank is the size of the entire training data set, the temperature is 0.07 and a momentum for memory update is 0.5. In the tumor classification training phase, all the models are trained for 100 epochs with a learning rate of 3e–3 and decayed by a factor of 10 at the 70th epoch. We use accuracy and weighted F1 score [16] to evaluate the models' performance.

Table 1. Linear classification performance of self-supervised pretext tasks. The table shows that the representation obtained by our self-supervised method performs superior on the given datasets.

Dataset	Self-supervised method	F1 score (weighted) (%)
Thyroid nodule	Autoencoder [21]	68.9
	Jigsaw [15]	71.6
	SimCLR [1]	78.7
	Ours	82.9
Kidney tumor	Autoencoder [21]	46.4 ± 5.3
	Jigsaw puzzles [15]	45.0 ± 3.5
	SimCLR [1]	50.0 ± 4.5
	Ours	52.0 ± 2.8

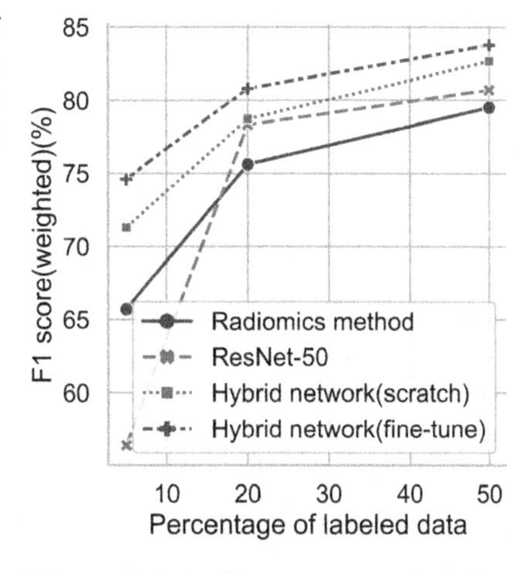

Fig. 2. Label-efficient image classification with the pre-trained model by radiomics-deep feature correspondence self-supervised learning.

Table 2. Evaluations of tumor classification using different models. The table shows that radiomics handcrafted features are useful in tumor diagnosis and fine-tuning our self-supervised model can improve the performance.

Dataset	Method	Accuracy (%)	F1 score (weighted) (%)
Thyroid nodule	Radiomics method [4]	81.3	80.7
	ResNet-50 [6]	82.6	82.2
	Hybrid network (scratch)	83.4	83.1
	Hybrid network (fine-tune)	84.9	84.4
Kidney tumor	Radiomics method [4]	56.2 ± 6.4	56.8 ± 7.0
	ResNet-50 [6]	50.0 ± 5.3	49.3 ± 7.6
	Hybrid network (scratch)	52.8 ± 5.7	52.0 ± 5.6
	Hybrid network (fine-tune)	64.3 ± 5.9	63.7 ± 6.9

3.3 Linear Classification on Self-supervised Model

Linear Classification is a general benchmark to evaluate unsupervised image representations' quality. We fix the pre-trained network as shown in Fig. 1 and add a fully connected layer behind the generated embeddings for evaluation. In addition, we compare other self-supervised methods using ResNet50. As Table 1 shows, F1 score of radiomics-deep feature correspondence network is higher than models that perform other pretext tasks. This indicates that the self-supervised task which incorporates prior knowledge can improve model's representations for tumors. Furthermore, the linear evaluation results of our model are comparable to the results of supervised models in Table 2. This implies that the embeddings obtained by our pretext task are discriminative and useful representations of tumors, although the distribution of medical image data is complicated.

3.4 Benefits of Self-supervised Pre-training

We now investigate the question of whether radiomics-deep feature correspondence method can improve the performance of image classification, even in the situation of small datasets, i.e., the kidney tumor dataset and a small part of the thyroid nodule dataset.

Supervised Baseline. We conduct three benchmark experiments. First, we use a conventional radiomics method [4] to analyze radiomics handcrafted features, that is, the pipeline of feature preprocessing, feature selecting, and classification. Second, we train ResNet-50 from scratch. Third, we train a hybrid network that combines radiomics and deep learning, namely the network in Sect. 3.3. But all the parameters in this model are random and all updated. The results are shown in Table 2. We can see that the radiomics method perform well on tumor classification, so some of the radiomics handcrafted features can indeed be correlated with tumor label. Moreover, the hybrid model that incorporates prior knowledge can improve performance compared to pure ResNet-50. In summary, quantitative radiomics handcrafted features can reflect the heterogeneity of tumors and increase the power of deep learning models for classification.

Transfer Learning Using All Labeled Data. We fine-tune the pre-trained hybrid network using our self-supervised task. We experimented with fine-tuning different numbers of layers. And we found that the best results were obtained without fine-tuning the last group of convolutions layers in ResNet-50 and the last two layers in fully connected network for processing radiomics handcrafted features. As Table 2 shows, evaluation scores for fine-tuning the pre-trained hybrid network are higher than training from scratch. It is noteworthy that fine-tuning the pre-trained hybrid network can achieve high scores in the kidney tumor classification task with less data and more types. This demonstrates the self-supervised approach can learn effective visual representations for tumor classification and the pre-trained model can generate more discriminative features for the target task by fine tuning.

Efficient Learning Using Less Labeled Data. We evaluate the performance of fine-tuning the pre-trained hybrid model as the size of the labeled dataset varies to 5%, 20%, 50%. The test dataset is always the same as the default setting. We only train these experiments on the thyroid nodule dataset, because the kidney tumor dataset is too small to do that. For comparison, the radiomics method, randomly-initialized ResNet-50 and hybrid network are trained with the above settings. The weighted F1 scores from models with different training datasets are displayed in Fig. 2. Compared to the three train-from-scratch models, the network with pre-trained weights always maintains high performance and brings a more remarkable gain with decreasing amounts of labeled data. These results reveal that our approach alleviates the current situation of insufficient medical labeled data to some extent. By the radiomics-deep feature correspondence pre-training method, the network learns rich visual features from a large amount of unlabeled data and can get free performance improvements with zero manual annotations.

4 Conclusion

In this paper, the radiomics-deep feature correspondence self-supervised learning approach is proposed to boost the model's predictive power of the tumor types. Contrasted by the radiomics view which contains phenotypic characteristics and domain knowledge, the hybrid network can learn good visual representation and mitigate overfitting when solving the target task with insufficient labeled data. Our method is validated on two public medical image datasets to demonstrate its label-efficient classification performance. Future works include enhancement of pre-trained models with more data in public medical image datasets and applications of pre-trained models used in other medical tasks.

Acknowledgements. This research was supported by National Natural Science Foundation under grants (31571001, 61828101). We thank the Big Data Center of Southeast University for providing the GPUs to support the numerical calculations in this paper.

References

1. Chen, T., Kornblith, S., Norouzi, M., Hinton, G.: A simple framework for contrastive learning of visual representations. In: International Conference on Machine Learning, pp. 1597–1607. PMLR (2020)
2. Chopra, S., Hadsell, R., LeCun, Y.: Learning a similarity metric discriminatively, with application to face verification. In: 2005 IEEE Computer Society Conference on Computer Vision and Pattern Recognition (CVPR'05), vol. 1, pp. 539–546. IEEE (2005)
3. Esteva, A., et al.: A guide to deep learning in healthcare. Nat. Med. **25**(1), 24–29 (2019)
4. Gillies, R.J., Kinahan, P.E., Hricak, H.: Radiomics: images are more than pictures, they are data. Radiology **278**(2), 563–577 (2016)
5. Gutmann, M., Hyvärinen, A.: Noise-contrastive estimation: a new estimation principle for unnormalized statistical models. In: Proceedings of the Thirteenth International Conference on Artificial Intelligence and Statistics, pp. 297–304. JMLR Workshop and Conference Proceedings (2010)
6. He, K., Zhang, X., Ren, S., Sun, J.: Deep residual learning for image recognition. In: Proceedings of the IEEE Conference on Computer Vision and Pattern Recognition, pp. 770–778 (2016)
7. He, T., Zhang, Z., Zhang, H., Zhang, Z., Xie, J., Li, M.: Bag of tricks for image classification with convolutional neural networks. In: Proceedings of the IEEE Conference on Computer Vision and Pattern Recognition, pp. 558–567 (2019)
8. Henaff, O.: Data-efficient image recognition with contrastive predictive coding. In: International Conference on Machine Learning, pp. 4182–4192. PMLR (2020)
9. Jamaludin, A., Kadir, T., Zisserman, A.: Self-supervised learning for spinal MRIs. In: Cardoso, M. et al. (eds.) Deep Learning in Medical Image Analysis and Multimodal Learning for Clinical Decision Support. LNCS, vol. 10553, pp. 294–302. Springer, Cham (2017). https://doi.org/10.1007/978-3-319-67558-9_34
10. Jiao, J., Cai, Y., Alsharid, M., Drukker, L., Papageorghiou, A.T., Noble, J.A.: Self-supervised contrastive video-speech representation learning for ultrasound. In: Martel, A.L. et al. (eds.) International Conference on Medical Image Computing and Computer-Assisted Intervention. MICCAI 2020. LNCS, vol. 12263, pp. 534–543. Springer, Cham (2020). https://doi.org/10.1007/978-3-030-59716-0_51
11. Lambin, P., et al.: Radiomics: the bridge between medical imaging and personalized medicine. Nat. Rev. Clin. Oncol. **14**(12), 749 (2017)
12. Larsson, G., Maire, M., Shakhnarovich, G.: Colorization as a proxy task for visual understanding. In: Proceedings of the IEEE Conference on Computer Vision and Pattern Recognition, pp. 6874–6883 (2017)
13. Nathan Mundhenk, T., Ho, D., Chen, B.Y.: Improvements to context based self-supervised learning. In: Proceedings of the IEEE Conference on Computer Vision and Pattern Recognition, pp. 9339–9348 (2018)
14. Ni, D.: Thyroid nodule segmentation and classification in ultrasound images (tn-scui2020) (2020). https://tn-scui2020.grand-challenge.org/

15. Noroozi, M., Favaro, P.: Unsupervised learning of visual representations by solving Jigsaw puzzles. In: Leibe, B., Matas, J., Sebe, N., Welling, M. (eds.) European Conference on Computer Vision. ECCV 2016. LNCS, vol. 9910, pp. 69–84. Springer, Cham (2016). https://doi.org/10.1007/978-3-319-46466-4_5
16. Pedregosa, F., et al.: Scikit-learn: machine learning in Python. J. Mach. Learn. Res. **12**, 2825–2830 (2011)
17. Poole, B., Ozair, S., Van Den Oord, A., Alemi, A., Tucker, G.: On variational bounds of mutual information. In: International Conference on Machine Learning, pp. 5171–5180. PMLR (2019)
18. Tao, X., Li, Y., Zhou, W., Ma, K., Zheng, Y.: Revisiting Rubik's cube: self-supervised learning with volume-wise transformation for 3D medical image segmentation. In: Martel, A.L. et al. (eds.) International Conference on Medical Image Computing and Computer-Assisted Intervention. MICCAI 2020. LNCS, vol. 12264, pp. 238–248. Springer, Cham (2020). https://doi.org/10.1007/978-3-030-59719-1_24
19. Tian, Y., Krishnan, D., Isola, P.: Contrastive multiview coding. In: Vedaldi, A., Bischof, H., Brox, T., Frahm, J.M. (eds.) Computer Vision – ECCV 2020. LNCS, vol. 12356, pp. 776–794. Springer, Cham (2020). https://doi.org/10.1007/978-3-030-58621-8_45
20. Van Griethuysen, J.J., et al.: Computational radiomics system to decode the radiographic phenotype. Cancer Res. **77**(21), e104–e107 (2017)
21. Vincent, P., Larochelle, H., Bengio, Y., Manzagol, P.A.: Extracting and composing robust features with denoising autoencoders. In: Proceedings of the 25th International Conference on Machine Learning, pp. 1096–1103 (2008)
22. Weight, C.: The 2019 kidney and kidney tumor segmentation challenge (KiTS19) (2019). https://kits19.grand-challenge.org/
23. Wu, Z., Xiong, Y., Yu, S.X., Lin, D.: Unsupervised feature learning via non-parametric instance discrimination. In: Proceedings of the IEEE Conference on Computer Vision and Pattern Recognition, pp. 3733–3742 (2018)

Learning 4D Infant Cortical Surface Atlas with Unsupervised Spherical Networks

Fenqiang Zhao[1], Zhengwang Wu[1], Li Wang[1], Weili Lin[1], Shunren Xia[2], Gang Li[1](✉), and the UNC/UMN Baby Connectome Project Consortium

[1] Department of Radiology and BRIC, University of North Carolina at Chapel Hill, Chapel Hill, NC, USA
gang_li@med.unc.edu

[2] Key Laboratory of Biomedical Engineering of Ministry of Education, Zhejiang University, Hangzhou, China

Abstract. Spatiotemporal (4D) cortical surface atlas during infancy plays an important role for surface-based visualization, normalization and analysis of the dynamic early brain development. Conventional atlas construction methods typically rely on classical group-wise registration on sub-populations and ignore longitudinal constraints, thus having *three* main issues: 1) constructing templates at discrete time points; 2) resulting in longitudinal inconsistency among different age's atlases; and 3) taking extremely long runtime. To address these issues, in this paper, we propose a fast *unsupervised learning-based surface atlas construction framework* incorporating *longitudinal constraints* to enforce the within-subject temporal correspondence in the atlas space. To well handle the difficulties of learning large deformations, we propose a *multi-level multi-modal spherical registration network* to perform cortical surface registration in a coarse-to-fine manner. Thus, only small deformations need to be estimated at each resolution level using the registration network, which further improves registration accuracy and atlas quality. Our constructed 4D infant cortical surface atlas based on 625 longitudinal scans from 291 infants is temporally *continuous*, in contrast to the state-of-the-art UNC 4D Infant Surface Atlas, which only provides the atlases at a few *discrete sparse* time points. By evaluating the intra- and inter-subject spatial normalization accuracy after alignment onto the atlas, our atlas demonstrates more detailed and fine-grained cortical patterns, thus leading to higher accuracy in surface registration.

Keywords: Surface registration · Infant cortical surface atlas

1 Introduction

Cortical surface atlases play an important role in neuroimaging studies by providing a common space for normalizing, comparing, and analyzing brain structure and function across different individuals and studies [14,19]. Considering

M. de Bruijne et al. (Eds.): MICCAI 2021, LNCS 12902, pp. 262–272, 2021.
https://doi.org/10.1007/978-3-030-87196-3_25

the rapid growth of the cerebral cortex during early brain development, spatiotemporal (4D) surface atlases are essential to characterize and model such dynamic development of infant brains [22] and are typically constructed on the spherical space by taking advantage of the intrinsic spherical typology of the cortex [9]. Conventional 4D atlas construction methods [1,9,10,12,16,22] generally perform several rounds of template estimation and individual-to-template registration, and finally average (based on Euclidean distance [9], Wasserstein distance [2], or sparse representation [17,22]) the cortical features in each age group to obtain the atlas for each age. Although these methods and their constructed infant cortical surface atlases are popular and widely used to investigate early brain development in the past decade, they have *three main issues. First*, these methods require an extremely long runtime due to the iterative group-wise registration process. *Second*, they create the atlas at each age based on a subset of the data without exploiting the rich information from the whole dataset, leading to suboptimal atlases. Accordingly, the resulting 4D atlases are constructed at discrete, sparse time points based on subpopulations scanned by predefined concentrated ages, e.g., 1, 3, 6, 9, 12, 18 and 24 months of age. These atlases are thus temporally discontinuous and cannot cover many important ages during infancy. Although the kernel regression approach [1,12,16] could be applied to interpolate the uncovered ages' atlases with specific kernel designs, it was not fully investigated and developed for infants, due to the difficulties in modeling the complicated, regionally-heterogeneous and nonlinear early brain development. *Third*, the methods [1,16] constructing atlases for each age independently based on the subset of the data may cause longitudinal inconsistency among different ages' atlases when longitudinal subjects are involved. The reason is that, as revealed by previous works [9,12,22], without considering the longitudinal information, the longitudinally-corresponding anatomical points at different ages of the same subject are independently deformed to different locations in the atlas space, consequently leading to a longitudinally-inconsistent 4D atlas.

To address these issues, we propose an unsupervised learning framework to learn a continuous 4D infant cortical surface atlas, inspired by the recent spherical networks developed on cortical surfaces [25] and an unsupervised atlas construction method [3]. Our approach jointly learns an *age-conditional atlas synthesis network* and an *unsupervised registration network* from the entire dataset without manually partitioning the dataset into subgroups, thus is computationally more efficient and powerful in learning rich, useful information from the whole dataset than conventional methods. Moreover, since learning large deformations is difficult [13,28], we propose a *deep multi-level multi-modal spherical registration network (MM-SRegNet)* to decompose the large deformations into small ones at each resolution step and perform cortical surface registration in a coarse-to-fine manner. Further considering the subject-specific longitudinal information of longitudinal subjects frequently seen in large-scale studies, we design a novel *longitudinal constraint (LC)* loss to enforce the within-subject temporal correspondence after registration to the atlas space, which is impor-

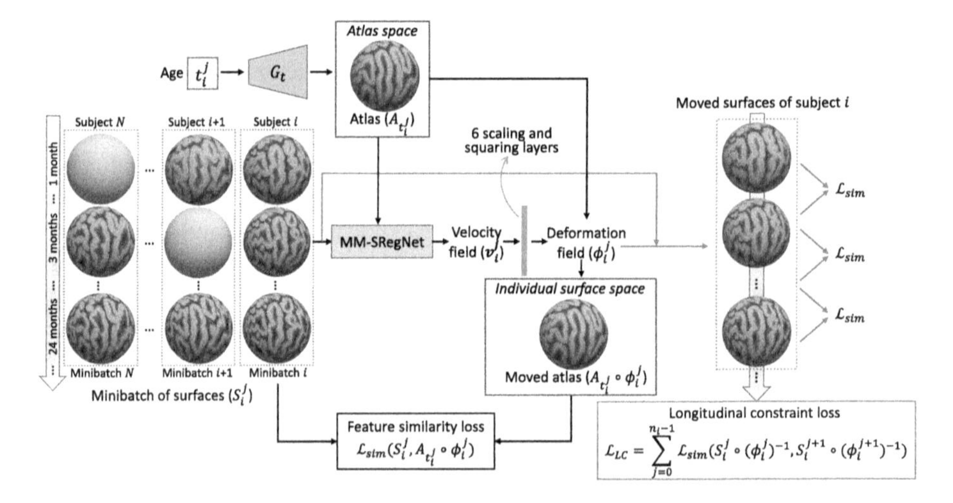

Fig. 1. Overview of our unsupervised learning framework for 4D infant cortical surface atlas construction. The atlas synthesis network G_t takes age attributes as input and outputs the conditional atlases, which are then registered to each surface in the input minibatch by MM-SRegNet (Fig. 2). Cortical surfaces of the same subject are grouped together to build the longitudinal constraint loss highlighted in orange color. The blank spheres indicate unavailable data at those time points. (Color figure online)

tant for establishing longitudinally consistent 4D atlases [9]. To sum up, our contributions in this paper are:

1. We propose an *unsupervised learning-based 4D cortical surface atlas construction framework* to jointly learn an *atlas synthesis network* and *MM-SRegNet* for generating the age-conditional atlas and the deformation field registering the cortical surfaces to the age-matched atlases simultaneously. Besides, our MM-SRegNet enables flexible and accurate cortical surface registration using multiple cortical features in a coarse-to-fine manner.
2. We efficiently and effectively *incorporate longitudinal constraints into our atlas learning framework*, which is essential for establishing within-subject longitudinally-consistent correspondences, thus obtaining longitudinally more consistent atlas with higher quality.
3. Leveraging our proposed method, we construct the *first longitudinally-consistent temporally-continuous 4D infant cortical surface atlas* based on 625 longitudinal scans from 291 infants and show that our generated 4D atlas preserves more details of cortical folding patterns compared to the state-of-the-art [22].

2 Method

As shown in Fig. 1, our goal is to jointly train a generative model G_t that can synthesize atlas given on-demand ages, and a registration model that registers the synthesized atlas to each input surface, thus avoiding the expensive time cost of conventional multi-round registration and iterative template refinement process. Let $S = \{S_1^1, \ldots, S_1^{n_1}, \ldots, S_i^j, \ldots, S_N^{n_N}\}$ denote a longitudinal cortical surface dataset with N subjects and each subject i ($i = 1, \ldots, N$) has n_i longitudinal scans, S_i^j denotes the jth time point cortical surface of subject i and t_i^j is the age of subject i at jth time point. Note that the number of longitudinal scans and the time points of each subject are not necessarily to be the same. Then a ***baseline*** framework aims to minimize the loss $\mathcal{L}_{BL}$ on the whole dataset:

$$\mathcal{L}_{BL} = \mathcal{L}_{sim}(S_i^j, A_{t_i^j} \circ \phi_i^j) + \lambda_c \|\overline{u}\|^2 + \lambda_d \sum_{i,j} \|u_i^j\|^2 + \lambda_s \sum_{i,j} \|\nabla_s u_i^j\|^2 \quad (1)$$

where $\mathcal{L}_{sim}(S_i^j, A_{t_i^j} \circ \phi_i^j) = 1 - \frac{cov(S_i^j, A_{t_i^j} \circ \phi_i^j)}{\sqrt{\sigma_{S_i^j} \cdot \sigma_{A_{t_i^j} \circ \phi_i^j}}}$ is the correlation coefficient loss for enforcing the similarity between the moved atlas and individual cortical surface maps, $A_{t_i^j} = G_t(t_i^j)$ is the synthesized atlas at age t_i^j, ϕ_i^j is the deformation field aligning atlas to jth surface of subject i and $\phi = exp(v)$ is computed using 6 "scaling and squaring" layers as in [24], which is an effective extension of diffeomorphic deformation from Euclidean space to spherical space. $u = \frac{\phi(x)}{x \cdot \phi(x)} - x$ is the tangent displacement vector on a unit sphere from x to $\phi(x)$. Hence, the rest terms in Eq. 1 regularize the unbiasedness, extent and smoothness of u, respectively. Note that we use the same ∇_s operator as in [26] on the spherical surface to approximate the tangent displacement field's gradients and accordingly penalize it to encourage a smooth deformation field on the spherical surface.

2.1 Longitudinal Constraint

Since the above baseline framework ignores the within-subject longitudinal consistency constraint when involving longitudinal subjects, we are motivated to incorporate it into our framework. To achieve this, we first group the longitudinal surfaces from the same subject in one minibatch, as shown in Fig. 1. Then we can explicitly enforce the longitudinal correspondence among the surfaces of the same subject after moving them to the atlas space, which is formulated as the longitudinal constraint (LC) loss:

$$\mathcal{L}_{LC} = \sum_i \sum_{j=0}^{n_i - 1} \mathcal{L}_{sim}(S_i^j \circ (\phi_i^j)^{-1}, S_i^{j+1} \circ (\phi_i^{j+1})^{-1}) \quad (2)$$

where $(\phi)^{-1}$ denotes the inverse deformation of ϕ, which can be conveniently computed using the method in [23]. This novel $\mathcal{L}_{LC}$ is very efficient to integrate, as it adds 0 parameters and only a small overhead over computation to the baseline framework.

2.2 Multi-level Multi-modal Spherical Registration Network

In unsupervised registration framework, it is known that large deformations are difficult to learn [13,28]. To address this issue, a typical method is stacking multiple networks in a pyramid fashion and perform the registration in a

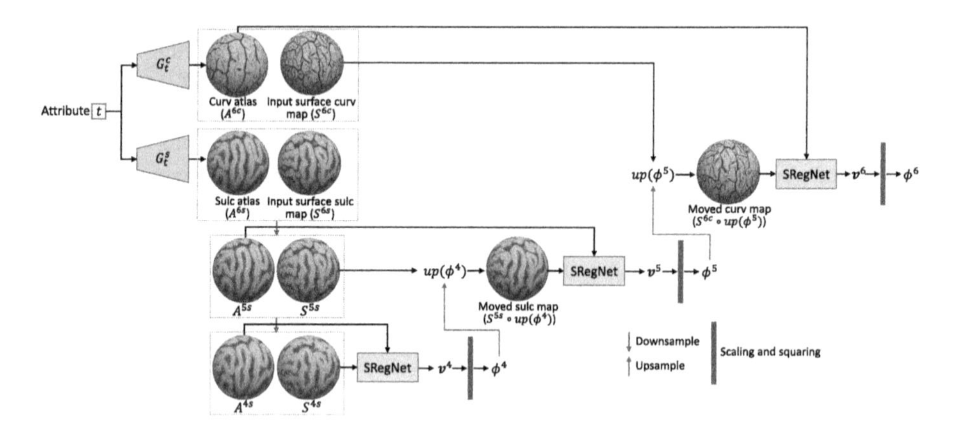

Fig. 2. Illustration of our MM-SRegNet. The superscript (6c) indicates the level 6 and the feature ('s' for sulc, 'c' for curv). 'A' denotes the atlas surface and 'S' is the individual surface. Note that this only provides a brief overview of our method, which can be flexibly extended to other levels and cortical features.

coarse-to-fine manner [13,20,28]. Inspired by these networks, we propose the MM-SRegNet to perform the coarse-to-fine cortical surface registration. A recent unsupervised cortical surface registration framework [26] employs three Spherical U-Nets [27] to predict spherical deformation fields for three equatorial regions at three orthogonal orientations, thus addressing the polar-distortion issue in spherical surface registration. We build on this method and incorporate the novel spherical "scaling and squaring" layers [23] to ensure diffeomorphic (invertible and topology-preserving) deformations to construct SRegNet and further stack it to construct MM-SRegNet, as shown in Fig. 2. Note that multimodal cortical features (such as geometry, myelin or functional gradient) are imperfectly correlated [15], therefore our MM-SRegNet also offers much flexibility in the choice of cortical features and thus can provide on-demand atlas flexibly.

In this paper, to obtain 4D atlas with fine cortical folding patterns, we follow the popular registration configuration [15,23] to align geometric features, i.e., 'sulc' (average convexity) and 'curv' (mean curvature), measuring cortical folding in a coarse and fine view, respectively. We stack 4 SRegNets at 4 levels (4, 5, 6, 6, with 2,562, 10,242, 40,962, 40,962 vertices, respectively) for aligning sulc, sulc, sulc, curv, respectively. Finally, we deform the generated sulc atlas to the final curv atlas space using the learned deformation field at the last level.

2.3 Network and Implementation Details

The generative model G_t for synthesizing the atlas consists of a fully connected layer with 10242*3 neurons, a reshape layer and an upsampling layer followed by 3 "batch normalization [7] + spherical 1-ring convolution [25] + ReLU" blocks each with [16, 32, 1] channels. Herein, G_t takes age t_i^j as a scalar value input and outputs the age-conditional atlas (40962×1). For the registration network, since we stack multiple SRegNets to perform the coarse-to-fine registration, we create a smaller version of Spherical U-Net [25] used in SRegNet, with 3 resolution steps and [8, 16, 32] channels, resulting in only 0.5M parameters in each SRegNet.

We implemented our method using PyTorch. We trained the atlas synthesis network G_t and registration network MM-SRegNet jointly by minimizing the full objective $\mathcal{L}_{BL} + \lambda_{LC}\mathcal{L}_{LC}$ in a minibatch style. We trained the network using Adam optimizer with learning rate 5e-4 for 50 epochs. $\lambda_c = 1.0$, $\lambda_d = 0.2$, $\lambda_s =$ 1.5, 2.5 and 4.0 at 4th, 5th and 6th level, respectively, and $\lambda_{LC} = 1.0$.

Note that although looking bigger, our final model is compact with only a small number of parameters, G_t with 66K, and MM-SRegNet with 6M parameters in total. Taking about 10 h to fully train the networks on a PC with an NVIDIA RTX2080 Ti GPU and an Intel Core i7-9700K CPU, it only needs *5 s* for one forward inference, including atlas generation and individual-to-atlas registration, which is much faster than conventional atlas construction and registration methods, where the whole process can take several days.

3 Experiments

3.1 Dataset and Preprocessing

We used a longitudinal infant dataset [6] with 625 scans from 291 subjects, with ages ranging from 0 to 5 years. The distribution of the number of longitudinal scans is, 44% infants with only 1 scan, 24% with 2 scans, 14% with 3, 10% with 4 and 8% with 5 or more scans. Cortical surfaces were reconstructed via iBEAT V2.0 Cloud (http://www.ibeat.cloud/) [8,11,18,21] and then mapped onto the sphere using FreeSurfer [5]. Each surface is then represented as a spherical map with 2 features at each vertex, i.e., 'sulc' and 'curv', and a parcellation map is also generated based on [4]. We randomly split the dataset into a training set with 418 scans from 185 subjects, and a test set with 207 scans from 106 subjects. In testing stage, we generate the age-matched atlas for each test surface and the corresponding deformation field aligning it to the age-specific atlas simultaneously.

To quantitatively validate the atlas, we calculate the widely-used within-group spatial normalization accuracy, which is measured by the Pearson correlation coefficient (PCC) of feature maps and Dice overlap metric of parcellation maps as in [9,15,22,26]. Generally, sharper atlases lead to higher registration accuracy and thus better spatial normalization performance. Therefore, a higher PCC and Dice indicate better within-group alignment and thus a better atlas.

3.2 Ablation Studies

MM-SRegNet. To demonstrate the effectiveness of our MM-SRegNet, we trained two baseline models on the 6th level, named **B-Sulc**, **B-Curv** for generating two baseline sulc and curv atlas with single level SRegNet. Since they use single-modal feature, the most critical problem of these two baselines is that the constructed atlases may not be in the same space, which causes trouble in neuroimaging analysis. As shown in Table 1, the **B-M** model using our MM-SRegNet as the registration network leads to a higher correlation and Dice for both within-subject and across-subject alignment. This is important for building atlases with other cortical features accurately and flexibly in the future study.

Table 1. Within-subject and inter-subject alignment performance of different models.

	Within-subject alignment			Inter-subject alignment		
	PCC_{sulc}	PCC_{curv}	Dice (%)	PCC_{sulc}	PCC_{curv}	Dice (%)
B-Sulc	0.879 ± 0.051	-	83.45 ± 5.76	0.672 ± 0.062	-	70.92 ± 4.40
B-Curv	–	0.557 ± 0.088	83.31 ± 5.79	–	0.337 ± 0.053	69.39 ± 6.16
B-M	0.907 ± 0.044	0.677 ± 0.100	85.02 ± 5.32	0.728 ± 0.060	0.484 ± 0.046	74.91 ± 5.46
B-M+LC	**0.927 ± 0.040**	**0.743 ± 0.067**	**86.52 ± 5.10**	**0.751 ± 0.052**	**0.528 ± 0.057**	**76.12 ± 4.85**

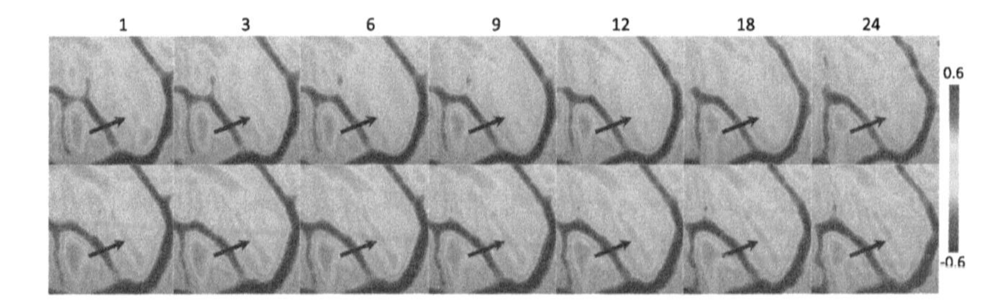

Fig. 3. Comparison of the learned 4D surface curv atlases using B-M (upper row) and B-M+LC (lower row) at discrete ages (months) in the first two postnatal years.

Longitudinal Constraint (LC). As shown in Table 1, adding LC loss to the baseline model (named **B-M+LC**) significantly improve the performance, which is reasonable for correlation metrics, since we directly add the within-subject correlation as an additional loss to train the model. As a more unbiased metric, the Dice improvement is a compelling evidence that incorporating LC strengthens the within-subject temporal correspondence and thus enhances the longitudinal consistency and the quality of the synthesized 4D atlas. Figure 3 provides a typical example of the curv atlases learned without LC and with LC. As we can see, the LC encourages longitudinally more consistent 4D cortical surface atlases along with much shaper cortical folding patterns, than the model without LC.

3.3 Comparison with State-of-the-Art

Visual Inspection. We used the B-M+LC model to generate the infant cortical surface atlases at corresponding discrete ages with the state-of-the-art UNC 4D infant cortical surface atlas [22] for comparison. Figure 4 shows the two atlases only in the first two postnatal years because of the most dynamic postnatal development during this period [11]. We can see that our atlas preserves more

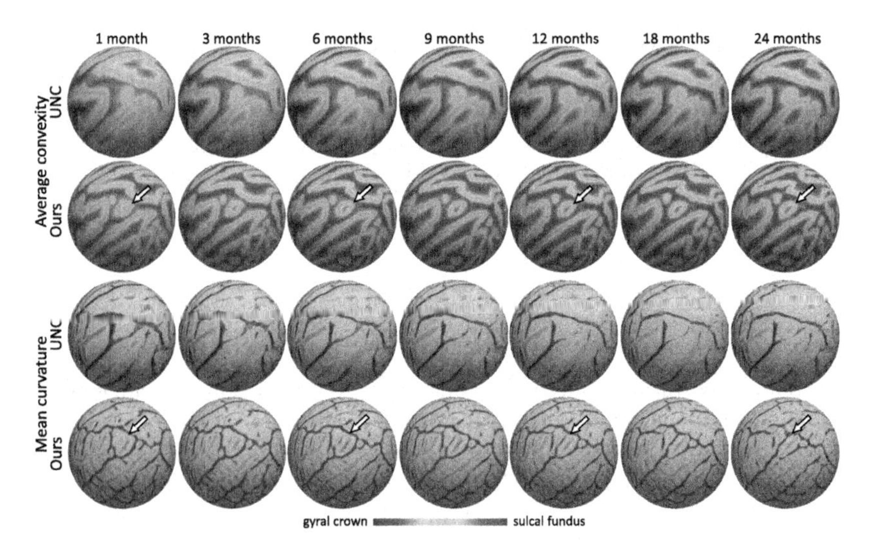

Fig. 4. Comparison of UNC 4D surface atlas and ours. Note that though limited time points are shown, our method offers *continuous* 4D atlas at arbitrary time points from 0 to 5 years of age.

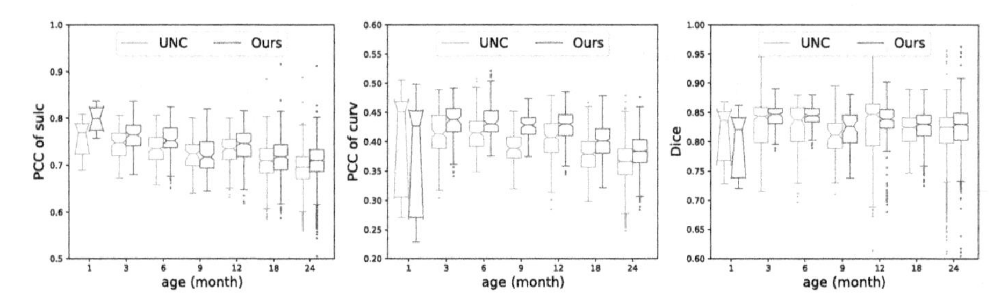

Fig. 5. Spatial normalization accuracy at each age group based on different atlases.

detailed cortical folding patterns than UNC atlas in most regions, e.g., insula, inferior parietal and lateral temporal cortex. More importantly, as indicated by arrows, minor cortical folds, which are still developing rapidly during infancy, are well captured by our atlas, but absent in UNC 4D atlas.

Quantitative Evaluation. We used a third-party registration tool, Spherical Demons [23], a conventional method solving registration problem using Gauss-Newton method, to fairly register the hold-out test surfaces to our atlas and UNC atlas, respectively. Note that we choose the age-matched UNC atlas that are closest to the individual surface's age for each individual surface (11 instances available in UNC atlas, i.e., 1, 3, 6, 9, 12, 18, 24, 36, 48, 60, and 72 months), while our method generates on-demand age-specific atlas at arbitrary time points and thus can provide more accurate and matched atlas to guide the registration. The average PCC of sulc, PCC of curv, and Dice is 0.706 ± 0.039, 0.373 ± 0.063,

0.812 ± 0.049 for UNC atlas, and 0.720 ± 0.043, 0.382 ± 0.058, 0.821 ± 0.030 for our atlas. Detailed results in the first two years are shown in Fig. 5. We can see that our atlas outperforms UNC atlas in most age groups and shows no significantly differences if not better, except PCC of curv and Dice at 1 month likely because of the limited samples in this group, demonstrating the practical usefulness of our higher-quality atlas in improving surface registration and parcellation accuracy.

4 Conclusion

In this paper, we present three methodological contributions to efficiently and accurately build the *first longitudinally-consistent temporally-continuous* 4D infant cortical surface atlas. *First*, we propose an unsupervised learning framework for simultaneously atlas synthesis and individual-to-atlas registration. Taking advantage of the deep learning ability of neural networks, our method can efficiently learn the atlas from the whole dataset and rapidly warp the new surfaces to the generated atlas. *Second*, our novel MM-SRegNet can accurately perform cortical surface registration in a coarse-to-fine manner flexibly using multiple cortical features. *Third*, incorporating the longitudinal constraint further improves within-subject longitudinal consistency and thus the quality of 4D atlas. Compared with the state-of-the-art UNC 4D atlas both visually and quantitatively, our atlas exhibits sharper and much more detailed patterns for more accurately mapping infant brain development. In future, we will release our model and atlas to the community to advance the neuroimaging studies of infants.

Acknowledgements. This work was partially supported by NIH grants (MH116 225, MH117943, MH109773, MH123202). This work also utilizes approaches developed by an NIH grant (1U01MH110274) and the efforts of the UNC/UMN Baby Connectome Project Consortium.

References

1. Boze, J., et al.: Construction of a neonatal cortical surface atlas using multimodal surface matching in the developing human connectome project. NeuroImage **179**, 11–29 (2018)
2. Chen, Z., et al.: Construction of 4D neonatal cortical surface atlases using Wasserstein distance. In: 2019 IEEE 16th International Symposium on Biomedical Imaging (ISBI 2019), pp. 995–998. IEEE (2019)
3. Dalca, A.V., Rakic, M., Guttag, J., Sabuncu, M.R.: Learning conditional deformable templates with convolutional networks. In: Proceedings of the 33rd International Conference on Neural Information Processing Systems, pp. 806–818 (2019)
4. Desikan, R.S., et al.: An automated labeling system for subdividing the human cerebral cortex on MRI scans into gyral based regions of interest. Neuroimage **31**(3), 968–980 (2006)

5. Fischl, B., Sereno, M.I., Dale, A.M.: Cortical surface-based analysis: II: inflation, flattening, and a surface-based coordinate system. Neuroimage **9**(2), 195–207 (1999)
6. Howell, B.R., et al.: The UNC/UMN baby connectome project (BCP): an overview of the study design and protocol development. NeuroImage **185**, 891–905 (2019)
7. Ioffe, S., Szegedy, C.: Batch normalization: Accelerating deep network training by reducing internal covariate shift. In: International Conference on Machine Learning, pp. 448–456. PMLR (2015)
8. Li, G., et al.: Measuring the dynamic longitudinal cortex development in infants by reconstruction of temporally consistent cortical surfaces. Neuroimage **90**, 266–279 (2014)
9. Li, G., Wang, L., Shi, F., Gilmore, J.H., Lin, W., Shen, D.: Construction of 4D high-definition cortical surface atlases of infants: methods and applications. Med. Image Anal. **25**(1), 22–36 (2015)
10. Li, G., Wang, L., Shi, F., Lin, W., Shen, D.: Constructing 4D infant cortical surface atlases based on dynamic developmental trajectories of the cortex. In: Golland, P., Hata, N., Barillot, C., Hornegger, J., Howe, R. (eds.) MICCAI 2014. LNCS, vol. 8675, pp. 89–96. Springer, Cham (2014). https://doi.org/10.1007/978-3-319-10443-0_12
11. Li, G., et al.: Computational neuroanatomy of baby brains: a review. NeuroImage **185**, 906–925 (2019)
12. Liao, S., Jia, H., Wu, G., Shen, D.: A novel longitudinal atlas construction framework by groupwise registration of subject image sequences. In: Székely, G., Hahn, H.K. (eds.) IPMI 2011. LNCS, vol. 6801, pp. 283–295. Springer, Heidelberg (2011). https://doi.org/10.1007/978-3-642-22092-0_24
13. Mok, T.C.W., Chung, A.C.S.: Large deformation diffeomorphic image registration with Laplacian pyramid networks. In: Martel, A.L., et al. (eds.) MICCAI 2020. LNCS, vol. 12263, pp. 211–221. Springer, Cham (2020). https://doi.org/10.1007/978-3-030-59716-0_21
14. Oishi, K., Chang, L., Huang, H.: Baby brain atlases. Neuroimage **185**, 865–880 (2019)
15. Robinson, E.C., et al.: MSM: a new flexible framework for multimodal surface matching. Neuroimage **100**, 414–426 (2014)
16. Serag, A., et al.: Construction of a consistent high-definition spatio-temporal atlas of the developing brain using adaptive kernel regression. Neuroimage **59**(3), 2255–2265 (2012)
17. Shi, F., et al.: Neonatal atlas construction using sparse representation. Hum. Brain Mapp. **35**(9), 4663–4677 (2014)
18. Sun, L., et al.: Topological correction of infant white matter surfaces using anatomically constrained convolutional neural network. NeuroImage **198**, 114–124 (2019)
19. Van Essen, D.C., Dierker, D.L.: Surface-based and probabilistic atlases of primate cerebral cortex. Neuron **56**(2), 209–225 (2007)
20. de Vos, B.D., Berendsen, F.F., Viergever, M.A., Sokooti, H., Staring, M., Išgum, I.: A deep learning framework for unsupervised affine and deformable image registration. Med. Image Anal. **52**, 128–143 (2019)
21. Wang, L., et al.: Volume-based analysis of 6-month-old infant brain MRI for autism biomarker identification and early diagnosis. In: Frangi, A.F., Schnabel, J.A., Davatzikos, C., Alberola-López, C., Fichtinger, G. (eds.) MICCAI 2018. LNCS, vol. 11072, pp. 411–419. Springer, Cham (2018). https://doi.org/10.1007/978-3-030-00931-1_47

22. Wu, Z., Wang, L., Lin, W., Gilmore, J.H., Li, G., Shen, D.: Construction of 4D infant cortical surface atlases with sharp folding patterns via spherical patch-based group-wise sparse representation. Hum. Brain Mapp. **40**(13), 3860–3880 (2019)
23. Yeo, B.T., Sabuncu, M.R., Vercauteren, T., Ayache, N., Fischl, B., Golland, P.: Spherical demons: fast diffeomorphic landmark-free surface registration. IEEE Trans. Med. Imaging **29**(3), 650–668 (2009)
24. Zhao, F., et al.: S3Reg: superfast spherical surface registration based on deep learning. IEEE Trans. Med. Imaging (2021)
25. Zhao, F., et al.: Spherical deformable U-Net: application to cortical surface parcellation and development prediction. IEEE Trans. Med. Imaging **40**(4), 1217–1228 (2021)
26. Zhao, F., et al.: Unsupervised learning for spherical surface registration. In: Liu, M., Yan, P., Lian, C., Cao, X. (eds.) MLMI 2020. LNCS, vol. 12436, pp. 373–383. Springer, Cham (2020). https://doi.org/10.1007/978-3-030-59861-7_38
27. Zhao, F., et al.: Spherical U-Net on cortical surfaces: methods and applications. In: Chung, A.C.S., Gee, J.C., Yushkevich, P.A., Bao, S. (eds.) IPMI 2019. LNCS, vol. 11492, pp. 855–866. Springer, Cham (2019). https://doi.org/10.1007/978-3-030-20351-1_67
28. Zhao, S., Dong, Y., Chang, E.I., Xu, Y., et al.: Recursive cascaded networks for unsupervised medical image registration. In: Proceedings of the IEEE/CVF International Conference on Computer Vision, pp. 10600–10610 (2019)

Multimodal Representation Learning via Maximization of Local Mutual Information

Ruizhi Liao[1(✉)], Daniel Moyer[1], Miriam Cha[2], Keegan Quigley[2], Seth Berkowitz[3], Steven Horng[3], Polina Golland[1], and William M. Wells[1,4]

[1] CSAIL, Massachusetts Institute of Technology, Cambridge, MA, USA
ruizhi@mit.edu
[2] MIT Lincoln Laboratory, Lexington, MA, USA
[3] Beth Israel Deaconess Medical Center, Harvard Medical School, Boston, MA, USA
[4] Brigham and Women's Hospital, Harvard Medical School, Boston, MA, USA

Abstract. We propose and demonstrate a representation learning approach by maximizing the mutual information between local features of images and text. The goal of this approach is to learn *useful* image representations by taking advantage of the rich information contained in the free text that describes the findings in the image. Our method trains image and text encoders by encouraging the resulting representations to exhibit high local mutual information. We make use of recent advances in mutual information estimation with neural network discriminators. We argue that the sum of local mutual information is typically a lower bound on the global mutual information. Our experimental results in the downstream image classification tasks demonstrate the advantages of using local features for image-text representation learning.

Keywords: Multimodal representation learning · Local feature representations · Mutual information maximization

1 Introduction

We present a novel approach for image-text representation learning by maximizing the mutual information between local features of the images and the text. In the context of medical imaging, the images could be, for example, radiographs and the text could be radiology reports that capture radiologists' impressions of the images. A large number of such image-text pairs are generated in the clinical workflow every day [7,13]. Jointly learning from images and raw text can support a leap in the quality of medical vision models by taking advantage of existing expert descriptions of the images.

Learning to extract *useful* feature representations from training data is an essential objective of a deep learning model. The definition of *usefulness* is task-specific [3,5,26]. In this work, we aim to learn image representations that improve

M. de Bruijne et al. (Eds.): MICCAI 2021, LNCS 12902, pp. 273–283, 2021.
https://doi.org/10.1007/978-3-030-87196-3_26

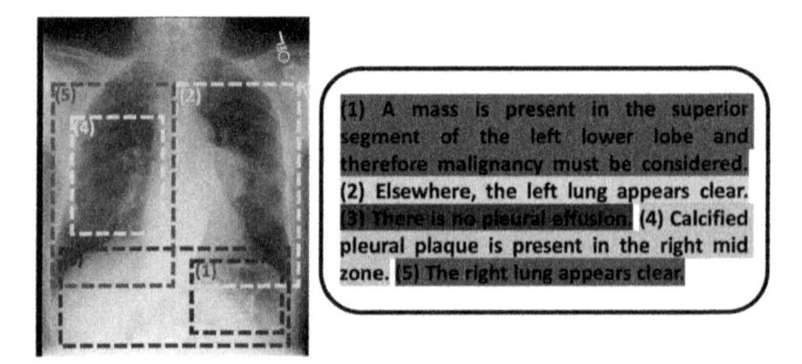

Fig. 1. An example image-text pair (a chest radiograph and its associated radiology report). Each sentence describes the image findings in a particular region of the image. This figure is best viewed in color. (Color figure online)

classification tasks, such as pathology detection, by making use of the rich information contained in the raw text that describe the findings in the image.

We exploit mutual information (MI) to learn useful image representations jointly with text. MI quantifies statistical dependencies between two random variables. Prior work has estimated and optimized MI across images for image registration [20,30], and MI between images and image features for unsupervised learning [6,10,24]. Since the text usually describes image findings that are relevant for downstream image classification tasks, it is sensible to encourage the image and text representations to exhibit high MI.

We propose to learn an image encoder and a text encoder by maximizing the MI of their resulting image and text representations. Moreover, we estimate and optimize the MI between local image features and sentence-level text representations. Figure 1 shows an example image-text pair, where the image is a chest radiograph and the document is the associated radiology report [13]. Each sentence in the report describes a local region in the image. A sentence is usually a minimal and complete semantic unit [25,33]. The findings described in that semantic unit are usually captured in a local region of the image [8].

Prior work in image-text joint learning has leveraged image-based text generation as an auxiliary task during the image model training [22,28,32], or has blended image and text features for downstream inference tasks [23]. Other work has leveraged contrastive learning, an approach to maximize a lower bound on MI to learn image and text representations jointly [4,33]. To the best of our knowledge, this work represents the first attempt to exploit the image spatial structure and sentence-level text features with MI maximization to learn image and text representations that are *useful* for subsequent analysis of images. In our experimental results, we demonstrate that the maximization of local MI yields the greatest improvement in the downstream image classification tasks.

This paper is organized as follows. In Sect. 2, we derive our approach for image-text representation learning by maximizing local MI. Section 3 discusses the relationship between the sum of local MIs and the global MI. This is followed by empirical evaluation in Sect. 4, where we describe the implementation details of our algorithms in application to chest radiographs and radiology reports.

2 Methods

Let x^{I} be an image and x^{R} be the associated free text such as a radiology report or a pathology report that describes findings in the image. The objective is to learn useful latent image representations $z^{\mathrm{I}}(x^{\mathrm{I}})$ and text representations $z^{\mathrm{R}}(x^{\mathrm{R}})$ from image-text data $\mathcal{X} = \{\mathrm{x}_j\}_{j=1}^{N}$, where $\mathrm{x}_j = (\mathrm{x}_j^{\mathrm{I}}, \mathrm{x}_j^{\mathrm{R}})$. We construct an image encoder and a text encoder parameterized by $\theta_{\mathrm{E}}^{\mathrm{I}}$ and $\theta_{\mathrm{E}}^{\mathrm{R}}$, respectively, to generate the representations $z^{\mathrm{I}}(x^{\mathrm{I}}; \theta_{\mathrm{E}}^{\mathrm{I}})$ and $z^{\mathrm{R}}(x^{\mathrm{R}}; \theta_{\mathrm{E}}^{\mathrm{R}})$.

Mutual Information Maximization. We seek such image and text encoders and learn their representations by maximizing MI between the image representation and the text representation:

$$I(z^{\mathrm{I}}, z^{\mathrm{R}}) \triangleq \mathbb{E}_{p(z^{\mathrm{I}}, z^{\mathrm{R}})}\left[\log \frac{p(z^{\mathrm{I}}, z^{\mathrm{R}})}{p(z^{\mathrm{I}})p(z^{\mathrm{R}})}\right]. \tag{1}$$

We employ MI as a statistical measure that captures dependency between images and text in the joint representation space. Maximizing MI between image and text representations is equivalent to maximizing the difference of the entropy and the conditional entropy of image representation given text: $I(z^{\mathrm{I}}, z^{\mathrm{R}}) = H(z^{\mathrm{I}}) - H(z^{\mathrm{I}}|z^{\mathrm{R}})$. This criterion encourages the model to learn feature representations where the information from one modality reduces the entropy of the other data modality, which is a better choice than solely minimizing the conditional entropy, where the image encoder could generate identical features for all data to achieve the conditional entropy minimum.

Stochastic Optimization of MI. Estimating mutual information between high-dimensional continuous variables from finite data samples is challenging. We leverage the recent advances that employ neural network discriminators for MI estimation and maximization [2,18,24,27]. The key idea is to construct a discriminator $f(\mathrm{z}_i^{\mathrm{I}}, \mathrm{z}_j^{\mathrm{R}}; \theta_{\mathrm{D}})$, parameterized by θ_{D}, that estimates the likelihood (or the likelihood ratio) of whether a sample pair $(\mathrm{z}_i^{\mathrm{I}}, \mathrm{z}_j^{\mathrm{R}})$ is sampled from the joint distribution $p(z^{\mathrm{I}}, z^{\mathrm{R}})$ or from the product of marginals $p(z^{\mathrm{I}})p(z^{\mathrm{R}})$. The discriminator is commonly found by maximizing the lower bound of the MI approximated by the likelihood ratio in Eq. (1) [2,24].

We train the discriminator $f(\mathrm{z}_i^{\mathrm{I}}, \mathrm{z}_j^{\mathrm{R}}; \theta_{\mathrm{D}})$ jointly with image and text encoders $z^{\mathrm{I}}(x^{\mathrm{I}}; \theta_{\mathrm{E}}^{\mathrm{I}})$ and $z^{\mathrm{R}}(x^{\mathrm{R}}; \theta_{\mathrm{E}}^{\mathrm{R}})$ via MI maximization:

$$\hat{\theta}_{\mathrm{E}}^{\mathrm{I}}, \hat{\theta}_{\mathrm{E}}^{\mathrm{R}}, \hat{\theta}_{\mathrm{D}} = \arg \max_{\theta_{\mathrm{E}}^{\mathrm{I}}, \theta_{\mathrm{E}}^{\mathrm{R}}, \theta_{\mathrm{D}}} \hat{I}(z^{\mathrm{I}}(x^{\mathrm{I}}; \theta_{\mathrm{E}}^{\mathrm{I}}), z^{\mathrm{R}}(x^{\mathrm{R}}; \theta_{\mathrm{E}}^{\mathrm{R}}); \theta_{\mathrm{D}}), \tag{2}$$

where $\hat{I}(z^{\mathrm{I}}, z^{\mathrm{R}}; \theta_{\mathrm{D}})$ is a lower bound on $I(z^{\mathrm{I}}, z^{\mathrm{R}})$. We consider two MI lower bounds: Mutual Information Neural Estimation (MINE) [2] and Contrastive Predictive Coding (CPC) [24]. In our experiments, we empirically show that our method is not sensitive to the choice of the lower bound. MINE estimates the MI lower bound by approximating the log likelihood ratio in Eq. (1), using the Donsker-Varadhan (DV) variational formula of the KL divergence between the joint distribution and the product of the marginals, which yields the lower bound

$$\hat{I}^{(\mathrm{MINE})}_{\theta^{\mathrm{I}}_{\mathrm{E}},\theta^{\mathrm{R}}_{\mathrm{E}},\theta_{\mathrm{D}}}(z^{\mathrm{I}},z^{\mathrm{R}}) = \mathbb{E}_{p(z^{\mathrm{I}},z^{\mathrm{R}})}\left[f(z^{\mathrm{I}},z^{\mathrm{R}};\theta_{\mathrm{D}})\right] - \log \mathbb{E}_{p(z^{\mathrm{I}})p(z^{\mathrm{R}})}\left[e^{f(z^{\mathrm{I}},z^{\mathrm{R}};\theta_{\mathrm{D}})}\right]. \quad (3)$$

CPC computes the MI lower bound by approximating the likelihood of an image-text feature pair being sampled from the joint distribution over the product of marginals, which leads to the objective function

$$\hat{I}^{(\mathrm{CPC})}_{\theta^{\mathrm{I}}_{\mathrm{E}},\theta^{\mathrm{R}}_{\mathrm{E}},\theta_{\mathrm{D}}}(z^{\mathrm{I}},z^{\mathrm{R}}) = \mathbb{E}_{p(z^{\mathrm{I}},z^{\mathrm{R}})}\left[f(z^{\mathrm{I}},z^{\mathrm{R}};\theta_{\mathrm{D}})\right] - \mathbb{E}_{p(z^{\mathrm{I}})p(z^{\mathrm{R}})}\left[\log \sum_{\hat{z}^{\mathrm{R}}_j \in z^{\mathrm{R}}} e^{f(z^{\mathrm{I}},\hat{z}^{\mathrm{R}}_j;\theta_{\mathrm{D}})}\right]. \quad (4)$$

Both methods sample from the matched image-text pairs and from shuffled pairs (to approximate the product of marginals), and train the discriminator to differentiate between these two types of sample pairs.

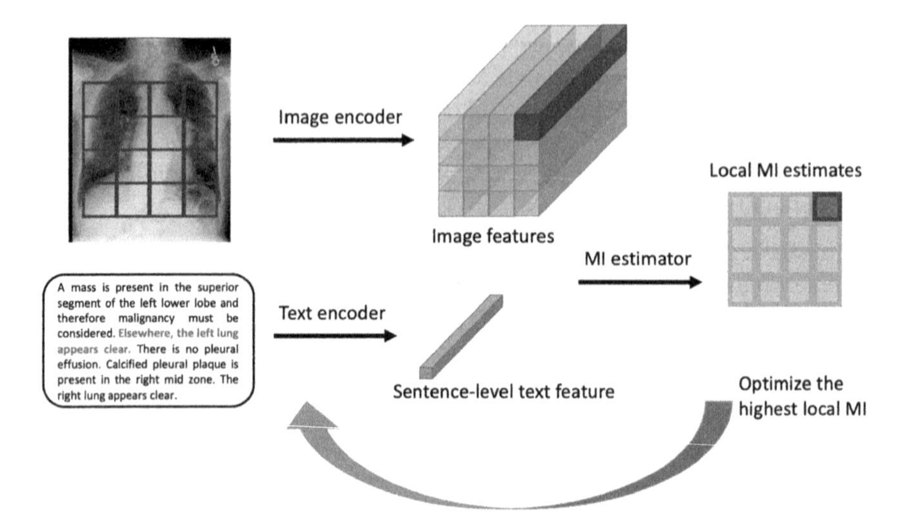

Fig. 2. Local MI maximization. First, we randomly select a sentence in the text and encode the sentence into a sentence-level feature. The corresponding image is encoded into a M × M × D feature block. We estimate the MI values between all local image features and the sentence feature. Note that the MI estimation needs shuffled image-text data, which is not illustrated in this diagram. We select the local image feature with the highest MI and update the image encoder, text encoder, and the MI discriminator such that the local MI between that image feature and the sentence feature is maximized.

Local MI Maximization. We propose to maximize MI between local features of images and sentence-level features from text. Given a sentence-level feature in the text, we estimate the MI values between all local image features and this sentence, select the image feature with the highest MI, and maximize the MI between that image feature and the sentence feature (Fig. 2). We train the image and text encoders, as well as the MI discriminator based on all the image-text data:

$$\hat{\theta}_{\mathrm{E}}^{\mathrm{I}}, \hat{\theta}_{\mathrm{E}}^{\mathrm{R}}, \hat{\theta}_{\mathrm{D}} = \arg \max_{\theta_{\mathrm{E}}^{\mathrm{I}}, \theta_{\mathrm{E}}^{\mathrm{R}}, \theta_{\mathrm{D}}} \sum_{j} \sum_{m} \max_{n} \hat{I}(\mathrm{z}_{j,(n)}^{\mathrm{I}}, \mathrm{z}_{j,(m)}^{\mathrm{R}}), \tag{5}$$

where $\mathrm{z}_{j,(n)}^{\mathrm{I}}$ is the n-th local feature in image $\mathrm{x}_j^{\mathrm{I}}$, and $\mathrm{z}_{j,(m)}^{\mathrm{R}}$ is the m-th sentence feature in text $\mathrm{x}_j^{\mathrm{R}}$. We use this *one-way* maximum, because in image captioning, every sentence was written to describe some finding in the corresponding image. In contrast, not every region in the image has a related sentence in the text that describes it.

3 Local MI vs Global MI

To provide further insight into the theoretical motivation for local mutual information, we show that the sum of local MIs between two variables is the lower bound of the global MI under a Markov condition. We consider MI between an image and two *halves* in its caption: $I(z^{\mathrm{I}}, z_{(1)}^{\mathrm{R}})$ and $I(z^{\mathrm{I}}, z_{(2)}^{\mathrm{R}})$, and also the global MI between this image and the entire caption: $I(z^{\mathrm{I}}, z^{\mathrm{R}})$, where $z^{\mathrm{R}} = (z_{(1)}^{\mathrm{R}}, z_{(2)}^{\mathrm{R}})$. We have:

$$I(z^{\mathrm{I}}, z_{(1)}^{\mathrm{R}}) + I(z^{\mathrm{I}}, z_{(2)}^{\mathrm{R}}) = I(z^{\mathrm{I}}, (z_{(1)}^{\mathrm{R}}, z_{(2)}^{\mathrm{R}})) + I(z^{\mathrm{I}}, z_{(1)}^{\mathrm{R}}, z_{(2)}^{\mathrm{R}}), \tag{6}$$

where $I(z^{\mathrm{I}}, z_{(1)}^{\mathrm{R}}, z_{(2)}^{\mathrm{R}})$ is an interaction information between the three variables [21]. We expect that, typically, since the two *halves* in the caption text both describe aspects of the same image, they form a Markov chain: $z_{(1)}^{\mathrm{R}} \leftrightarrow z^{\mathrm{I}} \leftrightarrow z_{(2)}^{\mathrm{R}}$. Under this Markov relationship, the interaction information item is non-negative and the sum of the local MIs is the lower bound of the global MI:

$$I(z^{\mathrm{I}}, z_{(1)}^{\mathrm{R}}) + I(z^{\mathrm{I}}, z_{(2)}^{\mathrm{R}}) \leq I(z^{\mathrm{I}}, z^{\mathrm{R}}). \tag{7}$$

Therefore, maximizing the local MIs is essentially maximizing a lower bound on the global MI, where the local MI optimization is usually an easier task given its lower dimension and more training samples available in the same dataset. The utility of our strategy is supported by our experimental results.

4 Experiments

Data and Model Evaluation. We demonstrate our approach on the MIMIC-CXR dataset v2.0 [13] that includes around 250K frontal-view chest radiographs with their associated radiology reports. We evaluate our representation learning methods on two downstream classification tasks:

Table 1. The number of images in the (labeled) training sets and the test sets.

–	Support devices	Cardiomegaly	Consolidation	Edema	Lung opacity
Training	76,492	65,129	20,074	56,203	58,105
Test	286	404	95	373	318
–	Pleural effusion	Pneumonia	Pneumothorax	Atelectasis	Edema severity
Training	86,871	43,951	56,472	50,416	7,066
Test	451	195	191	262	141

- **Pathology9.** Detecting 9 pathologies from the chest radiographs against the labels that were extracted from the corresponding radiology reports using a radiology report labeler CheXpert [12,14,15]. Note that there are 14 findings available in the repository [14]. We only train and evaluate 9 for which there are more than around 100 images available in the test set.
- **EdemaSeverity.** Assessing pulmonary edema severity from chest radiographs against the labels that were annotated by radiologists on the images [11,17,19,29]. The severity level ranges from 0 to 3 with a higher score indicating higher risk.

The test sets provided in MIMIC-CXR with CheXpert labels [14] and with edema severity labels [17] are used to evaluate our methods. The patients that are in either of the two test sets are excluded from the model training. Table 1 summarizes the size of the (labeled) training data and test data.

Experimental Design. Our goal is to learn representations that are useful for downstream classification tasks. Therefore, we use a fully supervised image model trained on the chest radiographs with available training labels as our benchmark. We compare two ways to use our image representations when *re-training* the image classifier: 1) freezing the image encoder; 2) fine-tuning the image encoder. In either case, the image encoder followed by a classifier is trained on the same training set that is used to train the fully supervised image model.

We compare our MI maximization approach on local features with the global MI maximization. We test both MINE [2] and CPC [24] as MI estimators. To summarize, we evaluate the variants of our model and training regimes as follows:

- **image-only-supervised:** An image-only model trained on the training data provided in [14,17].
- **global-mi-mine, global-mi-cpc:** Representation learning on the chest radiographs and the radiology reports using global MI maximization.
  - **encoder-frozen, encoder-tuned:** Once representation learning is completed, the image encoder followed by a classifier is *re-trained* on the labeled training image data, with the encoder frozen or fine-tuned.
- **local-mi-mine, local-mi-cpc:** Representation learning using local MI maximization in Eq. (5).
  - **encoder-frozen, encoder-tuned:** The resulting image encoder followed by a classifier is *re-trained*, with the encoder frozen or fine-tuned.

At the image model training or *re-training* time, all variants are trained on the same training sets. Note that the **local-mi** approach makes use of lower level image features. To make the **encoder-frozen** experiments comparable between **local-mi** and **global-mi**, we only freeze the same lower level feature extractor in both encoders.

Implementation Details. Chest radiographs are downsampled to 256×256. We use a 5-block resnet [9] as the image encoder in the local MI approach and the image feature representation z^{I} is 16×512 ($4 \times 4 \times 512$) feature vectors. We use a 6-block resnet as the image encoder for the global MI maximization, where the image representation z^{I} from this encoder is a 768-dimensional feature vector. We use the clinical BERT model [1] as the text encoder for both report-level and sentence-level feature extraction. The `[CLS]` token is used as the text feature z^{R}, which is a 768-dimensional vector. The MI discriminator for both MINE and CPC is a 1280→1024→512→1 multilayer perceptron to estimate local MI and a 1536→1024→512→1 multilayer perceptron to estimate global MI. The image feature and the text feature are concatenated to construct the input for the discriminator for MI estimation. The image models in all training variants at the image training or *re-training* time have the same architecture (6-block resnet followed by a fully connected layer).

The AdamW [31] optimizer is employed for the BERT encoder and the Adam [16] optimizer is used for the other parts of the model. The initial learning rate is $5 \cdot 10^{-4}$. The representation learning phase is trained for 5 epochs and the image model *re-training* phase is trained for 50 epochs. The fully supervised image model is trained for 100 epochs. Data augmentation including random rotation, translation, and cropping is performed on the images during training.

Results. In Table 2 and Table 3, we present the area under the receiver operating characteristic curve (AUC) statistics for the variants of our algorithms on the **EdemaSeverity** classification task and the **Pathology9** binary classification tasks. For most classification tasks, the local MI approach with encoder tuning performs the best and significantly improves the performance over solely supervised learning on labeled images. The local MI approach brings in noteworthy improvement compared to global MI. Both CPC and MINE perform similar in most tasks. Remarkably, the classification results from the frozen encoders approach the fully supervised learning results in many tasks, suggesting that the unsupervised learning captures useful features for image classification tasks even before supervision is provided.

The local MI offers substantial improvement in performance when the features are fine-tuned with the downstream model, while its performance is comparable with global MI if the features are frozen for the subsequent classification. In our experiments, training jointly with the downstream classifier (fine-tuning) typically improves performance of all tasks, with greater benefits for local MI. This suggests that local MI yields more flexible representations that adjust better for the downstream task. Our results are also supported by the analysis in Sect. 3 that shows that under the Markov assumption, the sum of local MI values is the lower bound to the global MI.

Table 2. The AUCs on the **EdemaSeverity** ordinal regression task. The average AUC score of **tuned local-mi** is 0.88 (±0.05); The average AUC score of **tuned global-mi** is 0.85 (±0.06).

Method	*Re-train* encoder?	Level 0 vs 1, 2, 3		Level 0, 1 vs 2, 3		Level 0, 1, 2 vs 3	
–	–	CPC	MINE	CPC	MINE	CPC	MINE
image-only	N/A	0.80		0.71		0.90	
global-mi	**frozen**	0.81	0.83	0.77	0.78	0.93	0.89
global-mi	**tuned**	0.81	0.82	0.79	0.81	0.93	0.93
local-mi	**frozen**	0.77	0.76	0.72	0.76	0.75	0.86
local-mi	**tuned**	**0.87**	0.83	0.83	**0.85**	**0.97**	0.93

Table 3. The AUCs on the **Pathology9** binary classification tasks. The average AUC score of **tuned local-mi** is 0.84 (±0.05); The average AUC score of **tuned global-mi** is 0.81 (±0.05).

Method	*Re-train* encoder?	Atelectasis		Cardiomegaly		Consolidation	
–	–	CPC	MINE	CPC	MINE	CPC	MINE
image-only	N/A	0.76		0.71		0.78	
global-mi	**frozen**	0.65	0.63	0.79	0.79	0.67	0.65
global-mi	**tuned**	0.74	0.77	0.81	0.81	0.81	0.82
local-mi	**frozen**	0.74	0.61	0.73	0.77	0.65	0.65
local-mi	**tuned**	0.73	**0.86**	0.82	**0.84**	**0.83**	**0.83**
–	–	Edema		Lung opacity		Pleural effusion	
–	–	CPC	MINE	CPC	MINE	CPC	MINE
image-only	N/A	**0.89**		0.86		0.69	
global-mi	**frozen**	0.81	0.81	0.69	0.68	0.74	0.74
global-mi	**tuned**	0.87	0.88	0.83	0.84	0.90	0.90
local-mi	**frozen**	0.78	0.80	0.66	0.69	0.69	0.72
local-mi	**tuned**	**0.89**	**0.89**	0.82	**0.88**	**0.92**	**0.92**
–	–	Pneumonia		Pneumothorax		Support devices	
–	–	CPC	MINE	CPC	MINE	CPC	MINE
image-only	N/A	0.75		0.65		0.72	
global-mi	**frozen**	0.71	0.70	0.65	0.66	0.70	0.68
global-mi	**tuned**	0.75	0.76	0.75	0.77	0.77	0.79
local-mi	**frozen**	0.61	0.66	0.70	0.67	0.72	0.74
local-mi	**tuned**	0.78	**0.79**	**0.79**	0.76	**0.87**	0.81

5 Conclusion

In this paper, we proposed a multimodal representation learning framework for images and text by maximizing the mutual information between their local features. The advantages of the local MI approach are tri-fold: 1) better fit to image-text structure: each sentence is typically a minimal and complete semantic unit that describes a local image region (Fig. 1) and therefore learning at the level of sentences and local image regions is more efficient than learning global descriptors; 2) better optimization landscape: the dimensionality of the representation is lower and every training image-report pair provides more samples of image-text descriptor pairs; 3) better representation fit to downstream tasks: as demonstrated in prior work, image classification usually relies on local features (e.g., pleural effusion detection based on the appearance of the region below the lungs) [10] and thus by learning local representations local MI improves classification performance.

By encouraging sentence-level features in the text to exhibit high MI with local image features, the image encoder learns to extract *useful* feature representations for subsequent image analysis. We provided further insight into local MI by showing that, under a Markov condition, maximizing local MI is equivalent to maximizing global MI. Our experimental results demonstrate that the local MI approach offers the greatest improvement for the downstream image classification tasks, and is not sensitive to the choice of the MI estimator.

Acknowledgments. This work was supported in part by NIH NIBIB NAC P41EB015902, Wistron, IBM Watson, MIT Deshpande Center, MIT J-Clinic, MIT Lincoln Lab, and US Air Force.

References

1. Alsentzer, E., et al.: Publicly available clinical BERT embeddings. arXiv preprint arXiv:1904.03323 (2019)
2. Belghazi, M.I., et al.: MINE: mutual information neural estimation. arXiv preprint arXiv:1801.04062 (2018)
3. Bojanowski, P., Joulin, A.: Unsupervised learning by predicting noise. In: International Conference on Machine Learning, pp. 517–526. PMLR (2017)
4. Chauhan, G., et al.: Joint Modeling of Chest Radiographs and Radiology Reports for Pulmonary Edema Assessment. In: Martel, A.L., et al. (eds.) MICCAI 2020. LNCS, vol. 12262, pp. 529–539. Springer, Cham (2020). https://doi.org/10.1007/978-3-030-59713-9_51
5. Chen, R.T., Li, X., Grosse, R., Duvenaud, D.: Isolating sources of disentanglement in variational autoencoders. arXiv preprint arXiv:1802.04942 (2018)
6. Chen, X., Duan, Y., Houthooft, R., Schulman, J., Sutskever, I., Abbeel, P.: InfoGAN: interpretable representation learning by information maximizing generative adversarial nets. In: Advances in Neural Information Processing Systems, pp. 2172–2180 (2016)
7. Demner-Fushman, D., et al.: Preparing a collection of radiology examinations for distribution and retrieval. J. Am. Med. Inform. Assoc. **23**(2), 304–310 (2016)

8. Harwath, D., Recasens, A., Surís, D., Chuang, G., Torralba, A., Glass, J.: Jointly discovering visual objects and spoken words from raw sensory input. In: Proceedings of the European Conference on Computer Vision (ECCV), pp. 649–665 (2018)
9. He, K., Zhang, X., Ren, S., Sun, J.: Deep residual learning for image recognition. In: Proceedings of the IEEE Conference on Computer Vision and Pattern Recognition, pp. 770–778 (2016)
10. Hjelm, R.D., et al.: Learning deep representations by mutual information estimation and maximization. arXiv preprint arXiv:1808.06670 (2018)
11. Horng, S., Liao, R., Wang, X., Dalal, S., Golland, P., Berkowitz, S.J.: Deep learning to quantify pulmonary edema in chest radiographs. Radiol. Artif. Intell. e190228 (2021)
12. Irvin, J., et al.: CheXpert: a large chest radiograph dataset with uncertainty labels and expert comparison. arXiv preprint arXiv:1901.07031 (2019)
13. Johnson, A.E., et al.: MIMIC-CXR, a de-identified publicly available database of chest radiographs with free-text reports. Sci. Data **6**(1), 1–8 (2019)
14. Johnson, A.E., et al.: MIMIC-CXR-JPG - chest radiographs with structured labels (version 2.0.0). PhysioNet (2019). https://doi.org/10.13026/8360-t248
15. Johnson, A.E., et al.: MIMIC-CXR-JPG, a large publicly available database of labeled chest radiographs. arXiv preprint arXiv:1901.07042 (2019)
16. Kingma, D.P., Ba, J.: Adam: a method for stochastic optimization. arXiv preprint arXiv:1412.6980 (2014)
17. Liao, R., Chauhan, G., Golland, P., Berkowitz, S.J., Horng, S.: Pulmonary edema severity grades based on MIMIC-CXR (version 1.0.1). PhysioNet (2021). https://doi.org/10.13026/rz5p-rc64
18. Liao, R., Moyer, D., Golland, P., Wells, W.M.: DEMI: discriminative estimator of mutual information. arXiv preprint arXiv:2010.01766 (2020)
19. Liao, R., et al.: Semi-supervised learning for quantification of pulmonary edema in chest x-ray images. arXiv preprint arXiv:1902.10785 (2019)
20. Maes, F., Collignon, A., Vandermeulen, D., Marchal, G., Suetens, P.: Multimodality image registration by maximization of mutual information. IEEE Trans. Med. Imaging **16**(2), 187–198 (1997)
21. McGill, W.: Multivariate information transmission. Trans. IRE Prof. Group Inf. Theory **4**(4), 93–111 (1954)
22. Moradi, M., Guo, Y., Gur, Y., Negahdar, M., Syeda-Mahmood, T.: A cross-modality neural network transform for semi-automatic medical image annotation. In: Ourselin, S., Joskowicz, L., Sabuncu, M.R., Unal, G., Wells, W. (eds.) MICCAI 2016. LNCS, vol. 9901, pp. 300–307. Springer, Cham (2016). https://doi.org/10.1007/978-3-319-46723-8_35
23. Moradi, M., Madani, A., Gur, Y., Guo, Y., Syeda-Mahmood, T.: Bimodal network architectures for automatic generation of image annotation from text. In: Frangi, A.F., Schnabel, J.A., Davatzikos, C., Alberola-López, C., Fichtinger, G. (eds.) MICCAI 2018. LNCS, vol. 11070, pp. 449–456. Springer, Cham (2018). https://doi.org/10.1007/978-3-030-00928-1_51
24. van den Oord, A., Li, Y., Vinyals, O.: Representation learning with contrastive predictive coding. arXiv preprint arXiv:1807.03748 (2018)
25. Reimers, N., Gurevych, I.: Sentence-BERT: sentence embeddings using siamese BERT-networks. arXiv preprint arXiv:1908.10084 (2019)
26. Rifai, S., Bengio, Y., Courville, A., Vincent, P., Mirza, M.: Disentangling factors of variation for facial expression recognition. In: Fitzgibbon, A., Lazebnik, S., Perona, P., Sato, Y., Schmid, C. (eds.) ECCV 2012. LNCS, vol. 7577, pp. 808–822. Springer, Heidelberg (2012). https://doi.org/10.1007/978-3-642-33783-3_58

27. Song, J., Ermon, S.: Understanding the limitations of variational mutual information estimators. In: International Conference on Learning Representations (2019)
28. Wang, X., Peng, Y., Lu, L., Lu, Z., Summers, R.M.: TienNet: text-image embedding network for common thorax disease classification and reporting in chest X-rays. In: Proceedings of the IEEE Conference on Computer Vision and Pattern Recognition, pp. 9049–9058 (2018)
29. Wang, X., et al.: Pulmonary edema severity estimation in chest radiographs using deep learning. In: International Conference on Medical Imaging with Deep Learning-Extended Abstract Track (2019)
30. Wells III, W.M., Viola, P., Atsumi, H., Nakajima, S., Kikinis, R.: Multi-modal volume registration by maximization of mutual information. Med. Image Anal. **1**(1), 35–51 (1996)
31. Wolf, T., et al.: Huggingface's transformers: state-of-the-art natural language processing. arXiv-1910 (2019)
32. Xue, Y., Huang, X.: Improved disease classification in chest X-rays with transferred features from report generation. In: Chung, A.C.S., Gee, J.C., Yushkevich, P.A., Bao, S. (eds.) IPMI 2019. LNCS, vol. 11492, pp. 125–138. Springer, Cham (2019). https://doi.org/10.1007/978-3-030-20351-1_10
33. Zhang, Y., Jiang, H., Miura, Y., Manning, C.D., Langlotz, C.P.: Contrastive learning of medical visual representations from paired images and text. arXiv preprint arXiv:2010.00747 (2020)

Inter-regional High-Level Relation Learning from Functional Connectivity via Self-supervision

Wonsik Jung[1], Da-Woon Heo[2], Eunjin Jeon[1], Jaein Lee[1], and Heung-Il Suk[1,2(✉)]

[1] Department of Brain and Cognitive Engineering, Korea University, Seoul, Republic of Korea
{ssikjeong1,eunjinjeon,wodls9212,hisuk}@korea.ac.kr
[2] Department of Artificial Intelligence, Korea University, Seoul, Republic of Korea
daheo@korea.ac.kr

Abstract. In recent studies, we have witnessed the applicability of deep learning methods on resting-state functional magnetic resonance image (rs-fMRI) analysis and its use for brain disease diagnosis, *e.g.*, autism spectrum disorder (ASD). However, it still remains challenging to learn discriminative representations from raw BOLD signals or functional connectivity (FC) with a limited number of samples. In this paper, we propose a simple but efficient representation learning method for FC in a self-supervised learning manner. Specifically, we devise a proxy task of estimating the randomly masked seed-based functional networks from the remaining ones in FC, to discover the complex high-level relations among brain regions, which are not directly observable from an input FC. Thanks to the random masking strategy in our proxy task, it also has the effect of augmenting training samples, thus allowing for robust training. With the pretrained feature representation network in a self-supervised manner, we then construct a decision network for the downstream task of ASD diagnosis. In order to validate the effectiveness of our proposed method, we used the ABIDE dataset that collected subjects from multiple sites and our proposed method showed superiority to the comparative methods in various metrics.

Keywords: Resting-state functional magnetic resonance imaging · Autism spectrum disorder · Deep learning · Representation learning · Self-supervised learning · Multi-site fMRI

1 Introduction

Resting-state functional magnetic resonance imaging (rs-fMRI) is a non-invasive technique that demonstrates spatio-temporal scales of regional brain activations by measuring blood-oxygen-level-dependent (BOLD) signals [7]. It is widely used for biomarkers identification, computer-aided diagnosis for diverse brain diseases,

M. de Bruijne et al. (Eds.): MICCAI 2021, LNCS 12902, pp. 284–293, 2021.
https://doi.org/10.1007/978-3-030-87196-3_27

such as schizophrenia [15], Alzheimer's disease [4,23], and autism spectrum disorder (ASD) [1,10]. In general, rs-fMRI based diagnosis methods employed a raw time signal [13,14] or a functional connectivity (FC) [6,27]. Specifically, FC statistically describes the temporal correlation between spatially distant brain regions [3,17]. Due to their useful properties, many existing methods favor the employment of FC obtained by rs-fMRI [1,10,21] for disease diagnosis.

With recent advances in machine learning, deep learning methods have gained significant attention as one of the major tools for neuroimaging in terms of both representation learning and classification tools. There have been efforts to learn a spatio-temporal representation or functional characteristics through unsupervised deep learning models (*e.g.*, stacked auto-encoder (SAE) [10,21,22] or restricted Boltzmann machines [11,12]). Although deep learning-based studies showed promising performance without any manual feature extraction, they required a large number of data to obtain a reliable model. Without sufficient amount of data samples, it is inevitable to face an overfitting problem that has difficulty in generalization for unseen samples. To address the aforementioned limitation, several studies devised a representation learning with random noise-added FC based on denoising autoencoder (dAE) [10] or SAEs [21] for ASD diagnosis. In other words, [10,21] learned only inter-regional relationships of FC in a low level.

In this paper, we propose a simple but efficient deep neural network for ASD diagnosis using rs-fMRI by formulating a self-supervised learning in an ROI level to make the network better utilize inter-regional relations. Concisely, we introduce an ROI level masking method and then train SAE and a classifier with the masked ROIs. Our proposed network learns not only inter-regional relations but also more higher-level relations in the overall inherent FCs. That is, by masking the connections between all regions from a specific seed region, our method generates more training data and encourages to learn more ROI-related information.

The main contributions of our methods are summarized as follows: (1) We devise a novel representation learning method for rs-fMRI data within a self-supervised learning framework, (2) Our model learns a feature representation in two levels: the first-order FC, which implies the relation between ROIs, and the high-level FC, which represents the relation between FC networks, (3) Our proposed method outperformed comparative methods in all metrics for ASD diagnosis.

2 Dataset and Preprocessing

In this study, we used preprocessed rs-fMRI datasets collected from the publicly available Autism Brain Imaging Database Exchange (ABIDE)[1] dataset [8]. From a total of 17 international sites[2], 1,046 samples are obtained, including

[1] http://fcon_1000.projects.nitrc.org/indi/abide/.

[2] {UM, NYU, MAX MUN, OHSU, SBL, OLIN, SDSU, CALTECH, TRINITY, YALE, PITT, LEUVEN, UCLA, USM, STANFORD, CMU, and KKI}.

510 from ASD and 536 from typical controls (TCs). The mean age of each class, *i.e.*, ASD and TC, is 16.8 ± 8.1 and 16.5 ± 7.1, respectively.

The collected data was preprocessed according to Data Processing Assistant for Resting-State fMRI [5] pipeline[3]. After the preprocessing, the brain space was parcellated into 116 ROIs using Automated Anatomical Labeling (AAL) template [24]. For each subject, we took an average of the signals in each ROI and then calculated FC by using the Pearson correlation coefficient between pairs of ROI based time series (normalized to z score using Fisher transformation). We then used the FC matrix as the input of our method.

3 Proposed Method

Our proposed method comprises of three phases: (i) an ROI connections masking, (ii) an autoencoder training, and (iii) a diagnostic classifier training, as illustrated in Fig. 1. Inspired by [20], prior to learning a class-discriminative feature representation for ASD diagnosis, we first randomly select a few ROIs and remove their region-connectivities (*i.e.*, masking) from the input, and regard those as target values to estimate from the remaining values in an input in the concept of self-supervised learning.

Specifically, the masked input data (*i.e.*, first-order FC) is fed into an SAE to learn a feature representation that reflects on the high-level FCs. The SAE is trained by employing greedy layer-wise training [2] so that it can better exploit the relatively small number of training samples to learn a feature representation called the high-level relations among region-based connectivities. Here, high-level features include inter-regional relations between the masked relations on FC and the remaining ones. With regards to the ASD diagnosis, a diagnostic classifier takes the high-level features as input and then outputs a predicted class probability.

3.1 ROI Connections Masking

Given an input matrix $\mathbf{X} \in \mathbb{R}^{R \times R}$, we generate a masking matrix for ROIs. To define such a mask matrix $\mathbf{M}_\gamma \in \mathbb{R}^{R \times R}$ denoted as the ROI connections mask, we first define a function $\phi_\gamma(\cdot)$ which randomly selects γ ratio of ROIs as follows:

$$\phi_\gamma(R) = \begin{cases} 0 & \text{if } R \text{ element is selected,} \\ 1 & \text{otherwise} \end{cases} \tag{1}$$

where $R \in \mathbb{R}^{116}$ means ROI indices. Note that we remove all connections associated with the selected ROIs to generate $\mathbf{M}_\gamma$. Thus, $\mathbf{M}_\gamma$ is composed of 0 for the removed connections and 1 for the remaining ones. To mask the input FC matrix, we first multiply the FC matrix with the mask matrix. Subsequently, we

[3] http://preprocessed-connectomes-project.org/abide/Pipelines.html.

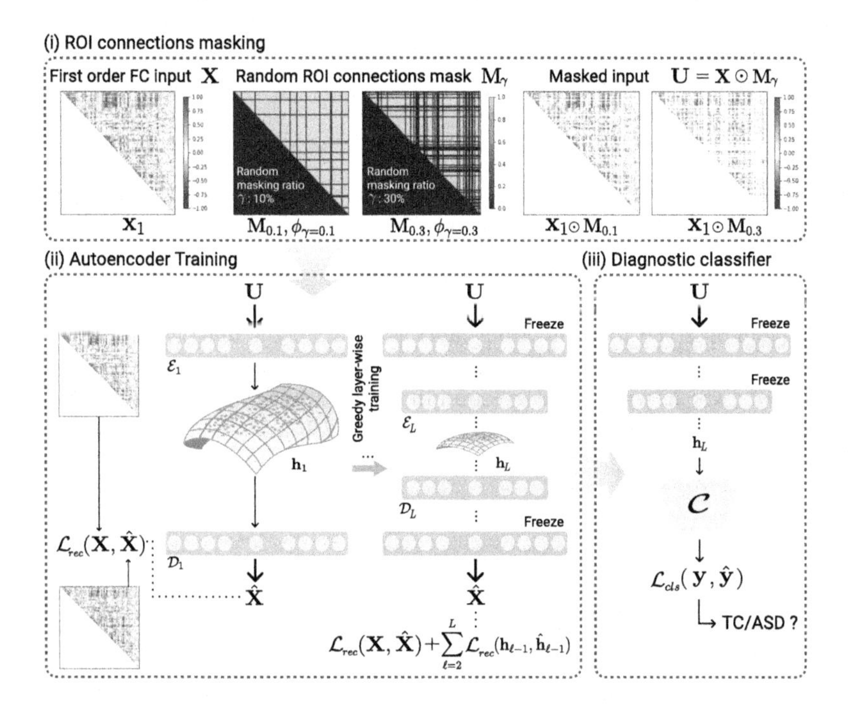

Fig. 1. Overview of our proposed method including three phases: (i) ROI connections masking which generates a masked input for ROIs, (ii) an autoencoder training to learn a more high-level feature representation among region-based connectivities, (iii) and a diagnostic classifier training to predict the corresponding labels.

flatten the masked FC matrix $\mathbf{U} \in \mathbb{R}^D$ where $D \in \mathbb{R}^{6670}$ denotes the flattened upper triangle of FC matrix as follows:

$$\mathbf{U} = \mathbf{M}_\gamma \odot \mathbf{X}, \tag{2}$$

$$\bar{\mathbf{U}} = (1 - \mathbf{M}_\gamma) \odot \mathbf{X} \tag{3}$$

where $\odot$ denotes an element-wise multiplication operator and $\bar{\mathbf{U}} \in \mathbb{R}^D$ represents the discarded region-connectivities. During every iteration in training phase, these masks are arbitrarily generated for each input. By doing so, we are able to augment training samples and also prevent the overfitting. Further, since we remove the specific ROI connections, it helps enhance the quality of feature representations from the neuroscientific perspectives.

3.2 Autoencoder Training

With the masked FCs (*i.e.*, $\mathbf{U}$) as input, we train our SAE described in the following such that it reconstructs the inputs $\mathbf{X}$ which consist of values in both $\mathbf{U}$ and the discarded region-connectivities $\bar{\mathbf{U}}$. To our best knowledge, this new learning strategy helps our neural network to discover the high-level relations among ROIs inherent in the FCs indirectly, thus potentially enriching the feature representations for predictive model building. To do this, we exploit SAE trained with greedy layer-wise training strategy [2].

High-Level Relation Representation. Initially, the masked input $\mathbf{U} \in \mathbb{R}^{D}$ is embedded to a hidden space with $\mathbf{h}_1 \in \mathbb{R}^{D_1}$ as follows:

$$\mathbf{h}_1 = \mathcal{E}_1(\mathbf{U}) = \tanh(\mathbf{W}_1\mathbf{U} + \mathbf{b}_1), \tag{4}$$

where $\mathcal{E}_1$ denotes the first encoding layer with a weight matrix ($\mathbf{W}_1 \in \mathbb{R}^{D_1 \times D}$) and a bias vector ($\mathbf{b}_1 \in \mathbb{R}^{D_1}$). The first hidden feature $\mathbf{h}_1$ is trained to reconstruct the original FCs $\mathbf{X}$ with the corresponding decoding layer $\mathcal{D}_1 = \tanh(\mathbf{W}'_1\mathbf{h}_1 + \mathbf{b}'_1) = \hat{\mathbf{X}}$ by minimizing a reconstruction loss as follows:

$$\min_{\mathbf{W}_1,\mathbf{b}_1,\mathbf{W}'_1,\mathbf{b}'_1} \mathcal{L}_{\text{rec}}(\mathbf{X}, \hat{\mathbf{X}}) = \sum_{i=1}^{N} ||\mathbf{X}^{(i)} - \mathcal{D}_1(\mathbf{h}_1^{(i)})|| \tag{5}$$

where N is the total number of training samples, $\hat{\mathbf{X}}$ is a reconstructed sample which is an output of $\mathcal{D}_1$, $\mathbf{W}'_1 \in \mathbb{R}^{D \times D_1}$ is a weight matrix, and $\mathbf{b}'_1 \in \mathbb{R}^{D}$ is a bias.

Subsequently, we train the next layer in the encoder $\mathcal{E}_\ell$ and decoder $\mathcal{D}_\ell$ pair to learn high-level feature representation of FCs where $\ell \in \{2, \ldots, L\}$. While we freeze the previous encoder $\mathcal{E}_{\ell-1}$ and decoder $\mathcal{D}_{\ell-1}$, we can estimate the ℓ-th level representation of FC input $\mathbf{h}_\ell$, which basically corresponds to the $(\ell+1)$-th level relations[4] among ROIs, and it can be fed into the $\mathcal{E}_\ell$ as follows:

$$\mathbf{h}_\ell = \mathcal{E}_\ell(\mathbf{h}_{\ell-1}) = \tanh(\mathbf{W}_\ell\mathbf{h}_{\ell-1} + \mathbf{b}_\ell) \tag{6}$$

where $\mathbf{W}_\ell \in \mathbb{R}^{D_\ell \times D_{\ell-1}}$ and $\mathbf{b}_\ell \in \mathbb{R}^{D_\ell}$. The encoder $\mathcal{E}_\ell$ and the subsequent decoder $\mathcal{D}_\ell = \tanh(\mathbf{W}'_\ell\mathbf{h}_\ell + \mathbf{b}'_\ell)$ are trained by minimizing the sum of each reconstruction loss of $\mathbf{h}_\ell$ and $\mathbf{X}$ as follows:

$$\begin{aligned} &\min_{\mathbf{W}_\ell,\mathbf{b}_\ell,\mathbf{W}'_\ell,\mathbf{b}'_\ell} \mathcal{L}_{\text{rec}}(\mathbf{X}, \hat{\mathbf{X}}) + \mathcal{L}_{\text{rec}}(\mathbf{h}_{\ell-1}, \hat{\mathbf{h}}_{\ell-1}) \\ &= \sum_{n=1}^{N} ||\mathbf{X}^{(i)} - \mathcal{D}_1(\mathbf{h}_1^{(i)})|| + \sum_{n=1}^{N}\sum_{\ell=2}^{L} ||\mathbf{h}_{\ell-1}^{(i)} - \mathcal{D}_\ell(\mathbf{h}_\ell^{(i)})||. \end{aligned} \tag{7}$$

Note that the target $\mathbf{X}$ in reconstruction loss Eq. (5) and (7) is an original complete before masking data. Specifically, it is defined as the sum of the masked input $\mathbf{U}$ (removed ROI connections) and the remaining connections $\bar{\mathbf{U}}$ (not included in $\mathbf{U}$). In this regard, our method is trained to reconstruct $\mathbf{U}$ and estimate $\bar{\mathbf{U}}$. Through this SAE training, we enforce the hidden layers to learn enriched representations inherent high-level relations among ROIs and the first-order connections for rs-fMRI.

[4] Note that the input FC $\mathbf{X}$ represents first-order relation among ROIs, and the ℓ-th hidden layer finds $(\ell+1)$-th relations among ROIs.

3.3 A Diagnostic Classifier

Once we obtain the high-level feature $\mathbf{h}_L$ from the encoder, it is fed into a diagnostic classifier composed of two linear fully-connected layers to make it be discriminative for ASD classification. The diagnostic classifier is trained to predict the clinical status $\hat{\mathbf{y}}$ by minimizing the cross-entropy loss as follows:

$$\mathcal{L}_{\text{cls}} = -\sum_{i=1}^{N} \mathbf{y}^i \log(\hat{\mathbf{y}}^i). \tag{8}$$

Table 1. Performance on the classification between ASD and TC. Note that γ denotes a ratio of the masking. ($*$: $p < 0.05$, w/ mask: with mask, w/o mask: without mask, G: Gaussian noise, FC-M: random FC connection mask)

	Models	γ	AUC	ACC (%)	SEN (%)	SPEC (%)
w/o mask	AE	×	$0.676 \pm 0.043^*$	$62.11 \pm 3.63^*$	59.41 ± 7.26	64.67 ± 2.42
	dAE [25]	×	$0.678 \pm 0.044^*$	$62.58 \pm 3.95^*$	59.22 ± 7.32	65.79 ± 4.70
	SAE	×	$0.691 \pm 0.053^*$	$63.64 \pm 3.35^*$	59.80 ± 4.90	67.29 ± 8.67
w/mask	AE ($\mathbf{M}_\gamma$)	0.1	$0.685 \pm 0.034^*$	$62.78 \pm 2.56^*$	54.71 ± 11.50	70.47 ± 11.07
	dAE (G) [25]	0.7	$0.689 \pm 0.046^*$	$63.16 \pm 2.55^*$	60.39 ± 4.19	65.79 ± 3.89
	SAE (G) [26]	0.2	$0.691 \pm 0.043^*$	$62.20 \pm 3.25^*$	58.24 ± 4.58	65.98 ± 8.45
	SAE (FC-M)	0.2	$0.693 \pm 0.039^*$	$63.54 \pm 2.65^*$	54.71 ± 8.30	71.96 ± 7.62
	Proposed: SAE ($\mathbf{M}_\gamma$)	0.6	$\mathbf{0.760 \pm 0.045}$	$\mathbf{69.19 \pm 3.70}$	$\mathbf{64.71 \pm 12.30}$	$\mathbf{73.46 \pm 13.23}$

4 Experiment and Analysis

4.1 Experimental Settings

For a fair comparison, we conducted a stratified five-fold cross validation, where one fold for validation set, another one for test set, and the remaining folds for train set from all multi-site samples in ABIDE dataset and then reported the averaged results in terms of four metrics: Area under the receiver operating characteristic curve (AUC), accuracy (ACC), sensitivity (SEN), and specificity (SPEC). We compared our method to seven competing methods, namely, basic autoencoder (AE) with/without $\mathbf{M}_\gamma$, dAE [25] with/without Gaussian noise, SAE without mask, SAE with Gaussian noise [26], or random FC connection mask [20], respectively. We compared our method with basic AE, dAE, and SAE without any masking methods using the same architecture. Further, we also compared to AE with $\mathbf{M}_\gamma$, dAE and SAE with Gaussian noise [26]. In order to validate the effectiveness of ROI-level masking, we trained SAE with randomly FC connection masking, inspired by [20].

Regarding our method, the SAE consisted of two (L) fully-connected layers with the units of $\{6670 \times 1.5, 6670 \times 0.3\}$ and the classifier was composed of two fully-connected layers with $\{10, 2\}$ units. We trained the model using an Adam optimizer [16] with a learning rate of 10^{-4} and a mini-batch size of 48. In addition, we applied an l_2 regularization by setting the corresponding hyper-parameter as 5×10^{-5}. We implemented all methods including the competing methods with PyTorch and they were trained with GPU GTX 1080 TI.

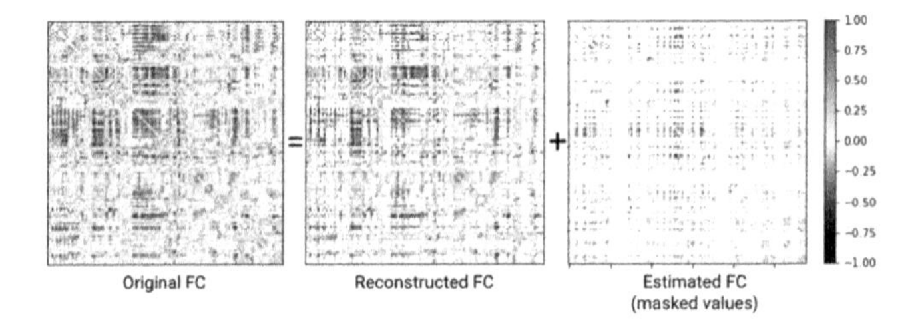

Fig. 2. Visualization of the reconstruction results from a randomly selected test samples. Note that original FC, reconstructed FC, and estimated FC denotes $\mathbf{X}, \mathbf{U}$, and $\bar{\mathbf{U}}$, respectively.

4.2 Experimental Results

Classification. We reported the averaged binary classification performance for all sites in the ABIDE dataset in Table 1. Note that we performed every experiment on all samples from whole sites. With regards to the making ratio (γ), we selected the best optimal γ based on the validation results for all methods. We observed that our method achieved the best performances for four metrics and showed a statistical significance of $p < 0.05$ in AUC and ACC.

Reconstruction. We randomly selected a test sample and removed a few connectivities including the randomly masked ROIs ($\gamma = 0.6$) in the sample. With the masked test sample, we illustrated its reconstructed output from trained model in Fig. 2. The experimental results showed that our proposed method well represented the input FC, by reconstructing the masked elements, regardless of the index of the masked ROI.

4.3 Analysis

We analyzed our method from two aspects: the feature representation and the neuroscientific explanation.

Feature Representation. We used t-SNE [18] to visualize the trained features $\mathbf{h}_2 = \mathcal{E}_2(\mathcal{E}_1(\mathbf{X}))$ from test samples in all sites, as shown in Fig. 3. We observed that our method learned the separable features between two classes regardless of sites. In other words, our method is more proficient in extracting common features among multi-sites, compared to other comparative methods.

Neuroscientific Analyses. We analyzed our model on the multi-site dataset, class-specific activated regions by using layer-wise relevance propagation (LRP) scores [19] for functional connectivities between TC and ASD groups as illustrated in Fig. 4. In order for better explanation, we summed up the extracted LRP scores for each group, then applied a threshold in the LRP scores smaller

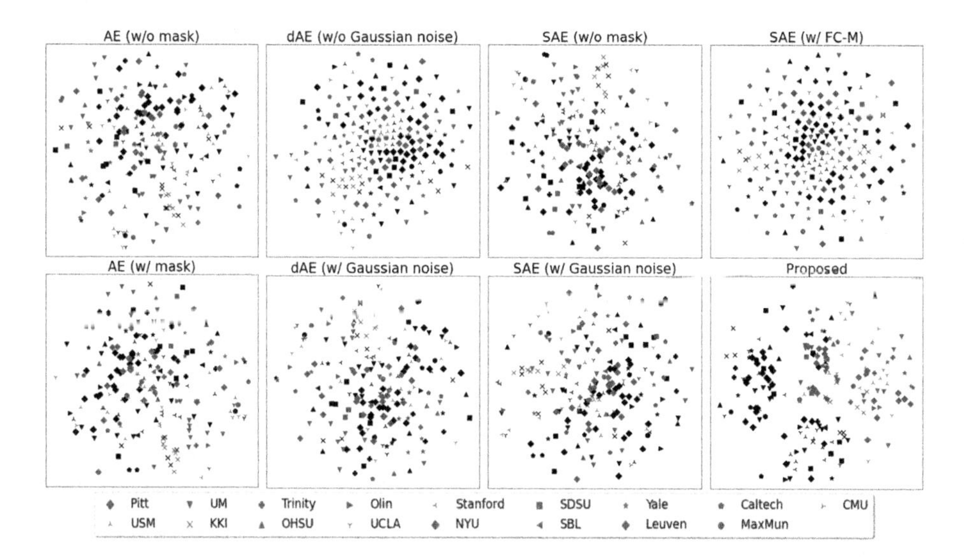

Fig. 3. Visualization of t-SNE [18] for the representation from the encoder in ABIDE dataset. We visualized features of the trained encoders from both our proposed method and competing methods, respectively. Note that the color and marker represent the class and site of the samples, respectively. (The red and blue color denotes the class of ASD and TC, respectively.) (Color figure online)

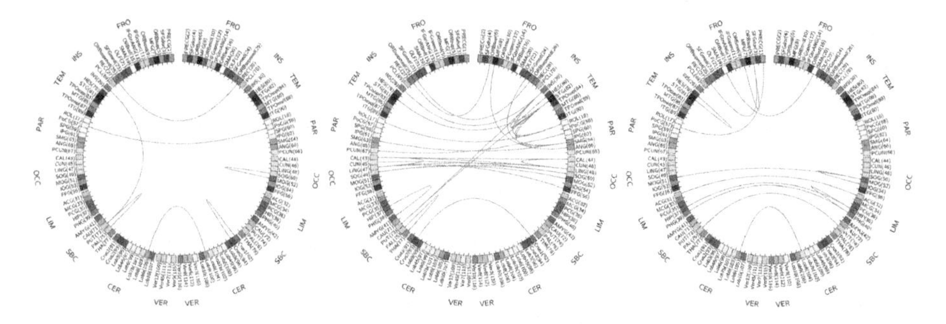

Fig. 4. Activated connections obtained from the layer-wise relevance propagation (LRP) [19] scores for FCs between TC and ASD groups over the dataset. Note that the common activated ROI connections between TC and ASD groups (left), highly activated ROI connections in TC group (center) and ASD group (right).

than 0.1. We observed the activated ROI connections in both groups as well as the highly activated ROI connections for each group. Several highly activated connections and their corresponding regions are reported in [9] as core findings in ASD group. We believe that these various connections can be used to explore multifaceted symptoms of ASD via further analysis.

5 Conclusion

In this work, we proposed a novel representation learning method that learns a latent feature that reflects both the first-order FC and the high-level FC, *i.e.*, inter-regional relations between disregarding and remaining of inherent FCs. Specifically, we utilized the ROI connections masking based on a neuroscientific viewpoint in a self-supervised manner. In our experiments on the ABIDE dataset, the proposed method showed superiority to the comparative methods in various metrics. Lastly, to support our assumption, we analyzed our method from two aspects: feature representation and neuroscientific explanation.

Acknowledgement. This work was supported by National Research Foundation of Korea (NRF) grant funded by the Korea government (MSIT) (No. 2019R1A2C1006543) and by Institute of Information & communications Technology Planning & Evaluation (IITP) grant funded by the Korea government(MSIT) (No. 2019-0-00079, Artificial Intelligence Graduate School Program (Korea University)).

References

1. Abraham, A., et al.: Deriving reproducible biomarkers from multi-site resting-state data: an autism-based example. NeuroImage **147**, 736–745 (2017)
2. Bengio, Y.: Learning Deep Architectures for AI. Now Publishers Inc, Hanover (2009)
3. Biswal, B.B.: Resting state fMRI: a personal history. Neuroimage **62**(2), 938–944 (2012)
4. Brier, M.R., et al.: Loss of intranetwork and internetwork resting state functional connections with alzheimer's disease progression. J. Neurosci. **32**(26), 8890–8899 (2012)
5. Chao-Gan, Y., Yu-Feng, Z.: DPARSF: a MATLAB toolbox for "pipeline" data analysis of resting-state fMRI. Front. Syst. Neurosci. **4** (2010)
6. Chen, X., et al.: High-order resting-state functional connectivity network for MCI classification. Hum. Brain Map. **37**(9), 3282–3296 (2016)
7. Cribben, I., Haraldsdottir, R., Atlas, L.Y., Wager, T.D., Lindquist, M.A.: Dynamic connectivity regression: determining state-related changes in brain connectivity. Neuroimage **61**(4), 907–920 (2012)
8. Di Martino, A., et al.: The autism brain imaging data exchange: towards a large-scale evaluation of the intrinsic brain architecture in autism. Mol. Psychiatry **19**(6), 659–667 (2014)
9. Dichter, G.S.: Functional magnetic resonance imaging of autism spectrum disorders. Dial. Clin. Neurosci. **14**(3), 319 (2012)
10. Heinsfeld, A.S., Franco, A.R., Craddock, R.C., Buchweitz, A., Meneguzzi, F.: Identification of autism spectrum disorder using deep learning and the abide dataset. NeuroImage Clin. **17**, 16–23 (2018)
11. Hjelm, R.D., Calhoun, V.D., Salakhutdinov, R., Allen, E.A., Adali, T., Plis, S.M.: Restricted Boltzmann machines for neuroimaging: an application in identifying intrinsic networks. NeuroImage **96**, 245–260 (2014)
12. Hu, X., et al.: Latent source mining in fMRI via restricted Boltzmann machine. Hum. Brain Map. **39**(6), 2368–2380 (2018)

13. Jeon, E., Kang, E., Lee, J., Lee, J., Kam, T.-E., Suk, H.-I.: Enriched representation learning in resting-state fMRI for early MCI diagnosis. In: Martel, A.L., et al. (eds.) MICCAI 2020. LNCS, vol. 12267, pp. 397–406. Springer, Cham (2020). https://doi.org/10.1007/978-3-030-59728-3_39
14. Kang, E., Suk, H.-I.: Probabilistic source separation on resting-state fMRI and its use for early MCI identification. In: Frangi, A.F., Schnabel, J.A., Davatzikos, C., Alberola-López, C., Fichtinger, G. (eds.) MICCAI 2018. LNCS, vol. 11072, pp. 275–283. Springer, Cham (2018). https://doi.org/10.1007/978-3-030-00931-1_32
15. Kim, J., Calhoun, V.D., Shim, E., Lee, J.H.: Deep neural network with weight sparsity control and pre-training extracts hierarchical features and enhances classification performance: evidence from whole-brain resting-state functional connectivity patterns of schizophrenia. Neuroimage **124**, 127–146 (2016)
16. Kingma, D.P., Ba, J.: Adam: a method for stochastic optimization. arXiv preprint arXiv:1412.6980 (2014)
17. Lee, L., Harrison, L.M., Mechelli, A.: A report of the functional connectivity workshop, Dusseldorf 2002. Neuroimage **19**(2), 457–465 (2003)
18. Van der Maaten, L., Hinton, G.: Visualizing data using t-SNE. J. Mach. Learn. Res. **9**(11) (2008)
19. Montavon, G., Lapuschkin, S., Binder, A., Samek, W., Müller, K.R.: Explaining nonlinear classification decisions with deep Taylor decomposition. Pattern Recogn. **65**, 211–222 (2017)
20. Pathak, D., Krahenbuhl, P., Donahue, J., Darrell, T., Efros, A.A.: Context encoders: feature learning by inpainting. In: Proceedings of the IEEE Conference on Computer Vision and Pattern Recognition, pp. 2536–2544 (2016)
21. Rakić, M., Cabezas, M., Kushibar, K., Oliver, A., Lladó, X.: Improving the detection of autism spectrum disorder by combining structural and functional MRI information. NeuroImage Clin. **25**, 102181 (2020)
22. Suk, H.I., Wee, C.Y., Lee, S.W., Shen, D.: State-space model with deep learning for functional dynamics estimation in resting-state fMRI. NeuroImage **129**, 292–307 (2016)
23. Supekar, K., Menon, V., Rubin, D., Musen, M., Greicius, M.D.: Network analysis of intrinsic functional brain connectivity in Alzheimer's disease. PLoS Comput. Biol. **4**(6), e1000100 (2008)
24. Tzourio-Mazoyer, N., et al.: Automated anatomical labeling of activations in SPM using a macroscopic anatomical parcellation of the MNI MRI single-subject brain. NeuroImage **15**(1), 273–289 (2002)
25. Vincent, P., Larochelle, H., Bengio, Y., Manzagol, P.A.: Extracting and composing robust features with denoising autoencoders. In: Proceedings of the 25th International Conference on Machine Learning, pp. 1096–1103 (2008)
26. Vincent, P., Larochelle, H., Lajoie, I., Bengio, Y., Manzagol, P.A., Bottou, L.: Stacked denoising autoencoders: learning useful representations in a deep network with a local denoising criterion. J. Mach. Learn. Res. **11**(12) (2010)
27. Zhao, F., Chen, Z., Rekik, I., Lee, S.W., Shen, D.: Diagnosis of autism spectrum disorder using central-moment features from low-and high-order dynamic resting-state functional connectivity networks. Front. Neurosci. **14** (2020)

Machine Learning - Semi-Supervised Learning

Semi-supervised Left Atrium Segmentation with Mutual Consistency Training

Yicheng Wu[1,2], Minfeng Xu[1], Zongyuan Ge[3,4], Jianfei Cai[2](✉), and Lei Zhang[1]

[1] DAMO Academy, Alibaba Group, Hangzhou 311121, China
[2] Department of Data Science and AI, Faculty of Information Technology, Monash University, Melbourne, VIC 3800, Australia
jianfei.cai@monash.edu
[3] Monash-Airdoc Research, Monash University, Melbourne, VIC 3800, Australia
[4] Monash Medical AI, Monash eResearch Centre, Melbourne, VIC 3800, Australia

Abstract. Semi-supervised learning has attracted great attention in the field of machine learning, especially for medical image segmentation tasks, since it alleviates the heavy burden of collecting abundant densely annotated data for training. However, most of existing methods underestimate the importance of challenging regions (e.g. small branches or blurred edges) during training. We believe that these unlabeled regions may contain more crucial information to minimize the uncertainty prediction for the model and should be emphasized in the training process. Therefore, in this paper, we propose a novel Mutual Consistency Network (MC-Net) for semi-supervised left atrium segmentation from 3D MR images. Particularly, our MC-Net consists of one encoder and two slightly different decoders, and the prediction discrepancies of two decoders are transformed as an unsupervised loss by our designed cycled pseudo label scheme to encourage mutual consistency. Such mutual consistency encourages the two decoders to have consistent and low-entropy predictions and enables the model to gradually capture generalized features from these unlabeled challenging regions. We evaluate our MC-Net on the public Left Atrium (LA) database and it obtains impressive performance gains by exploiting the unlabeled data effectively. Our MC-Net outperforms six recent semi-supervised methods for left atrium segmentation, and sets the new state-of-the-art performance on the LA database.

Keywords: Semi-supervised learning · Mutual consistency · Cycled pseudo label

1 Introduction

Accurately segmenting the organs or tissues is a fundamental and significant step to construct a computer-aided diagnosis (CAD) system, which plays a critical

Electronic supplementary material The online version of this chapter (https://doi.org/10.1007/978-3-030-87196-3_28) contains supplementary material, which is available to authorized users.

M. de Bruijne et al. (Eds.): MICCAI 2021, LNCS 12902, pp. 297–306, 2021.
https://doi.org/10.1007/978-3-030-87196-3_28

role for medical image quantitative analysis. Most of existing segmentation methods rely on abundant labeled data for training, where collecting a large number of labeled data is labor-intensive and time-consuming. Considering collecting unlabeled data is much easier, it is highly desirable to develop semi-supervised segmentation methods to effectively exploit rich unlabeled data.

A few recent semi-supervised models [9,11] have been proposed to study the consistency regularization. For example, Sohn et al. [11] applied weak or strong perturbations to augment data and constrained the model to output invariant results over different perturbations. Lee et al. [5] employed pseudo labels to encourage a low-density separation between classes as the entropy regularization for training. These semi-supervised methods have achieved promising progresses.

For medical tasks, there are also several semi-supervised works to segment human organs. For instance, Yu et al. [17] designed a teacher-student model to segment left atrium. Enforcing the consistency between the student model and the teacher model can facilitate the model learning. Li et al. [6] introduced an adversarial loss to encourage that the feature spaces of labeled data and unlabeled data are close. Luo et al. [7] studied the relation between semantic segmentation and shape regression to leverage unlabeled data. Xie et al. [15] utilized attention mechanisms to calculate the semantic similarities between labeled data and unlabeled data. Fang et al. [1] used a co-training framework to boost each sub-model and also adopted an adversarial loss to further improve the performance. However, these deep models either require extra components to obtain performance gains or underestimate the importance of some challenging regions (e.g. small branches or adhesive edges around the target) in the training process.

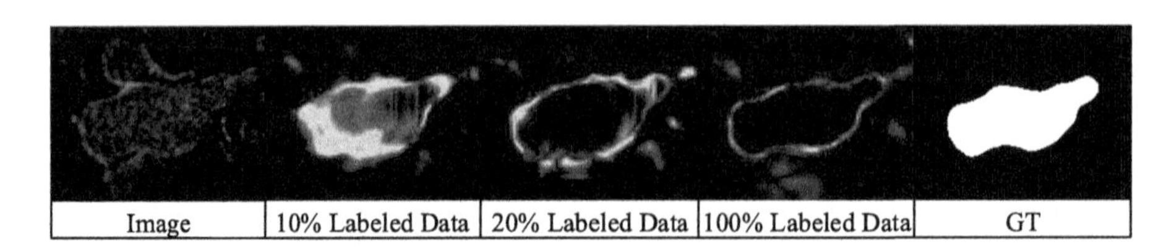

Fig. 1. Example MR image (1st column), the uncertainty of deep models trained with 10% labeled data (2nd column), 20% labeled data (3rd column) and all labeled data (4th column), and corresponding ground truth (5th column) on the LA database.

Specifically, due to the limitation of training data, deep models often have ambiguous predictions in some complex or blurred areas. For example, Fig. 1 gives the comparisons of uncertainty of deep models trained with 10% labeled data, 20% labeled data and all labeled data on the Left Atrium (LA) database [16]. First, we observe that ***(1) these ambiguous predictions mainly locate at challenging areas***. [17] claimed that these ambiguous targets may lead to meaningless and unreliable guidance. However, we believe that these challenging regions contain more crucial information and should be emphasized during training since hard examples can make training more efficient [10] and most of other regions can be correctly segmented by deep models even trained with 10%

labeled data. Second, Fig. 1 indicates that ***(2) deep models trained by more labeled data tend to output less ambiguous predictions***. The ambiguous predictions can be represented by the model-based epistemic uncertainty [3]. We deem that the epistemic uncertainty is able to evaluate the model's generalization ability since the most generalized model, i.e. the one trained with all labeled data, has the minimum uncertainty (see the fourth column in Fig. 1). Hence, based on the observations ***(1)*** and ***(2)***, we hypothesize that such uncertainty information can be regarded as additional supervision signals to boost model training. In other words, these most challenging and valuable regions should be explored to train the segmentation model even without labels.

Therefore, in this paper, we propose a novel Mutual Consistency model (MC-Net, see Fig. 2) for semi-supervised left atrium segmentation from 3D MR images. Our MC-Net is composed of one encoder and two slightly different decoders and the discrepancy of two outputs is used to capture the uncertainty information. Then, we employ a sharpening function [11] to generate soft pseudo labels from the probability outputs as the entropy regularization [5]. Afterwards, we design a new cycled pseudo label scheme to leverage the uncertainty information to learn more critical features by encouraging mutual consistency [18]. Such mutual consistency constrains MC-Net to generate consistent and low-entropy results of the same input so that the model-based epistemic uncertainty is reduced, which enables the model to learn generalized feature representation from these unlabeled challenging regions [9,11]. We evaluate our proposed MC-Net against six recent state-of-the-art (SOTA) methods on the popular LA database [16]. The experiment results reveal that the designed MC-Net outperforms all other existing semi-supervised segmentation methods. The ablation studies further demonstrate the effectiveness of each component design.

The contributions of our model include: (1) we explore the model-based uncertainty information to emphasize the unlabeled challenging regions during training; (2) we design a novel cycled pseudo label scheme to facilitate the model training by encouraging mutual consistency; (3) to our best knowledge, the proposed MC-Net has achieved the new state-of-the-art performance in the semi-supervised left atrium segmentation task on the LA database.

2 Method

As shown in Fig. 2, our MC-Net has two unique attributes. First, we embed an extra slightly different decoder into the original V-Net [8] and utilize the discrepancy of the two decoder outputs to capture the uncertainty information. Second, we design the cycled pseudo label scheme to transform such uncertainty as an unsupervised loss to encourage mutual consistency for training.

2.1 Model Architecture

There are several representative methods to measure the epistemic uncertainty. For example, the Monte Carlo dropout [3] is a popular one. Given a 3D input

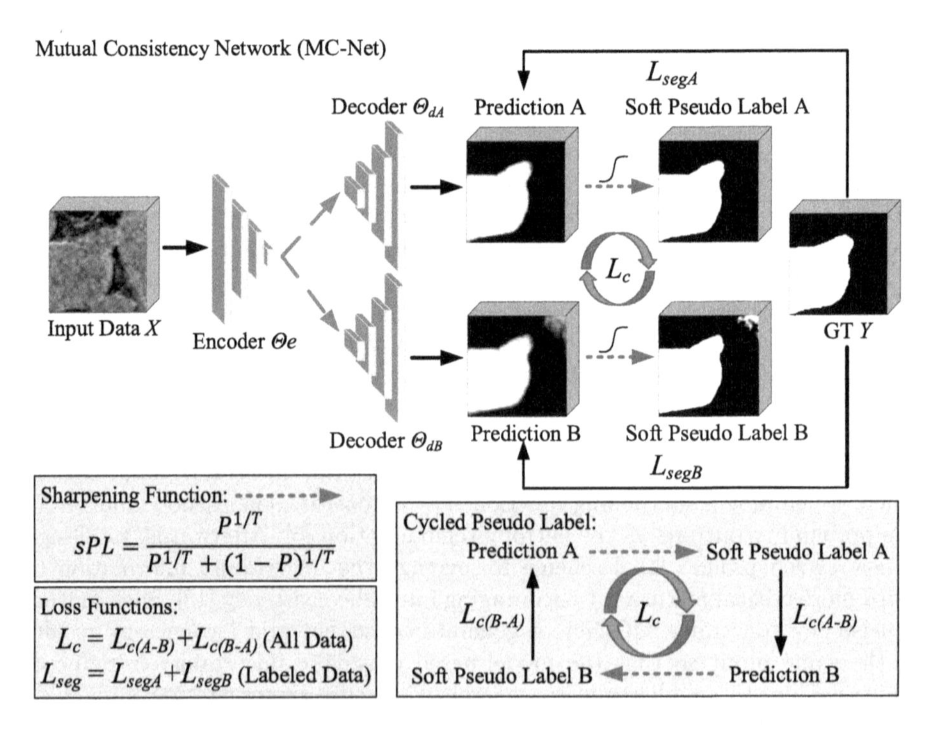

Fig. 2. Diagram of our proposed MC-Net.

sample $X \in \mathbb{R}^{H \times W \times D}$, we can perform N stochastic forward passes with random dropout, where the dropout layer is able to sample sub-models θ_n from the original model θ. In this way, the deep model θ outputs a set of probability vector: $\{P_n\}_{n=1}^N$. The uncertainty u can be approximated by the statistics of the predictions of all sub-models θ_n. For instance, a related work [17] employs the Monte Carlo dropout to estimate the uncertainty u as:

$$\mu_c = \frac{1}{N} \sum_n P_n^c, \quad u = -\sum_C \mu_c log \mu_c \tag{1}$$

where P_n^c represents the output of the c-th class in the n-th time, μ_c is the mean of N predictions, C is the number of classes and the uncertainty $u \in \mathbb{R}^{H \times W \times D}$ is essentially the voxel-wise entropy. One of the issues of such a method is that it requires multiple inferences, e.g. eight stochastic forward passes every iteration in [17], to estimate the uncertainty, which brings more computational costs.

Therefore, inspired by [20], we simplify the Monte Carlo dropout via introducing an auxiliary classifier, which reduces both training and inference costs. Specifically, we design two slightly different decoders to approximate the epistemic uncertainty. One decoder θ_{dA} employs original transposed convolution for up-sampling like V-Net and another decoder θ_{dB} uses tri-linear interpolation to expand the feature maps as an auxiliary classifier. Other modules are identical to V-Net. The two decoders receive the same deep features F_e from encoder θ_e and then generate two features F_A and F_B. This process can be formulated as

$$F_e = f_{\theta_e}(X), \quad F_A = f_{\theta_{dA}}(F_e), \quad F_B = f_{\theta_{dB}}(F_e) \tag{2}$$

where probability outputs P_A and P_B can be obtained from deep features F_A and F_B, respectively, with a Sigmoid activation function. Such two slightly different decoders can increase the diversity of segmentation models, and the diversified features in different sub-models are able to reduce over-fitting and improve the performance [13]. Meanwhile, for fair comparisons, we do not further introduce any other additional designs to enhance the V-Net backbone [8].

Compared to Monte Carlo dropout [3,17], the sub-models θ_n of our MC-Net are fixed and do not require additional perturbations in the training process. As shown in Fig. 2, the model-based epistemic uncertainty is approximated by the discrepancy between the two decoder outputs P_A and P_B.

2.2 Cycled Pseudo Label

Based on such model design, we then transform the prediction discrepancies as auxiliary supervision signals to boost training. First, we use a sharpening function [14] to transform probability outputs P_A and P_B into soft pseudo labels sPL_A and $sPL_B \in [0,1]^{H \times W \times D}$. The sharpening function is defined as

$$sPL = \frac{P^{1/T}}{P^{1/T} + (1-P)^{1/T}} \tag{3}$$

where T is a constant to control the temperature of sharpening. The soft pseudo labels make contribution to entropy regularization for training [5]. Compared with pseudo labels generating by a fixed threshold, soft pseudo labels are able to eliminate the impacts of some mislabeled training data [14].

Afterwards, we employ sPL_A to supervise P_B and then employ sPL_B to supervise P_A to achieve mutual consistency [18]. In this way, the two decoders can learn from each other and be trained in an 'end-to-end' manner. Via this design, the predictions P_A and P_B are encouraged to be consistent and of low entropy. Such consistency and entropy regularization facilitate the model to pay more attention to the unlabeled challenging and uncertain areas [9,11].

Hence, our MC-Net can be trained via the weighted sum of a segmentation loss L_{seg} and a consistency loss L_c. The total loss can be written as

$$loss = \underbrace{Dice(P_A, Y) + Dice(P_B, Y)}_{L_{seg}} + \lambda \times \underbrace{(L_2(P_A, sPL_B) + L_2(P_B, sPL_A))}_{L_c} \tag{4}$$

where $Dice$ represents the $Dice$ loss, L_2 is the Mean Squared Error (MSE) loss, Y is the ground truth and λ is a weight to balance L_{seg} and L_c. Note that, L_{seg} is only calculated from labeled data and L_c is unsupervised, which is used to supervise all training data.

3 Experiment and Results

3.1 Database

We evaluate the proposed MC-Net on the LA database [16] from the 2018 Atrial Segmentation Challenge[1]. The LA database consists of 100 gadolinium-enhanced MR imaging scans, with a fixed split of 80 samples for training and 20 samples for validation. The isotropic resolution is $0.625 \times 0.625 \times 0.625$ mm. We report the performance on the validation set for fair comparisons as [2,6,7,12,17].

3.2 Implementation Details

For pre-processing, we first cropped the 3D MR images with enlarged margins according to the targets. These scans were further normalized as zero mean and unit variance. For training, we randomly cropped 3D patches of size $112 \times 112 \times 80$ and applied the 2D rotation and flip operations as the data augmentation. Then, we set the batch size as 4 and each batch included two labeled patches and two unlabeled patches. The temperature constant T was set as 0.1 and the weight λ was set as a time-dependent Gaussian warming-up function [4]. The proposed MC-Net was trained by a SGD optimizer for 6K iterations, with an initial learning rate 0.01 decayed by 10% every 2.5K iterations. Most of the experiment settings were set same as the recent SOTA methods [6,7,17].

For testing, we employed a sliding window of size $112 \times 112 \times 80$ with a fixed stride ($18 \times 18 \times 4$) to extract patches. Then we recomposed the patch-based predictions as entire results. Note that, we used the mean of P_A and P_B as the final output during testing. All experiments were conducted on the same environments with fixed random seeds (Hardware: Intel Xeon E5-2682 CPU, NVIDIA Tesla P100 GPU; Software: Pytorch 1.7.0+cu110, and Python 3.8.5).

3.3 Results

Figure 3 shows the results obtained by UA-MT [17], SASSNet [6], DTC [7], our MC-Net, and the corresponding ground truth on the LA database from left to right. There are two popular semi-supervised settings that the first uses 10% labeled data with 90% unlabeled data and the second uses 20% labeled data with 80% unlabeled data. The corresponding results are also shown in Fig. 3. It can be seen that our model generates more complete left atrium than all other existing SOTA methods in either 3D or 2D view. Note that, we do not use any morphology algorithms as the post-processing module to refine the results. The MC-Net naturally generates better results in some challenging areas (see yellow circles) and eliminates most of isolated regions on the LA database. Such ability is essential for further medical quantitative analysis.

Table 1 shows the quantitative results in terms of the Dice, Jaccard, 95% Hausdorff Distance (95HD) and average surface distance (ASD). It also gives the

[1] http://atriaseg2018.cardiacatlas.org.

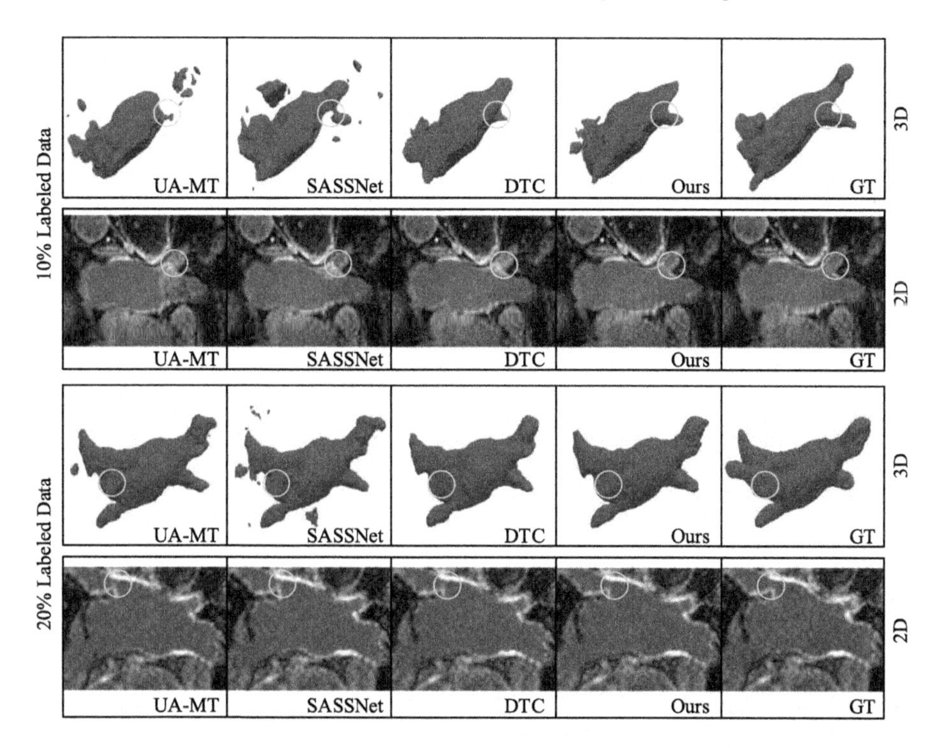

Fig. 3. Two segmentation results obtained by UA-MT [17] (1st column), SASSNet [6] (2nd column), DTC [7] (3rd column), our proposed MC-Net (4th column), and the corresponding ground truth (5th column) on the LA database. The comparisons using 10% or 20% labeled data are showed on the top or bottom two rows, respectively.

results of V-Net under fully supervised settings (with 10%, 20% and all labeled data) as the reference. By exploiting unlabeled data effectively, our proposed MC-Net produces impressive performance gains and our method with only 20% labeled training data obtain comparable results, e.g. 90.34% vs. 91.14% of Dice, with the upper bound (V-Net with 100% labeled training data). It can also be seen from Table 1 that our MC-Net significantly outperforms the six recent SOTA methods on the LA database in all semi-supervised settings.

3.4 Ablation Study

The ablation studies (see Table 2) were conducted to show the effectiveness of each component on the LA database. We can see from Table 2 that, in each semi-supervised setting, the one with two slightly different decoders (V2d-Net) generates better results than the one with two identical decoders (V2-Net). This suggests that increasing the diversity of the segmentation model can improve the performance. Moreover, encouraging consistent results, i.e. minimizing the differences between sPL_A and sPL_B (labeled as +sPL), in general leads to the performance gains and applying the entropy regularization, i.e. using the cycled

Table 1. Comparisons with six state-of-the-art methods on the LA database.

Method	# Scans used		Metrics			
	Labeled	Unlabeled	Dice (%)	Jaccard (%)	95HD (voxel)	ASD (voxel)
V-Net	8 (10%)	0	79.99	68.12	21.11	5.48
V-Net	16 (20%)	0	86.03	76.06	14.26	3.51
V-Net	80 (All)	0	91.14	83.82	5.75	1.52
DAP [19] (MICCAI'19)	8 (10%)	72	81.89	71.23	15.81	3.80
UA-MT [17] (MICCAI'19)	8 (10%)	72	84.25	73.48	13.84	3.36
SASSNet [6] (MICCAI'20)	8 (10%)	72	87.32	77.72	9.62	2.55
LG-ER-MT [2] (MICCAI'20)	8 (10%)	72	85.54	75.12	13.29	3.77
DUWM [12] (MICCAI'20)	8 (10%)	72	85.91	75.75	12.67	3.31
DTC [7] (AAAI'21)*	8 (10%)	72	86.57	76.55	14.47	3.74
MC-Net (Ours)	8 (10%)	72	**87.71**	**78.31**	**9.36**	**2.18**
DAP [19] (MICCAI'19)	16 (20%)	64	87.89	78.72	9.29	2.74
UA-MT [17] (MICCAI'19)	16 (20%)	64	88.88	80.21	7.32	2.26
SASSNet [6] (MICCAI'20)	16 (20%)	64	89.54	81.24	8.24	2.20
LG-ER-MT [2] (MICCAI'20)	16 (20%)	64	89.62	81.31	7.16	2.06
DUWM [12] (MICCAI'20)	16 (20%)	64	89.65	81.35	7.04	2.03
DTC [7] (AAAI'21)	16 (20%)	64	89.42	80.98	7.32	2.10
MC-Net (Ours)	16 (20%)	64	**90.34**	**82.48**	**6.00**	**1.77**

* Since the results of DTC model using 10% labeled data with 90% unlabeled data were not given, we conducted the experiments on the LA database as [7].

pseudo label (labeled as +CPL), is able to further improve the performance. Note that, our MC-Net can be easily combined with other shape-constrained models [6,7] to enhance the segmentation results.

Table 2. Ablation studies of our proposed MC-Net on the LA database.

Method	# Scans used		Metrics			
	Labeled	Unlabeled	Dice (%)	Jaccard (%)	95HD (voxel)	ASD (voxel)
V2-Net	8 (10%)	72	84.95	74.27	15.21	4.56
V2d-Net	8 (10%)	72	85.79	75.41	14.45	3.83
V2-Net+sPL	8 (10%)	72	86.61	76.65	13.39	3.93
V2d-Net+sPL	8 (10%)	72	86.84	76.97	14.16	4.05
V2-Net+CPL	8 (10%)	72	86.69	76.74	12.66	3.20
V2d-Net+CPL	8 (10%)	72	**87.71**	**78.31**	**9.36**	**2.18**
V2-Net	16 (20%)	64	88.14	79.10	14.22	3.78
V2d-Net	16 (20%)	64	88.98	80.36	7.61	2.25
V2-Net+sPL	16 (20%)	64	88.55	79.64	14.06	3.58
V2d-Net+sPL	16 (20%)	64	90.15	82.15	6.20	1.89
V2-Net+CPL	16 (20%)	64	89.66	81.36	11.14	2.86
V2d-Net+CPL	16 (20%)	64	**90.34**	**82.48**	**6.00**	**1.77**

4 Conclusion

In this paper, we have presented a mutual consistency network (MC-Net) for semi-supervised left atrium segmentation. Our key idea is that the unlabeled challenging regions should play a critical role in semi-supervised learning. Hence, via the designed cycled pseudo label scheme, our model is encouraged to generate consistent and low-entropy predictions so that more generalized features can be captured from these crucial areas to boost the model training. The proposed MC-Net achieves, to our best knowledge, the most accurate semi-supervised left atrium segmentation performance on the LA database.

Acknowledgements. This work was done during an internship at Alibaba Group and was partially supported by Monash FIT Start-up Grant. We also appreciate the efforts devoted to collect and share the LA database [16] and several available repositories [6,7,17].

References

1. Fang, K., Li, W.-J.: DMNet: difference minimization network for semi-supervised segmentation in medical images. In: Martel, A.I., et al. (eds.) MICCAI 2020. LNCS, vol. 12261, pp. 532–541. Springer, Cham (2020). https://doi.org/10.1007/978-3-030-59710-8_52
2. Hang, W., et al.: Local and global structure-aware entropy regularized mean teacher model for 3D left atrium segmentation. In: Martel, A.L., et al. (eds.) MICCAI 2020. LNCS, vol. 12261, pp. 562–571. Springer, Cham (2020). https://doi.org/10.1007/978-3-030-59710-8_55
3. Kendall, A., Gal, Y.: What uncertainties do we need in Bayesian deep learning for computer vision? arXiv preprint arXiv:1703.04977 (2017)
4. Laine, S., Aila, T.: Temporal ensembling for semi-supervised learning. arXiv preprint arXiv:1610.02242 (2016)
5. Lee, D.H., et al.: Pseudo-label: the simple and efficient semi-supervised learning method for deep neural networks. In: ICML, vol. 3, no. 2 (2013)
6. Li, S., Zhang, C., He, X.: Shape-aware semi-supervised 3D semantic segmentation for medical images. In: Martel, A.L., et al. (eds.) MICCAI 2020. LNCS, vol. 12261, pp. 552–561. Springer, Cham (2020). https://doi.org/10.1007/978-3-030-59710-8_54
7. Luo, X., et al.: Semi-supervised medical image segmentation through dual-task consistency. arXiv preprint arXiv:2009.04448 (2020)
8. Milletari, F., Navab, N., Ahmadi, S.A.: V-net: fully convolutional neural networks for volumetric medical image segmentation. In: 3DV 2016, pp. 565–571 (2016)
9. Ouali, Y., Hudelot, C., Tami, M.: Semi-supervised semantic segmentation with cross-consistency training. CVPR **2020**, 12674–12684 (2020)
10. Shrivastava, A., Gupta, A., Girshick, R.: Training region-based object detectors with online hard example mining. CVPR **2016**, 761–769 (2016)
11. Sohn, K., et al.: Fixmatch: Simplifying semi-supervised learning with consistency and confidence. In: NeurIPS 2020. vol. 33, pp. 596–608 (2020)
12. Wang, Y., et al.: Double-uncertainty weighted method for semi-supervised learning. In: Martel, A.L., et al. (eds.) MICCAI 2020. LNCS, vol. 12261, pp. 542–551. Springer, Cham (2020). https://doi.org/10.1007/978-3-030-59710-8_53

13. Xia, Y., et al.: 3D semi-supervised learning with uncertainty-aware multi-view co-training. WACV **2020**, 3646–3655 (2020)
14. Xie, Q., Dai, Z., Hovy, E., Luong, M.T., Le, Q.V.: Unsupervised data augmentation for consistency training. arXiv preprint arXiv:1904.12848 (2019)
15. Xie, Y.: Pairwise relation learning for semi-supervised gland segmentation. In: Martel, A.L., et al. (eds.) MICCAI 2020. LNCS, vol. 12265, pp. 417–427. Springer, Cham (2020). https://doi.org/10.1007/978-3-030-59722-1_40
16. Xiong, Z., et al.: A global benchmark of algorithms for segmenting the left atrium from late gadolinium-enhanced cardiac magnetic resonance imaging. Med. Image Anal. **67**, 101832 (2021)
17. Yu, L., Wang, S., Li, X., Fu, C.-W., Heng, P.-A.: Uncertainty-aware self-ensembling model for semi-supervised 3D left atrium segmentation. In: Shen, D., et al. (eds.) MICCAI 2019. LNCS, vol. 11765, pp. 605–613. Springer, Cham (2019). https://doi.org/10.1007/978-3-030-32245-8_67
18. Zhang, Y., Xiang, T., Hospedales, T.M., Lu, H.: Deep mutual learning. CVPR **2018**, 4320–4328 (2018)
19. Zheng, H., et al.: Semi-supervised segmentation of liver using adversarial learning with deep atlas prior. In: Shen, D., et al. (eds.) MICCAI 2019. LNCS, vol. 11769, pp. 148–156. Springer, Cham (2019). https://doi.org/10.1007/978-3-030-32226-7_17
20. Zheng, Z., Yang, Y.: Rectifying pseudo label learning via uncertainty estimation for domain adaptive semantic segmentation. Int. J. Comput. Vis. pp. 1–15 (2021)

Semi-supervised Meta-learning with Disentanglement for Domain-Generalised Medical Image Segmentation

Xiao Liu[1(✉)], Spyridon Thermos[1], Alison O'Neil[1,3], and Sotirios A. Tsaftaris[1,2]

[1] School of Engineering, University of Edinburgh, Edinburgh, EH9 3FB, UK
{Xiao.Liu,SThermos,S.Tsaftaris}@ed.ac.uk
[2] The Alan Turing Institute, London, UK
[3] Canon Medical Research Europe Ltd., Edinburgh, UK
Alison.Oneil@mre.medical.canon

Abstract. Generalising deep models to new data from new centres (termed here domains) remains a challenge. This is largely attributed to shifts in data statistics (domain shifts) between source and unseen domains. Recently, gradient-based meta-learning approaches where the training data are split into meta-train and meta-test sets to simulate and handle the domain shifts during training have shown improved generalisation performance. However, the current fully supervised meta-learning approaches are not scalable for medical image segmentation, where large effort is required to create pixel-wise annotations. Meanwhile, in a low data regime, the simulated domain shifts may not approximate the true domain shifts well across source and unseen domains. To address this problem, we propose a novel semi-supervised meta-learning framework with disentanglement. We explicitly model the representations related to domain shifts. Disentangling the representations and combining them to reconstruct the input image allows unlabeled data to be used to better approximate the true domain shifts for meta-learning. Hence, the model can achieve better generalisation performance, especially when there is a limited amount of labeled data. Experiments show that the proposed method is robust on different segmentation tasks and achieves state-of-the-art generalisation performance on two public benchmarks. Code is publicly available at: https://github.com/vios-s/DGNet.

Keywords: Domain generalisation · Disentanglement · Medical image segmentation

Electronic supplementary material The online version of this chapter (https://doi.org/10.1007/978-3-030-87196-3_29) contains supplementary material, which is available to authorized users.

M. de Bruijne et al. (Eds.): MICCAI 2021, LNCS 12902, pp. 307–317, 2021.
https://doi.org/10.1007/978-3-030-87196-3_29

1 Introduction

Despite recent progress in medical image segmentation [3,8,18], inference performance on unseen datasets, acquired from distinct scanners or clinical centres, is known to decrease [5,38]. Such reduction is mainly caused by shifts in data statistics between different clinical centres i.e. *domain shifts* [45], due to variation in patient populations, scanners, and scan acquisition settings [42]. The variation in population impacts the underlying anatomy and pathology due to factors such as gender, age, ethnicity, which may differ for patients in different locations [28,39,43]. The variation in scanners and scan acquisition settings impacts the characteristics of the acquired image, such as brightness and contrast [45].

The naive approach to handling domain shift is to acquire and label as many and diverse data as possible, the cost implications and difficulties of which are known to this community. Alternatively one can train a model on source domains to generalise for a target domain with some information on the target domain available i.e. domain adaptation [4] such as cross-site MRI harmonisation [37] to enforce the source and target domains to share similar image-specific characteristics [9]. A more strict alternative is to *not use any* information for the target domain, known as *domain generalisation* [22]. Herein, we focus on this more challenging and more widely applicable approach.

In domain generalisation, the overarching goal is to identify suitable representations that encode information about the task at hand whilst being insensitive to domain-specific information. There are several active research directions aiming to address this goal, including: direct augmentation of the source domain data [45], feature space regularisation [6,14,27,35,46], alignment of the source domain features or output distributions [25], and learning domain-invariant features with gradient-based meta-learning [11,24,29]. Of the above, gradient-based meta-learning methods have the advantage of not overfitting to dominant source domains which account for the more populous data in the training dataset [11]. Gradient-based meta-learning [22,26] exploits an episodic training paradigm [23] by splitting the source domains into meta-train and meta-test sets at each iteration. The model is trained to handle domain shift by simulating it during training. By using constraints to implicitly eliminate the information related to the simulated domain shifts, the model can learn to extract domain-invariant features. Previous work introduced different constraints in a fully supervised setting e.g. global class alignment and local sample clustering [11], shape-aware constraints [29] or simply the task objective [20,22], where [20] extends [22] to medical image segmentation.[1] These approaches do not scale in medical image segmentation as pixel-wise annotation is time-consuming, laborious, and requires expert knowledge. Meanwhile, in a low data regime where centres only provide a few labeled data samples, these methods may only learn to extract domain-invariant features from an under-represented data distribution [20,45]. In other

[1] With the exception of [41] which clusters unlabeled data to generate pseudo labels, but unfortunately is not applicable to segmentation.

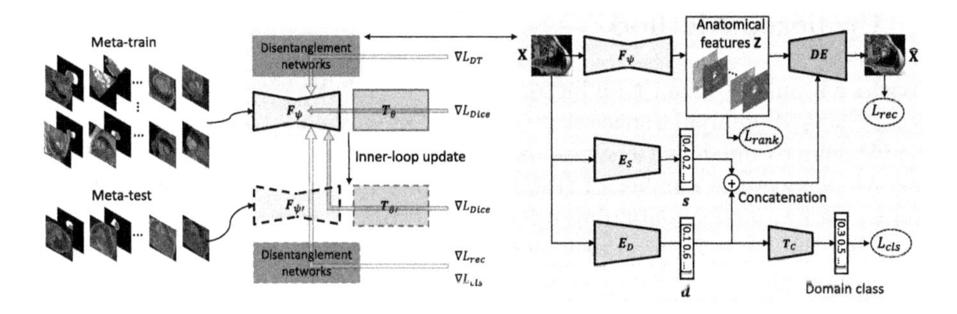

Fig. 1. At each iteration, the training dataset is split into meta-train and meta-test sets including labeled and unlabeled data. A feature network $\boldsymbol{F}_\psi$ extracts features $\mathbf{Z}$ for a task network $\boldsymbol{T}_\theta$ to predict segmentation masks. The model is trained in a semi-supervised setting, where $\mathcal{L}_{DT}$, $\mathcal{L}_{rec}$ and $\mathcal{L}_{cls}$ do not require pixel-wise annotation. In the inner-loop update, ψ' and θ' are computed for the meta-test step (see Eq. 1). Finally, all the gradients are computed to update $\boldsymbol{F}_\psi$ and $\boldsymbol{T}_\theta$ as in Eq. 2. The disentanglement networks decompose image $\mathbf{X}$ to common $\mathbf{s}$ and specific to the domain $\mathbf{d}$ representations to be disentangled with $\mathbf{Z}$ for meta-train and meta-test sets with the constraints ($\mathcal{L}_{DT}$ and $\mathcal{L}_{rec}$ and $\mathcal{L}_{cls}$). See Sect. 2 for loss definitions.

words, the simulated domain shifts may not well approximate the true domain shifts between source and unseen domains.

To address this problem, we propose to explicitly disentangle the representations related to domain shifts for meta-learning as we illustrate in Fig. 1. Learning these complete and sufficient representations [1] via reconstruction brings the benefit of unsupervised learning, thus we can better simulate the domain shifts by also using unlabeled data from any of the source domains. We consider two sources of shifts: one due to scanner and scan acquisition setting variation, and one due to population variation. Because our task is segmentation, we want to be sensitive to changes in anatomy but insensitive to changes in imaging characteristics be it some common across domains or domain-specific. We use spatial (grid-like) features as a representation of anatomy ($\mathbf{Z}$) and two vectors ($\mathbf{s}$,$\mathbf{d}$) to encode common or domain-specific imaging characteristics. We apply specific design and learning biases to disentangle the above. For example, a spatial $\mathbf{Z}$ is equivariant to segmentation and this has been shown to improve performance [7,16]. We further encourage $\mathbf{Z}$ to be disentangled from $\mathbf{s}$ and $\mathbf{d}$ by exploiting also low-rank regularisation [25]. Gradient-based meta-learning also encourages $\mathbf{Z}$, $\mathbf{s}$, and $\mathbf{d}$ to generalise well to unseen domains whilst at the same time improves (implicitly) their disentanglement. Our **contributions** are summarised as follows:

1. We propose the first, to the best of our knowledge, semi-supervised domain-generalisation framework combining meta-learning and disentanglement.
2. Use of low-rank regularisation as a learning bias to encourage better disentanglement and hence improved generalisation performance.
3. Extensive experiments on cardiac and gray matter datasets show improved performance over several baselines especially for the limited annotated data case.

2 Proposed Method

Consider a multi-domain training dataset $\mathcal{D} = \{\mathbf{X}_i^k, \mathbf{Y}_i^k\}_{i=1}^{N_k}, k \in \{1, 2, \cdots, K\}$ that is defined on a joint space $\mathcal{X} \times \mathcal{Y}$, where $\mathbf{X}_i^k$ is the ith training datum from the kth source domain with corresponding ground-truth segmentation mask $\mathbf{Y}_i^k$, and N_k denotes the number of training samples in the kth source domain. We aim to learn a model containing a feature network $\boldsymbol{F}_\psi : \mathcal{X} \rightarrow \mathcal{Z}$ to extract the anatomical features $\mathbf{Z}$ and a task network $\boldsymbol{T}_\theta : \mathcal{Z} \rightarrow \mathcal{Y}$ to predict the segmentation masks, where ψ and θ denote the network parameters.

2.1 Gradient-Based Meta-Learning for Domain Generalisation

In gradient-based meta-learning for domain generalisation, the domain shift is simulated by training the model on a sequence of episodes [23,24]. Specifically, the meta-train set $\mathcal{D}_{tr}$ and the meta-test set $\mathcal{D}_{te}$ are constructed by randomly splitting the source domains $\mathcal{D}$ for each iteration of training. Each iteration comprises a meta-train step followed by a meta-test step. For the meta-train step, the parameters ψ and θ of $\boldsymbol{F}_\psi$ and $\boldsymbol{T}_\theta$ are calculated by optimising the meta-train loss $\mathcal{L}_{meta-train}$ with data from $\mathcal{D}_{tr}$ (inner-loop update), as defined by:

$$(\psi', \theta') = (\psi, \theta) - \alpha \nabla_{\psi,\theta} \mathcal{L}_{meta-train}(\mathcal{D}_{tr}; \psi, \theta), \tag{1}$$

where α is the learning rate for the meta-train update step. Typically, $\mathcal{L}_{meta-train}$ is the task objective, e.g. the Dice loss [10] for a segmentation task. This step rewards accurate predictions on the meta-train source domains. For the meta-test step, the meta-test source domains $\mathcal{D}_{te}$ are processed by the updated parameters (ψ', θ') and the model is expected to contain certain properties quantified by the $\mathcal{L}_{meta-test}$ loss. $\mathcal{L}_{meta-test}$ is computed using the updated parameters (ψ', θ'), whilst the gradients are computed towards the original parameters (ψ, θ). The final objective is defined as:

$$\underset{\psi,\theta}{\operatorname{argmin}} \, \mathcal{L}_{meta-train}(\mathcal{D}_{tr}; \psi, \theta) + \mathcal{L}_{meta-test}(\mathcal{D}_{te}; \psi', \theta'). \tag{2}$$

The intuition behind this scheme is that the model should not only perform well on the source domains, but its future updates should also generalise well to unseen domains. Below, we will describe our meta-train and meta-test objectives but first we present how we disentangle representations related to domain shifts.

2.2 Learning Disentangled Representations

To model appearance in a single-domain setting, typically a single vector-based variational representation is used [7]. Here, due to our multi-domain setting, inspired by [17,44], we separately encode domain-specific imaging characteristics as an additional vector-based variational representation. Hence, we aim to learn two independent vector representations, where one ($\mathbf{s}$) captures common imaging characteristics across domains and the other one ($\mathbf{d}$) captures specific imaging characteristics for each domain. In addition, we encode spatial anatomy information in a separate representation $\mathbf{Z}$, which we encourage to be disentangled from $\mathbf{s}$ and $\mathbf{d}$.

In particular, the input image $\mathbf{X}$ is first encoded in a common (appearance) representation $\mathbf{s} = \boldsymbol{E_S}(\mathbf{X})$, and a domain representation $\mathbf{d} = \boldsymbol{E_D}(\mathbf{X})$ that is followed by a shallow domain classifier $\boldsymbol{T_C}(\mathbf{d})$ which predicts the source domain ($\hat{\mathbf{c}}$) label of $\mathbf{X}$. Then, a decoder $\boldsymbol{DE}$ combines the extracted features $\mathbf{Z} = \boldsymbol{F_\psi}(\mathbf{X})$ and the representations $\mathbf{s}$ and $\mathbf{d}$ to reconstruct the input image, i.e. $\hat{\mathbf{X}} = \boldsymbol{DE}(\mathbf{Z}, \mathbf{s}, \mathbf{d})$. Note that $\boldsymbol{DE}$ combines $\mathbf{Z}$ and $\mathbf{s}, \mathbf{d}$ using adaptive instance normalisation (AdaIN) layers [15]. As shown in [16], AdaIN improves disentanglement and encourages $\mathbf{Z}$ to encode spatially equivariant information, i.e. anatomical information useful for segmentation, and $\mathbf{s}, \mathbf{d}$ to only encode common or domain-specific appearance.

To achieve such "triple" disentanglement we consider several losses: **1)** KL divergences $\mathcal{L}_{KL}(\mathbf{s}, N(0,1)), \mathcal{L}_{KL}(\mathbf{d}, N(0,1))$ to induce a Gaussian $N(0,1)$ prior in $\mathbf{s}$ and $\mathbf{d}$, encouraging the representations to be robust on unseen domains [13]; **2)** Hilbert-Schmidt Independence Criterion (HSIC) loss $\mathcal{L}_{HSIC}(\mathbf{s}, \mathbf{d})$, to force $\mathbf{s}$ and $\mathbf{d}$ to be independent from each other [34]; **3)** a classification loss $\mathcal{L}_{cls}(\mathbf{c}, \hat{\mathbf{c}})$ such that the domain representation $\mathbf{d}$ is highly correlated with the domain-specific information [17]; and **4)** a reconstruction loss $\mathcal{L}_{rec}(\mathbf{X}, \hat{\mathbf{X}})$, defined as the ℓ_1 distance between $\mathbf{X}$ and $\hat{\mathbf{X}}$, to learn representations without supervision [7,16].

We further encourage the extracted features $\mathbf{Z}$ to be equivariant across the meta-train source domains and improve disentanglement between $\mathbf{Z}$ and $\mathbf{s}$, $\mathbf{d}$ by applying rank regularisation [25]. Specifically, consider a batch $\{\mathbf{X}^1_{i_1}, \mathbf{X}^2_{i_2}, \cdots, \mathbf{X}^{K_{tr}}_{i_{K_{tr}}}\}$ from K_{tr} meta-train source domains, and K_{tr} features $\{\mathbf{Z}^1_{i_1}, \mathbf{Z}^2_{i_2}, \cdots, \mathbf{Z}^{K_{tr}}_{i_{K_{tr}}}\}$ extracted using the feature network $\boldsymbol{F_\psi}$. By flattening and concatenating these features, we end up with a matrix $\mathbb{Z}$ with dimensions $[C, K_{tr} \times H \times W]$, where C, H, W denote the number of channels, height, and width of $\mathbb{Z}$. Then, by forcing the rank of $\mathbb{Z}$ to be m (i.e. the number of the segmentation classes), $\mathbf{Z}$ is encouraged to encode only globally-shared information across K_{tr} source domains in order to predict the segmentation mask as discussed in [25]. We achieve that by minimising the $(m+1)^{th}$ singular value σ_{m+1} of $\mathbb{Z}$, i.e. $\mathcal{L}_{rank} = \sigma_{m+1}$. Overall, $\mathcal{L}_{DT}$ is defined as:

$$\begin{aligned}\mathcal{L}_{DT} = &\lambda_{rank}\mathcal{L}_{rank}(\mathbf{Z}) + \lambda_{KL}(\mathcal{L}_{KL}(\mathbf{s}, N(0,1)) + \mathcal{L}_{KL}(\mathbf{d}, N(0,1))) \\ &+ \lambda_{rec}\mathcal{L}_{rec}(\mathbf{X}, \hat{\mathbf{X}}) + \lambda_{HSIC}\mathcal{L}_{HSIC}(\mathbf{s}, \mathbf{d}) + \lambda_{cls}\mathcal{L}_{cls}(\mathbf{c}, \hat{\mathbf{c}}),\end{aligned} \tag{3}$$

where $\mathbf{c}$ is the domain label. We adopt hyperparameter values according to our extensive early experiments and discussion from [7,25] as $\lambda_{rank} = 0.1$, $\lambda_{KL} = 0.1$, $\lambda_{rec} = 1$ and $\lambda_{cls} = 1$. Note that all the losses do not need ground-truth masks. The domain class label is available, as we know the centre where the data belong.

2.3 Meta-train and Meta-test Objectives

Our *meta-train* objective contains two components:

$$\mathcal{L}_{meta-train} = \lambda_{Dice}\mathcal{L}_{Dice}(\mathbf{Y}, \hat{\mathbf{Y}}) + \mathcal{L}_{DT}, \tag{4}$$

where $\lambda_{Dice} = 5$ when labeled data are available.

For the *meta-test* step, the model is expected to: **1)** accurately predict segmentation masks (by applying the task objective), and **2)** disentangle $\mathbf{Z}$ and $\mathbf{s}, \mathbf{d}$ to the same level as meta-train sets. A naive strategy for the latter is to use $\mathcal{L}_{DT}$ for meta-test sets. However, as analysed in [2,29], the meta-test step is unstable to train: the gradients from the meta-test loss are second-order statistics of ψ and θ. Our experiments revealed that including the unsupervised losses $\mathcal{L}_{KL}$ and $\mathcal{L}_{HSIC}$ make training even more unstable (even leading to model collapse). In addition, we use one domain for meta-test in experiments, while $\mathcal{L}_{rank}$ requires multiple domains. According to [31,33], considering fixed learning and design biases, the level of disentanglement can be proxied by the reconstruction quality (with ground-truth image $\mathbf{X}$) and the domain classification accuracy (with ground-truth label $\mathbf{c}$). Hence, we adopt as the meta-test loss:

$$\mathcal{L}_{meta-test} = \lambda_{Dice}\mathcal{L}_{Dice}(\mathbf{Y}, \hat{\mathbf{Y}}) + \lambda_{rec}\mathcal{L}_{rec}(\mathbf{X}, \hat{\mathbf{X}}) + \lambda_{cls}\mathcal{L}_{cls}(\mathbf{c}, \hat{\mathbf{c}}). \quad (5)$$

Note that for unlabeled data, $\mathcal{L}_{rec}$ and $\mathcal{L}_{cls}$ do not need ground-truth masks.

3 Experiments

3.1 Tasks and Datasets

Multi-centre, Multi-vendor & Multi-disease Cardiac Image Segmentation (M&Ms) Dataset [5]**:** The M&Ms challenge dataset contains 320 subjects. Subjects were scanned at 6 clinical centres in 3 different countries using 4 different magnetic resonance scanner vendors (Siemens, Philips, GE, and Canon) i.e. domains A, B, C and D. For each subject, only the end-systole and end-diastole phases are annotated. Voxel resolutions range from $0.85 \times 0.85 \times 10$ mm to $1.45 \times 1.45 \times 9.9$ mm. Domain A contains 95 subjects. Domain B contains 125 subjects. Both domains C and D contain 50 subjects.

Spinal Cord Gray Matter Segmentation (SCGM) Dataset [38]**:** The data from SCGM [38] are collected from 4 different medical centres with different MRI systems (Philips Achieva, Siemens Trio, Siemens Skyra) i.e. domains 1, 2, 3 and 4. The voxel resolutions range from $0.25 \times 0.25 \times 2.5$ mm to $0.5 \times 0.5 \times 5$ mm. Each domain has 10 labeled subjects and 10 unlabelled subjects.

3.2 Baseline Models

nnUNet [19]**:** is a self-adapting framework based on 2D and 3D U-Nets [40] which does not specifically target domain generalisation. Given a labelled training dataset, nnUNet automatically adapts its model design and hyperparameters to obtain optimal performance. In the M&Ms challenge, methods based on nnUNet achieved the top performance [5].

SDNet+Aug. [30]**:** disentangles the input image to a spatial anatomy and a non-spatial modality factors. Here we use intensity- and resolution- augmented

Table 1. Dice (%) results and the standard deviations on M&Ms dataset. For "SDNet+Aug." and our method, the training data contain all the unlabeled data and 2% or 5% of labeled data from source domains. The other models are trained by 2% or 5% labeled data only. Bold numbers denote the best performance.

Source		Target	nnUNet	SDNet+Aug.	LDDG	SAML	Ours
2%	B,C,D	A	52.87_{19}	54.48_{18}	59.47_{12}	56.31_{13}	$\mathbf{66.01_{12}}$
	A,C,D	B	64.63_{17}	67.81_{14}	56.16_{14}	56.32_{15}	$\mathbf{72.72_{10}}$
	A,B,D	C	72.97_{14}	76.46_{12}	68.21_{11}	$75.70_{8.7}$	$\mathbf{77.54_{10}}$
	A,B,C	D	73.27_{11}	74.35_{11}	68.56_{10}	$69.94_{9.8}$	$\mathbf{75.14_{8.4}}$
	Average		$65.94_{8.3}$	$68.28_{8.6}$	$63.16_{5.4}$	$64.57_{8.5}$	$\mathbf{72.85_{4.3}}$
5%	B,C,D	A	65.30_{17}	71.21_{13}	$66.22_{9.1}$	67.11_{10}	$\mathbf{72.40_{12}}$
	A,C,D	B	79.73_{10}	77.31_{10}	$69.49_{8.3}$	$76.35_{7.9}$	$\mathbf{80.30_{9.1}}$
	A,B,D	C	78.06_{11}	$81.40_{8.0}$	$73.40_{9.8}$	$77.43_{8.3}$	$\mathbf{82.51_{6.6}}$
	A,B,C	D	$81.25_{8.3}$	$79.95_{7.8}$	$75.66_{8.5}$	$78.64_{5.8}$	$\mathbf{83.77_{5.1}}$
	Average		$76.09_{6.3}$	$77.47_{3.9}$	$71.29_{3.6}$	$74.88_{4.6}$	$\mathbf{79.75_{4.4}}$

data in a semi-supervised setting. Compared to our method, "SDNet+Aug." only poses disentanglement to the latent features without meta-learning.

LDDG [25]**:** is the latest state-of-the-art model for domain-generalised medical image analysis. It also uses a rank loss and when applied in a fully supervised setting, LDDG achieved the best generalisation performance on SCGM.

SAML [29]**:** is another gradient-based meta-learning approach. SAML proposed to constraint the compactness and smoothness properties of segmentation masks across meta-train and meta-test sets in a fully supervised setting.

3.3 Implementation Details

Models are trained using the Adam optimiser [21] with a learning rate of $2e^{-5}$ for 50K iterations using batch size 4. Images are cropped to 224×244 for M&Ms and 144×144 for SCGM. $\boldsymbol{F}_\psi$ is a 2D UNet [40] to extract $\mathbf{Z}$ features with 8 channels of same height and width as input image. We follow the designs of SDNet [7] for $\boldsymbol{E}_S$, $\boldsymbol{T}_\theta$ and $\boldsymbol{DE}$. $\boldsymbol{E}_D$ has the same architecture as $\boldsymbol{E}_S$. Both $\mathbf{s}$ and $\mathbf{d}$ have 8 dimensions. $\boldsymbol{T}_C$ is a single fully-connected layer. All models are implemented in PyTorch [36] and are trained using an NVidia 2080 Ti GPU. In the semi-supervised setting, we use specific percentages of the subjects as labeled data and the rest as unlabeled data. We use Dice (%) and Hausdorff Distance [12] (in Appendix) as the evaluation metrics.

3.4 Results and Discussion

Tables 1 and 2 show that we consistently achieve the best generalisation performance on cardiac and gray matter segmentation. Particularly in the low data

Table 2. Dice (%) results and the standard deviations on SCGM dataset. For "SDNet+Aug." and our method, the training data contain all the unlabeled data and 20% or 100% of labeled data from source domains. The other models are trained by 20% or 100% of labeled data only. Bold numbers denote the best performance.

Source		Target	nnUNet	SDNet+Aug.	LDDG	SAML	Ours
20%	2,3,4	1	59.07_{21}	83.07_{16}	$77.71_{9.1}$	78.71_{25}	$\mathbf{87.45_{6.3}}$
	1,3,4	2	69.94_{12}	$80.01_{5.2}$	44.08_{12}	75.58_{12}	$\mathbf{81.05_{5.2}}$
	1,2,4	3	$60.25_{7.2}$	58.57_{10}	$48.04_{5.5}$	$54.36_{7.6}$	$\mathbf{61.85_{7.3}}$
	1,2,3	4	$70.13_{4.3}$	$85.27_{2.2}$	$83.42_{2.7}$	$85.36_{2.8}$	$\mathbf{87.96_{2.1}}$
	Average		$64.85_{5.2}$	76.73_{11}	63.31_{17}	73.50_{12}	$\mathbf{79.58_{11}}$
100%	2,3,4	1	$75.27_{8.3}$	$\mathbf{90.25_{4.5}}$	$88.21_{4.9}$	$90.22_{5.6}$	$90.01_{4.9}$
	1,3,4	2	$76.32_{2.9}$	$84.13_{4.2}$	$83.76_{3.1}$	$\mathbf{86.65_{3.5}}$	$85.48_{2.3}$
	1,2,4	3	$62.59_{6.9}$	62.18_{10}	$56.11_{9.3}$	$58.27_{9.4}$	$\mathbf{64.23_{9.7}}$
	1,2,3	4	$71.87_{2.5}$	$88.93_{1.9}$	$89.08_{2.7}$	$88.66_{2.6}$	$\mathbf{89.26_{2.5}}$
	Average		$71.51_{5.4}$	81.37_{11}	79.29_{13}	80.95_{13}	$\mathbf{82.25_{11}}$

regime we improve Dice by ≈5% on M&Ms and ≈3% on SCGM compared to the best performing baseline. In Appendix, we include the results on Hausdorff Distance, where similar conclusion can be drawn. For 100% annotations in M&Ms (see Appendix), our model still outperforms the baselines. We also show the qualitative results in Appendix, where the improved performance is visually observed.

M&Ms: Compared to "SDNet+Aug." which can also use (due to disentanglement) unlabeled data, our model performs consistently better. The results agree with the conclusion in [32]: without specific designs tuned to the tasks, disentanglement can not provide guaranteed generalisation ability. For LDDG and SAML, the generalisation performance significantly drops with small amounts of labeled data. Note that nnUNet adapts the model design per each run/training set. However, adapting the model design for different training data limits the scalability of nnUNet. In the Appendix, we also show that nnUNet possibly overfits the source domains in some cases.

SCGM: We obtain consistent improvements also on SCGM, demonstrating application in other organs. Our model benefits from the additional 10 unlabeled subjects of each domain leading to better performance overall.

Ablation Study. Here we conduct ablations on key losses crucial to disentanglement and the extraction of good anatomical features for good generalisation performance. We omit ablations on the KL losses as [13,17] showcase that variational encoding helps to learn robust vector representation for better generalisation. To illustrate that $\mathcal{L}_{rank}$ helps to disentangle $\mathbf{Z}$ to $(\mathbf{s}, \mathbf{d})$, and improves performance, we use Distance Correlation (DC) [31] to measure disentanglement (lower DC means a higher level of disentanglement). For M&Ms 5% cases, without $\mathcal{L}_{rank}$, the average DC on the test dataset between $\mathbf{Z}$ and $(\mathbf{s}, \mathbf{d})$ is 0.22 (an increase compared to 0.19 with $\mathcal{L}_{rank}$), and the average Dice is 78.54% (a decrease compared to 79.75% with $\mathcal{L}_{rank}$). We also ablate $\mathcal{L}_{cls}$ and $\mathcal{L}_{HSIC}$. The proposed model on M&Ms 5% cases had an average Dice 79.75% but without $\mathcal{L}_{cls}$, average Dice drops to 77.45% and without $\mathcal{L}_{HSIC}$, average Dice drops to 77.86%.

4 Conclusion

We have presented a novel semi-supervised meta-learning framework for domain generalisation. Using disentanglement our approach models domain shifts, and thanks to our reconstruction approach to disentanglement, our model can be trained also with unlabeled data. By applying the designed constraints (including the low-rank regularisation) to the gradient-based meta-learning approach, the model extracts robust anatomical features useful for predicting segmentation masks in a semi-supervised manner. Extensive quantitative results, especially when insufficient annotated data are available, indicate remarkable improvements compared to previous state-of-the-art approaches. The performance of our method might improve with the use of additional unlabeled data from other domains, which we leave as future work.

Acknowledgement. This work was supported by the University of Edinburgh, the Royal Academy of Engineering and Canon Medical Research Europe by a PhD studentship to Xiao Liu. This work was partially supported by the Alan Turing Institute under the EPSRC grant EP/N510129/1. We thank Nvidia for donating a Titan-X GPU. S.A. Tsaftaris acknowledges the support of Canon Medical and the Royal Academy of Engineering and the Research Chairs and Senior Research Fellowships scheme (grant RCSRF1819\8\25).

References

1. Achille, A., Soatto, S.: Emergence of invariance and disentanglement in deep representations. JMLR **19**(1), 1947–1980 (2018)
2. Antoniou, A., Edwards, H., Storkey, A.: How to train your MAML. In: Proceedings of the ICLR (2019)
3. Bernard, O., Lalande, A., Zotti, C., Cervenansky, F., Yang, X., et al.: Deep learning techniques for automatic MRI cardiac multi-structures segmentation and diagnosis: is the problem solved? IEEE TMI **37**(11), 2514–2525 (2018)
4. Bian, C., Yuan, C., Wang, J., Li, M., et al.: Uncertainty-aware domain alignment for anatomical structure segmentation. MedIA **64**, 101732 (2020)
5. Campello, V.M., Gkontra, P., Izquierdo, C., et al.: Multi-centre, multi-vendor and multi-disease cardiac segmentation: the M&MS challenge. IEEE Trans. Med. Imag. (2021)
6. Carlucci, F.M., D'Innocente, A., Bucci, S., Caputo, B., Tommasi, T.: Domain generalisation by solving jigsaw puzzles. In: Proceedings of the CVPR, pp. 2229–2238 (2019)

7. Chartsias, A., Joyce, T., Papanastasiou, G., et al.: Disentangled representation learning in cardiac image analysis. Med. Image Anal. **58**, 101535 (2019)
8. Chen, C., Qin, C., Qiu, H., et al.: Deep learning for cardiac image segmentation: a review. Front. Cardiovasc. Med. **7**(25), 1–33 (2020)
9. Dewey, B.E., et al.: A disentangled latent space for cross-site MRI harmonization. In: Martel, A.L., et al. (eds.) MICCAI 2020. LNCS, vol. 12267, pp. 720–729. Springer, Cham (2020). https://doi.org/10.1007/978-3-030-59728-3_70
10. Dice, L.R.: Measures of the amount of ecologic association between species. Ecology **26**(3), 297–302 (1945)
11. Dou, Q., Castro, D.C., Kamnitsas, K., Glocker, B.: Domain generalisation via model-agnostic learning of semantic features. Proc, NeurIPS (2019)
12. Dubuisson, M.P., Jain, A.K.: A modified hausdorff distance for object matching. In: Proceedings of the ICPR, vol. 1, pp. 566–568. IEEE (1994)
13. Higgins, I., Matthey, L., Pal, A., et al.: beta-VAE: learning basic visual concepts with a constrained variational framework In: Proceedings of the ICLR (2016)
14. Huang, J., Guan, D., Xiao, A., Lu, S.: FSDR: frequency space domain randomization for domain generalization. In: Proceedings of the CVPR (2021)
15. Huang, X., Belongie, S.: Arbitrary style transfer in real-time with adaptive instance normalization. In: Proceedings of the ICCV, pp. 1501–1510 (2017)
16. Huang, X., Liu, M.Y., Belongie, S., Kautz, J.: Multimodal unsupervised image-to-image translation. In: Proceedings of the ECCV, pp. 172–189 (2018)
17. Ilse, M., Tomczak, J.M., Louizos, C., Welling, M.: Diva: Domain invariant variational autoencoders. In: Proceedings of the MIDL, pp. 322–348. PMLR (2020)
18. Isensee, F., et al.: Automatic cardiac disease assessment on cine-MRI via time-series segmentation and domain specific features. In: Pop, M., et al. (eds.) STACOM 2017. LNCS, vol. 10663, pp. 120–129. Springer, Cham (2018). https://doi.org/10.1007/978-3-319-75541-0_13
19. Isensee, F., Jaeger, P.F., Kohl, S.A., Petersen, J., Maier-Hein, K.H.: NNU-net: a self-configuring method for deep learning-based biomedical image segmentation. Nat. Methods **18**(2), 203–211 (2021)
20. Khandelwal, P., Yushkevich, P.: Domain generalizer: a few-shot meta learning framework for domain generalization in medical imaging. In: Albarqouni, S., et al. (eds.) DART/DCL -2020. LNCS, vol. 12444, pp. 73–84. Springer, Cham (2020). https://doi.org/10.1007/978-3-030-60548-3_8
21. Kingma, D.P., Ba, J.: Adam: a method for stochastic optimization. In: Proceedings of the ICLR (2015)
22. Li, D., Yang, Y., Song, Y.Z., Hospedales, T.: Learning to generalise: meta-learning for domain generalisation. In: Proceedings of the AAAI (2018)
23. Li, D., Zhang, J., Yang, Y., Liu, C., Song, Y.Z., Hospedales, T.M.: Episodic training for domain generalisation. In: Proceedings of the ICCV, pp. 1446–1455 (2019)
24. Li, H., Pan, S.J., Wang, S., Kot, A.C.: Domain generalization with adversarial feature learning. In: Proceedings of the CVPR, pp. 5400–5409 (2018)
25. Li, H., Wang, Y., Wan, R., et al.: Domain generalisation for medical imaging classification with linear-dependency regularization. In: Proceedings of the NeurIPS (2020)
26. Li, X., et al.: Difficulty-aware meta-learning for rare disease diagnosis. In: Martel, A.L., et al. (eds.) MICCAI 2020. LNCS, vol. 12261, pp. 357–366. Springer, Cham (2020). https://doi.org/10.1007/978-3-030-59710-8_35
27. Li, Y., Tian, X., Gong, M., et al.: Deep domain generalization via conditional invariant adversarial networks. In: Proceedings of the ECCV, pp. 624–639 (2018)

28. Li, Y., Chen, J., Xie, X., Ma, K., Zheng, Y.: Self-loop uncertainty: a novel pseudo-label for semi-supervised medical image segmentation. In: Proceedings of the MICCAI, pp. 614–623. Springer, Cham (2020)
29. Liu, Q., Dou, Q., Heng, P.A.: Shape-aware meta-learning for generalising prostate MRI segmentation to unseen domains. In: Proceedings of the MICCAI, pp. 475–485. Springer, Cham (2020)
30. Liu, X., Thermos, S., Chartsias, A., et al.: Disentangled representations for domain-generalised cardiac segmentation. In: International Workshop on STACOM (2020)
31. Liu, X., Thermos, S., Valvano, G., et al.: Metrics for exposing the biases of content-style disentanglement. arXiv preprint arXiv:2008.12378 (2020)
32. Llera Montero, M., Ludwig, C.J.H., Ponte Costa, R., Malhotra, G., Bowers, J.: The role of disentanglement in generalisation. In: Proceedings of the ICLR (2021)
33. Locatello, F., et al.: Challenging common assumptions in the unsupervised learning of disentangled representations. In: Proceeding of the ICML, pp. 4114–4124. PMLR (2019)
34. Ma, W.D.K., Lewis, J., Kleijn, W.B.: The HSIC Bottleneck: Deep Learning without Back-Propagation. In: Proceedings of the AAAI, pp. 5085–5092 (2020)
35. Muandet, K., Balduzzi, D., Schölkopf, B.: Domain generalisation via invariant feature representation. In: Proceedings of the ICML, pp. 10–18. PMLR (2013)
36. Paszke, A., Gross, S., Massa, F., Lerer, A., et. al.: PyTorch: an imperative style, high-performance deep learning library. In: Proceedings of the NeurIPS, pp. 8026–8037 (2019)
37. Pomponio, R., Erus, G., et al.: Harmonization of large MRI datasets for the analysis of brain imaging patterns throughout the lifespan. NeuroImage **208**, 116450 (2020)
38. Prados, F., Ashburner, J., Blaiotta, C., Brosch, T., Carballido-Gamio, J., Cardoso, M.J., Conrad, B.N., Datta, E., Dávid, G., De Leener, B., et al.: Spinal cord grey matter segmentation challenge. Neuroimage **152**, 312–329 (2017)
39. Puyol-Anton, E., Ruijsink, B., Piechnik k., S., Neubauer, S., et al.: Fairness in cardiac MR image analysis: an investigation of bias due to data imbalance in deep learning based segmentation. arXiv preprint arXiv:2106.12387 (2021)
40. Ronneberger, O., Fischer, P., Brox, T.: U-Net: convolutional networks for biomedical image segmentation. In: Navab, N., Hornegger, J., Wells, W.M., Frangi, A.F. (eds.) MICCAI 2015. LNCS, vol. 9351, pp. 234–241. Springer, Cham (2015). https://doi.org/10.1007/978-3-319-24574-4_28
41. Sharifi-Noghabi, H., Asghari, H., Mehrasa, N., Ester, M.: Domain generalisation via semi-supervised meta learning. arXiv preprint arXiv:2009.12658 (2020)
42. Tao, Q., Yan, W., Wang, Y., Paiman, E.H., Shamonin, et al.: Deep learning-based method for fully automatic quantification of left ventricle function from cine MR images: a multivendor, multicenter study. Radiology **290**(1), 81–88 (2019)
43. Wang, J., Zhou, S., Fang, C., Wang, L., Wang, J.: Meta corrupted pixels mining for medical image segmentation. In: Proceedings of the MICCAI, pp. 335–345. Springer (2020)
44. Yu, X., Chen, Y., Li, T., Liu, S., Li, G.: Multi-mapping image-to-image translation via learning disentanglement. In: Proceedings of the NeurIPS (2019)
45. Zhang, L., Wang, X., Yang, D., Sanford, T., et al.: Generalizing deep learning for medical image segmentation to unseen domains via deep stacked transformation. IEEE Trans, Med. Image **39**(7), 2531–2540 (2020)
46. Zhao, S., Gong, M., Liu, T., Fu, H., Tao, D.: Domain generalization via entropy regularization. In: Proceedings of the NeurIPS, vol. 33 (2020)

Efficient Semi-supervised Gross Target Volume of Nasopharyngeal Carcinoma Segmentation via Uncertainty Rectified Pyramid Consistency

Xiangde Luo[1], Wenjun Liao[2], Jieneng Chen[3], Tao Song[4], Yinan Chen[4,5], Shichuan Zhang[6], Nianyong Chen[2], Guotai Wang[1](✉), and Shaoting Zhang[1,4]

[1] School of Mechanical and Electrical Engineering, University of Electronic Science and Technology of China, Chengdu, China
guotai.wang@uestc.edu.cn
[2] Department of Radiation Oncology, West China Hospital, Sichuan University, Chengdu, China
[3] College of Electronics and Information Technology, Tongji University, Shanghai, China
[4] SenseTime Research, Shanghai, China
[5] West China Hospital-SenseTime Joint Lab, West China Biomedical Big Data Center, Sichuan University West China Hospital, Chengdu, China
[6] Department of Radiation Oncology, Sichuan Cancer Hospital and Institute, University of Electronic Science and Technology of China, Chengdu, China

Abstract. Gross Target Volume (GTV) segmentation plays an irreplaceable role in radiotherapy planning for Nasopharyngeal Carcinoma (NPC). Despite that Convolutional Neural Networks (CNN) have achieved good performance for this task, they rely on a large set of labeled images for training, which is expensive and time-consuming to acquire. In this paper, we propose a novel framework with Uncertainty Rectified Pyramid Consistency (URPC) regularization for semi-supervised NPC GTV segmentation. Concretely, we extend a backbone segmentation network to produce pyramid predictions at different scales. The pyramid predictions network (PPNet) is supervised by the ground truth of labeled images and a multi-scale consistency loss for unlabeled images, motivated by the fact that prediction at different scales for the same input should be similar and consistent. However, due to the different resolution of these predictions, encouraging them to be consistent at each pixel directly has low robustness and may lose some fine details. To address this problem, we further design a novel uncertainty rectifying module to enable the framework to gradually learn from meaningful and reliable consensual regions at different scales. Experimental results on a dataset with 258 NPC MR images showed that with only 10% or 20% images labeled, our method largely improved the segmentation performance by

Electronic supplementary material The online version of this chapter (https://doi.org/10.1007/978-3-030-87196-3_30) contains supplementary material, which is available to authorized users.

M. de Bruijne et al. (Eds.): MICCAI 2021, LNCS 12902, pp. 318–329, 2021.
https://doi.org/10.1007/978-3-030-87196-3_30

leveraging the unlabeled images, and it also outperformed five state-of-the-art semi-supervised segmentation methods. Moreover, when only 50% labeled images, URPC achieved an average Dice score of 82.74% that was close to fully supervised learning. Code is available at: https://github.com/HiLab-git/SSL4MIS.

Keywords: Semi-supervised learning · Uncertainty rectifying · Pyramid consistency · Gross target volume · Nasopharyngeal carcinoma

1 Introduction

Nasopharyngeal Carcinoma (NPC) is one of the most common cancers in southern China, Southeast Asia, the Middle East, and North Africa [5]. The mainstream treatment strategy for NPC is radiotherapy, thus accurate target delineation plays an irreplaceable role for precise and effective radiotherapy. However, manual nasopharyngeal tumor contouring is tedious and laborious, since both the nasopharynx gross tumor volume (GTVnx) and lymph node gross tumor volume (GTVnd) need to be accurately delineated [13]. Recently, with a large amount of labeled data, deep learning has shown the potential for accurate GTV segmentation [13]. However, collecting a large labeled dataset for network training is difficult, as both time and expertise are needed to produce accurate annotation. In contrast, collecting a large set of unlabeled data is easier, which inspired us to develop a semi-supervised approach for NPC GTV segmentation by leveraging unlabeled data. What's more, semi-supervised learning (SSL) can largely reduce the workload of annotators for the development of deep learning models.

Recently, SSL has been widely used for medical image computing to reduce the annotation efforts [9,12,16,19]. Bai et al. [1] developed an iterative framework where in each iteration, pseudo labels for unannotated images are predicted by the network and refined by a Conditional Random Field (CRF), then the new pseudo labels are used to update the network. After that, the perturbation-based methods have achieved increasing attention in semi-supervised learning [2,4,12,19]. These methods add small perturbations to unlabeled samples and enforce the consistency between the model's predictions on the original data and the perturbed data. Meanwhile, the mean teacher-based [21] self-ensembling methods [7,9,26,27] were introduced for semi-supervised medical image segmentation. Following [7], some recent works [9,26,27] used uncertainty map to guide the student model to learn more stably. In [11,17,29], an adversarial training strategy was used as regularization for SSL, which aims to minimize the adversarial loss to encourage the prediction of unlabeled data is anatomical plausible. Luo et al. [15] proposed a dual-task consistency framework for SSL by representing segmentation as a pixel-wise classification task and a level set regression task simultaneously, the difference between which was minimized during training. Despite their higher performance than learning from available labeled images only, existing methods are limited by high computational cost and complex training strategies in practice. For example, the co-training-based methods need to train several models at the same time [20], and the uncertainty estimation-based frameworks need multiple forward passes [27]. Self-training-based approaches need to select and refine pseudo labels and update models' parameters in several rounds [1], which is time-consuming.

In this work, we propose a novel efficient semi-supervised learning framework for the segmentation of GTVnx and GTVnd by further utilizing the unlabeled data. Inspired by deep supervision [10], Our method leverages a network that gives a pyramid (i.e., multi-scale) prediction, and encourages the predictions at multiple scales to be consistent for a given input, which is a simple yet efficient idea for SSL. A standard supervised loss at multiple scales is used for learning from labeled images. For unlabeled images, we encourage the multi-scale predictions to be consistent, which serves as a regularization. Since the ground truth of unlabeled images is unknown, the model may produce some unreliable prediction or noise which may cause the model to collapse and lose details. To overcome these problems, some existing works [3,27] have leveraged model uncertainty to boost the stability of training and obtain better results. However, they typically estimate the uncertainty of each target prediction with Monte Carlo sampling [8], which needs massive computational costs as it requires multiple forward passes to obtain the uncertainty in each iteration. Differently from these methods, we estimate the uncertainty via the prediction discrepancy among multi-scale predictions, which just needs a single forward pass. With the guidance of the estimated uncertainty, we automatically emphasize the reliable predictions (low uncertainty) and weaken the unreliable ones (high uncertainty) when calculating the multi-scale consistency. Meanwhile, we introduce the uncertainty minimization [30] to reduce the prediction variance during training. Therefore, the proposed framework has high efficiency for semi-supervised segmentation by taking advantage of the unlabeled images. Our method was extensively evaluated on a clinical Nasopharyngeal Carcinoma dataset. Results show our method largely improved the segmentation performance by leveraging the unlabeled images, and it outperformed five state-of-the-art semi-supervised segmentation methods. Moreover, when only half of the training images are labeled, URPC achieves a very close result compared with fully supervised learning (the mean of Dice was 82.74% vs 83.51%).

2 Methods

The proposed URPC for semi-supervised segmentation is illustrated in Fig. 1. We add a pyramid prediction structure at the decoder of a backbone network and refer to it as PPNet. PPNet learns from the labeled data by minimizing a typical supervised segmentation loss directly. In addition, the PPNet is regularized by a multi-scale consistency between the pyramid predictions to deal with unlabeled data. The PPNet naturally leads to uncertainty estimation in a single forward pass by measuring the discrepancy between these predictions, and we propose to use this uncertainty to rectify the pyramid consistency considering the different spatial resolutions in the pyramid. To describe this work precisely, we first define

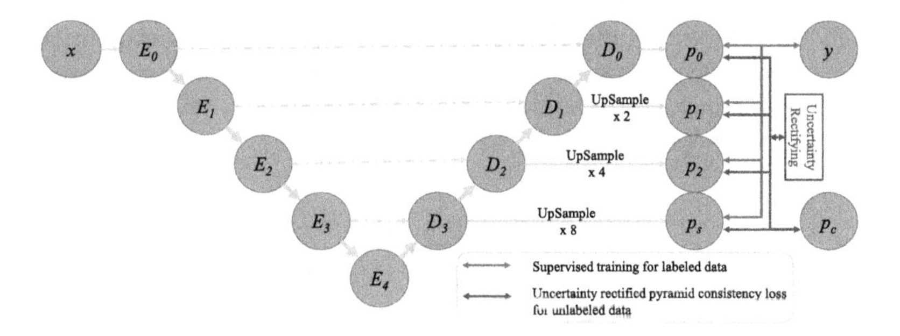

Fig. 1. Overview of the proposed Uncertainty Rectified Pyramid Consistency framework, which consists of a pyramid prediction network (PPNet) and an uncertainty rectifying module. It is based on a backbone of 3D UNet [6], where Es and Ds are the blocks in the encoder and decoder of 3D UNet respectively. p_s is the prediction as scale s. The URPC is optimized by minimizing the supervised loss on the labeled data and the pyramid consistency loss on the unlabeled data. In addition, an uncertainty rectifying module is designed to reduce the impact of noise in the pyramid consistency and boost the stability of training.

some mathematical terms. Let D_l, D_u and $f_\phi(x)$ be the labeled set, unlabeled set and the PPNet's parameters, respectively. Let $D = D_l \cup D_u$ be the whole provided dataset. We denote an unlabeled image as $x_i \in D_u$ and a labeled image pair as $(x_i, y_i) \in D_l$, where y_i is ground truth.

2.1 Multi-scale Prediction Network with Pyramid Consistency

To better exploit the prediction discrepancy of a single model at different scales, we firstly introduce the PPNet for the segmentation task, which can produce predictions with different scales. In this work, we employ 3D UNet [6] as a backbone and modify it to produce pyramid predictions by adding a prediction layer after each upsampling block in the decoder, where the prediction layer is implemented by $1 \times 1 \times 1$ convolution followed by a softmax layer. To introduce more perturbations in the network, a dropout layer and a feature-level noise addition layer are inserted before each of these prediction layers. For an input image x_i, PPNet $f_\phi(x)$ produces a set of multi-scale predictions $[p'_0, p'_1, ..., p'_s, ..., p'_{S-1}]$, where the p'_s is the prediction at scale s, and a smaller s corresponds to a higher resolution in the decoder, as shown in Fig. 1. S is the number of scales in the pyramid prediction. Then, we rescale these multi-scale predictions to the input size, and the corresponding results are denoted as $[p_0, p_1, ..., p_s, ..., p_{S-1}]$. For the labeled data, we use a supervised loss that is a combination of Dice and cross-entropy loss at multiple scales:

$$\mathcal{L}_{sup} = \frac{1}{S} \sum_{s=0}^{S-1} \frac{\mathcal{L}_{dice}(p_s, y_i) + \mathcal{L}_{ce}(p_s, y_i)}{2} \tag{1}$$

where y_i, $\mathcal{L}_{dice}$, $\mathcal{L}_{ce}$ denote the ground truth of input x_i, the Dice loss and the cross entropy loss, respectively.

To efficiently leverage unlabeled data, we introduce a regularization by encouraging the multi-scale predictions of PPNet to be consistent. Concretely, we design a pyramid consistency loss to minimize the discrepancy (i.e., variance) among the predictions at different scales. First, we denote the average prediction across these scales as:

$$p_c = \frac{1}{S}\sum_{s=0}^{S-1} p_s \tag{2}$$

Then, the pyramid consistency loss is defined as:

$$\mathcal{L}_{pyc} = \frac{1}{S}\sum_{s=0}^{S-1} \|p_s - p_c\|_2 \tag{3}$$

where we encourage a minimized L_2 distance between the prediction at each scale and the average prediction.

2.2 Uncertainty Rectified Pyramid Consistency Loss

As the pyramid prediction at a range of scales has different spatial resolutions, even they can be resampled to the same resolution as the input, the resampled results still have different spatial frequencies, i.e., the prediction at the lowest resolution captures the low-frequency component of the segmentation, and the prediction at the highest resolution obtains more high-frequency components. Directly imposing a voxel-level consistency among these predictions can be problematic due to the different frequencies, such as loss of fine details or model collapse. Inspired by existing works [3,24,25,27,30], we introduce an uncertainty-aware method to address these problems. Unlike existing methods, our uncertainty estimation is a scale-level approach and only requires a single forward pass, which needs less computational cost and running time than exiting methods.

Efficient Uncertainty Estimation Based on Pyramid Predictions. As our PPNet obtains multiple predictions in a single forward pass, uncertainty estimation can be obtained efficiently by justing measuring their discrepancy without extra efforts. To be specific, we use the KL-divergence between the average prediction and the prediction at scale s as the uncertainty measurement:

$$\mathcal{D}_s \approx \sum_{j=0}^{C} p_s^j \cdot \log \frac{p_s^j}{p_c^j} \tag{4}$$

where p_s^j is the j th channel of p_s, and C is the class (i.e., channel) number. The approximated uncertainty shows the difference between the p_s and p_c. Note that for a given voxel in $\mathcal{D}_s$, a larger value indicates the prediction for that pixel at scale s is far from the other scales, i.e., with high uncertainty. As result, we obtain a set of uncertainty maps $\mathcal{D}_0, \mathcal{D}_1, ...\mathcal{D}_{S-1}$, where $\mathcal{D}_s$ corresponds to uncertainty of p_s.

Uncertainty Rectifying. Based on the estimated uncertainty maps $\mathcal{D}_0, \mathcal{D}_1, ...$ $\mathcal{D}_{S-1}$, we further extend the pyramid consistency $\mathcal{L}_{pyc}$ to emphasize reliable parts and ignore unreliable parts of the predictions for stable unsupervised training. Specifically, for unlabeled data, we use the estimated uncertainty to automatically select reliable voxels for loss calculation. The rectified pyramid consistency loss is formulated as:

$$\mathcal{L}_{unsup} = \underbrace{\frac{1}{S}\frac{\sum_{s=0}^{S-1}\sum_{v}(p_s^v - p_c^v)^2 \cdot w_s^v}{\sum_{s=0}^{S-1}\sum_{v} w_s^v}}_{uncertainty\ rectification} + \underbrace{\frac{1}{S}\sum_{s=0}^{S-1}||\mathcal{D}_s||_2}_{uncertainty\ minimization} \tag{5}$$

where p_s^v and $\mathcal{D}_s^v$ are the corresponding prediction and uncertainty values for voxel v. The consistency loss consists of two terms: the first is an uncertainty rectification (UR) term and the second is uncertainty minimization (UM) term. For a more stable training, we follow the policy in [30] and we use a voxel- and scale-wise weight w_s^v to automatically rectify the MSE loss rather than the threshold-based cut-off approaches [3,27], as the threshold is hard to determine. The weight for a voxel v at scale s is defined as: $w_s^v = e^{-\mathcal{D}_s^v}$, it corresponds to voxel-wise exponential operation for - $\mathcal{D}_s$. According to this definition, for a given voxel at scale s, a higher uncertainty leads to a lower weight automatically. In addition, to encourage the PPNet to produce more consistent predictions at different scales, we use the uncertainty minimization term as a constraint directly. With this uncertainty rectified consistency loss, the PPNet can learn more reliable knowledge, which can then reduce the overall uncertainty of the model and produce more consistent predictions.

2.3 The Overall Loss Function

The proposed URPC framework learns from both labeled data and unlabeled data by minimizing the following combined objective function:

$$\mathcal{L}_{total} = \mathcal{L}_{sup} + \lambda \cdot \mathcal{L}_{unsup} \tag{6}$$

where $\mathcal{L}_{sup}$, $\mathcal{L}_{unsup}$ are defined in Eq. 1 and Eq. 5, respectively. λ is a widely-used time-dependent Gaussian warming up function [21,27] to control the balance between the supervised loss and unsupervised consistency loss, which is defined as $\lambda(t) = w_{max} \cdot e^{(-5(1-\frac{t}{t_{max}})^2)}$, where w_{max} means the final regularization weight, t denotes the current training step and t_{max} is the maximal training step.

3 Experiments and Results

3.1 Dataset and Implementations

The NPC dataset used in this work was collected from a local cancer center. A total number of 258 T1-weighted MRI images from 258 patients of NPC before radiotherapy were acquired on several 3T Siemens scanners. The mean resolution of the dataset was 1.23 mm × 1.23 mm × 1.10 mm, and the mean dimension was 176 × 286 × 245. The ground truth for GTVnx and GTVnd were obtained from manual segmentation by two experienced radiologists using ITK-SNAP [28]. The dataset was randomly split into 180 cases for training, 20 cases for validation, and 58 cases for testing. For the training images, only 18 (i.e., 10%) were used as labeled and the remaining 162 scans were used as unlabeled. For pre-processing, we just normalize each scan to zero mean and unit variance. In the evaluation stage, following existing work [13], we used the commonly-adopted Dice Similarity Coefficient (DSC) and the Average Surface Distance (ASD) as segmentation quality evaluation metrics.

The framework was implemented in PyTorch [18], using a node of a cluster with 8 TiTAN 1080TI GPUs. We used the SGD optimizer (weight decay = 0.0001, momentum = 0.9) with Eq. 6 for training our method. During the training processing, the poly learning rate strategy was used for learning rate decay, where the initial learning rate 0.1 was multiplied by $(1.0 - \frac{t}{t_{max}})^{\gamma}$ with $\gamma = 0.9$ and $t_{max} = 60k$. The batch size was set to 4, and each batch consists of two annotated images and two unannotated images. We randomly cropped 112×112×112 sub-volumes as the network input and employed data augmentation to enlarge the dataset and avoid over-fitting, including random cropping, random flipping, and random rotation. The final segmentation results were obtained by using a sliding window strategy. Following [27], the w_{max} was set to 0.1 for all experiments. (Details of the NPC dataset is presented in supplementary materials.)

Table 1. Ablation study of the proposed URPC framework on the NPC MRI dataset, where 18 labeled and 162 unlabeled images were used for training. UR and UM denote the uncertainty rectification term and uncertainty minimization term, respectively.

Method	GTVnx		GTVnd		Mean	
	DSC (%)	ASD (voxel)	DSC (%)	ASD (voxel)	DSC (%)	ASD (voxel)
Baseline (S = 1)	71.94 ± 11.60	2.42 ± 1.65	66.27 ± 14.62	3.60 ± 3.12	69.10 ± 10.15	3.01 ± 1.76
S = 2	79.88 ± 6.91	1.79 ± 1.27	72.82 ± 15.55	2.85 ± 2.54	76.35 ± 9.48	2.32 ± 1.46
S = 3	79.09 ± 5.82	**1.76 ± 0.97**	75.08 ± 13.22	**2.25 ± 2.27**	77.09 ± 7.85	**2.05 ± 1.24**
S = 4	**80.13 ± 6.37**	1.82 ± 1.30	**75.83 ± 12.93**	2.65 ± 2.77	**77.98 ± 8.00**	2.24 ± 1.53
S = 5	79.10 ± 6.53	1.84 ± 1.14	75.73 ± 13.71	2.29 ± 2.61	77.42 ± 8.27	2.06 ± 1.43
S = 4 + UR	**80.99 ± 5.50**	1.70 ± 1.12	75.22 ± 13.86	3.05 ± 3.16	78.11 ± 8.06	2.38 ± 1.65
S = 4 + UR + UM	80.76 ± 5.72	**1.69 ± 1.06**	**75.95 ± 12.74**	**2.20 ± 2.07**	**78.36 ± 7.66**	**1.95 ± 1.18**

3.2 Evaluation of Our Proposed URPC on the NPC Dataset

Ablation Study. Firstly, to investigate the impact of different numbers of scales in the pyramid prediction of PPNet, as shown in Fig. 1, we set S of PPNet to 2, 3, 4, and 5, respectively, and UR and UM were not used at this stage. They were compared with the baseline of 3D UNet [6] without multi-scale predictions and therefore it only learns from labeled data. In contrast, the PPNet learns

Table 2. Comparison between our method and existing methods on the NPC MRI dataset, when using 10% labeled data. * denotes p-value < 0.05 when comparing the proposed with the others.

Method	GTVnx		GTVnd		Mean		*T-T (h)*
	DSC (%)	*ASD* (voxel)	*DSC* (%)	*ASD* (voxel)	*DSC* (%)	*ASD* (voxel)	
SL (10%)	$71.94 \pm 11.60^*$	$2.42 \pm 1.65^*$	$66.27 \pm 14.62^*$	$3.60 \pm 3.12^*$	$69.10 \pm 10.15^*$	$3.01 \pm 1.76^*$	73
SL (100%)	$83.93 \pm 4.77^*$	$1.35 \pm 0.73^*$	$83.10 \pm 9.05^*$	$1.48 \pm 1.73^*$	$83.51 \pm 5.35^*$	$1.41 \pm 0.94^*$	61
MT [21]	$79.80 \pm 6.74^*$	1.70 ± 1.17	$69.78 \pm 16.34^*$	$2.81 \pm 2.57^*$	$74.79 \pm 9.15^*$	$2.25 \pm 1.40^*$	76
ICT [22]	80.58 ± 6.23	$1.58 \pm 1.02^*$	$72.62 \pm 13.47^*$	$2.72 \pm 2.61^*$	$76.59 \pm 7.98^*$	$2.15 \pm 1.38^*$	78
EM [23]	$79.85 \pm 6.32^*$	1.66 ± 1.06	$69.93 \pm 15.09^*$	$3.14 \pm 2.82^*$	$74.89 \pm 8.85^*$	$2.40 \pm 1.54^*$	74
UAMT [27]	$79.62 \pm 7.16^*$	1.67 ± 1.05	$71.98 \pm 15.66^*$	$2.55 \pm 2.58^*$	$75.78 \pm 9.67^*$	$2.11 \pm 1.39^*$	95
DAN [29]	80.47 ± 5.73	$\mathbf{1.56 \pm 0.81}^*$	$74.62 \pm 12.83^*$	$2.74 \pm 2.62^*$	77.55 ± 7.39	$2.15 \pm 1.26^*$	104
Ours	$\mathbf{80.76 \pm 5.72}$	1.69 ± 1.06	$\mathbf{75.95 \pm 12.74}$	$\mathbf{2.20 \pm 2.07}$	$\mathbf{78.36 \pm 7.66}$	$\mathbf{1.95 \pm 1.18}$	**74**

from both labeled data and unlabeled data. The quantitative results are shown in Table 1. It can be found that when S increases from 2 to 4, the performance of the proposed URPC improves gradually. However, we found that $S = 5$ achieved a lower performance than $S = 4$. That is because the resolution of p_4 is too small to preserve more details. It conforms to common sense and indicates that not more multi-scales predictions are better. Therefore, we used $S = 4$ for SSL in the following experiments. Secondly, to measure the contribution of the uncertainty rectifying module, we then turn on the UR term and UM term with $S = 4$ for training. From the last section of Table 1, we can see that both uncertainty rectifying (UR) term and uncertainty minimization (UM) term boost the model performance. What's more, combining all sub-modules into a unified framework results in a better gain where the mean *DSC* and *ASD* were improved by 9.26% and 1.06 voxels than the baseline, demonstrating their effectiveness for semi-supervised segmentation.

Comparison with Other Semi-supervised Methods. We compared our method with only using 18 annotated images for supervised learning with 3D UNet, which is denoted as SL (10%). Similarly, SL (100%) denote supervised learning with all the training images annotated, which gives the performance upper bound. In addition, we further compared our methods with five state-of-the-art semi-supervised segmentation methods, including Mean Teacher (MT) [21], Interpolation Consistency Training (ICT) [22], Entropy Minimization (EM) [23], Uncertainty Aware Mean Teacher (UAMT) [27] and Deep Adversarial Network (DAN) [29]. Note that, for a fair comparison, all these methods were implemented by using 3D UNet [6] as the backbone and they are online available [14]. Table 2 shows the quantitative comparison of these methods. It can be found that compared with SL (10%), all semi-supervised methods improve the segmentation performance by a large margin, as they can learn from the unannotated data by a regularization loss during the training, and the DAN [29] achieve the best results among existing methods. Our framework (URPC) achieves better performance than these semi-supervised methods when using 10% labeled data. These results show that our URPC has the capability to capture the rich information from the unlabeled data in addition to labeled data. What's more, our method is more efficient than existing methods and requires less training time (T-T) and computational cost, as it just needs to pass an input image once in an iteration. In Fig. 2(a), we visualize some 2D and 3D results of the supervised and semi-supervised methods when using 10% labeled data. Compared with supervised learning (SL) baseline and DAN [29], our method has a higher overlap ratio with the ground truth and reduces the false negative in both slice level and volume level, especially in GTVnd segmentation. We further visualized the estimated uncertainty ($\mathcal{D}_0$ in Eq. 4) in the last column of Fig. 2(a), showing that the uncertain region is mainly distributed near the boundary. We further performed a study on the data utilization efficiency of the URPC. Figure 2(b) shows the evolution curve of mean DSC of GTVnx and GTVnd segmentation obtained by SL, DAN [29] and URPC when using different numbers of labeled data. It can be found that URPC consistently outperforms SL and DAN [29], and when increasing the labeled ratio to 50%, URPC achieves the mean DSC of 82.74% which is very close to 83.51% obtained by SL (100%). These results demonstrate that the URPC has the capability to utilize the unlabeled data to bring performance gains. More results on 20% labeled data presented in supplementary materials showed that our method also outperforms these existing methods.

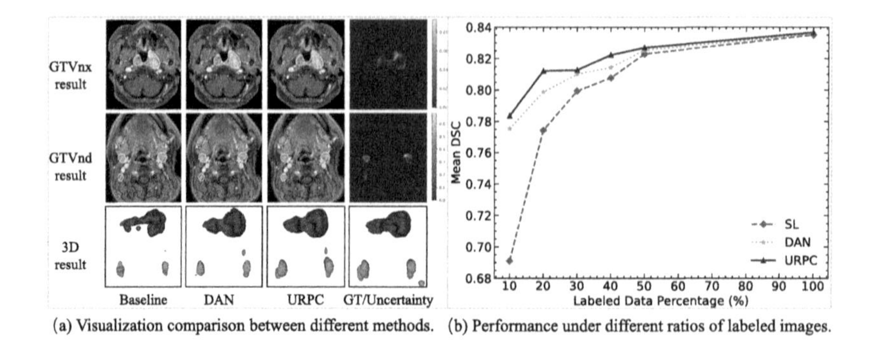

(a) Visualization comparison between different methods. (b) Performance under different ratios of labeled images.

Fig. 2. Comparison between different methods.

4 Conclusion

In this paper, we proposed a novel efficient semi-supervised learning framework URPC for medical image segmentation. A pyramid prediction network is employed to learn from the unlabeled data by encouraging to produce consistent predictions at multiple scales. An uncertainty rectifying module is designed to improve the stability of learning from unlabeled images and further boost model performance, where the uncertainty estimation can be obtained with a single forward pass efficiently. We applied the proposed method to the segmentation of GTVnx and GTVnd, and the results demonstrated the effectiveness and generalization of URPC and also indicated the promising potential of our proposed approach for further clinical use. In the future, we will evaluate this framework on other segmentation tasks.

Acknowledgment. This work was supported by the National Natural Science Foundations of China [81771921, 61901084], and also by key research and development project of Sichuan province, China [20ZDYF2817]. We thank M.D, Mengwan Wu and Yuanyuan Shen from the Sichuan Provincial People's Hospital for the data annotation and checking.

References

1. Bai, W., et al.: Semi-supervised learning for network-based cardiac MR image segmentation. In: Descoteaux, M., Maier-Hein, L., Franz, A., Jannin, P., Collins, D.L., Duchesne, S. (eds.) MICCAI 2017. LNCS, vol. 10434, pp. 253–260. Springer, Cham (2017). https://doi.org/10.1007/978-3-319-66185-8_29
2. Bortsova, G., Dubost, F., Hogeweg, L., Katramados, I., de Bruijne, M.: Semi-supervised medical image segmentation via learning consistency under transformations. In: Shen, D., et al. (eds.) MICCAI 2019. LNCS, vol. 11769, pp. 810–818. Springer, Cham (2019). https://doi.org/10.1007/978-3-030-32226-7_90
3. Cao, X., Chen, H., Li, Y., Peng, Y., Wang, S., Cheng, L.: Uncertainty aware temporal-ensembling model for semi-supervised abus mass segmentation. TMI **40**(1), 431–443 (2020)
4. Chaitanya, K., Karani, N., Baumgartner, C.F., Becker, A., Donati, O., Konukoglu, E.: Semi-supervised and task-driven data augmentation. In: Chung, A.C.S., Gee, J.C., Yushkevich, P.A., Bao, S. (eds.) IPMI 2019. LNCS, vol. 11492, pp. 29–41. Springer, Cham (2019). https://doi.org/10.1007/978-3-030-20351-1_3
5. Chen, W., et al.: Cancer statistics in China, 2015. CA: A Cancer J. Clin. **66**(2), 115–132 (2016)
6. Çiçek, Ö., Abdulkadir, A., Lienkamp, S.S., Brox, T., Ronneberger, O.: 3D U-Net: learning dense volumetric segmentation from sparse annotation. In: Ourselin, S., Joskowicz, L., Sabuncu, M.R., Unal, G., Wells, W. (eds.) MICCAI 2016. LNCS, vol. 9901, pp. 424–432. Springer, Cham (2016). https://doi.org/10.1007/978-3-319-46723-8_49
7. Cui, W., et al.: Semi-supervised brain lesion segmentation with an adapted mean teacher model. In: Chung, A.C.S., Gee, J.C., Yushkevich, P.A., Bao, S. (eds.) IPMI 2019. LNCS, vol. 11492, pp. 554–565. Springer, Cham (2019). https://doi.org/10.1007/978-3-030-20351-1_43
8. Gal, Y., Ghahramani, Z.: Dropout as a Bayesian approximation: representing model uncertainty in deep learning. In: ICML, pp. 1050–1059 (2016)
9. Hang, W., et al.: Local and global structure-aware entropy regularized mean teacher model for 3D left atrium segmentation. In: Martel, A.L., et al. (eds.) MICCAI 2020. LNCS, vol. 12261, pp. 562–571. Springer, Cham (2020). https://doi.org/10.1007/978-3-030-59710-8_55
10. Lee, C.Y., Xie, S., Gallagher, P., Zhang, Z., Tu, Z.: Deeply-supervised nets. In: Artificial Intelligence and Statistics, pp. 562–570. PMLR (2015)

11. Li, S., Zhang, C., He, X.: Shape-aware semi-supervised 3D semantic segmentation for medical images. In: Martel, A.L., et al. (eds.) MICCAI 2020. LNCS, vol. 12261, pp. 552–561. Springer, Cham (2020). https://doi.org/10.1007/978-3-030-59710-8_54
12. Li, X., Yu, L., Chen, H., Fu, C.W., Xing, L., Heng, P.A.: Transformation-consistent self-ensembling model for semisupervised medical image segmentation. TNNLS **32**(2), 523–534 (2020)
13. Lin, L., et al.: Deep learning for automated contouring of primary tumor volumes by MRI for nasopharyngeal carcinoma. Radiology **291**(3), 677–686 (2019)
14. Luo, X.: SSL4MIS (2020). https://github.com/HiLab-git/SSL4MIS
15. Luo, X., Chen, J., Song, T., Wang, G.: Semi-supervised medical image segmentation through dual-task consistency. In: AAAI, vol. 35, no. 10, pp. 8801–8809 (2021)
16. Ma, J., et al.: Active contour regularized semi-supervised learning for COVID-19 CT infection segmentation with limited annotations. Phys. Med. Biol. **65**(22), 225034 (2020)
17. Nie, D., Gao, Y., Wang, L., Shen, D.: ASDNet: attention based semi-supervised deep networks for medical image segmentation. In: Frangi, A.F., Schnabel, J.A., Davatzikos, C., Alberola-López, C., Fichtinger, G. (eds.) MICCAI 2018. LNCS, vol. 11073, pp. 370–378. Springer, Cham (2018). https://doi.org/10.1007/978-3-030-00937-3_43
18. Paszke, A., et al.: PyTorch: an imperative style, high-performance deep learning library. In: NeurIPS, pp. 8026–8037 (2019)
19. Peng, J., Pedersoli, M., Desrosiers, C.: Mutual information deep regularization for semi-supervised segmentation. In: MIDL, pp. 601–613. PMLR (2020)
20. Qiao, S., Shen, W., Zhang, Z., Wang, B., Yuille, A.: Deep co-training for semi-supervised image recognition. In: ECCV, pp. 135–152 (2018)
21. Tarvainen, A., Valpola, H.: Mean teachers are better role models: weight-averaged consistency targets improve semi-supervised deep learning results. In: NeurIPS, pp. 1195–1204 (2017)
22. Verma, V., Lamb, A., Kannala, J., Bengio, Y., Lopez-Paz, D.: Interpolation consistency training for semi-supervised learning. In: IJCAI, pp. 3635–3641 (2019)
23. Vu, T.H., Jain, H., Bucher, M., Cord, M., Pérez, P.: ADVENT: adversarial entropy minimization for domain adaptation in semantic segmentation. In: CVPR, pp. 2517–2526 (2019)
24. Wang, G., Aertsen, M., Deprest, J., Ourselin, S., Vercauteren, T., Zhang, S.: Uncertainty-guided efficient interactive refinement of fetal brain segmentation from stacks of MRI slices. In: Martel, A.L., et al. (eds.) MICCAI 2020. LNCS, vol. 12264, pp. 279–288. Springer, Cham (2020). https://doi.org/10.1007/978-3-030-59719-1_28
25. Wang, G., Li, W., Aertsen, M., Deprest, J., Ourselin, S., Vercauteren, T.: Aleatoric uncertainty estimation with test-time augmentation for medical image segmentation with convolutional neural networks. Neurocomputing **338**, 34–45 (2019)
26. Wang, Y., et al.: Double-uncertainty weighted method for semi-supervised learning. In: Martel, A.L., et al. (eds.) MICCAI 2020. LNCS, vol. 12261, pp. 542–551. Springer, Cham (2020). https://doi.org/10.1007/978-3-030-59710-8_53
27. Yu, L., Wang, S., Li, X., Fu, C.-W., Heng, P.-A.: Uncertainty-aware self-ensembling model for semi-supervised 3D left atrium segmentation. In: Shen, D., et al. (eds.) MICCAI 2019. LNCS, vol. 11765, pp. 605–613. Springer, Cham (2019). https://doi.org/10.1007/978-3-030-32245-8_67

28. Yushkevich, P.A., et al.: User-guided 3D active contour segmentation of anatomical structures: significantly improved efficiency and reliability. Neuroimage **31**(3), 1116–1128 (2006)
29. Zhang, Y., Yang, L., Chen, J., Fredericksen, M., Hughes, D.P., Chen, D.Z.: Deep adversarial networks for biomedical image segmentation utilizing unannotated images. In: Descoteaux, M., Maier-Hein, L., Franz, A., Jannin, P., Collins, D.L., Duchesne, S. (eds.) MICCAI 2017. LNCS, vol. 10435, pp. 408–416. Springer, Cham (2017). https://doi.org/10.1007/978-3-319-66179-7_47
30. Zheng, Z., Yang, Y.: Rectifying pseudo label learning via uncertainty estimation for domain adaptive semantic segmentation. IJCV **129**(4), 1106–1120 (2021). https://doi.org/10.1007/s11263-020-01395-y

Few-Shot Domain Adaptation with Polymorphic Transformers

Shaohua Li[1], Xiuchao Sui[1], Jie Fu[2], Huazhu Fu[3], Xiangde Luo[4], Yangqin Feng[1], Xinxing Xu[1(✉)], Yong Liu[1], Daniel S. W. Ting[5], and Rick Siow Mong Goh[1]

[1] Institute of High Performance Computing, A*STAR, Singapore, Singapore
[2] Mila, University of Montreal, Montreal, Canada
xuxinx@ihpc.a-star.edu.sg
[3] Inception Institute of Artificial Intelligence, Abu Dhabi, United Arab Emirates
[4] University of Electronic Science and Technology of China, Chengdu, China
[5] Singapore Eye Research Institute, Singapore, Singapore

Abstract. Deep neural networks (DNNs) trained on one set of medical images often experience severe performance drop on unseen test images, due to various domain discrepancy between the training images (source domain) and the test images (target domain), which raises a domain adaptation issue. In clinical settings, it is difficult to collect enough annotated target domain data in a short period. Few-shot domain adaptation, i.e., adapting a trained model with a handful of annotations, is highly practical and useful in this case. In this paper, we propose a Polymorphic Transformer (*Polyformer*), which can be incorporated into any DNN backbones for few-shot domain adaptation. Specifically, after the polyformer layer is inserted into a model trained on the source domain, it extracts a set of prototype embeddings, which can be viewed as a "basis" of the source-domain features. On the target domain, the polyformer layer adapts by only updating a projection layer which controls the interactions between image features and the prototype embeddings. All other model weights (except BatchNorm parameters) are frozen during adaptation. Thus, the chance of overfitting the annotations is greatly reduced, and the model can perform robustly on the target domain after being trained on a few annotated images. We demonstrate the effectiveness of Polyformer on two medical segmentation tasks (i.e., optic disc/cup segmentation, and polyp segmentation). The source code of Polyformer is released at https://github.com/askerlee/segtran.

Keywords: Transformer · Domain adaptation · Few-shot

Electronic supplementary material The online version of this chapter (https://doi.org/10.1007/978-3-030-87196-3_31) contains supplementary material, which is available to authorized users.

M. de Bruijne et al. (Eds.): MICCAI 2021, LNCS 12902, pp. 330–340, 2021.
https://doi.org/10.1007/978-3-030-87196-3_31

1 Introduction

Deep neural networks (DNNs) are notoriously fragile when being used on a domain not seen before. For example, it is common to witness 10–20% drop of accuracy on images captured with a device different from the training images. The training images and unseen test images are referred to as the *source domain* and the *target domain*, respectively. Domain adaptation (DA), i.e., modifying an existing model trained on the source domain, so that it performs well on the target domain, is important for deploying DNNs for medical image tasks.

Domain adaptation is trivial if a large set of annotated data exist in the target domain. However, such annotations are usually expensive to acquire, especially for segmentation tasks. Though, it is still cheap and feasible to obtain a handful of annotations. This work focuses on doing DA on a handful of annotations, or *few-shot* domain adaptation. It is a special case of semi-supervised learning. A large body of literature focuses on reducing the domain discrepancy by minimizing a domain adversarial loss [3,5,9,10,18,20]. Such methods can be used both in unsupervised and semi-supervised settings. As shown in our experiments, they are helpful for DA, and are complementary to our method.

In practice, a common approach to DA is retraining the model on the mixed source and target domain data [21]. However, it may be suboptimal in the few-shot scenario, as the joint dataset is dominated by the source domain. Another popular approach is fine-tuning the model weights on the target-domain annotated data. In the few-shot scenario, however, updating the whole pretrained model could easily overfit the limited target-domain annotations [13]. A remedy is to minimize the modification to the pretrained weights. For instance, we could freeze the feature extractor and just fine-tune the task head. Another scheme is to introduce adaptive modules into existing models, such as the DAM module [5], and freeze the pretrained weights.

Along the line of adaptive module-based methods, we propose a *polymorphic transformer* (polyformer). It can be inserted into a pretrained model, to take the responsibility of DA. It first extracts a set of *prototype embeddings* from the source domain, which is a condensed representation of the source-domain features. On the target domain, by attending with the prototype embeddings, the polyformer dynamically transforms the target-domain features. Thanks to the projection mechanism of transformers [19], after merely fine-tuning a projection layer, the transformed target-domain features can be made semantically compatible with the source domain. Hence, it can achieve good DA performance even in the few-shot scenario. As a proof-of-concept, we demonstrate the effectiveness of the polyformer on a vanilla U-Net model [17], evaluated on two cross-domain segmentation tasks: optic disc/cup area segmentation in fundus images, and polyp segmentation in colonoscopy images.

2 Related Work

Few-Shot Learning. Few-shot learning (FSL) is closely related, but different from few-shot DA. Typically, FSL is to adapt a pretrained model, so that it

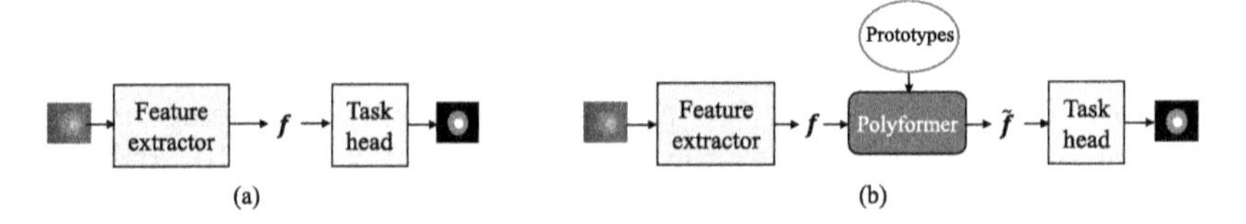

Fig. 1. (a) The original model pipeline consists of a feature extractor and a task head. (b) A modified pipeline for domain adaptation with an inserted polyformer. "Prototypes" are a set of prototype embeddings. For adaptation, the polyformer converts features $\boldsymbol{f}$ to $\tilde{\boldsymbol{f}}$. The weights of the feature extractor and the task head are frozen.

performs well on *novel tasks* (e.g. new classes) for which training examples are scarce [4]. In contrast, in few-shot DA, the model performs the same task on the source and target domains. A recent line of FSL research [22,23] is to first extract prototype embeddings (prototypes) from the support (training) images of each class, and then match them with query (test) image features using a distance metric. The prototypes are a set of "weak classifiers" [7] that vote to make the final prediction. Note that prototypes in FSL and in the polyformer serve different roles: in FSL the prototypes are used to make the final prediction, and in the polyformer they are used to transform target-domain features.

3 The Polymorphic Transformer

The polymorphic transformer (polyformer) is designed to bridge the gap between different domains. A pretrained network can offload DA onto a polyformer layer, so that it keeps all weights (except BatchNorms) frozen, and still performs robustly on a new domain. Adapting a polyformer layer only requires fine-tuning a projection layer, and thus a few annotated images are sufficient. In theory, the polyformer can be incorporated with any backbone networks, such as U-Net [17], DeepLabV3+ [2] or transformer-based models. In this work, we choose U-Net to illustrate how a polyformer performs DA on segmentation tasks.

Figure 1 illustrates how a polyformer layer is incorporated into an existing model. In Fig. 1(a), suppose a model M splits into a feature extractor M_1 and a task head M_2 (A similar formulation is found in [14]). For example, a U-Net can split into the encoder-decoder (M_1) and the segmentation head (M_2). On an input image x, the feature extractor generates feature maps $\boldsymbol{f}$, which are fed into M_2 to make predictions. On the target domain, due to domain discrepancy, the feature maps $\boldsymbol{f}$ follow different distributional properties, and thus M_2 is prone to make wrong predictions.

To bridge the domain gap, a polyformer layer is inserted between M_1 and M_2, as shown in Fig. 1(b). Now M_2 processes transformed features $\tilde{\boldsymbol{f}}$. After training the polyformer layer on the source domain and fine-tuning it on the target domain, $\tilde{\boldsymbol{f}}$ should "look more familiar" to M_2, leading to improved prediction accuracy. Instead of seeking adaptation between the myriads of features in the source and target domains, we propose to first find a condensed representation of

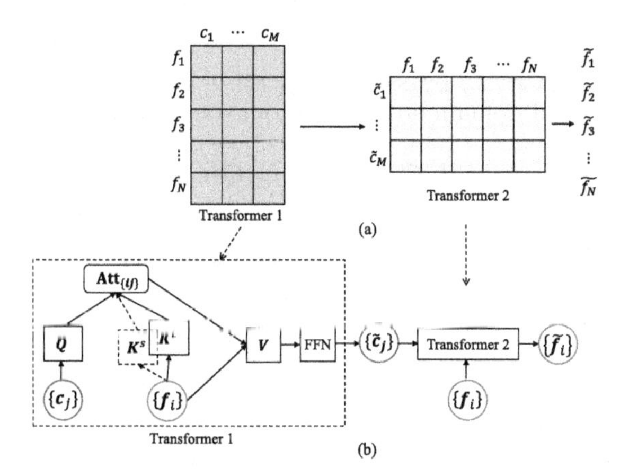

Fig. 2. A polyformer layer consists of two sub-transformers 1 and 2. (a) Schematic of the polyformer attention. In Transformer 1, the input feature vectors $f_1, \cdots, f_N$ attend with the prototypes $\boldsymbol{c}_1, \cdots, \boldsymbol{c}_M$, yielding $\tilde{\boldsymbol{c}}_1, \cdots, \tilde{\boldsymbol{c}}_M$, which then attend back with input $f_1, \cdots, f_N$ in Transformer 2 to generate the output features $\tilde{f}_1, \cdots, \tilde{f}_N$. (b) Zoom-in of Transformer 1. $\boldsymbol{K}^s$ and $\boldsymbol{K}^t$ are the key projections used on the source and target domains, respectively.

the source-domain features, namely a set of *prototype embeddings* (prototypes). Adaptation becomes much easier on a much smaller set of prototypes, making the model suited for few-shot scenarios.

3.1 Polyformer Architecture

Different designs can be chosen to implement the polyformer layer, and here we adopt the Squeeze-and-Expansion Transformer proposed in [12]. Figure 2 presents (a) the schematic of the two-step attention by the two sub-transformers, and (b) the zoomed-in architecture, especially the structure of Transformer 1.

A polyformer layer consists of two sub-transformer layers, denoted as Transformer 1 and 2. The prototypes are a set of M persistent embeddings $\boldsymbol{c}_1, \cdots, \boldsymbol{c}_M$, independent of the input. The input features $\boldsymbol{f}$ consist of a set of feature vectors $\{\boldsymbol{f}_1, \cdots, \boldsymbol{f}_N\}$, which are obtained by spatially flattening the input feature maps. Transformer 1 performs cross-attention between the prototypes and the input features, yielding intermediate features $\tilde{\boldsymbol{C}} = \tilde{\boldsymbol{c}}_1, \cdots, \tilde{\boldsymbol{c}}_M$. They attend back with $\boldsymbol{f}$ in Transformer 2, outputting the adapted features $\tilde{\boldsymbol{f}} = \{\tilde{\boldsymbol{f}}_1, \cdots, \tilde{\boldsymbol{f}}_N\}$:

$$\tilde{\boldsymbol{C}} = \text{Transformer1}(\boldsymbol{f}, \boldsymbol{C}), \tag{1}$$

$$\tilde{\boldsymbol{f}} = \text{Transformer2}(\tilde{\boldsymbol{C}}, \boldsymbol{f}) + \boldsymbol{f}, \tag{2}$$

where in Eq. (2), a residual connection builds upon the original representations. In a Squeeze-and-Expansion transformer, sub-transformers 1 and 2 are two Expanded Attention Blocks [12]. In each sub-transformer, the attention matrix is of $N \times M$.

Typically, the number of feature points $N > 10^4$, and we chose $M = 256$. Thus, the huge feature space is compressed using a set of 256 prototypes.

For the purpose of domain adaptation, our focus is on Transformer 1:

$$\text{Att\_weight}(\boldsymbol{f}, \boldsymbol{C}) = \sigma(\boldsymbol{K}(\boldsymbol{f}), \boldsymbol{Q}(\boldsymbol{C})) \in \mathbb{R}^{N \times M}, \tag{3}$$

$$\text{Attention}(\boldsymbol{f}, \boldsymbol{C}) = \text{Att\_weight}(\boldsymbol{f}, \boldsymbol{C}) \cdot \boldsymbol{V}(\boldsymbol{f}), \tag{4}$$

$$\tilde{\boldsymbol{C}} = \text{FFN}(\text{Attention}(\boldsymbol{f}, \boldsymbol{C})), \tag{5}$$

where $\boldsymbol{K}, \boldsymbol{Q}, \boldsymbol{V}$ are key, query, and value projections, respectively. σ is softmax after dot product. Att_weight$(\boldsymbol{f}, \boldsymbol{C})$ is a pairwise attention matrix, whose i, j-th element defines how much $\boldsymbol{f}_i$ contributes to the fused features of prototype j. FFN is a feed-forward network that transforms the fused features into $\tilde{\boldsymbol{c}}_j$.

In Eq. (3), $\boldsymbol{K}$ controls which subspace input features f_i are projected to, and influences the attention between $\boldsymbol{f}$ and $\boldsymbol{C}$. It leads to the following hypothesis:

Hypothesis 1. *In the target domain, by properly updating $\boldsymbol{K} : \boldsymbol{K}^s \rightarrow \boldsymbol{K}^t$, the polyformer layer will project input features $\boldsymbol{f}$ (which follow a different distribution) to a subspace similar to that in the source domain. Consequently, the output features from the polyformer are semantically compatible with the target-domain output feature space. As a result, the model will perform better on the target domain without updating the original model weights.*

According to Hypothesis 1, to adapt a trained model on a new domain, we can share $\boldsymbol{Q}, \boldsymbol{V}$ and FFN across domains, and update $\boldsymbol{K}$ only. This scheme inherites most of the representation powers from the source domain.

When the polyformer layer is an ordinary transformer without prototypes, the attention matrix is a huge $N \times N$ matrix ($N > 10^4$). Then cross-domain semantic alignment becomes much more harder, and hypothesis 1 may not satisfy.

3.2 Training and Adaptation of the Polyformer Layer

Training on the Source Domain. After a polyformer layer is inserted in a model M trained on the source domain, we need to train the polyformer layer to make the new pipeline M' keep similar performance on the source domain as the original pipeline. This is achieved by training again on the source-domain data. Specifically, all the weights (including BatchNorms) of M are frozen, and only the polyformer weights are to be updated. The same training protocol is performed on the source-domain training data $\{(x_1^s, y_1^s), \cdots, (x_n^s, y_n^s)\}$. After training, the prototypes are compressed representations of the source domain features.

Adapting to the Target Domain. On the target domain, all the weights (excluding BatchNorms) of M are frozen, and only the $\boldsymbol{K}$ projection weights and BatchNorm parameters are to be updated. The training is performed on the few-shot target-domain training data $(X^t, Y^t) = \{(x_1^t, y_1^t), \cdots, (x_m^t, y_m^t)\}$.

Note that traditional domain adversarial losses [9,18] could be incorporated to improve the adaptation, as shown in our ablation studies (Sect. 4.1):

$$\mathcal{L}_{adapt}(X^s, X^t, Y^t) = \mathcal{L}_{sup}(X^t, Y^t) + \mathcal{L}_{adv}(X^s, X^t). \tag{6}$$

There are two common choices for the domain adversarial loss: the discriminator could try to discriminate either 1) the features of a source vs. a target domain image, or 2) the predicted masks on a source vs. a target domain image.

Table 1. The dice scores on Fundus and Polyp target domains RIM-One and CVC-300, by five ablated Polyformer models and the standard "$\mathcal{L}_{sup} + \mathcal{L}_{adv} + K$". The U-Net and Polyformer trained on the source-domain were includes as references.

	RIM-One		CVC-300	Avg.
	Disc	Cup		
Trained on source domain				
U-Net	0.819	0.708	0.728	0.752
Polyformer	0.815	0.717	0.724	0.752
Adapted to target domain				
$\mathcal{L}_{adv} + K$	0.828	0.731	0.779	0.779
$\mathcal{L}_{sup} + K$, w/o BN	0.823	0.741	0.760	0.775
$\mathcal{L}_{sup} + K$	0.900	0.753	0.830	0.828
$\mathcal{L}_{sup} + \mathcal{L}_{adv}$ + All weights	0.892	0.741	0.826	0.820
$\mathcal{L}_{sup} + \mathcal{L}_{adv}$(mask) $+ K$	0.909	**0.763**	**0.836**	**0.836**
$\mathcal{L}_{sup} + \mathcal{L}_{adv} + K$ (standard setting)	**0.913**	0.758	0.834	0.835

4 Experiments

Different methods were evaluated on two medical image segmentation tasks:

Optic Disc/Cup Segmentation. This task does segmentation of the optic disc and cup in fundus images, which are 2D images of the rear of the eyes. The source domain was the 1200 training images provided in the REFUGE challenge [15]. The target domain, the RIM-One dataset [8], contains 159 images.

Polyp Segmentation. This task does polyp (fleshy growths inside the colon lining) segmentation in colonoscopy images. The source domain was a combination of two datasets: CVC-612 (612 images) [1] and Kvasir (1000 images) [16]. The target domain was the CVC-300 dataset (60 images) [6].

Number of Shots. For each task, **five** annotated images were randomly selected from the target domain to do few-shot supervised training. Each method was evaluated on the remaining target-domain images. Results with 10, 15 and 20 shots can be found in the supplementary file.

4.1 Ablation Studies

A standard Polyformer and five ablations were evaluated on the two tasks:

- $\mathcal{L}_{sup}+\mathcal{L}_{adv}+K$ **(standard setting)**, i.e., fine-tuning only the $\boldsymbol{K}$ projection, with both the few-shot supervision and the domain adversarial learning on features. It is the standard setting from which other ablations are derived;
- $\mathcal{L}_{adv}+K$, i.e., fine-tuning the $\boldsymbol{K}$ projection using the unsupervised domain adversarial loss on features, without using the few-shot supervision;
- $\mathcal{L}_{sup}+K$, w/o BN, i.e., freezing the BatchNorm affine parameters, but still updating the mean/std statistics on the target domain;
- $\mathcal{L}_{sup}+K$, i.e., fine-tuning the $\boldsymbol{K}$ projection using the few-shot supervision only, without the domain adversarial loss;
- $\mathcal{L}_{sup}+\mathcal{L}_{adv}$ + All weights, i.e., fine-tuning the whole polyformer layer, instead of only the $\boldsymbol{K}$ projection;
- $\mathcal{L}_{sup}+\mathcal{L}_{adv}(\text{mask})+K$, i.e., doing domain adversarial learning on the predicted masks, instead of on the extracted features.

Table 1 presents the results of the standard setting "$\mathcal{L}_{sup}+\mathcal{L}_{adv}+K$", as well as five ablated models. Without the few-shot supervision, the domain adversarial loss only marginally improved the target-domain performance ($0.752 \rightarrow 0.779$). Freezing the BatchNorm affine parameters greatly restricts adaptation ($0.752 \rightarrow 0.775$). Fine-tuning the whole polyformer layer led to worse performance than fine-tuning the $\boldsymbol{K}$ projection only (0.820 vs. 0.835), probably due to catastrophic forgetting [13] of the source-domain semantics encoded in the prototypes. Incorporating the domain adversarial complemented and helped the few-shot supervision obtain better performance ($0.828 \rightarrow 0.835$). The domain adversarial loss on features led to almost the same results as on masks.

4.2 Compared Methods

Two settings of Polyformer, as well as ten popular baselines, were evaluated:

- **U-Net (source)**, trained on the source domain without adaptation;
- $\mathcal{L}_{sup}$, fine-tuning U-Net (source) on the five target-domain images;
- $\mathcal{L}_{sup}$**(source + target)**, trained on a mixture of all source-domain images and the five target-domain images;
- **CycleGAN +** $\mathcal{L}_{sup}$**(source)** [11,24][1]. The CycleGAN was trained for 200 epochs to convert between the source and the target domains. The converted source-domain images were used to train a U-Net from scratch;
- **RevGrad** ($\mathcal{L}_{sup}+\mathcal{L}_{adv}$) [9], which fine-tunes U-Net (source), by optimizing the domain adversarial loss on the features with a gradient reversal layer;
- **ADDA** ($\mathcal{L}_{sup}+\mathcal{L}_{adv}$) [18], which uses inverted domain labels to replace the gradient reversal layer in RevGrad for more stable gradients;

[1] CycleGAN is the core component for DA in [11], but [11] is more than CycleGAN.

- **DA-ADV (tune whole model)** [3] also named as *p*OSAL in [20], which fine-tunes the whole U-Net (source) by discriminating whether the masks are generated on the source or the target domain images using RevGrad;
- **DA-ADV (tune last two layers)**, DA-ADV training that only fine-tunes the last two layers and all BatchNorm parameters of U-Net (source);
- **CellSegSSDA** ($\mathcal{L}_{sup}+\mathcal{L}_{adv}$(mask) + $\mathcal{L}_{recon}$) [10], which combines RevGrad on predicted masks, an image reconstruction loss and few-shot supervision;
- **Polyformer** ($\mathcal{L}_{sup}+K$), by fine-tuning the $\boldsymbol{K}$ projection in the polyformer layer, with the few-shot supervision only;
- **Polyformer** ($\mathcal{L}_{sup}+\mathcal{L}_{adv}+K$), i.e., the standard Polyformer training setting, which enhances Polyformer ($\mathcal{L}_{sup}+K$) with RevGrad on features;
- $\mathcal{L}_{sup}$ **(50% target)**, by fine-tuning U-Net (source) on 1/2 of the target-domain images, and tested on the remaining 1/2 images. This serves as an empirical upper-bound of all methods[2].

The domain adversarial methods RevGrad, ADDA and DA-ADV were combined with the few-shot supervision to do semi-supervised learning. All the methods were trained with a batch size of 4, and optimized with the AdamW optimizer at an initial learning rate of 0.001. The supervised training loss was the average of the pixel-wise cross-entropy loss and the dice loss.

4.3 Results

Table 2 presents the segmentation performance of different methods on the two target domains, measured in dice scores. The domain adversarial loss effectively reduced the performance gap between the source and target domains. "CycleGAN + $\mathcal{L}_{sup}$(source)" performed even worse than U-Net (source), as CycleGAN does not guarantee semantic alignment when doing conversion [24]. Without the domain adversarial loss, Polyformer ($\mathcal{L}_{sup}+K$) has already achieved higher average dice scores than all the baseline methods. By incorporating the domain adversarial loss RevGrad on predicted masks, Polyformer ($\mathcal{L}_{sup}+\mathcal{L}_{adv}+K$) achieved higher performance than Polyformer ($\mathcal{L}_{sup}+K$), showing that Polyformer is complementary with the traditional domain adversarial loss.

To gain an intuitive understanding of how different methods performed, Fig. 3 presents a fundus image from RIM-One, the ground-truth segmentation mask and the predicted masks by selected methods. In addition, a REFUGE image is presented in the left to visualize the source/target domain gap. Without adaptation, U-Net (source) was unable to find most of the disc/cup areas, as the image is much darker than typical source-domain REFUGE images. The mask predicted by "CycleGAN + $\mathcal{L}_{sup}$(source)" largely deviates from the ground-truth.

[2] However, the performance of $\mathcal{L}_{sup}$ (50% target) on CVC-300 was lower than Polyformer and other baseline methods with more shots, partly because CVC-300 is small (60 images) and sensitive to randomness. See the supplementary file for discussions and more experiments.

Table 2. Dice scores on Fundus and Polyp target domains RIM-One and CVC-300.

	RIM-One		CVC-300	Avg.
	Disc	Cup		
U-Net (source) [17]	0.819	0.708	0.728	0.752
$\mathcal{L}_{sup}$ [17]	0.871	0.665	0.791	0.776
$\mathcal{L}_{sup}$ (source + target) [17]	0.831	0.715	0.808	0.785
CycleGAN + $\mathcal{L}_{sup}$ (source) [11,24]	0.747	0.690	0.709	0.715
RevGrad ($\mathcal{L}_{sup} + \mathcal{L}_{adv}$) [9]	0.860	0.732	0.813	0.802
ADDA ($\mathcal{L}_{sup} + \mathcal{L}_{adv}$) [18]	0.874	0.726	**0.836**	0.812
DA-ADV (tune whole model) [3,20]	0.885	0.725	0.830	0.813
DA-ADV (tune last two layers) [3,20]	0.872	0.730	0.786	0.796
CellSegSSDA ($\mathcal{L}_{sup} + \mathcal{L}_{adv}$(mask) + $\mathcal{L}_{recon}$) [10]	0.869	0.756	0.805	0.810
Polyformer ($\mathcal{L}_{sup} + K$)	0.900	0.753	0.830	0.828
Polyformer ($\mathcal{L}_{sup} + \mathcal{L}_{adv} + K$)	**0.913**	**0.758**	0.834	**0.835**
$\mathcal{L}_{sup}$ (50% target) [17]	0.959	0.834	0.834	0.876

The mask from Polyformer was significantly improved by the domain adversarial loss, in that the artifacts were eliminated and the mask became much closer to the ground-truth. For comparison purposes, another example is presented in the supplementary file where all the major methods failed.

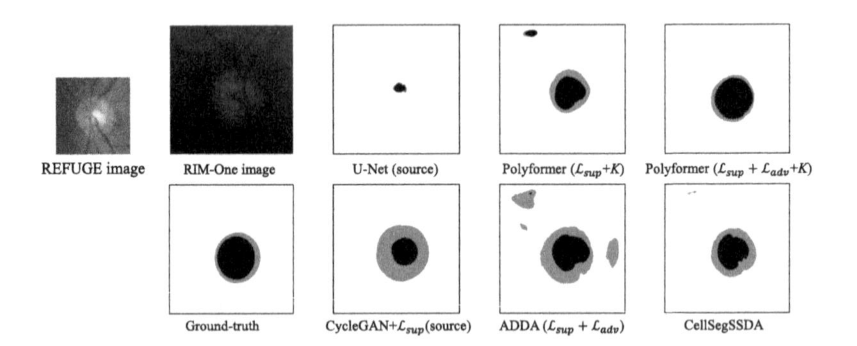

Fig. 3. The segmentation masks predicted by different methods on a RIM-One image.

5 Conclusions

In this work, we proposed a plug-and-play module called the polymorphic transformer (Polyformer) for domain adaptation. It can be plugged into a pretrained model. On a new domain, only fine-tuning a projection layer within Polyformer is sufficient to achieve good performance. We demonstrated the effectiveness of

Polyformer on two segmentation tasks, where it performed consistently better than strong baselines in the challenging few-shot learning setting.

Acknowledgements. We are grateful for the help and support of Wei Jing. This research is supported by A*STAR under its Career Development Award (Grant No. C210112016), and its Human-Robot Collaborative AI for Advanced Manufacturing and Engineering (AME) programme (Grant No. A18A2b0046).

References

1. Bernal, J., Tajkbaksh, N., et al.: Comparative validation of polyp detection methods in video colonoscopy: results from the MICCAI 2015 endoscopic vision challenge. IEEE Trans. Med. Imaging **36**(6), 1231–1249 (2017)
2. Chen, L.-C., Zhu, Y., Papandreou, G., Schroff, F., Adam, H.: Encoder-decoder with atrous separable convolution for semantic image segmentation. In: Ferrari, V., Hebert, M., Sminchisescu, C., Weiss, Y. (eds.) ECCV 2018. LNCS, vol. 11211, pp. 833–851. Springer, Cham (2018). https://doi.org/10.1007/978-3-030-01234-2_49
3. Dong, N., Kampffmeyer, M., Liang, X., Wang, Z., Dai, W., Xing, E.: Unsupervised domain adaptation for automatic estimation of cardiothoracic ratio. In: Frangi, A.F., Schnabel, J.A., Davatzikos, C., Alberola-López, C., Fichtinger, G. (eds.) MICCAI 2018. LNCS, vol. 11071, pp. 544–552. Springer, Cham (2018). https://doi.org/10.1007/978-3-030-00934-2_61
4. Dong, N., Xing, E.P.: Domain adaption in one-shot learning. In: Berlingerio, M., Bonchi, F., Gärtner, T., Hurley, N., Ifrim, G. (eds.) ECML PKDD 2018. LNCS (LNAI), vol. 11051, pp. 573–588. Springer, Cham (2019). https://doi.org/10.1007/978-3-030-10925-7_35
5. Dou, Q., Ouyang, C., Chen, C., Chen, H., Heng, P.A.: Unsupervised cross-modality domain adaptation of convnets for biomedical image segmentations with adversarial loss. In: IJCAI (2018)
6. Fan, D.-P., et al.: PraNet: parallel reverse attention network for polyp segmentation. In: Martel, A.L., et al. (eds.) MICCAI 2020. LNCS, vol. 12266, pp. 263–273. Springer, Cham (2020). https://doi.org/10.1007/978-3-030-59725-2_26
7. Freund, Y., Schapire, R.E.: A decision-theoretic generalization of on-line learning and an application to boosting. J. Comput. Syst. Sci. **55**, 119–139 (1997)
8. Fumero, F., Alayon, S., Sanchez, J.L., Sigut, J., et al.: RIM-ONE: an open retinal image database for optic nerve evaluation. In: 24th International Symposium on CBMS (2011)
9. Ganin, Y., Lempitsky, V.: Unsupervised domain adaptation by backpropagation. In: ICML (2015)
10. Haq, M.M., Huang, J.: Adversarial domain adaptation for cell segmentation. In: MIDL (2020)
11. Li, K., Wang, S., Yu, L., Heng, P.-A.: Dual-teacher: integrating intra-domain and inter-domain teachers for annotation-efficient cardiac segmentation. In: Martel, A.L., et al. (eds.) MICCAI 2020. LNCS, vol. 12261, pp. 418–427. Springer, Cham (2020). https://doi.org/10.1007/978-3-030-59710-8_41
12. Li, S., Sui, X., Luo, X., Xu, X., Liu, Y., Goh, R.: Medical image segmentation using squeeze-and-expansion transformers. In: IJCAI (2021)
13. Li, Z., Hoiem, D.: Learning without forgetting. In: Leibe, B., Matas, J., Sebe, N., Welling, M. (eds.) ECCV 2016. LNCS, vol. 9908, pp. 614–629. Springer, Cham (2016). https://doi.org/10.1007/978-3-319-46493-0_37

14. Motiian, S., Jones, Q., Iranmanesh, S.M., Doretto, G.: Few-shot adversarial domain adaptation. In: NeurIPS (2017)
15. Orlando, J.I., et al.: REFUGE challenge: a unified framework for evaluating automated methods for glaucoma assessment from fundus photographs. Med. Image Anal. **59**, 101570 (2020)
16. Pogorelov, K., Randel, K.R., et al.: KVASIR: a multi-class image dataset for computer aided gastrointestinal disease detection. In: Proceedings of ACM on Multimedia Systems Conference, MMSys 2017, pp. 164–169. ACM, New York (2017)
17. Ronneberger, O., Fischer, P., Brox, T.: U-Net: convolutional networks for biomedical image segmentation. In: Navab, N., Hornegger, J., Wells, W.M., Frangi, A.F. (eds.) MICCAI 2015. LNCS, vol. 9351, pp. 234–241. Springer, Cham (2015). https://doi.org/10.1007/978-3-319-24574-4_28
18. Tzeng, E., Hoffman, J., Saenko, K., Darrell, T.: Adversarial discriminative domain adaptation. In: CVPR (2017)
19. Vaswani, A., Shazeer, N., Parmar, N., Uszkoreit, J., Jones, L., Gomez, A.N., et al.: Attention is all you need. In: NeurIPS (2017)
20. Wang, S., Yu, L., Yang, X., Fu, C.W., Heng, P.A.: Patch-based output space adversarial learning for joint optic disc and cup segmentation. IEEE Trans. Med. Imaging **38**, 2485–2495 (2019)
21. Yao, L., Prosky, J., Covington, B., Lyman, K.: A strong baseline for domain adaptation and generalization in medical imaging. In: MIDL - Extended Abstract Track (2019)
22. Ye, H.J., Hu, H., Zhan, D.C., Sha, F.: Few-shot learning via embedding adaptation with set-to-set functions. In: CVPR, June 2020
23. Zhang, C., Cai, Y., Lin, G., Shen, C.: DeepEMD: few-shot image classification with differentiable earth mover's distance and structured classifiers. In: CVPR (2020)
24. Zhu, J., Park, T., Isola, P., Efros, A.A.: Unpaired image-to-image translation using cycle-consistent adversarial networks. In: ICCV (2017)

Lesion Segmentation and RECIST Diameter Prediction via Click-Driven Attention and Dual-Path Connection

Youbao Tang[1](✉), Ke Yan[1], Jinzheng Cai[1], Lingyun Huang[2], Guotong Xie[2], Jing Xiao[2], Jingjing Lu[3], Gigin Lin[4], and Le Lu[1]

[1] PAII Inc., Bethesda, MD, USA
[2] Ping An Technology, Shenzhen, People's Republic of China
[3] Beijing United Family Hospital, Beijing, People's Republic of China
[4] Chang Gung Memorial Hospital, Linkou, Taiwan, ROC

Abstract. Measuring lesion size is an important step to assess tumor growth and monitor disease progression and therapy response in oncology image analysis. Although it is tedious and highly time-consuming, radiologists have to work on this task by using RECIST criteria (Response Evaluation Criteria In Solid Tumors) routinely and manually. Even though lesion segmentation may be the more accurate and clinically more valuable means, physicians can not manually segment lesions as now since much more heavy laboring will be required. In this paper, we present a prior-guided dual-path network (PDNet) to segment common types of lesions throughout the whole body and predict their RECIST diameters accurately and automatically. Similar to [23], a click guidance from radiologists is the only requirement. There are two key characteristics in PDNet: 1) Learning lesion-specific attention matrices in parallel from the click prior information by the proposed prior encoder, named click-driven attention; 2) Aggregating the extracted multi-scale features comprehensively by introducing top-down and bottom-up connections in the proposed decoder, named dual-path connection. Experiments show the superiority of our proposed PDNet in lesion segmentation and RECIST diameter prediction using the DeepLesion dataset and an external test set. PDNet learns comprehensive and representative deep image features for our tasks and produces more accurate results on both lesion segmentation and RECIST diameter prediction.

1 Introduction

Assessing lesion growth across multiple time points is a major task for radiologists and oncologists. The sizes of lesions are important clinical indicators for monitoring disease progression and therapy response in oncology. A widely-used guideline is RECIST (Response Evaluation Criteria In Solid Tumors) [6], which requires users to first select an axial slice where the lesion has the largest spatial extent, then measure the longest diameter of the lesion (long axis), followed by its longest perpendicular diameter (short axis). This process is highly tedious

M. de Bruijne et al. (Eds.): MICCAI 2021, LNCS 12902, pp. 341–351, 2021.
https://doi.org/10.1007/978-3-030-87196-3_32

and time-consuming. More importantly, it is prone to inconsistency between different observers [21], even with considerable clinical knowledge. Segmentation masks may be another quantitative and meaningful metric to assess lesion sizes, which is arguably more accurate/precise than RECIST diameters and avoids the subjectivity of selecting long and short axes. However, it is impractical and infeasible for radiologists to manually delineate the contour of every target lesion on a daily basis due to the heavy work load that would require.

Automatic lesion segmentation has been extensively studied by researchers. Most existing works focused on tumors of specific types, such as lung nodules [24], liver tumors [12,22], and lymph nodes [18,27]. However, radiologists often encounter different types of lesions when reading an image. Universal lesion segmentation [1,3,20,23] and measurement [21,23] have drawn attentions in recent years, aiming at learning from a large-scale dataset to handle a variety of lesions in one algorithm. These work leverage NIH DeepLesion dataset [25], which contains the RECIST annotations of over 30K lesions of various types. Among them, [21] requires users to draw a box around the lesion to indicate the lesion of interest. It first employs a spatial transform network to normalize the lesion region, then adapts a stacked hourglass network to regress the four endpoints of the RECIST diameters. [23] requests users to only click a point on or near the lesion, more convenient and efficient than [21]. It uses an improved mask R-CNN to detect the lesion region and subsequently performs segmentation and RECIST diameter prediction [23]. User click information is fed into the model as the input together with the image. This strategy treats lesions with diverse sizes and shapes in the same way. Therefore, the previous work [23] has two major limitations: 1) the detection based LOI (Lesion of Interest) extraction method can produce erroneous LOIs when multiple lesions are spatially close to each other; 2) the click prior information is treated equally for all lesions with different scales, which may not be optimal at locating the lesion region precisely.

In this paper, we propose a novel framework named prior-guided dual-path network (PDNet) to precisely alleviate these limitations. Following [23], given a 2D computed tomography (CT) slice and a click guidance in a lesion region, our goal is to segment the lesion and predict its RECIST diameters automatically and reliably. To achieve this goal, we adopt a two-stage framework. The first stage extracts the lesion of interest (LOI) by segmentation rather than detection in [23], since sometimes the detection results do not cover the lesions that we click in. The second stage obtains the lesion segmentation and RECIST diameter prediction results from the extracted LOI. We propose a novel prior encoder to encode the click prior information into attention maps, which can deal with considerable size and shape variations of the lesions. We also design a scale-aware attention block with dual-path connection to improve the decoder. PDNet is evaluated on manually-labeled lesion masks and RECIST diameters in DeepLesion dataset [25]. To prove the generalizability of our method, we additionally collected an external test set from 6 public lesion datasets of 5 organs. Experimental results show that PDNet outperforms the previous state-of-the-art method [23] and a strong baseline nnUNet [10] on two test sets, for both lesion segmentation and RECIST diameter prediction tasks.

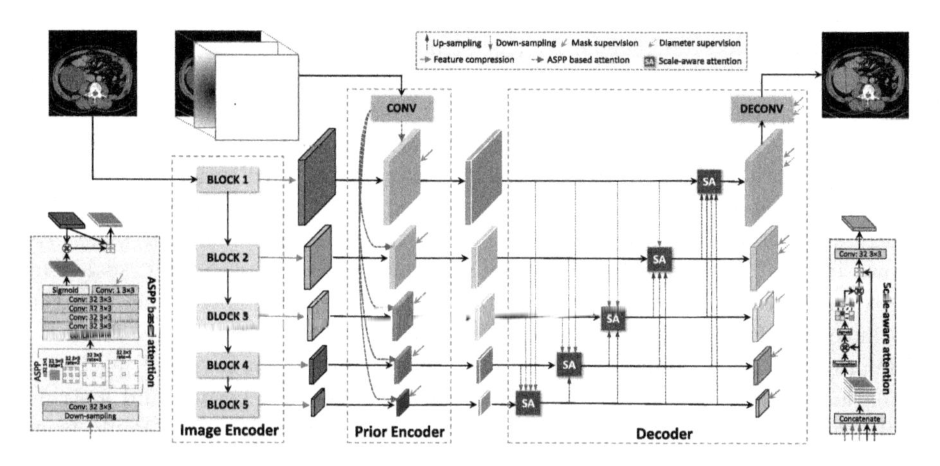

Fig. 1. Overview of the proposed prior-guided dual-path network (PDNet). (Color figure online)

2 Methodology

Our framework includes two stages. The first stage extracts the lesion of interest (LOI) by segmentation; the second stage performs lesion segmentation and RECIST diameter prediction from the extracted LOI. A prior-guided dual-path network (PDNet) is proposed for the tasks at both stages. Figure 1 shows the overview of the proposed PDNet. It consists of three components: an image encoder, a prior encoder with click-driven attention, and a decoder with dual-path connection.

Image Encoder: The image encoder aims to extract highly discriminative features from an input CT image. Also, representing features at multiple scales is of great importance for our tasks. Recently, Zhang *et al.* [26] present a split-attention block and stack several such blocks in ResNet style [8] to create a split-attention network, named ResNeSt. ResNeSt is able to capture cross-channel feature correlations by combining the channel-wise attention with multi-path network layout. Extensive experiments demonstrate that it universally improves the learned feature representations to boost performance across numerous vision tasks. Therefore, this work utilizes ResNeSt-50 [26] as backbone to extract highly discriminative multi-scale features in the image encoder. As shown in Fig. 1, ResNeSt-50 has five blocks that output multi-scale features with different channels. To relieve the computation burden, they are compressed to 32 channels using a convolutional layer with 32 3×3 kernels.

Prior Encoder with Click-Driven Attention: Given a click guidance, a click image and a distance transform image are generated and considered as prior information following [23], as shown in Fig. 1. In [23], the prior information is integrated into the model by directly treating it as input for feature extraction. We argue that the representation ability of features extracted from image encoder may be weaken by this strategy. That is because the sizes and shapes of different lesions are highly diverse, but their prior information generated using this strategy are the same. To avoid this, we separately build a prior encoder (PE) with click-driven attention, which is able to learn lesion-specific attention matrices by effectively exploring the click prior information. With them, the representation ability of the extracted multi-scale features from image encoder will be enhanced to improve the performance of our tasks. As shown in Fig. 1, the prior encoder takes as input the compressed multi-scale features and a 3-channel image (the original CT image, the click image, and the distance transform image), and outputs attention enhanced multi-scale features. The prior encoder includes five atrous spatial pyramid pooling (ASPP) [4] based attention modules and a convolutional layer with 32 3×3 kernels and a stride of 2. The detailed structure of ASPP based attention module can be found from the purple box in Fig. 1, where 5 side outputs (the pink solid arrows) are added to introduce the deep mask supervision to learn the attention matrices.

Decoder with Dual-Path Connection: It is known that the low-level scale features focus on fine-grained lesion parts (*e.g.*, edges) but are short of global contextual information, while the high-level scale features are capable of segmenting the entire lesion regions coarsely but at the cost of losing some detailed information. With this inspiration, unlike UNet [15] where the decoder only considers current scale features and its neighbouring high-level scale features gradually, we build a new decoder that can aggregate the attention enhanced multi-scale features more comprehensively. Specifically, each scale features are reasonably interacted with all lower-level and higher-level scale features in the decoder, which is accomplished by using dual-path connection (*i.e.*, top-down connection and bottom-up connection). The top-down connection (T2D) adopts a bilinear interpolation operation on the high-level scale features for up-sampling followed by a convolutional layer with 32 3×3 kernels for smoothing. The bottom-up connection (B2U) performs a convolution operation with 32 3×3 kernels and a large stride for down-sampling. Then the current scale features are concatenated with all up-sampled and down-sampled features from other scales in the channel dimension, suggesting that each concatenated features can represent the global contextual and local detail information of the lesion. The concatenated features can be directly used for lesion segmentation or RECIST diameter prediction with a convolutional layer of 1 or 4 3×3 kernels. But before that, to further improve the feature representations, we build a scale-aware attention module (SA) based on the channel attention mechanism of DANet [7], which selectively emphasizes interdependent channel features by integrating associated features among all feature channels. The SA's structure is shown in the red box of Fig. 1. Different lesions have different scales, but SA is able to adaptively select suitable scale or channel features for them for better accuracy in our tasks. To get a full-size prediction, a deconvolutional layer with 32 4×4 kernels and a stride of 2 is attached to the last SA. Also, 6 and 3 side outputs are added in the decoder to introduce the deep mask supervision (the pink solid arrows) and deep diameter supervision (the pink dotted arrows), respectively. The deep diameter supervision is only used for high-resolution side outputs, because a high-quality RECIST diameter prediction requires large spatial and detailed information.

Model Optimization: Following [21,23], we also convert the RECIST diameter prediction problem into a key point regression problem. It means that the model will predict four key point heatmaps to locate the four endpoints of RECIST diameters. For both tasks, a mean squared error loss (l_{mse}) is used to compute the errors between predictions and supervisions. As a pixel-wise loss, it will be affected by imbalanced foreground and background pixels. Unfortunately, lesion and non-lesion regions are highly imbalanced at stage 1 of our framework. To deal with this problem, an additional IOU loss [14] (l_{iou}) is introduced for the lesion segmentation task, which handles the global structures of lesions instead of every single pixel. As described above, 11 side outputs with deep mask supervision and 3 side outputs with deep diameter supervision are used in PDNet. Therefore, the loss is $l_{seg} = \sum_{i=1}^{11} [l_{mse}^i + l_{iou}^i]$ for lesion segmentation and $l_{dp} = \sum_{i=1}^{3} l_{mse}^i$ for RECIST diameter prediction. The final loss is $l = \lambda l_{seg} + (1 - \lambda) l_{dp}$, where λ is set to 0.01 to balance the magnitude of the two losses. Two PDNet models used in two stages are trained separately.

For the segmentation task, we do not have manual lesion masks in DeepLesion. Therefore, we first construct an ellipse from a RECIST annotation following [19]. Then the morphological snake (MS) algorithm [13] is used to refine the ellipse to get a pseudo mask with good quality, serving as the mask supervision. For l_{dp} update, we generate four 2D Gaussian heatmaps with a standard deviation of σ from four endpoints of each RECIST annotation, serving as the diameter supervision. We set $\sigma = 3$ at stage 1, $\sigma = 7$ at stage 2. We also apply an iterative refining strategy. When the training is done, we run the model over all training data to get their lesion segmentation results, and then use the MS algorithm to refine them. With an ellipse and a refined segmentation result, we can update the pseudo mask by setting their intersections as foreground, their differences as uncertain regions that will be ignored for loss computation during training, and the rest as background. The new pseudo masks can be used to retrain the models. The final models are obtained after three training iterations.

3 Experiments

Datasets and Evaluation Criteria: The DeepLesion dataset [25] contains 32,735 CT lesion images with RECIST diameter annotations from 10,594 studies of 4,459 patients. Various lesions throughout the whole body are included, such as lung nodules, bone lesions, liver tumors, enlarged lymph nodes, and so on. Following [3,23], 1000 lesion images from 500 patients with manual segmentations serve as a test set. The rest patient data are used for training. An external test set with 1350 lesions from 900 patients is built for external validation by collecting lung, liver, pancreas, kidney tumors, and lymph nodes from multiple public datasets, including Decathlon-Lung [17] (50), LIDC [2] (200), Decathlon-HepaticVessel [17] (200), Decathlon-Pancreas [17] (200), KiTS [9] (150), and

NIH-Lymph Node [16] (100), specifically. Each lesion has a 3D mask. To make it suitable for evaluation, we select an axial slice for each lesion where the lesion has the largest spatial extent based on its 3D mask. The long and short diameters calculated from the 2D lesion mask of the selected slice are treated as the ground truths of the RECIST diameters. We utilize the same criteria as in [23] to compute the quantitative results. The pixel-wise precision, recall, and dice coefficient (Dice) are used for lesion segmentation. The mean and standard deviation of differences between the diameter lengths (mm) of the predictions and manual annotations are used for RECIST diameter prediction.

Implementation Details: PDNet is implemented in PyTorch. The image encoder is initialized with the ImageNet [5] pre-trained weights. At both stages, we train PDNet using Adam optimizer [11] with an initial learning rate of 0.001 for 120 epochs and decay it by 0.1 after 60 and 90 epochs. During training, all CT images are resized to 512×512 first. Then the input images are generated by randomly rotating by $\theta \in [-10^\circ, 10^\circ]$ and cropping a square sub-image whose size is $s \in [480, 512]$ at stage 1 and 1.5 to 3.5 times as large as the lesion's long side with random offsets at stage 2. They are resized to 512×512 and 256×256 for both stages, respectively. For testing, input images are generated by resizing to 512×512 at stage 1 or cropping a square sub-image whose size is 2.5 times the long side of lesion segmentation result produced by the first PDNet model at stage 2. To mimic the clicking behavior of a radiologist, a point is randomly selected from a region obtained by eroding the ellipse to half of its size.

Experimental Results: As a powerful segmentation framework, nnUNet [10] built based on UNets [15] has been successfully used in many medical image segmentation tasks, thus it can serve as a strong baseline in this task. We train an nnUNet model for each stage by taking as input the 3-channel image and using the same setting as PDNet. At stage 1, nnUNet produced poor segmentation performance, *e.g.*, the DICE score is about 0.857 on the DeepLesion test set, suggesting that the LOI extracted from it is not good enough to serve as the input of stage 2. Therefore, the 2nd nnUNet takes as input the LOIs extracted by the 1st PDNet, achieving a DICE of 0.911.

Figure 2 shows five visual examples of the results produced by nnUNet and PDNet. We can see that **1)** the 1st PDNet can segment the lesion region (Fig. 2(c)), even if they are small (the 4th row), heterogeneous (the 2nd row), or have blurry boundaries (the 5th row), irregular shapes (the 3rd row), etc. It indicates that the LOIs can be extracted reliably at stage 1 (Fig. 2(b)). **2)** The lesion segmentation results can be improved significantly by the 2nd PDNet (Fig. 2(e)), but a part of them become worse when using the 2nd nnUNet (*e.g.*, the 1st and 3rd rows in Fig. 2(d)). **3)** The RECIST diameters predicted by PDNet are much closer to the references than nnUNet. The qualitative results validate that the proposed framework can segment the lesions and predict their RECIST diameters reliably using only a click guidance. It may struggle when lesions have highly blurry boundaries or irregular shapes (rows 3 and 5), in which all methods will fail to segment them well.

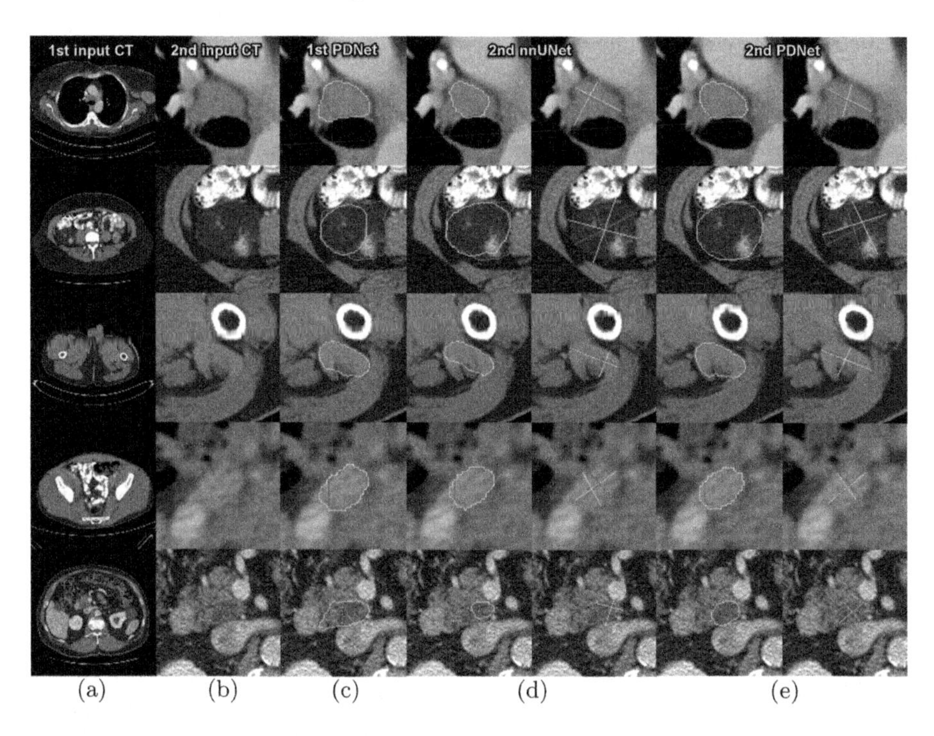

Fig. 2. Visual examples of results on the DeepLesion test set (the first three rows) and the external test set (the last two rows), where the pink and green curves/crosses are the manual annotations and automatic results. Given a CT image (a) and a click guidance (red spot), the 1st PDNet produces an initial lesion segmentation result (c) at stage 1, based on which a LOI (b) is extracted and taken as input of stage 2. The final results of lesion segmentation (left) and RECIST diameter prediction (right) are obtained by the 2nd nnUNet (d) and 2nd PDNet (e). Best viewed in color. (Color figure online)

Table 1. Results of lesion segmentation and RECIST diameter prediction on two test sets. The mean and standard deviation of all metrics are reported.

Method	Lesion segmentation			RECIST diameter prediction	
	Precision	Recall	Dice	Long axis	Short axis
DeepLesion test set					
Cai *et al.* [3]	0.893 ± 0.111	0.933 ± 0.095	0.906 ± 0.089	–	–
Tang *et al.* [21]	–	–	–	1.893 ± 2.185	1.614 ± 1.874
Tang *et al.* [23]	0.883 ± 0.057	$\mathbf{0.947 \pm 0.074}$	0.912 ± 0.039	1.747 ± 1.983	1.555 ± 1.808
nnUNet [10]	$\mathbf{0.977 \pm 0.033}$	0.852 ± 0.086	0.907 ± 0.050	2.108 ± 1.997	1.839 ± 1.733
PDNet	0.961 ± 0.044	0.898 ± 0.077	$\mathbf{0.924 \pm 0.045}$	$\mathbf{1.733 \pm 1.470}$	$\mathbf{1.524 \pm 1.374}$
External test set					
nnUNet [10]	$\mathbf{0.946 \pm 0.062}$	0.815 ± 0.099	0.870 ± 0.054	2.334 ± 1.906	1.985 ± 1.644
PDNet	0.927 ± 0.074	$\mathbf{0.857 \pm 0.093}$	$\mathbf{0.885 \pm 0.049}$	$\mathbf{2.174 \pm 1.437}$	$\mathbf{1.829 \pm 1.339}$

Table 2. Category-wise results in terms of segmentation Dice and the prediction error of diameter lengths on the external test set.

Method	Lung	Liver	Pancreas	Kidney	Lymph node
Lesion segmentation (Dice)					
nnUNet [10]	0.853 ± 0.054	0.876 ± 0.057	0.877 ± 0.055	0.890 ± 0.057	0.865 ± 0.050
PDNet	0.876 ± 0.046	0.893 ± 0.051	0.886 ± 0.050	0.911 ± 0.050	0.876 ± 0.045
RECIST diameter prediction (long axis)					
nnUNet [10]	2.396 ± 2.004	2.862 ± 2.090	2.655 ± 2.048	2.493 ± 1.963	1.958 ± 1.639
PDNet	2.435 ± 1.461	2.378 ± 1.463	2.220 ± 1.536	2.603 ± 1.533	1.897 ± 1.293
RECIST diameter prediction (short axis)					
nnUNet [10]	2.223 ± 1.404	2.383 ± 1.808	2.242 ± 1.637	2.342 ± 1.854	1.712 ± 1.440
PDNet	2.243 ± 1.333	2.168 ± 1.405	1.977 ± 1.359	2.362 ± 1.488	1.486 ± 1.174

Table 3. Results of different settings of our method in terms of Dice and the prediction error of diameter lengths on the DeepLesion test set.

Settings					Stage 1	Stage 2		
Base model	PE	T2D	B2U	SA	Dice	Dice	Long axis	Short axis
					0.871 ± 0.123	0.909 ± 0.068	1.961 ± 2.278	1.704 ± 1.948
	✓				0.890 ± 0.089	0.915 ± 0.055	1.861 ± 1.934	1.617 ± 1.684
	✓	✓			0.900 ± 0.067	0.919 ± 0.054	1.809 ± 1.731	1.577 ± 1.508
	✓	✓	✓		0.905 ± 0.070	0.921 ± 0.050	1.758 ± 1.696	1.544 ± 1.470
	✓	✓	✓	✓	0.911 ± 0.060	0.924 ± 0.045	1.733 ± 1.470	1.524 ± 1.374

Table 1 lists the quantitative results of different methods on two test sets. It can be seen that **1)** compared to the best previous work [23], PDNet boosts the Dice score by a large margin of 1.2% (from 0.912 to 0.924), and also gets smaller diameter errors on the DeepLesion test set. It means that PDNet can simultaneously segment the lesions accurately and produce reliable RECIST diameters close to the radiologists' manual annotations. **2)** Compared to the strong baseline nnUNet, PDNet gets much better results on both test sets. This is because PDNet is able to extract more comprehensive multi-scale features to better represent the appearances of different kinds of lesions. **3)** Compared to the DeepLesion test set, the performance drops for both nnUNet and PDNet on the external test set, *e.g.*, the Dice score of PDNet decreases from 0.924 to 0.885. In the external test set, some lesion masks are not well annotated, thus the generated ground-truth RECIST diameters will also be affected. The 4th row of Fig. 2 shows an unsatisfactory annotation, where the manual annotation is larger than the lesion's actual size. Meanwhile, the segmentation results produced by PDNet are better aligned to the lesion boundaries visually.

Table 2 lists the category-wise results on the external test set. PDNet achieves better performance in terms of all metrics and categories except the RECIST diameter prediction on lung and kidney tumors. After investigation, a potential reason is that a part of lung and kidney tumors have highly irregular shapes, whose diameters generated from manual masks are very likely to be larger, and nnUNet tends to predict larger diameters than PDNet in these cases. These results evidently demonstrate the effectiveness and robustness of our method.

Ablation Studies: To investigate the contributions of PDNet's components, *i.e.*, prior encoder (PE), top-down connection (T2D), bottom-up connection (B2U), and scale-aware attention module (SA), we configure different models by sequentially adding them into the base model that includes the image encoder with input of the 3-channel image and a UNet-style decoder. Table 3 outlines their quantitative comparisons. As can be seen, **1)** each added component improves the performance at both stages, demonstrating that the proposed strategies contribute to learning more comprehensive features for our tasks. **2)** The largest improvement gain is brought by introducing PE, especially for stage 1, demonstrating that PE can effectively explore the click prior information to learn lesion-specific attention matrices which heavily enhances the extracted multi-scale features for performance improvement.

4 Conclusions

This paper proposes a novel deep neural network architecture, prior-guided dual-path network (PDNet), for accurate lesion segmentation and RECIST diameter prediction. It works in a two-stage manner. Providing very simple human guide information, an LOI can be extracted precisely by segmentation at stage 1 and its segmentation and RECIST diameters can be predicted accurately at stage 2. As such, it offers a useful tool for radiologists to get reliable lesion size measurements (segmentation and RECIST diameters) with greatly reduced time and labor. It can potentially provide high positive clinical values.

References

1. Agarwal, V., Tang, Y., Xiao, J., Summers, R.M.: Weakly-supervised lesion segmentation on CT scans using co-segmentation. In: Medical Imaging: Computer-Aided Diagnosis, vol. 11314, p. 113141J (2020)
2. Armato III, S.G., et al.: The lung image database consortium (LIDC) and image database resource initiative (IDRI): a completed reference database of lung nodules on CT scans. Med. Phys. **38**(2), 915–931 (2011)
3. Cai, J., et al.: Accurate weakly-supervised deep lesion segmentation using large-scale clinical annotations: slice-propagated 3D mask generation from 2D RECIST. In: Frangi, A.F., Schnabel, J.A., Davatzikos, C., Alberola-López, C., Fichtinger, G. (eds.) MICCAI 2018. LNCS, vol. 11073, pp. 396–404. Springer, Cham (2018). https://doi.org/10.1007/978-3-030-00937-3_46
4. Chen, L.C., Papandreou, G., Kokkinos, I., Murphy, K., Yuille, A.L.: DeepLab: semantic image segmentation with deep convolutional nets, atrous convolution, and fully connected CRFs. IEEE Trans. Pattern Anal. Mach. Intell. **40**(4), 834–848 (2018)
5. Deng, J., Dong, W., Socher, R., Li, L., Li, K., Li, F.: ImageNet: a large-scale hierarchical image database. In: CVPR, pp. 248–255 (2009)

6. Eisenhauer, E.A., et al.: New response evaluation criteria in solid tumours: revised RECIST guideline (version 1.1). Eur. J. Cancer **45**(2), 228–247 (2009)
7. Fu, J., et al.: Dual attention network for scene segmentation. In: CVPR, pp. 3146–3154 (2019)
8. He, K., Zhang, X., Ren, S., Sun, J.: Deep residual learning for image recognition. In: CVPR, pp. 770–778 (2016)
9. Heller, N., et al.: The KiTS19 challenge data: 300 kidney tumor cases with clinical context, CT semantic segmentations, and surgical outcomes. arXiv preprint arXiv:1904.00445 (2019)
10. Isensee, F., et al.: nnU-Net: self-adapting framework for U-Net-based medical image segmentation. arXiv preprint arXiv:1809.10486 (2018)
11. Kingma, D.P., Ba, J.: Adam: a method for stochastic optimization. arXiv preprint arXiv:1412.6980 (2014)
12. Li, X., Chen, H., Qi, X., Dou, Q., Fu, C.W., Heng, P.A.: H-DenseUNet: hybrid densely connected UNet for liver and tumor segmentation from CT volumes. IEEE Trans. Med. Imaging **37**(12), 2663–2674 (2018)
13. Marquez-Neila, P., Baumela, L., Alvarez, L.: A morphological approach to curvature-based evolution of curves and surfaces. IEEE Trans. Pattern Anal. Mach. Intell. **36**(1), 2–17 (2013)
14. Rahman, M.A., Wang, Y.: Optimizing intersection-over-union in deep neural networks for image segmentation. In: Bebis, G., et al. (eds.) ISVC 2016. LNCS, vol. 10072, pp. 234–244. Springer, Cham (2016). https://doi.org/10.1007/978-3-319-50835-1_22
15. Ronneberger, O., Fischer, P., Brox, T.: U-Net: convolutional networks for biomedical image segmentation. In: Navab, N., Hornegger, J., Wells, W.M., Frangi, A.F. (eds.) MICCAI 2015. LNCS, vol. 9351, pp. 234–241. Springer, Cham (2015). https://doi.org/10.1007/978-3-319-24574-4_28
16. Roth, H.R., et al.: A new 2.5D representation for lymph node detection using random sets of deep convolutional neural network observations. In: Golland, P., Hata, N., Barillot, C., Hornegger, J., Howe, R. (eds.) MICCAI 2014. LNCS, vol. 8673, pp. 520–527. Springer, Cham (2014). https://doi.org/10.1007/978-3-319-10404-1_65
17. Simpson, A.L., et al.: A large annotated medical image dataset for the development and evaluation of segmentation algorithms. arXiv preprint arXiv:1902.09063 (2019)
18. Tang, Y.B., Oh, S., Tang, Y.X., Xiao, J., Summers, R.M.: CT-realistic data augmentation using generative adversarial network for robust lymph node segmentation. In: SPIE Medical Imaging, vol. 10950, p. 109503V (2019)
19. Tang, Y.B., Yan, K., Tang, Y.X., Liu, J., Xiao, J., Summers, R.M.: ULDor: a universal lesion detector for CT scans with pseudo masks and hard negative example mining. In: ISBI, pp. 833–836 (2019)
20. Tang, Y., et al.: Weakly-supervised universal lesion segmentation with regional level set loss. arXiv preprint arXiv:2105.01218 (2021)
21. Tang, Y., Harrison, A.P., Bagheri, M., Xiao, J., Summers, R.M.: Semi-automatic RECIST labeling on CT scans with cascaded convolutional neural networks. In: Frangi, A.F., Schnabel, J.A., Davatzikos, C., Alberola-López, C., Fichtinger, G. (eds.) MICCAI 2018. LNCS, vol. 11073, pp. 405–413. Springer, Cham (2018). https://doi.org/10.1007/978-3-030-00937-3_47
22. Tang, Y., Tang, Y., Zhu, Y., Xiao, J., Summers, R.M.: E^{2}Net: an edge enhanced network for accurate liver and tumor segmentation on CT scans. In: Martel, A.L., et al. (eds.) MICCAI 2020. LNCS, vol. 12264, pp. 512–522. Springer, Cham (2020). https://doi.org/10.1007/978-3-030-59719-1_50

23. Tang, Y., Yan, K., Xiao, J., Summers, R.M.: One click lesion RECIST measurement and segmentation on CT scans. In: Martel, A.L., et al. (eds.) MICCAI 2020. LNCS, vol. 12264, pp. 573–583. Springer, Cham (2020). https://doi.org/10.1007/978-3-030-59719-1_56
24. Wang, S., et al.: Central focused convolutional neural networks: developing a data-driven model for lung nodule segmentation. Med. Image Anal. **40**, 172–183 (2017)
25. Yan, K., Wang, X., Lu, L., Summers, R.M.: DeepLesion: automated mining of large-scale lesion annotations and universal lesion detection with deep learning. J. Med. Imaging **5**(3), 036501 (2018)
26. Zhang, H., et al.: ResNeSt: split-attention networks. arXiv preprint arXiv:2004.08955 (2020)
27. Zhu, Z., et al.: Lymph node gross tumor volume detection and segmentation via distance-based gating using 3D CT/PET imaging in radiotherapy. In: Martel, A.L., et al. (eds.) MICCAI 2020. LNCS, vol. 12267, pp. 753–762. Springer, Cham (2020). https://doi.org/10.1007/978-3-030-59728-3_73

Reciprocal Learning for Semi-supervised Segmentation

Xiangyun Zeng[1,2,3], Rian Huang[1,2,3], Yuming Zhong[1,2,3], Dong Sun[1,2,3], Chu Han[4], Di Lin[5], Dong Ni[1,2,3], and Yi Wang[1,2,3](✉)

[1] National-Regional Key Technology Engineering Laboratory for Medical Ultrasound, Guangdong Provincial Key Laboratory of Biomedical Measurements and Ultrasound Imaging, School of Biomedical Engineering, Health Science Center, Shenzhen University, Shenzhen, China

[2] Medical UltraSound Image Computing (MUSIC) Lab, Shenzhen University, Shenzhen, China

[3] Marshall Laboratory of Biomedical Engineering, Shenzhen University, Shenzhen, China

onewang@szu.edu.cn

[4] Department of Radiology, Guangdong Provincial People's Hospital, Guangdong Academy of Medical Sciences, Guangzhou, Guangdong, China

[5] College of Intelligence and Computing, Tianjin University, Tianjin, China

Abstract. Semi-supervised learning has been recently employed to solve problems from medical image segmentation due to challenges in acquiring sufficient manual annotations, which is an important prerequisite for building high-performance deep learning methods. Since unlabeled data is generally abundant, most existing semi-supervised approaches focus on how to make full use of both limited labeled data and abundant unlabeled data. In this paper, we propose a novel semi-supervised strategy called reciprocal learning for medical image segmentation, which can be easily integrated into any CNN architecture. Concretely, the reciprocal learning works by having a pair of networks, one as a student and one as a teacher. The student model learns from pseudo label generated by the teacher. Furthermore, the teacher updates its parameters autonomously according to the reciprocal feedback signal of how well student performs on the labeled set. Extensive experiments on two public datasets show that our method outperforms current state-of-the-art semi-supervised segmentation methods, demonstrating the potential of our strategy for the challenging semi-supervised problems. *The code is publicly available at* https://github.com/XYZach/RLSSS.

Keywords: Semi-supervised learning · Reciprocal learning · Segmentation · Meta learning · Deep learning

X. Zeng, R. Huang and Y. Zhong—Contribute equally to this work.

M. de Bruijne et al. (Eds.): MICCAI 2021, LNCS 12902, pp. 352–361, 2021.
https://doi.org/10.1007/978-3-030-87196-3_33

1 Introduction

Accurate and robust segmentation of organs or lesions from medical images is of great importance for many clinical applications such as disease diagnosis and treatment planning. With a large amount of labeled data, deep learning has achieved great success in automatic image segmentation [7,10]. In medical imaging domain, especially for volumetric images, reliable annotations are difficult to obtain as expert knowledge and time are both required. Unlabeled data, on the other hand, are easier to acquire. Therefore, semi-supervised approaches with unlabeled data occupying a large portion of the training set are worth exploring.

Bai et al. [1] introduced a self-training-based method for cardiac MR image segmentation, in which the segmentation prediction for unlabeled data and the network parameters were alternatively updated. Xia et al. [14] utilized co-training for pancreas and liver tumor segmentation tasks by exploiting multi-viewpoint consistency of 3D data. These methods enlisted more available training sources by creating pseudo labels, however, they did not consider the reliability of the pseudo labels which may leads to meaningless guidance. Some approaches to semi-supervised learning were inspired by the success of self-ensembling method. For example, Li et al. [5] embedded the transformation consistency into Π-model [3] to enhance the regularization for pixel-wise predictions. Yu et al. [16] designed an uncertainty-aware mean teacher framework, which can generate more reliable predictions for student to learn. To exploit the structural information for prediction, Hang et al. [2] proposed a local and global structure-aware entropy regularized mean teacher for left atrium segmentation. In general, most teacher-student methods update teacher's parameters using exponential moving average (EMA), which is an useful ensemble strategy. However, the EMA focuses on weighting the student's parameters at each stage during training process, without evaluating the quality of parameters explicitly. It is more expected that the teacher model could purposefully update the parameters through a parameter evaluation strategy, so as to generate more reliable pseudo-labels.

In this paper, we design a novel strategy named reciprocal learning for semi-supervised segmentation. Specifically, we make better use of the limited labeled data by using reciprocal learning strategy so that the teacher model can update its parameters with gradient descent algorithm and generate more reliable annotations for unlabeled set as the number of reciprocal learning step increases. We evaluate our approach on the pancreas CT dataset and the Atrial Segmentation Challenge dataset with extensive comparisons to existing methods. The results demonstrate that our segmentation network consistently outperforms the state-of-the-art method in respect to the evaluation metrics of Dice Similarity (Dice), Jaccard Index (Jaccard), 95% Hausdorff Distance (95 HD) and Average Symmetric Surface Distance (ASD). Our main contributions are three folds:

- We present a simple yet efficient reciprocal learning strategy for segmentation to reduce the labeling efforts. Inspired by the idea from *learning to learn*, we design a feedback mechanism for teacher network to generate more reliable pseudo labels by observing how pseudo labels would affect the student. In our implementation, the feedback signal is the performance of the student on the labeled set. By reciprocal learning strategy, the teacher can update its parameters autonomously.

- The proposed reciprocal learning strategy can be utilized directly in any CNN architecture. Specifically, any segmentation network can be used as the backbone, which means there are still opportunities for further enhancements.
- Experiments on two public datasets show our proposed strategy can further raise semi-supervised segmentation quality compared with existing methods.

2 Methods

Figure 1 illustrates our reciprocal learning framework for semi-supervised segmentation. We deploy a meta-learning concept for teacher model to generate better pseudo labels by observing how pseudo labels would affect the student. Specifically, the teacher and student are trained in parallel: the student learns from pseudo labels generated by the teacher, and the teacher learns from the feedback signal of how well the student performs on the labeled set.

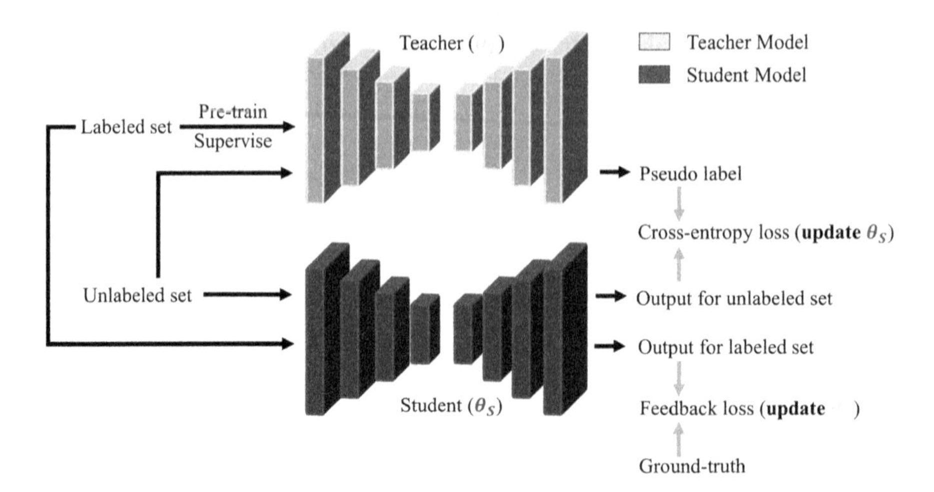

Fig. 1. The schematic illustration of our reciprocal learning framework for semi-supervised segmentation. In this paper, V-Net is used as the backbone. Again, we emphasize any segmentation network could be used as the backbone in our framework.

2.1 Notations

We denote the labeled set as (x_l, y_l) and the unlabeled set as x_u, where x is the input volume and y is the ground-truth segmentation. Let T and S respectively be the teacher model and the student model, and let their corresponding parameters be θ_T and θ_S. We denote the soft predictions of teacher network on the x_u as $T(x_u; \theta_T)$ and likewise for the student.

2.2 Reciprocal Learning Strategy

Figure 1 shows the workflow of our proposed reciprocal learning strategy. Firstly, the teacher model should be well pre-trained on labeled set (x_l, y_l) in a supervised manner. We use cross-entropy loss (CE) as loss function:

$$\mathcal{L}_{pre-train} = CE(y_l, T(x_l; \theta_T)). \tag{1}$$

Then we use the teacher's prediction on unlabeled set as pseudo labels $\widehat{y}_u$ to train the student model. Specifically, Pseudo Labels (PL) trains the student model to minimize the cross entropy loss on unlabeled set x_u:

$$\widehat{y}_u \sim T(x_u; \theta_T), \tag{2}$$

$$\theta_S^{\text{PL}} = \arg\min_{\theta_S} CE(\widehat{y}_u, S(x_u; \theta_S)). \tag{3}$$

After the student model updated, it's expected to perform well on the labeled set and achieve a low cross-entropy loss, i.e. $CE(y_l, S(x_l; \theta_S^{\text{PL}}))$. Notice that the optimal student parameters θ_S^{PL} always depend on the teacher parameters θ_T via the pseudo labels (see Eq. (2) and (3)). Therefore, we express the dependency as $\theta_S^{\text{PL}}(\theta_T)$ and further optimize $\mathcal{L}_{feedback}$ with respect to θ_T:

$$\min_{\theta_T} \quad \mathcal{L}_{feedback}(\theta_S^{\text{PL}}(\theta_T)) = CE(y_l, S(x_l; \theta_S^{\text{PL}}(\theta_T))). \tag{4}$$

For each reciprocal learning step (including one update for the student using Eq. (3) and one update for the teacher using Eq. (4) respectively), however, solving Eq. (3) to optimize θ_S until complete convergence is inefficient, as computing the gradient $\nabla_{\theta_T}\mathcal{L}_{feedback}(\theta_S^{\text{PL}}(\theta_T))$ requires unrolling the entire student training process. Instead, a meta-learning approach [6] is utilized to approximate θ_S^{PL} with one-step gradient update of θ_S:

$$\theta_S^{\text{PL}} \approx \theta_S - \eta_S \nabla_{\theta_S} CE(\widehat{y}_u, S(x_u; \theta_S)), \tag{5}$$

where η_S is the learning rate. In this way, the student model and the teacher model have an alternating optimization:

(1) Draw a batch of unlabeled set x_u, then sample $T(x_u; \theta_T)$ from the teacher model, and optimize with stochastic gradient descent (SGD):

$$\theta_S' = \theta_S - \eta_S \nabla_{\theta_S} CE(\widehat{y}_u, S(x_u; \theta_S)). \tag{6}$$

(2) Draw a batch of labeled set (x_l, y_l), and reuse the student's update to optimize with SGD:

$$\theta_T' = \theta_T - \eta_T \nabla_{\theta_T} \mathcal{L}_{feedback}(\theta_S'). \tag{7}$$

Optimize θ_S with Eq. (6) can be simply computed via back-propagation. We now present the derivation for optimizing θ_T. Firstly, by the chain rule, we have

$$\begin{aligned} \frac{\partial \mathcal{L}_{feedback}(\theta_S')}{\partial \theta_T} &= \frac{\partial CE(y_l, S(x_l; \theta_S'))}{\partial \theta_T} \\ &= \left.\frac{\partial \text{CE}\left(y_l, S\left(x_l; \theta_S'\right)\right)}{\partial \theta_S}\right|_{\theta_S = \theta_S'} \cdot \frac{\partial \theta_S'}{\partial \theta_T} \end{aligned} \tag{8}$$

We focus on the second term in Eq. (8)

$$\begin{aligned}\frac{\partial \theta_S'}{\partial \theta_T} &= \frac{\partial}{\partial \theta_T}[\theta_S - \eta_S \nabla_{\theta_S} CE(\widehat{y}_u, S(x_u; \theta_S))] \\ &= \frac{\partial}{\partial \theta_T}\left[-\eta_S \cdot \left(\frac{\partial CE(\widehat{y}_u, S(x_u; \theta_S))}{\partial \theta_S}\bigg|_{\theta_S=\theta_S} \right)^{\top}\right]\end{aligned} \tag{9}$$

To simplify notations, we define *the gradient*

$$g_S(\widehat{y}_u) = \left(\frac{\partial CE(\widehat{y}_u, S(x_u; \theta_S))}{\partial \theta_S}\bigg|_{\theta_S=\theta_S} \right)^{\top} \tag{10}$$

Since $g_S(\widehat{y}_u)$ has dependency on θ_T via $\widehat{y}_u$, we apply the REINFORCE equation [13] to achieve

$$\begin{aligned}\frac{\partial \theta_S'}{\partial \theta_T} &= -\eta_S \cdot \frac{\partial g_S(\widehat{y}_u)}{\partial \theta_T} \\ &= -\eta_S \cdot g_S(\widehat{y}_u) \cdot \frac{\partial \log P\left(\widehat{y}_u | x_u; \theta_T\right)}{\partial \theta_T} \\ &= \eta_S \cdot g_S(\widehat{y}_u) \cdot \frac{\partial CE(\widehat{y}_u, T(x_u; \theta_T))}{\partial \theta_T}\end{aligned} \tag{11}$$

Finally, we obtain the gradient

$$\begin{aligned}\nabla_{\theta_T} \mathcal{L}_{feedback}(\theta_S') = & \eta_S \cdot \left(\nabla_{\theta_S'} CE(y_l, S(x_l; \theta_S'))\right)^{\top} \cdot \\ & \nabla_{\theta_S} CE(\widehat{y}_u, S(x_u; \theta_S)) \cdot \nabla_{\theta_T} CE(\widehat{y}_u, T(x_u; \theta_T)).\end{aligned} \tag{12}$$

However, it might lead to overfitting if we rely solely on the student's performance to optimize the teacher model. To overcome this, we leverage labeled set to supervise teacher model throughout the course of training. Therefore, the ultimate optimal equation of the teacher model can be summarized as: $\theta_T' = \theta_T - \eta_T \nabla_{\theta_T}[\mathcal{L}_{feedback}(\theta_S') + \lambda CE(y_l, T(x_l; \theta_T))]$, where λ is the weight to balance the importance of different losses.

3 Experiments

3.1 Materials and Pre-processing

To demonstrate the effectiveness of our proposed method, experiments were carried on two different public datasets.

The first dataset is the pancreas dataset [11] obtained using Philips and Siemens MDCT scanners. It includes 82 abdominal contrast enhanced CT scans, which have resolutions of 512×512 pixels with varying pixel sizes and slice thickness between 1.5–2.5 mm. We used the soft tissue CT window range of [−125, 275] HU, and cropped the images centering at pancreas regions based on the ground truth with enlarged margins (25 voxels)[1] after normalizing them as zero mean and unit variance. We used 62 scans for training and 20 scans for validation.

The second dataset is the left atrium dataset [15]. It includes 100 gadolinium-enhanced MR images, which have a resolution of $0.625 \times 0.625 \times 0.625\,\text{mm}^3$. We cropped centering at heart regions and normalized them as zero mean and unit variance. We used 80 scans for training and 20 scans for validation.

In this work, we report the performance of all methods trained with 20% labeled images and 80% unlabeled images as the typical semi-supervised learning experimental setting.

3.2 Implementation Details

Our proposed method was implemented with the popular library Pytorch, using a TITAN Xp GPU. In this work, we employed V-Net [9] as the backbone. More importantly, it's flexible that any segmentation network can be the backbone. We set $\lambda = 1$. Both the teacher model and the student model share the same architecture but have independent weights. Both networks were trained by the stochastic gradient descent (SGD) optimizer for 6000 iterations, with an initial learning rate $\eta_T = \eta_S = 0.01$, decayed by 0.1 every 2500 iterations. To tackle the issues of limited data samples and demanding 3D computations cost, we randomly cropped $96 \times 96 \times 96$ (pancreas dataset) and $112 \times 112 \times 80$ (left atrium dataset) sub-volumes as the network input and adopted data augmentation for training. In the inference phase, we only utilized the student model to predict the segmentation for the input volume and we used a sliding window strategy to obtain the final results, with a stride of $10 \times 10 \times 10$ for the pancreas dataset and $18 \times 18 \times 4$ for the left atrium dataset.

3.3 Segmentation Performance

We compared results of our method with several state-of-the-art semi-supervised segmentation methods, including mean teacher self-ensembling model (MT) [12], uncertainty-aware mean teacher model (UA-MT) [16], shape-aware adversarial network (SASSNet) [4], uncertainty-aware multi-view co-training (UMCT) [14] and transformation-consistent self-ensembling model (TCSM) [5]. Note that we used the official code of MT, UA-MT, SASSNet, TCSM and reimplemented the UMCT which didn't release the official code. For a fair comparison, we obtained the results of our competitors by using the same backbone (V-Net) and retraining their networks to obtain the best segmentation results on the Pancreas dataset and the Left Atrium dataset.

[1] This study mainly focused on the challenging problem of semi-supervised learning for insufficient annotations. Several semi-supervised segmentation studies used cropped images for validations, e.g., UAMT [16] used cropped left atrium images, and [8] used cropped pancreas images. We followed their experimental settings.

The metrics employed to quantitatively evaluate segmentation include Dice, Jaccard, 95 HD and ASD. A better segmentation shall have larger values of Dice and Jaccard, and smaller values of other metrics.

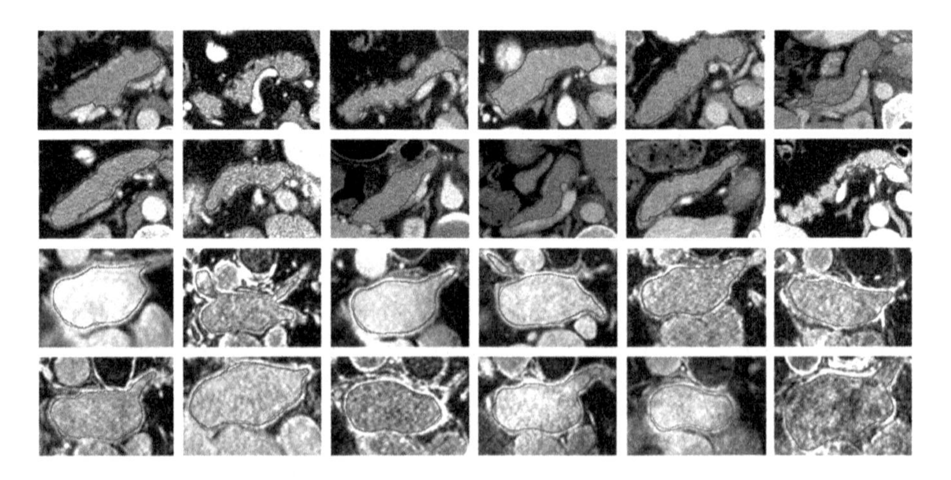

Fig. 2. 2D visualization of our proposed semi-supervised segmentation method under 20% labeled images. The first two rows are the segmentation results of pancreas and the last two rows are the segmentation results of left atrium. Red and blue colors show the ground truths and the predictions, respectively. (Color figure online)

We first evaluated our proposed method on pancreas dataset. The first two rows of Fig. 2 visualize 12 slices of the pancreas segmentation results. Apparently, our method consistently obtained similar segmented boundaries to the ground truths. Table 1 presents the quantitative comparison of several state-of-the-art semi-supervised segmentation methods. Compared with using only 20% annotated images (the first row), all semi-supervised segmentation methods achieved greater performance proving that they could both utilize unlabeled images. Notably, our method improved the segmentation by 9.76% Dice and 12.00% Jaccard compared with the fully supervised baseline's results. Furthermore, our method achieved the best performance over the state-of-the-art semi-supervised methods on all metrics. Compared with other methods, our proposed method utilized the limited labeled data in a better way by using reciprocal learning strategy so that the teacher model could update its parameters autonomously and generate more reliable annotations for unlabeled data as the number of reciprocal learning step increases. The first two rows of Fig. 3 visualize the pancreas segmentation results of different semi-supervised segmentation methods in 3D. Compared with other methods, our method produced less false positive predictions especially in the case as shown in the first row in Fig. 3.

We also evaluated our method on the left atrium dataset, which is a widely-used dataset for semi-supervised segmentation. The last two rows of Fig. 2 visualize 12 segmented slices. Obviously, our results can successfully infer the ambigu-

Table 1. Quantitative comparison between our method and other semi-supervised methods on the pancreas CT dataset.

Method	# Scans used		Metrics			
	Labeled	Unlabeled	Dice [%]	Jaccard [%]	ASD [voxel]	95HD [voxel]
V-Net	12	0	72.16	57.67	2.80	12.34
V-Net	62	0	84.89	73.89	1.22	4.44
MT	12	50	79.14	66.04	2.21	8.71
SASS	12	50	79.55	66.32	2.29	8.53
UMCT	12	50	79.74	66.59	4.13	14.01
UAMT	12	50	80.04	67.52	2.99	10.96
TCSM	12	50	78.17	64.95	5.06	17.52
Ours	12	50	**81.92**	**69.67**	**1.82**	**6.36**

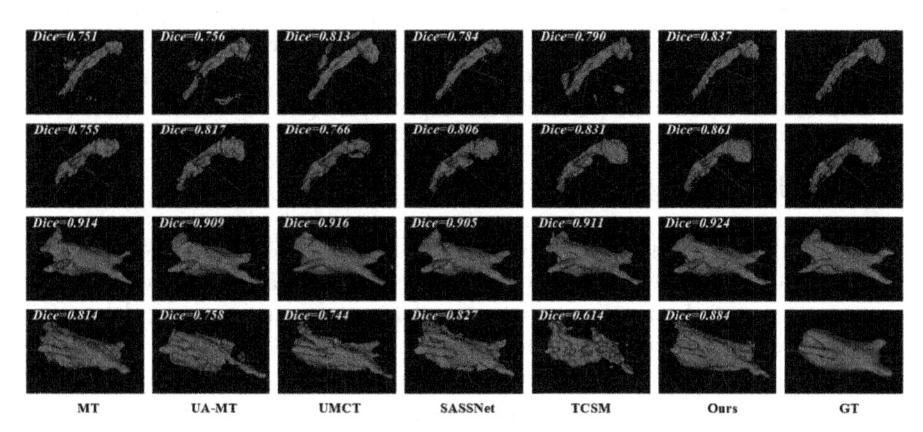

Fig. 3. Four cases of 3D visualization of different semi-supervised segmentation methods under 20% labeled images. The first two rows are the results of pancreas segmentation and the last two rows are the results of left atrium segmentation.

Table 2. Quantitative comparison between our method and other semi-supervised methods on the Left Atrium MRI dataset.

Method	# scans used		Metrics			
	Labeled	Unlabeled	Dice [%]	Jaccard [%]	ASD [voxel]	95HD [voxel]
V-Net	16	0	84.41	73.54	5.32	19.94
V-Net	80	0	91.42	84.27	1.50	5.15
MT	16	64	88.12	79.03	2.65	10.92
SASS	16	64	89.27	80.82	3.13	8.83
UMCT	16	64	89.36	81.01	2.60	7.25
UAMT	16	64	88.88	80.21	2.26	7.32
TCSM	16	64	86.26	76.56	2.35	9.67
Ours	16	64	**90.06**	**82.01**	**2.13**	**6.70**

ous boundaries and have a high overlap ratio with the ground truths. A quantitative comparison is shown in Table 2. Compared with using only 20% labeled images (the first row), our method improved the segmentation by 5.65% Dice and 8.47% Jaccard, which were very close to using 100% labeled images (the second row). In addition, it can be observed that our method achieved the best performance than the state-of-the-art semi-supervised methods on all evaluation metrics, corroborating that our reciprocal learning strategy has the fully capability to utilize the limited labeled data. The last two rows of Fig. 3 visualize the left atrium segmentation results of different semi-supervised segmentation methods in 3D. Compared with other methods, our results were close to the ground truths and preserved more details and produced less false positives, which demonstrates the efficacy of our proposed reciprocal learning strategy.

We further conducted an ablation study to demonstrate the efficacy of the proposed reciprocal learning strategy. Specifically, we discarded our reciprocal learning strategy by fixing teacher model after it was well pretrained. The results degraded to 73.82%/86.82% Dice, 59.38%/77.27% Jaccard, 4.62/3.69 ASD and 17.78/12.29 95HD on pancreas/left atrium datasets, which shows our reciprocal learning contributes to the performance improvement.

4 Conclusion

This paper develops a novel reciprocal learning strategy for semi-supervised segmentation. Our key idea is to fully utilize the limited labeled data by updating parameters of the teacher and the student model in a reciprocal learning way. Meanwhile, our strategy is simple and can be used directly in existing state-of-the-art network architectures, where the performance can be effectively enhanced. Experiments on two public datasets demonstrate the effectiveness, robustness and generalization of our proposed method. In addition, our proposed reciprocal learning strategy is a general solution and has the potential to be used for other image segmentation tasks.

Acknowledgements. This work was supported in part by the National Key R&D Program of China (No. 2019YFC0118300), in part by the National Natural Science Foundation of China under Grants 62071305, 61701312 and 81971631, in part by the Guangdong Basic and Applied Basic Research Foundation (2019A1515010847), in part by the Medical Science and Technology Foundation of Guangdong Province (B2019046), in part by the Natural Science Foundation of Shenzhen University (No. 860-000002110129), and in part by the Shenzhen Peacock Plan (No. KQTD2016053112051497).

References

1. Bai, W., et al.: Semi-supervised learning for network-based cardiac MR image segmentation. In: Descoteaux, M., Maier-Hein, L., Franz, A., Jannin, P., Collins, D.L., Duchesne, S. (eds.) MICCAI 2017. LNCS, vol. 10434, pp. 253–260. Springer, Cham (2017). https://doi.org/10.1007/978-3-319-66185-8_29

2. Hang, W., et al.: Local and global structure-aware entropy regularized mean teacher model for 3D left atrium segmentation. In: Martel, A.L., et al. (eds.) MICCAI 2020. LNCS, vol. 12261, pp. 562–571. Springer, Cham (2020). https://doi.org/10.1007/978-3-030-59710-8_55
3. Laine, S., Aila, T.: Temporal ensembling for semi-supervised learning. arXiv preprint arXiv:1610.02242 (2016)
4. Li, S., Zhang, C., He, X.: Shape-aware semi-supervised 3D semantic segmentation for medical images. In: Martel, A.L., et al. (eds.) MICCAI 2020. LNCS, vol. 12261, pp. 552–561. Springer, Cham (2020). https://doi.org/10.1007/978-3-030-59710-8_54
5. Li, X., Yu, L., Chen, H., Fu, C.W., Xing, L., Heng, P.A.: Transformation-consistent self-ensembling model for semisupervised medical image segmentation. TNNLS **32**(2), 523–534 (2020)
6. Liu, H., Simonyan, K., Yang, Y.: DARTS: differentiable architecture search. In: ICLR (2018)
7. Long, J., Shelhamer, E., Darrell, T.: Fully convolutional networks for semantic segmentation. In: CVPR, pp. 3431–3440 (2015)
8. Luo, X., Chen, J., Song, T., Wang, G.: Semi-supervised medical image segmentation through dual-task consistency. In: AAAI Conference on Artificial Intelligence (2021)
9. Milletari, F., Navab, N., Ahmadi, S.A.: V-Net: fully convolutional neural networks for volumetric medical image segmentation. In: 3DV, pp. 565–571. IEEE (2016)
10. Ronneberger, O., Fischer, P., Brox, T.: U-Net: convolutional networks for biomedical image segmentation. In: Navab, N., Hornegger, J., Wells, W.M., Frangi, A.F. (eds.) MICCAI 2015. LNCS, vol. 9351, pp. 234–241. Springer, Cham (2015). https://doi.org/10.1007/978-3-319-24574-4_28
11. Roth, H.R., et al.: DeepOrgan: multi-level deep convolutional networks for automated pancreas segmentation. In: Navab, N., Hornegger, J., Wells, W.M., Frangi, A.F. (eds.) MICCAI 2015. LNCS, vol. 9349, pp. 556–564. Springer, Cham (2015). https://doi.org/10.1007/978-3-319-24553-9_68
12. Tarvainen, A., Valpola, H.: Mean teachers are better role models: weight-averaged consistency targets improve semi-supervised deep learning results. In: NIPS, pp. 1195–1204 (2017)
13. Williams, R.J.: Simple statistical gradient-following algorithms for connectionist reinforcement learning. Mach. Learn. **8**(3–4), 229–256 (1992). https://doi.org/10.1007/BF00992696
14. Xia, Y., et al.: 3D semi-supervised learning with uncertainty-aware multi-view co-training. In: WACV, pp. 3646–3655 (2020)
15. Xiong, Z., et al.: A global benchmark of algorithms for segmenting the left atrium from late gadolinium-enhanced cardiac magnetic resonance imaging. Med. Image Anal. **67**, 101832 (2021)
16. Yu, L., Wang, S., Li, X., Fu, C.-W., Heng, P.-A.: Uncertainty-aware self-ensembling model for semi-supervised 3D left atrium segmentation. In: Shen, D., et al. (eds.) MICCAI 2019. LNCS, vol. 11765, pp. 605–613. Springer, Cham (2019). https://doi.org/10.1007/978-3-030-32245-8_67

Disentangled Sequential Graph Autoencoder for Preclinical Alzheimer's Disease Characterizations from ADNI Study

Fan Yang[1(✉)], Rui Meng[2], Hyuna Cho[3], Guorong Wu[4], and Won Hwa Kim[1,3(✉)]

[1] University of Texas at Arlington, Arlington, USA
[2] Lawrence Berkeley National Laboratory, Berkeley, USA
[3] Pohang University of Science and Technology, Pohang, South Korea
wonhwa@postech.ac.kr
[4] University of North Carolina, Chapel Hill, Chapel Hill, USA

Abstract. Given a population longitudinal neuroimaging measurements defined on a brain network, exploiting temporal dependencies within the sequence of data and corresponding latent variables defined on the graph (i.e., network encoding relationships between regions of interest (ROI)) can highly benefit characterizing the brain. Here, it is important to distinguish time-variant (e.g., longitudinal measures) and time-invariant (e.g., gender) components to analyze them individually. For this, we propose an innovative and ground-breaking Disentangled Sequential Graph Autoencoder which leverages the Sequential Variational Autoencoder (SVAE), graph convolution and semi-supervising framework together to learn a latent space composed of time-variant and time-invariant latent variables to characterize disentangled representation of the measurements over the entire ROIs. Incorporating target information in the decoder with a supervised loss let us achieve more effective representation learning towards improved classification. We validate our proposed method on the longitudinal cortical thickness data from Alzheimer's Disease Neuroimaging Initiative (ADNI) study. Our method outperforms baselines with traditional techniques demonstrating benefits for effective longitudinal data representation for predicting labels and longitudinal data generation.

1 Introduction

Representation learning is at the core of Image Analysis. Lots of recent attentions are at a disentangled representation of data, as the individual disentangled representations are highly sensitive to a specific factor whereas indifferent to others [2,10,13,23,30]. A typical disentangling method would find a low-dimensional

F. Yang and R. Meng are joint first authors.

M. de Bruijne et al. (Eds.): MICCAI 2021, LNCS 12902, pp. 362–372, 2021.
https://doi.org/10.1007/978-3-030-87196-3_34

latent space for high-dimensional data whose individual latent dimensions correspond to independent disentangling factors. For longitudinal data, one can expect to decompose the longitudinal data into time-invariant factors and time-variant factors by obtaining the "disentangled" representation as longitudinal observations are affected by both time-variant and static variables [12,19,31]. In the context of neuroimaging studies, the disentangled representation would be able to separate time-independent concepts (e.g. anatomical information) from dynamical information (e.g. modality information) [25], which may offer effective ways of compression, conditional data generation, classification and others.

Recent advances in variational autoencoders (VAE) [16] have made it possible to learn various representations in an unsupervised manner for neuroimaging analysis [1,30]. Moreover, various vibrant of autoencoders are also proposed to model temporal data; for example, [12] introduced the factorised hierarchical variational auto-encoder (FHVAE) for unsupervised learning of disentangled representation of time series. Sequential variational autoencoder was proposed in [19] benefiting from the usage of the hierarchical prior. It disentangles latent factors by factorizing them into time-invariant and time-dependent parts and applies an LSTM sequential prior to keep a sequential consistency for sequence generation. [31] modeled the time-varying variables via LSTM in both encoder and decoder for dynamic consistency.

There are two major issues with current approaches. First, while these methods can deal with temporal nature of the data, they do not necessarily introduce supervision at all. Moreover, from a neuroscience perspective, the domain knowledge tells us that the regions of interest (ROIs) in the brain are highly associated to each other both functionally and structurally [7,17,18,20]. This association provides a prior knowledge on connection between the ROIs as a graph; for example, structural brain connectivity from tractography on Diffusion Tensor Imaging (DTI) provides a path for anisotropic variation and diffusion of structural changes in the brain such as atrophy of cortical thickness. Most of the existing methods do not consider this arbitrary topology of variables, if there is any, into account, which can provide significant benefit for downstream tasks. To summarize, learning with (either full or partial) supervision on longitudinal neuroimaging measurements on a brain network is still **under-explored**.

Given longitudinal observations (e.g., cortical thickness) on specific ROIs in the brain and a structural brain network characterized by bundles of neuron fiber tracts, our aim is to develop a framework to learn a latent disentangled representation of the observations that are composed of time-variant and time-invariant latent variables. For this, we propose an innovative Semi-supervised Sequential Graph Autoencoder model which leverages ideas from the sequential variational autoencoder (SVAE), graph convolution and semi-supervising framework. The core idea is to incorporate target information as a supervision in the decoder with a supervised loss, which let us achieve more effective representation for downstream tasks by balancing extraction of underlying structure as well as accurately predicting class labels.

Our proposed framework learns a latent disentangled representation composed of time-variant and time-invariant latent variables to characterize the longitudinal measurements over the entire structural brain network that consists of ROIs. Our **contributions** are as summarized follows: our model can 1) learn an ideal disentangled representation which separates time-independent content or anatomical information from dynamical or modality information and conditionally generate synthetic sequential data; 2) perform semi-supervised tasks which can jointly incorporate supervised and unsupervised data for classification tasks; 3) leverage graph structure to robustly learn the disentangling latent structure. Using our framework, we analyzed longitudinal cortical thickness measures on brain networks with diagnostic labels of Alzheimer's Disease (AD) from Alzheimer's Disease Neuroimaging Initiative (ADNI) study. As AD is a progressive neurodegenerative condition characterized by neurodegeneration in the brain caused by synthetic factors [6,14,22,27,28], it is important to effectively characterize early symptoms of the disease. We expect that disentangling ROI measures with time-variant and static components can provide unique insights.

2 Background

Our proposed framework involves two important concepts: 1) graph convolutions and 2) SVAE. Hence, we begin with brief reviews of their basics.

Graph Convolutions. Let $G = \{\mathbb{V}, \mathbb{E}, A\}$ be an undirected graph, where $\mathbb{V}$ is a set of nodes with $|\mathbb{V}| = n$, $\mathbb{E}$ is a set of edges and A is an adjacent matrix that specify connections between the nodes. Graph Fourier analysis relies on the spectral decomposition of graph Laplacian defined as $\mathcal{L} = D - A$, where D is a diagonal degree matrix with $D_{i,i} = \sum_j A_{i,j}$. The normalized Laplacian is defined as $L = I_n - D^{-1/2}AD^{-1/2}$, where I_n is the identity matrix. Since L is real and positive semi-definite, it has a complete set of orthonormal eigenvectors $U = (\boldsymbol{u}_1, \ldots, \boldsymbol{u}_n)$ with corresponding non-negative real eigenvalues $\{\lambda_l\}_{l=1}^n$. Eigenvectors associated with smaller eigenvalues carry slow varying signals, indicating that connected nodes share similar values. In contrast, eigenvectors associated with larger values carry faster varying signals across the connected nodes. We are interested in the smallest eigenvalues due to the negation used to compute the Laplacian matrix in terms of the Euclidean Commute Time Distance [26]. Let $\boldsymbol{x} \in \mathbb{R}^n$ be a signal defined on the vertices of the graph. The graph Fourier transform of $\boldsymbol{x}$ is defined as $\hat{\boldsymbol{x}} = U^T\boldsymbol{x}$, with inverse operation given by $\boldsymbol{x} = U\hat{\boldsymbol{x}}$. The graphical convolution operation between signal $\boldsymbol{x}$ and filter $\boldsymbol{g}$ is

$$\boldsymbol{g} * \boldsymbol{x} = U((U^T\boldsymbol{g}) \odot (U^T\boldsymbol{x})) = U\hat{G}U^T\boldsymbol{x}. \tag{1}$$

Here, $U^T\boldsymbol{g}$ is replaced by a filter $\hat{G} = \text{diag}(\boldsymbol{\theta})$ parameterized by $\boldsymbol{\theta} \in \theta^n$ in Fourier domain. Unfortunately, eigendecomposition of $\boldsymbol{L}$ and matrix multiplication with U are expensive. Motivated by the Chebyshev polynomials approximation in [7,9] introduced a Chebyshev polynomial paramterization for ChebyNet that offers fast localized spectral filtering. Later, [17] provided a simplified version of ChebyNet by considering a second order approximation such that $\boldsymbol{g} * \boldsymbol{x} \approx \theta(I_n + D^{-1/2}AD^{-1/2})\boldsymbol{x}$ and illustrate promising model performance in graph-based semi-supervising learning tasks, and GCN is deeply studied in [18]. Then,

FastGCN was proposed in [4] which approximates the original convolution layer by Monte Carlo sampling, and recently, [29] leveraged graph wavelet transform to address the shortcomings of spectral graph convolutional neural networks.

Sequential Variational Autoencoder. Variational autoencoder (VAE), initially introduced in [16] as a class of deep generative mode, employs a reparameterized gradient estimator for a evidence lower bound (ELBO) while applying amortized variational inference to an autoencoder. It simultaneously trains both a probabilistic encoder and decoder for elements of a data set $\mathcal{D} = (\boldsymbol{x}_1, \ldots, \boldsymbol{x}_M)$ with latent variable $\boldsymbol{z}$. Sequential variational autoencoders (SVAEs) extend VAE to sequential data $\mathcal{D}$, where each data are $\boldsymbol{x}_{1:T} = (x_1, \ldots, x_T)$ [19,31]. SVAEs factorize latent variables into two disentangled variables: the time-invariant variable $\boldsymbol{f}$ and time-varying variable $\boldsymbol{z}_{1:T} = (\boldsymbol{z}_1, \ldots, \boldsymbol{z}_T)$. Accordingly, decoder is casted as a conditional probabilistic density $p_{\boldsymbol{\theta}}(\boldsymbol{x}|\boldsymbol{f}, \boldsymbol{z}_{1:T})$ and encoder is used to approximate the posterior distribution $p_{\boldsymbol{\theta}}(\boldsymbol{f}, \boldsymbol{z}_{1:T}|\boldsymbol{x})$ as $q_{\boldsymbol{\phi}}(\boldsymbol{f}, \boldsymbol{z}_{1:T}|\boldsymbol{x})$ that is referred to as an "inference network" or a "recognition network". $\boldsymbol{\theta}$ refer to the model parameters of generator and $\boldsymbol{\phi}$ refer to the model parameters of encoder. SVAEs are trained to maximize the following ELBO:

$$\mathcal{L}(\boldsymbol{\theta}, \boldsymbol{\phi}; \mathcal{D}) = \mathbb{E}_{\hat{p}(\boldsymbol{x}_{1:T})}\Big[\mathbb{E}_{q_{\boldsymbol{\phi}}(\boldsymbol{z}_{1:T}, \boldsymbol{f}|\boldsymbol{x}_{1:T})} \ln p_{\boldsymbol{\theta}}(\boldsymbol{x}_{1:T}|\boldsymbol{f}, \boldsymbol{z}_{1:T}) - \mathrm{KL}(q_{\boldsymbol{\phi}}(\boldsymbol{f}, \boldsymbol{z}_{1:T}|\boldsymbol{x}_{1:T}), p_{\boldsymbol{\theta}}(\boldsymbol{f}, \boldsymbol{z}_{1:T}))\Big], \tag{2}$$

where $\hat{p}(\boldsymbol{x}_{1:T})$ is the empirical distribution with respect to the data set $\mathcal{D}$, $q_{\boldsymbol{\phi}}(\boldsymbol{f}, \boldsymbol{z}_{1:T}|\boldsymbol{x}_{1:T})$ is the variational posterior, $p_{\boldsymbol{\theta}}(\boldsymbol{x}_{1:T}|\boldsymbol{f}, \boldsymbol{z}_{1:T})$ is the conditional likelihood and $p_{\boldsymbol{\theta}}(\boldsymbol{f}, \boldsymbol{z}_{1:T})$ is prior over the latent variables.

3 Proposed Model

Let us first formalize the problem setting. Consider a dataset consists of shared graph G, and M unsupervised data points $\mathcal{D} = \{\boldsymbol{X}_i\}_{i=1}^{M}$ and M^{sup} supervised data points $\mathcal{D}^{sup} = \{\boldsymbol{X}_i, y_i\}_{i=1}^{M^{sup}}$ as pairs. $\boldsymbol{X}_i = (X_{i,1}, \ldots X_{i,T_i})$ refer to the i-th sequential observations on N nodes of a graph G with C input channels, i.e., $X_{i,t} \in \mathbb{R}^{N \times C}$, and y_i is the corresponding class label such as diagnostic labels.

We propose a semi-supervised sequential variational autoencoder model, and for convenience we omit the index i whenever it is clear that we are referring to terms associated with a single data point and treat individual data as $(\boldsymbol{X}, y)$.

Objective Function. Typical semi-supervised learning pipelines for deep generative models, e.g., [15,24], define an objective function for optimization as

$$\mathcal{L}(\boldsymbol{\theta}, \boldsymbol{\phi}; \mathcal{D}, \mathcal{D}^{sup}) = \mathcal{L}(\boldsymbol{\theta}, \boldsymbol{\phi}; \mathcal{D}) + \tau \mathcal{L}^{sup}(\boldsymbol{\theta}, \boldsymbol{\phi}; \mathcal{D}^{sup}). \tag{3}$$

Similarly, our approach jointly models unsupervised and supervised collections of terms over $\mathcal{D}$ and $\mathcal{D}^{sup}$. The formulation in Eq. 3 introduces a constant τ to control the relative strength of the supervised term. As the unsupervised term in Eq. 3 is exactly same as that of Eq. 2, we focus on the supervised term $\mathcal{L}^{sup}$ in Eq. 3 expanded below. Incorporating a weighted component as in [15],

$$\mathcal{L}^{sup}(\boldsymbol{\theta}, \boldsymbol{\phi}; \mathcal{D}^{sup}) = \mathbb{E}_{\hat{p}(\boldsymbol{X}, y)}\left[\mathbb{E}_{q_{\boldsymbol{\phi}}(\boldsymbol{f}, \boldsymbol{z}|\boldsymbol{X}, y)}\left[\ln \frac{p_{\boldsymbol{\theta}}(\boldsymbol{X}, y, \boldsymbol{f}, \boldsymbol{z})}{q_{\boldsymbol{\phi}}(\boldsymbol{f}, \boldsymbol{z}|\boldsymbol{X}, y)}\right] + \alpha \ln q_{\boldsymbol{\phi}}(y|\boldsymbol{X})\right] \tag{4}$$

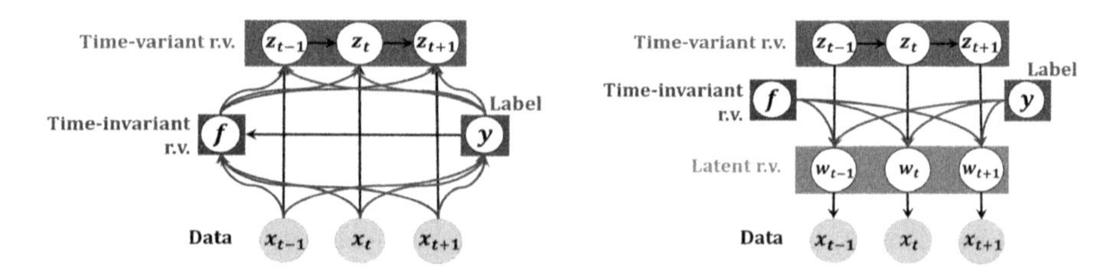

Fig. 1. A graphical model visualisation of the encoder (left) and decoder (right). In the encoder, label y is inferred by data $\boldsymbol{x}$ and time-invariant r.v. $\boldsymbol{f}$ are inferred by label y and data $\boldsymbol{x}$, and time-varying r.v. $\boldsymbol{z}$ are sequentially inferred by label y, time-invariant r.v. $\boldsymbol{f}$ and data $\boldsymbol{x}$. In the decoder, data are sequentially generated from time-invariant random variable (r.v.) $\boldsymbol{f}$, time-varying r.v. $\boldsymbol{z}$ and label y via latent r.v. $\boldsymbol{w}$.

where α balances the classification performance and reconstruction performance. Discussions on generative and inference model will continue in the later sections.

Generative Model. This section discusses modeling conditional probabilistic density $p_{\boldsymbol{\theta}}(\boldsymbol{X}|\boldsymbol{f}, \boldsymbol{z}, y)$ with its corresponding prior. We incorporate the topology information of the graph G into the generative process using a graph convolution. Specifically, we assume that sequences $\boldsymbol{X}$ are generated from P-dimensional latent vectors $\boldsymbol{W} = (W_1, \ldots, W_T)$ and $W_t \in \mathbb{R}^{N \times P}$ via

$$X_t = \hat{A} W_t \Theta, \tag{5}$$

where $\hat{A} = \tilde{D}^{-1/2} \tilde{A} \tilde{D}^{-1/2}$, $\tilde{A} = A + I$ and $\tilde{D}_{ii} = \sum_j \tilde{A}_{ij}$. A is the adjacent matrix for the graph G and Θ is the trainable weight matrix. Then we assume the latent variables $\boldsymbol{W}$ are generated from two disentangled variables: the time-invariant (or static) variable $\boldsymbol{f}$ and the time-varying (or dynamic) variables $\boldsymbol{z}$, as well as label $\boldsymbol{y}$, as shown in Fig. 1. A joint for the generative model is given as

$$p_{\boldsymbol{\theta}}(\boldsymbol{X}, y, \boldsymbol{z}, \boldsymbol{f}) = p_{\boldsymbol{\theta}}(\boldsymbol{f}) p_{\boldsymbol{\theta}}(y) \prod_{t=1}^{T} p_{\boldsymbol{\theta}}(\boldsymbol{z}_t | \boldsymbol{z}_{<t}) p_{\boldsymbol{\theta}}(X_t | y, \boldsymbol{f}, \boldsymbol{z}_t). \tag{6}$$

The prior of $\boldsymbol{f}$ is defined as a Gaussian distribution: $\boldsymbol{f} \sim \mathcal{N}(\boldsymbol{0}, I)$. Time-varying latent variables $\boldsymbol{z}_{1:T}$ follow a sequential prior $\boldsymbol{z}_t | \boldsymbol{z}_{t-1} \sim \mathcal{N}(\boldsymbol{\mu}_t, \text{diag}(\boldsymbol{\sigma}_t^2))$, where $[\boldsymbol{\mu}_t, \boldsymbol{\sigma}_t]$ are estimated by a recurrent network, such as LSTM [11] or GRU [5], in which the hidden states are updated temporally. The generating distribution of W_t is conditional on y, $\boldsymbol{f}$ and $\boldsymbol{z}_t$: $\text{vec}(W_t)|y, \boldsymbol{f}, \boldsymbol{z}_t \sim \mathcal{N}(\boldsymbol{\mu}_{w,t}, \text{diag}(\boldsymbol{\sigma}_{w,t}^2))$, where $[\boldsymbol{\mu}_{w,t}, \boldsymbol{\sigma}_{w,t}] = \psi^{Decoder}(y, \boldsymbol{f}, \boldsymbol{z}_t)$. This decoder $\psi^{Decoder}$ can be any flexible neural network such as multilayer perceptron (MLP). The $\boldsymbol{f}$ will be capable of modelling global aspects of the whole sequences which are time-invariant, while $\boldsymbol{z}_{1:T}$ will model time-varying features. As mentioned in [19], to separate the static and dynamic information, smaller dimension of $\boldsymbol{z}_t$ is preferred. In the context of ADNI study, $\boldsymbol{z}_t$ would encode how ROIs at timestamp t is morphed into those at timestamp $t+1$. In the context of generative model, we employ LSTM as the prior for $\boldsymbol{z}$ and use MLP for the conditional probabilistic density, and we set the dimension $P = 1$.

Inference Model. The developed SVAE within our framework proposes a recognition model $q_\phi(y, \boldsymbol{f}, \boldsymbol{z}|\boldsymbol{X}) = q_\phi(y|\boldsymbol{X})q_\phi(\boldsymbol{f}, \boldsymbol{z}|y, \boldsymbol{X})$ to approximate the posterior $p_\theta(y, \boldsymbol{f}, \boldsymbol{z}|\boldsymbol{X})$. The recognition model is formulated as

$$y \sim \mathrm{Cat}(\mathrm{Softmax}(\boldsymbol{p}_y)), \quad \boldsymbol{f} \sim \mathcal{N}(\boldsymbol{\mu}_f, \mathrm{diag}(\boldsymbol{\sigma}_f^2)), \quad \boldsymbol{z}_t \sim \mathcal{N}(\boldsymbol{\mu}_t, \mathrm{diag}(\boldsymbol{\sigma}_t^2)), \tag{7}$$

where $\boldsymbol{p}_y = \psi_y^{Encoder}(X_{1:T})$, $[\boldsymbol{\mu}_f, \boldsymbol{\sigma}_f] = \psi_f^{Encoder}(y, X_{1:T})$ and $[\boldsymbol{\mu}_t, 2\log \boldsymbol{\sigma}_t] = \psi_R^{Encoder}(y, X_{\leq t})$. It implies that the label y and the time-invariant variable $\boldsymbol{f}$ are conditional on the whole sequence via $\psi_y^{Encoder}$ and $\psi_f^{Encoder}$, while the time-dependent variable $\boldsymbol{z}_t$ is inferred by the sequence before time t, $X_{\leq t}$. The inference model is visualized in Fig. 1 and is factorized as

$$q_\phi(y, \boldsymbol{z}_{1:T}, \boldsymbol{f}|X_{1:T}) = q_\phi(y|X_{1:T})q_\phi(\boldsymbol{f}|y, X_{1:T})\prod_{t=1}^{T} q_\phi(\boldsymbol{z}_t|y, X_{\leq t}). \tag{8}$$

In the context of our inference model, we employ three independent LSTMs for three conditional probabilistic densities of y, $\boldsymbol{f}$ and $\boldsymbol{z}$.

4 Experimental Results

We conducted experiments on structural brain connectivity from DTI in ADNI. DTI images were processed by tractography, which extracted neuron fiber tracts and longitudinal cortical thickness measures registered at Destrieux atlas [8] with 148 ROIs. The dataset had five labels; we merged control (CN), Significant Memory Concern (SMC) and Early Mild Cognitive Impairment (EMCI) groups as Pre-clinical AD group, and Late Mild Cognitive Impairment (LMCI) and Alzheimer's Disease (AD) as Prodromal AD group to ensure sufficient sample

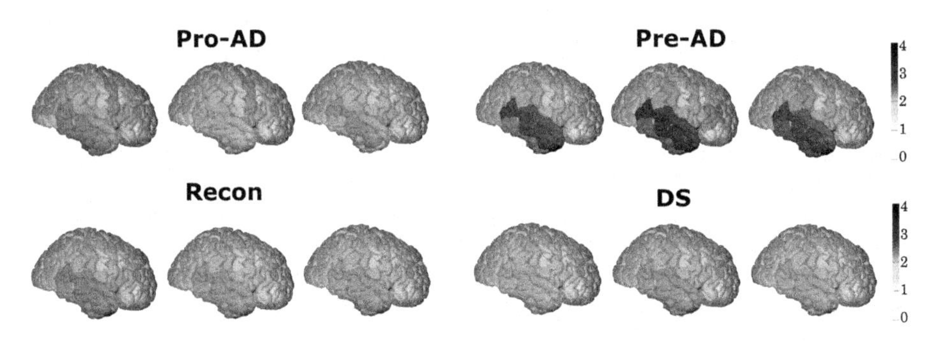

Fig. 2. Top panel shows the true brain surfaces at timestamp t_0, t_1 and t_2 for subject 1 (Pro-AD) and subject 2 (Pre-AD), respectively. Bottom panel shows the reconstructed brain surfaces for subject 1 (Recon) and subject 1's brain surfaces through the dynamic swapping (DS). Drawings generated using BrainPainter [21].

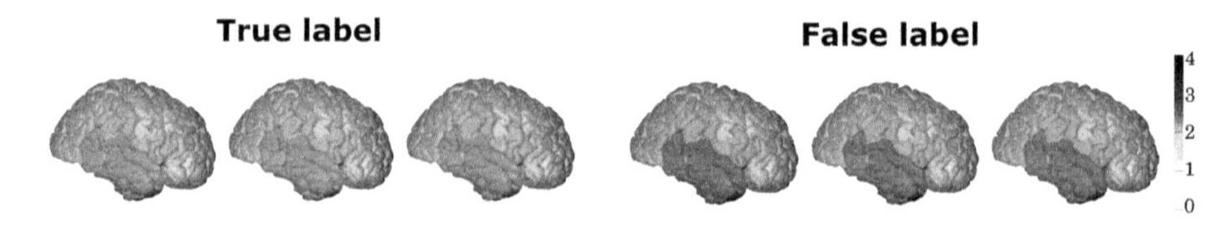

Fig. 3. Label swapping task. Left panel shows generated brain surfaces for subject 1 (Pro-AD) based on the true label at timestamp t_0, t_1 and t_2, respectively. Right panel shows generated brain surfaces for the same subject 1 but based on the false label.

size. The dataset included N = 140 subjects with the Pre-AD group (93 subjects/330 records) and the Pro-AD group (47 subjects/170 records). The mean (std) of ages and sex ratio (Male:Famale) in Pre-AD group and Pro-AD group are 74.02(6.72)/(185:145) and 74.87(6.92)/(95:75), respectively. An overall graph was obtained by taking the average of the adjacency matrices. Experiments for disentangle representation and quantitative analysis were performed given below.

4.1 Disentangled Representation

In this experiment, we randomly took 100 subjects' records for training, 20 subjects' records for validation and the other 20 subjects' records for testing. We set the dimension size of $\boldsymbol{f}$ as 8 and the dimension size of $\boldsymbol{z}$ as 32. We also set the size of hidden states in LSTMs as 32.

We randomly selected two subjects with more than three records (i.e., timepoints), where subject 1 belongs to Prodromal AD group and subject 2 belongs to Pre-clinical AD group. Suppose that the two subjects' sequential records are given for anatomical information and modality information denoted by R_1 and R_2. Our method performs the reconstruction task and the dynamic swapping task in which the record generation is based on the true y, $\boldsymbol{f}$ from R_1 and $\boldsymbol{z}$ from R_2 as in Fig. 2. It shows that the reconstruction captures both anatomical information and modality information, and figures generated from the dynamic swapping task illustrate that time-varying latent variables $\boldsymbol{z}$ succeed to learning the modality information.

In Fig. 3, we show results from the label swapping task on subject 1, where we generate cortical thickness based on the $\boldsymbol{f}$ from R_1, $\boldsymbol{z}$ from R_1 and true/false labels y. Comparing the generated measures of subject 1 with the true measures in Fig. 2, we found that generated measures based on the true label are more similar to the true measures and that based on the false label has totally different patterns but similar to the true measures of subject 2 in Fig. 2. It suggests that the decoder in our model correctly learns the label.

To understand the disentangled representation on the time invariant latent variable $\boldsymbol{f}$, we carry out latent traversals in $\boldsymbol{f}$ as in [3]. Specifically, we first computed the average Kullback–Leibler divergence for $\boldsymbol{f}$ with its prior. Then we selected the two dimensions in $\mathbf{f}$ with the largest two values (the 1st and 3rd elements), which refer to the two most informative dimensions and then

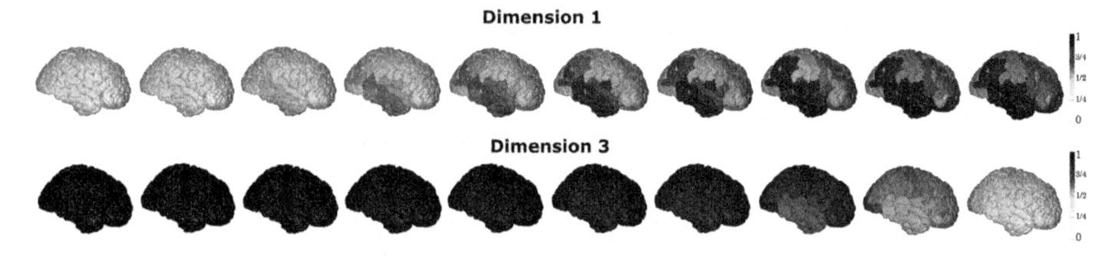

Fig. 4. Latent traversals task. Top: the latent brain surfaces for dim-1 on subject 1 (pro-AD). Bottom: the latent brain surfaces for dim-3 on subject 1.

traverse a single latent dimension on 10 equally spaced grids on $[-3, 3]$. For better visualization, we chose the first image as baseline and subtracted the baseline of image from all generated images. Then we normalized those images in a unit region $[0, 1]$ shown in Fig. 4.

4.2 Quantitative Analysis

We carry out 7-fold cross validation (CV) in which we take six folds for training (one fold for validation from the training set) and one fold for testing. We set the dimension size of $\boldsymbol{f}$ as 8 and the dimension size of $\boldsymbol{z}$ as 4. We also set the size of hidden states in LSTMs as 8. We compared our model with S3VAE model [31], which has a generator as in Fig. 1 but without a probabilistic model on label y. As S3VAE is unsupervised, we cannot directly compare our model with it. Instead, we tackle the classification task via a two-stage approach. Specifically, we train S3VAE to obtain latent $\boldsymbol{f}$ and train a naive neural network for the label classification. As for testing, we first get $\boldsymbol{f}$ from trained S3VAE and classify $\boldsymbol{f}$. Also, we propose a supervised loss based on the latent time-invariant f for S3VAE as one competitor. The generative model is modified as

$$p_\theta(\boldsymbol{X}, y, \boldsymbol{z}, \boldsymbol{f}) = p_\theta(\boldsymbol{f}) p_\theta(y|\boldsymbol{f}) \prod_{t=1}^{T} p_\theta(z_t|\boldsymbol{z}_{<t}) p_\theta(X_t|\boldsymbol{f}, z_t), \tag{9}$$

where we employ a fully connected network following a softmax activation function for $p_\theta(y|\boldsymbol{f})$. We treat the pro-AD as positive result and then report three classification measures, accuracy, precision and recall. We also report root mean

Table 1. Mean (Std) reconstruction and classification performance with 7-fold cross validation.

	RMSE	Accuracy	Precision	Recall
Our model ($\alpha = 1$)	0.257 (0.041)	0.657 (0.168)	0.416 (0.367)	0.446 (0.349)
Our model ($\alpha = 10$)	0.258 (0.046)	0.736 (0.151)	0.541 (0.346)	0.492 (0.337)
S3VAE (Supervised)	0.263 (0.042)	0.664 (0.164)	0.000 (0.000)	0.000 (0.000)
S3VAE (Two stages)	0.254 (0.043)	0.664 (0.164)	0.000 (0.000)	0.000 (0.000)

square error (RMSE) as a reconstruction measure for testing data in Table 1. As for our proposed model, we consider the regularization weights $\alpha = 1$ and $\alpha = 10$. We find that our model has a better reconstruction performance in comparison to the supervised S3VAE model and performs similarly to the two-stage S3VAE. As for classification, our model with $\alpha = 10$ outperforms other models. We note that S3VAE based methods always categorize patients into pre-AD group, suffering from the imbalance classification issue. Our model resolves this issue and obtains a significantly better classification result according to both higher precision and recall scores. Finally, we note that to get better reconstruction or prediction results, properly tuning the hyperparameter α is important.

5 Conclusion

In summary, we propose a novel Sequential Autoencoder model. It incorporates the graph information via graph convolution operation, and it jointly models supervised and unsupervised data. Our model is flexible for data generation and it can conditionally generate sequential data based on label, disentangled time-invariant and time-varying latent variables. Quantitatively, we show that this model has competitive classification and reconstruction performance compared with two modified state-of-the-art S3VAE models.

Acknowledgement. This work was supported by GAANN Doctoral Fellowships in Computer Science and Engineering at UTA sponsored by the U.S. Department of Education, NSF IIS CRII 1948510, NIH RF1 AG059312, NIH R03 AG070701, and IITP-2019-0-01906 funded by MSIT (AI Graduate School Program at POSTECH).

References

1. Baur, C., Wiestler, B., Albarqouni, S., Navab, N.: Deep autoencoding models for unsupervised anomaly segmentation in brain MR images. In: Crimi, A., et al. (eds.) BrainLes 2018. LNCS, vol. 11383, pp. 161–169. Springer, Cham (2019). https://doi.org/10.1007/978-3-030-11723-8_16
2. Bengio, Y., Courville, A., Vincent, P.: Representation learning: a review and new perspectives. IEEE Trans. Pattern Anal. Mach. Intell. **35**(8), 1798–1828 (2013)
3. Burgess, C.P., et al.: Understanding disentangling in beta-VAE. arXiv preprint arXiv:1804.03599 (2018)
4. Chen, J., Ma, T., Xiao, C.: Fastgcn: fast learning with graph convolutional networks via importance sampling. arXiv preprint arXiv:1801.10247 (2018)
5. Cho, K., et al.: Learning phrase representations using RNN encoder-decoder for statistical machine translation. In: EMNLP, pp. 1724–1734. ACL (2014)
6. Cho, Y., Seong, J.K., Jeong, Y., Shin, S.Y.: ADNI: individual subject classification for Alzheimer's disease based on incremental learning using a spatial frequency representation of cortical thickness data. Neuroimage **59**(3), 2217–2230 (2012)
7. Defferrard, M., Bresson, X., Vandergheynst, P.: Convolutional neural networks on graphs with fast localized spectral filtering. Adv. Neural Inf. Process. Syst. **29**, 3844–3852 (2016)

8. Destrieux, C., Fischl, B., Dale, A., Halgren, E.: Automatic parcellation of human cortical gyri and sulci using standard anatomical nomenclature. Neuroimage **53**(1), 1–15 (2010)
9. Hammond, D.K., Vandergheynst, P., Gribonval, R.: Wavelets on graphs via spectral graph theory. Appl. Comput. Harm. Anal. **30**(2), 129–150 (2011)
10. Higgins, I., et al.: beta-VAE: Learning basic visual concepts with a constrained variational framework (2016)
11. Hochreiter, S., Schmidhuber, J.: Long short-term memory. Neural Comput. **9**(8), 1735–1780 (1997)
12. Hsu, W.N., Zhang, Y., Glass, J.: Unsupervised learning of disentangled and interpretable representations from sequential data. In: NeurIPS, pp. 1878–1889 (2017)
13. Kim, H., Mnih, A.: Disentangling by factorising. In: International Conference on Machine Learning, pp. 2649–2658. PMLR (2018)
14. Kim, W.H., Racine, A.M., Adluru, N., et al.: Cerebrospinal fluid biomarkers of neurofibrillary tangles and synaptic dysfunction are associated with longitudinal decline in white matter connectivity: a multi-resolution graph analysis. NeuroImage Clin. **21**, 101586 (2019)
15. Kingma, D.P., Mohamed, S., Rezende, D.J., Welling, M.: Semi-supervised learning with deep generative models. In: Advances in Neural Information Processing Syst. pp. 3581–3589 (2014)
16. Kingma, D.P., Welling, M.: Auto-encoding variational bayes. CoRR abs/1312.6114 (2014)
17. Kipf, T.N., Welling, M.: Semi-supervised classification with graph convolutional networks. In: 5th International Conference on Learning Representations, ICLR 2017, Toulon, France, April 24–26, 2017, Conference Track Proceedings. OpenReview.net (2017)
18. Li, Q., Han, Z., Wu, X.M.: Deeper insights into graph convolutional networks for semi-supervised learning. In: AAAI, vol. 32 (2018)
19. Li, Y., Mandt, S.: Disentangled sequential autoencoder. arXiv preprint arXiv:1803.02991 (2018)
20. Ma, X., Wu, G., Hwang, S.J., Kim, W.H.: Learning multi-resolution graph edge embedding for discovering brain network dysfunction in neurological disorders. In: Feragen, A., Sommer, S., Schnabel, J., Nielsen, M. (eds.) IPMI 2021. LNCS, vol. 12729, pp. 253–266. Springer, Cham (2021). https://doi.org/10.1007/978-3-030-78191-0_20
21. Marinescu, R.V., Eshaghi, A., Alexander, D.C., Golland, P.: Brainpainter: A software for the visualisation of brain structures, biomarkers and associated pathological processes. In: Multimodal Brain Image Analysis and Mathematical Foundations of Computational Anatomy, pp. 112–120. Springer (2019), https://doi.org/10.1007/978-3-030-33226-6
22. McKhann, G.M., et al.: The diagnosis of dementia due to Alzheimer's disease: recommendations from the national institute on aging-Alzheimer's association workgroups on diagnostic guidelines for Alzheimer's disease. Alzheimer's Sementia **7**(3), 263–269 (2011)
23. Meng, R., Bouchard, K.: Bayesian inference in high-dimensional time-series with the orthogonal stochastic linear mixing model. arXiv preprint arXiv:2106.13379 (2021)
24. Siddharth, N., et al.: Learning disentangled representations with semi-supervised deep generative models. In: Advances in Neural Information Processing Systems, vol. 30. Curran Associates, Inc. (2017)

25. Ouyang, J., Adeli, E., Pohl, K.M., Zhao, Q., Zaharchuk, G.: Representation Disentanglement for Multi-modal MR Analysis. arXiv e-prints arXiv:2102.11456 (Feb 2021)
26. Saerens, M., Fouss, F., Yen, L., Dupont, P.: The principal components analysis of a graph, and its relationships to spectral clustering. In: Boulicaut, J.-F., Esposito, F., Giannotti, F., Pedreschi, D. (eds.) ECML 2004. LNCS (LNAI), vol. 3201, pp. 371–383. Springer, Heidelberg (2004). https://doi.org/10.1007/978-3-540-30115-8_35
27. Thompson, P.M., Hayashi, K.M., Sowell, E.R., Gogtay, N., Giedd, J.N., Rapoport, J.L., De Zubicaray, G.I., Janke, A.L., Rose, S.E., Semple, J., et al.: Mapping cortical change in Alzheimer's disease, brain development, and schizophrenia. Neuroimage **23**, S2–S18 (2004)
28. Wolz, R., et al.: Multi-method analysis of MRI images in early diagnostics of Alzheimer's disease. PLoS ONE **6**(10), e25446 (2011)
29. Xu, B., Shen, H., Cao, Q., Qiu, Y., Cheng, X.: Graph wavelet neural network. In: International Conference on Learning Representations (2019)
30. Zhao, Q., Adeli, E., Honnorat, N., Leng, T., Pohl, K.M.: Variational autoencoder for regression: application to brain aging analysis. In: Shen, D., Liu, T., Peters, T.M., Staib, L.H., Essert, C., Zhou, S., Yap, P.-T., Khan, A. (eds.) MICCAI 2019. LNCS, vol. 11765, pp. 823–831. Springer, Cham (2019). https://doi.org/10.1007/978-3-030-32245-8_91
31. Zhu, Y., Min, M.R., Kadav, A., Graf, H.P.: S3VAE: self-supervised sequential VAE for representation disentanglement and data generation. In: CVPR, pp. 6538–6547 (2020)

POPCORN: Progressive Pseudo-Labeling with Consistency Regularization and Neighboring

Reda Abdellah Kamraoui[1(✉)], Vinh-Thong Ta[1], Nicolas Papadakis[2], Fanny Compaire[1,2], José V. Manjon[3], and Pierrick Coupé[1]

[1] Univ. Bordeaux, Bordeaux INP, CNRS, LaBRI, UMR5800, PICTURA, 33400 Talence, France
reda-abdellah.kamraoui@u-bordeaux.fr
[2] Univ. Bordeaux, Bordeaux INP, CNRS, IMB, UMR5251, 33400 Talence, France
[3] ITACA, Universitat Politécnica de Valéncia, 46022 Valencia, Spain

Abstract. Semi-supervised learning (SSL) uses unlabeled data to compensate for the scarcity of annotated images and the lack of method generalization to unseen domains, two usual problems in medical segmentation tasks. In this work, we propose POPCORN, a novel method combining consistency regularization and pseudo-labeling designed for image segmentation. The proposed framework uses high-level regularization to constrain our segmentation model to use similar latent features for images with similar segmentations. POPCORN estimates a proximity graph to select data from easiest ones to more difficult ones, in order to ensure accurate pseudo-labeling and to limit confirmation bias. Applied to multiple sclerosis lesion segmentation, our method demonstrates competitive results compared to other state-of-the-art SSL strategies.

Keywords: Semi-supervised learning · Pseudo-labeling · Consistency regularization · MS lesion segmentation

1 Introduction

Semi-Supervised Learning (SSL) is a promising field which aims to exploit unlabeled data in order to enhance the performance achieved using only labeled data. SSL is explored to mitigate both problems of the limited availability of labeled data and the lack of model generalization to unseen domains. Among SSL works proposed for medical image segmentation tasks, we can distinguish three main categories:

Consistency Regularization (CR) constrains the model to give consistent predictions for the same unlabeled input under different perturbations. Bortsova *et al.* [4] constrained the model to produce similar segmentation when applying different elastic transformations to the same unlabeled images. Similarly, Perone *et al.* [18] used a mean teacher strategy where the consistency loss constrained

M. de Bruijne et al. (Eds.): MICCAI 2021, LNCS 12902, pp. 373–382, 2021.
https://doi.org/10.1007/978-3-030-87196-3_35

teacher and student predictions to be consistent. Orbes *et al.* [16] designed an adversarial loss to minimize the amount of information for a specific domain and to maximize segmentation consistency. CR offers interesting consistency properties on the learned features but it is usually trained under unrealistic scenarios (e.g., using the same input data under different perturbations). Such oversimplification does not guarantee a good generalization of the learned features. Besides, some works showed that consistency regularization using perturbation on input data is not adapted for segmentation [10,17].

Pseudo-Labeling (PL) strategies automatically assign labels to unlabeled data in order to use them during training in combination with labeled data. Pseudo-labels are generally assigned by a model trained on labeled data. Uncertainty can be used to measure the confidence of the predictions. For example, Sedai *et al.* [19] employed prediction uncertainty for estimating segmentation confidence on soft labels. Cao *et al.* [5] considered an uncertainty aware temporal ensembling strategy. Xia *et al.* [20] used uncertainty-weighted mechanism for the pseudo-label fusion of multiple networks predictions. PL is a simple way to use unlabeled data. PL is nevertheless prone to confirmation bias (*i.e.*, error propagation) [2]. So far, this is the main limitation of PL.

Auxiliary Tasks (AT) are secondary objectives combined with the main segmentation task which do not require ground truth annotations. Using unlabeled data, in such a way, implicitly extracts relevant features for the primary segmentation task. Li *et al.* [14] proposed the prediction of surface distance maps to capture more effectively shape-aware features. Kervadec *et al.* [12] predicted the size of the target segmentation as an intermediate task. Alternatively, Chen *et al.* [7] combined supervised segmentation and unsupervised input reconstruction. Finally, Luo *et al.* [15] proposed to predict geometry-aware level set representation of the transformed ground truth annotations. AT demonstrated good performance, but the choice of the AT highly depends on the addressed problem which limits the method generalization for other segmentation tasks.

In this work, our main contribution is threefold:

- We propose a novel framework that combines consistency regularization and pseudo-labeling for segmentation.
- We propose a consistency regularization strategy that ensures proximity in latent space of images with similar segmentations. This allows us to produce meaningful feature representation and accurate predictions.
- We propose a new pseudo-labeling strategy which selects progressively unlabeled samples according to their similarity with training data, in order to limit confirmation bias.

2 Method

2.1 Method Overview

The proposed strategy is a PrOgressive Pseudo-labeling with COnsistency Regularization and Neighboring (POPCORN) for semi-supervised learning in segmentation (see Fig. 1). First, the training is performed with a new CR ensuring

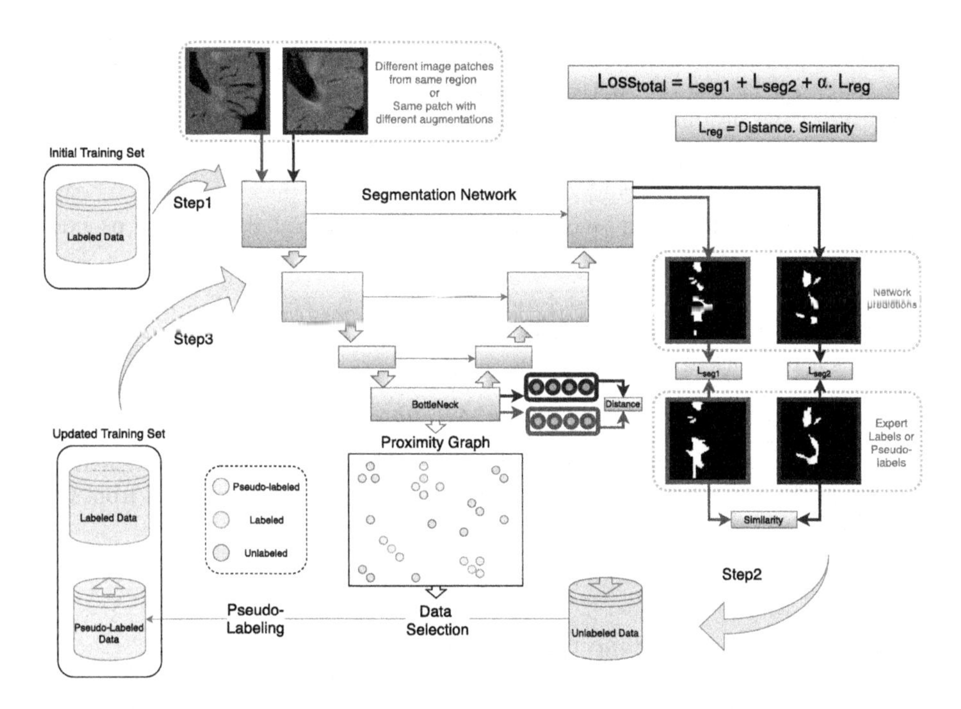

Fig. 1. The training process of POPCORN

that: *i*) augmented versions of the same image have identical feature maps, and *ii*) images with similar segmentations have similar feature maps. Second, PL of the unlabeled data is performed gradually. At each selection step, the proximity graph is used to select new unlabeled samples. The pseudo-labels of the chosen data-points are estimated with the current segmentation model and incorporated in the training set.

The main intuition is that our segmentation model is able to produce more accurate pseudo-labels for images similar to our training set. Since our CR ensures close features for similar data, features extracted from the model are used to select new samples.

2.2 Bottleneck Consistency Regularization

In POPCORN, the model architecture is based on 3D U-Net composed of an encoder and a decoder, linked by a bottleneck and skip connections at different scales (see Fig. 1). For an image X, $F(X)$ represents the prediction of the segmentation network, and $h(X)$ represents the latent features of X extracted at the bottleneck level. Our method is based on a dual/hybrid loss ensuring segmentation quality and consistency relevance.

Segmentation Loss: As traditionally done in supervised learning, we use the Dice similarity loss as the first element of our global loss. This loss ensures the similarity of the produced output with the expected one.

$$L_{seg}(F(X), y) = 1 - Dice(F(X), y), \tag{1}$$

where y is either the expert segmentation of X when available, or the pseudo-label otherwise.

Consistency Regularization Loss: Alongside L_{seg}, a regularization loss on the bottleneck is used:

$$L_{reg}[\,(X_i, y_i), (X_j, y_j)\,] = Distance(\,h(X_i),\, h(X_j)\,) \times Similarity(y_i, y_j), \tag{2}$$

where X_i, X_j are image patches randomly selected as either different augmented versions of the same patch, or patches from different images extracted from the same region and with the same orientation. Moreover, let us define:

$$Similarity(y_i, y_j) = e^{-mse(y_i, y_j)}, \tag{3}$$

$$Distance(h_i, h_j) = \frac{2\,||h_i - h_j||^2}{||h_i||^2 + ||h_j||^2}, \tag{4}$$

where mse is the mean squared error, and $||\,.\,||$ is the euclidean distance.

The total loss is a combination of (1) and (2) with a weighting coefficient α:

$$\begin{aligned} L_{total}[(X_i, y_i), (X_j, y_j)] = L_{seg}(F(X_i), y_i) + L_{seg}(F(X_j), y_j) \\ + \alpha.L_{reg}[(X_i, y_i), (X_j, y_j)]\,. \end{aligned} \tag{5}$$

2.3 Pseudo-Labeling Data Selection

Curriculum learning [3] showed that presenting data with an increasing difficulty can lead to a better learning process. We consider unlabeled data close to the training data as easy examples to be incorporated first in the training process, whereas distant samples are considered more challenging. Indeed, the latent distance between unlabeled and training data can be viewed as a measure of similarity. Thus, pseudo-labeling using the trained model is more accurate for unlabeled samples which are similar to training data.

Our data selection is performed in three steps (see Fig. 1). Step1: the training set is limited to labeled data. Once the segmentation model is trained until convergence, it is used to extract latent space representation for each unlabeled datapoint. Step2: the proximity graph is used to select K unlabeled data that guarantee a smooth learning (as described in Sect. 2.4). For each selected unlabeled datapoint, a pseudo-label (segmentation) is assigned by the trained model. Step3: the model is trained for N epochs with the new training set containing both labeled and pseudo-labeled data. Step two and three are repeated every N epochs by picking each time K new data points, their respective pseudo-labels are being computed with the newly updated segmentation model. The process is maintained until all unlabeled data are integrated into the training set.

2.4 Proximity Graph

The proximity graph represents the euclidean distance between the training and the unlabeled samples latent representations:

$$P_{i\in U,j\in T} = ||h(X_i) - h(X_j)||^2 \tag{6}$$

where T and U represent respectively the training set (labeled and pseudo-labeled), and unlabeled data. For data selection, we propose the following criteria to select K elements of U close to T. For each datapoint of U, the proximity with T is defined by the sum of the p closest elements of T to the datapoint:

$$[i_1^\star, i_2^\star \ldots, i_K^\star] = Argmin^K(\sum_{j\in Argmin^p(P_i)} P_{i,j}), \tag{7}$$

where $[i_1^\star, i_2^\star \ldots, i_K^\star]$ represent the K indices of the data selected. $Argmin^k(V)$ returns the indices of the k smallest values of the vector V.

3 Experiments

3.1 Dataset

Labeled Data: For labeled data, the ISBI training dataset [6] is used. It consists of 21 longitudinal multimodal images (including FLAIR modality) from only five different subjects with Multiple Sclerosis (MS). The images have been acquired on the same MRI scanner. MS lesions were delineated by two expert raters. This dataset has limited image quality diversity (all the images were acquired with the same protocol on a single site) and inter-subject variability (only 5 subjects).

Unlabeled Data: The unlabeled dataset consists of 2901 FLAIR MRI (large inter-subject variability) with white matter hyperintensities. It does not only contain MS, which increases pathology diversity. MRI have been collected across multiple acquisition sites based on different manufacturers, 1.5T and 3T scanners, 2D and 3D sequences. This dataset covers a large diversity in terms of image quality, pathology and inter-subject variability.

Testing Data: For assessing our results, the dataset described in [9] is used. It contains 3D multimodal MRI from 43 subjects diagnosed with MS. The images have been acquired with three different scanners and different acquisition protocols. Consequently this dataset proposes a larger diversity than the labeled dataset. Lesion masks have been obtained by expert manual delineation.

All images have been pre-processed using the same pipeline [9].

3.2 Reference Methods

POPCORN is compared to state-of-the-art strategies [7,19] and [4]. The following strategies have been implemented based on their published works and adapted to MS lesion segmentation. First, the multi-task attention-based SSL

[7] is an AT strategy. It combines supervised segmentation and unsupervised reconstruction objectives. The reconstruction task uses an attention mechanism to predict input image regions of different classes. Second, the uncertainty guided pseudo-labeling [19] is a PL strategy. The teacher model, trained only on labeled data, generates soft segmentation (pseudo-labels) and uncertainty maps for all the unlabeled data at once. The uncertainty is used for estimating segmentation confidence of the generated segmentation when training the student model. Finally, the semi-supervised transformation consistency [4] is based on CR. In addition to the primary loss, a consistency loss ensures that the prediction of the same images under transformations are consistent.

3.3 Implementation Details

The method hyperparameters were chosen empirically according to the size of labeled and unlabeled datasets. First, 200 from the $M = 2901$ unlabeled images were chosen after each training cycle that ran for 2 epochs ($K = 200$, $N = 2$) to limit computational burden. Second, the number of neighbors $p = 5$ was selected considering the initial training data of 21 labeled images. We suggest that this value is a good compromise in order to consider relevant near neighbors while avoiding far neighbors which mislead data selection.

In addition, we used the architecture proposed by [11] with a patch size of $[64 \times 64 \times 64]$ and a threshold of 0.5 to obtain the binary segmentation. Moreover, image quality data augmentation was used to introduce realistic perturbations, where blur, edge enhancement, and other augmentations simulated image quality heterogeneity [11]. Furthermore, the coefficients for the regularization part of the loss have been set to 0.2 ($\alpha = 0.2$). Finally, the experiments have been performed with Keras 2.2.4 [8] and Tensorflow 1.12.0 [1] on Python 3.6. The model was optimized with Adam [13] using a learning rate of 0.0001 and a momentum of 0.9.

3.4 Statistical Analysis

To assert the advantage of a technique obtaining the highest average score, we conducted a Wilcoxon test over the lists of Dice scores measured at image level. The significance of the test is established for a p-value below 0.05.

4 Results

4.1 Ablation Study

To evaluate our contributions, we compare POPCORN with other versions of our strategy when isolating key elements. As shown in Table 1, our full method achieves the highest Dice and the second best result in terms of precision. First, when comparing POPCORN without consistency regularization (corresponds to $\alpha = 0$ in (5)) and our full method, we notice a decrease in both precision and

Table 1. The table details the impact of each contribution: the consistency regularization (CR), the proximity graph, and using labeled/pseudo-labeled (lab/pseudo) data. Best result is displayed in bold, and the second best result is underlined.

Strategy	Trained on	CR on	Dice	Precision	Sensitivity
Our method	Lab + Pseudo	Lab + Pseudo	**73,09%**	<u>73,33%</u>	74,29%
Ours with half selection steps ($M = 1400$)	Lab + Pseudo	Lab + Pseudo	<u>70,59%</u>	68,26%	**75,91%**
Ours without CR	Lab + Pseudo	None	69,13%	70,49%	70,58%
Ours without proximity graph	Lab + Pseudo	Lab + Pseudo	68,06%	65,14%	<u>74,40%</u>
Baseline with CR	Lab	Lab	68,08%	**77,77%**	61,01%
Baseline	Lab	None	64,41%	61,80%	69,70%

sensitivity. This suggests that without CR, the latent space is less meaningful for our selection process of unlabeled data. Second, to underline the impact of the proximity graph, we consider another progressive PL strategy where pseudo labels are randomly selected. Although the strategy without proximity graph is slightly more sensitive, we observe an important drop in both Dice and precision compared to our full method. This demonstrates that the proposed progressive selection based on image proximity in latent space is more robust to confirmation bias than random selection. Next, when running only half the selection steps ($M = 1400$), our method obtained the second best Dice score. This shows that POPCORN with nearly half unlabeled data can achieve better performance than the other variations and methods with full dataset (see also 4.2). Finally, when combining the proposed CR (on labeled data only) with the baseline (supervised learning), the precision is considerably improved. This shows the importance of our CR on segmentation accuracy, beyond data selection. Overall, the statistical analysis shows that our full method has a significantly higher Dice than the baseline, the version without CR, baseline with CR, and Ours without proximity graph.

4.2 Comparison with State-of-the-Art Approaches

Table 2 shows the results of POPCORN compared to the reference methods presented in Sect. 3.2. First, all the SSL strategies obtain a significantly better Dice scores compared to the baseline. Second, POPCORN obtains the highest Dice followed by Uncertainty guided Pseudo-labeling [19]. Next, the multi-task attention-based SSL [7] and the semi-supervised transformation consistency [4] respectively obtain the highest precision and sensitivity rates. Finally, POPCORN obtains the best balance between precision and sensitivity, as opposed to the other strategies which are more prone to FP [4,19] and FN [7]. Overall, POPCORN has a significantly higher Dice compared to the other methods according to our Wilcoxon test.

Figure 2 shows image segmentations produced by POPCORN and the compared strategies. A, B, and C are images from the testing dataset, specifically

Table 2. The table represents results of POPCORN (our method) compared to other state-of-the-art strategies on the testing dataset (see 4.2 for complementary details).

Strategy	Dice	Precision	Sensitivity
POPCORN	**73,09%**	73,33%	74,29%
Multi-task Attention-based SSL [7]	67,23%	**75,72%**	61,99%
Uncertainty guided Pseudo-labeling [19]	68,31%	67,93%	71,95%
Semi-supervised transformation consistency [4]	66,75%	61,52%	**78,79%**
Baseline (labeled data only)	64,41%	61,80%	69,70%

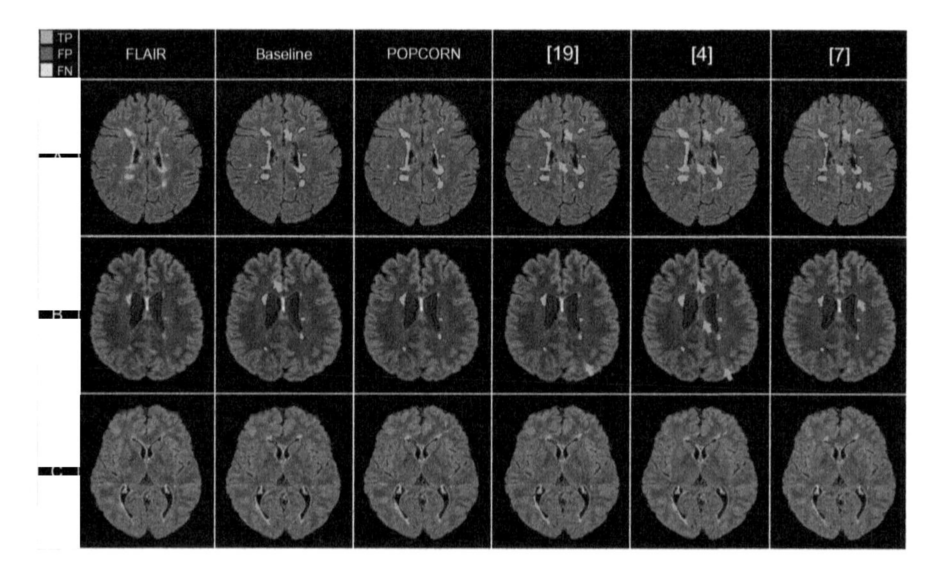

Fig. 2. Comparison of POPCORN, Uncertainty guided Pseudo-labeling [19], multi-task attention-based SSL [7], and the semi-supervised transformation consistency [4] lesion segmentations. Orange arrows indicate key segmentation differences.

chosen to showcase acquisition and lesion diversity. For A, we observe that POPCORN segmentation is the most accurate. On the contrary, [7,19] are the least sensitive with high volumes of false negative. Similarly, the segmentations obtained with the baseline and [4] do not cover all lesions. On image B, the segmentation provided by [4] contains several false positive lesions, compared to the other strategies. Both the baseline and [19] only include one or two false detections. POPCORN proposes an accurate segmentation. Last, the method [7] misses a small lesion. For C, we notice that [4,19] and the baseline detect many false positive lesions. POPCORN and [7] produce fewer false detection on this challenging sample. To conclude, our strategy segments accurately most lesions while minimizing false detection. Compared to the other strategies, POPCORN maintains the best balance between the sensitivity and the precision of lesion segmentation.

5 Conclusion

We propose a novel strategy for SSL segmentation. Our method combines consistency regularization and pseudo-labeling. POPCORN progressively selects unlabeled samples with an increasing difficulty using a proximity graph. Overall, we have shown the improvement of using POPCORN compared to other state-of-the-art strategies, as well as the impact of each of our contributions.

Acknowledgements. This work benefited from the support of the project DeepvolBrain of the French National Research Agency (ANR-18-CE45-0013). This study was achieved within the context of the Laboratory of Excellence TRAIL ANR-10-LABX-57 for the BigDataBrain project. Moreover, we thank the Investments for the future Program IdEx Bordeaux (ANR-10-IDEX-03-02, HL-MRI Project), Cluster of excellence CPU and the CNRS/INSERM for the DeepMultiBrain project. This study has been also supported by the DPI2017-87743-R grant from the Spanish Ministerio de Economia, Industria Competitividad. The authors gratefully acknowledge the support of NVIDIA Corporation with their donation of the TITAN Xp GPU used in this research.

References

1. Abadi, M., et al.: TensorFlow: a system for large-scale machine learning. In: 12th USENIX Symposium on Operating Systems Design and Implementation (OSDI 2016), pp. 265–283 (2016)
2. Arazo, E., Ortego, D., Albert, P., O'Connor, N.E., McGuinness, K.: Pseudo-labeling and confirmation bias in deep semi-supervised learning. In: 2020 International Joint Conference on Neural Networks (IJCNN), pp. 1–8. IEEE (2020)
3. Bengio, Y., Louradour, J., Collobert, R., Weston, J.: Curriculum learning. In: Proceedings of the 26th Annual International Conference on Machine Learning, pp. 41–48 (2009)
4. Bortsova, G., Dubost, F., Hogeweg, L., Katramados, I., de Bruijne, M.: Semi-supervised medical image segmentation via learning consistency under transformations. In: Shen, D., et al. (eds.) MICCAI 2019. LNCS, vol. 11769, pp. 810–818. Springer, Cham (2019). https://doi.org/10.1007/978-3-030-32226-7_90
5. Cao, X., Chen, H., Li, Y., Peng, Y., Wang, S., Cheng, L.: Uncertainty aware temporal-ensembling model for semi-supervised abus mass segmentation. IEEE Trans. Med. Imag. **40**(1), 431–443 (2020)
6. Carass, A., et al.: Longitudinal multiple sclerosis lesion segmentation: resource and challenge. NeuroImage **148**, 77–102 (2017)
7. Chen, S., Bortsova, G., García-Uceda Juárez, A., van Tulder, G., de Bruijne, M.: Multi-task attention-based semi-supervised learning for medical image segmentation. In: Shen, D., et al. (eds.) MICCAI 2019. LNCS, vol. 11766, pp. 457–465. Springer, Cham (2019). https://doi.org/10.1007/978-3-030-32248-9_51
8. Chollet, F., et al.: Keras (2015). https://keras.io
9. Coupé, P., Tourdias, T., Linck, P., Romero, J.E., Manjón, J.V.: LesionBrain: an online tool for white matter lesion segmentation. In: Bai, W., Sanroma, G., Wu, G., Munsell, B.C., Zhan, Y., Coupé, P. (eds.) Patch-MI 2018. LNCS, vol. 11075, pp. 95–103. Springer, Cham (2018). https://doi.org/10.1007/978-3-030-00500-9_11

10. French, G., Aila, T., Laine, S., Mackiewicz, M., Finlayson, G.: Semi-supervised semantic segmentation needs strong, high-dimensional perturbations (2020)
11. Kamraoui, R.A., Ta, V.T., Tourdias, T., Mansencal, B., Manjon, J.V., Coupé, P.: Towards broader generalization of deep learning methods for multiple sclerosis lesion segmentation. arXiv preprint arXiv:2012.07950 (2020)
12. Kervadec, H., Dolz, J., Granger, É., Ben Ayed, I.: Curriculum semi-supervised segmentation. In: Shen, D., et al. (eds.) MICCAI 2019. LNCS, vol. 11765, pp. 568–576. Springer, Cham (2019). https://doi.org/10.1007/978-3-030-32245-8_63
13. Kingma, D.P., Ba, J.: Adam: a method for stochastic optimization. arXiv preprint arXiv:1412.6980 (2014)
14. Li, S., Zhang, C., He, X.: Shape-aware semi-supervised 3D semantic segmentation for medical images. In: Martel, A.L., et al. (eds.) MICCAI 2020. LNCS, vol. 12261, pp. 552–561. Springer, Cham (2020). https://doi.org/10.1007/978-3-030-59710-8_54
15. Luo, X., Chen, J., Song, T., Chen, Y., Wang, G., Zhang, S.: Semi-supervised medical image segmentation through dual-task consistency. arXiv preprint arXiv:2009.04448 (2020)
16. Orbes-Arteaga, M., et al.: Multi-domain adaptation in brain MRI through paired consistency and adversarial learning. In: Wang, Q., et al. (eds.) DART/MIL3ID -2019. LNCS, vol. 11795, pp. 54–62. Springer, Cham (2019). https://doi.org/10.1007/978-3-030-33391-1_7
17. Ouali, Y., Hudelot, C., Tami, M.: Semi-supervised semantic segmentation with cross-consistency training. In: Proceedings of the IEEE/CVF Conference on Computer Vision and Pattern Recognition, pp. 12674–12684 (2020)
18. Perone, C.S., Cohen-Adad, J.: Deep semi-supervised segmentation with weight-averaged consistency targets. In: Stoyanov, D., et al. (eds.) DLMIA/ML-CDS - 2018. LNCS, vol. 11045, pp. 12–19. Springer, Cham (2018). https://doi.org/10.1007/978-3-030-00889-5_2
19. Sedai, S., et al.: Uncertainty guided semi-supervised segmentation of retinal layers in OCT images. In: Shen, D., et al. (eds.) MICCAI 2019. LNCS, vol. 11764, pp. 282–290. Springer, Cham (2019). https://doi.org/10.1007/978-3-030-32239-7_32
20. Xia, Y., et al.: 3D semi-supervised learning with uncertainty-aware multi-view co-training. In: The IEEE Winter Conference on Applications of Computer Vision, pp. 3646–3655 (2020)

3D Semantic Mapping from Arthroscopy Using Out-of-Distribution Pose and Depth and In-Distribution Segmentation Training

Yaqub Jonmohamadi[1(✉)], Shahnewaz Ali[1], Fengbei Liu[2], Jonathan Roberts[1], Ross Crawford[1], Gustavo Carneiro[2], and Ajay K. Pandey[1]

[1] School of Electrical Engineering and Robotics, Science and Engineering Faculty, Queensland University of Technology, Brisbane, Australia
[2] Australian Institute for Machine Learning, School of Computer Science, University of Adelaide, Adelaide, Australia

Abstract. Minimally invasive surgery (MIS) has many documented advantages, but the surgeon's limited visual contact with the scene can be problematic. Hence, systems that can help surgeons navigate, such as a method that can produce a 3D semantic map, can compensate for the limitation above. In theory, we can borrow 3D semantic mapping techniques developed for robotics, but this requires finding solutions to the following challenges in MIS: 1) semantic segmentation, 2) depth estimation, and 3) pose estimation. In this paper, we propose the first 3D semantic mapping system from knee arthroscopy that solves the three challenges above. Using out-of-distribution non-human datasets, where pose could be labeled, we jointly train depth+pose estimators using self-supervised and supervised losses. Using an in-distribution human knee dataset, we train a fully-supervised semantic segmentation system to label arthroscopic image pixels into femur, ACL, and meniscus. Taking testing images from human knees, we combine the results from these two systems to automatically create 3D semantic maps of the human knee. The result of this work opens the pathway to the generation of intra-operative 3D semantic mapping, registration with pre-operative data, and robotic-assisted arthroscopy. Source code: https://github.com/YJonmo/EndoMapNet.

Keywords: 3D semantic mapping · Endoscopy · Deep learning

1 Introduction

Minimally invasive surgery (MIS) is a surgical procedure where the operation is conducted via a few incision holes. It is favorable over open surgery due to its clinical benefits such as small scars, lower chances of bleeding and infection, and shorter recovery time. However, MIS forces the surgeon to lose direct eye contact with the scene and, consequently, to rely on endoscopic video for the whole

M. de Bruijne et al. (Eds.): MICCAI 2021, LNCS 12902, pp. 383–393, 2021.
https://doi.org/10.1007/978-3-030-87196-3_36

surgery. The limited field of view (FoV) and 2D nature of endoscopic images are challenges that surgeons face which quite often result in surgeons failing to identify the tissue structures and recourse to visual surveying by moving the camera around. In arthroscopy, this phenomenon happens repeatedly during surgery, which could prolong the operation time and lead to unintentional damage to critical tissue. According to a survey on knee arthroscopy [10,28], about 50% of the surgeons admitted to damage a knee once every 10 operations.

Computer vision could assist surgeons by augmenting the reality produced by the endoscopic image with the creation of a 3D semantic map of the scene, color-coded to represent the different anatomical structures and surgical tools. To produce a 3D semantic map in arthroscopy, we need to solve the following challenges: semantic segmentation, depth estimation, and pose estimation. In other types of MIS, like laparascopy and sinus endoscopy, techniques such as simultaneous localization and mapping (SLAM) [8] and structure from motion (SfM) [12] have been applied successfully. However, such techniques will fail in arthroscopy, due to poor texture, lack of photometric constancy across the frames, and assumptions regarding the camera motion.

In this paper, we introduce the first method to produce 3D semantic maps from arthroscopy. To achieve this goal, we create two datasets: 1) out-of-distribution (OOD) datasets containing non-human knees that have camera pose annotation, and 2) an in-distribution (ID) dataset containing human knees that have semantic segmentation annotation of femur, Articular Cartilage Ligament (ACL), and meniscus. We train a system with the OOD datasets to estimate depth+pose using self-supervised view synthesis loss + supervised pose loss. We also train a method to produce semantic segmentation using the ID dataset in [11]. We then combine the pose, depth, and semantic segmentation of both systems and use the method in [30] to produce 3D semantic maps of testing images from the ID dataset. Quantitative results of the pose estimation and qualitative visual results from the 3D semantic maps suggest that our approach can be reliably used for mapping human knees, even though part of the training was based on OOD training sets. To the best of our knowledge, this is the first method that can estimate the depth and pose from arthroscopy and the first to create 3D semantic maps in clinical endoscopy.

2 Related Work

Deep learning has shown impressive results in complex computer vision tasks such as segmentation, depth perception, and pose estimation [7,26,31]. These approaches work well on feature rich datasets like road scenes but perform poorly for environments such as medical endoscopy as shown in [24]. This is because of poor texture information and the lack of photometric constancy between frames in endoscopy due to the joint motion between the camera and light source [14]. Recently, depth and pose estimation methods above have been adapted for colonoscopy [3,4,17] and sinus endoscopy [14–16]. Compared to the original work on self-supervised estimation of depth and pose shown in [7,26,31], a key aspect to these proposed methods is the incorporation of supervision for depth+pose estimation. For example, [3,14–17] used structure from motion (SfM) [25] to create sparse depth frames from the training images and used them for supervision of the depth+pose training. Arthroscopy images have little texture information due to the smooth bone surfaces. Furthermore, the problem of over and under illumination in arthroscopy is a frequent occurrence that will impact the approaches above [2]. As a result, feature tracking based techniques such as SfM, cannot create reliable feature maps in arthroscopy as has been shown in [18].

Hence, we advocate the use of pose annotation acquired from images from non-human environments to supervise the training of depth+pose using a self-supervised+supervised loss function. We also trained a novel supervised model for semantic segmentation with the method in [1] that extends the semantic segmentation in [11] based on the use of multi-spectral frame reconstruction [20]. By considering that the biological compositions of each tissue type namely bone, ACL, and meniscus are intrinsically different, the RGB arthroscopic frames are transformed into 36 spectral bands and then the spatial features of anatomical structures are used at wavelengths from 380–740 nm with 10 nm of intervals as a preprocessing step. A segmentation network extracts spatial characteristics at these 36 spectral bands and subsequently learns the location along with its label.

3 Methods

The aim of depth + pose network, for a given source image at time t, $\mathbf{I}_t$, and source frames, $\mathbf{I}_S$, is to estimate the pixel level depth $\hat{\mathbf{D}}_t = f_D(\mathbf{I}_t)$ and the ego motion $\mathbf{X}^l_{t\rightarrow t+\Delta t} = [Rot\ Trl]$, where $Rot = [\alpha, \beta, \gamma]$ and $Trl = [x, y, z]$ refer to the 6 degree of freedom, rotation and translation, in the Euler coordinates We achieve this by training the depth+pose network on the self-supervised plus supervised objectives. In our case, with a stereo endoscope, the source images $\mathbf{I}_S$ are the left image at time $t+1 : \mathbf{I}^l_{t+1}$ and the right image at time $t : \mathbf{I}^r_t$, while the target image is the left image at time $t : \mathbf{I}^l_t$.

Self-supervised Objective minimizes a photometric reprojection error between the synthesized target image, $\hat{\mathbf{I}}_t$ and the target image, $\mathbf{I}_t$ as shown in [6,31] and edge-aware smoothing term as shown in [7,29]:

$$L_{self}(\mathbf{I}_t, \hat{\mathbf{I}}_t) = L_{phot}(\mathbf{I}_t, \mathbf{I}_S) \odot \mathbf{M}_{phot} + \lambda_{smoo} L_{smoo}(\hat{\mathbf{D}}) \tag{1}$$

where $L_{phot}(\mathbf{I}_t, \hat{\mathbf{I}}_t)$ is the pixel level photometric reprojection error shown in [6] and consist of structural similarity (SSIM) term [27] and a $L1$ loss:

$$L_{phot}(\mathbf{I}_t, \hat{\mathbf{I}}_t) = \alpha\ \frac{1 - SSIM(\mathbf{I}_t, \hat{\mathbf{I}}_t)}{2} + (1-\alpha)||\mathbf{I}_t - \hat{\mathbf{I}}_t||, \tag{2}$$

where $\alpha = 0.85$. Similar to [7], the minimum reprojection error is used to minimize the effect of the pixels which are not visible in some of the source images compared with the target image due to ego motion or occlusion:

$$L_{phot}(\mathbf{I}_t, \mathbf{I}_s) = \min_{\mathbf{I}_S} L_{phot}(\mathbf{I}_t, \hat{\mathbf{I}}_t). \tag{3}$$

The minimum reprojection loss is particularly helpful in reducing the edge artifacts of the depth. The auto masking term $\mathbf{M}_{phot}$ is a binary mask to reject the pixels with no change in appearance between frames such as static scenes and the moving objects at the same velocity and orientation of the camera [7]:

$$\mathbf{M}_{phot} = \min_{\mathbf{I}_S} L_{phot}(\mathbf{I}_t, \mathbf{I}_s) > \min_{\mathbf{I}_S} L_{phot}(\mathbf{I}_t, \hat{\mathbf{I}}_t). \tag{4}$$

Similar to [6] the weighted edge-aware term $\lambda_{smoo} L_{smoo}(\hat{\mathbf{D}})$ is used to regularize the depth on low texture areas:

$$L_{smoo}(\hat{\mathbf{D}}) = |\delta_x \hat{\mathbf{D}}_t| e^{-|\delta_x \mathbf{I}_t|} + |\delta_y \hat{\mathbf{D}}_t| e^{-|\delta_y \mathbf{I}_t|}|, \tag{5}$$

where δ refers to the gradient function is 1e-3.

Supervised Objective is to minimize the error between the estimated ego motion by the pose network $\hat{\mathbf{X}}_{t \to t+1} = [\hat{Rot}\ \hat{Trl}]$ and the groundtruth relative camera pose $\mathbf{X}_{t \to t+1} = [Rot\ Trl]$:

$$\begin{aligned} L_{pose}(\mathbf{X}_{t \to t+1}, \hat{\mathbf{X}}_{t \to t+1}) = L_{Trl} + L_{Ang}, \quad and \\ L_{Trl} = L'_{Trl} + L''_{Trl}, \ \ L'_{Trl} = ||Trl' - \hat{Trl}'||, \ \ L''_{Trl} = ||Trl - \hat{Trl}||, and \\ L_{Ang} = L'_{Ang} + L''_{Ang}, \ \ L'_{Ang} = ||Ang' - \hat{Ang}'||, \ \ L''_{Ang} = ||Ang - \hat{Ang}||. \end{aligned} \tag{6}$$

The term L'_{Trl} is calculated on the normalized translations, i.e., $Trl' = Trl/||Trl||$ and $\hat{Trl}' = \hat{Trl}/||\hat{Trl}||$. Similarly, $Ang' = Ang/||Ang||$ and $\hat{Ang}' = \hat{Ang}/||\hat{Ang}||$. This is because the relative displacement from frame to frame could substantially change for the endoscopic sequences with some frames having more than 20 times change in translation or angle compared with other frames. Without the L'_{Trl}, the network performs poorly for the frames with small changes in motion.

The final loss equation for training the depth+pose network in this work is:

$$L_{tot}(\mathbf{I}_t, \hat{\mathbf{I}}_t) = L_{self}(\mathbf{I}_t, \hat{\mathbf{I}}_t) + L_{pose}(\mathbf{X}_{t \to t+1}, \hat{\mathbf{X}}_{t \to t+1}) \tag{7}$$

Since most of the variation in the camera pose is in at the x and y axes of the translation, the weighting of [0.5, 0.5, 1] was applied to the L_{Trl}. Figure 1 shows the pipeline for training the depth+pose and segmentation networks.

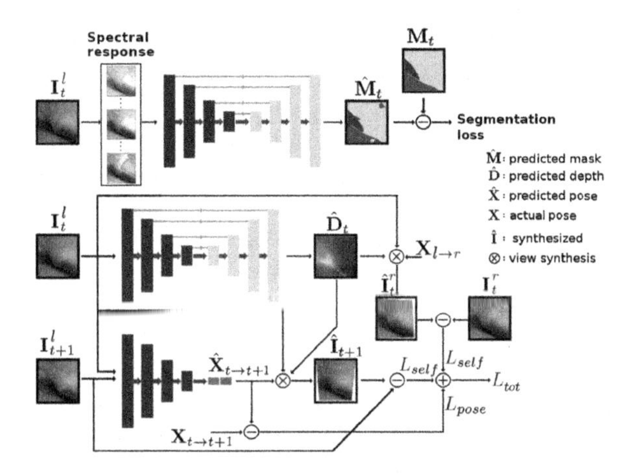

Fig. 1. The training pipeline for the segmentation, depth and pose estimation. The upper network performs semantic segmentation in a supervised manner. The second and third networks are the depth+pose networks being trained jointly using the supervise+self-supervised approach.

The method in [30] was used to fuse n depth frames and create the 3D maps in chunks. Since in the actual arthroscopy, the endoscope movement is limited to areas surrounding the incision holes, the sequences used for fusion are typically 3 to 8 s long ($n \in \{70, ..., 200\}$ frames), which correspond to sweep in translation to cover a certain part of the knee accessible via the incision hole.

4 Experimental Setup

4.1 Depth+Pose Training

Training + validation data was recorded using the stereo endoscope with 384×384 resolution, 1.52 mm baseline, 87.5° FoV, and 25 fps. Stereo images were rectified and downsampled to 256×256. The groundtruth poses of the camera tip were recorded by attaching an NDI magnetic sensor.

Liu et al. [13] showed that training a self-supervised depth network directly on the arthroscopy videos failed due to the poor texture and pretraining on texture rich frames is crucial. Therefore, for the training, a 3D printed model of the knee was placed in a water tank and recorded while the magnetic sensor was attached to the camera tip to provide the groundtruth camera poses. The images of the 3D printed knee provided rich texture and are ideal for the training phase. Another data was recorded from a sheep joint with the corresponding groundtruth camera poses. Similar to the human knee, the animal joint video frames suffer from the poor texture problem. During the training and validation, the images from the 3D printed knee and the animal joint were used at the same time.

In total 28500 frames from the 3D printed knee and 10000 frames from the animal experiment were recorded, out of which 25000 from the 3D printed and

8800 from the animal experiment were used for training and the rest for validation. The Absolute Trajectory Error (ATE) was used to quantify the error between the network estimated and the groudtruth poses.

The testing data was obtained from a cadaveric experiment in which multiple sequences were recorded from a left knee. About 9 sequences were recorded with varying lengths, few seconds to a few minutes. This resulted in approximately 12000 images. No groundtruth camera pose was available for these sequences.

The depth network architecture was similar to Disp-Net [19], but also uses skip connection [22] from the encoder's activation blocks, leading to higher resolution details [6] with sigmoids at the output. For the pose, only the encoder was used. The ResNet50 [9] was used as the encoder for both depth and pose networks. The weights were pretrained on ImageNet [23]. The training augmentations was used with 50% possibility of random brightness, contrast, saturation, and hue jitter with respective ranges of ± 0.2, ± 0.2, ± 0.2, and ± 0.1 [7]. The model was trained on 30 epochs, using Adam [5] optimizer, with a batch size of 18. Since most of the variation in the camera pose is in at the x and y axes of the translation, the weighting of [0.5, 0.5, 1] was applied to the translation loss.

4.2 Segmentation Training

Data from four cadaveric experiments were used for training and testing. Data from the last experiment were used as the test data for the depth+pose networks [11]. There were 2868 images from the first experiment and 1524 from the last experiment (two sequences among nine) that were used for training. The remaining images from the last experiment (3460 frames) were used for testing along with the other three cadaver experiments. We test on two sets: i) high quality of images, and ii) all remaining cadaver datasets, excluding saturated and bad frames. It has been confirmed that the accuracy of the proposed method can be improved if high quality imaging system and sufficient information about the irregular knee geometrical structures are provided. More details are available on [1]. The training data was augmented 6 times using shift and rotation (with angles 90, 180, 270), flip vertical and horizontal, and brightness changes.

The model in [1] is a U-Net with the contraction layer containing two successive convolution layers and a 3×3 kernel. The spatial context map is downsampled by max pooling operation with pool size 2×2. Padding 'same' is used to get the same resolution of input and output images. Kernel initializer is used to set initialize weights of the convolution layer during training. The dilation rate is set to 2 which provides a wider field of view so that it can avoid adjacent pixels having the same reflectance. The softmax is used at the final layer. Categorical cross-entropy loss function and Stochastic gradient descent optimizer are used for training. The Tensorflow [21] was used to implement all the models.

5 Results

The training and validation losses of the depth+pose networks are shown in Fig. 2. For comparison between the supervised+self-supervised depth+pose estimation using images at time t + 1 and the stereo pair, we included the plot

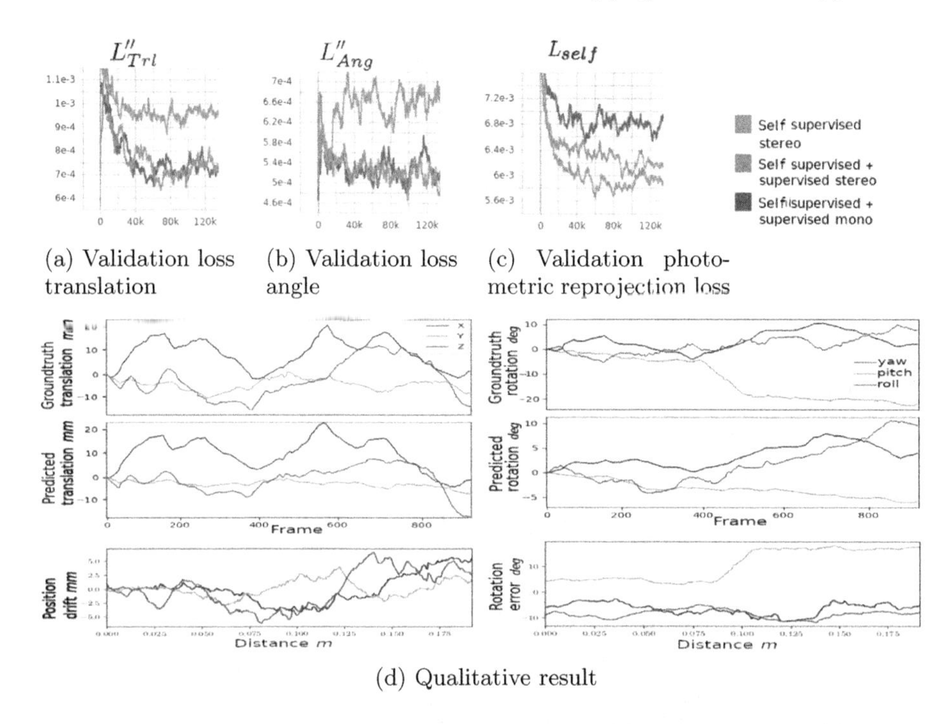

(d) Qualitative result

Fig. 2. The validation losses are shown in subfigure (a) for translation, (b) for the angle, and (c) for photometric reprojection loss. The qualitative results are showing the actual camera poses for sheep joint in (d). The first row of (d) is the groundtruth, the middle row is the corresponding network predictions (supervised+self-supervised stereo) and the third row is the ATE. The left column is the translation and the right column is the angle (rotation).

for the self-supervised counterpart as well as supervised+self-supervised using image at time t + 1 only, i.e., mono supervised+self-supervised. In this way, it is possible to evaluate the impact of the pose supervision and stereo versus mono scenario.

According to the plots on the training and validation data, Fig. 2(a) and (b) respectively, the photometric reprojection loss (which is an indication of the combined accuracy of camera pose and the depth estimation) is lowest for the self-supervised network. It is closely followed by the supervised+self-supervised stereo networks. The supervised+self-supervised mono has the poorest outcome with the photometric reprojection loss. On the other hand, the supervised+self-supervised networks outperformed the self-supervised network on the camera pose estimation for the validation data as shown in Fig. 2(c) and (d). These results from Fig. 2 indicate that the supervised+self-supervised networks using the stereo pairs and time t + 1 as reference images have higher accuracy in depth and slightly higher on camera pose estimation compared with the supervised+self-supervised mono. Hence, this model was considered for 3D mapping of the scene.

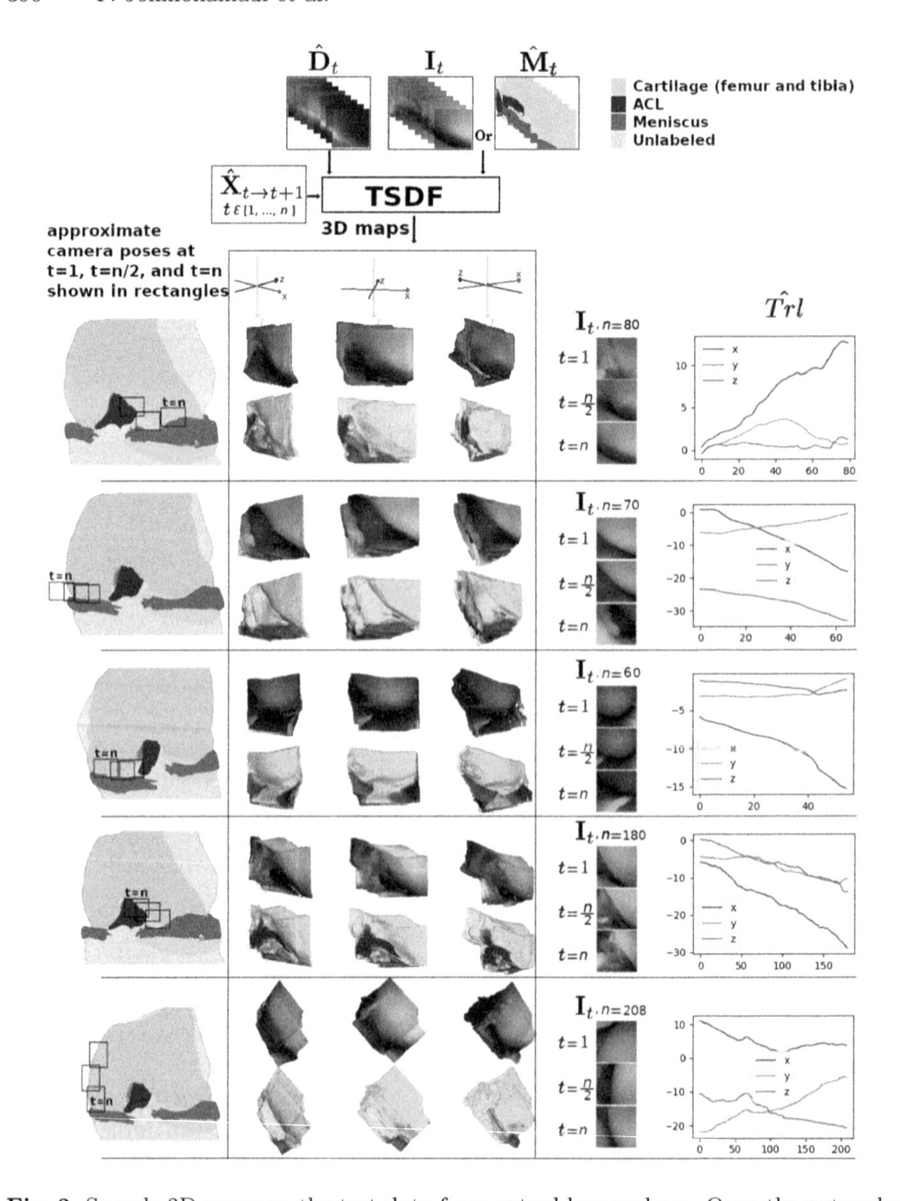

Fig. 3. Sample 3D maps on the test data from actual human knee. Once the networks are trained, the arthroscopic frames are provided as input to them and their output of depth $\hat{\mathbf{D}}_t$ and camera pose, $\hat{\mathbf{X}}_{t\rightarrow t+1}$, will be given as inputs to the TSDF function [30] to create the extended map. Either the image $\hat{\mathbf{D}}_t$ or the corresponding semantic label, $\hat{\mathbf{M}}_t$, can be provided as the 3rd input to the TSDF function. The variable n refers to the number of the frames used to create the corresponding map. Since most of the variation in $\hat{\mathbf{X}}$ is due to translation, only the $\hat{\mathbf{Trl}}$ was shown in the figure. The knee models on the left hand side of the figure with dark squares, show the approximate locations of the camera with respect to the actual knee.

The actual pose network predictions on the validation data (animal experiment) are shown in Fig. 2(d). The groundtruth is shown in the first row while the prediction is on the second row and the corresponding ATE on the 3rd row. In general, the changes in rotation proved to be harder for the network to predict than translation. Overall, the prediction of the camera rotation was more difficult than the translation. Figure 3 shows sample 3D maps obtained by fusing chunks of arthroscope frames from a human cadaver knee. The number of frames to create the maps is indicated by n. The corresponding camera translations are shown in the plots on the right side of each map. For every map, the semantic map is also provided with green being the cartilage (femur and tibia), meniscus in red, and ACL in blue. The cyan refers to other structures such as typically floating fat and skin. The segmentations appear correct except for a minor error in the third map where the segmentation network falsely detects meniscus (red) on the right hand side of the map.

6 Conclusion

In this work for the first time, we presented a pipeline to perform 3D semantic mapping in arthroscopy. To the best of our knowledge, this has not been done in any medical endoscopy before. To achieve these, we used the deep learning approaches for semantic segmentation, depth perception and camera pose estimation. The proposed domain adaptive approach produced superior accuracy in camera pose estimation and comparable depth accuracy in comparison with the self-supervised counterpart. Furthermore, we used a segmentation tool to semantically segment the images into cartilage, ACL, and meniscus. The segmentation approach utilizes the multi-spectral properties of surgical tissue in the images rather than merely the geometrical cues as was shown in [11,13].

Acknowledgements. Supported by AISRF53820 and Australian Research Council through grants DP180103232 and FT190100525.

References

1. Ali, S., Jonmohamadi, Y., Roberts, J., Crawford, R., Carneiro, G., Pandey, A.K.: Arthroscopic multi-spectral scene segmentation using deep learning. arXiv preprint arXiv:2001.05566 (2021)
2. Ali, S., Jonmohamadi, Y., Takeda, Y., Roberts, J., Crawford, R., Pandey, A.K.: Supervised scene illumination control in stereo arthroscopes for robot assisted minimally invasive surgery. IEEE Sens. J. **4**(10), 11577–11587 (2020)
3. Bae, G., Budvytis, I., Yeung, C.-K., Cipolla, R.: Deep multi-view stereo for dense 3D reconstruction from monocular endoscopic video. In: Martel, A.L., et al. (eds.) MICCAI 2020. LNCS, vol. 12263, pp. 774–783. Springer, Cham (2020). https://doi.org/10.1007/978-3-030-59716-0_74
4. Chen, R.J., Bobrow, T.L., Athey, T., Mahmood, F., Durr, N.J.: Slam endoscopy enhanced by adversarial depth prediction. arXiv preprint arXiv:1907.00283 (2019)

5. Da, K.: A method for stochastic optimization. arXiv preprint arXiv:1412.6980 (2014)
6. Godard, C., Mac Aodha, O., Brostow, G.J.: Unsupervised monocular depth estimation with left-right consistency. In: Proceedings of the IEEE Conference on Computer Vision and Pattern Recognition, pp. 270–279 (2017)
7. Godard, C., Mac Aodha, O., Firman, M., Brostow, G.J.: Digging into self-supervised monocular depth estimation. In: Proceedings of the IEEE International Conference on Computer Vision, pp. 3828–3838 (2019)
8. Grasa, O.G., Bernal, E., Casado, S., Gil, I., Montiel, J.M.M.: Visual SLAM for handheld monocular endoscope. IEEE Trans. Med. Imag. **33**(1), 135–146 (2013)
9. He, K., Zhang, X., Ren, S., Sun, J.: Deep residual learning for image recognition. In: Proceedings of the IEEE Conference on Computer Vision and Pattern Recognition, pp. 770–778 (2016)
10. Jaiprakash, A., et al.: Orthopaedic surgeon attitudes towards current limitations and the potential for robotic and technological innovation in arthroscopic surgery. J. Orthop. Surg. **25**(1), 2309499016684993 (2017)
11. Jonmohamadi, Y., et al.: Automatic segmentation of multiple structures in knee arthroscopy using deep learning. IEEE Access **8**, 51853–51861 (2020)
12. Leonard, S., et al.: Evaluation and stability analysis of video-based navigation system for functional endoscopic sinus surgery onin vivoclinical data. IEEE Trans. Med. Imag. **37**(10), 2185–2195 (2018)
13. Liu, F., Jonmohamadi, Y., Maicas, G., Pandey, A.K., Carneiro, G.: Self-supervised Depth Estimation to Regularise Semantic Segmentation in Knee Arthroscopy. In: International Conference on Medical Image Computing and Computer-Assisted Intervention. pp. 594–603. Springer (2020). https://doi.org/10.1007/10704282
14. Liu, X., et al.: Dense depth estimation in monocular endoscopy with self-supervised learning methods. IEEE Trans. Med. Imag. **39**(5), 1438–1447 (2019)
15. Liu, X., et al.: Self-supervised Learning for Dense Depth Estimation in Monocular Endoscopy. arXiv pp. arXiv-1806 (2018)
16. Liu, X., et al.: Reconstructing sinus anatomy from endoscopic video-towards a radiation-free approach for quantitative longitudinal assessment. arXiv preprint arXiv:2003.08502 (2020)
17. Ma, R., Wang, R., Pizer, S., Rosenman, J., McGill, S.K., Frahm, J.-M.: Real-time 3D reconstruction of colonoscopic surfaces for determining missing regions. In: Shen, D., et al. (eds.) MICCAI 2019. LNCS, vol. 11768, pp. 573–582. Springer, Cham (2019). https://doi.org/10.1007/978-3-030-32254-0_64
18. Marmol, A., Banach, A., Peynot, T.: Dense-arthroSLAM: Dense intra-articular 3-D reconstruction with robust localization prior for arthroscopy. IEEE Robot. Autom. Lett. **4**(2), 918–925 (2019)
19. Mayer, N., et al.: A large dataset to train convolutional networks for disparity, optical flow, and scene flow estimation. In: 2016 IEEE Conference on Computer Vision and Pattern Recognition (CVPR). pp. 4040–4048. IEEE, Las Vegas, NV, USA, June 2016. /DOIurl 0.1109/CVPR.2016.438, http://ieeexplore.ieee.org/document/7780807/
20. Otsu, H., Yamamoto, M., Hachisuka, T.: Reproducing spectral reflectances from tristimulus colours. Comput. Graph. Forum (2018). https://doi.org/10.1111/cgf.13332
21. Paszke, A., et al.: Gross automatic differentiation in pytorch (2017)

22. Ronneberger, O., Fischer, P., Brox, T.: U-Net: convolutional networks for biomedical image segmentation. In: Navab, N., Hornegger, J., Wells, W.M., Frangi, A.F. (eds.) MICCAI 2015. LNCS, vol. 9351, pp. 234–241. Springer, Cham (2015). https://doi.org/10.1007/978-3-319-24574-4_28
23. Russakovsky, O., et al.: Imagenet large scale visual recognition challenge. Int. J. Comput. Vis. **115**(3), 211–252 (2015)
24. Sharan, L., et al.: Domain gap in adapting self-supervised depth estimation methods for stereo-endoscopy. Curr. Direct. Biomed. Eng. **6**(1) (2020)
25. Ullman, S.: The interpretation of structure from motion. Proc. R. Soc. Lond. Ser B Biol. Sci. **203**(1153), 405–426 (1979)
26. Vijayanarasimhan, S., Ricco, S., Schmid, C., Sukthankar, R., Fragkiadaki, K.: Sfm-net: learning of structure and motion from video. arXiv preprint arXiv:1704.07804 (2017)
27. Wang, Z., Bovik, A.C., Sheikh, H.R., Simoncelli, E.P.: Image quality assessment: from error visibility to structural similarity. IEEE Trans. Image Process. **13**(4), 600–612 (2004)
28. Wu, L., et al.: Robotic and image-guided knee arthroscopy. In: Handbook of Robotic and Image-Guided Surgery, pp. 493–514. Elsevier, Amesterdam (2020)
29. Yang, Z., Wang, P., Wang, Y., Xu, W., Nevatia, R.: Lego: learning edge with geometry all at once by watching videos. In: Proceedings of the IEEE Conference on Computer Vision and Pattern Recognition, pp. 225–234 (2018)
30. Zach, C., Pock, T., Bischof, H.: A globally optimal algorithm for robust tv-l 1 range image integration. In: 2007 IEEE 11th International Conference on Computer Vision, pp. 1–8. IEEE (2007)
31. Zhou, T., Brown, M., Snavely, N., Lowe, D.G.: Unsupervised learning of depth and ego-motion from video. In: Proceedings of the IEEE Conference on Computer Vision and Pattern Recognition, pp. 1851–1858 (2017)

Semi-Supervised Unpaired Multi-Modal Learning for Label-Efficient Medical Image Segmentation

Lei Zhu[1(✉)], Kaiyuan Yang[1], Meihui Zhang[2], Ling Ling Chan[3], Teck Khim Ng[1], and Beng Chin Ooi[1]

[1] National University of Singapore, Singapore, Singapore
e203764@u.nus.edu

[2] Beijing Institute of Technology, Beijing, China

[3] Singapore General Hospital, Singapore, Singapore

Abstract. Multi-modal learning using unpaired labeled data from multiple modalities to boost the performance of deep learning models on each individual modality has attracted a lot of interest in medical image segmentation recently. However, existing unpaired multi-modal learning methods require a considerable amount of labeled data from both modalities to obtain satisfying segmentation results which are not easy to obtain in reality. In this paper, we investigate the use of unlabeled data for label-efficient unpaired multi-modal learning, with a focus on the scenario when labeled data is scarce and unlabeled data is abundant. We term this new problem as Semi-Supervised Unpaired Multi-Modal Learning and thereupon, propose a novel deep co-training framework. Specifically, our framework consists of two segmentation networks, where we train one of them for each modality. Unlabeled data is effectively applied to learn two image translation networks for translating images across modalities. Thus, labeled data from one modality is employed for the training of the segmentation network in the other modality after image translation. To prevent overfitting under the label scarce scenario, we introduce a new semantic consistency loss to regularize the predictions of an image and its translation from the two segmentation networks to be semantically consistent. We further design a novel class-balanced deep co-training scheme to effectively leverage the valuable complementary information from both modalities to boost the segmentation performance. We verify the effectiveness of our framework with two medical image segmentation tasks and our framework outperforms existing methods significantly.

Keywords: Semi-supervised learning · Unpaired multi-modal learning · Deep co-training · Segmentation

---

Electronic supplementary material The online version of this chapter (https://doi.org/10.1007/978-3-030-87196-3_37) contains supplementary material, which is available to authorized users.

M. de Bruijne et al. (Eds.): MICCAI 2021, LNCS 12902, pp. 394–404, 2021.
https://doi.org/10.1007/978-3-030-87196-3_37

1 Introduction

Multi-Modal Learning is a well-researched area with various applications in medical image analysis, where it leverages complementary information from different modalities to learn a better performing model compared with single modal learning. However, Multi-Modal Learning often requires paired and spatially well-aligned images from different modalities, which is expensive and sometimes even infeasible to collect. Recently, Unpaired Multi-Modal Learning [2,3,14] has been proposed, which applies unpaired labeled data from multiple modalities to boost the performance of deep learning models on each individual modality and has attracted a lot of interest.

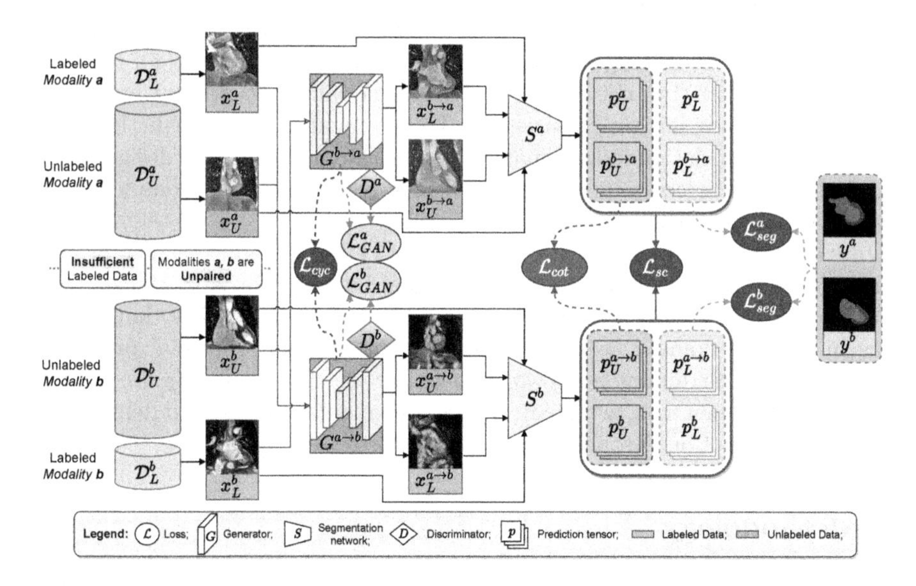

Fig. 1. Architecture and dataflow of our proposed deep co-training framework. Note the scarcity of labeled data for both modalities, and modalities a and b are unpaired.

Valindria et al. [14] are the first to systematically investigate different dual-stream convolutional neural networks for unpaired MRI/CT multi-organ segmentation, where they find an "X"-shaped model is more effective for multi-modal learning compared to other network architectures they investigated. Recently, Dou et al. [3] propose a compact "Chilopod"-shaped model and introduce a novel knowledge distillation loss for joint training of unpaired MRI/CT data for both cardiac substructure and abdominal multi-organ segmentation. Chen et al. [2] propose a deep class-specific affinity-guided convolutional network, which shares the same "Chilopod"-shaped architecture but with a newly proposed class-specific affinity loss to distill between-layer relationships across modalities for unpaired multi-modal learning. While existing unpaired multi-modal learning methods can effectively leverage complementary information from different modalities to boost the segmentation performance, however, they all require a considerable amount of labeled data from both modalities, which are often impractical to obtain in reality. In this paper, we investigate the use of unlabeled data for label-efficient unpaired multi-modal learning, where we particularly focus on the scenario when labeled data is scarce and unlabeled data is abundant. We term this new problem as Semi-Supervised Unpaired Multi-Modal Learning and propose a novel deep co-training framework to address the problem. In a nutshell, our framework consists of two segmentation networks, where we train each segmentation network with labeled data from a corresponding modality. The main challenge faced by existing unpaired multi-modal learning methods when labeled data is scarce is that there may not exist sufficient supervision signal for the training of the segmentation networks. To alleviate the problem, our framework applies unlabeled data to learn two image translation networks so that labeled data from one modality could be employed for the training of the segmentation network in the other modality. To prevent overfitting, we introduce a new semantic consistency loss to regularize the predictions of an image and its translation to be semantically consistent. Finally, we design a novel class-balanced deep co-training scheme to effectively exploit the complementary information from both modalities.

In summary, our main contributions are: **(1)** We introduce a new problem termed Semi-Supervised Unpaired Multi-Modal Learning for label-efficient medical image segmentation, which is of great practical value in view that annotations in medical domain are difficult to obtain; **(2)** We propose a principally designed deep co-training framework to tackle the problem; **(3)** We conduct extensive experiments on two medical image segmentation tasks to demonstrate the effectiveness of our proposed framework.

2 Methodology

Denote $\mathcal{D}_L^a = \{(x_{L,i}^a, y_{L,i}^a)\}_{i=1}^{N^a}$ and $\mathcal{D}_L^b = \{(x_{L,i}^b, y_{L,i}^b)\}_{i=1}^{N^b}$ as two unpaired labeled dataset in modalities a and b, where $y_{L,i}$ is the corresponding segmentation mask for $x_{L,i}$. We assume there exist two additional unpaired unlabeled dataset $\mathcal{D}_U^a = \{x_{U,i}^a\}_{i=1}^{M^a}$ and $\mathcal{D}_U^b = \{x_{U,i}^b\}_{i=1}^{M^b}$ in the two modalities. We focus on the scenario in which N^a and N^b are extremely small and $N^a \ll M^a$, $N^b \ll M^b$. The task is to combine both labeled dataset and unlabeled dataset to learn accurate deep segmentation models for the two modalities. To the best of our knowledge, this is the first work that uses both unpaired labeled data and unpaired unlabeled data for multi-modal learning. Figure 1 presents an overview of our proposed deep co-training framework for this problem.

2.1 Segmentation

As illustrated in Fig. 1, our framework comprises two segmentation networks S^a and S^b for modalities a and b. We utilize the two labeled dataset $\mathcal{D}_L^a$ and $\mathcal{D}_L^b$ for the training of the two segmentation networks with the following loss:

$$\mathcal{L}^a_{seg}(S^a) = \mathbb{E}_{x^a_L \sim \mathcal{D}^a_L}[H(y^a_L, S^a(x^a_L)) + Dice(y^a_L, S^a(x^a_L))] \tag{1}$$

$$\mathcal{L}^b_{seg}(S^b) = \mathbb{E}_{x^b_L \sim \mathcal{D}^b_L}[H(y^b_L, S^b(x^b_L)) + Dice(y^b_L, S^b(x^b_L))] \tag{2}$$

where $H(\cdot)$ calculates the pixel-wise cross-entropy loss and $Dice(\cdot)$ calculates the *Dice* loss [4].

2.2 Image Translation and Semantic Consistency

The main challenge of unpaired multi-modal learning when labeled data is scarce is that there may not exist sufficient supervision signal for the training of deep segmentation networks. Due to their large capacity, deep segmentation networks will easily overfit to the labeled data and result in poor generalization performance when in deployment. To tackle this problem, we novelly apply the abundant unlabeled data to learn two image translation networks $G^{a\rightarrow b}$ and $G^{b\rightarrow a}$ for translating images across modalities. (Note the learning of the translation networks does not require data annotation.) Image translation reduces the distribution shift across modalities at pixel level [1], thus we utilize the translated labeled images to train each segmentation network. In this way, we increase the supervision signal for the training of each segmentation network. The segmentation loss with translated images is defined as follows:

$$\mathcal{L}^{b\rightarrow a}_{seg}(G^{b\rightarrow a}, S^a) = \mathbb{E}_{x^b_L \sim \mathcal{D}^b_L}[H(y^b_L, S^a(G^{b\rightarrow a}(x^b_L))) + Dice(y^b_L, S^a(G^{b\rightarrow a}(x^b_L)))] \tag{3}$$

$$\mathcal{L}^{a\rightarrow b}_{seg}(G^{a\rightarrow b}, S^b) = \mathbb{E}_{x^a_L \sim \mathcal{D}^a_L}[H(y^a_L, S^b(G^{a\rightarrow b}(x^a_L))) + Dice(y^a_L, S^b(G^{a\rightarrow b}(x^a_L)))] \tag{4}$$

Furthermore, we utilize the unlabeled data to regularize the training of the two segmentation networks together with labeled data. Our insight is that an image and its translation, assuming to be different views of the same object, should have the same prediction results when inputted into the respective segmentation network in each modality. Therefore, we introduce a new Semantic Consistency (SC) loss where we regularize the prediction result of a translated image so that it is consistent with the prediction result of the original image.

We define the SC loss with symmetric Kullback–Leibler (KL)-divergence as follows:

$$\begin{aligned}\mathcal{L}_{sc}(G^{a\rightarrow b}, G^{b\rightarrow a}, S^a, S^b) &= \mathcal{L}^a_{sc}(G^{a\rightarrow b}, S^b) + \mathcal{L}^b_{sc}(G^{b\rightarrow a}, S^a)\\ &= \mathbb{E}_{x^a \sim D^a_L \cup D^a_U}[Div(S^b(G^{a\rightarrow b}(x^a)), S^a(x^a))] + \mathbb{E}_{x^b \sim D^b_L \cup D^b_U}[Div(S^a(G^{b\rightarrow a}(x^b)), S^b(x^b))]\end{aligned} \tag{5}$$

where $Div(\cdot)$ measures the average pixel-wise symmetric KL-divergence between the two prediction masks. Note that with the SC regularization, only semantically consistent pair of segmentation networks S^a and S^b will be learned, thus effectively reducing the hypothesis space for the two segmentation networks and preventing overfitting.

2.3 Class-Balanced Deep Co-training

To further exploit the complementary information from both modalities, we design a novel class-balanced deep co-training scheme. Specifically, we assign pseudo-labels to unlabeled data with existing segmentation network in each modality in a class-balanced manner, and utilize the pseudo-labels to train the segmentation network in the other modality after image translation. Denote $\hat{y}_U^a = \text{argmax}\, S^a(x_U^a)$ and $\hat{y}_U^b = \text{argmax}\, S^b(x_U^b)$ as the segmentation masks of x_U^a and x_U^b respectively. We introduce two selection masks, denoted as M^a and M^b to selectively assign pseudo-labels to each pixel, where $M^a[i,j] = 1$ if the pixel of x^a at position (i,j) is labeled, otherwise $M^a[i,j] = 0$, similarly for M^b. Let "$\circ$" denote element-wise dot product, the co-training loss is defined as follows:

$$\mathcal{L}_{cot}(G^{a\to b}, G^{b\to a}, S^a, S^b) = \mathcal{L}_{cot}^a(G^{a\to b}, S^b) + \mathcal{L}_{cot}^b(G^{b\to a}, S^a)$$

$$= \mathbb{E}_{x^a\sim\mathcal{D}_U^a}[H(M^a \circ \hat{y}_U^a, S^b(G^{a\to b}(x^a)))] + \mathbb{E}_{x^b\sim\mathcal{D}_U^b}[H(M^b \circ \hat{y}_U^b, S^a(G^{b\to a}(x^b)))] \quad (6)$$

We assign pseudo-label to a pixel only if its predicted probability is among the top $\alpha\%$ in its predicted class within a training batch so that some large classes would not dominant the population of the pseudo-labels and the labeling error is reduced. In this way, the pseudo-labels are generated in a more class balanced manner because every class gets $\alpha\%$ of pixels which are predicted in the class labeled. We gradually increase α from 0 to an upper bound η along the training process so that we include more pseudo-labels when the segmentation networks become more accurate. With the co-training loss, confident knowledge from one modality is effectively transferred to the learning of the segmentation network in the other modality thus enhancing the segmentation performance mutually.

In summary, the overall objective for our framework is as follows:

$$\begin{aligned}\mathcal{L}_o = &\mathcal{L}_{seg}^a(S^a) + \mathcal{L}_{seg}^b(S^b) + \mathcal{L}_{seg}^{b\to a}(G^{b\to a}, S^a) + \mathcal{L}_{seg}^{a\to b}(G^{a\to b}, S^b) \\ &+ \lambda_{sc}\mathcal{L}_{sc}(G^{a\to b}, G^{b\to a}, S^a, S^b) + \lambda_{cot}\mathcal{L}_{cot}(G^{a\to b}, G^{b\to a}, S^a, S^b) \\ &+ \lambda_{adv}\mathcal{L}_{GAN}^a(G^{b\to a}, D^a) + \lambda_{adv}\mathcal{L}_{GAN}^b(G^{a\to b}, D^b) + \lambda_{cyc}\mathcal{L}_{cyc}(G^{a\to b}, G^{b\to a})\end{aligned} \quad (7)$$

where $\lambda_{sc} = 0.1$, $\lambda_{cot} = 1$, $\lambda_{adv} = 1$, $\lambda_{cyc} = 10$ by default. The $\mathcal{L}_{GAN}^a$, $\mathcal{L}_{GAN}^b$ and $\mathcal{L}_{cyc}$ losses in Eq. 7 are adopted for the training of the image translation networks. We introduce two discriminators D^a and D^b, where D^a discriminates whether an image is a translated image in modality a and the translation network $G^{b\to a}$ competes with D^a to generate more authentic translated images, similarly for D^b and $G^{a\to b}$. The image translation networks and the discriminators form two pairs of GANs and standard GAN loss $\mathcal{L}_{GAN}$ [5] are applied for the training. To preserve the shape and structure information during image translation, we also add a cycle consistency loss $\mathcal{L}_{cyc}$ [16] for the translation networks.

3 Experiments and Results

Datasets. We evaluate the effectiveness of our framework with datasets from three public medical image segmentation challenges, cardiac substructure segmentation [17] and abdominal multi-organ segmentation [7,8]. The cardiac dataset consists of 20 unpaired MRI and CT volumes with ground truth masks on four heart substructures: ascending aorta (AA), left atrium blood cavity (LAC), left ventricle blood cavity (LVC), and myocardium of the left ventricle (MYO). The abdominal dataset consists of 20 unpaired T2-SPIR MRI and 30 CT volumes collected from two public datasets with ground truth masks on four organs: spleen, right kidney, left kidney, and liver. In order to obtain similar field of view for all the volumes in each task, we manually crop the original scans to cover the structures/organs which we aim to segment. All the data are normalized with zero mean and unit variance and re-sampled into the size of 256 × 256. Coronal view of cardiac volumes and axial view of abdominal volume are used to train the 2D network. Each modality is randomly split with 80% scans for training and 20% for testing. We employ the Dice similarity coefficient (Dice), a commonly-used metric, to quantitatively evaluate the segmentation performance (Please refer to the supplementary material for experiment results on the average symmetric surface distance (ASD)). The evaluation is performed on the subject-level segmentation volume. Our experiment results show a significant performance increase over state-of-the-art methods.

Implementation. The architecture of our segmentation network consists of ten convolutional operation groups, two dilated convolutional groups [15] and one softmax layer. Image translation network follows the architecture in [16]. We train the entire framework end-to-end using Adam optimizer with learning rate of 2×10^{-4}, and we train the framework in two phases. In the initial phase, we warm up the image translation networks and segmentation networks without the co-training loss by setting $\lambda_{cot} = 0$ for $20k$ iterations. In the second phase, we add co-training loss by setting $\lambda_{cot} = 1$ and train the framework for another $20k$ iterations. The batch size is set to 8 with half labeled data and half unlabeled data on a NVIDIA GeForce GTX 1080p GPU. η is set to 0.5 in all experiments. The predictions from the two segmentation networks are ensembled for final prediction, where we take the average of their prediction probabilities.

Quantitative Comparison. We focus on unpaired multi-modal learning under the scenario when labeled data is scarce and unlabeled data is abundant. To this end, we conduct two experimental studies where we randomly select either 0.5% (approx. 48 images) or 2.5% (approx. 240 images) of data from the training set in accord with patient ID in both modalities as labeled dataset. For the remaining data, we exclude those with same patient ID in the labeled dataset and use the rest as unlabeled dataset. We compare our framework with state-of-the-art unpaired multi-modal learning methods with Y-shape architecture [14], X-shape architecture [14], and Ummkd [3]. We also compare with semi-supervised learning methods based on self-training [9], where we implement both Self-Training with single modality data, denoted as ST-single and Self-Training with joint modality data, denoted as ST-joint. For ST-joint, we apply modality specific batch normalization layers [6] in the network to reduce modal difference so that the network can effectively leverage shared cross-modality information for semi-supervised learning. (We provide additional experiment results on two cross-modality medical image segmentation methods, namely Dual-Teacher [10] and Omkd [11] in the supplementary material.) Finally, we implement the Single method, which only utilizes the labeled data from a single modality for learning as the baseline method. All comparison methods are implemented with the same network architecture for fair comparison.

Table 1 presents the quantitative performance of different methods. We have the following observations: 1). our framework significantly outperforms the comparison methods in both experimental studies and in both tasks and achieves high Dice score even under label scarce scenario. Specifically, our framework outperforms state-of-the-art unpaired multi-modal learning methods, demonstrating the important value of unlabeled data. For the abdominal multi-organ segmentation task with 0.5% of training set labeled, our framework has achieved a Dice score of 87.4, which has improved upon the second best by 7.4 points; 2). unpaired multi-modal learning methods generally outperform the baseline method via exploiting labeled data from both modalities; 3). semi-supervised learning methods which use additional unlabeled data for training exceed the baseline method in most cases; 4). ST-joint employs all available labeled and unlabeled data in both modalities for training. However, the simple ST-joint method cannot fully exploit the value of all data, while our principally designed method achieves significantly better performance; 5). finally, with the increase of the percentage of labeled data, all methods perform better.

Qualitative Comparison. Figure 2(a) shows the qualitative comparison results for cardiac substructure segmentation. As shown in the figure, the segmentation masks produced by our method are closer to the ground truth and contain fewer wrong semantic prediction results compared to other methods. (Please view the supplementary material for more qualitative comparison results.)

Ablation Study. Table 2 presents the ablation study of different components of our framework. With the gradual addition of Semantic Consistency (SC) and class-balanced deep co-training, our framework results in consistently improved performance, demonstrating the importance of each component in our framework.

Performance with Fully Labeled Data. We have mainly focused our study on the scenario when labeled data is scarce, but how will our method perform when labeled data become abundant? To answer this question, we conduct an experiment where we use the whole training set as labeled set and there is no unlabeled data. We compare our method with state-of-the-art method Ummkd [3]. We find that our framework even outperforms Ummkd in this scenario. As shown in Table 3, for abdominal multi-organ segmentation, our method has significantly improved the Dice score by 2 points for MRI segmentation and 3 points for CT segmentation. The experiment results verify the effectiveness of

Table 1. Segmentation performance of cardiac substructures (top) and abdominal multi-organ (bottom) tasks measured in Dice coefficients (%). Best and second best results are in red and blue bold respectively.

Cardiac segmentation (MRI/CT)										
	Labeled percentage = 0.5%					Labeled percentage = 2.5%				
Method	AA	LAC	LVC	MYO	Avg.	AA	LAC	LVC	MYO	Avg.
Single	68.0/60.9	64.0/78.2	84.6/82.5	64.9/71.6	70.4/73.3	81.4/73.7	81.0/88.0	90.4/89.7	74.5/83.5	81.8/83.8
Y-shape [14]	**70.2**/61.5	62.3/79.1	84.2/83.7	66.7/71.1	70.8/73.8	80.5/78.0	81.2/89.3	89.9/89.6	74.4/83.9	81.6/85.2
X-shape [14]	**70.2**/62.7	65.4/81.2	83.5/83.1	65.0/71.8	71.0/74.7	81.2/76.6	81.6/88.4	90.2/89.7	75.4/83.6	82.1/84.6
Ummkd [3]	69.7/70.1	**71.4/86.4**	88.3/84.7	65.6/71.9	73.8/78.2	**82.3**/78.8	**83.0**/91.2	92.3/91.2	77.9/85.3	**83.9/86.6**
ST-single [9]	70.0/63.8	67.2/80.5	82.7/83.4	67.0/73.2	71.7/75.2	80.9/74.4	80.6/88.2	90.0/90.0	75.1/83.7	81.7/84.1
ST-joint [9]	69.5/**79.5**	70.4/85.5	87.2/**84.9**	**69.1/77.2**	**74.0/81.8**	81.9/**83.7**	80.6/89.6	91.0/89.0	76.0/84.3	82.4/**86.6**
Ours	76.1/83.6	76.4/87.5	**87.7**/88.1	69.2/78.0	77.4/84.3	83.0/88.8	84.4/**90.1**	**92.0/90.7**	**77.0/85.0**	84.1/88.6

Abdominal Multi-organ segmentation (MRI/CT)										
	Labeled percentage = 0.5%					Labeled percentage = 2.5%				
Method	Spleen	R. Kidney	L. Kidney	Liver	Avg.	Spleen	R. Kidney	L. Kidney	Liver	Avg.
Single	65.7/58.1	73.5/68.5	71.4/69.4	80.9/70.7	72.9/69.4	79.3/63.8	84.3/75.3	77.7/70.5	81.4/72.9	80.7/71.6
Y-shape [14]	67.5/67.5	76.9/71.7	70.3/74.1	79.9/74.8	73.6/74.4	80.5/65.2	81.9/78.0	78.8/78.4	81.9/75.2	80.8/76.3
X-shape [14]	69.8/61.2	73.0/68.1	70.2/70.2	81.1/67.8	73.5/69.2	75.8/63.5	83.5/74.1	76.9/75.4	83.1/71.0	79.8/73.6
Ummkd [3]	72.0/70.3	86.3/80.3	76.4/75.8	**85.5/80.1**	**80.0**/77.3	76.1/74.5	89.4/81.7	80.4/76.4	85.6/80.0	**82.9**/79.1
ST-single [9]	71.2/62.2	82.4/71.7	73.7/73.6	60.8/78.0	72.0/72.0	71.2/**77.2**	**90.8**/79.2	86.3/74.8	62.6/73.7	77.7/76.8
ST-joint [9]	**77.5/76.4**	**90.0**/72.9	**85.5/81.9**	55.3/79.0	77.1/**78.9**	**83.4**/80.3	90.1/**80.3**	**86.8/78.5**	68.6/**80.6**	82.2/**80.4**
Ours	86.3/77.0	91.0/**80.1**	85.6/84.5	86.8/87.6	87.4/82.0	86.1/75.4	92.3/80.2	87.0/87.0	88.2/88.2	88.4/83.8

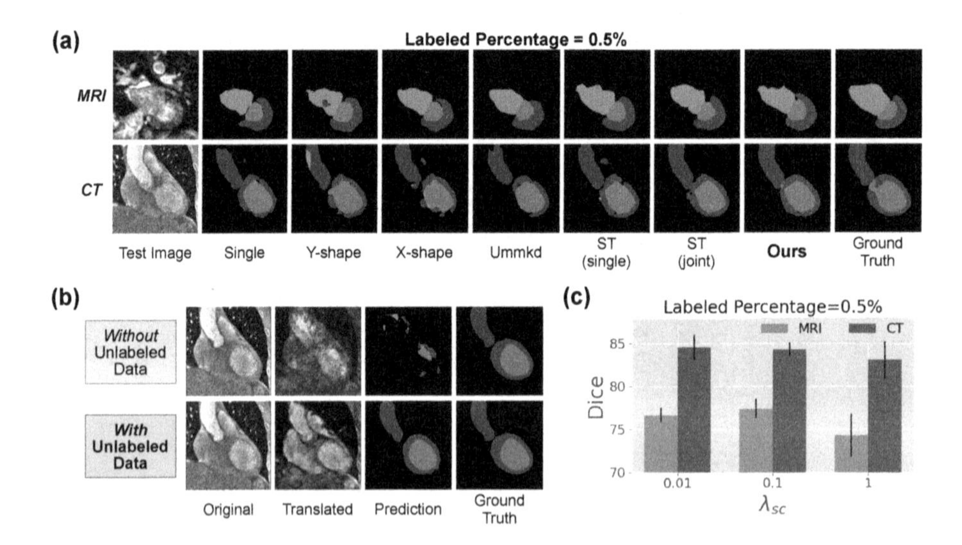

Fig. 2. (a) Visual comparisons of segmentation results on cardiac substructure segmentation. (b) Effect of unlabeled data on boosting segmentation performance. The cardiac substructure of AA, LAC, LVC and MYO are indicated in green, orange, purple, and blue masks respectively. (c) Sensitivity Analysis. (Color figure online)

Table 2. Ablation study for 0.5% labeled data in Dice score (%).

Cardiac task			
Modality	Plain	$+SC$	Ours
MRI	73.6	76.5	**77.4**
CT	81.2	83.4	**84.3**

Table 3. Comparison with Ummkd for fully labeled data in Dice score (%).

Modality	Cardiac task		Multi-organ task	
	Ummkd	Ours	Ummkd	Ours
MRI	85.6	**85.8**	92.4	**94.4**
CT	90.4	**90.9**	90.7	**93.7**

our framework design and demonstrates the wide applicability of our framework under various scenarios.

Effect of Unlabeled Data. We employ the abundant unlabeled data for the training of the two image translation networks. How does this help the performance of our framework? Figure 2(b) shows the translated images by the image translation networks with or without unlabeled data for training, and their prediction masks representing the amount of semantic information contained in the translated images. We find that training with unlabeled data enables translation networks to generate more authentic translated images and preserve more semantic information when compared to ground truth, as a consequence, translated images can better help in learning and regularizing the segmentation networks for better performance.

Sensitivity Analysis. We conduct sensitivity analysis of our framework on λ_{sc}. Figure 2(c) shows that our method is generally robust to the change of λ_{sc} and the default value of 0.1 gives the best results.

4 Conclusions

In this paper, we introduce Semi-Supervised Unpaired Multi-Modal Learning for label-efficient medical image segmentation. We then propose a novel deep co-training framework which significantly outperforms existing methods in two segmentation tasks. Our proposal serves to reduce time-consuming and expensive medical data annotation to facilitate wider adoption of learning solutions in healthcare. The proposed framework[1], which makes use of Apache SINGA [13] for distributed training, has been integrated into our MLCask [12] for handling healthcare images and analytics.

Acknowledgement. We would like to thank the reviewers for their helpful suggestions and comments. This research is supported by the National Research Foundation, Singapore under its AI Singapore Programme (AISG Award No: AISG-GC-2019-002). Meihui Zhang's work is supported by the National Natural Science Foundation of China (62050099). Ling Ling Chan's work is supported by the National Medical Research Council, Singapore (CSASI20nov-0008).

References

1. Bousmalis, K., Silberman, N., Dohan, D., Erhan, D., Krishnan, D.: Unsupervised pixel-level domain adaptation with generative adversarial networks. In: Proceedings of the IEEE Conference on Computer Vision and Pattern Recognition, pp. 3722–3731 (2017)
2. Chen, J., Li, W., Li, H., Zhang, J.: Deep class-specific affinity-guided convolutional network for multimodal unpaired image segmentation. In: Martel, A.L., et al. (eds.) MICCAI 2020. LNCS, vol. 12264, pp. 187–196. Springer, Cham (2020). https://doi.org/10.1007/978-3-030-59719-1_19
3. Dou, Q., Liu, Q., Heng, P.A., Glocker, B.: Unpaired multi-modal segmentation via knowledge distillation. IEEE Trans. Med. Imaging **39**(7), 2415–2425 (2020)
4. Dou, Q., Ouyang, C., Chen, C., Chen, H., Heng, P.A.: Unsupervised cross-modality domain adaptation of convnets for biomedical image segmentations with adversarial loss. In: Proceedings of the 27th International Joint Conference on Artificial Intelligence, IJCAI 2018, pp. 691–697. AAAI Press (2018)
5. Goodfellow, I., et al.: Generative adversarial nets. In: Advances in neural Information Processing Systems, pp. 2672–2680 (2014)
6. Ioffe, S., Szegedy, C.: Batch normalization: accelerating deep network training by reducing internal covariate shift. In: International Conference on Machine Learning, pp. 448–456. PMLR (2015)
7. Kavur, A.E., et al.: Chaos challenge-combined (CT-MR) healthy abdominal organ segmentation. Med. Image Anal. **69**, 101950 (2021)

[1] Source code can be found in https://github.com/nusdbsystem/SSUMML.

8. Landman, B., Xu, Z., Iglesias, J.E., Styner, M., Langerak, T.R., Klein, A.: Multi-atlas labeling beyond the cranial vault (2020)
9. Lee, D.H., et al.: Pseudo-label: the simple and efficient semi-supervised learning method for deep neural networks. In: Workshop on Challenges in Representation Learning, ICML, vol. 3 (2013)
10. Li, K., Wang, S., Yu, L., Heng, P.-A.: Dual-teacher: integrating intra-domain and inter-domain teachers for annotation-efficient cardiac segmentation. In: Martel, A.L., et al. (eds.) MICCAI 2020. LNCS, vol. 12261, pp. 418–427. Springer, Cham (2020). https://doi.org/10.1007/978-3-030-59710-8_41
11. Li, K., Yu, L., Wang, S., Heng, P.A.: Towards cross-modality medical image segmentation with online mutual knowledge distillation. Proceedings of the AAAI Conference on Artificial Intelligence, vol. 34, no. 01, pp. 775–783, April 2020
12. Luo, Z., et al.: MLCask: efficient management of component evolution in collaborative data analytics pipelines. In: 37th IEEE International Conference on Data Engineering (ICDE) (2021)
13. Ooi, B.C., et al.: SINGA: a distributed deep learning platform. In: Proceedings of the 23rd ACM International Conference on Multimedia, pp. 685–688 (2015)
14. Valindria, V.V., et al.: Multi-modal learning from unpaired images: application to multi-organ segmentation in CT and MRI. In: 2018 IEEE Winter Conference on Applications of Computer Vision (WACV), pp. 547–556. IEEE (2018)
15. Yu, F., Koltun, V., Funkhouser, T.: Dilated residual networks. In: Proceedings of the IEEE Conference on Computer Vision and Pattern Recognition, pp. 472–480 (2017)
16. Zhu, J.Y., Park, T., Isola, P., Efros, A.A.: Unpaired image-to-image translation using cycle-consistent adversarial networks. In: Proceedings of the IEEE International Conference on Computer Vision, pp. 2223–2232 (2017)
17. Zhuang, X., Shen, J.: Multi-scale patch and multi-modality atlases for whole heart segmentation of MRI. Med. Image Anal. **31**, 77–87 (2016)

Implicit Neural Distance Representation for Unsupervised and Supervised Classification of Complex Anatomies

Kristine Aavild Juhl[1(✉)], Xabier Morales[2], Ole de Backer[3], Oscar Camara[2], and Rasmus Reinhold Paulsen[1]

[1] Section for Image Computing, Technical University of Denmark, Kgs. Lyngby, Denmark
{kajul,rapa}@dtu.dk

[2] Physense, Department of Information and Communication Technologies, Universitat Pompeu Fabra, Barcelona, Spain
{xabier.morales,oscar.camara}@upf.edu

[3] Department of Cardiology, Rigshospitalet, University of Copenhagen, Copenhagen, Denmark
ole.de.backer@regionh.dk

Abstract. The task of 3D shape classification is closely related to finding a good representation of the shapes. In this study, we focus on surface representations of complex anatomies and on how such representations can be utilized for super- and unsupervised classification. We present a novel Implicit Neural Distance Representation based on unsigned distance fields (UDFs). The UDFs can be embedded into a low-dimensional latent space, which is optimized using only the shape itself. We demonstrate that this self-optimized latent space holds important global shape information useful for reconstructing the anatomies, but also that unsupervised clustering of the latent vectors successfully separates different anatomies (left atrium, left/right ear-canals and human faces). Finally, we show how the representation can be used to do gender classification of human face geometries, which is a notoriously hard problem.

Keywords: Implicit functions · Unsigned distance fields · Shape analysis

1 Introduction

Being able to describe and classify complex anatomical shapes are tasks inevitably linked to finding a good representation of the shape. In the human body, a variety of complex anatomies exists and often such anatomies are represented as a 3D mesh surface either as an isosurface in an image volume or directly from a 3D surface scanner. Representing such 3D meshes in a geometric deep learning framework that can extract global shape characteristics and use them for clustering or classification is not straightforward. Methods working directly

M. de Bruijne et al. (Eds.): MICCAI 2021, LNCS 12902, pp. 405–415, 2021.
https://doi.org/10.1007/978-3-030-87196-3_38

on surface meshes make use of templates or surface parameterization techniques [10,19], but often these methods either make strong assumptions on the topology, the number of vertices in the mesh or make use of local shape patches that are difficult to combine into a global shape descriptor. The simplest way of extending the ground-breaking convolutional neural networks (CNNs) from 2D images into 3D space is by representing the shape as a uniform grid, on which the main components of the CNN (convolution, pooling, etc.) scales naturally to 3D [20,25]. Regardless, whether the grid stores binary occupancies or truncated distance values, the 3D extension comes with a cubically growing memory requirement only allowing for shallow networks or limited resolution. Another popular shape representation is point clouds, where a variety of deep learning approaches exists [18]. Point cloud data are however not capable of capturing the topology, and is therefore not ideal for representing complex anatomies. Finally, a surface can also be represented implicitly, for example in the form of a distance field (DF) [11]. In theory, a DF is a continuous function, where the surface is implicitly represented as the zero-level isosurface. Where the traditional DF represent a single shape, recent work on neural distance representations allow for using deep neural networks to represent an entire class of shapes [14].

Neural signed distance fields (SDFs) have shown to be effective to reconstruct surfaces from partial or low-resolution data by learning the mapping between a point in space and the signed distance from that point to the surface [8,14]. This allows for learning a continuous function whose zero-level isosurface represents the surface at an theoretical arbitrary resolution. When using SDFs one can only describe closed surfaces, but many anatomies are of arbitrary topologies such as planes, open tubes, etc. Unsigned distance fields (UDFs) can handle such topologies, but requires more advanced methods to retrieve the zero-level isosurface [4,5,9,23]. Compared to [9] our method relies on automatically derived latent vectors as opposed to multiscale features, which allows us to derive a single latent space holding both global and local features. The proposed method borrows ideas from these closely related works, but distinguishes itself from those in their goal. Where previous work have solely focused on reconstruction of 3D shapes from a low resolution or noisy input, our paper focus on shape representation and classification using a self-optimized latent space.

We adopt the likelihood-based auto-decoder formulation from [14] to create a latent space that is optimized using no information except the shape itself. We show that the neural unsigned distance representation also can be used to reconstruct complex anatomical shapes of arbitrary topology, but more importantly that the learned latent space holds global feature descriptors that can be used for shape clustering and classification. All code is publicly available[1].

[1] https://github.com/kristineaajuhl/Implicit-Neural-Distance-Representation-of-Complex-Anatomie.

2 Methodology

An UDF is an implicit surface representation, where the magnitude of the field defines the distance from the given point to the nearest point on the surface. UDFs can represent surfaces of arbitrary topology, but are challenged by the discontinuity at the surface and the difficulty in retrieving the zero-level isosurface.

Self-optimized Latent Space Learning. In general terms, the aim is to learn a function f that, given any point in space $\mathbf{p} \in \mathrm{IR}^3$, outputs the distance $d \in \mathrm{IR}^+$ from $\mathbf{p}$ to the surface. To enable the function to describe multiple shapes, a unique latent vector $\mathbf{z}$ for each shape is introduced, and the function f can be written as:

$$f(\mathbf{z}, \mathbf{p}) \mapsto d, \qquad \mathbf{p} \in \mathrm{IR}^3, d \in \mathrm{IR}^+ \tag{1}$$

To represent the distance field with a deep neural network, we use the auto-decoder formulation from [14]. The aim of the auto-decoder is to jointly optimize the latent-space and the decoder parameters. An illustration of the training and testing setups can be seen in Fig. 1.

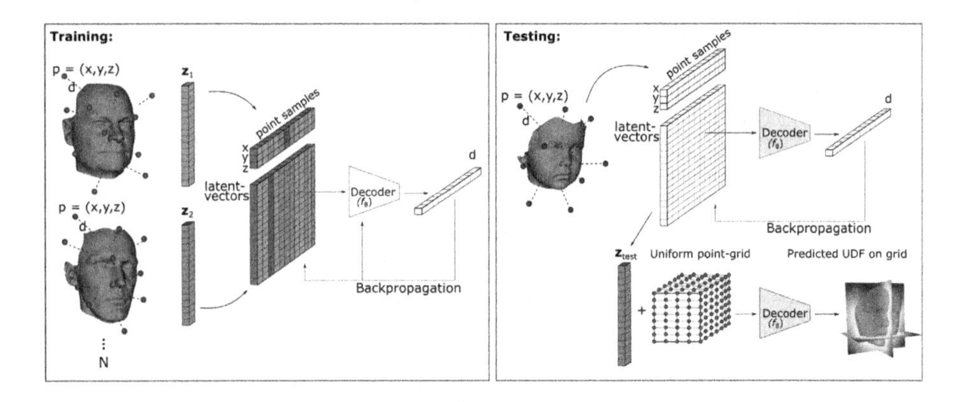

Fig. 1. During network training the decoder parameters θ and the individual latent space vectors $\{\mathbf{z}_i\}_{i=1}^N$ are updated by backpropagating the error between the predicted distance $\hat{d}$ and the true distance d. At testing time the decoder parameters are locked and the optimal vector $\mathbf{z}_{test}$ is found. Based on this vector the unsigned distance field (UDF) is predicted on a uniform grid and the zero-level isosurface can be extracted.

We prepare a set of points $\mathbf{p}_{1,..,j}$ and corresponding distances $d_{1,..,j}$ from each of the N shapes in the dataset. The point-coordinates are stacked with a randomly initialized latent vector and passed through the decoder. The decoder predicts a distance value $\hat{d}$ for each point, which is compared to the true distance. We minimize the following loss with respect to the individual latent vectors $\{\mathbf{z}_i\}_{i=1}^N$ and the neural network parameters θ:

$$\underset{\theta,\{\mathbf{z}_i\}_{i=1}^N}{\arg\min} \sum_{i=1}^{N}\sum_{j=1}^{K} \mathcal{L}(f_\theta(\mathbf{z}_i, \mathbf{p}_j), d_j) + \frac{1}{\sigma^2}||\mathbf{z}_i||_2^2, \tag{2}$$

where $\mathcal{L}(f_\theta(\mathbf{z}_i, \mathbf{p}_j), d_j) = |\text{clamp}(f_\theta(\mathbf{z}_i, \mathbf{p}_j), \delta) - \text{clamp}(d_j, \delta)|$ is the clamped L1-loss and $\frac{1}{\sigma^2}$ is a regularization term that encourages a compact latent space and is required to converge to good solutions. The clamping-parameter δ controls the distance from the surface, over which we expect to learn an accurate distance field, whereas σ^2 balances the L1-loss and the regularization.

After training the network, one can find the latent vector for an unseen shape by fixing the network parameters (θ) and minimizing the following loss:

$$\hat{\mathbf{z}} = \underset{\mathbf{z}}{\arg\min} \sum_{j=1}^{K} \mathcal{L}(f_\theta(\mathbf{z}, \mathbf{p}_j), d_j) + \frac{1}{\sigma^2}||\mathbf{z}||_2^2, \tag{3}$$

where $\mathbf{p}_j$ and d_j are point locations and distance samples from the test shape. We use a fully connected neural network with 8 hidden layers to approximate the function f_θ. The first seven layers consists of 512 units, before narrowing to one unit to output the distance. Since we only have positive distance values, we use the rectified linear unit (ReLU) activation for all layers. The point coordinate and latent vector are given as input to the first layer, and are reintroduced to the network in the fourth layer, where it is concatenated with the output of the fourth fully connected layer to yield a 512 vector.

Zero-Level Surface Extraction. In an SDF, the surface can be computed as the zero-crossing of the field. This can be done with a standard iso-surface extraction method like marching-cubes [12]. For an UDF we are looking for a minimum in the field, which traditional iso-surface extraction methods are not for. As opposed to many related works, our goal has not been to make accurate reconstructions of the inputs surfaces, but since an accurate reconstruction also indicates that important information is pertained in the latent space, we have decided to also evaluate the reconstruction abilities. Our approach is inspired by [9], but can be executed without access to any information about the normals. We start by computing a thin envelope using marching cubes with a small positive iso-value (in the order of the voxel spacing). Secondly, the vertices from the envelope are projected into the local minimum of the UDF by a bisection search along the vertex normal. The vertex normal is computed using the envelope mesh [6]. Finally, the mesh is iteratively optimised by the process described in [16] using a combination of edge flips, mesh smoothing, and projection to optimise triangle aspect ratios. The result is a double surface lying on the minimum of the distance field. For a manifold surface, a single surface can be generated by splitting sharp edges and keeping the largest connected surface patch.

3 Experiments

To conduct experiments on the deep neural implicit distance representation we used three distinct datasets of distinct anatomies and topologies. The three datasets are carefully chosen to demonstrate the core features of the method on tasks with a clear ground truth and objective. The three datasets present complex anatomical or biomedical data with the difficulties such datasets possess; few examples to learn from, different topologies, large number of vertices, noise, etc. Participants in all three datasets gave informed consent for use of their data.

- **ESOF** contains 3D face-scans of 394 humans with natural expressions acquired using a Canfield Vectra M3 scanner. Only the 3D geometry was used. For each scan the self-reported gender and age was noted (192/202 male/female, age 0–84 years). The raw scans were manually cropped to include the face-only region and pre-aligned using automatically annotated facial landmarks [17].
- **EARS** contains 3D scans of 571 ear canals from humans (259/312 right/left ear) acquired with a 3Shape S200 scanner. The ear canals were roughly aligned from the scanner and accurate pre-alignment was achieved using iterative closest point (ICP) [21]. The ear canals were aligned depending on side (left/right).
- **LA** contains 106 3D mesh models of the Left Atrium (LA) and the Left Atrial Appendage (LAA). The mesh models were created using manual segmentation of cardiac computed tomography angiographies (CCTA) supplied by the Department of Radiology, Copenhagen Central Hospital, Denmark. The segmentation was carried out in 3D Slicer[2] The CCTA-images were recorded for research purposes and fully anonymized. The LA surfaces were pre-aligned using ICP with centroid matching.

Experiment 1: Unsupervised Clustering of Complex Anatomies
In this experiment we aim to show that the self-optimized latent-space can be used to distinguish the three datasets of different topologies (ESOF/EARS/LA). We further show that the latent space contains information that can separate the left from the right EAR and male from female ESOF faces.

For each dataset independently, all surfaces were aligned to a randomly chosen template and scaled rigidly to lie within a unit sphere. For each shape we sampled approximately 250 000 samples randomly on the mesh and perturbed the points randomly with two zero-mean Gaussian noise kernels with variance 2.5/0.5/10 mm and 5/1/20 mm for the LA/EARS/ESOF examples respectively. An additional 25 000 points were uniformly sampled within the unit sphere. For each point sample the shortest distance to the surface was measured using an octree-based spatial search algorithm from the Visualization Toolkit [22].

The data were divided such that 86/511/345 LA/EARS/ESOF examples were included in the training set and 20/60/50 LA/EARS/ESOF examples were reserved for testing. We use 128 dimensions for the latent space and initialize the latent vectors randomly from $\mathcal{N}(0, 0.01^2)$. It is important to initialize the

[2] www.slicer.org.

latent vectors quite small to ensure that similar shapes do not diverge because of different initialization. We trained the network in batches of 10 shapes and subsampled 16 384 points per shape in the batch. The parameters were optimized using Adam optimization, with an initial learning rate of 0.0005 for the decoder parameters and 0.001 for the latent vectors. Both learning rates were dropped by a factor 0.5 every 500 epochs. We used $\delta = 0.1$ as the truncation distance and $\sigma = 0.0001$ as the regularization parameter. The network was trained with 20% dropout of the decoder parameters for 1000 epochs. It is implemented in PyTorch [15] and trained on a single Nvidia Titan X (12 GB).

We initialized the surface reconstruction algorithm with a marching cubes iso-surface at 0.01/0.02/0.02 normalized distance (1.129/0.538/1.317 mm) for ESOF/EARS/LA respectively. These values were chosen based on a trade-off between high surface accuracy and high surface completion. The reconstructed surfaces were evaluated using the average symmetric chamfer distance, mesh accuracy and mesh completion metrics. We used $\delta = 2/0.2/1$ mm for the ESOF/EARS/LA mesh completion measure. For more details on the hyper-parameter settings and evaluation methods the reader is referred to [14].

We investigated the clustering abilities in latent space using a simple K-means algorithm. We fitted the algorithm to the training samples and assigned each cluster to a class based on majority voting. We report the classification accuracy on dataset, ear-side and gender classification based on the test examples.

Results. Figure 2 shows examples of reconstructed surfaces from the three datasets on both training and testing examples. Table 1 shows the evaluation measures. It is evident that the decoder is capable of reconstructing distinct features for each of the three datasets, while also capturing individual traits. It is for example seen that the overall shape and size of the LA is preserved, even though some finer details on the LAA is missing. The decoder was also capable of reconstructing ear-canals from both the left and the right ear with individual shape characteristics such as the width and bend of the ear canal tube. Similarly, the faces were reconstructed in way such that individuals could be recognized. Based on the visual appearance and the evaluation methods, we argue that the network is capable of representing complex anatomies across different topologies with high accuracy on both seen and unseen examples.

Figure 3 (left) shows the first 2 principal components of the 128 dimensional latent space. It is evident that the three datasets are clearly separated in the latent space and a K-means algorithm with three clusters achieves 99.23% accuracy. To enable separation between all subdivisions we increase the number of clusters to six and thereby achieves 100% accuracy on dataset classification, 98.33% accuracy on ear side (EARS) and 72% accuracy on gender (ESOF). Two of the six clusters are assigned to the right ear, whereas the remaining clusters are unique to a class.

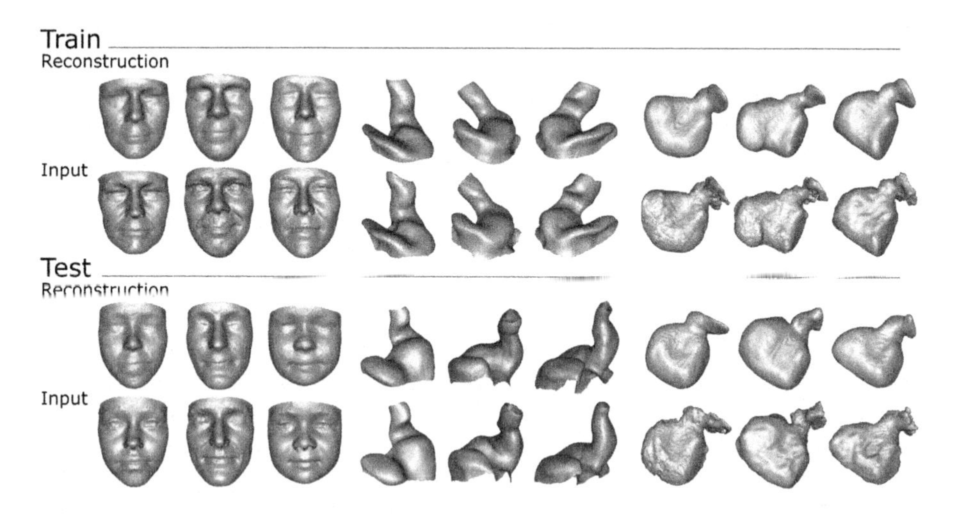

Fig. 2. Reconstructions of examples from training (above line) and testing (below line). The upper rows show the reconstructions and the original surface is shown below.

Table 1. Reconstruction evaluations measured as the average symmetric chamfer distance [mm] (CD), mesh accuracy [mm] (M. Acc.) and mesh completion [%] (M. compl.) for the three datasets on the train and test sets. Arrows indicate whether higher (↑) or lower (↓) is better.

	Train			Test		
	↓ CD	↓ M. Acc.	↑ M. Compl.	↓ CD	↓ M. Acc.	↑ M. Compl.
LA	0.326	0.688	93.75%	0.435	0.944	86.27%
EARS	0.291	0.607	93.05%	0.204	0.422	84.29%
ESOF	0.267	0.551	90.54%	0.574	1.179	89.27%

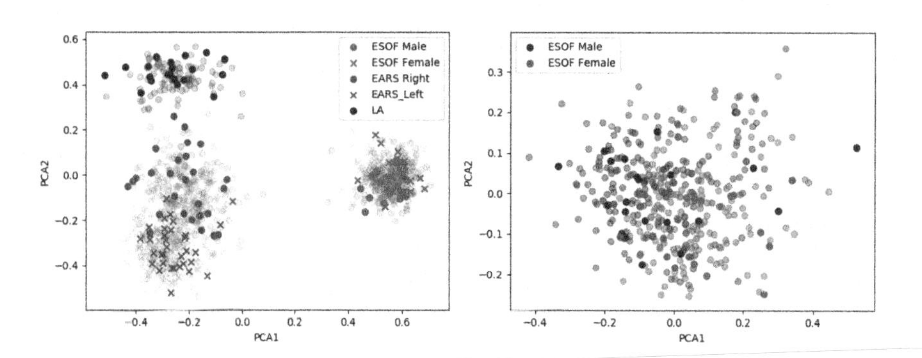

Fig. 3. Visualization of the latent space projected to 2D using principle component analysis (PCA) for the 128 dimensional latent space of the three datasets (left) and the ESOF-faces only (right). Opaque: training set, Solid: test set.

Experiment 2: Gender Classification in Latent Space Self-optimized for Face Reconstruction
The latent space was self-optimized to create accurate reconstructions as described above, this time using only the ESOF faces. Due to the smaller dataset we also reduced the latent space dimensions to 64.

We compared two methods for gender-classification based on the latent-vectors; a simple K-means algorithm with two clusters and a more advanced neural net classifier. The neural net classifier was a fully connected network that takes the 64 dimensional latent vector as input. The network consisted of two hidden layers with 32 and 16 nodes respectively, and an output layer with two nodes - male vs. female. The network was optimized using stochastic gradient descent with the initial learning rate set to 0.01 and dropped by a factor 0.1 for every 50 epochs. To prevent overfitting the network was trained using 20% dropout on the hidden layers and using early stopping. The network was trained on 300 training examples and validated on 41 examples. We report the classification accuracy for the same 50 test examples as Experiment 1.

Results. Figure 3 (right) shows the 64-dimensional latent space projected to 2D using PCA. Even though the decoder and latent-optimization are now able to focus only on this dataset, there is still no clear separation between the males and females in the 2D latent space, and the result of K-means clustering with two centroids achieves similar performance as before with 68% accuracy on the test-set. The neural net classifier are however able to learn more complex relations between the latent vector and the gender and achieves an accuracy of 88% on the test-set.

4 Discussion and Conclusion

Table 2 shows the performance of state-of-the-art geometric gender classification algorithms. The best performing methods makes use of anthropometric measurements between manually annotated landmarks. In our approach, we do not use manually defined features, but rely solely on the self-optimized latent space to contain the necessary information. The human face has well-defined features (such as nose, eyes, etc.) to serve as important guides for placing landmarks. This is not the case for the two other datasets (EARS and LA), where it is difficult to design a landmarking scheme suitable for describing the highly varying morphologies. The LAA morphology is known to correlate with thrombus risk in patients with heart arrythmias. The correlation is however hard to quantify due to the missing quantitative measures of shape. Future work aim to investigate this hypothesis further using the self-optimized latent features. The current model are not capable of reconstructing the finer details of the LAA, which is attributed the small dataset and the simple model, which needs to be addressed.

Because the proposed method relies solely on self-optimization, we cannot control which differences or similarities the latent space highlights. Since the representation is variant to similarity transforms, the latent space may enhance

Table 2. Comparison to state-of-the-art gender classification based on 3D geometry. NN: Neural Network Classifier.

Method	K-Means	NN	[1]	[2]	[7]	[24]	[13]
Accuracy	**68%**	**88%**	90%	92%	86%	69.7%	85.4%

differences in alignment above differences in shape, if the shapes are not sufficiently pre-aligned. Future extensions of this work may therefore include efforts into making it invariant to these transforms by for example including it as an extra learning objective [3]. A related avenue may be to encourage the latent space to enhance the relevant features for a given task. This includes different regularization terms aiming to promote the clustering objective, or the inclusion of auxiliary tasks, in which certain shape-characteristics are needed.

We have used UDFs to overcome the problem of open surfaces such as tubes or planes. Another commonly adopted solution is to artificially close the holes or extrapolate an SDF from the open surface. We argue, that a neural network aiming to learn a representation of an SDF from one of those methods wastes capacity on trying to represent areas without importance for the shape itself. As an example, the closing of a frontal face scan will inevitably introduce a large surface area that has no meaningful anatomical information.

To conclude, we have shown that UDFs can be represented by a neural network and that such representation is suited for representing complex anatomical shapes of arbitrary topology. A self-optimized latent space is learned from the shape itself, and can be used to describe and classify 3D shapes. The proposed method allows for unsupervised shape analysis for example in scenarios, where manually defined features or landmarks are not available.

Acknowledgements. This work was supported by a PhD grant from the Technical University of Denmark - Department of Applied Mathematics and Computer Science (DTU Compute) and the Spanish Ministry of Science, Innovation and Universities under the Retos I+D Programme (RTI2018-101193-B-I00).

References

1. Abbas, H., Hicks, Y., Marshall, D., Zhurov, A.I., Richmond, S.: A 3D morphometric perspective for facial gender analysis and classification using geodesic path curvature features. Comput. Visual Media **4**(10), 17–32 (2018). https://doi.org/10.1007/s41095-017-0097-1
2. Abbas, H.H., Altameemi, A.A., Farhan, H.R.: Biological landmark vs quasi-landmarks for 3D face recognition and gender classification. Int. J. Electr. Comput. Eng. (IJECE) **9**(5), 4069–4076 (2019)
3. Afolabi, O., Yang, A.Y., Sastry, S.S.: Extending DeepSDF for automatic 3D shape retrieval and similarity transform estimation (2020)
4. Atzmon, M., Lipman, Y.: SAL: sign agnostic learning of shapes from raw data. In: Proceedings of the IEEE Computer Society Conference on Computer Vision and Pattern Recognition pp. 2562–2571 (2020)

5. Atzmon, M., Lipman, Y.: SALD: sign agnostic learning with derivatives. In: International Conference on Learning Representations (ICLR) 2021 (2021)
6. Bærentzen, J.A., Aanaes, H.: Signed distance computation using the angle weighted pseudonormal. IEEE Trans. Vis. Comput. Graph. **11**(3), 243–253 (2005)
7. Ballihi, L., Ben Amor, B., Daoudi, M., Srivastava, A., Aboutajdine, D.: Boosting 3D-geometric features for efficient face recognition and gender classification. IEEE Trans. Inf. Forensics Secur. **7**(6), 1766–1779 (2012)
8. Chibane, J., Alldieck, T., Pons-Moll, G.: Implicit functions in feature space for 3D shape reconstruction and completion. In: Proceedings of the IEEE Computer Society Conference on Computer Vision and Pattern Recognition, pp. 6968–6979 (2020)
9. Chibane, J., Mir, A., Pons-Moll, G.: Neural unsigned distance fields for implicit function learning. In: Proceedings Neural Information Processing Systems (2020)
10. Defferrard, M., Bresson, X., Vandergheynst, P.: Convolutional neural networks on graphs with fast localized spectral filtering. In: Advances in Neural Information Processing Systems, pp. 3844–3852 (2016)
11. Hoppe, H., DeRose, T., Duchamp, T., McDonald, J., Stuetzle, W.: Surface reconstruction from unorganized points. In: Proceedings of the 19th Annual Conference on Computer Graphics and Interactive Techniques, pp. 71–78 (1992)
12. Lorensen, W.E., Cline, H.E.: Marching cubes: a high resolution 3D surface construction algorithm. ACM SIGGRAPH Comput. Graph. **21**(4), 163–169 (1987)
13. Lu, X., Chen, H., Jain, A.K.: Multimodal facial gender and ethnicity identification. In: Zhang, D., Jain, A.K. (eds.) ICB 2006. LNCS, vol. 3832, pp. 554–561. Springer, Heidelberg (2005). https://doi.org/10.1007/11608288_74
14. Park, J.J., Florence, P., Straub, J., Newcombe, R., Lovegrove, S.: DeepSDF: learning continuous signed distance functions for shape representation. In: The IEEE Conference on Computer Vision and Pattern Recognition (CVPR) (2019)
15. Paszke, A., et al.: Pytorch: an imperative style, high-performance deep learning library. Adv. Neural Inf. Process. Syst. **32**, 8024–8035 (2019)
16. Paulsen, R.R., Bærentzen, J.A., Larsen, R.: Markov random field surface reconstruction. IEEE Trans. Vis. Comput. Graph. **16**(4), 636–646 (2009)
17. Paulsen, R.R., Juhl, K.A., Haspang, T.M., Hansen, T., Ganz, M., Einarsson, G.: Multi-view consensus CNN for 3D facial landmark placement. In: Jawahar, C.V., Li, H., Mori, G., Schindler, K. (eds.) ACCV 2018. LNCS, vol. 11361, pp. 706–719. Springer, Cham (2019). https://doi.org/10.1007/978-3-030-20887-5_44
18. Qi, C.R., Yi, L., Su, H., Guibas, L.J.: PointNet++: deep hierarchical feature learning on point sets in a metric space. In: Advances in Neural Information Processing Systems, pp. 5100–5109 (2017)
19. Ranjan, A., Bolkart, T., Sanyal, S., Black, M.J.: Generating 3D faces using convolutional mesh autoencoders. In: Proceedings European Conference on Computer Vision, pp. 725–741 (2018)
20. Riegler, G., Ulusoy, A.O., Geiger, A.: OctNet: learning deep 3D representations at high resolutions. In: Proceedings Computer Vision and Pattern Recognition, pp. 6620–6629 (2016)
21. Rusinkiewicz, S., Levoy, M.: Efficient variants of the ICP algorithm. In: Proceedings Third International Conference on 3-D Digital Imaging and Modeling, pp. 145–152. IEEE (2001)
22. Schroeder, W., Martin, K., Lorensen, B.: The Visualization Toolkit-An Object-Oriented Approach to 3D Graphics, 4th edn. Kitware Inc., New York (2006)

23. Venkatesh, R., Sharma, S., Ghosh, A., Jeni, L., Singh, M.: DUDE: Deep Unsigned Distance Embeddings for Hi-Fidelity Representation of Complex 3D Surfaces (2020)
24. Wu, J., Smith, W.A.P., Hancock, E.R.: Supervised principal geodesic analysis on facial surface normals for gender classification. In: da Vitoria Lobo, N., et al. (eds.) SSPR /SPR 2008. LNCS, vol. 5342, pp. 664–673. Springer, Heidelberg (2008). https://doi.org/10.1007/978-3-540-89689-0_70
25. Wu, Z., et al.: 3D ShapeNets: a deep representation for volumetric shapes. In: Proceedings Computer Vision and Pattern Recognition, pp. 1912–1920 (2015)

3D Graph-S²Net: Shape-Aware Self-ensembling Network for Semi-supervised Segmentation with Bilateral Graph Convolution

Huimin Huang[1], Nan Zhou[1], Lanfen Lin[1](✉), Hongjie Hu[2](✉), Yutaro Iwamoto[3], Xian-Hua Han[3], Yen-Wei Chen[1,3,4](✉), and Ruofeng Tong[1,4]

[1] College of Computer Science and Technology, Zhejiang University, Hangzhou, China
llf@zju.edu.cn, chen@is.ritsumei.ac.jp

[2] Department of Radiology, Sir Run Run Shaw Hospital, Hangzhou, China
hongjiehu@zju.edu.cn

[3] College of Information Science and Engineering, Ritsumeikan University, Kyoto, Japan

[4] Research Center for Healthcare Data Science, Zhejiang Lab, Hangzhou, China

Abstract. Semi-supervised learning (SSL) algorithms have attracted much attentions in medical image segmentation due to challenge in acquiring pixel-wise annotations by using unlabeled data. However, most of existing SSLs neglected the geometric shape constraint in object, leading to unsatisfactory boundary and non-smooth of object. In this paper, we propose a shape-aware semi-supervised 3D medical image segmentation network, named 3D Graph-S²Net, which incorporates the flexible shape information and learns duality constraints between semantics and geometrics in the graph domain. Specifically, our method consists of two parts: a multi-task learning network (3D S²Net) and a graph-based cross-task module (3D BGCM). The 3D S²Net improves the existing self-ensembling model (i.e., Mean-Teacher model) by adding a signed distance map (SDM) prediction task, which encodes richer features of object shape and surface. Moreover, the 3D BGCM explores the co-occurrence relations between the semantics segmentation and SDM prediction task, so that the network learns stronger semantic and geometric correspondences from both labeled and unlabeled data. Experimental results on the Atrial Segmentation Challenge confirm that our 3D Graph-S²Net outperforms the state-of-the-arts in semi-supervised segmentation.

Keywords: Semi-supervised segmentation task · Signed distance map · 3D bilateral graph convolution

1 Introduction

Accurate medical image segmentation is an essential prerequisite for many clinical applications [1]. Recently, a variety of convolutional neural networks (CNNs) have been developed for segmentation tasks [33–35]. Though these methods achieved satisfactory results, they needed massive pixel-wise annotation and to be trained in fully supervision. In the medical field, however, sufficient labeled data is still unavailable as the manual

M. de Bruijne et al. (Eds.): MICCAI 2021, LNCS 12902, pp. 416–427, 2021.
https://doi.org/10.1007/978-3-030-87196-3_39

annotation is costly and time-consuming. To address this problem, semi-supervised learning (SSL) has been introduced into the field of segmentation, which provides the means to use both labeled data and arbitrary amounts of unlabeled data in training.

Recent SSL have focused on incorporating unlabeled data into training, which can be categorized as: self-training [2, 29], co-training [3, 4], GANs based methods [5–8, 28] and self-ensembling (Π model [9, 27] and Mean-Teacher model [10–12]). For example, Bai *et al.* [8] proposed a self-training-based method that updated the segmentation results of unlabeled data; while Xia *et al.* [3] achieved co-training by exploiting multi viewpoint consistency. Li *et al.* [8] designed a GANs based model, named SASSNet, which enforced the segmentation of unlabeled images to be similar to those of the labeled ones. Based on Mean-Teacher model [9], Yu *et al.* [12] proposed UA-MT that used an uncertainty map to guide the student network learning. Despite their promising performance, they often ignore the geometric information and/or the inherent semantic and geometric correspondences, which lead to unsatisfactory boundary and non-smooth of object since the ambiguity of structure boundary and heterogeneous texture. To alleviate this dilemma, we aim to utilize the flexible shape information in the semi-supervised segmentation network, as well as learn duality constraints between semantics and geometrics in the graph domain, which is motivated by the following two aspects.

Firstly, considering that the signed distance map (SDM) [8, 13, 30–32] encodes richer features of object shape and surface, we predict the SDM as an auxiliary task for the semi-supervised segmentation network. Specifically, we devise a 3D **S**hape-aware **S**elf-ensembling **Net**work (3D S^{2}Net), which is based on Mean-Teacher (MT) algorithm. It jointly predicts semantic segmentation and SDM with a shared network, and can be considered as our backbone. In this multi-task learning way, the semantic segmentation produces the probability of being the object, which provides the smoothness and continuity constraints; while the SDM prediction generates its distance from the closest boundary, which enforce a global shape and boundary constraints.

Secondly, as there exists the duality constraints between two tasks, it is obvious that semantic segmentation and SDM prediction can benefit each other by mutual interaction and promotion to boost the performance of semi-supervised segmentation. Inspired by this, we design a 3D Bilateral graph convolution module (3D BGCM) to explore co-occurrence relations and diffuse information between the semantics segmentation and SDM prediction. To establish the relationships effectively, we utilize graph convolution [15–23] to mine the intra-task and inter-task relations within and between two tasks. Specifically, the flexible intra-task reasoning can capture long-range dependencies over non-local regions and refine the visual features in separate tasks; while the inter-task reasoning can model the similar latent representations between two tasks and enable information propagation in a bidirectional way. In this way, our 3D Graph-S^{2}Net, that consists of the backbone 3D S^{2}Net and the cross-task module 3D BGCM, can be aware of the reciprocal relations between tasks and exhibits superior performance.

Our main contributions are four-fold: (**i**) We propose a 3D Graph-S^{2}Net to enforce semantic and geometric constraints in semi-supervised segmentation. It combines a multi-task learning network (3D S^{2}Net) and a graph-based module (3D BGCM) reasoning between tasks. (**ii**) We devise a 3D S^{2}Net that jointly predict the segmentation and SDM, which learns shape information from both labeled and unlabeled data. (**iii**) We design a 3D BGCM to enforce duality constraints between semantics and geometrics by using bilateral graph convolution, which globally mines the intra-task and inter-task relations. (**iv**) We conduct extensive experiments on the Atrial Segmentation Challenge dataset, where our 3D Graph-S^{2}Net surpasses the state-of-the-art methods.

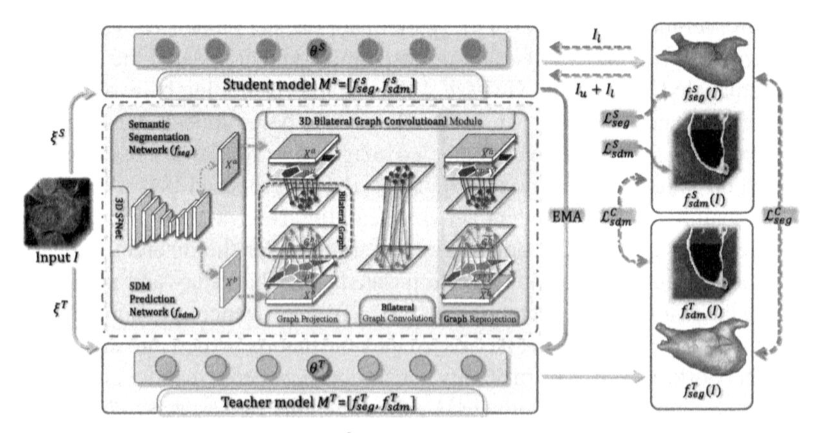

Fig. 1. An overview of our 3D Graph-S^{2}Net. The MT-based framework takes a 3D volume as input, and predicts a SDM and a segmentation after feature extraction by 3D S^{2}Net and feature enhancement by 3D BGCM. The student model and teacher model share the same architecture, and the student learns from the teacher by minimizing the supervised loss ($\mathcal{L}^S_{seg}$ and $\mathcal{L}^S_{sdm}$) on labeled data I_l and the consistency loss ($\mathcal{L}^C_{seg}$ and $\mathcal{L}^C_{sdm}$) on both unlabeled I_u and labeled I_l.

2 Method

Figure 1 illustrates an overview of our 3D Graph-S^{2}Net for semi-supervised medical image segmentation, which consists of two components: (**i**) 3D S^{2}Net that improves the existing MT algorithm by adding a SDM prediction task and works as our backbone; (**ii**) 3D BGCM that encodes co-occurrence relations and diffuses information between tasks.

The 3D S^{2}Net contains two branches: The semantic segmentation branch (f_{seg}) and SDM prediction branch (f_{sdm}), which share the network (e.g. V-Net [24]) but have the task-specific final decoder layer. During the training, f_{seg} and f_{sdm} learn the feature maps X^a and X^b by focusing on the semantics and geometrics, respectively. To explore the mutual information between tasks, our 3D BGCM firstly projects X^a and X^b in the coordinate domain into the fully-connected graphs $\mathcal{G}^a$ and $\mathcal{G}^b$ in the graph domain, where relational reasoning can be efficiently computed. After reasoning, relation-aware features are reversed back to the coordinate domain for further prediction.

2.1 3D Shape-Aware Self-ensembling Network (3D S^{2}Net)

Formally, let $D_L = \{(I_l, Y_l)\}_{l=1}^{N}$ denotes the labeled set and $D_U = \{(I_u)\}_{u=N+1}^{N+M}$ denotes the unlabeled set, where $Y_l = \left\{Y_l^a, Y_l^b\right\}$ is the segmentation ground truth (Y_l^a) and SDM ground truth (Y_l^b). In this work, our 3D S^{2}Net takes Mean-Teacher algorithm (MT) [10] as the basic framework, which consists of a student model M^S and a teacher model M^T sharing the same architecture but having individual parameters. To leverage the hidden information in D_U, we enforce the student to learn the knowledge from the teacher by encouraging prediction consistency under small perturbations ($\xi^S$and ξ^T). Besides the segmentation branch f_{seg}, our multi-task learning network, 3D S^{2}Net, incorporates an additional SDM prediction branch f_{sdm}, which makes the network more sensitive to boundaries. f_{seg} is trained with pixel-wise semantic annotations and yields segmentation mask $f_{seg}(I) \in [0, 1]$; while f_{sdm} is used to predict SDM of object $f_{sdm}(I) \in [-1, 1]$, which is activated by *tanh* function. Each element of $f_{sdm}(I)$ indicates the signed distance of a voxel to its closest surface point after normalization, where sign encodes whether the pixel is inside (negative) or outside (positive) of the boundary.

The student learns from the teacher by minimizing the supervised loss $\mathcal{L}^S$ on the labeled data and the unsupervised loss $\mathcal{L}^C$ with respect to the targets from the teacher on all data. To this end, our 3D S^{2}Net is optimized by the following objective function:

$$\min_{\theta^S} \sum_{i=1}^{N} \mathcal{L}^S\left(M^S\left(I_i; \theta^S\right), Y_i\right) + \lambda \sum_{i=1}^{N+M} \mathcal{L}^C\left(M^T\left(I_i; \theta^T, \xi^T\right), M^S\left(I_i; \theta^S, \xi^S\right)\right) \tag{1}$$

where (**i**) $\mathcal{L}^S$is consists of two losses: supervised segmentation loss $\mathcal{L}_{seg}^S$ and SDM loss $\mathcal{L}_{sdm}^S$. We use a joint cross-entropy loss and dice loss for $\mathcal{L}_{seg}^S$ and introduce the mean squared error (MSE) for $\mathcal{L}_{sdm}^S$. (**ii**) $\mathcal{L}^C$ measures the consistency of the same inputs under different perturbations. It contains unsupervised segmentation loss $\mathcal{L}_{seg}^C$ and SDM loss $\mathcal{L}_{sdm}^C$, which are calculated by using MSE. Here, θ^S and θ^T are the parameters weights of the student model and teacher model; while ξ^S and ξ^T are the different input perturbations (*e.g.* Gaussian noise) applied to the two models. λ is a ramp-up weighting factor that controls the balance between the supervised and unsupervised loss.

Instead of sharing the weights with the student model, the teacher model uses the Exponential Moving Average (EMA) weights of the student: $\theta_t^T = \alpha\theta_{t-1}^T + (1-\alpha)\theta_t^S$, where θ_t^T and θ_t^S are the teacher's and student's weights in $(t-1)$ step, and α controls the updating rate, which is set as 0.99 referring to the previous work [10].

2.2 3D Bilateral Graph Convolutional Module (3D BGCM)

Formally, we define a graph as $\mathcal{G} = (\mathcal{N}, \mathcal{A}, \mathcal{H})$, where $\mathcal{N}$ is a set of nodes, and $|\mathcal{N}|$ denotes the number of nodes. The adjacent matrix $\mathcal{A} \in \mathbb{R}^{|\mathcal{N}|\times|\mathcal{N}|}$ describes the edge weights and $\mathcal{H} \in \mathbb{R}^{|\mathcal{N}|\times K}$ is the feature matrix with the rows corresponding to the nodes in $\mathcal{N}$. Our 3D BGCM consists of three operations: Graph Projection, Bilateral Graph Reasoning and Graph Reprojection. Specifically, Graph Projection is the first step that maps the feature map X in the coordinate domain onto a set of node features $\mathcal{H}$ in the graph domain; while Graph Reprojection is the final step that reverse the updated graph features $\tilde{\mathcal{H}}$ back to $\tilde{X}$. Bilateral Graph Convolution is the critical step that models intra-task and inter-task relations and diffuses information between tasks in graph domain.

Graph Projection and Reprojection: We adopt the same strategies to project and reproject semantic-aware graph $\mathcal{G}^a$ and geometric-aware graph $\mathcal{G}^b$. For simplicity, we take $\mathcal{G}^a$ as an example. As seen in Fig. 2, we use a convolution, $\phi^a(.)$, with a kernel size of $1 \times 1 \times 1$ and stride of γ to reduce the size of X^a from $HWT \times C$ to $HWT/\gamma^3 \times L$, which not only enhances the capacity of the projection, but also reduces the computational cost. The next step is to obtain the anchors of the vertices. These anchors denote the centers of each region of pixels. Here, we use a convolution operation, $Anc^a(\cdot)$, with a kernel size of $1 \times 1 \times 1$ and stride of ε. Then, we adopt the multiplication, $\otimes$, of $\phi^a(X^a)$ and anchors to capture the similarity between anchors and each pixel. The range of the projection matrix, $\mathcal{P}^a$, is constrained to (0, 1) by using a softmax function:

$$\mathcal{P}^a = softmax\Big(Anc^a\big(\phi^a(X^a)\big) \otimes \phi^a(X^a)^T\Big) \tag{2}$$

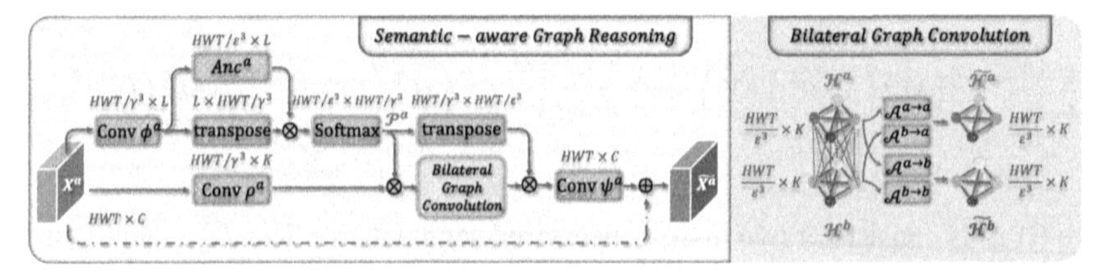

Fig. 2. Architecture of the 3D Bilateral Graph Convolution Module (3D BGCM).

Based on the projection matrix $\mathcal{P}^a$, the feature map X^a is then mapped into the graph domain as: $\mathcal{H}^a = \mathcal{P}^a \otimes \rho^a(X^a)$, where $\rho^a(\cdot)$ is a convolution with $1 \times 1 \times 1$ kernel and stride of γ to obtain a small-size feature map, leading to $\rho^a(X^a) \in \mathbb{R}^{HWT/\gamma^3 \times K}$. The projection process is formulated as a linear combination, which aggregates the pixels with similar features as an anchor to one node. This results in a semantic-aware graph feature $\mathcal{H}^a \in \mathbb{R}^{HWT/\varepsilon^3 \times K}$. Similarly, we can obtain a geometric-aware graph feature $\mathcal{H}^b$.

After the reasoning, we adopt linear reprojection given by $\widetilde{X^a} = (\mathcal{P}^a)^T \widetilde{\mathcal{H}^a}$. But small - sized $\widetilde{X^a} \in \mathbb{R}^{HWT/\gamma^3 \times K}$ is still inconsistent with the original feature map $X^a \in \mathbb{R}^{HWT \times C}$. Thus, we attach a $1 \times 1 \times 1$ Deconvolution layer $\psi^a(\cdot)$ with a stride of γ for size expansion, so that the output can seamlessly match the input to form a residual path:

$$\widetilde{X^a} = X^a + \psi^a\Big((\mathcal{P}^a)^T \widetilde{\mathcal{H}^a}\Big) \tag{3}$$

Bilateral Graph Reasoning: Given the $\mathcal{G}^a$ and $\mathcal{G}^b$, we adopt the graph convolution to diffuse information on the graphs. Here, we use a similar approach as in [11] to define graph convolution. We first define the augmentation form of a bilateral graph as:

$$\mathcal{H} = \left[(\mathcal{H}^a)^T, \left(\mathcal{H}^b\right)^T\right]^T, \mathcal{W} = \left[(\mathcal{W}^a)^T, \left(\mathcal{W}^b\right)^T\right]^T \tag{4}$$

where $\mathcal{H} \in \mathbb{R}^{(|\mathcal{N}^a|+|\mathcal{N}^b|)\times K}$, $\mathcal{W} \in \mathbb{R}^{2K\times K'}$ is the augmented form of bilateral node feature and weight matrix. $\mathcal{W}^a$ and $\mathcal{W}^b \in \mathbb{R}^{K\times K'} \mathcal{W}^b \in \mathbb{R}^{K\times K'}$ are two trainable weight matrixes that alter the node dimension of $\mathcal{H}^a$ and $\mathcal{H}^b$, respectively.

Instead of performing on a single graph as in [11], our bilateral graph convolution captures the co-occurrence relations over two graphs via the intra-graph and inter-graph reasoning. Specifically, the intra-graph reasoning models the non-local dependencies in each graph. This is performed on the semantic-to-semantic edges ($\mathcal{A}^{a\to a}$) and geometric-to-geometric edges ($\mathcal{A}^{b\to b}$). The inter-graph reasoning explores the mutual relations between the graphs hence it is applied to the semantic-to-geometric edges ($\mathcal{A}^{a\to b}$) and geometric-to-semantic edges ($\mathcal{A}^{b\to a}$). Based on the above, the adjacent matrix $\mathcal{A}$ is a combination of intra-graph matrix $\mathcal{A}^{intra}$ and inter-graph matrix $\mathcal{A}^{inter}$:

$$\mathcal{A} = \mathcal{A}^{intra} + \mathcal{A}^{inter} \tag{5}$$

$$\mathcal{A}^{intra} = \begin{pmatrix} \mathcal{A}^{a\to a} & 0 \\ 0 & \mathcal{A}^{b\to b} \end{pmatrix}, \mathcal{A}^{inter} = \begin{pmatrix} 0 & \mathcal{A}^{b\to a} \\ \mathcal{A}^{a\to b} & 0 \end{pmatrix} \tag{6}$$

where $\mathcal{A}^{a\to b} = \left\{\eta_{ij}^{a\to b}\right\} \in \mathbb{R}^{|\mathcal{N}^a|\times|\mathcal{N}^b|}$ assembles the correlation weight from j-the node of $\mathcal{G}^a$ to the i-the node of $\mathcal{G}^b$, and $\mathcal{A}^{a\to a}$, $\mathcal{A}^{b\to b}$, $\mathcal{A}^{b\to a}$ are explained similarly. The coefficients, η_{ij}, which indicate the importance of node j for node i, are obtained for every neighboring node pair using an attention mechanism [25]:

$$\eta_{ij} = \frac{\exp(\delta(W[h_i||h_j]))}{\sum_{z\in\mathcal{N}_i}\exp(\delta(W[h_i||h_z]))} \tag{7}$$

where the attention function is a single-layer neural network parameterized by a weight vector $W \in \mathbb{R}^{2K}$. $||$ is the concatenation and δ is LeakyReLU nonlinear. $\mathcal{N}_i$ is the neighborhood of node i, which contains all nodes in our fully-connected graph. Note that the graphs constructed here is a directional graph, as weight vector W is different in learning η_{ij} and η_{ji}. With the normalized adjacent matrix $\mathcal{A}$, augmented bilateral node feature $\mathcal{H}$ and weight matrix $\mathcal{W}$, a single graph convolution layer is designed as:

$$\tilde{\mathcal{H}} = \mathcal{F}(\mathcal{H}||\sigma(\mathcal{A}(\mathcal{H}\otimes\mathcal{W}))), \mathcal{H}\otimes\mathcal{W} = \left[(\mathcal{H}^a\mathcal{W}^a)^T, \left(\mathcal{H}^b\mathcal{W}^b\right)^T\right]^T \tag{8}$$

where $\mathcal{F}(\cdot)$ fuses the original features and updated features by using a 1×1 convolution.

3 Experiments and Results

3.1 Datasets

We evaluated our method on the Left Atrium (LA) dataset from Atrial Segmentation Challenge [26]. The dataset contains 100 3D gadolinium-enhanced MR imaging scans (GE-MRIs) and LA segmentation mask for training and validation, with an isotropic resolution of $0.625 \times 0.625 \times 0.625\ \text{mm}^3$. Following [12], we split the 100 scans into 80 scans for training and 20 scans for validation. All the scans were cropped centering at the heart region for better comparison of the segmentation performance on different methods, and normalized as zero mean and unit variance.

3.2 Implementation Details and Metrics

We adopted Stochastic Gradient Descent (SGD) to update the parameters, where the momentum was set to 0.9 and the weight decay to 1e−4. The initial learning rate was 1e−2, which was divided by 10 every 4000 iterations. We trained the models in 8000 iterations as the network had converged. We used a batch size of 2, including 1 labeled image and 1 unlabeled image. We randomly cropped 112 × 112 × 80 sub-volumes as the input and the final segmentation results were obtained using a sliding window strategy. Following [12], we performed the standard data augmentation to avoid overfitting, including randomly flipping, and rotating with 90°, 180° and 270° along the axial plane.

Here, we adopted V-Net [24] as the backbone shared by f_{seg} and f_{sdm}. The short residual connection in each convolution block was removed. After feature extraction, feature map X in the final decoder had the size of 112 × 112 × 80 × 16, which was taken as the input of 3D BGCM. Massive experiments were conducted for hyper-parameter optimization, and the best performance was achieved with node feature dimension $K = 8$, reduced dimension $L = 8$, stride $\gamma = 2$ and $\varepsilon = 16$. Following [12], we used a piecewise weight function $\lambda(t) = 0.1 \times e^{(-5(1-t/t_{max}))^2}$ in Eq. (1), where t was the current training step and t_{max} was the maximum training step. During testing, we used four standard evaluation metrics, including Dice coefficient (Dice), Jaccard Index (Jaccard), Average Symmetric Surface Distance (ASD) and 95% Hausdorff Distance (95HD).

Table 1. Ablation of 3D S²Net and 3D BGCM with 10% labeled data.

#	3D S²Net		3D BGCM				Dice [%]
	Seg	SDM	Intra-task		Inter-task		
			$\mathcal{G}^a \rightarrow \mathcal{G}^a$	$\mathcal{G}^b \rightarrow \mathcal{G}^b$	$\mathcal{G}^a \rightarrow \mathcal{G}^b$	$\mathcal{G}^b \rightarrow \mathcal{G}^a$	
1	√						83.04
2	√	√					85.36
3	√	√	√				86.32
4	√	√		√			85.83
5	√	√	√	√			86.75
6	√	√	√	√	√		87.02
7	√	√	√	√		√	87.48
8	√	√	√	√	√	√	**87.94**

3.3 Ablation Study

Ablation of 3D S²Net. First, we presented the ablation of our 3D S²Net backbone with 10% labeled data. We examined the impact of introducing the SDM prediction branch (f_{sdm}). As seen in Table 1 (#1 and # 2), the Dice was improved from 83.04% to 85.36%. This indicates the essential role of geometric information in segmentation.

Ablation of 3D BGCM. To validate the efficiency of our 3D BGCM, we further considered different graph reasoning directions. Regarding the single $\mathcal{G}^a$, the accuracy was improved because of the semantic-wise reasoning that considered the correlations among proposals. As for the single $\mathcal{G}^b$, the performance was increased from 85.36% to 85.83%. With applying the intra-task reasoning, an increment from 86.75% to 87.48% was achieved, where 0.27% and 0.73% improvements were produced by using a single direction of $\mathcal{G}^a \rightarrow \mathcal{G}^b$ (from $\mathcal{G}^a$ to $\mathcal{G}^b$) and $\mathcal{G}^b \rightarrow \mathcal{G}^a$ (from $\mathcal{G}^a$ to $\mathcal{G}^b$), respectively. The best performance was achieved at 87.94% dice accuracy by using bilateral reasoning, which learned the mutual relations from tasks and yielded clear boundaries.

3.4 Comparison with the State-of-the-Art Methods

We compared our method with eight state-of-the-art semi-supervised methods, including, DAP [6], ASDNet [5], TCSE [27], EM [28], CCT [29], SASSNet [8], MT [10] and UA-MT [12]. For fair comparison, we used the same V-Net backbone in these methods.

Quantitative Evaluation: Table 2 showed the quantitative results with four metrics, in which we first presented the upper-bound performance achieved in the fully-supervised mode, followed by two individual settings in the semi-supervised mode: (**i**) The first setting took 20% (i.e., 16) scans as labeled data and the remaining 64 scans as unlabeled data. As seen, the V-Net trained with only 20% labeled data already achieved good result (Dice: 86.03%). A possible explanation for the inferior performance of MT [10] and UA-MT [12] attributes to negligence of geometric constraint in object. SASSNet [8] achieved the top result in the previous methods by jointly predicting SDM, but the dual relationships between two tasks are not explicitly considered. Notably, our 3D Graph-S^{2}Net achieved a better performance over SASSNet, obtaining improvement in Dice: +0.56%, Jaccrad: +0.87%, ASD: -1.01 voxel and 95HD: -1.56 voxel. (**ii**) We also considered a more challenging setting with only 8 labeled images for training. Experimental results showed that our 3D Graph-S^{2}Net surpassed all the other methods in four metrics, especially outperforming MT [10] and UA-MT [12] with a large margin.

Qualitative Comparison: Figure 3(a) and (b) respectively exhibited 2D and 3D segmentation results of MT [10], UA-MT [12], SASSNet [8] and 3D Graph-S^{2}Net with 20% labeled training data. It can be observed that our 3D Graph-S^{2}Net produces more accurate boundaries than other methods despite the existence of complex background with challenging edges. We further illustrated the number of error pixels *vs.* their Euclidean distances to the boundaries on four methods in Fig. 3(c). We can observe that, our 3D Graph-S^{2}Net achieved accurate segmentation with high-quality edges, especially for the dominated error distributing within ~4 pixels width along the boundary.

Table 2. Comparison between our method (show in orange) and various methods.

Method	# Scan used		Metrics			
	Labeled	Unlabeled	Dice[%]	Jaccard[%]	ASD[voxel]	95HD[voxel]
V-Net	80	0	91.14	83.82	1.52	5.75
V-Net	16	0	86.03	76.06	3.51	14.26
DAP [6]	16	64	87.89	78.72	2.74	9.29
ASDNet [5]	16	64	87.90	78.85	**2.08**	9.24
TCSE [27]	16	64	88.15	79.20	2.44	9.57
EM [28]	16	64	88.45	79.51	3.72	14.14
CCT [29]	16	64	88.83	80.06	2.49	8.44
SASSNet [8]	16	64	89.27	80.82	3.13	8.24
MT [10]	16	64	88.32	79.37	2.76	10.50
UA-MT [12]	16	64	88.88	80.21	2.26	7.32
3D Graph- S^2Net	**16**	**64**	**89.83**	**81.69**	2.12	**6.68**
V-Net	8	0	79.99	68.12	5.48	21.11
DAP [6]	8	72	81.89	71.23	3.80	15.81
SASSNet [8]	8	72	86.81	76.92	3.94	12.54
MT [10]	8	72	83.04	72.55	3.85	15.21
UA-MT [12]	8	72	84.25	73.48	3.36	13.84
3D Graph- S^2Net	**8**	**72**	**87.94**	**78.90**	**2.32**	**8.99**

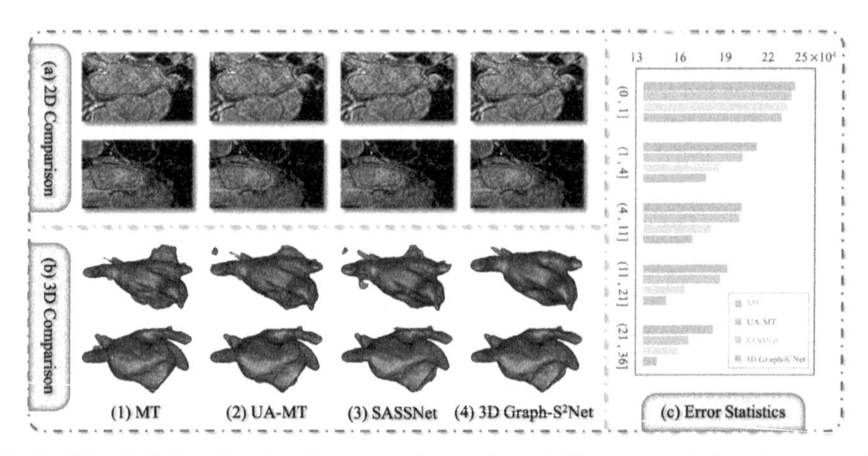

Fig. 3. 2D and 3D Visualization of two typical examples and Histogram statistics of errors on the whole validation dataset with 20% labeled data. In (a) 2D visualization, **blue edges** are ground truth boundaries and **red edges** the are predictions. In (b) 3D visualization, **pink areas** are true positives (TP), **gray areas** are false negatives, and **cyan areas** are false positives. (c) shows the number of error pixels *vs.* their Euclidean distances to the boundaries on four methods.

4 Conclusion

In this paper, we propose a novel 3D Graph-S^{2}Net for semi-supervised medical image segmentation, which consists of two novel components: backbone 3D S^{2}Net and cross-task module 3D BGCM. The Mean-Teacher based 3D S^{2}Net jointly conducts segmentation and SDM prediction in a multi-task learning way. Under this framework, the graph-based 3D BGCM interacts between tasks to mine intra- and inter-task relations, which enhances both two tasks. Experimental results on Atrial Segmentation Challenge confirm that our method surpasses all state-of-the-art approaches.

Acknowledgement. This work was supported in part by Major Scientific Research Project of Zhejiang Lab under Grant No. 2020ND8AD01, and in part by Grant-in Aid for Scientific Research from the Japanese Ministry for Education, Science, Culture and Sports (MEXT) under Grant No. 20KK0234, No. 21H03470 and No. 20K21821.

References

1. Lu, F., et al.: Automatic 3D liver location and segmentation via convolutional neural network and graph cut. Int. J. Comput. Assist. Radiol. Surg. **12**(2), 171–182 (2016). https://doi.org/10.1007/s11548-016-1467-3
2. Bai, W., et al.: Semi-supervised learning for network-based cardiac MR image segmentation. In: Descoteaux, M., Maier-Hein, L., Franz, A., Jannin, P., Collins, D.L., Duchesne, S. (eds.) MICCAI 2017. LNCS, vol. 10434, pp. 253–260. Springer, Cham (2017). https://doi.org/10.1007/978-3-319-66185-8_29
3. Xia, Y., et al.: 3D semi-supervised learning with uncertainty - aware multi-view co-training. arXiv:1811.12506 (2018)
4. Zhou, Y., et al.: Semi-supervised 3D abdominal multi-organ segmentation via deep multi-planar co-training. In: IEEE Winter Conference on Applications of Computer Vision (2019)
5. Nie, D., Gao, Y., Wang, L., Shen, D.: ASDNet: attention based semi-supervised deep networks for medical image segmentation. In: Frangi, A.F., Schnabel, J.A., Davatzikos, C., Alberola-López, C., Fichtinger, G. (eds.) MICCAI 2018. LNCS, vol. 11073, pp. 370–378. Springer, Cham (2018). https://doi.org/10.1007/978-3-030-00937-3_43
6. Zheng, H., et al.: Semi-supervised segmentation of liver using adversarial learning with deep atlas prior. In: Shen, D., et al. (eds.) MICCAI 2019. LNCS, vol. 11769, pp. 148–156. Springer, Cham (2019). https://doi.org/10.1007/978-3-030-32226-7_17
7. Huang, H., et al.: Medical image segmentation with deep atlas prior. IEEE Trans. Med. Imaging (2021)
8. Li, S., Zhang, C., He, X.: Shape-aware semi-supervised 3D semantic segmentation for medical images. In: Martel, A.L., et al. (eds.) MICCAI 2020. LNCS, vol. 12261, pp. 552–561. Springer, Cham (2020). https://doi.org/10.1007/978-3-030-59710-8_54
9. Laine, S., Aila, T.: Temporal ensembling for semi-supervised learning. In: 5th International Conference on Learning Representations, Toulon, France (2017)
10. Tarvainen, A., Valpola, H.: Mean teachers are better role models: weight-averaged consistency targets improve semi-supervised deep learning results. NIPS (2017)
11. Cui, W., et al.: Semi-supervised brain lesion segmentation with an adapted mean teacher model. In: Chung, A.C.S., Gee, J.C., Yushkevich, P.A., Bao, S. (eds.) IPMI 2019. LNCS, vol. 11492, pp. 554–565. Springer, Cham (2019). https://doi.org/10.1007/978-3-030-20351-1_43

12. Yu, L., Wang, S., Li, X., Fu, C.-W., Heng, P.-A.: Uncertainty-aware self-ensembling model for semi-supervised 3D left atrium segmentation. In: Shen, Dinggang, et al. (eds.) MICCAI 2019. LNCS, vol. 11765, pp. 605–613. Springer, Cham (2019). https://doi.org/10.1007/978-3-030-32245-8_67
13. Dangi, S., Linte, C.A., Yaniv, Z.: A distance map regularized CNN for cardiac cine MR image segmentation. Med. Phys. **46**, 5637–5651 (2019)
14. Kipf, T.N., Welling, M.: Semi-supervised classification with graph convolutional networks. In: International Conference on Learning Representations (2017)
15. Li, Y., Abhinav, G.: Beyond grids: Learning graph representations for visual recognition. Adv. Neural Inf. Process. Syst. **31**, 9225–9235 (2018)
16. Chen, Y., et al.: Graph-based global reasoning networks. In: Proceedings of the IEEE Conference on Computer Vision and Pattern Recognition (2019)
17. Huang, H., et al.: Graph-based pyramid global context reasoning with a saliency-aware projection for COVID-19 lung infections segmentation. In: ICASSP (2021)
18. Huang, S., et al.: Referring image segmentation via cross-modal progressive comprehension. In: Computer Vision and Pattern Recognition (2020)
19. Shin, S., Lee, S., Yun, I., Lee, K.: Deep vessel segmentation by learning graphical connectivity. Med. image Anal. **58**, 101556 (2019)
20. Li, X., et al.: Spatial pyramid based graph reasoning for semantic segmentation. In: Proceedings of the IEEE/CVF Conference on Computer Vision and Pattern Recognition (2020)
21. Zhang, L., Li, X., Arnab, A., Yang, K., Tong, Y., Torr, P.: Dual graph convolutional network for semantic segmentation. arXiv preprint arXiv:1909.06121 (2019)
22. Te, G., et al.: Edge-aware graph representation learning and reasoning for face parsing. arXiv preprint arXiv:2007.11240 (2020)
23. Wu, Y., Zhang, G., et al.: Bidirectional graph reasoning network for panoptic segmentation. In: Conference on Computer Vision and Pattern Recognition (2020)
24. Milletari, F., Navab, N., Ahmadi, S.A.: V-Net: fully convolutional neural networks for volumetric medical image segmentation. In: 3DV (2016)
25. Velickovic, P., Cucurull, G., Casanova, A., et al.: Graph attention networks. arXiv preprint arXiv:1710.10903 (2017)
26. Atrial Segmentation Challenge. http://atriaseg2018.cardiacatlas.org/
27. Li, X., et al.: Semi-supervised skin lesion segmentation via transformation consistent self-ensembling model. arXiv preprint arXiv:1808.03887 (2018)
28. Vu, T.-H., et al.: Advent: adversarial entropy minimization for domain adaptation in semantic segmentation. In: CVRP (2019)
29. Ouali, Y., et al.: Semi-supervised semantic segmentation with cross-consistency training. In: CVPR (2020)
30. Park, J.J., Florence, P., Straub, J., Newcombe, R., Lovegrove, S.: DeepSDF: learning continuous signed distance functions for shape representation. In: CVPR (2019)
31. Xue, Y., et al.: Shape-aware organ segmentation by predicting signed distance maps. arXiv preprint arXiv:1912.03849 (2019)
32. Perera, S., Barnes, N., He, X., Izadi, S., Kohli, P., Glocker, B.: Motion segmentation of truncated signed distance function based volumetric surfaces. In: WACV (2015)
33. Huang, H., et al.: UNet 3+: a full-scale connected UNet for medical image segmentation. In: ICASSP (2020)

34. Chen, L.-C., et al.: Semantic image segmentation with deep convolutional nets, atrous convolution, and fully connected CRFs. IEEE Trans. Pattern Anal. Mach. Intell. **40**(4), 834–848 (2018)
35. Ronneberger, O., Fischer, P., Brox, T.: U-Net: convolutional networks for biomedical image segmentation. In: Navab, N., Hornegger, J., Wells, W.M., Frangi, A.F. (eds.) MICCAI 2015. LNCS, vol. 9351, pp. 234–241. Springer, Cham (2015). https://doi.org/10.1007/978-3-319-24574-4_28

Duo-SegNet: Adversarial Dual-Views for Semi-supervised Medical Image Segmentation

Himashi Peiris[1(✉)], Zhaolin Chen[1,2], Gary Egan[2], and Mehrtash Harandi[1]

[1] Department of Electrical and Computer Systems Engineering, Monash University, Melbourne, Australia
{Edirisinghe.Peiris,Zhaolin.Chen,Mehrtash.Harandi}@monash.edu

[2] Monash Biomedical Imaging (MBI), Monash University, Melbourne, Australia
Gary.Egan@monash.edu

Abstract. Segmentation of images is a long-standing challenge in medical AI. This is mainly due to the fact that training a neural network to perform image segmentation requires a significant number of pixel-level annotated data, which is often unavailable. To address this issue, we propose a semi-supervised image segmentation technique based on the concept of multi-view learning. In contrast to the previous art, we introduce an adversarial form of dual-view training and employ a critic to formulate the learning problem in multi-view training as a min-max problem. Thorough quantitative and qualitative evaluations on several datasets, indicate that our proposed method outperforms state-of-the-art medical image segmentation algorithms consistently and comfortably. The code is publicly available at https://github.com/himashi92/Duo-SegNet.

Keywords: Deep learning · Semi-supervised learning · Medical image segmentation · Multi-view learning · Adversarial learning

1 Introduction

In this paper, we propose a semi-supervised technique based on the concept of multi-view learning [18] to segment medical images. Accurate segmentation of medical images is a key-step in developing Computer-Aided Diagnosis (CAD) and automating various clinical tasks such as image-guided interventions.

The prevailing idea in medical image segmentation is to employ an encoder-decoder structure (*e.g.*, UNet [16] and its variants [12,20]) and formulate the problem as a dense classification/regression problem, depending on the type of input and the desired output. Most of the existing algorithms are breeze through supervised setting when sizable, annotated datasets are available. However, annotating large-scale datasets for image segmentation is challenging and expensive. On one hand, most medical image modalities are hefty in size (*e.g.*, 3D volumes as in MRI and CT) and hence annotation is extremely laborious.

M. de Bruijne et al. (Eds.): MICCAI 2021, LNCS 12902, pp. 428–438, 2021.
https://doi.org/10.1007/978-3-030-87196-3_40

On the other hand, annotating medical images requires expert knowledge and cannot be crowd-sourced. Add to this the fact that medical images often contain low contrast slices and ambiguous regions, which in turn makes annotation very difficult. While attaining large annotated datasets is challenging, unlabeled data comes (almost) for free and is abundant.

Multi-view learning makes use of multiple distinct views of data and benefits from the resulting relationships to achieve accurate models. The principle of consensus [18] states that by minimizing the disagreement on multiple distinct hypotheses, the error of each hypothesis will be minimized. To be specific, suppose h^1 and h^2 are two distinct hypothesis defined on a distribution. Under some mild assumptions and as shown in [5]

$$P(h^1 \neq h^2) \geq \max\left\{P_{\text{err}}(h^1), P_{\text{err}}(h^2)\right\}.$$

That is, reducing the disagreement of the two hypotheses minimizes the error of each hypothesis. The consensus principle provides an efficient way with strong theoretical properties to benefit from unlabeled data as shown for example in the celebrated work of Blum and Mitchell [4]. Except a handful of studies [13,14], multi-view learning has been mostly studied under the classification regime [7,8]. This raises a natural question, *is multi-view learning beneficial when it comes to segmentation?* and *if yes, how it must be formulated?* Our work takes a step in this direction and provides a way to cultivate unlabeled data for image segmentation. In particular, we propose a dual-view UNet model and equip it with a critic network. Each UNet provides a view of the data distribution and will be trained by minimizing a supervised loss on the labeled data and a disagreement loss over the unlabeled data. The critic is used to facilitate two objectives, **1.** to ensure that the output of UNets resembles the ground-truth segmentation masks (in the ideal case, the critic is not able to distinguish whether its input is a ground-truth mask or a prediction mask from the UNets), and **2.** to identify confident parts of a prediction mask to enforce agreement across the views. We should stress that unlike classification where (dis)agreement can be formulated readily between predictions, in segmentation, we face a dense prediction problem, meaning a prediction mask includes tens of thousands, if not more, predictions at pixel-level. A naive treatment of (dis)agreement loss could lead to inferior results as simply the networks can overfit to agree on the background, which is the dominant part of the prediction mask in many cases. Therefore, to train the Duo-SegNet, we formulate the learning as a min-max problem by allowing a critic to stand in as a quantitative subjective referee. Thorough empirical evaluations demonstrate that our method performs well both qualitatively and quantitatively, utilizing small fraction of labeled data.

In short, we have made the following contributions in this work, **1.** We propose a dual-view learning scheme for semi-supervised medical image segmentation. **2.** We make use of a critic to regularize the training and identify confident parts of a prediction mask towards learning from unlabelled data.

2 Methodology

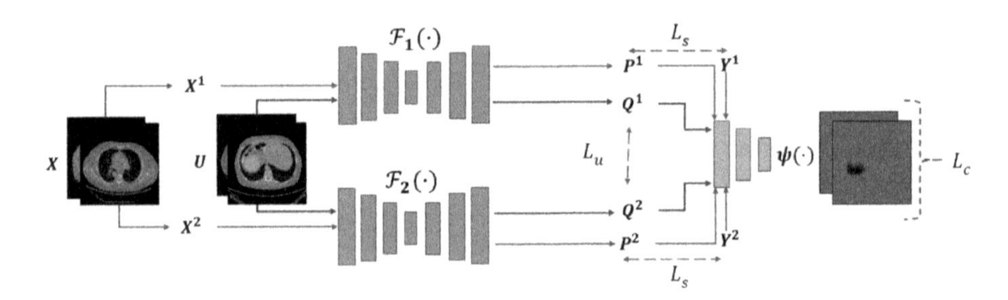

Fig. 1. Proposed Adversarial Dual View Network. $\{\mathcal{F}_i(\cdot)\}_{i=1}^{2}$ and $\psi(\cdot)$ denote Segmentation networks and Critic network. Here, Critic criticizes between prediction masks and the ground truth masks to perform the min-max game.

We start this section by providing an overview of the Duo-SegNet (see Fig. 1 for a conceptual illustration). Let $\mathcal{X} = \{(\mathbf{X}_1, \mathbf{Y}_1), \cdots, (\mathbf{X}_m, \mathbf{Y}_m)\}$ be a labeled set, where each pair $(\mathbf{X}_i, \mathbf{Y}_i)$ consists of an image $\mathbf{X}_i \in \mathbb{R}^{C\times H\times W}$ and its associated ground-truth mask $\mathbf{Y}_i \in \{0,1\}^{H\times W}$. Furthermore, let $\mathcal{U} = \{\mathbf{U}_i\}_{i=1}^{n}, \mathbf{U}_i \in \mathbb{R}^{C\times H\times W}$ be a set of n unlabelled images with $n \gg m$. The primary objective is to learn a segmentation model from $\mathcal{D} = \mathcal{X} \cup \mathcal{U}$.

The Duo-SegNet includes two basic modules, a dual view segmentation networks and a critic, shown by $\mathcal{F}_1$, $\mathcal{F}_2$, and ψ respectively in Fig. 1. Each leg in the dual-view network (*i.e.*, $\mathcal{F}_1$ and $\mathcal{F}_2$) benefits from an encoder-decoder design, and is structured as a UNet [16]. In this work, by incorporating a critic, we implicitly enforce segmentation networks to create predictions that are more similar to the desired masks holistically. The parameters of the model are the parameters of the dual-view network θ_1, θ_2 and the critic network θ_C. We collectively show the parameters of the dual-view network (*i.e.*, θ_1, θ_2) by $\mathbf{\Theta}$ to avoid cluttering the equations. To train the model, we propose optimizing the following min-max problem:

$$\min_{\mathbf{\Theta}} \max_{\theta_C} \mathcal{L}(\mathbf{\Theta}; \mathcal{D}). \tag{1}$$

The familiar min-max problem in (1) encourages the dual-view segmentation networks to yield segmentation masks that look like realistic ones by deceiving the critic. Similar to standard multi-view training, our approach adapted dual view training to train two segmentation models collaboratively. We propose to train dual view segmentation networks by minimizing a multitask loss function consists of three loss terms:

$$\mathcal{L}(\mathbf{\Theta}; \mathcal{D}) := \lambda_s \mathcal{L}_{\mathrm{s}}(\mathbf{\Theta}; \mathcal{X}) + \lambda_u \mathcal{L}_{\mathrm{u}}(\mathbf{\Theta}; \mathcal{U}) + \lambda_c \mathcal{L}_{\mathrm{c}}(\mathbf{\Theta}; \theta_C; \mathcal{D}), \tag{2}$$

where $\mathcal{L}_{\mathrm{s}}$, $\mathcal{L}_{\mathrm{u}}$, and $\mathcal{L}_{\mathrm{c}}$ denote the supervised, the unsupervised and the critic loss respectively. Furthermore, $\lambda_{\mathrm{s}}, \lambda_{\mathrm{u}}, \lambda_{\mathrm{c}} > 0$ are hyper-parameters of the algorithm, controlling the contribution of each loss term. We note that the supervised and unsupervised loss are only dependent on the dual-view networks while the critic loss is defined based on the parameters of the whole model. In practice, we find that tuning the hyper-parameters of the network is not difficult at all and the Duo-SegNet works robustly as long as these parameters are defined in a reasonable range. For example, in all our experiments in this paper, we set $\lambda_s = 1.0$, $\lambda_u = 0.3$ and $\lambda_c = 0.2$. To better understand Eq. (2), we start with the supervised loss. This loss makes use of the labeled data and can be defined as a cross-entropy (*i.e.*, Eq. (3)) or dice loss (*i.e.*, Eq. (4)) or even a linear combination of both, which is a common practice in segmentation [16].

$$\mathcal{L}_{\mathrm{ce}}(\theta_i; \mathcal{X}) = \mathbb{E}_{(\mathbf{X},\mathbf{Y})\sim\mathcal{X}}\left[\Big\langle \mathbf{Y}, \log\left(\mathcal{F}_i(\mathbf{X})\right)\Big\rangle\right] \tag{3}$$

$$\mathcal{L}_{\mathrm{dice}}(\theta_i; \mathcal{X}) = 1 - \mathbb{E}_{(\mathbf{X},\mathbf{Y})\sim\mathcal{X}}\left[\frac{2\langle \mathbf{Y}\ ,\ \mathcal{F}_i(\mathbf{X})\rangle}{\|\mathbf{Y}\|_1 + \|\mathcal{F}_i(\mathbf{X})\|_1}\right], \tag{4}$$

where we use $\langle \mathbf{A}, \mathbf{B}\rangle = \sum_{i,j} \mathbf{A}[i,j]\mathbf{B}[i,j]$ and $\|\mathbf{A}\|_1 = \sum_{i,j} |\mathbf{A}[i,j]|$.

We define the unsupervised loss as a means to realize the principle of consensus as;

$$\mathcal{L}_{\mathrm{u}}(\boldsymbol{\Theta}; \mathcal{U}) := \mathbb{E}_{\mathbf{U}\sim\mathcal{U}}\left[\Big\langle \mathcal{F}_1(\mathbf{U}), \log\left(\mathcal{F}_2(\mathbf{U})\right)\Big\rangle + \Big\langle \mathcal{F}_2(\mathbf{U}), \log\left(\mathcal{F}_1(\mathbf{U})\right)\Big\rangle\right]. \tag{5}$$

We identify Eq. (5) as a symmetric form of the cross-entropy loss. In essence, the unsupervised loss $\mathcal{L}_{\mathrm{u}}$ acts as an agreement loss during dual-view training. The idea is that the two segmentation networks should generate similar segmentation masks for unlabeled data. In the proposed method, diversity among two views are retrieved based on segmentation model predictions. Unlike in [13,14] where adversarial examples for a model are used to teach other model in the ensemble, we make use of a critic to approach inter diversity among segmentation models for both labeled and unlabeled distributions. When using the labeled data, two segmentation networks are supervised by both the supervised loss with the ground truth and the adversarial loss. We denote the functionality of the critic by $\Psi : [0,1]^{H\times W} \rightarrow [0,1]^{H\times W}$ and define the normalized loss of critic for labeled prediction distribution as:

$$\begin{aligned}\mathcal{L}_{adv1}(\boldsymbol{\Theta}; \theta_C; \mathcal{X}) := \mathbb{E}_{(\mathbf{X},\mathbf{Y})\sim\mathcal{X}}\Bigg[-\sum_{a\in H}\sum_{b\in W}\Big\{(1-\eta)\log\Big(\psi(\mathbf{Y})[a,b]\Big) \\ +\,\eta\log\Big(1-\psi(\mathcal{F}_i(\mathbf{X}))[a,b]\Big)\Big\}\Bigg],\end{aligned} \tag{6}$$

where $\eta = 0$ if the sample is generated by the segmentation network, and $\eta = 1$ if the sample is drawn from the ground truth labels. For unlabeled data, it is obvious that we cannot apply any supervised loss since there is no ground truth annotation available. However, adversarial loss can be applied as it only requires the knowledge about whether the mask is from the ground-truth labels or generated by the segmentation networks. The adversarial loss for the distribution of unlabeled predictions is defined as:

$$\mathcal{L}_{adv2}(\mathbf{\Theta};\theta_C;\mathcal{U}) := \mathbb{E}_{\mathbf{U}\sim\mathcal{U}}\left[-\sum_{a\in H}\sum_{b\in W}\left\{\log\left(1-\psi(\mathcal{F}_i(\mathbf{U}))[a,b]\right)\right\}\right], \quad (7)$$

Remark 1. The basis of having a critic is that a well trained critic can produce pixel-wise uncertainty map/confidence map which imposes a higher-order consistency measure of prediction and ground truth. So, it infers the pixels where the prediction masks are close enough to the ground truth distribution.

Therefore, the critic loss is defined as:

$$\mathcal{L}_{\mathrm{c}}(\mathbf{\Theta};\theta_C;\mathcal{D}) = \mathcal{L}_{\mathrm{adv1}}(\mathbf{\Theta};\theta_C;\mathcal{X}) + \mathcal{L}_{\mathrm{adv2}}(\mathbf{\Theta};\theta_C;\mathcal{U}). \quad (8)$$

With this aggregated adversarial loss, we co-train two segmentation networks to fool the critic by maximizing the confidence of the predicted segmentation being generated from the ground truth distribution. To train the critic, we use labeled prediction distribution along with given ground truth segmentation masks and calculate the normalized loss as in Eq. (6). The overall min-max optimization process is summarized in Algorithm 1.

Algorithm 1: Duo-SegNet (*training*)

Input: Define Segmentation networks $\{\mathcal{F}_i(\cdot)\}_{i=1}^2$, critic $\psi(\cdot)$, batch size $\mathcal{B}$, maximum epoch E_{max}, number of steps k_s and k_c for segmentation networks and critic, Labeled images $\mathcal{X} = \{(X_1, Y_1), ..., (X_m, Y_m)\}$, Unlabeled images $\mathcal{U} = \{U_1, ..., U_n\}$ and two labeled sets $\mathcal{X}^1; \mathcal{X}^2 \subset \mathcal{X}$;

Output: Network Parameters $\{\theta_i\}_{i=1}^2$ and θ_C;

Initialize Network Parameters $\{\theta_i\}_{i=1}^2$ and θ_C;

for *epoch* = 1, $\cdots$, E_{max} **do**
- **for** *batch* = 1, $\cdots$, $\mathcal{B}$ **do**
  - **for** k_s *steps* **do**
    - Generate predictions for labeled data $\mathcal{F}_1(x)$ for all $X_i \in \mathcal{X}^1$, $\mathcal{F}_2(x)$ for all $X_i \in \mathcal{X}^2$ and for unlabeled data $\mathcal{F}_1(x)$ and $\mathcal{F}_2(x)$ for all $U_i \in \mathcal{U}$;
    - Generate confidence maps for all predictions using $\psi(\cdot)$;
    - Let $\mathcal{L} = \mathcal{L}_s + \mathcal{L}_u + \mathcal{L}_c$, as defined in Eqs. (2)–(8);
    - Update $\{\theta_i\}_{i=1}^2$ by descending its stochastic gradient on $\mathcal{L}$;
  - **end**
  - **for** k_c *steps* **do**
    - Generate confidence maps for all labeled predictions and ground truth masks using $\psi(\cdot)$;
    - Let $\mathcal{L}_c = \mathcal{L}_{adv1}$, as defined in 6;
    - Update θ_C by ascending its stochastic gradient on $\mathcal{L}_c$;
  - **end**
- **end**

end

3 Related Work

We begin by briefly discussing alternative approaches to semi-supervised learning, with a focus on the most relevant ones to our approach, namely pseudo labelling [9], mean teacher model [17], Virtual Adversarial Training (VAT) [11], recently published deep co-training [13]. Our goal here is to discuss the resemblance and differences between our proposed approach and some of the methods that have been adopted for semi-supervised image segmentation.

In Pseudo Labelling, as the name implies, the model uses the predicted labels of the unlabeled data and treat it as ground-truth labels for training. The drawback here is, sometimes incorrect pseudo labels may diminish the generalization performance and weaken the training of deep neural networks. In contrast, Duo-SegNet by incorporating a critic, increases the tolerance of these incorrect pseudo labels which stabilizes the generalization performance. Similar to Duo-SegNet, the mean teacher approach benefits from two neural networks, namely the teacher and the student networks. While the student model is trained in a stochastic manner, the parameters of the teacher model are updated slowly by a form of moving averaging of the student's parameters. This, in turn, results in better robustness to prediction error as one could hope that averaging attenuates the effect of noisy gradient (as a result of incorrect pseudo labels). Unlike mean teacher model, Duo-SegNet simultaneously train both networks and models can learn from one another during training. VAT can be understood as an effective regularization method which optimizes generalization power of a model for unlabeled examples. This is achieved by generating adversarial perturbations to the input of the model, followed by making the model robust to the adversarial perturbations. In contrast to VAT, Duo-SegNet makes use of a critic to judge if predictions are from the same or different distribution compared to labeled examples. With this, segmentation networks are encouraged to generate similar predictive distribution for both labeled and unlabeled data. Similar to our work, in co-training two models are alternately trained on distinct views, while learning from each other is encouraged. Recently, Peng *et al.* introduced a deep co-training method for semi-supervised image segmentation task [13] based on the approach presented by Qiao *et al.* for image recognition task in [14]. In this approach diversity among views are achieved via adversarial examples following VAT [11]. We note that adversarial examples, from a theoretical point of view, cannot guarantee diversity, especially when unlabeled data is considered. That is, a wrong prediction can intensify even more once adversarial examples are constructed from it. In contrast, inspired by Generative Adversarial Networks (GANs) [6] and GAN based medical imaging applications including medical image segmentation [10], reconstruction [15] and domain adaptation [19], our proposed method graciously encloses the min-max formulation in dual-view learning for segmenting medical images where high-confidence predictions for unlabeled data are leveraged, which is simple and effective.

4 Experiments

Implementation Details: The proposed model is developed in PyTorch [3]. Training was done from scratch without using any pre-trained model weights. For training of segmentation network and critic, we use SGD optimizer (LR = 1e−02) and RMSProp optimizer (LR = 5e−05), respectively. We divide the original dataset into training (80%) and test set (20%). Experiments were conducted for 5%, 20% and 50% of labeled training sets.

Datasets: We use three medical image datasets for model evaluation covering three medical image modalities: 670 Fluorescence Microscopy (FM) images from Nuclei [1], 20 MRI volumes from Heart [2] and 41 CT volumes from Spleen [2]. For our experiments, 2D images are obtained by slicing the high-resolution MRI and CT volumes for Heart (2271 slices) and Spleen (3650 slices) datasets. Each slice is then resized to a resolution of 256×256.

Competing Methods and Evaluation Metrics: We compare our proposed method with Mean Teacher [17], Pseudo Labelling [9], VAT [11], Deep Co-training [13] and fully supervised U-Net [16]. For all baselines, we follow the same configurations as for our method. All approaches are evaluated using: 1) Dice Sørensen coefficient (DSC) and 2) Mean Absolute Error (MAE).

Performance Comparison: The qualitative results for the proposed and competing methods are shown in Fig. 3. The quantitative results comparison of the proposed method to the four state-of-the-art methods are shown in Table 1. The results reveal that the proposed method comfortably outperform other studied methods for smaller fractions of annotated data (*e.g.* Spleen 5%). The gap between the Duo-SegNet and other competitors decreases on the Nuclei dataset, when the amount of labeled data increases. That said, we can still observe a significant improvement on the Heart and Spleen dataset. The proposed network can produce both accurate prediction masks and confidence maps representing which regions of the prediction distribution are close to the ground truth label distribution. This is useful when training unlabeled data. Figure 2 shows the visual analysis of confidence maps.

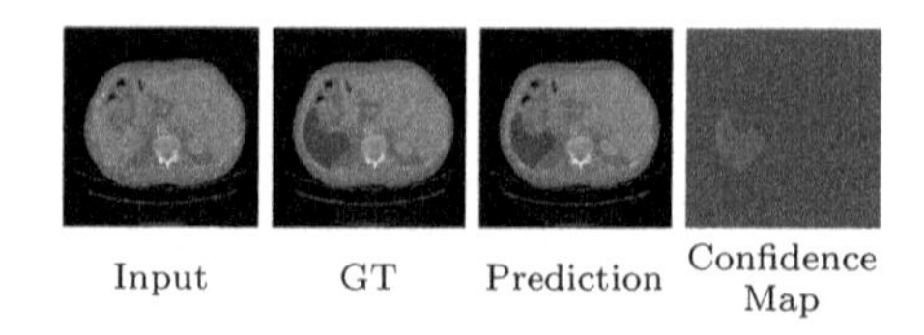

Fig. 2. Visual analysis of confidence map generated by the critic during training

Table 1. Comparison with state-of-the-art methods.

Dataset	Method	DSC			MAE		
Nuclei	Fully supervised	**91.36**			**2.25**		
		$l_a = 5\%$	$l_a = 20\%$	$l_a = 50\%$	$l_a = 5\%$	$l_a = 20\%$	$l_a = 50\%$
	Mean teacher	83.78	84.92	87.99	4.78	4.30	3.36
	Pseudo labeling	60.90	72.46	85.91	8.40	6.37	3.84
	VAT	85.24	86.43	88.45	4.09	3.77	3.26
	Deep co-training	85.83	87.15	89.20	4.08	3.80	3.08
	Duo-SegNet	**87.14**	**87.83**	**89.28**	**3.57**	**3.43**	**3.03**
Heart	Fully supervised	**97.17**			**0.02**		
		$l_a = 5\%$	$l_a = 20\%$	$l_a = 50\%$	$l_a = 5\%$	$l_a = 20\%$	$l_a = 50\%$
	Mean teacher	71.00	87.59	93.43	0.22	0.09	0.05
	Pseudo labeling	65.92	79.86	80.75	0.20	0.13	0.13
	VAT	85.33	91.60	94.83	0.11	0.06	0.04
	Deep co-training	85.96	91.54	94.55	**0.10**	0.06	0.04
	Duo-SegNet	**86.79**	**93.21**	**95.56**	**0.10**	**0.05**	**0.03**
Spleen	Fully supervised	**97.89**			**0.02**		
		$l_a = 5\%$	$l_a = 20\%$	$l_a = 50\%$	$l_a = 5\%$	$l_a = 20\%$	$l_a = 50\%$
	Mean teacher	75.44	90.76	92.98	0.20	0.08	0.06
	Pseudo labeling	67.70	68.81	84.81	0.24	0.21	0.12
	VAT	78.31	91.37	94.34	0.19	0.07	0.05
	Deep co-training	79.16	89.65	94.90	0.16	0.09	0.05
	Duo-SegNet	**88.02**	**92.19**	**96.03**	**0.10**	**0.07**	**0.03**

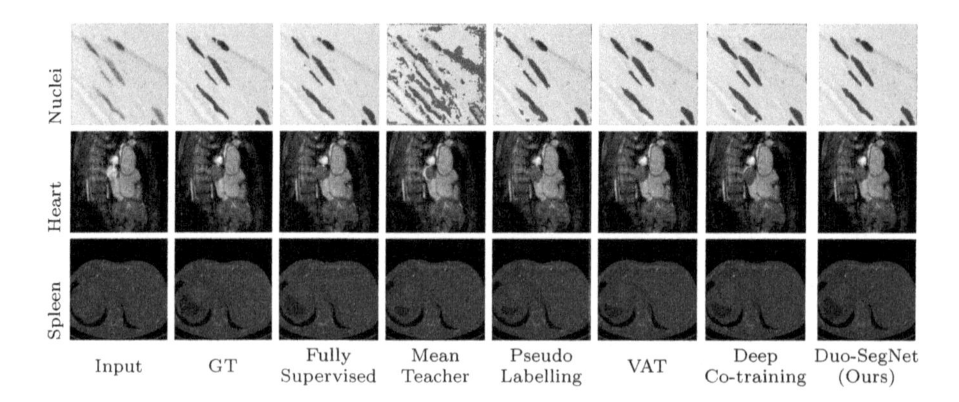

Fig. 3. Visual comparison of our method with state-of-the-art models. Segmentation results are shown for 5% of labeled training data.

Ablation Study: We also perform ablation studies to show the effectiveness of adding a critic in dual-view learning in semi-supervised setting and the importance of dual view network structure. In our algorithm, we benefit from unlabeled data via (1) criss-cross exchange of confident regions, (2) improving the critic which in essence minimizes an upper-bound of error. To justify this, we conducted an additional experiment without unlabeled data. It can be seen that there is an impact in the performance of segmentation model for varying values for λ_u. For our experiments we choose λ_u in the range of 0.3 to 0.4. All experiments in Table 2 are conducted for spleen dataset with 5% of annotated data.

Table 2. Ablation study

(a) Network Structure Analysis.

Experiment	DSC	MAE
Duo-SegNet	88.02	0.10
w/o Critic	77.69	0.19
w/o Unlabeled Data	76.67	0.17
One Segmentation Network	82.44	0.16

(b) Hyper-parameter Analysis for λ_u.

λ_u	0.1	0.2	0.3	0.4	0.5
DSC	83.58	85.62	88.02	87.14	78.89
MAE	0.15	0.12	0.10	0.11	0.20

5 Conclusion

We proposed an adversarial dual-view learning approach for semi-supervised medical image segmentation and demonstrated its effectiveness on publicly available three medical datasets. Our extensive experiments showed that employing a min-max paradigm into multi-view learning scheme sharpens boundaries between different regions in prediction masks and yield a performance close to full-supervision with limited annotations. The dual view training can still be improved by self-tuning mechanisms, which will be considered in our future works.

References

1. Data science bowl. https://www.kaggle.com/c/data-science-bowl-2018. Accessed 14 Feb 2021
2. Medical segmentation decathlon. http://medicaldecathlon.com/. Accessed 14 Feb 2021
3. Pytorch. https://pytorch.org/. Accessed 05 July 2021
4. Blum, A., Mitchell, T.: Combining labeled and unlabeled data with co-training. In: Proceedings of the Eleventh Annual Conference on Computational Learning Theory, pp. 92–100 (1998)
5. Dasgupta, S., Littman, M.L., McAllester, D.: PAC generalization bounds for co-training. In: Proceedings Advances in Neural Information Processing Systems (NeurIPS), pp. 375–382 (2002)
6. Goodfellow, I., et al.: Generative adversarial nets. In: Advances in Neural Information Processing Systems, pp. 2672–2680 (2014)
7. Kiritchenko, S., Matwin, S.: Email classification with co-training. In: Proceedings of the 2001 Conference of the Centre for Advanced Studies on Collaborative Research, p. 8. Citeseer (2001)
8. Kumar, A., Daumé, H.: A co-training approach for multi-view spectral clustering. In: Proceedings of the 28th International Conference on Machine Learning (ICML-11), pp. 393–400 (2011)
9. Lee, D.H., et al.: Pseudo-label: The simple and efficient semi-supervised learning method for deep neural networks
10. Mahmood, F., et al.: Deep adversarial training for multi-organ nuclei segmentation in histopathology images. IEEE Trans. Med. Imag. **39**, 3257 (2019)
11. Miyato, T., Maeda, S.I., Koyama, M., Ishii, S.: Virtual adversarial training: a regularization method for supervised and semi-supervised learning. IEEE Trans. Pattern Anal. Mach. Intell. **41**(8), 1979–1993 (2018)
12. Oktay, O., et al.: Attention u-net: Learning where to look for the pancreas. arXiv preprint arXiv:1804.03999 (2018)
13. Peng, J., Estrada, G., Pedersoli, M., Desrosiers, C.: Deep co-training for semi-supervised image segmentation. Pattern Recogn. **107**, 107269 (2020)
14. Qiao, S., Shen, W., Zhang, Z., Wang, B., Yuille, A.: Deep co-training for semi-supervised image recognition. In: Proceedings European Conference on Computer Vision (ECCV), pp. 135–152 (2018)
15. Quan, T.M., Nguyen-Duc, T., Jeong, W.K.: Compressed sensing MRI reconstruction with cyclic loss in generative adversarial networks. arXiv preprint arXiv:1709.00753 (2017)
16. Ronneberger, O., Fischer, P., Brox, T.: U-Net: convolutional networks for biomedical image segmentation. In: Navab, N., Hornegger, J., Wells, W.M., Frangi, A.F. (eds.) MICCAI 2015. LNCS, vol. 9351, pp. 234–241. Springer, Cham (2015). https://doi.org/10.1007/978-3-319-24574-4_28

17. Tarvainen, A., Valpola, H.: Mean teachers are better role models: weight-averaged consistency targets improve semi-supervised deep learning results. In: Proceedings Advances in Neural Information Processing Systems (NeurIPS), pp. 1195–1204 (2017)
18. Xu, C., Tao, D., Xu, C.: A survey on multi-view learning. arXiv preprint arXiv:1304.5634 (2013)
19. Zhang, Y., Miao, S., Mansi, T., Liao, R.: Task driven generative modeling for unsupervised domain adaptation: application to X-ray image segmentation. In: Frangi, A.F., Schnabel, J.A., Davatzikos, C., Alberola-López, C., Fichtinger, G. (eds.) MICCAI 2018. LNCS, vol. 11071, pp. 599–607. Springer, Cham (2018). https://doi.org/10.1007/978-3-030-00934-2_67
20. Zhou, Z., Rahman Siddiquee, M.M., Tajbakhsh, N., Liang, J.: UNet++: a nested U-Net architecture for medical image segmentation. In: Stoyanov, D., et al. (eds.) DLMIA/ML-CDS -2018. LNCS, vol. 11045, pp. 3–11. Springer, Cham (2018). https://doi.org/10.1007/978-3-030-00889-5_1

Neighbor Matching for Semi-supervised Learning

Renzhen Wang[1], Yichen Wu[1], Huai Chen[2], Lisheng Wang[2], and Deyu Meng[1,3](✉)

[1] School of Mathematics and Statistics, Xi'an Jiaotong University, Xi'an, Shaanxi, People's Republic of China
[2] Institute of Image Processing and Pattern Recognition, Department of Automation, Shanghai Jiao Tong University, Shanghai, People's Republic of China
[3] Faculty of Information Technology, Macau University of Science and Technology, Taipa, Macau
dymeng@mail.xjtu.edu.cn

Abstract. Consistency regularization has shown superiority in deep semi-supervised learning, which commonly estimates pseudo-label conditioned on each single sample and its perturbations. However, such a strategy ignores the relation between data points, and probably arises error accumulation problems once one sample and its perturbations are integrally misclassified. Against this issue, we propose Neighbor Matching, a pseudo-label estimator that propagates labels for unlabeled samples according to their neighboring ones (labeled samples with the same semantic category) during training in an online manner. Different from existing methods, for an unlabeled sample, our Neighbor Matching defines a mapping function that predicts its pseudo-label conditioned on itself and its local manifold. Concretely, the local manifold is constructed by a memory padding module that memorizes the embeddings and labels of labeled data across different mini-batches. We experiment with two distinct benchmark datasets for semi-supervised classification of thoracic disease and skin lesion, and the results demonstrate the superiority of our approach beyond other state-of-the-art methods. Source code is publicly available at https://github.com/renzhenwang/neighbor-matching.

Keywords: Semi-supervised learning · Pseudo-label estimation · Memory padding · Lesion classification

1 Introduction

Deep learning models have achieved sound performance in various medical image analysis tasks [13], which commonly attributes to large amounts of labeled training data for supervising the training phase of the models. However, collecting such labeled data is expensive in medical imaging, since it necessarily involves expert knowledge. A widely-used strategy to alleviate this challenge is semi-supervised learning (SSL), which exploits unlabeled data to help deep models improve performance so that reduce the demand for labeled data.

M. de Bruijne et al. (Eds.): MICCAI 2021, LNCS 12902, pp. 439–449, 2021.
https://doi.org/10.1007/978-3-030-87196-3_41

Among existing SSL methods, consistency regularization is an efficient strategy that adds a loss term on unlabeled data to encourage the model to generalize better to unseen data. More specifically, it enforces the model to produce the same prediction for an unlabeled data point when it is perturbed or augmented. This consistency regularization often involves pseudo-label prediction for unlabeled training data. For instance, Pseudo-Ensembles [1] and VAT [14] could be viewed as leveraging a perturbed teacher model to generate pseudo-labels for the training of the student model. Temporal Ensembling [12] used an exponential moving average (EMA) prediction of each unlabeled sample as its pseudo-label, and Mean Teacher [18] updated the teacher model's parameters with EMA for outputting more stable pseudo-label. In addition, pseudo-label prediction was a crucial step in MixMatch [3] and Global Latent Mixing (GLM) [8], where the models were trained with a mixup strategy [19,21]. All these approaches estimate the pseudo-labels along the line that each single sample together with its perturbations or augmentations have the same class semantics. However, this ignores the relation between data points, and perhaps arises error accumulation problems once one sample and its perturbations are integrally misclassified.

In fact, images with the same category usually lie in a low-dimensional manifold, where nearby points should have the same label predictions. A real-world scenario is that patients with the same disease often show similar symptoms, according to which physicians could make a diagnosis and the subsequent treatment. Inspired by this, we propose Neighbor Matching, a semi-supervised method that estimates the pseudo-label of an unlabeled data according to the nearby points lying in its local manifold, rather than its perturbations using in traditional SSL methods. Specifically, we use an existing neural network to project the labeled and unlabeled samples into an embedding space, and employ an attention mechanism to query the nearest neighbors (samples with the same category) from labeled samples for unlabeled samples, whose pseudo-labels are subsequently determined by the weighted average of the labels of labeled samples. By this paradigm, pseudo-labels could be adaptively predicted on the local manifold within each mini-batch during training. Considering that the labeled neighbors of unlabeled data may not be contained in a randomly sampled mini-batch, we further propose a memory padding module that uses a memory bank to preserve the embeddings of labeled samples across different mini-batches. This allows unlabeled data to match their neighbors from a larger pooling, and more accurate pseudo-labels will be predicted.

In summary, our contributions are mainly three-fold: First, a new pseudo-label estimation framework is proposed for SSL, through which the pseudo-labels of unlabeled data could be online propagated according to their local manifold. Second, a memory padding module is designed to dynamically update the local manifold around the unlabeled data, rather than their perturbations only, which results in a more accurate pseudo-label estimation during training. Third, the proposed pseudo-label estimator is plug-and-play so that no extra inference burden is introduced during testing, and we verify the superiority of our approach on two distinct benchmark datasets.

2 Related Work

Our method is mostly related to two lines of research, including semi-supervised learning and neural attention mechanism. In medical image analysis, most existing semi-supervised works mainly focus on enforcing consistency regularization to deep models. One typical approach aims to regularize the local smoothness of deep models by encouraging their prediction consensus among each single data point and its perturbations. For instance, a Siamese architecture is introduced in [4] for segmentation consistency between the predictions of the two branches, each of which input a differently transformed image. Similarly, a student network was required to predict the same label prediction with the teacher network, whose parameters were updated through a EMA of the student network's parameters [6]. In LSSE [9], a stochastic embedding was learned to randomly augment the input, and the models were encouraged to make consistent predictions for such perturbation samples. Another commonly used approach attempts to constrain deep models with global smoothness, such as filling the void in between data points through the mixup strategy [8], constructing an affinity graph according to the predictions to maintain intra-class clusters and inter-class margins [17]. Despite their sound performance, all the mentioned methods propagated the pseudo-label of unlabeled data points from their perturbations or ensemble predictions. Instead, our method introduces a novel pseudo-label estimator that directly infers the labels from themselves together with their local manifolds.

Neural attention mechanism has achieved vast success in modeling relevance by aiding models to dynamically pay attention to partial input. In this paper, we draw inspiration from some remarkable works, such as sequence to sequence with attention [2] in machine translation, matching network [20] and prototypical networks [16] in few-shot learning. Apart from application scenarios, our approach differs from the above methods by just adopting an attention-like mechanism to dynamically find the local manifold for unlabeled points, with the help of our designed memory padding module.

3 Methods

In SSL tasks, we are given a labeled training set $\mathcal{D}_l = \{\mathbf{x}_i, \mathbf{y}_i\}_{i=1}^{n}$ and an unlabeled set $\mathcal{D}_u = \{\mathbf{x}_j\}_{j=1}^{m}$, where n and m denote the number of $\mathcal{D}_l$ and $\mathcal{D}_u$. In deep learning, we aim to learn a deep network $f_\theta(\mathbf{x}) = g(h(\mathbf{x};\theta_1);\theta_2)$ parameterized with $\theta = \{\theta_1, \theta_2\}$, where $h(\cdot)$ is a feature extractor used to capture the high-level semantic information of input, and $g(\cdot)$ is a classifier. For an unlabeled sample $\mathbf{x}_u \in \mathcal{D}_u$, our method can be summarized to estimate its pseudo-label $\hat{\mathbf{y}}$ conditioned on the embedding $\hat{h}(\mathbf{x}_u) \in \mathcal{X}$ and a memory bank $\mathcal{M} = \{\mathbf{k}_i, \mathbf{y}_i\}_{i=1}^{t}$, where $\hat{h}$ could adopt a flexible form (e.g., partial or all layers of h, or an EMA of h) whose parameters don't need gradient back-propagation; $\mathbf{k}_i = \hat{h}(\mathbf{x}_i)$ is the embedding of labeled sample $\mathbf{x}_i$, $\mathbf{y}_i$ is the corresponding label of $\mathbf{x}_i$, and t is the size of $\mathcal{M}$. In the following subsections, we first mathematically formulate our Neighbor Matching that estimates the pseudo-labels for a batch of unlabeled

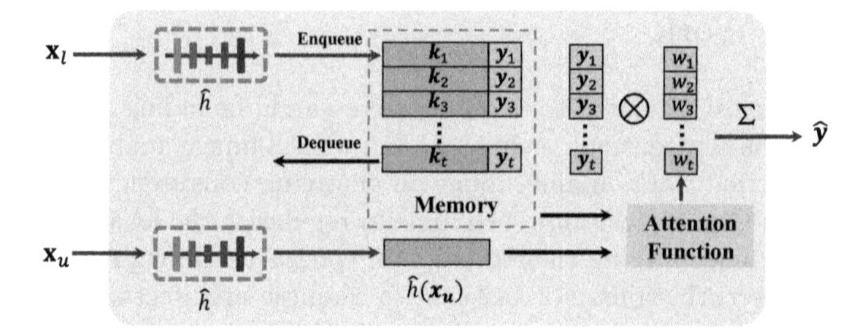

Fig. 1. The framework of Neighbor Matching. Two branches of encoders project a mini-batch of labeled and unlabeled samples into an imbedding space, and here we set the mini-batch size is 1 for better visualization. The embeddings and labels of labeled samples are memorized in a dynamic queue (Memory), and the embeddings of unlabeled ones are used to match their neighbors from the Memory through an attention mechanism. For each unlabeled sample, a weight vector is returned to estimate the pseudo-label as a linear combination of the labels in Memory.

samples during training. We then introduce the proposed memory padding module that dynamically updates the memory bank $\mathcal{M}$. Finally, we summarize the resulting SSL framework based on consistency regularization. Figure 1 shows the overall flowchart of the proposed Neighbor Matching method.

3.1 Neighbor Matching

For an unlabeled sample $\mathbf{x}_u$, Neighbor Matching defines a mapping function P that predicts the pseudo-label $\hat{\mathbf{y}}$ conditioned on its embedding $\hat{h}(\mathbf{x}_u)$ and the memory bank $\mathcal{M}$, and it formulates as

$$P(\hat{\mathbf{y}}|\hat{h}(\mathbf{x}_u), \mathcal{M}) = \sum_{i=1}^{t} w(\hat{h}(\mathbf{x}_u), \mathbf{k}_i)\mathbf{y}_i, \tag{1}$$

where $w(\cdot, \cdot)$ is an attention function that computes the similarities between $\hat{h}(\mathbf{x}_u)$ and the embedding $\mathbf{k}_i$ in $\mathcal{M}$. The higher output of w represents the larger similarities between the input pairs. This equation means Neighbor Matching outputs a soft pseudo-label, i.e., a linear combination of the value $\mathbf{y}_i$ in $\mathcal{M}$, and it enforces unlabeled data to have the same category semantics with their neighboring labeled data within the embedding space $\mathcal{X}$. As a comparison, in traditional pseudo-label estimators, the labels of unlabeled data are usually inherited from their perturbations or augmentations.

We define the attention function using a softmax function with a temperature parameter T, i.e.,

$$w(\hat{h}(\mathbf{x}_u), \mathbf{k}_i) = \frac{\exp(d(\hat{h}(\mathbf{x}_u), \mathbf{k}_i)/T)}{\sum_{j=1}^{t} \exp(d(\hat{h}(\mathbf{x}_u), \mathbf{k}_j)/T)}, \tag{2}$$

where $d(\cdot, \cdot)$ represents the similarity metric, such as cosine similarity and negative Euclidean distance. In this paper, we empirically select the former throughout all our experiments, which can guarantee a stable and sound performance.

3.2 Memory Padding

The memory bank $\mathcal{M}$ acts an important role in the pseudo-label estimation phase. As Eq. (1) shows, the pseudo-labels of unlabeled data are propagated from themselves together with their labeled neighbors in the embedding space. A plain idea is that saving the embeddings and labels of current mini-batch of labeled samples as $\mathcal{M}$. However, it is problematic because unlabeled samples would be misclassified, if the class they belong to was not memorized in $\mathcal{M}$, especially under a small mini-batch size. Another feasible idea is to compute the embeddings for all the labeled samples at each training iteration, which, however, is extremely time consuming.

To address the aforementioned issues, we draw inspiration from physicians who continually accumulate experience from past cases so that they can make a confident diagnosis for patients suffering from the same diseases. Accordingly, we propose a novel memory padding module satisfying two updating rules: 1) **dynamic updating** and 2) **unbiased updating**. For dynamic updating, as shown in Fig. 1, the memory bank $\mathcal{M}$ is updated through a queue, where we enqueue the embeddings and labels from the current mini-batch, and dequeue the earliest embedding-label pairs from the memory bank $\mathcal{M}$. As for unbiased updating, which satisfies that the number of each class in $\mathcal{M}$ should be always kept equal, and only correctly classified samples whose embedding-label pairs can enqueue into $\mathcal{M}$. Assume a labeled sample $(\mathbf{x}, \mathbf{y}) \in \mathcal{D}_l$ that belongs to the c-th class, then we have

$$(\hat{h}(\mathbf{x}), \mathbf{y}) \rightarrow \mathcal{M} \iff \delta[f(\mathbf{x})] = c \tag{3}$$

where $\rightarrow$ is enqueue operation, and $\delta[\mathbf{v}]$ denotes the index where vector $\mathbf{v}$ achieves the largest element. This tailored memory padding module makes a more accurate pseudo-label estimation and thus largely alleviates the error accumulation problems during training.

3.3 Loss Function

The standard semi-supervised methods based on consistency regularization commonly minimize a combined loss with supervised and unsupervised loss terms, which is formulated as

$$\mathcal{L} = \mathcal{L}_l + \lambda_u \mathcal{L}_u, \tag{4}$$

where λ_u is the hyper-parameter of unsupervised term. The supervised loss term $\mathcal{L}_l$ is usually the cross-entropy between labels and predictions of labeled data. For unlabeled data, we simply define $\mathcal{L}_u$ as the cross-entropy between their pseudo-labels estimated by Eq. (1) and the predictions predicted by f. The training algorithm is listed in Algorithm 1, where $\mathrm{H}(\cdot, \cdot)$ denotes cross-entropy function.

Algorithm 1. Mini-batch training of SSL with Neighbor Matching

Input: Labeled training data $\mathcal{D}_l$, Unlabeled training data $\mathcal{D}_u$, batch size s, the size t of memory bank $\mathcal{M}$, max iteration T

Output: Parameters θ of the recognition network f

1: Initialize the memory bank $\mathcal{M}$.
2: **for** $t = 1$ **to** T **do**
3: $\hat{\mathcal{D}}_l = \{\mathbf{x}_i, \mathbf{y}_i\}_{i=1}^{s} \leftarrow \text{SampleMiniBatch}(\mathcal{D}_l, s)$.
4: $\hat{\mathcal{D}}_u = \{\mathbf{x}_j\}_{j=1}^{s} \leftarrow \text{SampleMiniBatch}(\mathcal{D}_u, s)$.
5: Forward to compute the supervised loss: $\mathcal{L}_l = \frac{1}{|\hat{\mathcal{D}}_l|} \sum_{\mathbf{x}_i, \mathbf{y}_i \in \hat{\mathcal{D}}_l} \text{H}(\mathbf{y}_i, f(\mathbf{x}_i))$.
6: Update the memory bank $\mathcal{M}$ according to Eq. (3).
7: Estimate pseudo-labels for $\hat{\mathcal{D}}_u$ by Eq. (1).
8: Forward to compute the unsupervised loss: $\mathcal{L}_u = \frac{1}{|\hat{\mathcal{D}}_u|} \sum_{\mathbf{x}_j \in \hat{\mathcal{D}}_u} \text{H}(\hat{\mathbf{y}}_j, f(\mathbf{x}_j))$.
9: Backward according to Eq. (4) to update the parameters θ.
10: **end for**

4 Experiments

We experiment with two publicly available medical datasets, ISIC 2018 Skin dataset [5] and Chexpert [10], to evaluate the proposed methods.

4.1 Implementation Details

Chexpert [10] is a large chest X-ray dataset for thoracic disease recognition, in which 224,316 chest radiographs were collected from 65,240 patients, and 14 pathology categories were extracted from radiology reports as labels. We follow the preprocessing strategies and augmentation strategy presented in [8], where all uncertain and lateral view samples are removed, and the public data splits [9] are used for the labeled training set (under five different settings with samples ranging from 100 to 500), unlabeled training set, validation set, and test set. Each image is resized to 128×128, rotated in the range of $(-10°, 10°)$ and translated in the range of (0.1, 0.1) fraction of the image.

ISIC 2018 skin dataset [5] was provided by the ISIC2018 Challenge with 7 disease categories, in which 10015 dermoscopic images and associated labels are contained. Since serious data imbalance are suffered from different classes, we randomly sample 50/10 samples from each class as test/validation sets, and the remainder serve as training data to further splits for labeled and unlabeled training sets. Similar to [8], we choose a subset of examples (considering three different settings with 350, 800, and 1200 labeled samples, respectively) from the training data to act as the labeled training set, then the rest acts as the unlabeled training set. The same resizing and augmentation strategies as Chexpert dataset are adopted for each image.

For fair comparison, we follow [8] to use AlexNet [11] as the baseline to perform all our experiments. In this paper, we set the penultimate fully-connected layer as 256 neurons to reduce the computational cost of neighbor matching phase presented in Eq. (1). Following the setup in [8], all the models are trained by SGD optimizer with an initial learning rate of 1e−4, a mini-batch size of 128 for a total of 256 epochs, and the learning rate is decayed by a factor of 10 at the $50th$ and $125th$ epochs. We linearly ramp up the weight of unsupervised loss term from 0 to its final value λ_u, which is set as 1 throughout all our experiments, and we empirically set the temperature hyper-parameter T as 100. For the size t of the memory bank $\mathcal{M}$, we initially set it as the nearest integer to our mini-batch size, i.e., $t = 14 * \lfloor 128/14 \rfloor$ for Chexpert [10] and $t = 7 * \lfloor 128/7 \rfloor$ for ISIC 2018 skin dataset [5]. Note that the class number is an aliquot part of the mini-batch size for satisfying the unbiased updating rule mentioned in Sect. 3.2. All experiments are implemented with the PyTorch platform [15].

4.2 Skin Lesion Recognition Results

Table 1. Ablation analysis on ISIC 2018 skin dataset [5] with 800 labeled training samples. Note that t is the memory bank size.

Methods	AUC	MCA
Baseline	0.7835	0.4168
EntMin [7]	0.7916	0.4127
Ours (w/o Unbiased Updating, w/t = 126)	0.8147	0.4184
Ours (w/Unbiased Updating, w/t = 7)	0.8184	0.4524
Ours (w/Unbiased Updating, w/t = 126)	0.8271	0.4846
Ours (w/Unbiased Updating, w/t = 252)	**0.8303**	**0.4933**
Ours (w/Unbiased Updating, w/t = 504)	0.8178	0.4701

For better understanding of the proposed Neighbor Matching method, we first conduct an ablation study on ISIC 2018 skin dataset [5] with 800 labeled training samples. Evaluation metrics include mean AUC (area under the ROC curve) and mean class accuracy (MCA, i.e., average recall over all classes). We firstly quantify the effectiveness of our pseudo-label estimator compared with two base models: 1) Baseline, no unlabeled data is used during training; 2) EntMin [7], the predictions of the network directly taken as pseudo-labels of unlabeled data. As shown in Table 1, all our models consistently outperform the two base models, improving AUC and MCA by around 3.9% and 7.7%, respectively, which verifies the effectiveness of Neighbor Matching as a pseudo-label estimator.

We further study the effect of the memory padding module, including the updating rules and the memory bank size. Without unbiased updating rule, the memory bank is updated by the dynamic updating rule only, and the performance (0.8271 AUC and 0.4846 MCA) largely drops compared with the standard model updated by both updating rules. This is mainly because an unlabeled sample may fail to find its labeled neighbors from such a class-imbalanced memory bank, which manifests the superiority of the proposed memory padding module. As for the memory bank size t, the results show that increasing the size from 7 to 252, i.e., 1 to 36 samples per class, is helpful for the semi-supervised task. However, it yields a negative effect when t continues to increase, since the overlarge memory bank will violate the unbiased updating rule.

Table 2. The performance on ISIC 2018 skin dataset [10].

Method	350		800		1200	
	AUC	MCA	AUC	MCA	AUC	MCA
Baseline	0.7692	0.3951	0.7835	0.4168	0.8340	0.4935
GLM [8]	0.7870	0.3856	0.8479	0.4515	0.8754	0.5034
Ours (w/o mixup)	0.7743	0.4145	0.8303	0.4933	0.8547	0.5539
Ours (w/mixup)	**0.8176**	**0.4729**	**0.8603**	**0.5186**	**0.8864**	**0.5887**

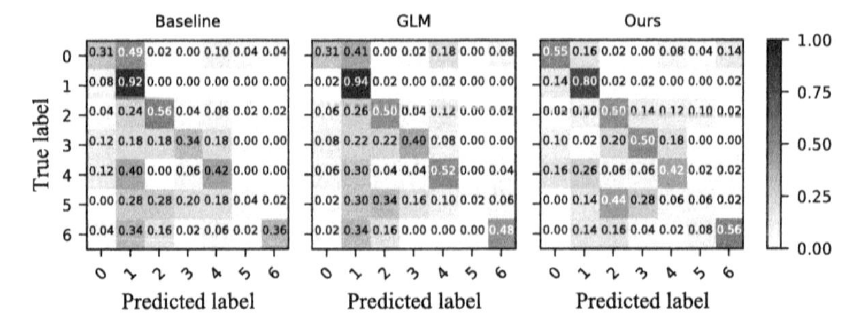

Fig. 2. Confusion matrices of Baseline, GLM [8], and ours on ISIC 2018 skin dataset [10] with 800 labeled training samples. Under this setting, the number of seven categories in labeled training set, in turn, is: 89, 535, 41, 26, 88, 9, 12.

Furthermore, we compare our method with Baseline and current state-of-the-art GLM [8] under varying number of labeled data. As GLM is mixup-based method, we report our results with/without mixup augmentation [21] for fair comparison. As listed in Table 2, our method significantly outperforms GLM when mixup is applied. Even without mixup augmentation, we still achieve a higher MCA score than GLM. A further qualitative comparison through the confusion matrices is shown in Fig. 2. It is clear that the head class (*2nd* column) easily dominate the training procedure, as many samples in other classes are misclassified into this one. Instead, our methods largely alleviate this challenge issue by propagating more accurate labels for unlabeled data using an unbiased memory bank. Moreover, diagonal elements of these confusion matrices present the recall of each class, and an obvious finding is that our method achieves a relatively class-balanced recall for all the seven classes.

4.3 Thoracic Disease Recognition Results

Table 3. The performance (mean AUC) on Chexpert dataset [10].

Method	Size of labeled data set (k)				
	100	200	300	400	500
Baseline	0.5576	0.6166	0.6208	0.6343	0.6353
LSSE [9]	0.6200	0.6386	0.6484	0.6637	0.6697
GLM [8]	0.6512	0.6641	0.6739	0.6796	0.6847
Ours	**0.6515**	**0.6711**	**0.6830**	**0.6889**	**0.6934**

Chexpert dataset [10] is a public dataset for multi-label thoracic disease recognition. Aside from Baseline, we also compared with two state-of-the-art methods, i.e., LSSE [9] and GLM [8]. Since all are experimental settings follow [8], we directly refer to the results reported in [8]. The mean AUC results are reported as the average of five random seeds runs. Note that mixup augmentation is not applied here, since it brings little performance gain in multi-label recognition. As shown in Table 3, for varying number of labeled data, our method consistently attains the best performance among all comparison methods.

5 Conclusion

In this work we have proposed a pseudo-label estimation framework to address the semi-supervised recognition problem. Our key insight is to predict the pseudo-label of an unlabeled sample conditioned on itself and its local manifold, through which the pseudo-label could be adaptively propagated from its neighboring labeled points during training. This is inspired by the prior that patients with the same disease often show similar symptoms, according to which physicians could make a diagnosis and follow-up treatment. To construct the manifold for each unlabeled point, we further introduce a memory padding module to memorize the structural information of labeled points across different mini-batches. Experimental results on two benchmark datasets demonstrate the superiority of our method in semi-supervised lesion recognition. In future, applying the method in other practical problems will be investigated.

Acknowledgments. This research was supported by National Key R&D Program of China (2020YFA0713900), the Macao Science and Technology Development Fund under Grant 061/2020/A2, the China NSFC projects (62076196, 11690011, 61721002, U1811461).

References

1. Bachman, P., Alsharif, O., Precup, D.: Learning with pseudo-ensembles. In: Advances in Neural Information Processing Systems, pp. 3365–3373 (2014)
2. Bahdanau, D., Cho, K., Bengio, Y.: Neural machine translation by jointly learning to align and translate. In: International Conference on Learning Representations (2015)
3. Berthelot, D., Carlini, N., Goodfellow, I., Papernot, N., Oliver, A., Raffel, C.A.: Mixmatch: a holistic approach to semi-supervised learning. In: Advances in Neural Information Processing Systems. pp. 5049–5059 (2019)
4. Bortsova, G., Dubost, F., Hogeweg, L., Katramados, I., de Bruijne, M.: Semi-supervised medical image segmentation via learning consistency under transformations. In: Shen, D., et al. (eds.) MICCAI 2019. LNCS, vol. 11769, pp. 810–818. Springer, Cham (2019). https://doi.org/10.1007/978-3-030-32226-7_90
5. Codella, N., et al.: Skin lesion analysis toward melanoma detection 2018: a challenge hosted by the international skin imaging collaboration (ISIC). arXiv preprint arXiv:1902.03368 (2019)
6. Cui, W., et al.: Semi-supervised brain lesion segmentation with an adapted mean teacher model. In: Chung, A.C.S., Gee, J.C., Yushkevich, P.A., Bao, S. (eds.) IPMI 2019. LNCS, vol. 11492, pp. 554–565. Springer, Cham (2019). https://doi.org/10.1007/978-3-030-20351-1_43
7. Grandvalet, Y., Bengio, Y.: Semi-supervised learning by entropy minimization. Adv. Neural Inf. Process. Syst. **17**, 529–536 (2004)
8. Gyawali, P.K., Ghimire, S., Bajracharya, P., Li, Z., Wang, L.: Semi-supervised medical image classification with global latent mixing. In: Martel, A.L., et al. (eds.) MICCAI 2020. LNCS, vol. 12261, pp. 604–613. Springer, Cham (2020). https://doi.org/10.1007/978-3-030-59710-8_59
9. Gyawali, P.K., Li, Z., Ghimire, S., Wang, L.: Semi-supervised learning by disentangling and self-ensembling over stochastic latent space. In: Shen, D., et al. (eds.) MICCAI 2019. LNCS, vol. 11769, pp. 766–774. Springer, Cham (2019). https://doi.org/10.1007/978-3-030-32226-7_85
10. Irvin, J., et al.: Chexpert: a large chest radiograph dataset with uncertainty labels and expert comparison. In: Proceedings of the AAAI Conference on Artificial Intelligence, vol. 33, pp. 590–597 (2019)
11. Krizhevsky, A., Sutskever, I., Hinton, G.E.: ImageNet classification with deep convolutional neural networks. Commun. ACM **60**(6), 84–90 (2017)
12. Laine, S., Aila, T.: Temporal ensembling for semi-supervised learning. In: International Conference on Learning Representations (2017)
13. Litjens, G., et al.: A survey on deep learning in medical image analysis. Med. Image Anal. **42**, 60–88 (2017)
14. Miyato, T., Maeda, S.I., Koyama, M., Ishii, S.: Virtual adversarial training: a regularization method for supervised and semi-supervised learning. IEEE Trans. Pattern Anal. Mach. Intell. **41**(8), 1979–1993 (2018)
15. Paszke, A., Gross, S., Chintala, S., et al.: Automatic differentiation in PyTorch. In: Advances in Neural Information Processing Systems Workshop Autodiff, pp. 1–4 (2017)
16. Snell, J., Swersky, K., Zemel, R.: Prototypical networks for few-shot learning. In: Advances in Neural Information Processing Systems, pp. 4077–4087 (2017)

17. Su, H., Shi, X., Cai, J., Yang, L.: Local and global consistency regularized mean teacher for semi-supervised nuclei classification. In: Shen, D., et al. (eds.) MICCAI 2019. LNCS, vol. 11764, pp. 559–567. Springer, Cham (2019). https://doi.org/10.1007/978-3-030-32239-7_62
18. Tarvainen, A., Valpola, H.: Mean teachers are better role models: weight-averaged consistency targets improve semi-supervised deep learning results. In: Advances in Neural Information Processing Systems, pp. 1195–1204 (2017)
19. Verma, V., et al.: Manifold mixup: better representations by interpolating hidden states. In: International Conference on Machine Learning, pp. 6438–6447. PMLR (2019)
20. Vinyals, O., Blundell, C., Lillicrap, T., Wierstra, D., et al.: Matching networks for one shot learning. In: Advances in Neural Information Processing Systems, pp. 3630–3638 (2016)
21. Zhang, H., Cisse, M., Dauphin, Y.N., Lopez-Paz, D.: Mixup: beyond empirical risk minimization. In: International Conference on Learning Representations (2018)

Tripled-Uncertainty Guided Mean Teacher Model for Semi-supervised Medical Image Segmentation

Kaiping Wang[1], Bo Zhan[1], Chen Zu[2], Xi Wu[3], Jiliu Zhou[1,3], Luping Zhou[4], and Yan Wang[1](✉)

[1] School of Computer Science, Sichuan University, Chengdu, China
[2] Department of Risk Controlling Research, JD.COM, Chengdu, China
[3] School of Computer Science, Chengdu University of Information Technology, Chengdu, China
[4] School of Electrical and Information Engineering, University of Sydney, Sydney, Australia

Abstract. Due to the difficulty in accessing a large amount of labeled data, semi-supervised learning is becoming an attractive solution in medical image segmentation. To make use of unlabeled data, current popular semi-supervised methods (e.g., temporal ensembling, mean teacher) mainly impose data-level and model-level consistency on unlabeled data. In this paper, we argue that in addition to these strategies, we could further utilize auxiliary tasks and consider task-level consistency to better leverage unlabeled data for segmentation. Specifically, we introduce two auxiliary tasks, i.e., a foreground and background reconstruction task for capturing semantic information and a signed distance field (SDF) prediction task for imposing shape constraint, and explore the mutual promotion effect between the two auxiliary and the segmentation tasks based on mean teacher architecture. Moreover, to handle the potential bias of the teacher model caused by annotation scarcity, we develop a tripled-uncertainty guided framework to encourage the three tasks in the teacher model to generate more reliable pseudo labels. When calculating uncertainty, we propose an uncertainty weighted integration (UWI) strategy for yielding the segmentation predictions of the teacher. Extensive experiments on public 2017 ACDC dataset and PROMISE12 dataset have demostrated the effectiveness of our method. Code is available at https://github.com/DeepMedLab/Tri-U-MT.

Keywords: Semi-supervised segmentation · Mean teacher · Multi-task learning · Tripled-uncertainty

1 Introduction

Segmentation is a basic yet essential task in the realm of medical image processing and analysis. To enable clinical efficiency, recent deep learning frameworks [1, 2] have made a quantum leap in automatic segmentation with sufficient annotations. However, such

K. Wang and B. Zhan—The authors contribute equally to this work.

M. de Bruijne et al. (Eds.): MICCAI 2021, LNCS 12902, pp. 450–460, 2021.
https://doi.org/10.1007/978-3-030-87196-3_42

annotations are hard to acquire in real world due to their expensive and time-consuming nature. To alleviate annotation scarcity, a feasible approach is to take semi-supervised learning [3, 4] which leverages both labeled and unlabeled data to train the deep network effectively.

Considerable efforts have been devoted to the semi-supervised segmentation community, which can be broadly categorized into two groups. The first group refers to those methods trying to predict pseudo labels on unlabeled images and mixing them with ground truth labels to provide additional training information [5–7]. However, the segmentation result of such self-training based method is susceptible due to the uneven quality of the predicted pseudo labels. The second group of semi-supervised segmentation methods lies in the consistency regularization, that is, encouraging the segmentation predictions to be consistent under different perturbations for the same input. A typical example is Π-model [8] which minimizes the distance between the results of two forward passes with different regularization strategies. To improve the stability, a temporal ensembling model [8] is further proposed based on Π-model, which aggregates the exponential moving average (EMA) predictions and encourages the consensus between the ensembled predictions and current predictions for unlabeled data. To accelerate the training and enable the online learning, mean teacher [9] further improves the temporal ensembling model by enforcing prediction consistency between the current training model (i.e., the student model) and the corresponding EMA model (i.e., the teacher model) in each training step. Nevertheless, unreliable results from the teacher model may mislead the student model, thus deteriorating the whole training. Researchers therefore incorporated uncertainty map to the mean teacher model, forcing the student to learn high confidence predictions from the teacher model [10, 11]. Our research falls in the second group of semi-supervised approaches.

On the other hand, different from the above semi-supervised segmentation methods which mainly focus on the consistency under the disturbances at data level or model level, there are also research works [12–14] that explore to improve segmentation from another perspective: multi-task learning. These methods jointly train multiple tasks to boost segmentation performance. For instance, Chen et al. [13] applied a reconstruction task to assist the segmentation of medical images. Li et al. [14] proposed a shape-aware semi-supervised model by incorporating a signed distance map generation task to enforce a shape constraint on the segmentation result. By exploring the relationship between the main and the auxiliary tasks, the learned segmentation model could bypass the over-fitting problem and learn more general representations.

In this paper, inspired by the success of semi-supervised learning and multi-task learning, we propose a novel end-to-end semi-supervised mean teacher model guided by tripled-uncertainty maps from three highly related tasks. Concretely, apart from the segmentation task, we bring in two additional auxiliary tasks, i.e., a foreground and background reconstruction task and a signed distance field (SDF) prediction task. The reconstruction task can help the segmentation network capture more semantic information, while the predicted SDF describes the signed distance of a corresponding pixel to its closest boundary point after normalization, thereby constraining the global geometric shape of the segmentation result. Following the spirit of mean teacher architecture, we

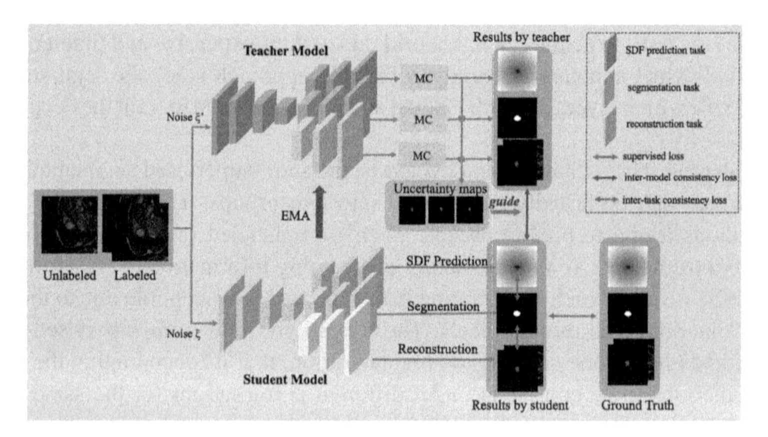

Fig. 1. Overview of our tripled-uncertainty guided mean teacher model.

build a teacher model and a student model, each targeting at all three tasks above. Additionally, to tackle the unreliability and noise in teacher model predictions, we impose uncertainty constraints on all the three tasks, expecting the student model can learn as accurate information as possible.

Our main contributions are three-fold: (1) We inject the spirit of multi-task learning into mean teacher architecture, so that the segmentation task could also benefit from the enhanced semantic and geometric shape information by extracting the correlations among the segmentation task, the reconstruction task, and the SDF prediction task. In this manner, our mean teacher model simultaneously takes account of the data-, model- and task-level consistency to better leverage unlabeled data for segmentation. (2) We impose the uncertainty estimation on all tasks and develop a tripled-uncertainty to guide the student model to learn more reliable predictions from the teacher model. (3) Current approaches tend to generate uncertainty maps by averaging the results from multiple Monte Carlo (MC) samplings, neglecting the discrepancy of different results. In contrast, we propose an uncertainty weighted integration (UWI) strategy to assign different weights for different sampling results, generating a more accurate segmentation prediction.

2 Methodology

An overview of our proposed network is illustrated in Fig. 1, consisting of a teacher model and a student model, following the idea of mean teacher. The student model is the trained model, and it assigns the exponential moving average of its weights to the teacher model at each step of training. On the other hand, the predictions of the teacher model would be viewed as additional supervisions for the student model to learn. These two models share a same encoder-decoder structure, where the encoder is shared among different tasks while the decoders are task-specific. In our problem setting, we are given a training set containing N labeled data and M unlabeled data samples, where $N \ll M$. The labeled set is defined as $\mathcal{D}^l = \left\{\mathbf{X}^i, \mathbf{Y}^i\right\}_{i=1}^{N}$ and the unlabeled set as $\mathcal{D}^u = \left\{\mathbf{X}^i\right\}_{i=N+1}^{N+M}$, where

$\mathbf{X}^i \in \mathbb{R}^{H\times W}$ is the intensity image, $\mathbf{Y}^i \in \{0,1\}^{H\times W}$ is the corresponding segmentation label. For labeled data, we can also obtain the ground truth of SDF $\mathbf{Z}^i \in \mathbb{R}^{H\times W}$ from $\mathbf{Y}^i$ via the SDF function in [17] for the SDF prediction task. Similarly, for the reconstruction task, the ground truths of foreground and background $\mathbf{G}^i \in \mathbb{R}^{2\times H\times W}$ can be obtained by $\mathbf{Y}^i \odot \mathbf{X}^i$ and $\left(1-\mathbf{Y}^i\right) \odot \mathbf{X}^i$ where $\odot$ refers to element-wise multiplication. Given $\mathcal{D}^l$ and $\mathcal{D}^u$, the student model is optimized by minimizing 1) the supervised segmentation loss $\mathcal{L}_s$ on labeled data $\mathcal{D}^l$, 2) the inter-model consistency loss $\mathcal{L}_{cons}^{model}$ between the student model and teacher model on both $\mathcal{D}^l$ and $\mathcal{D}^u$, and 3) the inter-task consistency loss $\mathcal{L}_{cons}^{task}$ among different tasks on $\mathcal{D}^l$ and $\mathcal{D}^u$. On the other hand, the tripled-uncertainty maps with respect to the three tasks generated by the teacher model guide the student model to learn more reliable predictions from the teacher model. Moreover, to enhance the robustness of our mean teacher model, different perturbations ξ and ξ' are fed into the student and teacher model, respectively. More details will be introduced in subsequent sections.

2.1 Student Model

Our student model employs U-net [2] as backbone with respect to three tasks, i.e., segmentation task, reconstruction task and SDF prediction task. Note that the encoder is shared by the three tasks with the same parameters while the parameters in the task-specific decoders are different to fit the different tasks. In this manner, the encoder is forced to capture features related to the semantic information and geometric shape information, leading to a low disparity between the output segmentation result and the ground truth. Concretely, the encoder consists of three down-sampling blocks with 3 × 3 convolutional layers while the decoders take similar structures with three up-sampling blocks but differ in the last task head layers. The segmentation head is equipped with softmax activation while the reconstruction head and the SDF head utilize sigmoid activation. Following [13], we drop the skip connection in the reconstruction task. Given an input image $\mathbf{X}^i \in \mathcal{D}^l$, these tasks can generate the segmentation result $\tilde{\mathbf{Y}}_S^i$, the reconstruction result $\tilde{\mathbf{G}}_S^i$, and the SDF result $\tilde{\mathbf{Z}}_S^i$, as follows:

$$\tilde{\mathbf{Y}}_S^i = f_{seg}\left(\mathbf{X}^i;\theta_{seg},\xi\right),\ \tilde{\mathbf{G}}_S^i = f_{rec}\left(\mathbf{X}^i;\theta_{rec},\xi\right),\ \tilde{\mathbf{Z}}_S^i = f_{sdf}\left(\mathbf{X}^i;\theta_{sdf},\xi\right), \quad (1)$$

where $f_{seg}, f_{rec}, f_{sdf}$ represent the segmentation network, the reconstruction network and the SDF prediction network with corresponding parameters θ_{seg}, θ_{rec}, θ_{sdf}, and ξ is the noise perturbation of the student model.

2.2 Teacher Model

The network of our teacher model is reproduced from the student model, yet they have different ways for updating parameters. The student model updates its parameters $\boldsymbol{\theta} = \{\theta_{seg},\theta_{rec},\theta_{sdf}\}$ by gradient descent while the teacher model updates its parameters $\boldsymbol{\theta}' = \left\{\theta'_{seg},\theta'_{rec},\theta'_{sdf}\right\}$ as the EMA of the student model parameters $\boldsymbol{\theta}$ in different training steps. In particular, at training step t, the parameters of the teacher model, i.e., $\boldsymbol{\theta}_t'$, are updated according to:

$$\boldsymbol{\theta}_t' = \tau\boldsymbol{\theta}_{t-1}' + (1-\tau)\boldsymbol{\theta}_t, \quad (2)$$

where τ is the coefficient of EMA decay to control the updating rate.

Moreover, as there is no label on $\mathcal{D}^u$, the results of the teacher model may be biased. To relieve such unreliability, we bring in the uncertainty estimation in the teacher model. Specifically, we perform K times forward passes with Monte Carlo (MC) dropout, thus obtaining K preliminary results of all the tasks with regard to the input $\mathbf{X}^i$, i.e., $\left\{\tilde{\mathbf{Y}}_T^{ij}\right\}_{j=1}^K$, $\left\{\tilde{\mathbf{G}}_T^{ij}\right\}_{j=1}^K$, and $\left\{\tilde{\mathbf{Z}}_T^{ij}\right\}_{j=1}^K$.

Please note that, for the main segmentation task, we design an uncertainty weighted integration (UWI) strategy to assign different weights for different sampling results, instead of averaging the K preliminary results simply. Specifically, we derive the uncertainty maps U_{seg}^{ij} for each preliminary result by calculating the entropy $-\sum_{c \in C} \tilde{\mathbf{Y}}_T^{ijc} \log_C \tilde{\mathbf{Y}}_T^{ijc}$, where the base of the log function, C, is the number of classes to be segmented and set to 2 here. By doing so, the value range of uncertainty maps is between 0 and 1, and a larger value represents a higher degree of uncertainty. Since entropy reflects the uncertainty degree of information, we use 1-entropy to measure the confidence level for each preliminary result, leading to K confidence maps while each pixel corresponds to a vector with length of K. The values of the vector are further normalized to [0, 1] by applying a softmax operation. Afterwards, we can regard the softmax probability map at channel j, i.e., $softmax\left\{1 - U_{\text{seg}}^{ij}\right\}_{j=1}^K$, as a weight map $\mathbf{W}_j$ which guides the teacher model implicititly to heed the areas with higher confidence during aggregation. Thus, the aggregated prediction of segmentation $\tilde{\mathbf{Y}}_T^i$ can be formulated as $\tilde{\mathbf{Y}}_T^i = \sum_{j=1}^K \mathbf{W}_j \odot \tilde{\mathbf{Y}}_T^{ij}$.

As for the other two auxiliary tasks, it is noteworthy that they predict the real regression values rather than the probabilistic values as the segmentation task. Accordingly, the entropy is unsuitable for the uncertainty estimation for them. Therefore, we obtain the aggregated results directly by averaging operation, that is, $\tilde{\mathbf{G}}_T^i = \frac{1}{K}\sum_{j=1}^K \tilde{\mathbf{G}}_T^{ij}$, $\tilde{\mathbf{Z}}_T^i = \frac{1}{K}\sum_{j=1}^K \tilde{\mathbf{Z}}_T^{ij}$. For the same reason, we utilize the variance instead of entropy as the uncertainty of the aggregated results of these two auxiliary tasks by following [20]. To sum up, leveraging the aggregated results of three tasks, we can acquire tripled-uncertainty maps of all the tasks by:

$$U_{\text{seg}} = -\sum\nolimits_{c \in C} \tilde{\mathbf{Y}}_T^{ic} \log_C \tilde{\mathbf{Y}}_T^{ic}, U_{\text{rec}} = \frac{1}{K}\sum\nolimits_{j=1}^K \left(\tilde{\mathbf{G}}_T^{ij} - \tilde{\mathbf{G}}_T^i\right)^2, U_{\text{sdf}} = \frac{1}{K}\sum\nolimits_{j=1}^K \left(\tilde{\mathbf{Z}}_T^{ij} - \hat{\mathbf{Z}}_T^i\right)^2 \quad (3)$$

With the tripled-uncertainty guidance, the student model can avoid the misleading information from the teacher model and learn more trustworthy knowledge.

2.3 Objective Functions

As aforementioned, the objective function is composed of three aspects: 1) Supervised loss $\mathcal{L}_s$ on labeled data $\mathcal{D}^l$; 2) Inter-model consistency loss $\mathcal{L}_{\text{cons}}^{\text{model}}$ between the student model and teacher model on both $\mathcal{D}^l$ and $\mathcal{D}^u$; 3) Inter-task consistency loss $\mathcal{L}_{\text{cons}}^{\text{task}}$ among different tasks in the student model on $\mathcal{D}^l$ and $\mathcal{D}^u$.

Specifically, $\mathcal{L}_s$ is the weighted sum of the supervised losses on three tasks, i.e., $\mathcal{L}_s^{seg}$, $\mathcal{L}_s^{rec}$, $\mathcal{L}_s^{sdf}$, and can be formulated as:

$$\mathcal{L}_s = \mathcal{L}_s^{seg} + \alpha_1 \mathcal{L}_s^{rec} + \alpha_2 \mathcal{L}_s^{sdf}, \tag{4}$$

where $\mathcal{L}_s^{seg}$ uses Dice loss following [18], $\mathcal{L}_s^{rec}$ and $\mathcal{L}_s^{sdf}$ use mean squared error (MSE) loss, α_1 and α_2 are coefficients for balancing the loss terms.

For the same input from $\mathcal{D}^l$ or $\mathcal{D}^u$, since the teacher model is an ensembling of the student model, the outputs of both models on three tasks should be identical. Therefore, we employ the inter-model consistency loss $\mathcal{L}_{cons}^{model}$ to constrain this condition as follows:

$$\begin{aligned}
\mathcal{L}_{cons}^{model} &= \mathcal{L}_{cons}^{seg} + \mu_1 \mathcal{L}_{cons}^{rec} + \mu_2 \mathcal{L}_{cons}^{sdf}, \\
\mathcal{L}_{cons}^{seg} &= \frac{1}{N+M} \sum_{i=1}^{N+M} exp(-U_{seg}) \odot \left(\tilde{\mathbf{Y}}_S^i - \tilde{\mathbf{Y}}_T^i\right)^2, \\
\mathcal{L}_{cons}^{rec} &= \frac{1}{N+M} \sum_{i=1}^{N+M} exp(-U_{rec}) \odot \left(\tilde{\mathbf{G}}_S^i - \tilde{\mathbf{G}}_T^i\right)^2, \\
\mathcal{L}_{cons}^{sdf} &= \frac{1}{N+M} \sum_{i=1}^{N+M} exp(-U_{sdf}) \odot \left(\tilde{\mathbf{Z}}_S^i - \tilde{\mathbf{Z}}_T^i\right)^2,
\end{aligned} \tag{5}$$

where the tripled-uncertainty U_{seg}, U_{rec}, U_{sdf} are used as weight maps to encourage the student model learning meaningful information from the teacher model, and μ_1, μ_2 are balancing coefficients.

Similarly, owing to the shared encoder, the results of three tasks are supposed to have consistent semantic information for the same input. Based on this, we devise the inter-task consistency loss $\mathcal{L}_{cons}^{task}$ to narrow the gap between $\tilde{\mathbf{Y}}_S^i$ and $\tilde{\mathbf{G}}_S^i$, $\tilde{\mathbf{Z}}_S^i$. Accordingly, $\mathcal{L}_{cons}^{task}$ is formulated as:

$$\mathcal{L}_{cons}^{task} = \frac{1}{N+M} \sum_{i=1}^{N+M} \left(\left(\tilde{\mathbf{Z}}_S^i - SDF\left(\tilde{\mathbf{Y}}_S^i\right)\right)^2 + \left(\tilde{\mathbf{G}}_S^i - Mask\left(\tilde{\mathbf{Y}}_S^i, \mathbf{X}^i\right)\right)^2\right), \tag{6}$$

where $SDF\left(\tilde{\mathbf{Y}}_S^i\right)$ converts $\tilde{\mathbf{Y}}_S^i$ to the domain of SDF following the function in [17], and $Mask\left(\tilde{\mathbf{Y}}_S^i, \mathbf{X}^i\right)$ is the concatenation of $\tilde{\mathbf{Y}}_S^i \odot \mathbf{X}^i$ and $\left(1 - \tilde{\mathbf{Y}}_S^i\right) \odot \mathbf{X}^i$.

Finally, the total objective function $\mathcal{L}_{total}$ can be summarized as:

$$\mathcal{L}_{total} = \mathcal{L}_s + \lambda_1 \mathcal{L}_{cons}^{model} + \lambda_2 \mathcal{L}_{cons}^{task}, \tag{7}$$

where λ_1 and λ_2 are the ramp-up weighting coefficients for balancing the supervised loss and the two consistency losses.

2.4 Training Details

Our network is implemented by Pytorch and trained with two NVIDIA GeForce 2070SUPER GPUs with total 16 GB memory. We utilize Adam optimizer to train the whole network for 100 epochs with learning rate of 1e−4 and batchsize of 2. To achieve a balance betweem the training efficiency and uncertainty map quality, we perform $K = 8$ times MC dropout in the teacher model. While in testing phase, we turn off the dropout to generate the estimation directly. For updating Eq. (2), we set τ as 0.99 according to [9]. Based on our trial studies, the hyper-parameters α_1, α_2 in Eq. (4) are set to 1, μ_1, μ_2 in Eq. (5) are set to 0.2 and 1, respectively. As for λ_1 and λ_2 in Eq. (7), following [9], we set them equally as a time-dependent Gaussian warming-up function $\lambda(t) = 0.1 * e^{(-5(1-t/t_{max})^2)}$ where t and t_{max} indicate the current training step and total training steps, respectively. Note that, only the student model is retained for generating segmentation predictions in the test phase.

3 Experiment and Analysis

Table 1. Quantitative comparison results on 2017 ACDC dataset. * means our method is significantly better than the compared method with $p < 0.05$ via paired t-test.

n/m	5/70		10/65		20/55	
	Dice [%]	JI [%]	Dice [%]	JI [%]	Dice [%]	JI [%]
U-net [2]	60.1(24.7)*	47.3(23.4)*	70.9(23.6)*	53.1(28.5)*	90.0(7.7)	82.5(11.3)
Curriculum [3]	67.5(9.9)*	51.8(10.5)*	69.2(14.6)*	50.0(14.5)*	86.6(9.0)*	77.5(13.0)*
Mean Teacher [9]	52.1(18.7)*	37.3(16.2)*	80.0(14.3)*	68.8(17.5)*	88.8(9.3)*	80.9(13.4)*
MASSL [13]	77.4(16.7)*	66.0(18.5)*	86.0(14.9)	77.8(18.5)	90.6(8.8)	84.0(12.6)
Shape-aware [17]	81.4(14.2)*	70.8(16.9)*	85.0(12.2)*	75.6(15.8)*	91.0(7.8)	84.3(11.5)
UA-MT [10]	70.7(14.1)*	56.4(15.9)*	80.6(17.8)*	70.7(21.8)*	88.7(10.5)*	81.2(14.7)*
MI-SSS [19]	81.2(20.9)*	72.4(23.4)*	84.7(15.2)	75.8(18.6)	91.2(5.6)	84.3(8.9)
Proposed	**84.6(13.9)**	**75.6(17.6)**	**87.1(11.5)**	**78.8(15.5)**	**91.3(7.6)**	**84.9(11.6)**

Dataset and Evaluation. We evaluate our method on the public datasets of 2017 ACDC challenge [15] for cardiac segmentation and PROMISE12 [16] for prostate segmentation. The 2017 ACDC dataset contains 100 subjects, where 75 subjects are assigned to training set, 5 to validating set and 20 to testing set. The PROMISE12 dataset has 50 transversal T2-weighted magnetic resonance imaging (MRI) images, from which we randomly selected 35 samples as training set, 5 as validating set and 10 as testing set. In the training set, the partition of the labeled set and the unlabeled set is denoted as n/m, where n and m are the numbers of labeled and unlabeled samples, respectively. To quantitatively assess the performance, we use two standard evaluation metrics, i.e., Dice coefficient (Dice $= \frac{2*|X \cap Y|}{|X|+|Y|}$) and Jaccard Index (JI $= \frac{|X \cap Y|}{|X \cup Y|}$). Higher scores indicate better segmentation performance.

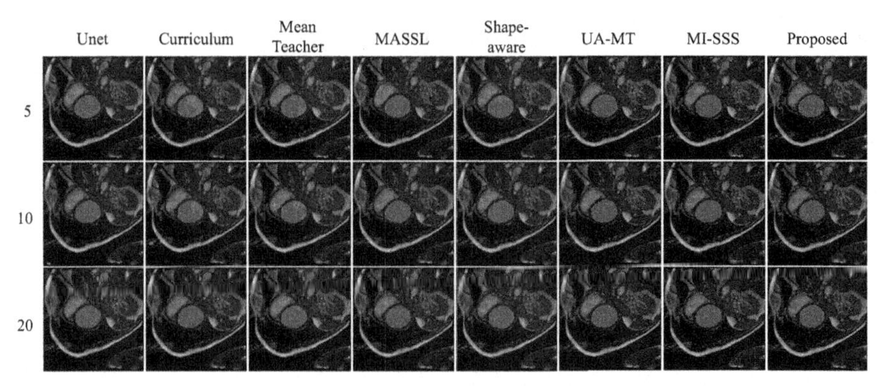

Fig. 2. Visual comparison results on 2017 ACDC dataset. Orange indicates the correct segmented area, green the unidentified and yellow the miss-identified. (Color figure online)

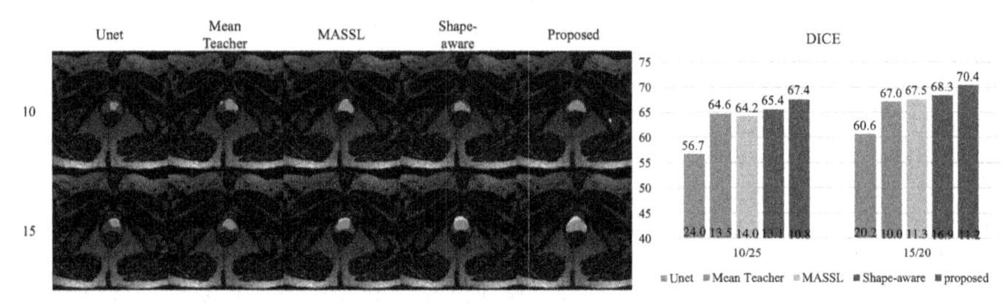

Fig. 3. Visual comparison results on PROMISE12 dataset. Averages and standard deviations are provided above and below the bars, respectively.

Comparison with Other Methods. To demonstrate the superiority of the proposed method in leveraging unlabeled and labeled data, we compare our method with several methods including U-net [2], Curriculum [3], Mean Teacher [9], MASSL [13], Shape-aware [17], UA-MT [10] and MI-SSS [19]. It is worth noting that only U-net is trained in a fully supervised manner while others are semi-supervised. Table 1 is a summary of quantitative results on 2017 ACDC dataset in different n/m settings. As observed, our proposed method outperforms all the compared methods with the highest Dice and JI values in all n/m settings. Specifically, for the fully supervised U-net, our method can leverage the unlabeled data and largely improve Dice and JI from 60.1%, 47.3% to 84.6%, 75.6%, respectively, when $n = 5$. Besides, when n is small, the improvement of our method is statistically significant with $p < 0.05$ via paired t-test. With more labeled data available, all the methods show an upward trend with a narrowing gap, but our method still ranks first. We also give a qualitative comparison result on 2017 ACDC dataset in Fig. 2. It can be seen that the target area is better sketched by our method with more accurate boundaries. On the other hand, our method produces the fewest false positive predictions, thereby generating the most similar segmentation results to ground truth over all of the compared methods. Figure 3 provides comparison results on PROMISE12 dataset. Clearly, our methods achieves the best performance in both qualitative and quantitative measures.

Ablation Study. To investigate the contributions of key components of our method, we further conduct a series of experiments in different model settings on 2017 ACDC dataset. First, to validate the effectiveness of auxiliary tasks, we compare the models of (1) the segmentation task alone (Seg), (2) the segmentation task and the SDF prediction task (Seg+SDF), and (3) all the three tasks (Seg+SDF+Rec). The quantitative results are shown in the upper part in Table 2, from where we can see that the Seg model exhibits the worst performance. With the SDF prediction task and the reconstruction task joining, Dice and JI are improved to varying degrees. Especially when n is small, the improvement by SDF is significant, revealing its large contribution to the utilization of unlabeled data.

Table 2. Ablation study of our method on 2017 ACDC dataset. * means our method is significant better than compared method with $p < 0.05$ via paired t-test.

n/m	5/70		10/65		20/55	
	Dice [%]	JI [%]	Dice [%]	JI [%]	Dice [%]	JI [%]
Seg	60.1(24.7)*	47.3(23.4)*	70.9(23.6)*	53.1(28.5)*	90.0(7.7)	82.5(11.3)
Seg+SDF	81.0(16.7)*	70.8(19.5)*	83.4(17.2)*	74.5(20.2)*	90.2(8.5)	83.1(12.5)
Seg+SDF+Rec	81.2(13.8)*	70.9(17.8)*	84.2(15.1)*	75.4(18.6)*	90.7(7.2)	83.8(10.8)
S+T	52.1(18.7)*	37.3(16.2)*	80.0(14.3)*	68.8(17.5)*	88.8(9.3)*	80.9(13.4)*
S+T+UncA	70.8(22.4)*	58.8(22.6)*	82.8(20.5)*	74.7(23.0)*	90.2(8.2)	83.1(12.0)
S+T+UncW	79.3(23.4)*	70.5(25.3)*	83.2(18.7)*	74.7(21.8)*	90.7(7.8)	83.8(11.6)
Proposed w/o. Tri-U	82.6(13.0)	72.4(16.6)	86.0(12.2)	77.1(16.0)	90.9(8.4)	84.2(12.4)
Proposed	**84.6(13.9)**	**75.6(17.6)**	**87.1(11.5)**	**78.8(15.5)**	**91.3(7.7)**	**84.9(11.6)**

Second, we regard the Seg model as the student model (S) and construct variant models by incorporating (1) the teacher model (T), (2) the uncertainty estimation with averaging (UncA), and (3) the uncertainty estimation with the proposed UWI (UncW). The middle part of Table 2 presents the detailed results. We can find that the S+T model decreases Dice and JI by 8.0% and 10% compared with S only when $n = 5$. This may be explained as that the teacher model is susceptible to noise when labeled data is few, thus degrading the performance. However, by considering the uncertainty estimation, the S+T+UncA model rises Dice and JI remarkably by 18.7% and 21.5% for $n = 5$, and the proposed S+T+UncW model yields higher indicator values, proving their effectiveness.

Third, to verify the guiding role of the tripled-uncertainty, we compare the proposed model and that without the tripled uncertainty, i.e., proposed w/o. tri-U, and display the results in the Table 2. As observed, our complete model gains better performance, demonstrating the promotion effect of the devised tripled-uncertainty.

4 Conclusion

In this paper, we propose a tripled-uncertainty guided semi-supervised model for medical image segmentation. Based on a mean teacher architecture, our model explores the relationship among the segmentation task, the foreground and background reconstruction task and the SDF prediction task. To eliminate the possible misdirection caused by the noisy unlabeled data, we employ uncertainty estimation on all three tasks in the teacher model. In contrast to the common uncertainty averaging integration strategy, we consider the differences of each sampling and develop a novel uncertainty weighted integration strategy. The experimental results demonstrate the feasibility and superiority of our method.

Acknowledgement. This work is supported by National Natural Science Foundation of China (NFSC 62071314) and Sichuan Science and Technology Program (2021YFG0326, 2020YFG0079).

References

1. Chen, J., Yang, L., Zhang, Y., et al.: Combining fully convolutional and recurrent neural networks for 3D biomedical image segmentation. In: NIPS (2016)
2. Ronneberger, O., Fischer, P., Brox, T.: U-Net: convolutional networks for biomedical image segmentation. In: Navab, N., Hornegger, J., Wells, W.M., Frangi, A.F. (eds.) MICCAI 2015. LNCS, vol. 9351, pp. 234–241. Springer, Cham (2015). https://doi.org/10.1007/978-3-319-24574-4_28
3. Kervadec, H., Dolz, J., Granger, É., Ayed, I.B.: Curriculum semi-supervised segmentation. In: Shen, D., et al. (eds.) MICCAI 2019. LNCS, vol. 11765, pp. 568–576. Springer, Cham (2019). https://doi.org/10.1007/978-3-030-32245-8_63
4. Bortsova, G., Dubost, F., Hogeweg, L., Katramados, I., Bruijne, M.: Semi-supervised medical image segmentation via learning consistency under transformations. In: Shen, D., et al. (eds.) MICCAI 2019. LNCS, vol. 11769, pp. 810–818. Springer, Cham (2019). https://doi.org/10.1007/978-3-030-32226-7_90
5. Zheng, Z., et al.: Semi-supervised segmentation with self-training based on quality estimation and refinement. In: Liu, M., Yan, P., Lian, C., Cao, X. (eds.) MLMI 2020. LNCS, vol. 12436, pp. 30–39. Springer, Cham (2020). https://doi.org/10.1007/978-3-030-59861-7_4
6. Park, S., Hwang, W., Jung, K.H.: Integrating reinforcement learning to self- training for pulmonary nodule segmentation in chest x-rays. arXiv preprint arXiv:1811.08840 (2018)
7. Zheng, H., et al.: Cartilage segmentation in high-resolution 3D micro-CT images via uncertainty-guided self-training with very sparse annotation. In: Martel, A.L., et al. (eds.) MICCAI 2020. LNCS, vol. 12261, pp. 802–812. Springer, Cham (2020). https://doi.org/10.1007/978-3-030-59710-8_78
8. Laine, S., Aila, T.: Temporal ensembling for semi-supervised learning. arXiv preprint arXiv:1610.02242 (2016)
9. Tarvainen, A., Valpola, H.: Mean teachers are better role models: weight-averaged consistency targets improve semi-supervised deep learning results. In: Proceedings of the 31st International Conference on Neural Information Processing Systems, pp. 1195–1204 (2017)
10. Lequan, Y., Wang, S., Li, X., Chi-Wing, F., Heng, P.-A.: Uncertainty-aware self-ensembling model for semi-supervised 3D left atrium segmentation. In: Shen, D., et al. (eds.) MICCAI 2019. LNCS, vol. 11765, pp. 605–613. Springer, Cham (2019). https://doi.org/10.1007/978-3-030-32245-8_67
11. Wang, Y., et al.: Double-uncertainty weighted method for semi-supervised learning. In: Martel, A.L., et al. (eds.) MICCAI 2020. LNCS, vol. 12261, pp. 542–551. Springer, Cham (2020). https://doi.org/10.1007/978-3-030-59710-8_53

12. Luo, X., Chen, J., Song, T., et al.: Semi-supervised medical image segmentation through dual-task consistency. arXiv preprint arXiv:2009.04448 (2020)
13. Chen, S., Bortsova, G., Juárez, A.-U., Tulder, G., Bruijne, M.: Multi-task attention-based semi-supervised learning for medical image segmentation. In: Shen, D., et al. (eds.) MICCAI 2019. LNCS, vol. 11766, pp. 457–465. Springer, Cham (2019). https://doi.org/10.1007/978-3-030-32248-9_51
14. Li, S., Zhang, C., He, X.: Shape-aware semi-supervised 3d semantic segmentation for medical images. In: Martel, A.L., et al. (eds.) MICCAI 2020. LNCS, vol. 12261, pp. 552–561. Springer, Cham (2020). https://doi.org/10.1007/978-3-030-59710-8_54
15. Bernard, O., Lalande, A., Zotti, C., et al.: Deep learning techniques for automatic MRI cardiac multi-structures segmentation and diagnosis: is the problem solved? IEEE Trans. Med. Imaging **37**(11), 2514–2525 (2018)
16. Litjens, G., Toth, R., van de Ven, W., et al.: Evaluation of prostate segmentation algorithms for MRI: the PROMISE12 challenge. Med. Image Anal. **18**(2), 359–373 (2014)
17. Xue, Y., Tang, H., Qiao, Z., et al.: Shape-aware organ segmentation by predicting signed distance maps. In: Proceedings of the AAAI Conference on Artificial Intelligence, vol. 34, no. 7, pp. 12565–12572 (2020)
18. Milletari, F., Navab, N., Ahmadi, S.A.: V-Net: fully convolutional neural networks for volumetric medical image segmentation. In: 2016 Fourth International Conference on 3D Vision (3DV), pp. 565–571. IEEE (2016)
19. Peng, J., Pedersoli, M., Desrosiers, C., et al.: Boosting semi-supervised image segmentation with global and local mutual information regularization. arXiv preprint arXiv:2103.04813 (2021)
20. Kendall, A., Gal, Y.: What uncertainties do we need in bayesian deep learning for computer vision? arXiv preprint arXiv:1703.04977 (2017)

Learning with Noise: Mask-Guided Attention Model for Weakly Supervised Nuclei Segmentation

Ruoyu Guo, Maurice Pagnucco, and Yang Song(✉)

School of Computer Science and Engineering, University of New South Wales, Sydney, Australia
yang.song1@unsw.edu.au

Abstract. Deep convolutional neural networks have been highly effective in segmentation tasks. However, high performance often requires large datasets with high-quality annotations, especially for segmentation, which requires precise pixel-wise labelling. The difficulty of generating high-quality datasets often constrains the improvement of research in such areas. To alleviate this issue, we propose a weakly supervised learning method for nuclei segmentation that only requires annotation of the nuclear centroid. To train the segmentation model with point annotations, we first generate boundary and superpixel-based masks as pseudo ground truth labels to train a segmentation network that is enhanced by a mask-guided attention auxiliary network. Then to further improve the accuracy of supervision, we apply Confident Learning to correct the pseudo labels at the pixel-level for a refined training. Our method shows highly competitive performance of cell nuclei segmentation in histopathology images on two public datasets. Our code is available at: https://github.com/RuoyuGuo/MaskGA_Net.

Keywords: Nuclei segmentation · Weakly supervised learning · Noisy labels · Point annotations

1 Introduction

Identifying abnormalities in cell nuclei is crucial to cancer diagnosis. The traditional method requires pathologists to look at Hematoxylin and Eosin (H&E) stained histopathology images to identify those features, which is highly inefficient. Recently, the development of computer vision and Convolutional Neural Networks (CNNs) has enabled accurate localisation of cells, significantly reducing the burden of manual nuclei segmentation. For instance, CIA-Net [20] achieves excellent performance by incorporating an additional decoder for nuclei contour refinement. BiO-Net [15] applies skip connection in both forward and backward directions to enhance the semantic feature learning recursively feeds feature maps to increase performance. However, the performance of CNNs relies heavily on the size of datasets and the quality of annotations.

M. de Bruijne et al. (Eds.): MICCAI 2021, LNCS 12902, pp. 461–470, 2021.
https://doi.org/10.1007/978-3-030-87196-3_43

Unlike natural image datasets, it is difficult to obtain large-scale digital pathology datasets with high-quality annotations. Therefore, weakly-supervised learning has received increasing attention in recent years. In comparison to fully supervised learning, it is more efficient to provide weak annotations, such as points or bounding boxes. However, missing and noisy annotations are the major obstacles limiting the performance of weakly-supervised segmentation methods. To acquire satisfactory results, a well-designed training method is essential and pixel-level pseudo ground truths are commonly required.

To generate pixel-wise ground truth labels from point annotations, a distance transformation has been applied to generate the Voronoi boundary as negative labels so that supervised learning with binary cross entropy is possible [17]. They also integrate an auxiliary network to enforce edge detection during training. Another method utilises k-means clustering on input images and Voronoi distance transform to produce richer annotations to train the network, and employs the dense conditional random field (CRF) loss for model refinement in nuclei segmentation [11]. Similarly, a self-supervised iterative refinement framework based on Voronoi boundary is proposed for nuclei weakly-supervised segmentation [12]. The framework also includes a Sobel edge map to refine the contour. In addition, rather than learning with cross-entropy loss, an instance loss is designed to help the neural network to learn a continuous peak-shaped map from point annotations [3]. A multi-task learning method [2] that combines three different losses and learns from one of them at each iteration also shows accuracy gain. Nonetheless, these weakly-supervised methods all require an estimation of pixel-wise ground truth labels for training the network. Such pseudo ground truth can be highly noisy with lots of mislabelling, and this issue can significantly affect the network performance.

To improve the robustness of trained models with noisy ground truth labels, noise-robust training has attracted some attention in recent years. For instance, [5] presents a probabilistic model to learn the probability transformation from the true labels to noisy labels and estimate the true labels by using the Expectation Maximum (EM) algorithm. Another approach assigns a lower weight to the corrupted data at each training batch based on the evaluated label quality [4]. Additionally, a noise detectable teacher-student architecture is introduced [19] in which the teacher network is first trained with corrupted data and then the student network is trained with an adaptive loss after applying Confident Learning (CL) [10] to recognise incorrect ground truth. Such methods, however, have not been investigated in the context of weakly-supervised segmentation with point annotations.

To this end, we propose a novel method that can not only mitigate the heavy dependence of precise annotations in nuclei segmentation but also correct the noisy labelling in the generated pseudo ground truth. The pipeline of our method is shown in Fig. 1, which consists of a main network to learn different scales of features, an attention network to enhance the noise robustness of the main network and the CL technique for label refinement. Also, unlike aforementioned methods that use Voronoi boundaries as the pixel-level labels for segmentation, we use them for denoising the pseudo ground truth generated based on superpixels. Our contributions can be summarized as follows: (i) We design a weakly-supervised learning framework with label correction for nuclei segmentation based on point annotations. (ii) We introduce an attention network that can greatly reduce the distraction of noisy labels without introducing extra computational cost at test time. (iii) Our experimental results on two public datasets show that our method can effectively accommodate high levels of label noise and achieve highly effective segmentation of cell nuclei.

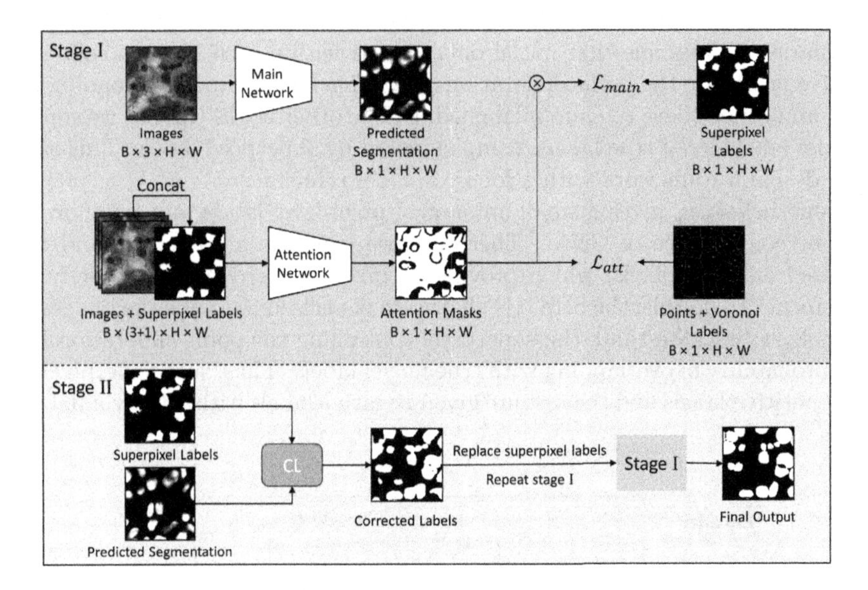

Fig. 1. Overview of our proposed weakly-supervised nuclei segmentation method. B, H and W represent batch size, height and width of input images.

2 Methods

As shown in Fig. 1, at Stage I, the main network and attention network are trained using the pseudo ground truth labels that are generated based on superpixel and Voronoi boundary information. At Stage II, the CL algorithm is adopted to refine labels, and we replace the superpixel labels with the revised labels and conduct the same training process again as Stage I. At test time, only the main network is used without involving the attention network.

2.1 Annotations

While the datasets provide full pixel-wise annotations, to experiment with weakly-supervised learning, we generate the point annotations automatically by calculating the centroids of nuclei based on the pixel-wise annotations. Subsequently, since the point annotations are incapable of providing pixel-level information for training an accurate segmentation network, we need to estimate the pixel-wise annotations from the point annotations. To do this, we generate two types of pseudo ground truth masks, namely y_v and y_s, which are **Points+Voronoi Labels** and **Superpixel Labels**, as shown in Fig. 1 respectively.

The first pseudo ground truth y_v is generated following [7], which aims to provide localisation information and boundary constraints. Specifically, we apply the Euclidean distance transform based on point annotations to obtain the Voronoi boundaries and assume that nuclei only appear within each Voronoi cell. Then, positive labels are the point annotations, negative labels are the Voronoi boundaries, and other pixels remain unlabelled. Unlike other works [12,17], we consider Voronoi labels to be true labels, complemented by superpixel labels. This enable us to design a framework with a focus on label refinement.

Nonetheless, y_v gives a small amount of pixel-level labels and this may lead to poor convergence of CNNs. Therefore, we generate a second ground truth y_s based on superpixels, which provides a pixel-wise ground truth for further refinement. To do this, the SLIC [1] algorithm is utilised to divide the images into superpixels first. We think the superpixel containing the point annotation has a high probability of overlapping with true nuclei labels. Thus, such superpixels are given positive labels and the rest are given negative labels without any unlabelled pixels.

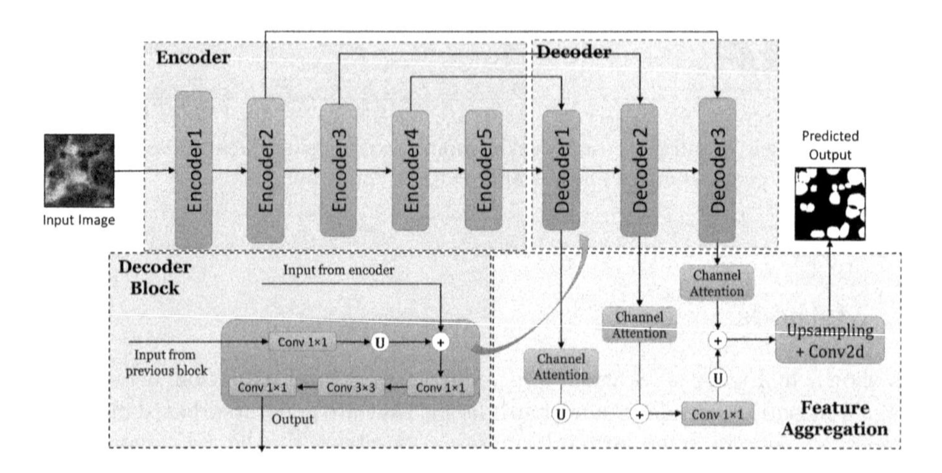

Fig. 2. Feature aggregation network architecture.

2.2 Main Network Architecture

In order to adapt to the complexity of nuclei, we design a feature aggregation network as our main segmentation network. The feature aggregation network follows an encoder-decoder architecture with an additional feature aggregation module. ResNet-50 is employed as the encoder, and only the last four blocks are connected to the decoder network because the spatial dimension of the first block is too large. Moreover, we adopt the dilation strategy in encoder5, with an output stride of 1/16 to preserve resolution while increasing the receptive fields [18]. In addition, since the feature maps from each decoder block have different scales and not all levels of features contribute equally to the final output, we equip an aggregation module to the encoder-decoder network. The aggregation module has a channel attention layer (CBAM [14]) to highlight useful information from the output of each decoder block for better aggregation. Besides, two 1×1 Conv2d layers is attached after addition. As shown in Fig. 2, the feature maps from three channel attention layers are merged by element-wise addition and fed through an upsampling operation to obtain the final predicted probability maps. For training the main network, we choose binary cross-entropy loss as $\mathcal{L}_{main}$ to optimise the main network. At Stage I, we use the superpixel ground truth y_s as the supervision for the main network only. The other ground truth y_v, which contains the location information of nuclei, is used to detect noise in y_s, as described in the next section.

2.3 Mask-Guided Attention

Noise is unavoidable in y_s because it is generated based on the superpixel information. While data re-weighting [4,8,13,16] can address this problem to a certain extent, a significant number of clean data are required in these methods, which is not suitable for our problem. Therefore, we introduce the mask-guided attention network, which can be considered a data-dependent denoising method. The attention mask generated by the attention network can help identify the true labels y_s and also guide the main network to focus on these regions. Specifically, as illustrated in Fig. 1, we first concatenate the input image and its corresponding superpixel labels to form a 4-channel input, which is then fed into an attention network to produce the spatial attention mask. The attention network employs the same feature aggregation network architecture but with a smaller ResNet-18 as the encoder.

To recognise the noise in y_s, the partial pixel-wise $\mathcal{L}_1$ loss is selected as $\mathcal{L}_{att}$ for loss calculation between the attention network output and the Voronoi ground truth y_v:

$$\mathcal{L}_{att}(g(c(I, y_s)), y_v) = \frac{\sum_{y \in y_v^+} ||1 - g(c(I, y_s))||_1}{|y_v^+|} + \beta \cdot \frac{\sum_{y \in y_v^-} ||1 - g(c(I, y_s))||_1}{|y_v^-|} \tag{1}$$

where g, y_v^+, y_v^-, $|y_v^+|$ and $|y_v^-|$ denote the attention network, positive labels and negative labels, and the total numbers of positive labels and negative labels of input image I in y_v. $c(I, y_s)$ is the concatenation operation. β is a constant to balance the positive and negative labels. Note that this loss ignores all the pixels that are not annotated in y_v. The effectiveness of $\mathcal{L}_{att}$ could be explained as follows: pixels that are mislabelled in y_s would not result in expected probabilities in the attention mask, which would then increase the value of $\mathcal{L}_{att}$.

Furthermore, the predicted segmentation of the main network is multiplied by the attention mask to compute $\mathcal{L}_{main}$ so that the segmentation by the main network is guided by the attention masks during training. To train the main network and attention network simultaneously, we define the total loss $\mathcal{L}$ as a linear combination of $\mathcal{L}_{att}$ and $\mathcal{L}_{main}$:

$$\mathcal{L} = \mathcal{L}_{main}(f(I) \odot g(c(I, y_s)), y_s) + \alpha \cdot \mathcal{L}_{att}(g(c(I, y_s)), y_v) \tag{2}$$

where $\odot$ is the element-wise multiplication and α is the hyper-parameter to indicate the reliability of the attention maps.

2.4 Pseudo Ground Truth Revision

Given the predicted segmentation masks, we apply the CL method to identify the noisy labels in the superpixel ground truth y_s by estimating the joint distribution $\mathbf{Q}_{y_s,y^*}$ between the pseudo ground truth y_s and real labels y^*.

To estimate $\mathbf{Q}_{y_s,y^*}$, we first build an $|m| \times |m|$ matrix, $\mathbf{C}_{y_s,y^*}$, where m is the set of unique class ({positive, negative} in our method and $|m| = 2$). Each entry $\mathbf{C}_{y_s=i,y^*=j}$ represents the number of pixels with pseudo label i and having $\hat{p}(y = j; x) > t_j$, and $\hat{p}(y = j; x)$ is the predicted probability of a pixel x belonging to class j. Here, j in $\mathbf{C}_{y_s=i,y^*=j}[i][j]$ is the class that has the highest predicted probability among all classes of that pixel, and threshold t_j is the averaged predicted probability of all pixels belonging to class j and having $y_s = j$. Given $\mathbf{C}_{y_s,y^*}$, we estimate $\mathbf{Q}_{y_s,y^*}$ by row normalising $\mathbf{C}_{y_s,y^*}$ first. Secondly, each row in the normalised $\mathbf{C}_{y_s,y^*}$ will be multiplied by the total number of pixels that belong to class i in y_s to get $\mathbf{C}'_{y_s,y^*}$. Finally, $\mathbf{C}'_{y_s,y^*}$ is divided by the sum of $\mathbf{C}'_{y_s,y^*}$ to obtain $\mathbf{Q}_{y_s,y^*}$.

The mislabelled pixels in y_s are selected from each off-diagonal entry in $\mathbf{Q}_{y_s,y^*}$. Specifically, we first compute the margin $\hat{p}(y = j; x) - \hat{p}(y = i; x)$ of each pixel in $\mathbf{Q}_{y_s=i,y^*=j,i \neq j}$, and the top $n \cdot \mathbf{Q}_{y_s=i,y^*=j}$ (n is the total number of pixels in an image) pixels within the margin are selected as mislabelled pixels in each entry. Finally, we generate a modified ground truth y_{re} by flipping these mislabelled pixels. Once the revised ground truth y_{re} is obtained, we train our whole framework again by replacing y_s with y_{re}.

3 Experiments and Results

The proposed method is evaluated on two public datasets, MoNuSeg [6] and TNBC [9]. MoNuSeg contains 30 images with a total of 21,623 nuclei from 7 different organs, and each image size is $1,000 \times 1,000$ pixels. TNBC has 50 images with a total of 4022 nuclei from 11 different patients with image resolution of 512×512 pixels. These datasets provide full annotations of nuclei. Therefore,

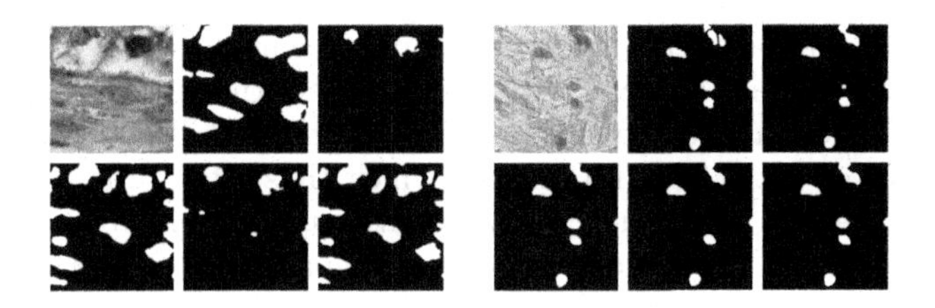

Fig. 3. Segmentation results using different architectures. For each block of six images, from left to right, top to bottom: input image, ground truth, segmentation result obtained by Main, Main+Att, Main+CL, Main+Att+CL. The left block is from MoNuSeg and the right one is from TNBC.

centroids of nuclei can be generated automatically and evaluation of segmentation performance on full masks is possible. We follow [17] to conduct 10-fold cross-validation with Intersection over Union (IoU) as the evaluation metric. In each iteration of cross-validation, the data are divided with an 8:1:1 split ratio for the training, validation and test sets.

Table 1. Segmentation results comparing weakly and fully supervised approaches.

Supervision	Method	MoNuSeg	TNBC
Point annotations	Laradji et al. [6]	0.5710 (±0.02)	0.5504 (±0.04)
	Qu et at. [11]	0.6099	–
	Yoo et al. [17]	0.6136 (±0.04)	0.6038 (±0.03)
	Tian et al. [12]	0.6239 (±0.03)	**0.6393 (±0.03)**
	Our method	**0.6401 (±0.03)**	0.6234 (±0.02)
Pixel-wise annotation	Yoo et al. [17]	0.6522 (±0.03)	0.6619 (±0.04)
	Xiang et al. [15]	**0.704**	0.651
	Tian et al. [12]	0.6494 (±0.04)	**0.6950 (±0.03)**
	Our method	0.6723 (±0.05)	0.6537 (±0.03)

Our method is implemented with PyTorch. We train the model for 60 epochs with an initial learning rate of 0.001 and a weight decay of 0.0005. The learning rate is halved when the loss per epoch on the validation set does not decrease for 4 epochs compared to the last minimal loss. For data augmentation, Gaussian blur, hue adjustment and saturation, affine transformations, horizontal and vertical flips are applied. During training, the batch size is set to 2 and the Adam optimiser is used. Every convolutional layer is followed by a batch normalization layer and ReLU layer. Bilinear upsampling is used for upsampling operations. α and β are set to 0.8 and 0.4 respectively. These settings are the same for all ablation studies.

Table 1 summarises our method's performance. Our method achieves higher performance than the recent weakly-supervised segmentation approaches

[6,11,17]. Our approach leads to 0.0162 improvement in IoU over the state-of-the-art method [12] on MoNuSeg. However, the performance on TNBC is not as great as [12]. The main reason is that the low quality of TNBC dataset images and its generated y_s (0.5118 IoU score), which thus affects the overall performance of our method. Even with poor annotations, our framework still shows its effectiveness on denoising. Table 2 shows the comparison of segmentation performance with initial weak labels and after refinement. Additionally, the F1-score of y_s is 0.73 on MoNuSeg and 0.69 on TNBC, which is increasing to 0.83 and 0.76 with Main+Att. Furthermore, as shown in Fig. 3, when the main network is equipped with the attention network and CL, more nuclei are detected and the nuclear shape is closer to the ground truth. We also experimented with replacing the point annotation with full pixel-wise ground truth masks while keeping our model unchanged. Our result on MoNuSeg still provides highly accurate segmentation with full supervision compared to other existing approaches [12,17]. Although [12] achieves higher performance on TNBC with point annotations than our method, the improvement of this method on MoNuSeg is moderate even with pixel-wise annotation. It is worth noting that our result on MoNuSeg with point annotations is close to [12] with pixel-wise annotation.

Table 2. Ablation study of our method components and comparison of different models with different noise levels in the point annotations.

Point location	Methods	MoNuSeg ($r = 5$)	TNBC ($r = 3$)
Centroid (r = 0)	Main	0.5580 (±0.07)	0.5329 (±0.05)
	Main+CL	0.5919 (±0.06)	0.5585 (±0.04)
	Main+Att	0.6222 (±0.05)	0.5788 (±0.04)
	Main+Att+CL	**0.6401 (±0.03)**	**0.6234 (±0.02)**
	Yoo et al. [17]	0.6173 (±0.05)	0.5907 (±0.08)
r = 3 or 5	Main	0.5437 (±0.09)	0.5196 (±0.08)
	Main+CL	0.5897 (±0.08)	0.5531 (±0.03)
	Main+Att	0.6296 (±0.07)	0.5776 (±0.04)
	Main+Att+CL	**0.6324 (±0.04)**	**0.6153 (±0.04)**
	Yoo et al. [17]	0.6011 (±0.08)	0.5893 (±0.07)

In addition, we consider that it might not be easy to precisely annotate the centroids of nuclei as the ground truth. Therefore, to comprehensively investigate our approach, we simulate the noisy point supervision at different levels. To do this, two extra sets of point annotations are obtained by randomly shifting the centroid points in a circle within a radius $r = 3$ and $r = 5$. We then conduct the ablation study of our model using 10-fold cross validation. For comparison, we have trained and tested the model [17] using both sets of annotations. The experimental results shown in Table 2 demonstrate that without CL, our attention network also shows competitive results compared to [17].

4 Conclusions

In this paper, we propose a mask-guided attention network model and a two-stage method for weakly-supervised nucleus segmentation with point annotations. The main network can appropriately aggregate features from different scales. The attention network takes the noisy ground truth and input images as input to highlight the correct labels by extracting information from point annotations. Many weakly-supervised methods could be equipped with our attention network. Besides, the CL algorithm is utilised to refine noisy labels. The experimental results demonstrate that our method is highly effective for cell nuclei segmentation on two histopathology image datasets. In future works, we will experiment the effects of different pseudo labels generation methods and label refinement methods that appropriate for nuclei domain.

References

1. Achanta, R., Shaji, A., Smith, K., Lucchi, A., Fua, P., Süsstrunk, S.: Slic superpixels compared to state-of-the-art superpixel methods. IEEE Tran. Pattern Anal. Mach. Intell. **34**, 2274–2282 (2012)
2. Chamanzar, A., Nie, Y.: Weakly supervised multi-task learning for cell detection and segmentation. In: 2020 IEEE 17th International Symposium on Biomedical Imaging (ISBI), pp. 513–516 (2020)
3. Dong, M., et al.: Towards neuron segmentation from macaque brain images: a weakly supervised approach. In: Martel, A.L., et al. (eds.) MICCAI 2020. LNCS, vol. 12265, pp. 194–203. Springer, Cham (2020). https://doi.org/10.1007/978-3-030-59722-1_19
4. Zhu, H., Shi, J., Wu, J.: Pick-and-Learn: automatic quality evaluation for noisy-labeled image segmentation. In: Shen, D., et al. (eds.) MICCAI 2019. LNCS, vol. 11769, pp. 576–584. Springer, Cham (2019). https://doi.org/10.1007/978-3-030-32226-7_64
5. Joseph, B., Jacob, G.: Training deep neural-networks based on unreliable labels. In: 2016 IEEE International Conference on Acoustics, Speech and Signal Processing (ICASSP), vol. 2016, pp. 2682–2686 (2016)
6. Kumar, N., Verma, R., Sharma, S., Bhargava, S., Vahadane, A., Sethi, A.: A dataset and a technique for generalized nuclear segmentation for computational pathology. IEEE Trans. Med. Imag. (TMI) **36**(7), 1550–1560 (2017)
7. Laradji, I.H., Rostamzadeh, N., Pinheiro, P.O., Vazquez, D., Schmidt, M.: Where are the blobs: counting by localization with point supervision. In: Ferrari, V., Hebert, M., Sminchisescu, C., Weiss, Y. (eds.) ECCV 2018. LNCS, vol. 11206, pp. 560–576. Springer, Cham (2018). https://doi.org/10.1007/978-3-030-01216-8_34
8. Mirikharaji, Z., Yan, Y., Hamarneh, G.: Learning to segment skin lesions from noisy annotations. In: Wang, Q., et al. (eds.) DART/MIL3ID -2019. LNCS, vol. 11795, pp. 207–215. Springer, Cham (2019). https://doi.org/10.1007/978-3-030-33391-1_24
9. Naylor, P., Laé, M., Reyal, F., Walter, T.: Segmentation of nuclei in histopathology images by deep regression of the distance map. IEEE Trans. Med. Imag. (TMI) **38**(2), 448–459 (2019)

10. Northcutt, C.G., Jiang, L., Chuang, I.L.: Confident learning: Estimating uncertainty in dataset labels. arXiv preprint arXiv:1911.00068 (2019)
11. Qu, H., et al.: Weakly supervised deep nuclei segmentation using points annotation in histopathology images. In: Proceedings of The 2nd International Conference on Medical Imaging with Deep Learning, pp. 390–400 (2019)
12. Tian, K., et al.: Weakly-Supervised nucleus segmentation based on point annotations: a coarse-to-fine self-stimulated learning strategy. In: Martel, A.L., et al. (eds.) MICCAI 2020. LNCS, vol. 12265, pp. 299–308. Springer, Cham (2020). https://doi.org/10.1007/978-3-030-59722-1_29
13. Wang, J., Zhou, S., Fang, C., Wang, L., Wang, J.: Meta corrupted pixels mining for medical image segmentation. In: Martel, A.L., et al. (eds.) MICCAI 2020. LNCS, vol. 12261, pp. 335–345. Springer, Cham (2020). https://doi.org/10.1007/978-3-030-59710-8_33
14. Woo, S., Park, J., Lee, J., Kweon, I.: CBAM: Convolutional block attention module. In: European Conference on Computer Vision (ECCV), pp. 3–19 (2018)
15. Xiang, T., Zhang, C., Liu, D., Song, Y., Huang, H., Cai, W.: BiO-Net: learning recurrent bi-directional connections for encoder-decoder architecture. In: Martel, A.L., et al. (eds.) MICCAI 2020. LNCS, vol. 12261, pp. 74–84. Springer, Cham (2020). https://doi.org/10.1007/978-3-030-59710-8_8
16. Xue, C., Dou, Q., Shi, X., Chen, H., Heng, P.: Robust learning at noisy labeled medical images: applied to skin lesion classification. In: 2019 IEEE 16th International Symposium on Biomedical Imaging (ISBI 2019), pp. 1280–1283 (2019)
17. Yoo, I., Yoo, D., Paeng, K.: PseudoEdgeNet: nuclei segmentation only with point annotations. In: Shen, D., et al. (eds.) MICCAI 2019. LNCS, vol. 11764, pp. 731–739. Springer, Cham (2019). https://doi.org/10.1007/978-3-030-32239-7_81
18. Yu, F., Koltun, V., Funkhouser, T.: Dilated residual networks. In: IEEE Conference on Computer Vision and Pattern Recognition (CVPR), pp. 636–644 (2017)
19. Zhang, M., et al.: Characterizing label errors: confident learning for noisy-labeled image segmentation. In: Martel, A.L., et al. (eds.) MICCAI 2020. LNCS, vol. 12261, pp. 721–730. Springer, Cham (2020). https://doi.org/10.1007/978-3-030-59710-8_70
20. Zhou, Y., Onder, O.F., Dou, Q., Tsougenis, E., Chen, H., Heng, P.-A.: CIA-Net: robust nuclei instance segmentation with contour-aware information aggregation. In: Chung, A.C.S., Gee, J.C., Yushkevich, P.A., Bao, S. (eds.) IPMI 2019. LNCS, vol. 11492, pp. 682–693. Springer, Cham (2019). https://doi.org/10.1007/978-3-030-20351-1_53

Order-Guided Disentangled Representation Learning for Ulcerative Colitis Classification with Limited Labels

Shota Harada[1(✉)], Ryoma Bise[1,2], Hideaki Hayashi[1], Kiyohito Tanaka[3], and Seiichi Uchida[1,2]

[1] Kyushu University, Fukuoka City, Japan
shota.harada@human.ait.kyushu-u.ac.jp
[2] National Institute of Informatics, Tokyo, Japan
[3] Kyoto Second Red Cross Hospital, Kyoto, Japan

Abstract. Ulcerative colitis (UC) classification, which is an important task for endoscopic diagnosis, involves two main difficulties. First, endoscopic images with the annotation about UC (positive or negative) are usually limited. Second, they show a large variability in their appearance due to the location in the colon. Especially, the second difficulty prevents us from using existing semi-supervised learning techniques, which are the common remedy for the first difficulty. In this paper, we propose a practical semi-supervised learning method for UC classification by newly exploiting two additional features, the location in a colon (e.g., left colon) and image capturing order, both of which are often attached to individual images in endoscopic image sequences. The proposed method can extract the essential information of UC classification efficiently by a disentanglement process with those features. Experimental results demonstrate that the proposed method outperforms several existing semi-supervised learning methods in the classification task, even with a small number of annotated images.

Keywords: Endoscopic image classification · Ulcerative colitis · Semi-supervised learning · Disentangled representation learning

1 Introduction

In the classification of ulcerative colitis (UC) using deep neural networks, where endoscopic images are classified into lesion and normal classes, it is difficult to collect a sufficient number of labeled images because the annotation requires significant effort by medical experts. UC is an inflammatory bowel disease that causes inflammation and ulcers in the colon. Specialist knowledge is required to annotate UC because texture features, such as bleeding, visible vascular patterns, and ulcers, should be captured among the image appearances that drastically vary depending on the location in the colon to detect UC.

M. de Bruijne et al. (Eds.): MICCAI 2021, LNCS 12902, pp. 471–480, 2021.
https://doi.org/10.1007/978-3-030-87196-3_44

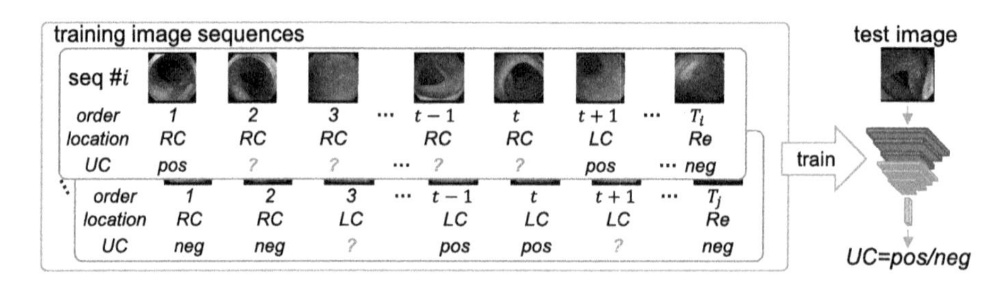

Fig. 1. Underlying concept for the proposed method. The objective of the study is to train an ulcerative colitis (UC) classifier with incomplete UC labels. The order and location are used as the guiding information (RC: right colon. LC: left colon. Re: rectum).

Semi-supervised learning methods [1,2,7,11] have been used to train classifiers based on a limited number of labeled images, involving the use of both labeled and unlabeled images. If a classifier with a moderate classification performance is obtained with few labeled data, the performance of a classifier can be further improved by applying these semi-supervised learning methods. However, existing semi-supervised learning methods do not show satisfactory performance for UC classification because they implicitly assume that the major appearance of images is determined by the classification target class, whereas the major appearance of UC images is determined by the location in the colon, not by the disease condition.

Incorporating domain-dependent knowledge can also compensate for the lack of labeled data. In endoscopic images, we can utilize two types of prior knowledge: location information and temporal ordering information, that is, the order in which the endoscopic images were captured. Location information can be obtained easily by tracking the movement of the endoscope during the examination [6,10], with the rough appearance of endoscopic images characterized by their location. Endoscopic images are acquired in sequence while the endoscope is moved through the colon. Therefore, the temporal ordering information is readily available, and temporally adjacent images tend to belong to the same UC label. If the above information can be incorporated into semi-supervised learning, more accurate and reliable networks for UC classification can be developed.

In this study, we propose a semi-supervised learning method for UC classification that utilizes location and temporal ordering information obtained from endoscopic images. Figure 1 shows the underlying concept for the proposed method. In the proposed method, a UC classifier is trained with incomplete UC labels, whereas the location and ordering information are available. By utilizing the location information, we aim to improve UC classification performance by simultaneously extracting the UC and location features from endoscopic images. We introduce disentangled representation learning [8,9] to effectively embed the UC and location features into the feature space separately. To compensate for the lack of UC-labeled data using temporal ordering information, we formulated the ordinal loss, which is an objective function that brings temporally adjacent images closer in the feature space.

The contributions of this study are as follows:

- We propose a semi-supervised learning method that utilizes the location and temporal ordering information for UC classification. The proposed method introduces disentangled representation learning using location information to extract UC classification features that are separated from the location features.
- We formulate an objective function for order-guided learning to utilize temporal ordering information of endoscopic images. Order-guided learning can obtain the effective feature for classifying UC from unlabeled images by considering the relationship between the temporally adjacent images.

2 Related Work

Semi-supervised learning methods that utilize unlabeled samples efficiently have been reported in the training of classifiers when limited labeled data are available [1,2,7,11]. Lee [7] proposed a method called Pseudo-Label, which uses the class predicted by the trained classifier as the ground-truth for unlabeled samples. Despite its simplicity, this method improves the classification performance in situations where labeled images are limited. Sohn *et al.* [11] proposed FixMatch, which improves the classification performance by making the predictions for weakly and strongly augmented unlabeled images closer during training. These semi-supervised learning methods work well when a classifier with a moderate classification performance has already been obtained using limited labels. However, in UC classification, which requires the learning of texture features from endoscopic images whose appearance varies depending on imaging location, it is difficult to obtain a classifier with a moderate classification performance using limited labeled endoscopic images, and applying these methods to UC classifications may not improve classification performance. Therefore, we propose a semi-supervised learning method that does not directly use the prediction results returned by a classifier trained by limited-labeled data, but utilizes two additional features: the location and the temporal ordering.

Several methods that utilize the temporal ordering information of images have been reported [3–5]. For example, Cao *et al.* [3] proposed Temporal-Cycle Consistency (TCC), which is a self-supervised learning method that utilizes temporal alignment between sequences. The TCC yields good image feature representation by maximizing the number of points where the temporal alignment matches. Dwibedi *et al.* [4] proposed a few-shot video classification method that utilizes temporal alignment between labeled and unlabeled video, then improved the video classification accuracy by minimizing the distance between temporally aligned frames. Moreover, a method for segmenting endoscopic image sequences has been proposed [5]. By utilizing the prior knowledge that temporally adjacent images tend to belong to the same class, this method segments an image sequence without requiring additional annotation. However, the methods proposed [3,4] are not suitable for our task, where involves a sequence with indefinite class transitions, because they assume that the class transitions in the sequence are the same. Furthermore, the method proposed in [5], which assumes segmentation of normal organ image sequences, is not suitable for our task where the target image sequence consists of images of both normal and inflamed organs. In the proposed method, temporal ordering information is used to implement order-guided learning, which brings together temporal adjacency images that tend to belong to the same UC class, thus obtaining a good feature representation for detecting UC in the feature space.

3 Order-Guided Disentangled Representation Learning for UC Classification with Limited Labels

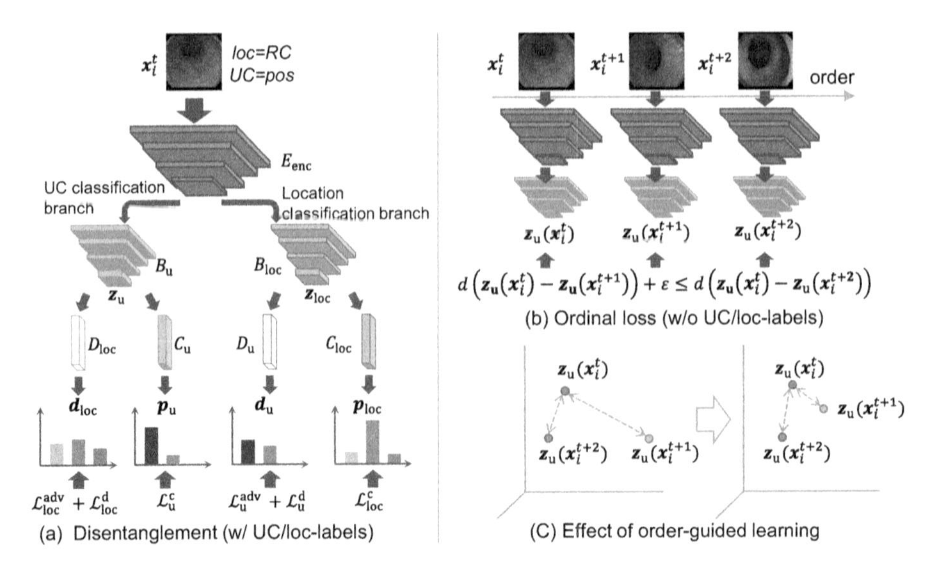

Fig. 2. Overview of the proposed method. (a) Disentanglement into the UC feature z_u and the location feature z_{loc}. (b) Ordinal loss for order-guided learning. (c) Effect of order-guided learning.

The classification of UC using deep neural networks trained by general learning methods is difficult for two reasons. First, the appearances of the endoscopic images vary dynamically depending on the location in the colon, whereas UC is characterized by the texture of the colon surface. Second, the number of UC-labeled images is limited because annotating UC labels to a large number of images requires significant effort by medical experts.

To overcome these difficulties, the proposed method introduces *disentangled representation learning* and *order-guided learning*. Figure 2 shows the overview of the proposed method. In disentangled representation learning using location information, we disentangle the image features into features for UC-dependent and location-dependent to mitigate the worse effect from the various appearance depending on the location. Order-guided learning utilizes the characteristics of an endoscopic image sequence in which temporally adjacent images tend to belong to the same class. We formulated an objective function that represents this characteristic and employs it during learning to address the limitation of the UC-labeled images.

3.1 Disentangled Representation Learning Using Location Information

Disentangled representation learning for the proposed method aims to separate the image features into UC and location-dependent features. These features are obtained via multi-task learning of UC and location classification. Along with the training of classifiers for UC and location classification tasks, the feature for one task is learned to fool the classifier for the other task; that is, the UC-dependent feature is learned to be non-discriminative with respect to location classification, and vice versa.

The network structure for learning disentangled representations is shown in Fig. 2(a). This network has a hierarchical structure in which a feature extraction module branches into two task-specific modules, each of which further branches into two classification modules. The feature extraction module E_{enc} extracts a common feature vector for UC and location classification from the input image. The task-specific modules B_{u} and B_{loc} extract the UC feature $\boldsymbol{z}_{\mathrm{u}}$ and the location feature $\boldsymbol{z}_{\mathrm{loc}}$, which are disentangled features for UC and location classification. Out of four classification modules, the modules C_{u} and C_{loc} are used for UC and location classification, respectively, whereas D_{u} and D_{loc} are used to learn the disentangled representations.

In the left branch of Fig. 2(a), the network obtains the prediction results for UC classes, $\boldsymbol{p}_{\mathrm{u}}$, as the posterior probabilities, based on the disentangled UC feature $\boldsymbol{z}_{\mathrm{u}}$ through learning. Hereinafter, we explain only the training of the left branch in detail because that of the right branch can be formulated by simply swapping the subscripts "loc" and "u" in the symbols for the left branch.

Given a set of N image sequences and corresponding location class labels $\{\boldsymbol{x}_i^{(1:T_i)}, \boldsymbol{l}_i^{(1:T_i)}\}_{i=1}^N$ and a set of limited UC class labels $\{\boldsymbol{u}_j^k \mid (j,k) \in \mathcal{U}\}$, where T_i is the number of images in the i-th image sequence and $\boldsymbol{u}_j^k$ is the UC class label corresponding to the j-th image in the k-th sequence, the training is performed based on three losses: classification loss $\mathcal{L}_{\mathrm{u}}^{\mathrm{c}}$, discriminative loss $\mathcal{L}_{\mathrm{loc}}^{\mathrm{d}}$, and adversarial loss $\mathcal{L}_{\mathrm{loc}}^{\mathrm{adv}}$. To learn the UC classification, we minimize the classification loss $\mathcal{L}_{\mathrm{u}}^{\mathrm{c}}$, which is computed by taking the cross-entropy between the UC class label $\boldsymbol{u}_i^t$ and the UC class prediction $\boldsymbol{p}_{\mathrm{u}}(\boldsymbol{x}_i^t)$ that is output from C_{u}. The discriminative loss $\mathcal{L}_{\mathrm{loc}}^{\mathrm{d}}$ and adversarial loss $\mathcal{L}_{\mathrm{loc}}^{\mathrm{adv}}$ are used to learn the disentangled representation, and are formulated as follows:

$$\mathcal{L}_{\mathrm{loc}}^{\mathrm{d}}(\boldsymbol{x}_i^t) = -\sum_{j=1}^{K_{\mathrm{loc}}} l_i^t \log d_{\mathrm{loc}}^j(\boldsymbol{x}_i^t),\ \mathcal{L}_{\mathrm{loc}}^{\mathrm{adv}}(\boldsymbol{x}_i^t) = \sum_{j=1}^{K_{\mathrm{loc}}} \log d_{\mathrm{loc}}^j(\boldsymbol{x}_i^t), \tag{1}$$

where $\boldsymbol{d}_{\mathrm{loc}}(\boldsymbol{x}_i^t)$ is the location class prediction estimated by D_{loc}. By minimizing the discriminative loss $\mathcal{L}_{\mathrm{loc}}^{\mathrm{d}}$, the classification module D_{loc} is trained to classify the location. In contrast, the minimization of the adversarial loss $\mathcal{L}_{\mathrm{loc}}^{\mathrm{adv}}$ results in the UC feature $\boldsymbol{z}_{\mathrm{u}}$ that is non-discriminative with respect to the location. Note that $\mathcal{L}_{\mathrm{loc}}^{\mathrm{d}}$ is back-propagated only to D_{loc}, whereas the parameters of D_{loc} are frozen during the back-propagation of $\mathcal{L}_{\mathrm{loc}}^{\mathrm{adv}}$. As mentioned above, some images are not labeled for UC classification in this problem. Therefore, the classification loss $\mathcal{L}_{\mathrm{u}}^{\mathrm{c}}$ and the disentangle losses $\mathcal{L}_{\mathrm{u}}^{\mathrm{adv}}$ and $\mathcal{L}_{\mathrm{u}}^{\mathrm{d}}$ are ignored for UC-unlabeled images.

3.2 Order-Guided Learning

Order-guided learning considers the relationship between temporally adjacent images, as shown in Fig. 2(b). Since an endoscopic image is more likely to belong to the same UC class as its temporally adjacent images than the UC class of temporally distant images, the UC-dependent features of temporally adjacent images should be close to each other. To incorporate this assumption into learning of the network, the ordinal loss for order-guided learning is formulated as:

$$\mathcal{L}_{\mathrm{seq}}(\boldsymbol{x}_i^t, \boldsymbol{x}_i^{t+1}, \boldsymbol{x}_i^{t+2}) = \left[||\boldsymbol{z}_{\mathrm{u}}(\boldsymbol{x}_i^t) - \boldsymbol{z}_{\mathrm{u}}(\boldsymbol{x}_i^{t+1})||_2^2 - ||\boldsymbol{z}_{\mathrm{u}}(\boldsymbol{x}_i^t) - \boldsymbol{z}_{\mathrm{u}}(\boldsymbol{x}_i^{t+2})||_2^2 + \varepsilon\right]_+, \tag{2}$$

where $\boldsymbol{z}_{\mathrm{u}}(\boldsymbol{x}_i^t)$ is a UC feature vector for the sample $\boldsymbol{x}_i^t$ and is extracted via E_{enc} and B_{u}, $[\cdot]_+$ is a function that returns zero for a negative input and outputs the input directly otherwise, and ε is a margin that controls the degree of discrepancy between two temporally separated samples.

The UC features of temporally adjacent samples get closer by updating the network with the order-guided learning, as shown in Fig. 2(c). This warping in the UC feature space functions as a regularization that allows the network to make more correct predictions because the temporally adjacent images tend to belong to the same UC class. The order-guided learning can be applied without the UC label, and therefore it is also effective for the UC-unlabeled images.

4 Experimental Results

We conducted the UC classification experiment to evaluate the validity of the proposed method. In the experiment, we used an endoscopic image dataset collected from the Kyoto Second Red Cross Hospital. Participating patients were informed of the aim of the study and provided written informed consent before participating in the trial. The experiment was approved by the Ethics Committee of the Kyoto Second Red Cross Hospital.

4.1 Dataset

The dataset consists of 388 endoscopic image sequences, each of which contains a different number of images, comprising 10,262 images in total. UC and location labels were attached to each image based on annotations by medical experts. Out of 10,262 images, 6,678 were labeled as UC (positive) and the remaining 3,584 were normal (negative). There were three classes for the location label: right colon, left colon, and rectum. In the experiments, the dataset was randomly split into image sequence units, and 7,183, 2,052, and 1,027 images were used as training, validation, and test set, respectively. To simulate the limitation of the UC-labeled images, the labeled image ratio R for the training set used by the semi-supervised learning methods was set to 0.1.

4.2 Experimental Conditions

We compared the proposed method with two semi-supervised learning methods. One is the Pseudo-Label [7], which is one of the famous semi-supervised learning methods. The other is FixMatch [11], which is the state-of-the-art semi-supervised learning method for the general image classification task. Since the distribution of data differs greatly between general and endoscopic images, we changed the details of FixMatch to maximize its performance for UC classification. Specifically, strong augmentation was changed to weak augmentation, and weak augmentation was changed to rotation-only augmentation for processing unlabeled images. We also compared the proposed method with two classifiers trained with only labeled images in the training set with the labeled image ratio $R = 0.1$ and 1.0.

In addition, we conducted an ablation study to evaluate the effectiveness of the location label, disentangled representation learning, and order-guided learning. The best network parameter for each method was determined based on the accuracy of the validation set. We used precision, recall, F1 score, specificity, and accuracy as the performance measures.

4.3 Results

Table 1 shows the result of the quantitative performance evaluation for each method. Excluding specificity, the proposed method achieved the best performance for all performance measures. Although the specificity of the proposed method was the third-best, it was hardly different from that of the fully supervised classification. Moreover, we confirmed that the proposed method improved all measures of the classifier trained using only UC-labeled images in the training set with $R = 0.1$. In particular, the improvement in recall was confirmed only in the proposed method. Therefore, disentangled representation learning and order-guided learning, which use additional information other than UC information, were effective for improving UC classification performance.

Table 1. Quantitative performance evaluation. Labeled image ratio R represents the ratio of the UC-labeled images in the training set.

Method	R	Precision	Recall	F1	Specificity	Accuracy
Supervised learning	1.0	80.52	84.89	82.64	90.23	88.51
	0.1	69.16	67.07	68.10	85.78	79.75
Pseudo-Label [7]	0.1	75.19	61.33	67.55	90.37	81.10
FixMatch [11]	0.1	75.24	46.82	57.73	**92.67**	77.90
Proposed	0.1	**77.56**	**73.11**	**75.27**	89.94	**84.52**

Table 2. Results of the ablation study with the location label (Location), and disentangled representation learning (Disentangle), and order-guided learning (Order)

Location	Disentangle	Order	Precision	Recall	F1	Specificity	Accuracy
			69.16	67.07	68.10	85.78	79.75
✓			**79.50**	67.98	73.29	**91.67**	84.03
✓	✓		72.02	**73.11**	72.56	86.49	82.18
✓	✓	✓	77.56	**73.11**	**75.27**	89.94	**84.52**

Table 2 shows the results of the ablation study. The results demonstrated that each element of the proposed method was effective for improving the UC classification. The location information was effective for improving the precision with keeping the recall. Moreover, since the recall and the specificity were improved using the order-guided learning, temporal ordering information was useful for incorporating the order-related feature that cannot be learned by only the annotation to individual images.

To demonstrate the effect of the order-guided learning, the examples of prediction results were shown in Fig. 3. In this figure, the prediction results from the proposed method with the order-guided learning for temporally adjacent images tend to belong to the same class. For example, the proposed method predicted the first and second images from the right in Fig. 3(b) as the same class, whereas the proposed method without the order-guided learning predicted them as different classes.

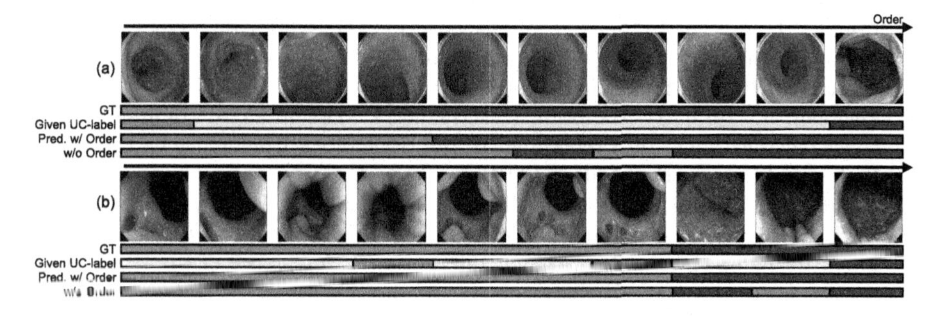

Fig. 3. Examples of the prediction results. Each bar represents the ground-truth labels, labels given during training, prediction results by the proposed method with and without order-guided learning. The red, blue, and gray bars represent UC, normal, and unlabeled images, respectively. (Color figure online)

5 Conclusion

We proposed a semi-supervised learning method for learning ulcerative colitis (UC) classification with limited UC-labels. The proposed method utilizes the location and temporal ordering information of endoscopic images to train the UC classifier. To obtain the features that separate the UC-dependent and location-dependent features, we introduced disentangled representation learning using location information. Moreover, to compensate for the limitation of UC-labeled data using temporal ordering information, we introduced order-guided learning, which considers the relationship between temporally adjacent images. The experimental results using endoscopic images demonstrated that the proposed method outperforms existing semi-supervised learning methods.

In future work, we will focus on extending the proposed framework to other tasks. Although this study applies the proposed method exclusively to UC classification, the proposed framework based on location and temporal ordering information can be applied to other tasks involving endoscopic images, such as the detection of polyps and cancer.

Acknowledgments. This work was supported by JSPS KAKENHI Grant Number JP20H04211 and AMED Grant Number JP20lk1010036h0002.

References

1. Arazo, E., Ortego, D., Albert, P., O'Connor, N.E., McGuinness, K.: Pseudo-labeling and confirmation bias in deep semi-supervised learning. In: Proceedings of the International Joint Conference on Neural Networks (2020)
2. Berthelot, D., Carlini, N., Goodfellow, I., Papernot, N., Oliver, A., Raffel, C.A.: MixMatch: a holistic approach to semi-supervised learning. In: Proceedings of the Advances in Neural Information Processing Systems, vol. 32 (2019)

3. Cao, K., Ji, J., Cao, Z., Chang, C.Y., Niebles, J.C.: Few-shot video classification via temporal alignment. In: Proceedings of the IEEE/CVF Conference on Computer Vision and Pattern Recognition (2020)
4. Dwibedi, D., Aytar, Y., Tompson, J., Sermanet, P., Zisserman, A.: Temporal cycle-consistency learning. In: Proceedings of the IEEE/CVF Conference on Computer Vision and Pattern Recognition (2019)
5. Harada, S., Hayashi, H., Bise, R., Tanaka, K., Meng, Q., Uchida, S.: Endoscopic image clustering with temporal ordering information based on dynamic programming. In: Proceedings of the Annual International Conference of the IEEE Engineering in Medicine and Biology Society, pp. 3681–3684 (2019)
6. Herp, J., Deding, U., Buijs, M.M., Kroijer, R., Baatrup, G., Nadimi, E.S.: Feature point tracking-based localization of colon capsule endoscope. Diagnostics **11**(2), 193 (2021)
7. Lee, D.H.: Pseudo-Label : the simple and efficient semi-supervised learning method for deep neural networks. In: Proceedings of the ICML 2013 Workshop: Challenges in Representation Learning (2013)
8. Liu, A.H., Liu, Y.C., Yeh, Y.Y., Wang, Y.C.F.: A unified feature disentangler for multi-domain image translation and manipulation. In: Proceedings of the Advances in Neural Information Processing Systems, vol. 31 (2018)
9. Liu, Y., Wei, F., Shao, J., Sheng, L., Yan, J., Wang, X.: Exploring disentangled feature representation beyond face identification. In: Proceedings of the IEEE/CVF Conference on Computer Vision and Pattern Recognition (2018)
10. Mori, K., et al.: A method for tracking the camera motion of real endoscope by epipolar geometry analysis and virtual endoscopy system. In: Proceedings of the Medical Image Computing and Computer-Assisted Intervention, pp. 1–8 (2001)
11. Sohn, K., et al.: FixMatch: simplifying semi-supervised learning with consistency and confidence. Proce. Adv. Neural Inf. Process. Syst. **33**, 596–608 (2020)

Semi-supervised Contrastive Learning for Label-Efficient Medical Image Segmentation

Xinrong Hu[1], Dewen Zeng[1], Xiaowei Xu[2], and Yiyu Shi[1](✉)

[1] University of Notre Dame, Notre Dame, IN 46556, USA
{xhu7,dzeng2,yshi4}@nd.edu
[2] Guangdong Provincial People's Hospital, Guangdong 510000, China

Abstract. The success of deep learning methods in medical image segmentation tasks heavily depends on a large amount of labeled data to supervise the training. On the other hand, the annotation of biomedical images requires domain knowledge and can be laborious. Recently, contrastive learning has demonstrated great potential in learning latent representation of images even without any label. Existing works have explored its application to biomedical image segmentation where only a small portion of data is labeled, through a pre-training phase based on self-supervised contrastive learning without using any labels followed by a supervised fine-tuning phase on the labeled portion of data only. In this paper, we establish that by including the limited label information in the pre-training phase, it is possible to boost the performance of contrastive learning. We propose a supervised local contrastive loss that leverages limited pixel-wise annotation to force pixels with the same label to gather around in the embedding space. Such loss needs pixel-wise computation which can be expensive for large images, and we further propose two strategies, downsampling and block division, to address the issue. We evaluate our methods on two public biomedical image datasets of different modalities. With different amounts of labeled data, our methods consistently outperform the state-of-the-art contrast-based methods and other semi-supervised learning techniques.

Keywords: Semi-supervised learning · Contrastive learning · Semantic segmentation · Label efficient learning

1 Introduction

Accurate semantic segmentation result is of great value to medical application, which provides physicians with the anatomical structural information for disease diagnosis and treatment. With the emergence of convolutional neural network,

Electronic supplementary material The online version of this chapter (https://doi.org/10.1007/978-3-030-87196-3_45) contains supplementary material, which is available to authorized users.

M. de Bruijne et al. (Eds.): MICCAI 2021, LNCS 12902, pp. 481–490, 2021.
https://doi.org/10.1007/978-3-030-87196-3_45

supervised deep learning has achieved state-of-the-art performance in many biomedical image segmentation tasks, including different organs and different modalities [6,13,15,16]. These methods all need abundant labeled data with the class of every pixel or voxel known to guide the training. However, it is difficult to collect densely labeled medical images, because labeling medical images require domain-specific knowledge and pixel-wise annotations can be time-consuming.

To deal with this issue, various methods have been proposed. One branch of them is data augmentation, which expands labeled dataset by generative adversarial networks (GAN) [9] for data synthesis [3,5,8] or by simple linear combination[17]. However, data augmentation only leads to limited improvement of segmentation performance due to the defect of artificially generated images and labels. As a better alternative, semi-supervised learning based approaches [2,12,14,18] can utilize both labeled data and unlabeled data efficiently. Recently, contrastive learning [7,10], by forcing the embedding features of similar images to be close in the latent space and those of dissimilar ones to be apart, achieved state-of-the-art in self-supervised classification problems. The powerful ability of extracting features that are useful for downstream tasks from unlabeled data makes it a great candidate for label efficient learning.

However, most existing contrastive learning methods target image classification tasks. Semantic segmentation, on the other hand, requires pixel-wise classification. [4] first attempted at using contrastive learning to improve segmentation accuracy when only part of the data is labeled. In the pre-training phase, a two-step self-supervised contrastive learning scheme is used to learn both global and local features from unlabeled data. Specifically, it first projects a 2D slice to latent space with the encoder path only and computes a global contrastive loss, similar to what has been used for image classification problem. Then based on the trained encoder, decoder blocks are added and the intermediate feature maps are used to perform local contrastive learning at pixel level. Only a small number of points are selected from fixed positions in the feature maps. Next, in the fine-tuning phase the pre-trained model from contrastive learning is trained with supervision on the labeled part of the data. In such a framework, the label information is never utilized in the pre-training phase. Yet intuitively, by including such information to provide certain supervision, the features can be better extracted by contrastive learning.

Following this philosophy, in this work we propose a semi-supervised framework consisting of self-supervised global contrast and supervised local contrast to take advantage of the available labels. Compared with the unsupervised local contrast in the literature, the supervised one can better enforce the similarity between features within the same class and discriminate those that are not. In addition, we use two strategies, downsampling and block division, to address the high computation complexity associated with the supervised local contrastive loss. We conduct experiments on two public medical image segmentation datasets of different modalities. We find that, for different amount of labeled data, our framework can always generate segmentation results with higher Dice than the state-of-the-art methods.

2 Methods

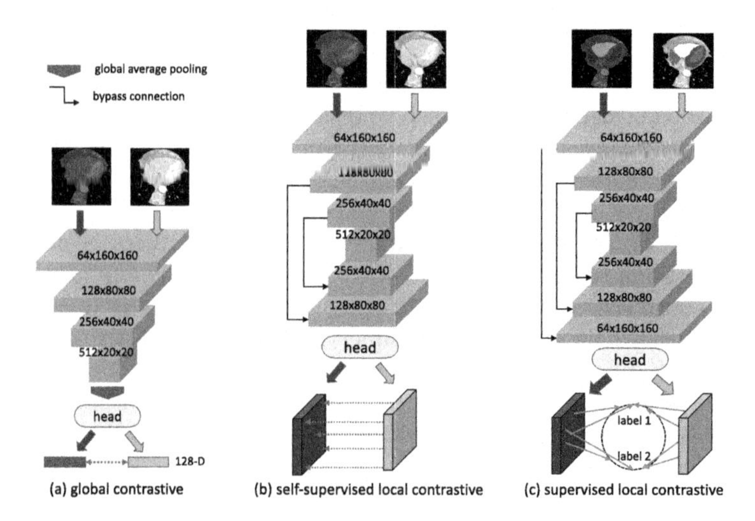

Fig. 1. Illustration of (a) self-supervised global contrastive loss, (b) self-supervised local contrastive loss used in existing works, and (c) our proposed supervised local contrastive loss.

2.1 Overview of the Semi-supervised Framework

The architecture backbone of a deep segmentation network usually consists of two parts, the encoder E and the decoder D. The encoder comprises several downsampling blocks to extract features from input images and the decoder path deploys multiple upsampling blocks to restore the resolution for pixel-wise prediction. In the pre-training stage, our framework contains a self-supervised global contrastive learning step which only uses unlabeled data to train E and captures image-level features, followed by a supervised local contrastive learning step which uses the limited labeled data to further train both E and D and captures pixel-level features.

The global contrastive learning is similar to what is used in the literature (e.g. SimCLR [7]) and for the sake of completeness we briefly describe it here with an illustration in Fig. 1(a). Given an input batch $B = \{x_1, x_2, ..., x_b\}$, in which x_i represents an input 2D slice, random combinations of image transformations $aug(\cdot)$ are performed twice on each image in B to form a pair of augmented images, same as the existing contrastive learning methods. These augmented images form an augmented image set A. Let $a_i, i \in I = \{1...2b\}$ represent an augmented image in A and let j(i) be the index of the other image in A that is derived from the same image in B as a_i, i.e., a_i and $a_{j(i)}$ is an augmented pair. The global contrastive loss then takes the following form

$$L_g = -\frac{1}{|A|}\sum_{i\in I} log\frac{exp(z_i \cdot z_{j(i)}/\tau)}{\sum_{k\in I-\{i\}} exp(z_i \cdot z_k)/\tau)}, \quad (1)$$

where z_i is the normalized output feature by adding a header $h_1(\cdot)$ to encoder E. The header is usually a multi-layer perceptron that enhances the representing ability of extracted features from E. Mathematically put, $z_i = |h_1(E(a_i))|$ and $z_{j(i)} = |h_1(E(a_{j(i)}))|$. τ is the temperature constant.

Only teaching the encoder to extract image-level features is not enough, since segmentation requires the prediction of all pixels in the input image. After the global contrastive learning, we attach the decoder path D to the encoder and retrain the whole network with supervised local contrastive loss. This is where our work differs from existing works and the details will be described in the following section. The two-step pre-training stage will eventually generate a model that can be used as an initialization for the subsequent fine-tuning stage on the labeled data to boost segmentation accuracy.

2.2 Supervised Local Contrastive Learning

The goal of local contrastive learning is to train the decoder to extract distinctive local representations. Let $f^l(\tilde{x}_i) = h_2(D_l(E(a_i)))$ be the output feature map of the l-th uppermost decoder block D_l for an augmented input a_i, where h_2 is a two-layer point-wise convolution. For feature map $f(a_i)$, the local contrastive loss is defined as

$$loss(a_i) = -\frac{1}{|\Omega|}\sum_{(u,v)\in\Omega}\frac{1}{|P(u,v)|} log\frac{\sum_{(u_p,v_p)\in P(u,v)} exp(f^l_{u,v} \cdot f^l_{u_p,v_p}/\tau)}{\sum_{(u',v')\in N(u,v)} exp(f^l_{u,v} \cdot f^l_{u',p'}/\tau)} \quad (2)$$

$f^l_{u,v} \in \mathbb{R}^c$, where c is the channel number, stands for the feature at the u-th column and v-th row of the feature map. Ω is the set of points in $f_{u,v}$ that are used to compute the loss. $P(u,v)$ and $N(u,v)$ denote the positive set and negative set of $f_{u,v}(a_i)$, respectively. Then the overall local contrastive loss is

$$L_l = \frac{1}{|A|}\sum_{a_i\in A} loss(a_i) \quad (3)$$

In self-supervised setting, [4] only selects 9 or 13 points from the feature map as illustrated by Fig. 2(b). Without annotation information, $P(u,v)$ only contains points from the same position at the feature maps of the paired images, that is $f_{u,v}(a_{j(i)})$, and the negative set is the union of Ω of all images in A minus the positive set.

According to SimCLR, when applying contrastive learning to image classification tasks, the data augmentation that boosts the accuracy most is the spatial transformation, such as crop and cutout. Yet in self-supervised setting, the transformation $aug(\cdot)$ can only include intensity transformation such as color distort, Gaussian noise and Gaussian blur. It is because after spatial transformation, features at the same position of $f(a_i$ and $f(a_{j(i)})$ could originate from different points or areas but would still be treated as similar when calculating (2). Moreover, without any supervision, chances are that the selected points in Ω might all correspond to the background area, in which case the model should fail to effectively learn local features.

Inspired by [11], since the dataset contains limited labeled data anyway to be used in the fine-tuning stage, we can make use of it in the pre-training stage through a supervised local contrastive learning. Observing that the last point-wise convolution in U-Net functions similarly as the fully connected layer of classification network, we view the feature maps $f^1 \in \mathbb{R}^{c \times h \times h}$ just before the 1×1 convolutional layer as the projections of all pixels into the latent space, where c equals the channel number that is also the dimension of latent space and h is the input image dimension. Given a point (u, v) in the feature map, the definition of positive set $P(u, v)$ is all features in the feature maps of A that share the same annotation as (u, v). The negative set is then all features in the feature maps of A that are labeled as different classes. In addition, two embedding features of background pixels make up the majority of positive pairs and the subsequent segmentation task benefits less from comparing features as these pairs. Hence, for supervised setting, we change the definition of Ω to only contain points with non-background annotation. Please note that background piexls are still included in the negative set $N(u, v)$. Then the supervised contrastive loss takes the same form as (2), except for the new meaning of $\Omega, P(u, v), N(u, v)$ as discussed above. With the guidance of labels, supervised contrastive loss provides additional information on the similarity of features derived from the same class and dissimilarity of inter-class features.

2.3 Downsampling and Block Division

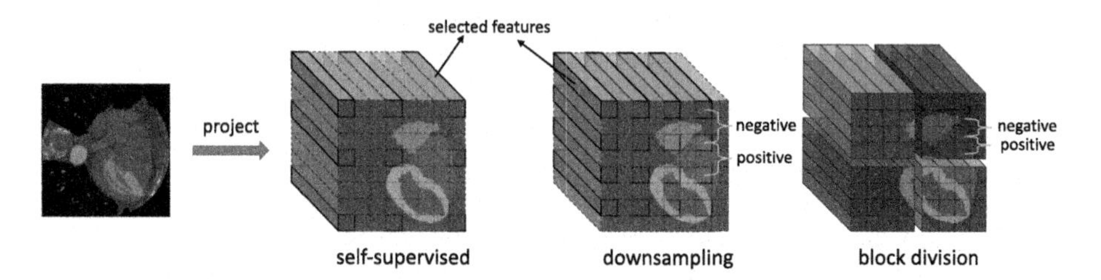

Fig. 2. Illustration of three strategies to select features from feature maps for local contrastive learning. Left: self-supervised as in [4]. Middle: downsampling where the contrastive loss is computed with features separated with fixed stride; Right: block division where contrastive loss is computed within each block and then averaged.

As we try to utilize the segmentation annotation, the feature map is of the same size as the input image. Though we only consider the points that do not belong to background, $|P(u,v)|$ can still be very large, that is $O(h^2)$ with input image dimension $h \times h$. The overall computation complexity for the superivsed local contrastive loss is then $O(h^4)$. For example, the image size in MMWHS is 512×512. Even after we reshape the slices to 160×160, the size of negative set is on the order of 10^4 and the total multiplications needed is on the order of 10^8 to evaluate the loss. To address this issue, we propose two strategies to reduce the size of the negative set as illustrated in Fig. 2. The first one is to downsample the feature map with fixed stride. The idea behind this method is that neighboring pixels contain redundant information and their embedded features appear quite similar in the latent space. With stride number s, the size of $P(u,v)$ and the computation complexity of the loss function are reduced by a factor of s^2 and s^4 respectively. As larger stride leads to fewer negative pairs and the number of negative pairs is important for contrastive learning [10]. In this work, we set the stride to 4, the minimum value that will not cause out-of-memory (OOM) issue on our experiment platform.

Alternatively, we can divide the feature maps into several blocks of the same size $h' \times h'$ as shown in Fig. 2. The local contrastive loss is calculated within each block only and then averaged across all the blocks. If all the pixels in a block are labeled as background, we will not compute its local contrastive loss. The size of $P(u,v)$ decreases to the number of features in a block $O(h'^2)$ and the total computation complexity is $O((h/h')^2 \cdot h'^4) = O(h^2 h'^2)$. Smaller block size will also lead to fewer negative pairs. Therefore the block size should also be determined based on the acceptable training time/memory consumption. In this paper, we set it as 16×16, the maximum value that will not cause OOM.

3 Experiments and Results

Dataset and Preprocessing. We evaluate our methods on two public medical image segmentation datasets of different modalities, MRI and CT. (1) Hippocampus dataset is from the MICCAI 2018 medical segmentation decathlon challenge[1]. There are 260 3D volumes of mono-modal MRI scans with the annotation of anterior and posterior hippocampus body. (2) MMWHS [19] comes from the MICCAI 2017 challenge, which consists of 20 3D cardiac CT volumes. The annotations include seven structures of heart: left ventricle, left atrium, right ventricle, right atrium, myocardium, ascending aorta, and pulmonary artery. Regarding preprocessing, we normalize the intensity for the whole volume. Then we resize the slices of all volumes in the same dataset to unify them using bilinear interpolation. The resolutions of the two reshaped dataset are 64×64 for Hippocampus and 160×160 for MMWHS.

Experiment Settings. Similar to [4], the segmentation network backbone we choose is 2D U-Net with three blocks for both encoder and decoder path, the implementation of which is based on the PyTorch library. For both sets, we divide the data into three parts, training set X_{tr}, validation set X_{vl}, and test set X_{ts}.

Table 1. Comparison between state-of-the-art methods and the proposed methods w.r.t. subsequent segmentation dice scores on two datasets. Ablation studies of our methods (w/o global contrast) are also included.

Methods	Hippocampus			MMWHS		
	% of data labeled in X_{tr}			% of data labeled in X_{tr}		
	5	10	20	10	20	40
Random	0.788	0.833	0.852	0.328	0.440	0.715
Global [7]	0.817	0.834	0.861	0.359	0.487	0.724
Global+local (self) [4]	0.808	0.843	0.858	0.367	0.490	0.730
Mixup [17]	0.818	0.847	0.861	0.365	0.541	0.755
TCSM [14]	0.796	0.838	0.855	0.347	0.489	0.733
Local(stride)	0.818	0.845	0.860	0.354	0.485	0.743
Local(block)	0.817	0.843	0.862	0.366	0.475	0.736
Global+local(stride)	0.822	**0.851**	0.863	**0.384**	0.525	0.758
Global+local(block)	**0.824**	0.849	**0.866**	0.382	**0.553**	**0.764**

The ratio $|X_{tr}|$:$|X_{vl}|$:$|X_{ts}|$ for Hippocampus is 3:1:1 and 2:1:1 for MMWHS. Then we randomly choose certain number of samples in X_{tr} as labeled volumes and the rest are viewed as unlabeled. We split the dataset four times independently to generate four different folds and the average dice score on X_{ts} is used as the metric for comparison. We choose Adam as the optimizer for both contrastive learning and the following segmentation training. The learning rate is 0.0001 for contrastive learning and is 0.00001 for segmentation training. The training epoch for contrastive learning is 70, and for segmentation training, it stops after 120 epochs. We run experiments of Hippocampus dataset on two NVIDIA GTX 1080 gpus and the experiments of MMWHS dataset on two NVIDIA Tesla P100 gpus.

We implement four variants of the proposed methods: the two-step global contrast and supervised local contrast with downsampling (global+local(stride)) and block division (global+local(block)); and for the purpose of ablation study, these two methods without global contrast (local(stride) and local(block)). For comparison, we implement two state-of-the-art contrast based methods: a global contrastive learning method [7] (global), and a self-supervised global and local contrastive learning method [4] (global+local(self)). Note that the entire training data is used but without any labels for global contrast and self-supervised local contrast, while only labeled training data is used for our supervised local contrast. We further implement three non-contrast baseline methods: training from scratch with random initialization (random), a data augmentation based method Mixup [17], and a state-of-the-art semi-supervised learning method TCSM [14].

Table 1 lists the average dice scores of all the methods. There are several important observations we can make from the table: First, all contrastive learning methods lead to higher dice scores compared to the random approach, which implies that contrastive learning is a feasible initialization when labels are limited. Second, we notice that the proposed supervised local contrast alone

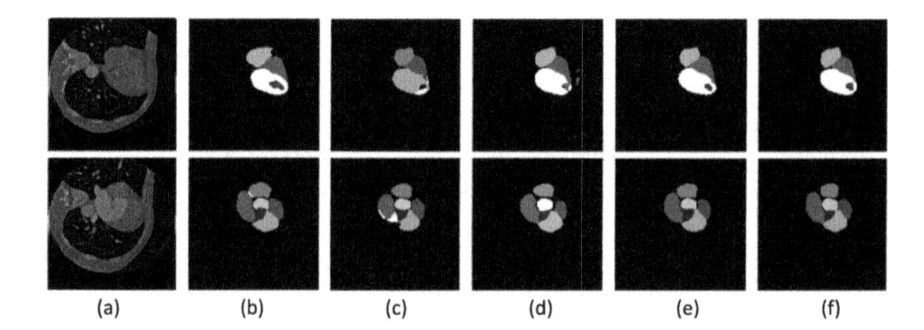

Fig. 3. Visualization of segmentation results on MMWHS with 40% labeled data. (a) and (b) represent input image and ground truth. (c)–(f) are the predictions of random, global+local(self), global+local(stride), and global+local(block). Different color represents different structure categories. (Better to view in colors)

(local(stride) and local(block)) generates segmentation results comparable with those from existing contrast based methods. Considering that the former only makes use of the labeled data while the latter leverage all volumes, the supervised contrastive loss efficiently teaches the model the pixel-wise representations shared by the same class. Lastly, combining the supervised local contrast with the global contrast, the semi-supervised setting gains a considerable improvement of segmentation performance over the state-of-the-art. It demonstrates that the image-level representation extracted from unsupervised global contrast is complementary to the pixel-level representation from the supervised local contrast.

Finally, the segmentation results on MMWHS with 40% labeled data from our methods, the state-of-the-art (global+local(self) [4]), and the random approach are shown in Fig. 3. From the figure we can clearly see that our approaches perform the best. With the same setting, the embedding features after applying t-SNE for these methods are visualized in Fig. 4. From the figure we can see that compared with the random approach, the features for the same class are tightly clustered and those for different classes are clearly separated for the global+local(block) method, while those from the state-of-the-art are more scattered. This aligns with the results in Table 1 where global+local(block) method achieves the highest Dice (rightmost column).

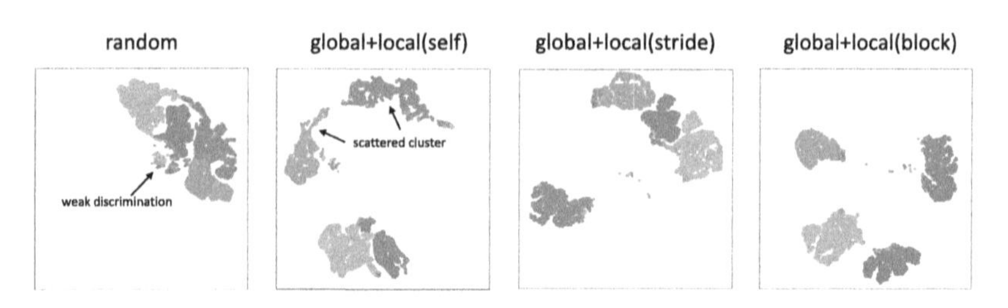

Fig. 4. Visualization of embedding features after applying t-SNE on MMWHS with 40% labeled data. Different color represents different classes. The features are all from the feature map before the last point-wise convolution. (Better to view in colors.)

4 Conclusion

The limited annotation is always a hurdle for the development of deep learning methods for medical image segmentation. Aiming at tackling this challenge, in this work, we propose a supervised local contrastive loss to learn the pixel-wise representation. Further, we put forward two strategies to reduce the computational complexity when computing the loss. Experiments on two public medical image datasets with only partial labels show that when combining the proposed supervised local contrast with global contrast, the resulting semi-supervised contrastive learning achieves substantially improved segmentation performance over the state-of-the-art.

Acknowledgement. This work is partially supported by NSF award IIS-2039538.

References

1. Medical segmentation decathlon challenge (2018). http://medicaldecathlon.com/index.html
2. Bai, W., et al.: Semi-supervised learning for network-based cardiac MR image segmentation. In: Descoteaux, M., Maier-Hein, L., Franz, A., Jannin, P., Collins, D.L., Duchesne, S. (eds.) MICCAI 2017. LNCS, vol. 10434, pp. 253–260. Springer, Cham (2017). https://doi.org/10.1007/978-3-319-66185-8_29
3. Bowles, C., et al.: GAN augmentation: augmenting training data using generative adversarial networks. arXiv preprint arXiv:1810.10863 (2018)
4. Chaitanya, K., Erdil, E., Karani, N., Konukoglu, E.: Contrastive learning of global and local features for medical image segmentation with limited annotations. arXiv preprint arXiv:2006.10511 (2020)
5. Chaitanya, K., Karani, N., Baumgartner, C.F., Becker, A., Donati, O., Konukoglu, E.: Semi-supervised and task-driven data augmentation. In: Chung, A.C.S., Gee, J.C., Yushkevich, P.A., Bao, S. (eds.) IPMI 2019. LNCS, vol. 11492, pp. 29–41. Springer, Cham (2019). https://doi.org/10.1007/978-3-030-20351-1_3
6. Chen, L.C., Papandreou, G., Kokkinos, I., Murphy, K., Yuille, A.L.: DeepLab: Semantic image segmentation with deep convolutional nets, atrous convolution, and fully connected CRFs. IEEE Trans. Pattern Anal. Mach. Intell. **40**(4), 834–848 (2017)
7. Chen, T., Kornblith, S., Norouzi, M., Hinton, G.: A simple framework for contrastive learning of visual representations. In: International Conference on Machine Learning, pp. 1597–1607. PMLR (2020)
8. Costa, P., et al.: End-to-end adversarial retinal image synthesis. IEEE Trans. Med. Imag. **37**(3), 781–791 (2017)
9. Goodfellow, I.J., et al.: Generative adversarial networks. arXiv preprint arXiv:1406.2661 (2014)
10. He, K., Fan, H., Wu, Y., Xie, S., Girshick, R.: Momentum contrast for unsupervised visual representation learning. In: Proceedings of the IEEE/CVF Conference on Computer Vision and Pattern Recognition, pp. 9729–9738 (2020)
11. Khosla, P., et al.: Supervised contrastive learning. arXiv preprint arXiv:2004.11362 (2020)

12. Li, S., Zhang, C., He, X.: Shape-aware semi-supervised 3D semantic segmentation for medical images. In: Martel, A.L., et al. (eds.) MICCAI 2020. LNCS, vol. 12261, pp. 552–561. Springer, Cham (2020). https://doi.org/10.1007/978-3-030-59710-8_54
13. Li, X., Chen, H., Qi, X., Dou, Q., Fu, C.W., Heng, P.A.: H-DenseUNet: hybrid densely connected UNet for liver and tumor segmentation from CT volumes. IEEE Trans. Med. Imag. **37**(12), 2663–2674 (2018)
14. Li, X., Yu, L., Chen, H., Fu, C.W., Xing, L., Heng, P.A.: Transformation-consistent self-ensembling model for semisupervised medical image segmentation. IEEE Trans. Neural Networks Learn. Syst. **42**, 523 (2020)
15. Milletari, F., Navab, N., Ahmadi, S.A.: V-Net: fully convolutional neural networks for volumetric medical image segmentation. In: 2016 Fourth International Conference on 3D Vision (3DV), pp. 565–571. IEEE (2016)
16. Ronneberger, O., Fischer, P., Brox, T.: U-Net: convolutional networks for biomedical image segmentation. In: Navab, N., Hornegger, J., Wells, W.M., Frangi, A.F. (eds.) MICCAI 2015. LNCS, vol. 9351, pp. 234–241. Springer, Cham (2015). https://doi.org/10.1007/978-3-319-24574-4_28
17. Zhang, H., Cisse, M., Dauphin, Y.N., Lopez-Paz, D.: Mixup: beyond empirical risk minimization. arXiv preprint arXiv:1710.09412 (2017)
18. Zhang, Y., Yang, L., Chen, J., Fredericksen, M., Hughes, D.P., Chen, D.Z.: Deep adversarial networks for biomedical image segmentation utilizing unannotated images. In: Descoteaux, M., Maier-Hein, L., Franz, A., Jannin, P., Collins, D.L., Duchesne, S. (eds.) MICCAI 2017. LNCS, vol. 10435, pp. 408–416. Springer, Cham (2017). https://doi.org/10.1007/978-3-319-66179-7_47
19. Zhuang, X., Shen, J.: Multi-scale patch and multi-modality atlases for whole heart segmentation of MRI. Med. Image Anal. **31**, 77–87 (2016)

Functional Magnetic Resonance Imaging Data Augmentation Through Conditional ICA

Badr Tajini, Hugo Richard, and Bertrand Thirion(✉)

Inria, CEA, Université Paris-Saclay, Gif-sur-Yvette, France
bertrand.thirion@inria.fr

Abstract. Advances in computational cognitive neuroimaging research are related to the availability of large amounts of labeled brain imaging data, but such data are scarce and expensive to generate. While powerful data generation mechanisms, such as Generative Adversarial Networks (GANs), have been designed in the last decade for computer vision, such improvements have not yet carried over to brain imaging. A likely reason is that GANs training is ill-suited to the noisy, high-dimensional and small-sample data available in functional neuroimaging. In this paper, we introduce Conditional Independent Components Analysis (Conditional ICA): a fast functional Magnetic Resonance Imaging (fMRI) data augmentation technique, that leverages abundant resting-state data to create images by sampling from an ICA decomposition. We then propose a mechanism to condition the generator on classes observed with few samples. We first show that the generative mechanism is successful at synthesizing data indistinguishable from observations, and that it yields gains in classification accuracy in brain decoding problems. In particular it outperforms GANs while being much easier to optimize and interpret. Lastly, Conditional ICA enhances classification accuracy in eight datasets without further parameters tuning.

Keywords: Conditional ICA · Data generation · Decoding studies

1 Introduction

As a non-invasive brain imaging technique, task fMRI records brain activity while participants are performing specific cognitive tasks. Univariate statistical methods, such as general linear models (GLMs) [7] have been successfully applied to identifying the brain regions involved in specific tasks. However such methods

B. Tajini and H. Richard—These authors contributed equally to this work.

Electronic supplementary material The online version of this chapter (https://doi.org/10.1007/978-3-030-87196-3_46) contains supplementary material, which is available to authorized users.

M. de Bruijne et al. (Eds.): MICCAI 2021, LNCS 12902, pp. 491–500, 2021.
https://doi.org/10.1007/978-3-030-87196-3_46

do not capture well correlations and interactions between brain-wide measurements. By contrast, classifiers trained to *decode* brain maps, i.e. to discriminate between specific stimulus or task types [15,25,27], take these correlations into account. The same framework is also popular for individual imaging-based diagnosis.

However, the large sample-complexity of these classifiers currently limits their accuracy. To tackle this problem, data generation is an attractive approach, as it could potentially compensate for the shortage of data. Generative Adversarial Networks (GANs) are promising generative models [8]. However, GANs are ill-suited to the noisy, high-dimensional and small-sample data available in functional neuroimaging. Furthermore the training of GANs is notoriously unstable and there are many hyper-parameters to tune.

In this work, we introduce Conditional ICA: a novel data augmentation technique using ICA together with conditioning mechanisms to generate surrogate brain imaging data and improve image classification performance. Conditional ICA starts from a generative model of resting state data (unconditional model), that is fine-tuned into a conditional model that can generate task data. This way, the generative model for task data benefits from abundant resting state data and can be trained with few labeled samples. We first show that the generative model of resting state data shipped in Conditional ICA produces samples that neither linear nor non-linear classifiers are able to distinguish. Then we benchmark Conditional ICA as a generative model of task data against various augmentation methods including GANs and conditional GANs on their ability to improve classification accuracy on a large task fMRI dataset. We find that Conditional ICA yields highest accuracy improvements. Lastly, we show on 8 different datasets that the use of Conditional ICA results in systematic improvements in classification accuracy ranging from 1% to 5%.

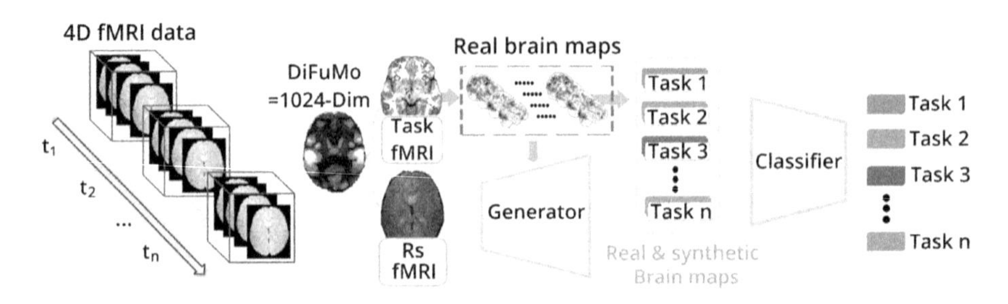

Fig. 1. Conditional ICA approach. Our method aims to generate surrogate data from Task and Rest fMRI data by synthesizing statistical maps that qualitatively fit the distribution of the original maps. These can be used to improve the accuracy of machine learning models that identify contrasts from the corresponding brain activity patterns.

2 Methods

Notations. We write matrices as bold capital letters, vectors as small bold letters. $\mathbf{X}^\dagger$ refers to the Moore-Penrose pseudo inverse of matrix $\mathbf{X}$, $\text{tr}(\mathbf{X})$ refers to the trace of matrix $\mathbf{X}$ and $\mathbf{I}_k$ refers to the identity matrix in $\mathbb{R}^{k,k}$. $\mathbf{0}_k$ refers to the null vector in $\mathbb{R}^k$.

Spatial Dimension Reduction. The outline of the proposed approach is presented in Fig. 1. While brain maps are high-dimensional, they span a smaller space than that of the voxel grid. For the sake of tractability, we reduce the dimension of the data by projecting the voxel values on the high-resolution version of the Dictionaries of Functional Modes *DiFuMo* atlas [3], i.e. with $p = 1024$ components. The choice of dimension reduction technique generally has an impact on the results. However we consider this question to be out of the scope of the current study and leave this to future work.

Unconditional Generative Models (Resting State Data Generation). Given a large-scale resting-state dataset $\mathbf{X}^{rest}$ in $\mathbb{R}^{p,n}$ where n is the number of images (samples) and $p = 1024$ the number of components in the atlas, let us consider how to learn its distribution. Assuming a Gaussian distribution is standard in this setting, yet, as shown later, it misses key distributional features. Moreover, we consider a model that subsumes the distribution of any type of fMRI data (task or rest): a linear mixture of $k \leq p$ independent temporal signals. We therefore use temporal ICA to learn a dimension reduction and unmixing matrix $\mathbf{W}^{rest} \in \mathbb{R}^{k,p}$ such that the k sources i.e. the k components of $\mathbf{S}^{rest} = \mathbf{W}^{rest}\mathbf{X}^{rest}$ are as independent as possible.

A straightforward method to generate new rest data would be to independently sample them from the distribution of the sources. This is easy because such distribution has supposedly independent marginals. We apply an invertible quantile transform q^{rest} to the sources S^{rest} so that the distribution of $\mathbf{z}^{rest} = q^{rest}(\mathbf{s}^{rest})$ has standardized Gaussian marginals. Since the distribution of $\mathbf{z}^{rest}$ has independent marginals, it is given by $\mathcal{N}(\mathbf{0}_k, I_k)$ from which we can easily sample. As shown later, this approach fails: such samples are still separable from actual rest data.

We hypothesize that this is because independence does not hold, and thus a latent structure among the marginals of the source distribution has to be taken into account. Therefore we assume that the distribution of $\mathbf{z}^{rest}$ is given by $\mathcal{N}(\mathbf{0}_k, \boldsymbol{\Lambda}^{rest})$ where $\boldsymbol{\Lambda}^{rest}$ is a definite positive matrix. $\boldsymbol{\Lambda}^{rest}$ can easily be learned from a standard shrunk covariance estimator: $\boldsymbol{\Lambda}^{rest} = \boldsymbol{\Sigma}^{rest}(1-\alpha) + \alpha \text{tr}(\boldsymbol{\Sigma}^{rest})\mathbf{I}_k$ where α is given by the Ledoit-Wolf formula [14] and $\boldsymbol{\Sigma}^{rest}$ is the empirical covariance of $\mathbf{Z}^{rest}$.

Our encoding model for rest data is therefore given by $\mathbf{Z}^{rest} = q^{rest}(\mathbf{W}^{rest}\mathbf{X}^{rest})$ and we assume that the distribution of $\mathbf{Z}^{rest}$ is $\mathcal{N}(\mathbf{0}_k, \boldsymbol{\Lambda}_k)$. The generative model is given by the pseudo inverse of the encoding model: $\tilde{\mathbf{X}}^{rest} = (\mathbf{W}^{rest})^\dagger (q^{rest})^{-1}(\boldsymbol{\varepsilon})$ where $\boldsymbol{\varepsilon} \sim \mathcal{N}(\mathbf{0}_k, \Lambda^{rest})$.

Conditional Generative Models (Generative Model for Task Data). While resting state datasets have a large number of samples ($10^4 \sim 10^5$), task datasets have a small number of samples ($10 \sim 10^2$). As a result, there are too few samples to learn high quality unmixing matrices. Therefore, using the unmixing matrix $\mathbf{W}^{rest}$ learned from the resting state data, we rely on the following nonlinear generative model for brain maps in a certain class c:

$$\mathbf{x}_c = (\mathbf{W}^{rest})^{\dagger} q^{-1}(\boldsymbol{\varepsilon}) \tag{1}$$

with $\boldsymbol{\varepsilon} \sim \mathcal{N}(\boldsymbol{\mu}_c, \boldsymbol{\Lambda})$.

In order to maximize the number of samples used to learn the parameters of the model, we assume that the quantile transform q and the latent covariance $\boldsymbol{\Lambda}$ do not depend on the class c. However, the mean $\boldsymbol{\mu}_c$, that can be learned efficiently using just a few tens of samples, depends on class c. An overview of our generative method is shown in Fig. 2.

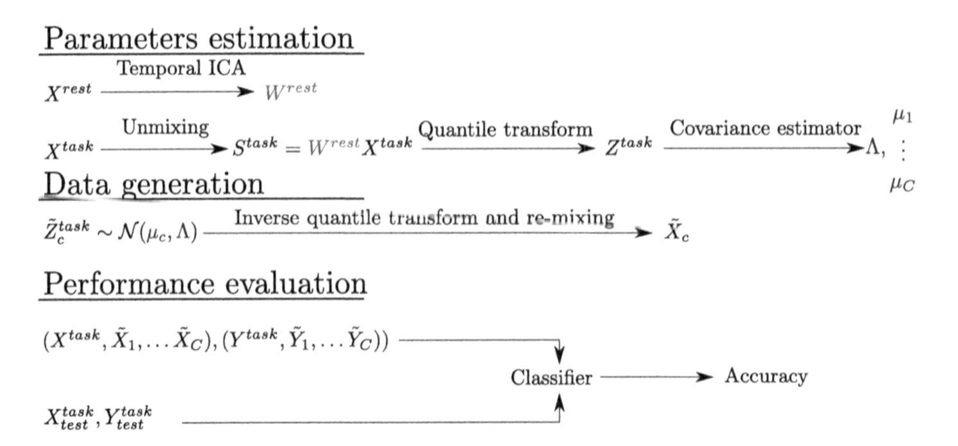

Fig. 2. Conditional ICA approach in depth. The approach proceeds by learning a temporal ICA of rest data $\mathbf{X}^{rest} \in \mathbb{R}^{p,n}$, resulting in independent sources and unmixing matrix $\mathbf{W}^{rest} \in \mathbb{R}^{k,p}$. Applying the unmixing matrix to the task data, we obtain samples in the source space $\mathbf{S}^{task} \in \mathbb{R}^{k,n}$. Afterwards, we map $\mathbf{S}^{task}$ to a normal distribution, yielding $\mathbf{Z}^{task} \in \mathbb{R}^{k,n}$. Then, we estimate the covariance $\boldsymbol{\Lambda} \in \mathbb{R}^{k,k}$ (all classes are assumed to have the same covariance) and the class-specific means $\boldsymbol{\mu}_1, \dots, \boldsymbol{\mu}_C \in \mathbb{R}^k$ according to Ledoit-Wolf's method. For each class c, we can draw random samples $\tilde{\mathbf{Z}}_c^{task} \in \mathbb{R}^{k,n_{\text{fakes}}}$ from the resulting multivariate Gaussian distribution $\mathcal{N}(\boldsymbol{\mu}_c, \boldsymbol{\Lambda})$ and obtain fake data $\tilde{\mathbf{X}}_c \in \mathbb{R}^{p,n_{\text{fakes}}}$ by applying the inverse quantile transform and re-mixing the data using the pseudo inverse of the unmixing matrix. We append these synthetic data to the actual data to create our new augmented dataset on which we train classifiers.

3 Related Work

In image processing, data augmentation is part of standard toolboxes and typically includes operations like cropping, rotation, translation. On fMRI data these methods do not make much sense as brain images are not invariant to such transformations. More advanced techniques [29] are based on generative models such as GANs or variational auto-encoders [12]. Although GAN-based method are powerful they are slow and difficult to train [2]. In appendix Table 7, we show that Conditional ICA is several orders of magnitude faster than GAN-based methods.

Our method is not an adversarial procedure. However it relates to other powerful generative models such as variational auto-encoders [12] with which it shares strong similarities. Indeed the analog of the encoding function in the variational auto-encoder is given by $e(\mathbf{x}) = \boldsymbol{\Lambda}^{-\frac{1}{2}} q(\mathbf{W}^{rest}\mathbf{x})$ in our model and the analog to the decoding function in the variational auto-encoder is given by $d(\mathbf{z}) = (\mathbf{W}^{rest})^{\dagger} q^{-1}(\boldsymbol{\Lambda}^{\frac{1}{2}}\mathbf{z})$ in our model. As in the variational auto-encoder, e approximately maps the distribution of the data to a standardized Gaussian distribution, while the reconstruction error defined by the difference in l2 norm $\|d(e(\mathbf{x})) - \mathbf{x}\|_2^2$ must remain small.

Another classical generative model related to ours is given by normalizing flows [13]. When $\mathbf{W}^{rest}$ is squared (no dimension reduction in ICA), the decoding operator d is invertible (its inverse is e) making our model an instance of normalizing flows. A great property is thus the simplicity and reduced cost of data generation.

Software Tools. We use the FastICA algorithm [11] to perform independent component analysis and use $k = 900$ components. We use Nilearn [1] for fMRI processing, Scikit-learn [19], Numpy [10] and Scipy [28] for data manipulation and machine learning tools.

Code Availability. Our code is available on the following git repository: https://github.com/BTajini/augfmri.

4 Experiments

4.1 Dataset, Data Augmentation Baselines and Classifiers Used

The unmixing matrices are learned on the rest HCP dataset [26] using 200 subjects. These data were used after standard preprocessing, including linear detrending, band-pass filtering ($[0.01, 0.1]$ Hz) and standardization of the time courses. The other 8 datasets [18,20–24,26] are obtained from the Neurovault repository [9]. The classes used in each dataset correspond to the activation maps related to the contrasts (such as "face vs tools") present in the set of tasks of each dataset. Details are available in appendix Table 3.

We consider 5 alternative augmentation methods: *ICA*, *Covariance*, *ICA + Covariance*, *GANs* and *CGANs*. When no augmentation method is applied we use the label *Original*.

The *ICA* method applies ICA to $\mathbf{X}^{task}$ to generate unmixing matrices $\mathbf{W}^{task}$ and sources $\mathbf{S}^{task} = \mathbf{W}^{task}\mathbf{X}^{task}$. To generate a sample $\tilde{\mathbf{x}}_c$ from class c, we sample independently from each source restricted to the samples of class c yielding $\tilde{\mathbf{s}}_c^{task}$ and mix the data: $\tilde{\mathbf{x}}_c = (\mathbf{W}^{task})^{\dagger}\tilde{\mathbf{s}}_c^{task}$.

The *Covariance* method generates a new sample of synthetic data in class c by sampling from a Multivariate Gaussian with mean $\boldsymbol{\mu}_c$ and covariance $\boldsymbol{\Sigma}$, where $\boldsymbol{\mu}_c$ is the class mean and $\boldsymbol{\Sigma}$ is the covariance of centered task data estimated using Ledoit-Wolf method. In brief, it assumes normality of the data per class.

The *ICA + Covariance* method combines the augmentation methods *ICA* and *Covariance*: samples are drawn following the ICA approach, but with some additive non-isotropic Gaussian noise. As in *ICA* we estimate $\mathbf{W}^{task}$ and $\mathbf{S}^{task}$ from $\mathbf{X}^{task}$ via ICA. Then we consider $\mathbf{R}_{task} = \mathbf{X}_{task} - \mathbf{W}_{task}\mathbf{S}_{task}$ and estimate the covariance $\boldsymbol{\Sigma}_R$ of $\mathbf{R}_{task}$ via LedoitWolf's method. We then generate a data sample $\tilde{\mathbf{x}}_c$ from class c as with ICA and add Gaussian noise $\tilde{\mathbf{n}} \sim \mathcal{N}(0, \boldsymbol{\Sigma}_R)$. Samples are thus generated as $\tilde{\mathbf{x}}_c + \tilde{\mathbf{n}}$.

The *GANs* (respectively *CGANs*) method use a GANs (respectively CGANs) to generate data. The generator and discriminator have a mirrored architecture with 2 fully connected hidden layer of size (256 and 512). The number of epochs, batch size, momentum and learning rate are set to 20k, 16, 0.9, 0.01 and we use the Leaky RELU activation function.

We evaluate the performance of augmentation methods through the use of classifiers: logistic regression (LogReg), linear discriminant analysis with Ledoit-Wold estimate of covariance (LDA) perceptron with two hidden layers (MLP) and random forrests (RF). The hyper-parameters in each classifier are optimized through an internal 5-Fold cross validation. We set the number of iterations in each classifier so that convergence is reached. The exact specifications are available in appendix Table 4.

4.2 Distinguish Fake from Real HCP Resting State Data

This experiment is meant to assess the effectiveness of the data augmentation scheme in producing good samples. Data augmentation methods are trained on 200 subjects taken from HCP rest fMRI dataset which amounts to $960k$ samples (4800 per individual). Then synthetic data corresponding to 200 synthetic subjects are produced, yielding $960k$ fake samples and various classifiers are trained to distinguish fake from real data using 5-Fold cross validation. The cross-validated accuracy is shown in Table 1. Interestingly, we observe a dissociation between linear models (LogReg and LDA) that fail to discriminate between generated and actual data, and non-linear models (MLP and RF) that can discriminate samples from alternative augmentation methods. By contrast, all classifiers are at chance when Conditional ICA is used.

4.3 Comparing Augmentation Methods Based on Classification Accuracy on Task HCP Dataset

In order to compare the different augmentation methods, we measure their relative benefit in the context of multi-class classification. We use 787 subjects from the HCP task dataset that contains 23 classes and randomly split the dataset into a train set that contains 100 subjects and a test set that contains 687 subjects. In each split we train augmentation methods on the train set to generate fake

Table 1. Distinguish fake from real HCP resting state data We use HCP resting state data from $n = 200$ subjects ($960k$ samples) and produce an equally large amount of fake data ($960k$ samples) using data augmentation methods. The table shows the 5-fold cross validated accuracy obtained with various classifiers. When Conditional ICA is used, all classifiers are at chance.

Models	LDA	LogReg	Random forest	MLP
ICA	0.493	0.500	0.672	0.697
Covariance	0.473	0.461	0.610	0.626
ICA + Covariance	0.509	0.495	0.685	0.706
GANs [8]	0.501	0.498	0.592	0.607
CGANs [17]	0.498	0.493	0.579	0.604
Conditional ICA	0.503	0.489	0.512	0.523

samples corresponding to 200 subjects. These samples are then appended to the train set, resulting in an augmented train set on which the classifiers are trained. Results, displayed in Table 2, show that Conditional ICA always yields a higher accuracy than when no augmentation method is applied. The gains are over 1% on all classifiers tested. By contrast, ICA+Covariance and ICA lead to a decrease in accuracy while the Covariance approach leads to non-significant gains. To further investigate the significance of differences between the proposed approach and other state-of-the-art methods, we perform a t-test for paired samples (see Table 5 in appendix). Most notably, the proposed method performs significantly better than other data augmentation techniques. Given the large size of the HCP task data evaluation set, this significance test would demonstrate that the gains are robust. Note that the random forest classifier is not used in this experiment as it leads to a much lower accuracy than other methods. For completeness, we display the results obtained with random forest in appendix Table 6.

Visualization of fake examples produced by GANs, CGANs and Conditional ICA are available in appendix Fig. 5.

4.4 Gains in Accuracy Brought by Conditional ICA on Eight Datasets

In this experiment we assess the gains brought by Conditional ICA data augmentation on eight different task fMRI dataset. The number of subjects, classes and the size of the training and test sets differ between dataset and are reported in appendix Table 4. The rest of the experimental pipeline is exactly the same as with the HCP task dataset. We report in Fig. 3 the cross-validated accuracy of classifiers with and without augmentation. We notice that the effect of data augmentation is consistent across datasets, classifiers and splits, with 1% to 5% net gains. An additional experiment studying the sensitivity of Conditional ICA to the number of components used is described in appendix Fig. 4.

Table 2. Comparing augmentation methods based on classification accuracy on task HCP dataset We compare augmentation methods based on the classification accuracy **(Acc)**, precision **(Pre)** and recall **(Rec)** obtained by 2 linear classifiers (LDA and LogReg) and one non-linear classifier trained on augmented datasets on HCP Task fMRI data. We report the mean accuracy across 5 splits.

Models	LDA			LogReg			MLP		
	Acc	Pre	Rec	Acc	Pre	Rec	Acc	Pre	Rec
Original	0.893	0.889	0.891	0.874	0.869	0.873	0.779	0.782	0.778
ICA	0.814	0.809	0.813	0.840	0.836	0.839	0.803	0.805	0.802
Covariance	0.895	0.894	0.895	0.876	0.877	0.875	0.819	0.823	0.820
ICA + Covariance	0.816	0.811	0.812	0.840	0.839	0.840	0.815	0.819	0.814
GANs [8]	0.877	0.871	0.870	0.863	0.865	0.864	0.771	0.779	0.775
CGANs [17]	0.874	0.875	0.872	0.874	0.872	0.875	0.726	0.731	0.725
Conditional ICA	**0.901**	**0.903**	**0.905**	**0.890**	**0.888**	**0.890**	**0.832**	**0.835**	**0.831**

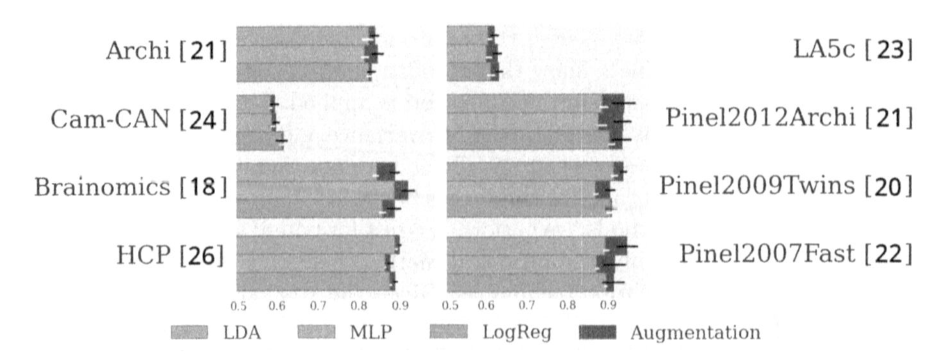

Fig. 3. Accuracy of models for eight multi-contrast datasets. Cross validated accuracy of two linear (LDA and LogReg) and one non-linear classifier (MLP) with or without using data augmentation. The improvement yielded by data augmentation is displayed in red. Black error bars indicate standard deviation across splits while white error bars indicate standard deviation across splits with no augmentation.

5 Discussion and Future Work

In this work we introduced Conditional ICA a fast generative model for rest and task fMRI data. It produces samples that cannot be distinguished from actual rest by linear as well as non-linear classifiers, showing that it captures higher-order statistics than naive ICA-based generators. When Conditional ICA is used as a data augmentation method, it yields consistent improvement in classification accuracy: on 8 tasks fMRI datasets, we observe increase in accuracy between 1% and 5% depending on the dataset and the classifier used. Importantly, this performance was obtained without any fine-tuning of the method, showing its reliability. One can also notice that our experiments cover datasets with different cardinalities, from tens to thousand, and different baseline prediction accuracy.

It is noteworthy that Conditional ICA is essentially a linear generative model with pointwise non-linearity, which makes it cheap, easy to instantiate on new data, and to introspect.

The speed and simplicity of Conditional ICA as well as the systematic performance improvement it yields, also make it a promising candidate for data augmentation in a wide range of contexts. Future work may focus on its applicability to other decoding tasks such as the diagnosis of Autism Spectrum Disorder (ASD) [4–6] or Attention-Deficit/Hyperactivity Disorder detection (ADHD) [16]. Other extensions of the present work concern the adaptation to individual feature prediction (e.g. age) where fMRI has shown some potential.

Acknowledgements. This project has received funding from the European Union's Horizon 2020 Framework Program for Research and Innovation under Grant Agreement No 945539 (Human Brain Project SGA3) and the KARAIB AI chair (ANR-20-CHIA-0025-01).

References

1. Abraham, A., et al.: Machine learning for neuroimaging with Scikit-learn. Front. Neuroinformatics **8**, 14 (2014)
2. Arjovsky, M., Chintala, S., Bottou, L.: Wasserstein GAN. arXiv:1701.07875 [cs, stat] (2017)
3. Dadi, K., et al.: Fine-grain atlases of functional modes for fMRI analysis. NeuroImage **221**, 117126 (2020)
4. Dvornek, N.C., Ventola, P., Pelphrey, K.A., Duncan, J.S.: Identifying autism from resting-state fMRI using long short-term memory networks. In: Wang, Q., Shi, Y., Suk, H.-I., Suzuki, K. (eds.) MLMI 2017. LNCS, vol. 10541, pp. 362–370. Springer, Cham (2017). https://doi.org/10.1007/978-3-319-67389-9_42
5. Eslami, T., Mirjalili, V., Fong, A., Laird, A.R., Saeed, F.: ASD-DiagNet: a hybrid learning approach for detection of autism spectrum disorder using fMRI data. Front. Neuroinformatics **13**, 70 (2019)
6. Eslami, T., Saeed, F.: Auto-ASD-network: a technique based on deep learning and support vector machines for diagnosing autism spectrum disorder using fMRI data. In: Proceedings of the 10th ACM International Conference on Bioinformatics, Computational Biology and Health Informatics, pp. 646–651 (2019)
7. Friston, K.J., et al.: Analysis of fMRI time-series revisited. Neuroimage **2**(1), 45–53 (1995)
8. Goodfellow, I., et al.: Generative adversarial nets. In: Advances in Neural Information Processing Systems, pp. 2672–2680 (2014)
9. Gorgolewski, K.J., et al.: Neurovault. org: a web-based repository for collecting and sharing unthresholded statistical maps of the human brain. Front. Neuroinformatics **9**, 8 (2015)
10. Harris, C.R., et al.: Array programming with Numpy. Nature **585**(7825), 357–362 (2020)
11. Hyvärinen, A.: Fast and robust fixed-point algorithms for independent component analysis. IEEE Trans. Neural Networks **10**(3), 626–634 (1999)
12. Kingma, D.P., Welling, M.: Auto-encoding variational bayes. arXiv preprint arXiv:1312.6114 (2013)

13. Kobyzev, I., Prince, S., Brubaker, M.: Normalizing flows: an introduction and review of current methods. In IEEE Trans. Pattern Anal. Mach. Intell. 1 (2020). https://doi.org/10.1109/TPAMI.2020.2992934
14. Ledoit, O., Wolf, M.: A well-conditioned estimator for large-dimensional covariance matrices. J. Multivariate Anal. **88**(2), 365–411 (2004)
15. Loula, J., Varoquaux, G., Thirion, B.: Decoding fMRI activity in the time domain improves classification performance. NeuroImage **180**, 203–210 (2018)
16. Mao, Z., et al.: Spatio-temporal deep learning method for ADHD fMRI classification. Inf. Sci. **499**, 1–11 (2019)
17. Mirza, M., Osindero, S.: Conditional generative adversarial nets. arXiv preprint arXiv:1411.1784 (2014)
18. Orfanos, D.P., et al.: The brainomics/localizer database. Neuroimage **144**, 309–314 (2017)
19. Pedregosa, F., et al.: Scikit-learn: machine learning in Python. J. Mach. Learn. Res. **12**, 2825–2830 (2011)
20. Pinel, P., Dehaene, S.: Genetic and environmental contributions to brain activation during calculation. Neuroimage **81**, 306–316 (2013)
21. Pinel, P., et al.: The functional database of the archi project: potential and perspectives. NeuroImage **197**, 527–543 (2019)
22. Pinel, P., et al.: Fast reproducible identification and large-scale databasing of individual functional cognitive networks. BMC Neurosci. **8**(1), 91 (2007)
23. Poldrack, R.A., et al.: A phenome-wide examination of neural and cognitive function. Sci. Data **3**(1), 1–12 (2016)
24. Shafto, M.A., et al.: The Cambridge centre for ageing and neuroscience (CAM-CAN) study protocol: a cross-sectional, lifespan, multidisciplinary examination of healthy cognitive ageing. BMC Neurol. **14**(1), 204 (2014)
25. Shirer, W.R., Ryali, S., Rykhlevskaia, E., Menon, V., Greicius, M.D.: Decoding subject-driven cognitive states with whole-brain connectivity patterns. Cerebral Cortex (New York, NY) **22**(1), 158–165 (2012)
26. Van Essen, D.C., et al.: The Wu-Minn human connectome project: an overview. Neuroimage **80**, 62–79 (2013)
27. Varoquaux, G., Thirion, B.: How machine learning is shaping cognitive neuroimaging. GigaScience **3**, 28 (2014)
28. Virtanen, P., et al.: SciPy 1.0 contributors: SciPy 1.0: fundamental algorithms for scientific computing in Python. Nat. Methods **17**, 261–272 (2020). https://doi.org/10.1038/s41592-019-0686-2
29. Zhuang, P., Schwing, A.G., Koyejo, O.:fMRI data augmentation via synthesis. In: 2019 IEEE 16th International Symposium on Biomedical Imaging (ISBI 2019), pp. 1783–1787. IEEE (2019)

Scalable Joint Detection and Segmentation of Surgical Instruments with Weak Supervision

Ricardo Sanchez-Matilla[1(✉)], Maria Robu[1], Imanol Luengo[1], and Danail Stoyanov[1,2]

[1] Digital Surgery, a Medtronic Company, London, UK
ricardo.sanchez-matilla@medtronic.com
[2] Wellcome/EPSRC Centre for Interventional and Surgical Sciences, University College London, London, UK

Abstract. Computer vision based models, such as object segmentation, detection and tracking, have the potential to assist surgeons intra-operatively and improve the quality and outcomes of minimally invasive surgery. Different work streams towards instrument detection include segmentation, bounding box localisation and classification. While segmentation models offer much more granular results, bounding box annotations are easier to annotate at scale. To leverage the granularity of segmentation approaches with the scalability of bounding box-based models, a multi-task model for joint bounding box detection and segmentation of surgical instruments is proposed. The model consists of a shared backbone and three independent heads for the tasks of classification, bounding box regression, and segmentation. Using adaptive losses together with simple yet effective weakly-supervised label inference, the proposed model use weak labels to learn to segment surgical instruments with a fraction of the dataset requiring segmentation masks. Results suggest that instrument detection and segmentation tasks share intrinsic challenges and jointly learning from both reduces the burden of annotating masks at scale. Experimental validation shows that the proposed model obtain comparable results to that of single-task state-of-the-art detector and segmentation models, while only requiring a fraction of the dataset to be annotated with masks. Specifically, the proposed model obtained 0.81 weighted average precision (wAP) and 0.73 mean intersection-over-union (IOU) in the Endovis2018 dataset with 1% annotated masks, while performing joint detection and segmentation at more than 20 frames per second.

Keywords: Instrument detection · Instrument segmentation · Multi-task learning · Semi-supervised learning

---

Electronic supplementary material The online version of this chapter (https://doi.org/10.1007/978-3-030-87196-3_47) contains supplementary material, which is available to authorized users.

M. de Bruijne et al. (Eds.): MICCAI 2021, LNCS 12902, pp. 501–511, 2021.
https://doi.org/10.1007/978-3-030-87196-3_47

1 Introduction

Detection of surgical instruments in minimally invasive surgery video frames allows automatic generation of offline surgical analytics, that can provide valuable information for improving surgical procedures [1]. Additionally, surgical instrument detection can provide real-time decision support during the surgery and notify preventable risks during computer assisted interventions [2].

Accurate models are required to successfully use decision support systems during surgical procedures. Current machine learning approaches typically estimate the location and type of surgical instruments via either bounding boxes *detection* [3,4] or semantic *segmentation* [5,6]. Tool detection models generally rely on annotated bounding boxes during training. This has a major limitation for instrument detection as the annotated bounding boxes include a high number of background pixels due to the elongated dimensions of the surgical instruments, which might impede a model from learning discriminative features of the instruments. Alternatively, segmentation models directly estimate the probability of each pixel to belong to a specific instrument type by relying on fine-grained pixel-wise segmentation mask annotations. While masks solve the aforementioned challenge faced by bounding boxes, the annotation cost significantly grows up to almost two orders of magnitude for annotating masks with respect to only annotating frame-level labels or bounding boxes [7]. In practice, the annotation of datasets with masks at scale could be unfeasible, which can prevent models from achieving the generalisation and robustness required to be applied in real-world applications.

To address some of the challenges above and leverage the strengths of both workstreams, a multi-task model is proposed that jointly learns to estimate bounding boxes and masks for surgical instruments. The model aggregates information from the multiple tasks by using a shared backbone as encoder, while having a head for each task: instrument classification, bounding box regression and segmentation. While the classification and regression heads allow to localise and classify surgical instruments using scalable annotations, the segmentation head achieves the detailed pixel-wise annotations. To alleviate the burden of expensive pixel-wise annotation on large datasets, we introduce a training framework that accounts for missing masks and uses a weakly-supervised loss computed on frame-level labels which can be freely obtained from the bounding box annotations. Experimental results show that our model achieves detection and segmentation performance on par with fully supervised alternatives, while requiring as little as 1% of the masks in training.

2 Related Work

Surgical tool identification and localisation is an active research field, which has resulted in multi-centre collaborations releasing novel surgical datasets to encourage the research community to design models to advance segmentation quality [8], model robustness and generalisation [9].

Proposed research directions to tackle surgical tool identification and localisation include semantic segmentation and tool detection. Segmentation models are able to segment instruments against background (binary segmentation) [10], tool types (semantic segmentation) [6] or tools instances (instance segmentation) [5]. Most relevant, Gonzalez et al. [5] proposed to segment entire instruments instances instead of pixel-wise segmentation to achieve state-of-the-art results. A comprehensive overview of the latest segmentation models is available in [9].

Recent detection models [3,4] are anchor-based and are composed of a convolutional backbone with ResNet architecture, that generates feature maps at different scales, and two task-specific heads that perform object classification and bounding box regression from the feature pyramid. This approach faces an extreme foreground-background class imbalance during training. This can be handled by using the focal loss [3], a variation of the cross-entropy loss function that down-weights the loss assigned to well-classified examples. EfficientDet [4] proposed to jointly scale up model width, depth and resolution to meet real-time requirements without sacrificing detection accuracy. The model computes a feature pyramid using EfficientNet [11] as backbone. EfficientDet proposed a weighted bi-directional feature pyramid network (BiFPN) to efficiently leverage the multi-scale feature information for object detection.

More complex approaches have also been proposed. Joint detection and segmentation can be learnt jointly [12]. A semi-supervised object segmentation model can rely on a single manual bounding box initialisation to produce class-agnostic object masks and rotated bounding boxes with a fully-convolutional siamese model [13]. Multi-task models using weak supervision can jointly detect and segment with a weakly-supervised cyclic policy can be used to complement the learning of both tasks simultaneously [14]. In the same line of work, other works show that a weakly-supervised convolutional model can be used to estimate the presence and localisation of surgical instruments using only frame-label annotations [15,16]. However, the performance of these weakly-supervised models is still far from that of the fully supervised ones.

3 Proposed Model

3.1 Joint Detection and Segmentation

Let $\mathbf{x} \in \{0, 255\}^{W,H,C}$ be an RGB image with width W, height H and $C = 3$ colour channels. Let $D(\cdot) : \mathbf{x} \rightarrow (\mathbb{B}^{N,4}, \mathbb{C}^{N}, \mathbb{M}^{W,H,M})$ be a joint detection and segmentation model that localises and classifies surgical instruments within $\mathbf{x}$ which outputs are a set of bounding boxes ($\mathbb{B}^{N,4}$), their corresponding estimated classes ($\mathbb{C}^{N}$), and a segmentation mask ($\mathbb{M}^{W,H,M}$), with N being the number of detected instruments and M the total number instrument types in the dataset.

The problem is formulated as a multi-task learning problem. The proposed model, depicted in Fig. 1, is composed of three main components, namely, backbone, feature fusion module, and three heads - one for each task: localisation,

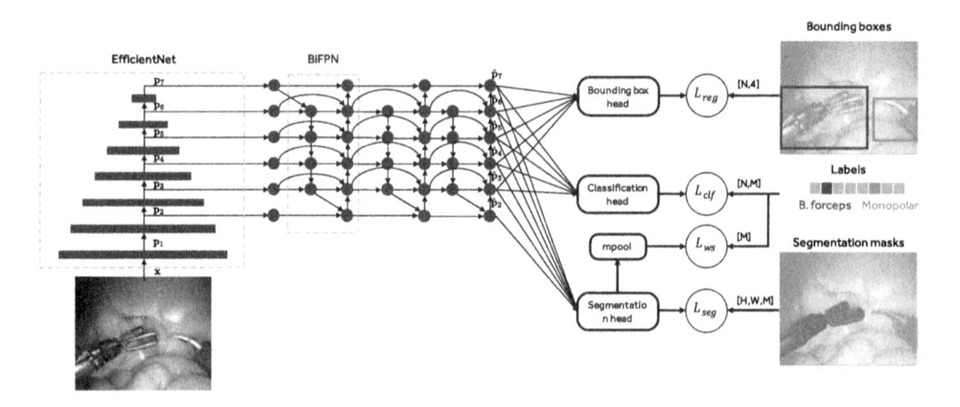

Fig. 1. Proposed multi-task model composed of an EfficientNet backbone, a set of bi-directional feature pyramid network (BiFPN), and the proposed three heads for the tasks of bounding box regression, bounding box classification, and segmentation. The model is trained using four loss functions named regression L_{reg}, classification L_{clf}, weak supervision L_{ws} and segmentation L_{seg}. The model requires bounding box and label annotations for every frame, and segmentation masks for a reduced number of frames. *mpool* is the global maxpool operation. The text on the most-right arrows indicates the shape of the annotations, where N is the number of instruments in a given frame, M is the total number instrument types in the dataset, and W, H are the dimensions of the input frame.

classification, and segmentation. The shared backbone acts as a joint representation learning module whose aim is to learn multi-level feature representations suitable for all the tasks. Having $\mathbf{x}$ as input, the backbone $\beta(\cdot) : \mathbf{x} \rightarrow \mathbb{P}$ generates a pyramid of features at S scales $\mathbb{P} = (p_s)_{s=1}^{s=S}$. The feature pyramid is fed to a bi-directional feature pyramid network (BiFPN) [4] that fuses the features across scales while maintaining their number and resolution $\gamma(\cdot) : \mathbb{P} \rightarrow \hat{\mathbb{P}}$. The heads guide the learning of the backbone and the feature fusion modules to learn more discriminative and complementary features to further improve the three tasks while adapting the generated features for task-specific problems. In our implementation, we use the localisation and classification heads proposed in [4].

The following subsections describe the proposed segmentation head as well as the mechanism that allows learning to segment with only a fraction of annotated masks via weak supervision.

3.2 Segmentation Head

The segmentation head aims to generate a mask $\mathbb{M}^{W,H,M}$ from the fused feature pyramid $\hat{\mathbb{P}}$. The segmentation head architecture (Fig. 2) is composed of three main steps: feature upsampling and concatenation, convolutional block, and upsampling.

To make use of the information contained in the multiple scales of the fused feature pyramid, $\hat{\mathbb{P}}$, the $S-2$ smallest feature maps $(\hat{p}_s)_{s=2}^{s=S}$ are first upsampled

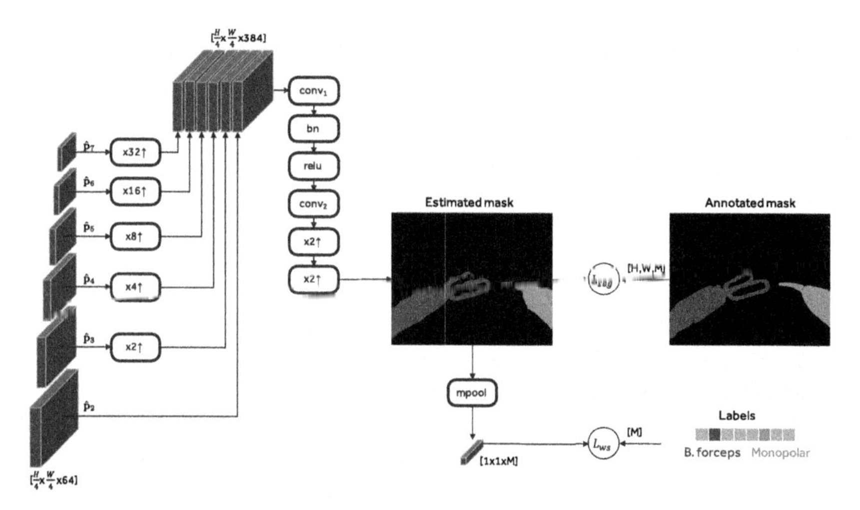

Fig. 2. Diagram of the segmentation head with the proposed weak-supervision module. L_{seg} is the cross entropy segmentation loss that is computed for a given frame when the mask is annotated. L_{ws} is the cross entropy weak supervision loss computed at every frame. *mpool* is the global maxpool operation. The text on the most-right arrows indicates the shape of the annotations, where W, H are the dimensions of the input frame, and M is the total number instrument types in the dataset.

to the resolution of $\hat{p}_2$ using bi-linear interpolation. Then, the $S-1$ feature maps are concatenated

$$\tilde{\mathbb{P}} = (\hat{p}_2, U_2(\hat{p}_3), U_2(\hat{p}_4), U_2(\hat{p}_5), U_2(\hat{p}_6), U_2(\hat{p}_7)) \tag{1}$$

where $U_2(\cdot)$ is the bilinear interpolation operation that upsamples the feature map to the resolution of $\hat{p}_2$, and $(\cdot, \ldots, \cdot)$ represents the concatenation operation. A convolutional block is then applied to achieve a feature map with M channels as

$$\tilde{\mathbb{M}} = conv_2(relu(bn(conv_1(\tilde{\mathbb{P}})))), \tag{2}$$

where $conv_1(\cdot)$ is a 2D convolution with kernel (1×1) and $S-1$x64 channels that fuses the features with different resolutions, $bn(\cdot)$ is a batch normalisation layer, $relu(\cdot)$ is the Rectified Linear Unit (ReLU) operation, and $conv_2(\cdot)$ is a 2D convolution with kernel (1×1) and M channels to reduce the number of channels to the number of instrument types in the dataset, M.

Finally, $\tilde{\mathbb{M}}$ is upsampled to generate masks with same dimensions than the input images

$$\mathbb{M} = U_0(\tilde{\mathbb{M}}), \tag{3}$$

where $U_0(\cdot)$ is the bilinear interpolation operation that upsamples the feature map to the resolution of the input image $\mathbf{x}$.

3.3 Semi-supervised Learning with Weak Supervision

When the annotated mask, $\bar{\mathbb{M}}$, is available for a given frame, we use the cross-entropy loss function ($L_{CE}(\cdot,\cdot)$) for training the segmentation head. However, as not all samples will have an annotated mask, in each batch we weight the cross-entropy loss by the ratio of samples with annotated masks A to the total number of samples within the batch B as

$$L_{seg} = \frac{A}{B} L_{CE}(\mathbb{M}, \bar{\mathbb{M}}). \tag{4}$$

Thus, batches with fewer annotated masks have a lower weight.

In addition, we enable the training of the segmentation head in the absence of annotated masks. We reduce the estimated mask with a global max pooling into a vector that is supervised with frame-level annotations (presence of the instruments in each frame) [15] as

$$\mathbb{O} = mpool(\mathbb{M}), \tag{5}$$

where $mpool(\cdot)$ is the 2D maxpool operation with kernel size $(H,\ W)$ that generates a vector $\mathbb{O} \in \mathbb{R}^{1,1,M}$. The information within $\mathbb{O}$ estimates the presence/absence of each instrument type within the frame. These outputs indicate the presence of a given instrument type within the frame. Note that these annotations, which are cheap to generate, are already available to the model within the bounding box annotations.

The weakly-supervised loss is the cross entropy between $\mathbb{O}$ and the instrument type frame-level multi-label annotations, $\bar{\mathbb{O}}$ as

$$L_{ws} = L_{CE}(\mathbb{O}, \bar{\mathbb{O}}). \tag{6}$$

Note that we compute $L_{ws}(\cdot)$ for all frames, regardless of whether their mask is provided or not.

In conclusion, the full loss used to train the backbone, BiFPN, and heads is

$$L = w_{reg} \cdot L_{reg} + w_{clf} \cdot L_{clf} + w_{seg} \cdot L_{seg} + w_{ws} \cdot L_{ws}, \tag{7}$$

where (L_{reg}, L_{clf}) is the focal loss [3], and w_{reg}, w_{clf}, w_{seg}, and w_{ws} are weights of regression, classification, segmentation and weak supervision losses that tune the contribution of each loss.

4 Experimental Setup

4.1 Dataset

We validate the performance of the proposed model in EndoVis2018 dataset released as part of the Robotic Scene Segmentation Challenge [8]. The dataset is composed of 15 sequences of 149 frames each. We use the annotated masks of the instrument type provided by [5]. We automatically generate bounding boxes from the masks by selecting the minimum and maximum values for the vertical and horizontal coordinates of each mask. We split the data in training and validation sets as done by [5]. Sequences 5, 9, and 15 compose the validation set and the remaining ones the training set. The sequence *seq2* is discarded from either of the sets as it contains frames with two instances of the same instrument type, and therefore, we cannot automatically generate bounding boxes.

4.2 Performance Metrics

We evaluate detection using the mean Average Precision weighted (wAP) by the number of samples in the validation set. As proposed by [5], segmentation is evaluated using IOU averaged over the number of classes, M, and over the number of images, K:

$$IOU = \frac{1}{K}\sum_{k=1}^{K}\left(\frac{1}{M}\sum_{m=1}^{M}\frac{\mathbb{M}_{k,m} \cap \bar{\mathbb{M}}_{k,m}}{\mathbb{M}_{k,m} \cup \bar{\mathbb{M}}_{k,m}}\right). \tag{8}$$

4.3 Implementation Details

As backbone, we use EfficientNet-D0 [11] pre-trained on ImageNet. We modify the BiFPN layer [4] to aggregate $S = 6$ feature scales instead of five for improved segmentation accuracy. The five smallest scales are used for both the regression and classification heads. Images are downscaled to 512×512 pixels and data augmentation that includes geometrical and colour transformations is used. A sampler to balance the number of different instruments types in each epoch is used. All models are trained for 150 epochs. We report the results on the validation set obtained in the last epoch. SGD optimiser with momentum and *1Cycle* learning scheduler [17] with cosine decay and a maximum learning rate of $5e^{-4}$ is used. Each batch contains 32 samples. The proposed loss (Eq. 7) weights are empirically set to: $w_{reg} = 1$, $w_{clf} = 5$, $w_{seg} = 700$, and $w_{ws} = 5$. These parameters encourage all losses to be in a similar range. These settings remain fixed for all the experiments.

4.4 Results

Ablation Study. We first study how the proposed joint detection and segmentation model compares against the only-detection and only-segmentation alternatives. Results in the supplementary material indicate that performing detection and segmentation jointly slightly improves detection while maintaining similar segmentation performance when 100% of the masks are available. However, the segmentation-only model performance rapidly degrades when fewer masks are available during training. For instance, the performance drops from an IOU of 0.821 to 0.651 when reducing the masks from 100% to 20%, and to 0.544 when further reduced to 1%. Secondly, we perform an ablation study to understand how the presence/absence of the weakly supervised loss impacts the performance

Table 1. Comparison of the proposed model against state-of-the-art detection and segmentation models. The proposed model performance is evaluated for a range of different availability of masks during training. When available, mean and standard deviation (mean ± std) over three trainings with different random seeds are reported. KEY: *, sequence seq2 is used in the validation set.

Model	Task		Weak superv.	% annotated masks	Detection	Segmentation
	Det.	Segm.			wAP	IOU
[4] EfficientDet	✓			–	0.808 ± 0.008	–
[18] TernausNet*		✓		100%	–	0.399
[19] MFTAPnet*		✓		100%	–	0.391
[5] ISINet*		✓		100%	–	0.710
[6] HRNet*		✓		100%	–	0.714 ± 0.019
[6] HRNet		✓		100%	–	0.738
Proposed	✓	✓	✓	100%	0.827 ± 0.007	0.822 ± 0.015
Proposed	✓	✓	✓	20%	0.817 ± 0.003	0.800 ± 0.014
Proposed	✓	✓	✓	5%	0.808 ± 0.026	0.791 ± 0.022
Proposed	✓	✓	✓	1%	0.813 ± 0.016	0.728 ± 0.006
Proposed	✓	✓	✓	0%	0.786 ± 0.021	0.345 ± 0.012

with limited availability of annotated segmentation masks. Results in the supplementary material show that when the weakly supervised loss is present the performance approximately remains stable, even when reducing the number of masks up to 1%. For instance, the performance between using 100% or 1% of the masks varies in 0.01, and 0.09 points for wAP and IOU, respectively.

Comparison Against State of the Art. The proposed model obtain competitive segmentation results against fully-supervised state-of-the-art alternatives while only requiring a 1% of annotated masks (Table 1). In terms of detection, the proposed model outperforms by up to 3% with respect to the detection-only model. We also study how reducing the number of segmentation masks available during training (100%, 20%, 10%, 5%, 1%, and 0% of the total number of training samples) affects the detection and segmentation performance of the proposed model. During the different training setups for different mask ratios, frames with masks are sampled equally spaced across the dataset. Results show that the proposed model outperforms the rest of the alternatives while only requiring 5% of masks. Even with only 1% of the masks available, the proposed model obtain competitive results when compared with fully-supervised alternatives. When no masks are available (0%), the model solely relies on the weakly-supervised module for learning to segment, and both detection and segmentation performance significantly drops (see the last column of Fig. 3). Three visual segmentation samples, one per each sequence of the validation set, are displayed in Fig. 3 for models trained using 100%, 20%, 5%, and 0% of annotated masks. The estimated masks maintain the quality even when the available masks are reduced to

5%. Some classification errors are observed in the second sequence when limited masks are used. When no masks are used during training (0%) the estimated masks tend to only focus on representative parts of the instrument. Additional visual examples are available in the supplementary material.

The proposed model has 4.01M parameters and can perform detection and segmentation simultaneously while requiring only 5% and 1% more parameters than the only-detection and only-segmentation models, respectively. The proposed model obtains an inference speed of 22.4 fps in an NVIDIA Quadro RTX 6000.

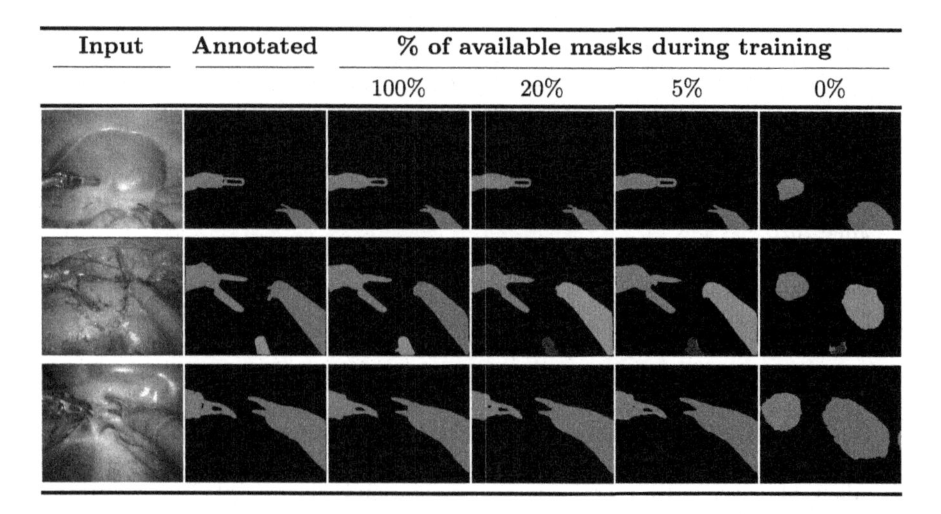

Fig. 3. Visual segmentation results of the proposed model with different percentage of available annotated masks during training. The colours encode the instrument type.

5 Conclusion

This work[1] proposed a multi-task model that jointly learns to detect and segment surgical instruments. A weakly-supervised adaptive loss is also proposed, that enables the learning of segmentation masks when only a fraction of masks are available during training by supervising the learning with frame-level annotations. Experimental results showed that the proposed model obtains comparable results to a fully-supervised alternative, while only requiring a 1% of the frames to have annotated masks.

[1] This work was supported by the Wellcome/EPSRC Centre for Interventional and Surgical Sciences (WEISS) at UCL (203145Z/16/Z), EPSRC (EP/P012841/1, EP/P027938/1, EP/R004080/1) and the H2020 FET (GA 863146). Danail Stoyanov is supported by a Royal Academy of Engineering Chair in Emerging Technologies (CiET18196) and an EPSRC Early Career Research Fellowship (EP/P012841/1).

Further investigation is required to understand how to effectively add temporal information and consistency to the model as well as how to further interrelate the learning of the multiple tasks.

References

1. Trehan, A., Barnett-Vanes, A., Carty, M.J., McCulloch, P., Maruthappu, M.: The impact of feedback of intraoperative technical performance in surgery: a systematic review. BMJ Open **5**(6) (2015)
2. Jo, K., Choi, Y., Choi, J., Chung, J.W.: Robust real-time detection of laparoscopic instruments in robot surgery using convolutional neural networks with motion vector prediction. Appl. Sci. **9**(14), 2865 (2019)
3. Lin, T.-Y., Goyal, P., Girshick, R., He, K., Dollar, P.: Focal loss for dense object detection. In: Proceedings of the IEEE International Conference on Computer Vision, October 2017
4. Tan, M., Pang, R., Le, Q.V.: EfficientDet: scalable and efficient object detection. In: Proceedings of the IEEE Conference on Computer Vision and Pattern Recognition, June 2020
5. González, C., Bravo-Sánchez, L., Arbelaez, P.: ISINet: an instance-based approach for surgical instrument segmentation. In: Martel, A.L., et al. (eds.) International Conference on Medical Image Computing and Computer Assisted Intervention, pp. 595–605 (2020)
6. Sun, K., et al.: High-resolution representations for labeling pixels and regions. CoRR, abs/1904.04514 (2019)
7. Bilen, H.: Weakly supervised object detection. In: Proceedings of the IEEE Conference on Computer Vision and Pattern Recognition (2018)
8. Allan, M., et al.: 2018 robotic scene segmentation challenge (2020)
9. Ross, T., et al.: Robust medical instrument segmentation challenge 2019 (2020)
10. García-Peraza-Herrera, L.C., et al.: Real-time segmentation of non-rigid surgical tools based on deep learning and tracking. In: Peters, T., et al. (eds.) Computer-Assisted and Robotic Endoscopy, pp. 84–95 (2017)
11. Tan, M., Le, Q.: EfficientNet: rethinking model scaling for convolutional neural networks. In: Proceedings of the International Conference on Machine Learning, pp. 6105–6114 (2019)
12. Cao, J., Pang, Y., Li, X.: Triply supervised decoder networks for joint detection and segmentation. In: Proceedings of the IEEE Conference on Computer Vision and Pattern Recognition, pp. 7384–7393 (2019)
13. Wang, Q., Zhang, L., Bertinetto, L., Hu, W., Torr, P.H.S.: Fast online object tracking and segmentation: a unifying approach. CoRR, abs/1812.05050 (2018)
14. Shen, Y., Ji, R., Wang, Y., Wu, Y., Cao, L.: Cyclic guidance for weakly supervised joint detection and segmentation. In: Proceedings of the IEEE Conference on Computer Vision and Pattern Recognition, June 2019
15. Vardazaryan, A., Mutter, D., Marescaux, J., Padoy, N.: Weakly-supervised learning for tool localization in laparoscopic videos. In: Stoyanov, D., et al. (eds.) Intravascular Imaging and Computer Assisted Stenting and Large-Scale Annotation of Biomedical Data and Expert Label Synthesis, pp. 169–179 (2018)
16. Nwoye, C.I., Mutter, D., Marescaux, J., Padoy, N.: Weakly supervised convolutional LSTM approach for tool tracking in laparoscopic videos. Int. J. Comput. Assist. Radiol. Surg. **14**(6), 1059–1067 (2019)

17. Smith, L.N., Topin, N.: Super-convergence: very fast training of residual networks using large learning rates. CoRR, abs/1708.07120 (2017)
18. Shvets, A.A., Rakhlin, A., Kalinin, A.A., Iglovikov, V.I.: Automatic instrument segmentation in robot-assisted surgery using deep learning. In: 2018 17th IEEE International Conference on Machine Learning and Applications (ICMLA), pp. 624–628 (2018)
19. Jin, Y., Cheng, K., Dou, Q., Heng, P.-A.: Incorporating temporal prior from motion flow for instrument segmentation in minimally invasive surgery video. In: Shen, D., et al. (eds.) MICCAI 2019. LNCS, vol. 11768, pp. 440–448. Springer, Cham (2019). https://doi.org/10.1007/978-3-030-32254-0_49

Machine Learning - Weakly Supervised Learning

Weakly-Supervised Universal Lesion Segmentation with Regional Level Set Loss

Youbao Tang[1](✉), Jinzheng Cai[1], Ke Yan[1], Lingyun Huang[2], Guotong Xie[2], Jing Xiao[2], Jingjing Lu[3], Gigin Lin[4], and Le Lu[1]

[1] PAII Inc., Bethesda, MD, USA
[2] Ping An Technology, Shenzhen, People's Republic of China
[3] Beijing United Family Hospital, Beijing, People's Republic of China
[4] Chang Gung Memorial Hospital, Linkou, Taiwan, ROC

Abstract. Accurately segmenting a variety of clinically significant lesions from whole body computed tomography (CT) scans is a critical task on precision oncology imaging, denoted as universal lesion segmentation (ULS). Manual annotation is the current clinical practice, being highly time-consuming and inconsistent on tumor's longitudinal assessment. Effectively training an automatic segmentation model is desirable but relies heavily on a large number of pixel-wise labelled data. Existing weakly-supervised segmentation approaches often struggle with regions nearby the lesion boundaries. In this paper, we present a novel weakly-supervised universal lesion segmentation method by building an attention enhanced model based on the High-Resolution Network (HRNet), named AHRNet, and propose a regional level set (RLS) loss for optimizing lesion boundary delineation. AHRNet provides advanced high-resolution deep image features by involving a decoder, dual-attention and scale attention mechanisms, which are crucial to performing accurate lesion segmentation. RLS can optimize the model reliably and effectively in a weakly-supervised fashion, forcing the segmentation close to lesion boundary. Extensive experimental results demonstrate that our method achieves the best performance on the publicly large-scale DeepLesion dataset and a hold-out test set.

1 Introduction

Basing on global cancer statistics, 19.3 million new cancer cases and almost 10.0 million cancer deaths occurred in 2020 [22]. Cancer is one of the critical leading causes of death and a notorious barrier to increasing life expectancy in every country of the world. To assess cancer progress and treatment responses, tumor size measurement in medical imaging and its follow-ups is one of the most widely accepted protocols for cancer surveillance [9]. In current clinical practice, most of these measurements are performed by doctors or radiology technicians [2]. It is time-consuming and often suffers from large inter-observer

M. de Bruijne et al. (Eds.): MICCAI 2021, LNCS 12902, pp. 515–525, 2021.
https://doi.org/10.1007/978-3-030-87196-3_48

variations, especially with the growing cancer incidence. Automatic or semi-automatic lesion size measurement approaches are in need to alleviate doctors from this tedious clinical load, and more importantly, to significantly improve assessment consistency [3,27]. In this work, we develop a new universal lesion segmentation (ULS) method to measure tumor sizes accurately on selected CT cross sectional images, as defined by RECIST guideline [9].

Many efforts have been developed for automating lesion size measurement. Specifically, deep convolutional neural networks are successfully applied to segment tumors in brain [11], lung [29,34], pancreas [32,36], liver [6,7,16,20,25,31], enlarged lymph node [17,23,35], *etc.* Most of these approaches are specifically designed for a certain lesion type, however, an effective and efficient lesion size measurement tool should be able to handle a variety of lesions in practice. Our ULS approach is proposed via leveraging a sophisticated network architecture and an effective weakly-supervised learning strategy. On one hand, more sophisticated network backbones allow ULS to have larger model capacities to cope with lesions with various appearances, locations and sizes. On the other hand, weakly-supervised learning strategy may drastically simplify the annotation complexity that permits large amounts of bookmarked cancer images (already stored in PACS system) to be used for model initialization. In weakly-supervised learning, we propose a new regional level set (RLS) loss as the key component to refine segment regions near lesion boundaries so as to improve the quality of segmentation supervisory signal.

We follow the literature to formulate lesion size measurement as a two dimensional region segmentation problem, which performs dense pixel-wise classification on RECIST-defined CT axial slices. Such region segmentation based tumor size, area or volume assessment should perform more accurately in measuring solid tumor's response than lesion diameters [1,9,24,26,27]. The main reason why diameter is adopted in [9] was due to the easier reproducibility by human raters. To precisely delineate the tumor boundary, we propose three main contributions in ULS: 1) an effective network architecture (AHRNet) based on HRNet [28] that renders rich high-resolution representations with strong position sensitivity, by being augmented with a decoder (DE) and a novel attention mechanism combining both dual attention (DA) [10] and scale attention (SA). 2) the RLS loss as a reformulated deep learning based level set loss [4] with specific modifications for lesion segmentation. 3) AHRNet and the RLS loss integrated within a simple yet effective weakly-supervised training strategy so that our AHR-Net model can be trained on large-scale PACS stored lesion databases, such as DeepLesion [30].

Our contributions result in a new state-of-the-art segmentation accuracy that outperforms nnUNet [13] by 2.6% in averaged Dice score and boosts the former best result [27] from 91.2% to 92.6% in Dice on the DeepLesion test set. Our AHRNet model is also more robust by improving the worst case of 78.9% Dice from 43.6% in [13]. It is trained with a large-scale database and generalizes well on a hold-out test set, outpacing nnUNet by 2.7% and achieving 88.2% Dice score. Over 92% of the testing lesions are segmented with >85% Dice scores,

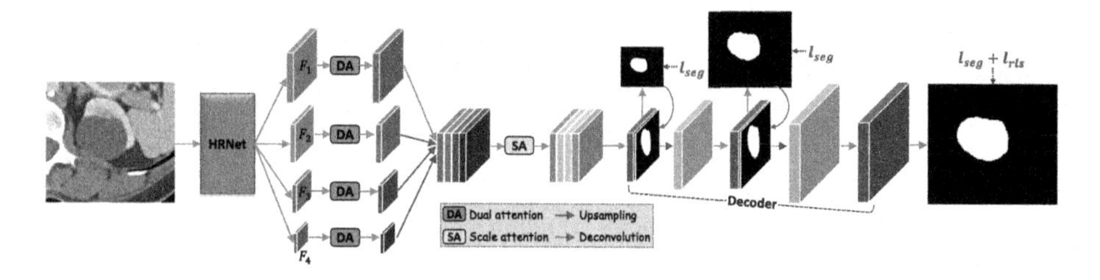

Fig. 1. Illustrated framework of our proposed AHRNet, where ℓ_{seg} is the segmentation loss defined in Sect. 2.3 that consists of a binary cross entropy loss and an IoU loss, ℓ_{rls} is the regional level set loss described in Sect. 2.2.

demonstrating that our AHRNet model is a reliable tool for lesion size measurement. The network components, *i.e.*, DE, DA, SA, and RLS, could work seamlessly with different network backbones including nnUNet, HRNet, and our AHRNet. We validate the effectiveness of each sub-module via ablation studies.

2 Methodology

This work aims to produce reliable and accurate lesion masks on the given lesion sub-images. Following previous work [3,27], we assume that the lesion sub-images have been obtained in the form of bounding boxes that could be either automatically generated by lesion detection algorithms or semi-automatically drawn by radiologists. Figure 1 illustrates the overall AHRNet framework.

2.1 AHRNet Architecture

HRNet has been demonstrated of achieving state-of-the-art performance in a wide range of computer vision applications [28], including semantic segmentation, object detection, and human pose estimation, suggesting that HRNet may be a strong versatile CNN backbone. It can connect high and low resolution convolutions in parallel, maintain high resolution through the whole process, and fuse multi-resolution representations repeatedly, rendering rich hierarchical, high-resolution representations with strong position sensitivity. These characteristics of HRNet are crucial for pixel-wise dense prediction tasks. We choose HRNet as the backbone to extract rich multi-scale features for lesion segmentation here. As shown in Fig. 1, given a CT image $I \in \mathbb{R}^{H\times W}$, HRNet produces stacked multi-scale image features $F = \{F_k \in \mathbb{R}^{2^{k+4}\times \frac{H}{2^{k+1}}\times \frac{W}{2^{k+1}}} \mid k \in \{1,2,3,4\}\}$.

A straightforward means of lesion segmentation is to upsample F_i to have the same resolution (*e.g.*, $\frac{1}{4}$ of the input image), concatenate them, and follow a convolutional layer with a 1×1 kernel to get the prediction, which serves as our baseline. The resolution of deep image features is important for accurate lesion segmentation, especially for small lesions. Thus to get more accurate predictions, we set up a small decoder (DE) to obtain higher resolution features. From Fig. 1,

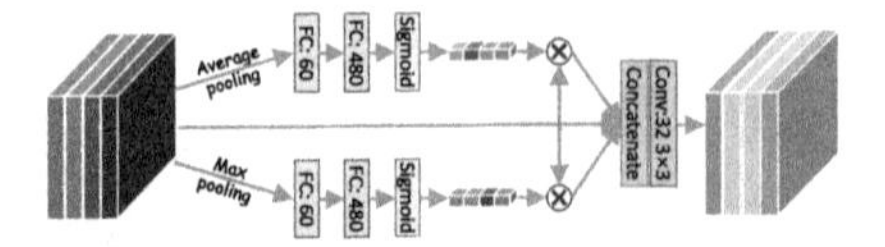

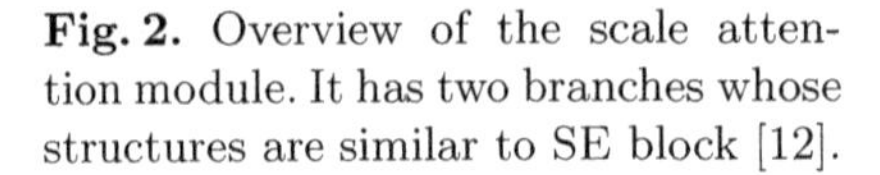

Fig. 2. Overview of the scale attention module. It has two branches whose structures are similar to SE block [12].

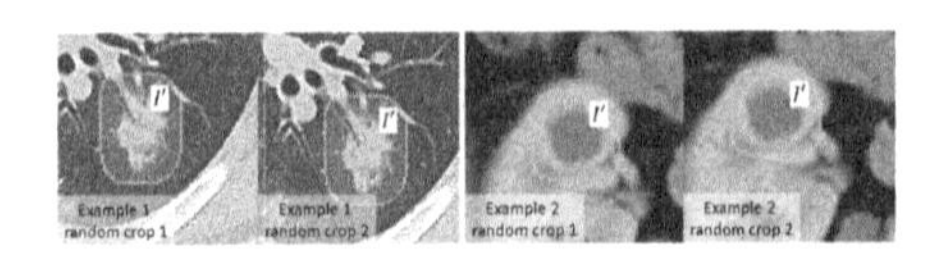

Fig. 3. Two examples of using lesion-adaptive regions I' for ℓ_{rls} computation defined in Eq. 2.

it contains two deconvolutional layers with 32 4×4 kernels and a stride of 2 and three convolutional layers with 32 3×3 kernels, where the dimensions of feature maps are $\frac{1}{4}$, $\frac{1}{2}$, and 1 of the input image, respectively. Another three convolutional layers with a 1×1 kernel are added to get the corresponding predictions. Each deconvolutional layer takes as input the features and the prediction.

As described above, we do not model the long-range dependencies of features in F_i for lesion segmentation. However, long-range contextual information can be crucial in obtain precise predictions. Fu *et al.* [10] present a dual attention (DA) module that can capture the long-range contextual information over local feature representations in spatial and channel dimensions respectively. In this work, we model the global contextual information in F_i by employing a DA module [10] to adaptively aggregate their rich long-range contextual dependencies in both spatial and channel dimensions, and enhancing feature representations to improve the performance of lesion segmentation. Since the studied lesion sizes are very diverse, to better address lesion segmentation under different scales, we introduce a scale attention (SA) module to effectively combine the multi-scale features by treating them input-specifically with learnable scale attention vectors. It contains two branches that are built upon SE block [12], as in Fig. 2.

2.2 Regional Level Set Loss

A classic level set method is proposed for image segmentation [4], treating segmentation as an energy minimization problem. The energy function is defined as:

$$\begin{aligned} E\left(c_1, c_2, \phi\right) = \mu \cdot \text{Length}(\phi) + \nu \cdot \text{Area}(\phi) + \lambda_1 \sum_{i \in I} \left|i - c_1\right|^2 H(\phi(i)) \\ + \lambda_2 \sum_{i \in I} |i - c_2|^2 (1 - H(\phi(i))), \end{aligned} \tag{1}$$

where μ, ν, λ_1 and λ_2 are the predefined non-negative hyper-parameters, i is the intensity of its corresponding image location, $\phi(\cdot)$ is the level set function, Length(ϕ) and Area(ϕ) are the regularization terms with respect to the length and the inside area of the contour, c_1 and c_2 represent the mean pixel intensity of inside and outside areas of the contour, and H is the Heaviside function: $H(\phi(i)) = 1$, if $\phi(i) \geq 0$; $H(\phi(i)) = 0$ otherwise.

Recently, researchers have studied to integrate this energy function into deep learning frameworks for semantic segmentation [14] and medical image segmentation [5,33]. Approaches [5,14] replace the original image I in Eq. 1 with a binary image that is reconstructed from the ground truth mask of each object. Method [33] computes a cross-entropy loss between the outputs of Eq. 1 when setting ϕ as the prediction and ground truth. Neither formulation applies for our task due to the lack of ground truth masks of lesions for training. To tackle this issue, based on Eq. 1, we introduce a regional level set (RLS) loss defined by

$$\ell_{rls} = \frac{1}{|I'|}\sum_{i \in I'}[\lambda_1 \cdot p(i) \cdot |i - c_1|^2 + \lambda_2 \cdot (1 - p(i)) \cdot |i - c_2|^2], \tag{2}$$

where $p(i)$ is the predicted probability map of pixel i, I' is the constrained region of the input image I and $|I'|$ is the number of pixels in I'. We experimentally set $\lambda_1 = 1$, $\lambda_2 = 3$. Here, terms of Length(ϕ) and Area(ϕ) in Eq. 1 have been removed because they are sensitive to object sizes (which vary greatly in our task).

During training, we first obtain a lesion pseudo mask g that is an ellipse for the given lesion image, fitted from four endpoints of its RECIST annotation, then construct the constrained region I' by dilating g to four times its size so that I' is lesion-adaptive. After testing the model using different sizes (2/3/4/5 times the size of pseudo mask) of the constrained region for RLS loss computation, we found that using 4 times provides the best performance and the performance slightly changes when using 3/4/5 times. Figure 3 shows two examples, where the region inside the red curve is the constrained region I'. As we can see, for the same lesion, the size of I' remains stable under different data augmentations, *e.g.*, randomly cropping and rotating.

2.3 Model Optimization

As in Fig. 1, AHRNet takes as input a CT image and outputs three probability maps (denoted as p_1, p_2, and p_3). Besides the regional level set loss ℓ_{rls}, we also use a segmentation loss (ℓ_{seg}) to compute the errors between the predicted probability maps and the pseudo masks (denoted as g_1, g_2, and g_3) for optimization. ℓ_{seg} is the summation of a binary cross entropy loss (ℓ_{bce}) and an IoU loss (ℓ_{iou}) [19], *i.e.*, $\ell_{seg} = \sum_{k=1}^{3}[\ell_{bce}(p_k, g_k) + \ell_{iou}(p_k, g_k)]$, which are defined as

$$\begin{aligned} \ell_{bce}(p, g) &= -\frac{1}{|I|}\sum_{i \in I}[g(i)\log(p(i)) + (1 - g(i))\log(1 - p(i))], \\ \ell_{iou}(p, g) &= 1 - \left(\sum_{i \in I} g(i)p(i)\right) / \left(\sum_{i \in I} g(i) + p(i) - g(i)p(i)\right), \end{aligned} \tag{3}$$

where we omit the subscript k of p and g for simplicity. Although as a pixel-wise loss, ℓ_{bce} does not consider the global structure of lesion, ℓ_{iou} can optimize the global structure of the segmented lesion rather than focusing on a single pixel.

In order to make ℓ_{rls} to provide effective gradients for back propagation, we do not add ℓ_{rls} for training until the model converges using only ℓ_{seg}. That means the model can produce a good-quality prediction at its early training stage with ℓ_{seg}, which could be considered as a good initialization for ℓ_{rls}. We then add ℓ_{rls}, which is reduced by a factor of 0.1, at the later training stage so that it can provide useful gradients for optimization, making the prediction closer to the lesion boundary.

The supervision for training is the constructed pseudo mask g and its quality directly affects the final lesion segmentation performance. However, our straightforward ellipse estimation is not guaranteed to always generating lesion masks with high fidelity. Therefore, based on the prediction p from the trained model and the fitted ellipse e, we further construct an updated pseudo mask g' by setting $p \cap e$ as the foreground, $p \cup e - p \cap e$ as the ignored region, and the rest as the background. With the updated pseudo masks, we can retrain the model using the same way described above. The training has converged after three rounds.

3 Experiments

Datasets and Evaluation Metrics. NIH DeepLesion dataset [30] has 32,735 CT lesion images from 4,459 patients, where a variety of lesions over the whole body parts are included, such as lung nodules, liver lesions, enlarged lymph nodes, etc. Each lesion has only a RECIST annotation that serves as weak supervision for model optimization. Following [3], 1,000 lesion images from 500 patients are manually segmented as a test set for quantitative evaluation. The rest patient data are used for training. Besides, we collect a hold-out test set from our collaborated *anonymous* hospital for external validation. It contains 470 lesions from 170 patients with pixel-wise manual masks, which also covers various lesion types over the whole body. The precision, recall, and Dice coefficient are used for performance evaluation.

Implementation Details. AHRNet is implemented in PyTorch [18] and its backbone is initialized with ImageNet [8] pre-trained weights [28]. It is trained using Adam optimizer [15] with an initial learning rate of 0.001 for 80 epochs reduced by 0.1 at epoch 40 and 60. The data augmentation operations include randomly scaling, cropping, rotating, brightness and contrast adjusting, and Gaussian blurring. After data augmentation, the long sides of all training images are randomly resized into a range of [128, 256]. For testing, an image is taken as input directly if its long side is in the range, otherwise, it is resized into the closest bound first.

Quantitative Results. nnUNet [13] is a robust and self-adapting framework on the basis of vanilla UNets [21]. It has been widely used and overwhelmingly successful in many medical image segmentation tasks, suggesting itself as a strong baseline for comparisons. For empirical comparisons, besides the existing work [3,27], we train three segmentation models, *i.e.*, nnUNet, HRNet, and the proposed AHRNet, with or without our proposed RLS loss. Table 1 reports the quantitative results of different methods/variations on two test sets. As can be seen, 1) our method "AHRNet+RLS" achieves the highest Dice of 92.6% surpassing the best previous work [27] by 1.4%, which demonstrates its effectiveness for weakly-supervised lesion segmentation. 2) When RLS is not used,

Table 1. Lesion segmentation results of different methods. The mean and standard deviation of pixel-wise recall, precision and Dice score are reported (%).

Method	DeepLesion test set			Hold-out test set		
	Precision	Recall	Dice	Precision	Recall	Dice
Cai *et al.* [3]	89.3 ± 11.1	93.3 ± 9.5	90.6 ± 8.9	–	–	–
Tang *et al.* [27]	88.3 ± 5.7	**94.7 ± 7.4**	91.2 ± 3.9	–	–	–
nnUNet [13]	95.5 ± 5.3	85.8 ± 8.8	90.0 ± 4.9	88.2 ± 12.3	85.5 ± 13.0	85.5 ± 8.7
nnUNet+RLS	96.8 ± 4.7	87.1 ± 8.6	91.4 ± 5.7	89.8 ± 10.9	85.8 ± 10.3	86.8 ± 6.9
HRNet [28]	**97.5 ± 3.2**	84.9 ± 8.6	90.5 ± 5.3	86.0 ± 13.9	88.7 ± 11.7	86.0 ± 9.4
HRNet+RLS	95.0 ± 5.8	89.7 ± 9.4	91.8 ± 6.2	86.9 ± 12.1	**90.3 ± 10.4**	87.6 ± 8.1
AHRNet	97.0 ± 3.7	87.0 ± 8.3	91.5 ± 5.1	88.5 ± 11.3	87.7 ± 11.7	86.8 ± 6.4
AHRNet+RLS	95.8 ± 4.5	90.2 ± 7.4	**92.6 ± 4.3**	**89.8 ± 10.0**	88.3 ± 9.6	**88.2 ± 6.0**

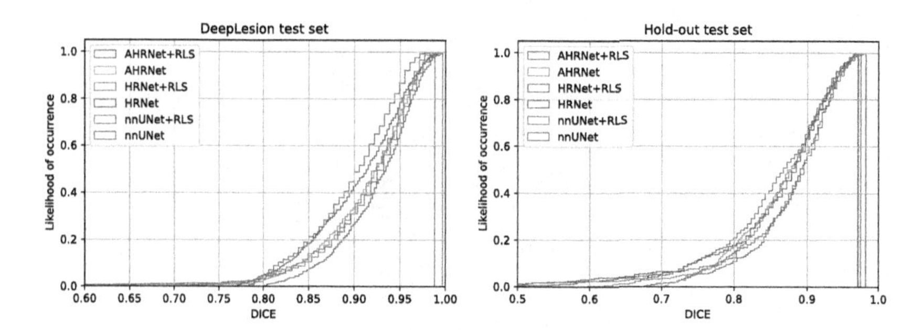

Fig. 4. Dice cumulative histograms of different methods on both test sets. The x axes are truncated for better visualization when drawing the figures.

the proposed AHRNet still has the best Dice score, indicating that the designed components are effective to enhance the feature representations for lesion segmentation. 3) For all three models, *i.e.*, nnUNet, HRNet, and AHRNet, the performance is consistently and remarkedly improved when using RLS; the Dice gains are 1.4%, 1.3%, and 1.1%, respectively. This shows that RLS is capable of making the segmentation outcome promisingly close to lesion boundaries and can be effectively optimized via a weakly-supervised fashion. Figure 4 shows the Dice cumulative histograms of different methods. Our method is observed with about 99% or 90% lesions having Dice ≥ 0.8 on the DeepLesion or hold-out test sets, respectively. Figure 4 evidently validates the overall improvements by our methods.

Figure 5 shows some **visual examples** of lesion segmentation results by different methods for qualitative comparisons. As can be seen, 1) our lesion segmentation results are closer to the manual annotations than others, suggesting that our AHRNet model has desirable capability to learn more comprehensive features for distinguishing pixels nearby the lesion boundaries. 2) When using RLS, the results produced by all methods are closer to the manual annotations than the ones without RLS. Through optimizing the regional level set loss, models can push or pull the segmentation results to improve the alignment with lesion boundaries. 3) When lesions have highly irregular shapes and blurry boundaries, all methods cannot segment them well, as shown in the last two columns of Fig. 5. Beyond the weakly-supervised learning means, using a large number of fully- annotated data for training may alleviate these limitations.

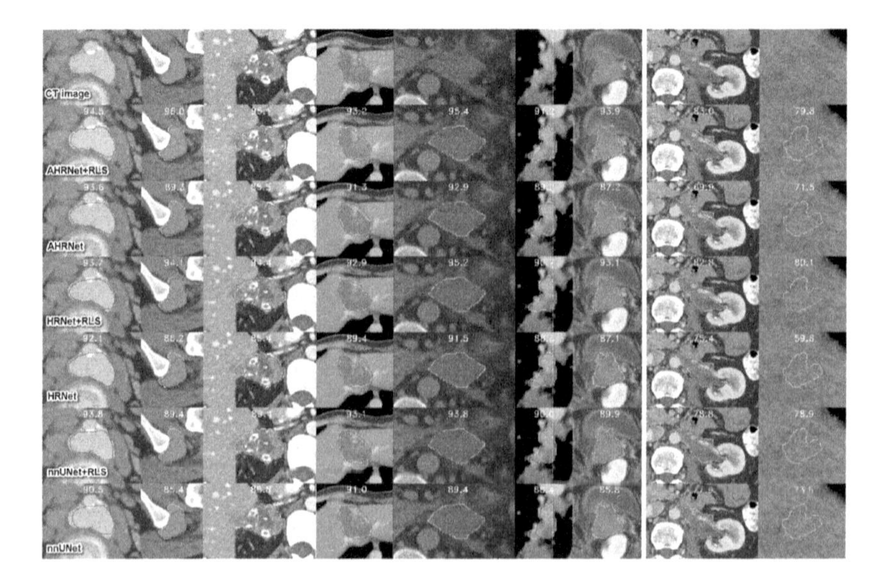

Fig. 5. Visual examples of results produced by different methods. The green and red curves represent the manual annotations and automatic segmentation results, respectively. The yellow digits indicate Dice scores. The last two columns give two failure cases. (Color figure online)

Table 2. Results of different configurations on the DeepLesion test set.

Configurations		Precision	Recall	Dice
(1)	Baseline (HRNet [28])	**97.5 ± 3.2**	84.9 ± 8.6	90.5 ± 5.3
(2)	+ DE	95.5 ± 5.8	87.7 ± 9.1	91.0 ± 5.9
(3)	+ DE + DA	95.1 ± 6.1	88.4 ± 8.1	91.3 ± 5.4
(4)	+ DE + DA + SA	97.0 ± 3.7	87.0 ± 8.3	91.5 ± 5.1
(5)	+ DE + DA + SA + LS	96.2 ± 4.3	89.4 ± 7.7	92.2 ± 4.6
(6)	+ DE + DA + SA + RLS	95.8 ± 4.5	**90.2 ± 7.4**	**92.6 ± 4.3**

Ablation Studies. Table 2 lists quantitative comparisons of using different configurations to construct models for lesion segmentation. From Table 2, when gradually introducing these components, *i.e.*, the decoder (DE), the dual-attention module (DA), and the scale attention module (SA), into the baseline

(HRNet) sequentially, the performance is also improved accordingly. This indicates that these design options are useful to learn more representative features for lesion segmentation. When adding our RLS loss for training, it brings the largest performance gain (seeing row 4 versus row 6), *e.g.*, the Dice score is improved from 91.5% to 92.6%. The importance of RLS in our entire framework is validated. We also compute ℓ_{rls} using the entire input image rather than the constrained region during training, denoted as LS. From row 5 and 6, RLS achieves better performance than LS, implying that using the constrained regions for ℓ_{rls} computation is more reliable and beneficial for model optimization.

4 Conclusions

In this paper, we present an attention enhanced high-resolution network (AHRNet) and a regional level set (RLS) loss for accurate weakly-supervised universal lesion segmentation. Instead of directly using the deep image features extracted by HRNet, AHRNet is able to learn more representative high-resolution features for lesion segmentation by integrating a decoder and attention mechanism. Assisted with the RLS loss, AHRNet model can further distinguish the pixels nearby the lesion boundaries more accurately. Extensive experimental results demonstrate that our proposed methods bring in better and more robust lesion segmentation results; specifically, RLS improves the performance significantly.

References

1. Agarwal, V., Tang, Y., Xiao, J., Summers, R.M.: Weakly-supervised lesion segmentation on CT scans using co-segmentation. In: Medical Imaging: Computer-Aided Diagnosis, vol. 11314, p. 113141J (2020)
2. Beaumont, H., et al.: Radiology workflow for RECIST assessment in clinical trials: can we reconcile time-efficiency and quality? Eur. J. Radiol. **118**, 257–263 (2019)
3. Cai, J., et al.: Accurate weakly-supervised deep lesion segmentation using large-scale clinical annotations: slice-propagated 3d mask generation from 2D RECIST. In: Frangi, A.F., Schnabel, J.A., Davatzikos, C., Alberola-López, C., Fichtinger, G. (eds.) MICCAI 2018. LNCS, vol. 11073, pp. 396–404. Springer, Cham (2018). https://doi.org/10.1007/978-3-030-00937-3_46
4. Chan, T.F., Vese, L.A.: Active contours without edges. IEEE Trans. Image Process. **10**(2), 266–277 (2001)
5. Chen, X., Williams, B.M., Vallabhaneni, S.R., Czanner, G., Williams, R., Zheng, Y.: Learning active contour models for medical image segmentation. In: CVPR, pp. 11632–11640 (2019)
6. Chlebus, G., Schenk, A., Moltz, J.H., van Ginneken, B., Hahn, H.K., Meine, H.: Automatic liver tumor segmentation in CT with fully convolutional neural networks and object-based postprocessing. Sci. Rep. **8**(1), 1–7 (2018)
7. Christ, P.F., et al.: Automatic liver and tumor segmentation of CT and MRI volumes using cascaded fully convolutional neural networks. arXiv preprint arXiv:1702.05970 (2017)
8. Deng, J., Dong, W., Socher, R., Li, L., Li, K., Li, F.: ImageNet: a large-scale hierarchical image database. In: CVPR, pp. 248–255 (2009)

9. Eisenhauer, E.A., et al.: New response evaluation criteria in solid tumours: revised RECIST guideline (version 1.1). Eur. J. Cancer **45**(2), 228–247 (2009)
10. Fu, J., et al.: Dual attention network for scene segmentation. In: CVPR, pp. 3146–3154 (2019)
11. Havaei, M., et al.: Brain tumor segmentation with deep neural networks. Med. Image Anal. **35**, 18–31 (2017)
12. Hu, J., Shen, L., Albanie, S., Sun, G., Wu, E.: Squeeze-and-excitation networks. IEEE Trans. Pattern Anal. Mach. Intell. **42**(8), 2011–2023 (2020)
13. Isensee, F., et al.: nnU-Net: self-adapting framework for U-Net-based medical image segmentation. arXiv preprint arXiv:1809.10486 (2018)
14. Kim, Y., Kim, S., Kim, T., Kim, C.: CNN-based semantic segmentation using level set loss. In: WACV, pp. 1752–1760 (2019)
15. Kingma, D.P., Ba, J.: Adam: a method for stochastic optimization. arXiv preprint arXiv:1412.6980 (2014)
16. Li, X., Chen, H., Qi, X., Dou, Q., Fu, C.W., Heng, P.A.: H-Denseunet: Hybrid densely connected UNet for liver and tumor segmentation from CT volumes. IEEE Trans. Med. Imaging **37**(12), 2663–2674 (2018)
17. Nogues, I., et al.: Automatic lymph node cluster segmentation using holistically-nested neural networks and structured optimization in CT images. In: Ourselin, S., Joskowicz, L., Sabuncu, M.R., Unal, G., Wells, W. (eds.) MICCAI 2016. LNCS, vol. 9901, pp. 388–397. Springer, Cham (2016). https://doi.org/10.1007/978-3-319-46723-8_45
18. Paszke, A., et al.: PyTorch: an imperative style, high-performance deep learning library. In: NeurIPS, pp. 8024–8035 (2019)
19. Rahman, M.A., Wang, Y.: Optimizing intersection-over-union in deep neural networks for image segmentation. In: ISVC, pp. 234–244 (2016)
20. Raju, A., et al.: Co-heterogeneous and adaptive segmentation from multi-source and multi-phase CT imaging data: a study on pathological liver and lesion segmentation. In: Vedaldi, A., Bischof, H., Brox, T., Frahm, J.-M. (eds.) ECCV 2020. LNCS, vol. 12368, pp. 448–465. Springer, Cham (2020). https://doi.org/10.1007/978-3-030-58592-1_27
21. Ronneberger, O., Fischer, P., Brox, T.: U-Net: convolutional networks for biomedical image segmentation. In: Navab, N., Hornegger, J., Wells, W.M., Frangi, A.F. (eds.) MICCAI 2015. LNCS, vol. 9351, pp. 234–241. Springer, Cham (2015). https://doi.org/10.1007/978-3-319-24574-4_28
22. Sung, H., et al.: Global cancer statistics 2020: GLOBOCAN estimates of incidence and mortality worldwide for 36 cancers in 185 countries. CA Cancer J. Clin. (2021)
23. Tang, Y.B., Oh, S., Tang, Y.X., Xiao, J., Summers, R.M.: CT-realistic data augmentation using generative adversarial network for robust lymph node segmentation. In: SPIE Med. Imaging **10950**, 109503V (2019)
24. Tang, Y., Harrison, A.P., Bagheri, M., Xiao, J., Summers, R.M.: Semi-automatic RECIST labeling on CT scans with cascaded convolutional neural networks. In: Frangi, A.F., Schnabel, J.A., Davatzikos, C., Alberola-López, C., Fichtinger, G. (eds.) MICCAI 2018. LNCS, vol. 11073, pp. 405–413. Springer, Cham (2018). https://doi.org/10.1007/978-3-030-00937-3_47
25. Tang, Y., Tang, Y., Zhu, Y., Xiao, J., Summers, R.M.: E^2Net: an edge enhanced network for accurate liver and tumor segmentation on CT scans. In: Martel, A.L., et al. (eds.) MICCAI 2020. LNCS, vol. 12264, pp. 512–522. Springer, Cham (2020). https://doi.org/10.1007/978-3-030-59719-1_50
26. Tang, Y., et al.: Lesion segmentation and RECIST diameter prediction via click-driven attention and dual-path connection. arXiv preprint arXiv:2105.01828 (2021)

27. Tang, Y., Yan, K., Xiao, J., Summers, R.M.: One click lesion RECIST measurement and segmentation on CT scans. In: Martel, A.L., et al. (eds.) MICCAI 2020. LNCS, vol. 12264, pp. 573–583. Springer, Cham (2020). https://doi.org/10.1007/978-3-030-59719-1_56
28. Wang, J., et al.: Deep high-resolution representation learning for visual recognition. IEEE Trans. Pattern Anal. Mach. Intell. (2020)
29. Wang, S., et al.: Central focused convolutional neural networks: developing a data-driven model for lung nodule segmentation. Med. Image Anal. **40**, 172–183 (2017)
30. Yan, K., Wang, X., Lu, L., Summers, R.M.: DeepLesion: automated mining of large-scale lesion annotations and universal lesion detection with deep learning. J. Med. Imaging **5**(3), 036501 (2018)
31. Zhang, D., Chen, B., Chong, J., Li, S.: Weakly-supervised teacher-student network for liver tumor segmentation from non-enhanced images. Med. Image Anal. **70**, 102005 (2021)
32. Zhang, L., et al.: Robust pancreatic ductal adenocarcinoma segmentation with multi-institutional multi-phase partially-annotated CT scans. In: Martel, A.L., et al. (eds.) MICCAI 2020. LNCS, vol. 12264, pp. 491–500. Springer, Cham (2020). https://doi.org/10.1007/978-3-030-59719-1_48
33. Zhang, M., Dong, B., Li, Q.: Deep active contour network for medical image segmentation. In: Martel, A.L., et al. (eds.) MICCAI 2020. LNCS, vol. 12264, pp. 321–331. Springer, Cham (2020). https://doi.org/10.1007/978-3-030-59719-1_32
34. Zhou, B., Crawford, R., Dogdas, B., Goldmacher, G., Chen, A.: A progressively-trained scale-invariant and boundary-aware deep neural network for the automatic 3D segmentation of lung lesions. In: WACV, pp. 1–10. IEEE (2019)
35. Zhu, Z., et al.: Lymph node gross tumor volume detection and segmentation via distance-based gating using 3D CT/PET imaging in radiotherapy. In: Martel, A.L., et al. (eds.) MICCAI 2020. LNCS, vol. 12267, pp. 753–762. Springer, Cham (2020). https://doi.org/10.1007/978-3-030-59728-3_73
36. Zhu, Z., Xia, Y., Xie, L., Fishman, E.K., Yuille, A.L.: Multi-scale coarse-to-fine segmentation for screening pancreatic ductal adenocarcinoma. In: Shen, D., et al. (eds.) MICCAI 2019. LNCS, vol. 11769, pp. 3–12. Springer, Cham (2019). https://doi.org/10.1007/978-3-030-32226-7_1

Bounding Box Tightness Prior for Weakly Supervised Image Segmentation

Juan Wang[1(✉)] and Bin Xia[2]

[1] Delta Micro Technology, Inc., Laguna Hills, CA 92653, USA
jwang134@hawk.iit.edu
[2] Shenzhen SiBright Co. Ltd., Shenzhen 518052, Guangdong, China
b.xia@sibionics.com

Abstract. This paper presents a weakly supervised image segmentation method that adopts tight bounding box annotations. It proposes generalized multiple instance learning (MIL) and smooth maximum approximation to integrate the bounding box tightness prior into the deep neural network in an end-to-end manner. In generalized MIL, positive bags are defined by parallel crossing lines with a set of different angles, and negative bags are defined as individual pixels outside of any bounding boxes. Two variants of smooth maximum approximation, i.e., α-softmax function and α-quasimax function, are exploited to conquer the numeral instability introduced by maximum function of bag prediction. The proposed approach was evaluated on two pubic medical datasets using Dice coefficient. The results demonstrate that it outperforms the state-of-the-art methods. The codes are available at https://github.com/wangjuan313/wsis-boundingbox.

Keywords: Weakly supervised image segmentation · Bounding box tightness prior · Multiple instance learning · Smooth maximum approximation · Deep neural networks

1 Introduction

In recent years, image segmentation has been made great progress with the development of deep neural networks in a fully-supervised manner [3,12,16]. However, collecting large-scale training set with precise pixel-wise annotation is considerably labor-intensive and expensive. To tackle this issue, there have been great interests in the development of weakly supervised image segmentation. All kinds of supervisions have been considered, including image-level annotations [1,19], scribbles [10], bounding boxes [5,15], and points [2]. This work focuses on image segmentation by employing supervision of bounding boxes.

In the literature, some efforts have been made to develop the weakly supervised image segmentation methods adopting the bounding box annotations. For example, Rajchl *et al.* [15] developed an iterative optimization method for image segmentation, in which a neural network classifier was trained from bounding

M. de Bruijne et al. (Eds.): MICCAI 2021, LNCS 12902, pp. 526–536, 2021.
https://doi.org/10.1007/978-3-030-87196-3_49

box annotations. Khoreva *et al.* [6] employed GrabCut [18] and MCG proposals [14] to obtain pseudo label for image segmentation. Hsu *et al.* [4] exploited multiple instance learning (MIL) strategy and mask R-CNN for image segmentation. Kervadec *et al.* [5] leveraged the tightness prior to a deep learning setting via imposing a set of constraints on the network outputs for image segmentation.

In this work, we present a generalized MIL formulation and smooth maximum approximation to integrate the bounding box tightness prior into the network in an end-to-end manner. Specially, we employ parallel crossing lines with a set of different angles to obtain positive bags and use individual pixels outside of any bounding boxes as negative bags. We consider two variants of smooth maximum approximation to conquer the numeral instability introduced by maximum function of bag prediction. The experiments on two public medical datasets demonstrate that the proposed approach outperforms the state-of-the-art methods.

2 Methods

2.1 Preliminaries

Problem Description. Suppose I denotes an input image, and $Y \in \{1, 2, \cdots, C\}$ is its corresponding pixel-level category label, in which C is the number of categories of the objects. The image segmentation problem is to obtain the prediction of Y, denoted as P, for the input image I.

In the fully supervised image segmentation setting, for an image I, its pixel-wise category label Y is available during training. Instead, in this study we are only provided its bounding box label B. Suppose there are M bounding boxes, then its bounding box label is $B = \{b_m, y_m\}, m = 1, 2, \cdots, M$, where the location label b_m is a 4-dimensional vector representing the top left and bottom right points of the bounding box, and $y_m \in \{1, 2, \cdots, C\}$ is its category label.

Deep Neural Network. This study considers deep neural networks which output the pixel-wise prediction of the input image, such as Unet [16], FCN [12], etc. Due to the possible overlaps of objects of different categories in images, especially in medical images, the image segmentation problem is formulated as a multi-label classification problem in this study. That is, for a location k in the input image, it outputs a vector $\mathbf{p}_k$ with C elements, one element for a category; each element is converted to the range of $[0, 1]$ by the sigmoid function.

2.2 MIL Baseline

MIL Definition and Bounding Box Tightness Prior. Multiple instance learning (MIL) is a type of supervised learning. Different from the traditional supervised learning which receives a set of training samples which are individually labeled, MIL receives a set of labeled bags, each containing many training samples. In MIL, a bag is labeled as negative if all of its samples are negative, a bag is positive if it has at least one sample which is positive.

Tightness prior of bounding box indicates that the location label of bounding box is the smallest rectangle enclosing the whole object, thus the object must touch the four sides of its bounding box, and does not overlap with the region outside its bounding box. The crossing line of a bounding box is defined as a line with its two endpoints located on the opposite sides of the box. In an image I under consideration, for an object with category c, any crossing line in the bounding box has at least one pixel belonging to the object in the box; any pixels outside of any bounding boxes of category c do not belong to category c. Hence pixels on a cross line compose a positive bag for category c, while pixels outside of any bounding boxes of category c are used for negative bags.

MIL Baseline. For category c in an image, the baseline approach simply considers all of the horizontal and vertical crossing lines inside the boxes as positive bags, and all of the horizontal and vertical crossing lines that do not overlap any bounding boxes of category c as negative bags. This definition is shown in Fig. 1(a) and has been widely employed in the literature [4,5].

Fig. 1. Demonstration of positive and negative bags, in which the red rectangle is the bounding box of the object. (a) MIL baseline. Examples of positive bags are denoted by green dashed lines, and examples of negative bags are marked by the blue dashed lines. (b) Generalized MIL. Examples of positive bags with different angles are shown ($\theta = 25°$ for the yellow dashed lines and $\theta = 0°$ for the green dashed lines). Individual pixels outside of the red box are negative bags, which are omitted here for simplicity. (Color figure online)

To optimize the network parameters, MIL loss with two terms are considered [4]. For category c, suppose its positive and negative bags are denoted by $\mathcal{B}_c^+$ and $\mathcal{B}_c^-$, respectively, then loss $\mathcal{L}_c$ can be defined as follows:

$$\mathcal{L}_c = \phi_c(P; \mathcal{B}_c^+, \mathcal{B}_c^-) + \lambda \varphi_c(P) \tag{1}$$

where ϕ_c is the unary loss term, φ_c is the pairwise loss term, and λ is a constant value controlling the trade off between the unary loss and the pairwise loss.

The unary loss ϕ_c enforces the tightness constraint of bounding boxes on the network prediction P by considering both positive bags $\mathcal{B}_c^+$ and negative bags $\mathcal{B}_c^-$. A positive bag contains at least one pixel inside the object, hence the pixel with highest prediction tends to be inside the object, thus belonging to category c. In contrast, no pixels in negative bags belong to any objects, hence even the pixel with highest prediction does not belong to category c. Based on these observations, the unary loss ϕ_c can be expressed as:

$$\phi_c = -\frac{1}{|\mathcal{B}_c^+| + |\mathcal{B}_c^-|}\left(\sum_{b\in\mathcal{B}_c^+} \log P_c(b) + \sum_{b\in\mathcal{B}_c^-} \log(1 - P_c(b))\right) \tag{2}$$

where $P_c(b) = \max_{k\in b}(p_{kc})$ is the prediction of the bag b being positive for category c, p_{kc} is the network output of the pixel location k for category c, and $|\mathcal{B}|$ is the cardinality of $\mathcal{B}$.

The unary loss is binary cross entropy loss for bag prediction $P_c(b)$, it achieves minimum when $P_c(b) = 1$ for positive bags and $P_c(b) = 0$ for negative bags. More importantly, during training the unary loss adaptively selects a positive sample per positive bag and a negative sample per negative bag based on the network prediction for optimization, thus yielding an adaptive sampling effect.

However, using the unary loss alone is prone to segment merely the discriminative parts of an object. To address this issue, the pairwise loss as follows is introduced to pose the piece-wise smoothness on the network prediction.

$$\varphi_c = \frac{1}{|\varepsilon|}\sum_{(k,k')\in\varepsilon} (p_{kc} - p_{k'c})^2 \tag{3}$$

where ε is the set containing all neighboring pixel pairs, p_{kc} is the network output of the pixel location k for category c.

Finally, considering all C categories, the MIL loss $\mathcal{L}$ is:

$$\mathcal{L} = \sum_{c=1}^{C} \mathcal{L}_c \tag{4}$$

2.3 Generalized MIL

Generalized Positive Bags. For an object of height H pixels and width W pixels, the positive bag definition in the MIL baseline yields only $H+W$ positive bags, a much smaller number when compared with the size of the object. Hence it limits the selected positive samples during training, resulting in a bottleneck for image segmentation.

To eliminate this issue, this study proposes to generalize positive bag definition by considering all parallel crossing lines with a set of different angles. An parallel crossing line is parameterized by an angle $\theta \in (-90°, 90°)$ with respect to the edges of the box where its two endpoints located. For an angle θ, two sets of parallel crossing lines can be obtained, one crosses up and bottom edges of the box, and the other crosses left and right edges of the box. As examples, in Fig. 1(b), we show positive bags of two different angles, in which those marked by yellow dashed colors have $\theta = 25°$, and those marked by green dashed lines are with $\theta = 0°$. Note the positive bag definition in MIL baseline is a special case of the generalized positive bags with $\theta = 0°$.

Generalized Negative Bags. The similar issue also exists for the negative bag definition in the MIL baseline. To tackle this issue, for a category c, we propose to define each individual pixel outside of any bounding boxes of category c as a negative bag. This definition greatly increases the number of negative bags, and forces the network to see every pixel outside of bounding boxes during training.

Improved Unary Loss. The generalized MIL definitions above will inevitably lead to imbalance between positive and negative bags. To deal with this issue, we borrow the concept of focal loss [17] and use the improved unary loss as follows:

$$\phi_c = -\frac{1}{N^+}\left(\sum_{b \in \mathcal{B}_c^+} \beta\left(1 - P_c(b)\right)^\gamma \log P_c(b) + \sum_{b \in \mathcal{B}_c^-} (1-\beta) P_c(b)^\gamma \log(1 - P_c(b))\right) \tag{5}$$

where $N^+ = \max(1, |\mathcal{B}_c^+|)$, $\beta \in [0, 1]$ is the weighting factor, and $\gamma \geq 0$ is the focusing parameter. The improved unary loss is focal loss for bag prediction, it achieves minimum when $P_c(b) = 1$ for positive bags and $P_c(b) = 0$ for negative bags.

2.4 Smooth Maximum Approximation

In the unary loss, the maximum prediction of pixels in a bag is used as bag prediction $P_c(b)$. However, the derivative $\partial P_c / \partial p_{kc}$ is discontinuous, leading to numerical instability. To solve this problem, we replace the maximum function by its smooth maximum approximation [8]. Let the maximum function be $f(\mathbf{x}) = \max_{i=1}^n x_i$, its two variants of smooth maximum approximation as follows are considered.

(1) *α-softmax function:*

$$S_\alpha(\mathbf{x}) = \frac{\sum_{i=1}^n x_i e^{\alpha x_i}}{\sum_{i=1}^n e^{\alpha x_i}} \tag{6}$$

where $\alpha > 0$ is a constant. The higher the α value is, the closer the approximation $S_\alpha(\mathbf{x})$ to $f(\mathbf{x})$. For $\alpha \to 0$, a soft approximation of the mean function is obtained.

(2) *α-quasimax function:*

$$Q_\alpha(\mathbf{x}) = \frac{1}{\alpha} \log\left(\sum_{i=1}^n e^{\alpha x_i}\right) - \frac{\log n}{\alpha} \tag{7}$$

where $\alpha > 0$ is a constant. The higher the α value is, the closer the approximation $Q_\alpha(\mathbf{x})$ to $f(\mathbf{x})$. One can easily prove that $Q_\alpha(\mathbf{x}) \leq f(\mathbf{x})$ always holds.

In real application, each bag usually has more than one pixel belonging to object segment. However, $\partial f/\partial x_i$ has value 0 for all but the maximum x_i, thus the maximum function considers only the maximum x_i during optimization. In contrast, the smooth maximum approximation has $\partial S_\alpha/\partial x_i > 0$ and $\partial Q_\alpha/\partial x_i > 0$ for all x_i, thus it considers every x_i during optimization. More importantly, in the smooth maximum approximation, large x_i has much greater derivative than small x_i, thus eliminating the possible adverse effect of negative samples in the optimization. In the end, besides the advantage of conquering numerical instability, the smooth maximum approximation is also beneficial for performance improvement.

3 Experiments

3.1 Datasets

This study made use of two public medical datasets for performance evaluation. The first one is the prostate MR image segmentation 2012 (PROMISE12) dataset [11] for prostate segmentation and the second one is the anatomical tracings of lesions after stroke (ATLAS) dataset [9] for brain lesion segmentation.

Prostate Segmentation: The PROMISE12 dataset was first developed for prostate segmentation in MICCAI 2012 grand challenge [11]. It consists of the transversal T2-weighted MR images from 50 patients, including both benign and prostate cancer cases. These images were acquired at different centers with multiple MRI vendors and different scanning protocols. Same as the study in [5], the dataset was divided into two non-overlapping subsets, one with 40 patients for training and the other with 10 patients for validation.

Brain Lesion Segmentation: The ATLAS dataset is a well-known open-source dataset for brain lesion segmentation. It consists of 229 T1-weighted MR images from 220 patients. These images were acquired from different cohorts and different scanners. The annotations were done by a group of 11 experts. Same as the study in [5], the dataset was divided into two non-overlapping subsets, one with 203 images from 195 patients for training and the other with 26 images from 25 patients for validation.

3.2 Implementation Details

All experiments were implemented using PyTorch in this study. Image segmentation was conducted on the 2D slices of MR images. The parameters in the MIL loss (1) were set as $\lambda = 10$ based on experience, and those in the improved unary loss (5) were set as $\beta = 0.25$ and $\gamma = 2$ according to the focal loss [17]. As indicated below, most experimental setups were set to be same as study in [5] for fairness of comparison.

For the PROMISE12 dataset, a residual version of UNet [16] was used for segmentation [5]. The models were trained with Adam optimizer [7] with the following parameter values: batch size = 16, initial learning rate = 10^{-4}, $\beta_1 = 0.9$, and $\beta_2 = 0.99$. To enlarge the set of images for training, an off-line data augmentation procedure [5] was applied to the images in the training set as follows: 1) mirroring, 2) flipping, and 3) rotation.

For the ATLAS dataset, ENet [13] was employed as a backbone architecture for segmentation [5]. The models were trained with Adam optimizer with the following parameter values: batch size = 80, initial learning rate = 5×10^{-4}, $\beta_1 = 0.9$, and $\beta_2 = 0.99$. No augmentation was performed during training [5].

3.3 Performance Evaluation

To measure the performance of the proposed approach, the Dice coefficient was employed, which has been applied as a standard performance metric in medical image segmentation. The Dice coefficient was calculated based on the 3D MR images by stacking the 2D predictions of the networks together.

To demonstrate the overall performance of the proposed method, we considered the baseline method in the experiments for comparison. Moreover, we further perform comparisons with state-of-the-art methods with bounding-box annotations, including deep cut [15] and global constraint [5].

4 Results

4.1 Ablation Study

Generalized MIL. To demonstrate the effectiveness of the proposed generalized MIL formulation, in Table 1 we show the performance of the generalized MIL for the two datasets. For comparison, we also show the results of the baseline method. It can be observed that the generalized MIL approach consistently outperforms the baseline method at different angle settings. In particular, the PROMISE12 dataset got best Dice coefficient of 0.878 at $\theta_{best} = (-40°, 40°, 20°)$ for the generalized MIL, compared with 0.859 for the baseline method. The ATLAS dataset achieved best Dice coefficient of 0.474 at $\theta_{best} = (-60°, 60°, 30°)$ for the generalized MIL, much higher than 0.408 for the baseline method.

Smooth Maximum Approximation. To demonstrate the benefits of the smooth maximum approximation, in Table 2 we show the performance of the MIL baseline method when the smooth maximum approximation was applied. As can be seen, for the PROMISE12 dataset, the better performance is obtained for α-softmax function with $\alpha = 4$ and 6 and for α-quasimax function with $\alpha = 6$ and 8. For the ATLAS dataset, the improved performance can also be observed for α-softmax function with $\alpha = 6$ and 8 and for α-quasimax function with $\alpha = 8$.

Table 1. Dice coefficients of the proposed generalized MIL for image segmentation, where $\theta = (\theta_1, \theta_2, \Delta)$ denotes evenly spaced angle values within interval (θ_1, θ_2) with step Δ. The standard deviation of Dice coefficients among different MR images are reported in the bracket. For comparison, results of the baseline method are also given.

Method	PROMISE12	ATLAS
MIL baseline	0.859 (0.038)	0.408 (0.249)
$\theta = (-40°, 40°, 10°)$	0.868 (0.031)	0.463 (0.278)
$\theta = (-40°, 40°, 20°)$	**0.878 (0.027)**	0.466 (0.248)
$\theta = (-60°, 60°, 30°)$	0.868 (0.047)	**0.474 (0.262)**

Table 2. Dice coefficients of the MIL baseline method when smooth maximum approximation was applied.

	Method	PROMISE12	ATLAS
α-softmax	$\alpha = 4$	0.861 (0.031)	0.401(0.246)
	$\alpha = 6$	0.861 (0.036)	0.424(0.255)
	$\alpha = 8$	0.859 (0.030)	0.414(0.264)
α-quasimax	$\alpha = 4$	0.856 (0.026)	0.405(0.246)
	$\alpha = 6$	0.873 (0.018)	0.371(0.240)
	$\alpha = 8$	0.869 (0.024)	0.414(0.256)

4.2 Main Experimental Results

In Table 3 the Dice coefficients of the proposed approach are given, in which $\alpha = 4, 6$, and 8 are considered in smooth maximum approximation functions and those with highest dice coefficient are reported. For comparison, the full supervision results are also shown in Table 3. As can be seen, the PROMISE12 dataset gets Dice coefficient 0.878 for α-softmax function and 0.880 for α-quasimax function, which are same as or higher than the results in the ablation study. More importantly, these values are close to 0.894 obtained by the full supervision. The similar trends are observed for the ATLAS dataset.

Table 3. Comparison of Dice coefficients for different methods.

Method	PROMISE12	ATLAS
Full supervision	0.894 (0.021)	0.512 (0.292)
θ_{best} + α-softmax	**0.878 (0.031)**	**0.494 (0.236)**
θ_{best} + α-quasimax	**0.880 (0.024)**	**0.488 (0.240)**
Deep cut [15]	0.827 (0.085)	0.375 (0.246)
Global constraint [5]	0.835 (0.032)	0.474 (0.245)

Furthermore, we also show the Dice coefficients of two state-of-the-art methods in Table 3. The PROMISE12 dataset gets Dice coefficient 0.827 for deep cut and 0.835 for global constraint, both of which are much lower than those of the proposed approach. Similarly, the proposed approach also achieves higher Dice coefficients compared with these two methods for the ATLAS dataset.

Finally, to visually demonstrate the performance of the proposed approach, qualitative segmentation results are depicted in Fig. 2. It can be seen that the proposed method achieves good segmentation results for both prostate segmentation task and brain lesion segmentation task.

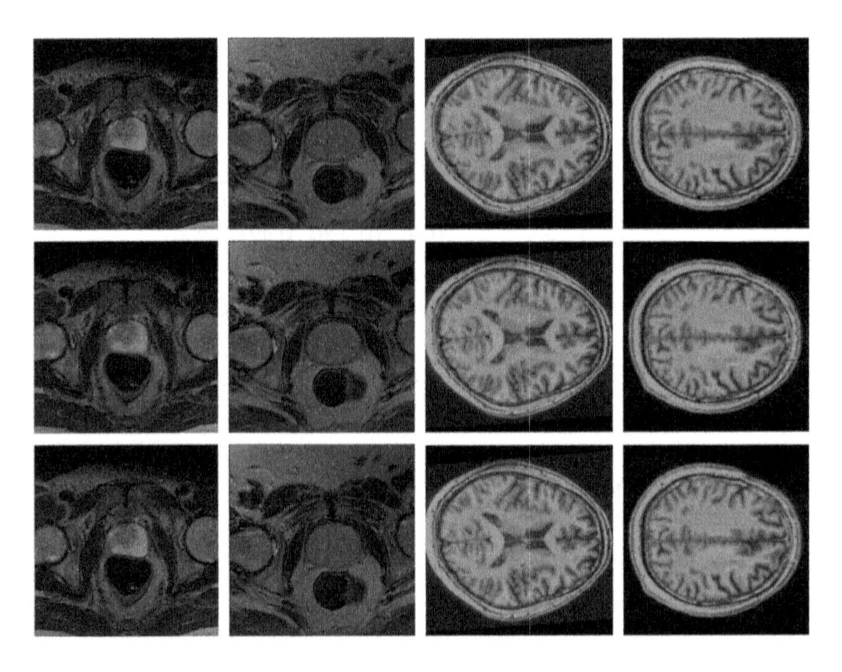

Fig. 2. Ground-truth segmentation (top row) and predicted segmentation results for the proposed approach with setting θ_{best} + α-softmax (middle row) and setting θ_{best} + α-quasimax (bottom row) on the validation set for PROMISE12 (first two columns) and ATLAS (last two columns) datasets.

5 Conclusion

This paper described a weakly supervised image segmentation method with tight bounding box supervision. It proposed generalized MIL and smooth maximum approximation to integrate the supervision into the deep neural network. The experiments demonstrate the proposed approach outperforms the state-of-the-art methods. However, there is still performance gap between the weakly supervised approach and the fully supervised method. In the future, it would be interesting to study whether using multi-scale outputs and adding auxiliary object detection task improve the segmentation performance.

References

1. Ahn, J., Cho, S., Kwak, S.: Weakly supervised learning of instance segmentation with inter-pixel relations. In: Proceedings of the IEEE/CVF Conference on Computer Vision and Pattern Recognition, pp. 2209–2218 (2019)
2. Bearman, A., Russakovsky, O., Ferrari, V., Fei-Fei, L.: What's the point: semantic segmentation with point supervision. In: Leibe, B., Matas, J., Sebe, N., Welling, M. (eds.) ECCV 2016. LNCS, vol. 9911, pp. 549–565. Springer, Cham (2016). https://doi.org/10.1007/978-3-319-46478-7_34
3. Chen, L.C., Zhu, Y., Papandreou, G., Schroff, F., Adam, H.: Encoder-decoder with atrous separable convolution for semantic image segmentation. In: Proceedings of the European Conference on Computer Vision, pp. 801–818 (2018)
4. Hsu, C.C., Hsu, K.J., Tsai, C.C., Lin, Y.Y., Chuang, Y.Y.: Weakly supervised instance segmentation using the bounding box tightness prior. In: Advances in Neural Information Processing Systems, vol. 32, pp. 6586–6597 (2019)
5. Kervadec, H., Dolz, J., Wang, S., Granger, E., Ayed, I.B.: Bounding boxes for weakly supervised segmentation: global constraints get close to full supervision. In: Medical Imaging with Deep Learning, pp. 365–381. PMLR (2020)
6. Khoreva, A., Benenson, R., Hosang, J., Hein, M., Schiele, B.: Simple does it: weakly supervised instance and semantic segmentation. In: Proceedings of the IEEE Conference on Computer Vision and Pattern Recognition, pp. 876–885 (2017)
7. Kingma, D.P., Ba, J.: Adam: a method for stochastic optimization. arXiv preprint arXiv:1412.6980 (2014)
8. Lange, M., Zühlke, D., Holz, O., Villmann, T., Mittweida, S.G.: Applications of LP-norms and their smooth approximations for gradient based learning vector quantization. In: ESANN, pp. 271–276 (2014)
9. Liew, S.L., et al.: A large, open source dataset of stroke anatomical brain images and manual lesion segmentations. Sci. Data **5**(1), 1–11 (2018)
10. Lin, D., Dai, J., Jia, J., He, K., Sun, J.: ScribbleSup: scribble-supervised convolutional networks for semantic segmentation. In: Proceedings of the IEEE Conference on Computer Vision and Pattern Recognition, pp. 3159–3167 (2016)
11. Litjens, G.: Evaluation of prostate segmentation algorithms for MRI: the PROMISE12 challenge. Med. Image Anal. **18**(2), 359–373 (2014)
12. Long, J., Shelhamer, E., Darrell, T.: Fully convolutional networks for semantic segmentation. In: Proceedings of the IEEE Conference on Computer Vision and Pattern Recognition, pp. 3431–3440 (2015)
13. Paszke, A., Chaurasia, A., Kim, S., Culurciello, E.: ENet: a deep neural network architecture for real-time semantic segmentation. arXiv preprint arXiv:1606.02147 (2016)
14. Pont-Tuset, J., Arbelaez, P., Barron, J.T., Marques, F., Malik, J.: Multiscale combinatorial grouping for image segmentation and object proposal generation. IEEE Trans. Pattern Anal. Mach. Intell. **39**(1), 128–140 (2016)
15. Rajchl, M., et al.: DeepCut: object segmentation from bounding box annotations using convolutional neural networks. IEEE Trans. Med. Imaging **36**(2), 674–683 (2016)
16. Ronneberger, O., Fischer, P., Brox, T.: U-Net: convolutional networks for biomedical image segmentation. In: Navab, N., Hornegger, J., Wells, W.M., Frangi, A.F. (eds.) MICCAI 2015. LNCS, vol. 9351, pp. 234–241. Springer, Cham (2015). https://doi.org/10.1007/978-3-319-24574-4_28

17. Ross, T.Y., Dollár, G.: Focal loss for dense object detection. In: Proceedings of the IEEE Conference on Computer Vision and Pattern Recognition, pp. 2980–2988 (2017)
18. Rother, C., Kolmogorov, V., Blake, A.: "GrabCut" interactive foreground extraction using iterated graph cuts. ACM Trans. Graph. **23**(3), 309–314 (2004)
19. Wang, Y., Zhang, J., Kan, M., Shan, S., Chen, X.: Self-supervised equivariant attention mechanism for weakly supervised semantic segmentation. In: Proceedings of the IEEE/CVF Conference on Computer Vision and Pattern Recognition, pp. 12275–12284 (2020)

OXnet: Deep Omni-Supervised Thoracic Disease Detection from Chest X-Rays

Luyang Luo[1(✉)], Hao Chen[2], Yanning Zhou[1], Huangjing Lin[1,3], and Pheng-Ann Heng[1,4]

[1] Department of Computer Science and Engineering, The Chinese University of Hong Kong, Hong Kong, China
lyluo@cse.cuhk.edu.hk

[2] Department of Computer Science and Engineering, The Hong Kong University of Science and Technology, Hong Kong, China
jhc@cse.ust.hk

[3] Imsight AI Research Lab, Shenzhen, China

[4] Guangdong-Hong Kong-Macao Joint Laboratory of Human-Machine Intelligence-Synergy Systems, Shenzhen Institutes of Advanced Technology, Chinese Academy of Sciences, Beijing, China

Abstract. Chest X-ray (CXR) is the most typical diagnostic X-ray examination for screening various thoracic diseases. Automatically localizing lesions from CXR is promising for alleviating radiologists' reading burden. However, CXR datasets are often with massive image-level annotations and scarce lesion-level annotations, and more often, without annotations. Thus far, unifying different supervision granularities to develop thoracic disease detection algorithms has not been comprehensively addressed. In this paper, we present OXnet, the first deep omni-supervised thoracic disease detection network to our best knowledge that *uses as much available supervision as possible* for CXR diagnosis. We first introduce supervised learning via a one-stage detection model. Then, we inject a global classification head to the detection model and propose dual attention alignment to guide the global gradient to the local detection branch, which enables learning lesion detection from image-level annotations. We also impose intra-class compactness and inter-class separability with global prototype alignment to further enhance the global information learning. Moreover, we leverage a soft focal loss to distill the soft pseudo-labels of unlabeled data generated by a teacher model. Extensive experiments on a large-scale chest X-ray dataset show the proposed OXnet outperforms competitive methods with significant margins. Further, we investigate omni-supervision under various annotation granularities and corroborate OXnet is a promising choice to mitigate the plight of annotation shortage for medical image diagnosis (Code is available at https://github.com/LLYXC/OXnet.).

L. Luo and H. Chen—Contributed equally.

Electronic supplementary material The online version of this chapter (https://doi.org/10.1007/978-3-030-87196-3_50) contains supplementary material, which is available to authorized users.

M. de Bruijne et al. (Eds.): MICCAI 2021, LNCS 12902, pp. 537–548, 2021.
https://doi.org/10.1007/978-3-030-87196-3_50

Keywords: Omni-supervised learning · Chest X-ray · Disease localization · Attention

1 Introduction

Modern object detection algorithms often require a large amount of supervision signals. However, annotating abundant medical images for disease detection is infeasible due to the high dependence of expert knowledge and expense of human labor. Consequently, many medical datasets are weakly labeled or, more frequently, unlabeled [22]. This situation especially exists for chest X-rays (CXR), which is the most commonly performed diagnostic X-ray examination. Apart from massive unlabeled data, CXR datasets often have image-level annotations that can be easily obtained by text mining from numerous radiological reports [9,27], while lesion-level annotations (e.g., bounding boxes) scarcely exist [7,28]. Therefore, efficiently leveraging available annotations to develop thoracic disease detection algorithms has significant practical value.

Omni-supervised learning [20] aims to leverage the existing fully-annotated data and other available data (e.g. unlabeled data) as much as possible, which could practically address the mentioned challenge. Distinguished from previous studies that only include extra unlabeled data [6,20,25], we target on utilizing *as much available supervision as possible* from data of various annotation granularities, i.e., fully-labeled data, weakly-labeled data, and unlabeled data, to develop a *unified framework* for thoracic disease detection on chest X-rays.

In this paper, we present OXnet, a unified deep framework for omni-supervised chest X-ray disease detection. To this end, we first leverage limited bounding box annotations to train a base detector. To enable learning from weakly labeled data, we introduce a dual attention alignment module that guides gradient from an image-level classification branch to the local lesion detection branch. To further enhance learning from global information, we propose a global prototype alignment module to impose intra-class compactness and inter-class separability. For unlabeled data, we present a soft focal loss to distill the soft pseudo-labels generated by a teacher model. Extensive experiments show that OXnet not only outperforms the baseline detector with a large margin but also achieves better thoracic disease detection performance than other competitive methods. Further, OXnet also show comparable performance to fully-supervised method with fewer fully-labeled data and sufficient weakly-labeled data. In summary, OXnet can effectively utilize all the available supervision signals, demonstrating a promisingly feasible and general solution to real-world applications.

2 Method

Let $\mathcal{D}_F$, $\mathcal{D}_W$, and $\mathcal{D}_U$ denote the fully-labeled data (with bounding box annotations), weakly-labeled data (with image-level annotations), and unlabeled data, respectively. As illustrated in Fig. 1, OXnet correspondingly consists of three main parts: a base detector to learn from $\mathcal{D}_F$, a global path to learn from $\mathcal{D}_W$, and an unsupervised branch to learn from $\mathcal{D}_U$. As the first part is a RetinaNet [15] backbone supervised by the focal loss and bounding box regression loss, we will focus on introducing the latter two parts in the following sections.

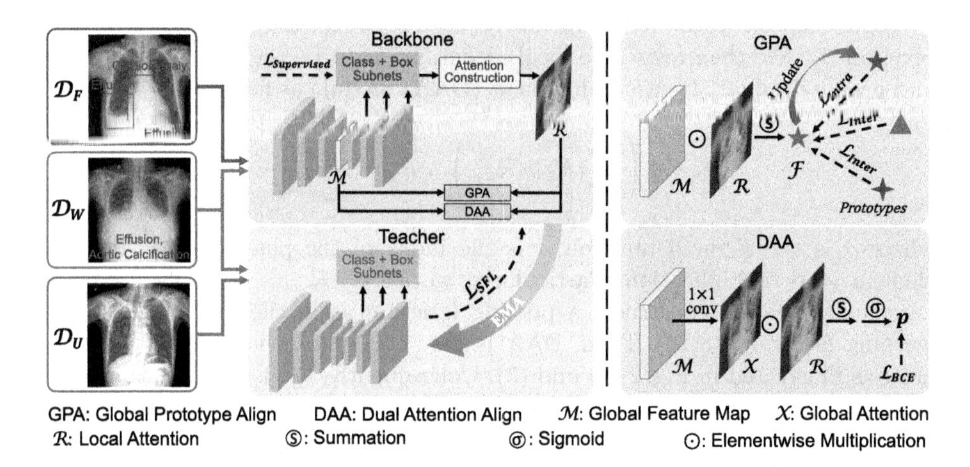

Fig. 1. Overview of OXnet. For $\mathcal{D}_F$, the model learns via the original supervised loss of RetinaNet. For $\mathcal{D}_W$ and $\mathcal{D}_F$, DAA aligns the global and local attentions and enables global information flowing to the local branch (Sect. 2.1). Further, GPA enhances the learning from global labels (Sect. 2.2). For learning $\mathcal{D}_U$ and $\mathcal{D}_W$, a teacher model would guide the RetinaNet with soft focal loss (Sect. 2.3).

2.1 Dual Attention Alignment for Weakly-Supervised Learning

To enable learning from $\mathcal{D}_W$, we first add a multi-label classification global head to the ResNet [5] part of RetinaNet. This global head consists of a 1×1 convolutional layer for channel reduction, a global average pooling layer, and a sigmoid layer. Then, the binary cross entropy loss is used as the objective:

$$\mathcal{L}_{\text{BCE}}(p, y) = -\sum_{c=1}^{N}[y_c \cdot \log(p_c) + (1 - y_c) \cdot \log(1 - p_c)] \tag{1}$$

where c is the index of total N categories, y_c is the image-level label, and p_c is the prediction. The 1×1 convolution gets input $\mathcal{M}$ and outputs feature map $\mathcal{X}$ with N channels, where $\mathcal{X}$ is exactly the global attention map, i.e., class activating map [32]. Meanwhile, the local classification head of RetinaNet conducts dense prediction and generates outputs of size Width $\times$ Height $\times$ Classes $\times$ Anchors for each feature pyramid level [14]. These classification maps can be directly used as the local attentions for each class and anchor. Hence, we first take the maximum on the anchor dimension to get the attentions $\mathcal{A}$ on number of M pyramid levels. Then, the final local attention for each class is obtained by:

$$\mathcal{A}_c = \text{argmax}_{\mathcal{A}_c}(\max(\mathcal{A}_c^1), \max(\mathcal{A}_c^2), \cdots, \max(\mathcal{A}_c^M)) \tag{2}$$

Essentially, we observed that the global and local attentions are often mismatched as shown in Fig. 2(1), indicating the two branches making decisions out of different regions of interests (ROIs). Particularly, the local attention usually covers more precise ROIs by learning from lesion-level annotations. Therefore, we argue that the local attention can be used to weigh the importance of each pixel on $\mathcal{X}_c$. We then resize the local attention $\mathcal{A}_c$ to be the same shape as $\mathcal{X}_c$ and propose a dual attention alignment (DAA) module as follows:

$$p_c = \sigma\left(\sum_i [\mathcal{X}_c \odot \mathcal{R}_c]_i\right), \mathcal{R}_c = \frac{\mathcal{A}_c}{\sum_i [\mathcal{A}_c]_i} \tag{3}$$

where σ is the sigmoid function, i is the index of the pixels, and $\odot$ denotes element-wise multiplication. Particularly, we let $\sum_i \mathcal{R}_c = 1$, so the element-wise multiplication constructs a pooling function as in the multiple instance learning literature [8,23]. Then, DAA is used to replace the mentioned global head as illustrated in Fig. 2 (2) and (3). Consequently, the local attention helps rectify the decision-making regions of the global branch, and the gradient from the global branch could thus flow to the local branch. To this stage, the local classification branch not only receives the strong yet limited supervision from $\mathcal{D}_F$ but also massive weak supervision from $\mathcal{D}_W$.

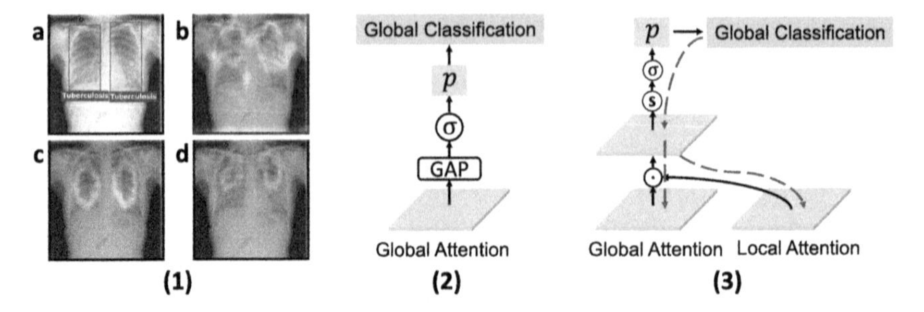

Fig. 2. **(1)**. (a) The ground truth of lesions; (b) The global attention without DAA; (c) The local attention map; (d) The global attention with DAA. **(2)**. The global classification head without DAA. **(3)**. The DAA module, where solid line and dashed line represent forward and backward flow, respectively.

2.2 Global Prototype Alignment for Multi-label Metric Learning

With DAA, any global supervision signals could flow to the local branch. To further enhance this learning process, we propose a multi-label metric learning loss that imposes intra-class compactness and inter-class separability to the network. It is worth noting that multi-label metric learning can be hard as the deep features often capture multi-class information [17]. Specially, in our case, we can obtain *category-specific* features under the guidance of local attention as follows:

$$\mathcal{F}_c = \sum_i \mathcal{M} \odot \mathcal{R}_c \tag{4}$$

where $\mathcal{M}$ is the feature map before the global 1×1 convolutional layer. Particularly, $\mathcal{R}_c$ is of shape $W \times H$ with W and H being width and height, respectively, i is the index of the feature vectors ($\mathcal{M}$ contains $\#W \times H$ vectors), and $\odot$ is the element-wise multiplication over the dimensions of W and H. Here, the local attention $\mathcal{R}_c$ highlights the feature vectors related to each category, leading to the aggregated category-specific feature vector $\mathcal{F}_c$. For better metric learning regularization, we generate global prototypes [29,31] for each category as follows:

$$\mathcal{P}_{t+1} = \beta \cdot \mathcal{P}_t + (1 - \beta) \cdot \frac{\sum_k^K p_k \cdot \mathcal{F}_k}{\sum_k^K p_k} \tag{5}$$

where β is 0 if $t = 0$ otherwise 0.7, and K is the number of data in a batch. Particularly, the prototype is updated with the weighted average of confidences [21,30]. We then align $\mathcal{F}_c$ with the prototypes as follows:

$$\begin{aligned} \mathcal{L}_{\text{Intra}} &= \frac{1}{N} \sum_{c=1}^{N} \|\mathcal{F}_c - \mathcal{P}_c\|_2^2 \\ \mathcal{L}_{\text{Inter}} &= \frac{1}{N(N-1)} \sum_{c=1}^{N} \sum_{0 \leq j \neq c \leq N}^{N} \max(0, \delta - \|\mathcal{F}_j - \mathcal{P}_c\|_2^2) \end{aligned} \tag{6}$$

where δ is a scalar representing the inter-class margin.

2.3 Soft Focal Loss for Unsupervised Knowledge Distillation

To learn from the unlabeled data, we first obtain a mean teacher model by exponential moving average [24]: $\tilde{\theta}_{t+1} = \lambda \cdot \tilde{\theta}_t + (1 - \lambda) \cdot \theta_{t+1}$, where θ and $\tilde{\theta}$ are the parameters of RetinaNet and the teacher model, respectively, λ is a decay coefficient, and t represents training step. Then, the student RetinaNet will learn from the teacher model's classification predictions, which are probabilities in range (0, 1). Particularly, the foreground-background imbalance problem [15] should also be taken care of. Therefore, we extend the focal loss to its soft form inspired by [12,26]. Specifically, the original focal loss is as follows:

$$\mathcal{L}_{\text{FL}}(q) = -\alpha \cdot (1 - q)^{\gamma} \cdot log(q) \tag{7}$$

where α is a pre-set weight for balancing foreground and background, q is the local classification prediction from the model (we eliminate the subscript c for simplicity), γ is a scalar controlling $(1-q)^{\gamma}$ to assign more weights onto samples which are less-well classified. We obtain the soft focal loss as follows:

$$\mathcal{L}_{\text{SFL}}(q, \tilde{q}) = -(\tilde{q}\alpha + \epsilon) \cdot \|\tilde{q} - q\|^{\gamma} \cdot \mathcal{L}_{\text{BCE}}(q, \tilde{q}) \tag{8}$$

Here, instead of assigning α according to whether an anchor is positive or negative, we assume its value changes linearly with the value of teacher model's prediction $\tilde{q}$ and modify it to be $\tilde{q}\alpha + \epsilon$, where ϵ is a constant. Meanwhile, we notice that $(1-q)^\gamma$ in Eq. (7) depends on how closer q is to its target ground truth. Therefore, we modify the focal weight to be $\|\tilde{q}-q\|^\gamma$ to give larger wights to the instances with higher disagreement between student-teacher models.

2.4 Unified Training for Deep Omni-Supervised Detection

The overall training loss sums up the supervised losses [15], weakly-supervised losses ($\mathcal{L}_{\text{BCE}}$, $\mathcal{L}_{\text{Intra}}$, and $\mathcal{L}_{\text{Inter}}$), and unsupervised loss ($\mathcal{L}_{\text{SFL}}$). Particularly, weakly-supervised losses are applied to $\mathcal{D}_W$ and $\mathcal{D}_F$, and unsupervised loss is applied to $\mathcal{D}_U$ and $\mathcal{D}_W$. Note that there were works on the NIH dataset [27] leveraging two types of annotations [13,16,18,34] for lesion localization. However, these methods localize lesions with attention maps, which cannot be easily compared to bounding box-based detection model that we aim to develop.

β is set to 0 for the first step otherwise 0.7, δ is set to 1, λ is set to 0.99 by tuning on the validation set. As suggested in [26], α is set to 0.9, ϵ is 0.05, and γ is set to 2. CXR images are resized to 512 $\times$ 512 without cropping. Data augmentation is done following [33]. Adam [10] is used as the optimizer. The learning rate is initially set to 1e−5 and divided by ten when validation mAP stagnate. All implementations use Pytorch [19] on an NVIDIA TITAN Xp GPU.

3 Experiments

3.1 Dataset and Evaluation Metrics

In total, 32,261 frontal CXR images taken from 27,253 patients were used. A cohort of 10 radiologists (4–30 years of experience) was involved in labeling the lesion bounding boxes, and each image was labeled by two with corresponding text report. If the initial labelers disagreed on the annotation, a senior radiologist ($\geq$20-year experience) would make the final decision. Nine thoracic diseases including aortic calcification, cardiomegaly, fracture, mass, nodule, pleural effusion, pneumonia, pneumothorax, and tuberculosis were finally chosen in this study by the consensus of the radiologists. The dataset was split into training, validation, and testing sets with 13,963, 5,118, and 13,180 images, respectively, without patients overlapping. Detailed numbers of images and annotations per set can be found in the supplementary.

We used Average Precision (AP) [3] as the evaluation metric. Specifically, we reported the mean AP (mAP) from AP^{40} to AP^{75} with an interval of 5, following the Kaggle Pneumonia Detection Challenge[1] and [4]. We also evaluated AP^{S}, AP^{M}, and AP^{L} for small, medium, and large targets, respectively, using the COCO API[2]. All reported statistics were averaged over the nine abnormalities.

[1] https://www.kaggle.com/c/rsna-pneumonia-detection-challenge.
[2] https://cocodataset.org/#detection-eval.

3.2 Comparison with Other Methods

To our best knowledge, few previous works simultaneously leveraged fully-labeled data, weakly-labeled data, and unlabeled data. Hence, we first trained a RetinaNet (with pre-trained Res101 weight from ImageNet [2]; a comparison of baselines could be found in the supplementary) on our dataset. Then, we implemented several state-of-the-art semi-supervised methods finetuned from RetinaNet, including RetinaNet [15], Π Model [11], Mean Teacher [24], MMT-PSM [33], and FocalMix [26]. We further enabled learning from the weak annotations by adding a global classification head to each semi-supervised method to construct multi-task (MT) models.

Quantitative Results. All semi-supervised methods (row 2 to 5 in Table 1) clearly outperform the RetinaNet baseline (1st row), demonstrating effective utilization of the unlabeled data. After incorporating the global classification heads, the four multi-task networks (row 6 to 9) get further improvement with about 0.2–0.4 points raising in mAP. This finding suggests that large image-level supervision can help learning abnormalities detection, but the benefits are still limited without proper design. On the other hand, OXnet achieves 22.3 in mAP and outperforms the multi-task models on various sizes of targets. The results corroborate that our proposed method can more effectively leverage the less well-labeled data for the thoracic disease detection task.

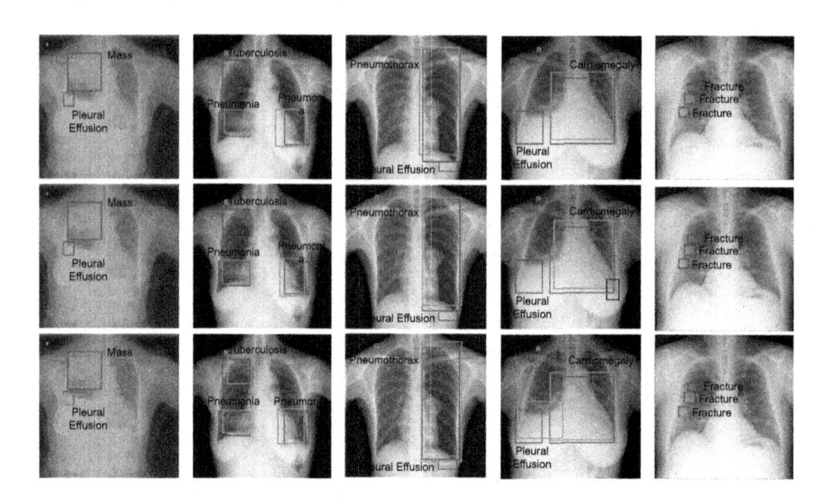

Fig. 3. Visualization of results generated by RetinaNet (first row), Mean Teacher + MT (second row), and our method (third row). Ground truth is in red, true positives are in green, and false positives are in blue. Best viewed in color. (Color figure online)

Table 1. Quantitative comparison of different methods.

Method	# images used			Metrics					
	Fully labeled	Weakly labeled	Unlabeled	mAP	AP^{40}	AP^{75}	AP^{S}	AP^{M}	AP^{L}
RetinaNet [15]	2725	0	0	18.4	27.7	7.5	8.0	16.7	25.4
Π Model [11]	2725	0	11238	20.0	29.3	9.3	9.2	20.8	27.0
Mean Teacher [24]	2725	0	11238	20.0	29.2	9.4	9.1	20.4	26.9
MMT-PSM [33]	2725	0	11238	19.1	28.4	8.4	8.8	19.3	26.5
FocalMix [26]	2725	0	11238	19.8	29.1	9.0	8.6	19.6	26.3
Π Model [11] + MT	2725	11238	0	20.2	29.6	9.2	9.4	21.3	27.6
Mean Teacher [24] + MT	2725	11238	0	20.4	29.6	9.4	9.3	20.6	27.1
MMT-PSM [33] + MT	2725	11238	0	19.3	28.4	8.8	8.4	17.9	26.8
FocalMix [26] + MT	2725	11238	0	20.1	29.7	8.8	8.5	18.3	27.4
SFL	2725	0	11238	20.4	29.7	9.4	9.3	21.6	29.8
DAA	2725	11238	0	21.2	31.2	9.4	9.8	20.7	28.0
DAA + GPA	2725	11238	0	21.4	31.4	9.5	**11.0**	20.9	27.1
OXnet (SFL + DAA + GPA)	2725	11238	0	**22.3**	**32.4**	**10.3**	9.6	**21.8**	**31.4**

Table 2. Results under different ratios of annotation granularities.

Method	# images used			Metrics					
	Fully labeled	Weakly labeled	Unlabeled	mAP	AP^{40}	AP^{75}	AP^{S}	AP^{M}	AP^{L}
RetinaNet	682	0	0	12.5	18.9	5.5	7.0	16.4	18.1
	1372	0	0	14.6	21.8	6.5	5.8	10.4	19.5
	2725	0	0	18.4	27.7	7.5	8.0	16.7	25.4
OXnet (ours)	682	13281	0	14.9	22.6	6.6	0.4	16.8	20.3
	1372	12591	0	17.8	26.9	8.0	6.8	16.2	24.4
	2725	11238	0	22.3	32.4	10.3	9.6	21.8	31.4
	2725	8505	2733	21.9	32.0	10.0	9.7	21.6	29.9
	2725	2733	8505	21.2	31.0	10.0	9.0	22.2	29.0
	2725	0	11238	20.7	30.3	9.4	8.3	20.7	28.3

Ablation Study. The last four rows in Table 1 report the ablation study of the proposed components, i.e., soft focal loss (SFL), dual attention alignment (DAA), and global prototype alignment (GPA). Our SFL achieves an mAP of 20.4 and outperforms other semi-supervised methods, demonstrating its better capability of utilizing the unlabeled data. On the other hand, incorporating only DAA reaches an mAP of 21.2, showing effective guidance from the weakly labeled data. Adding GPA to DAA improves the mAP to 21.4, demonstrating the effectiveness of the learned intra-class compactness and inter-class separability. By unifying the three components, OXnet reaches the best results in 5 out of 6 metrics, corroborating the complementarity of the proposed methods.

Qualitative Results. We also visualize the outputs generated by RetinaNet, Mean Teacher + MT (the best method among those compared with ours), and OXnet in Fig. 3. As illustrated, our model yields more accurate predictions for multiple lesions of different diseases in each chest X-ray sample. We also illustrates more samples of the attention maps in the supplementary.

3.3 Omni-Supervision Under Different Annotation Granularities

Efficient learning is crucial for medical images as annotations are extremely valuable and scarce. Thus, we investigate the effectiveness of OXnet given different annotation granularities. With results in Table 2, we find that: (1) Finer annotations always lead to better performance, and OXnet achieves consistent improvements as the annotation granularity becomes finer (row 1 to 9); (2) Increasing fully labeled data benefits OXnet more (mAP improvements are 2.9 from row 4 to 5 and 4.5 from row 5 to 6) than RetinaNet (mAP improvements are 2.1 from row 1 to 2 and 3.8 from row 2 to 3); and (3) With less fully-labeled data and more weakly-labeled data, OXnet can achieve comparable performance to RetinaNet (row 2 vs. row 4, row 3 vs. row 5). These findings clearly corroborate OXnet's effectiveness in utilizing as much available supervision as possible. Moreover, unlabeled data are easy to acquire without labeling burden, and weakly-labeled data can also be efficiently obtained with natural language processing methods [1,9,27]. Therefore, we believe OXnet could serve as a promisingly feasible and general approach to real-world clinic applications.

4 Conclusion

We present OXnet, a deep omni-supervised learning approach for thoracic disease detection from chest X-rays. The OXnet simultaneously utilizes well-annotated, weakly-annotated, and unlabeled data as a unified framework. Extensive experiments have been conducted and our OXnet has demonstrated superiority in effectively utilizing various granularities of annotations. In summary, the proposed OXnet has shown as a promisingly feasible and general solution to real-world applications by leveraging as much available supervision as possible.

Acknowledgement. This work was supported by Key-Area Research and Development Program of Guangdong Province, China (2020B010165004), Hong Kong Innovation and Technology Fund (Project No. ITS/311/18FP and Project No. ITS/426/17FP.), and National Natural Science Foundation of China with Project No. U1813204.

References

1. Bustos, A., Pertusa, A., Salinas, J.M., de la Iglesia-Vayá, M.: PadChest: a large chest x-ray image dataset with multi-label annotated reports. MedIA **66**, 101797 (2020)
2. Deng, J., Dong, W., Socher, R., Li, L.J., Li, K., Fei-Fei, L.: ImageNet: a large-scale hierarchical image database. In: CVPR, pp. 248–255. IEEE (2009)
3. Everingham, M., Van Gool, L., Williams, C.K., Winn, J., Zisserman, A.: The pascal visual object classes (VOC) challenge. IJCV **88**(2), 303–338 (2010)
4. Gabruseva, T., Poplavskiy, D., Kalinin, A.: Deep learning for automatic pneumonia detection. In: CVPR Workshops, pp. 350–351 (2020)
5. He, K., Zhang, X., Ren, S., Sun, J.: Deep residual learning for image recognition. In: CVPR, pp. 770–778 (2016)
6. Huang, R., Noble, J.A., Namburete, A.I.L.: Omni-supervised learning: scaling up to large unlabelled medical datasets. In: Frangi, A.F., Schnabel, J.A., Davatzikos, C., Alberola-López, C., Fichtinger, G. (eds.) MICCAI 2018. LNCS, vol. 11070, pp. 572–580. Springer, Cham (2018). https://doi.org/10.1007/978-3-030-00928-1_65
7. Huang, Y.J., Liu, W., Wang, X., Fang, Q., Wang, R., Wang, Y., et al.: Rectifying supporting regions with mixed and active supervision for rib fracture recognition. IEEE TMI **39**(12), 3843–3854 (2020)
8. Ilse, M., Tomczak, J., Welling, M.: Attention-based deep multiple instance learning. In: ICML, pp. 2127–2136. PMLR (2018)
9. Irvin, J., et al.: CheXpert: a large chest radiograph dataset with uncertainty labels and expert comparison. In: AAAI, vol. 33, pp. 590–597 (2019)
10. Kingma, D.P., Ba, J.: Adam: a method for stochastic optimization. In: ICLR (2015)
11. Laine, S., Aila, T.: Temporal ensembling for semi-supervised learning. In: ICLR (2017)

12. Li, X., et al.: Generalized focal loss: Learning qualified and distributed bounding boxes for dense object detection. In: NeurIPS (2020)
13. Li, Z., et al.: Thoracic disease identification and localization with limited supervision. In: CVPR, pp. 8290–8299 (2018)
14. Lin, T.Y., Dollár, P., Girshick, R., He, K., Hariharan, B., Belongie, S.: Feature pyramid networks for object detection. In: CVPR, pp. 2117–2125 (2017)
15. Lin, T.Y., Goyal, P., Girshick, R., He, K., Dollár, P.: Focal loss for dense object detection. In: ICCV, pp. 2980–2988 (2017)
16. Liu, J., Zhao, G., Fei, Y., Zhang, M., Wang, Y., Yu, Y.: Align, attend and locate: chest x-ray diagnosis via contrast induced attention network with limited supervision. In: CVPR, pp. 10632–10641 (2019)
17. Luo, L., Yu, L., Chen, H., Liu, Q., Wang, X., Xu, J., et al.: Deep mining external imperfect data for chest x-ray disease screening. IEEE TMI **39**(11), 3583–3594 (2020)
18. Ouyang, X., et al.: Learning hierarchical attention for weakly-supervised chest x-ray abnormality localization and diagnosis. IEEE TMI (2020)
19. Paszke, A., et al.: PyTorch: an imperative style, high-performance deep learning library. In: NeurIPS, vol. 32, pp. 8026–8037. Curran Associates, Inc. (2019)
20. Radosavovic, I., Dollár, P., Girshick, R., Gkioxari, G., He, K.: Data distillation: towards omni-supervised learning. In: CVPR, pp. 4119–4128 (2018)
21. Shi, Y., Yu, X., Sohn, K., Chandraker, M., Jain, A.K.: Towards universal representation learning for deep face recognition. In: CVPR, pp. 6817–6826 (2020)
22. Tajbakhsh, N., Jeyaseelan, L., Li, Q., Chiang, J.N., Wu, Z., Ding, X.: Embracing imperfect datasets: a review of deep learning solutions for medical image segmentation. MedIA **63**, 101693 (2020)
23. Tang, P., Wang, X., Bai, X., Liu, W.: Multiple instance detection network with online instance classifier refinement. In: CVPR, pp. 2843–2851 (2017)
24. Tarvainen, A., Valpola, H.: Mean teachers are better role models: weight-averaged consistency targets improve semi-supervised deep learning results. In: NeurIPS, vol. 30, pp. 1195–1204 (2017)
25. Venturini, L., Papageorghiou, A.T., Noble, J.A., Namburete, A.I.L.: Uncertainty estimates as data selection criteria to boost omni-supervised learning. In: Martel, A.L., et al. (eds.) MICCAI 2020. LNCS, vol. 12261, pp. 689–698. Springer, Cham (2020). https://doi.org/10.1007/978-3-030-59710-8_67
26. Wang, D., Zhang, Y., Zhang, K., Wang, L.: FocalMix: semi-supervised learning for 3D medical image detection. In: CVPR, pp. 3951–3960 (2020)
27. Wang, X., Peng, Y., Lu, L., Lu, Z., Bagheri, M., Summers, R.M.: ChestX-ray8: hospital-scale chest x-ray database and benchmarks on weakly-supervised classification and localization of common thorax diseases. In: CVPR, pp. 2097–2106 (2017)
28. Wang, Y., et al.: Knowledge distillation with adaptive asymmetric label sharpening for semi-supervised fracture detection in chest x-rays. In: IPMI (2021)
29. Wen, Y., Zhang, K., Li, Z., Qiao, Yu.: A discriminative feature learning approach for deep face recognition. In: Leibe, B., Matas, J., Sebe, N., Welling, M. (eds.) ECCV 2016. LNCS, vol. 9911, pp. 499–515. Springer, Cham (2016). https://doi.org/10.1007/978-3-319-46478-7_31
30. Xu, M., Wang, H., Ni, B., Tian, Q., Zhang, W.: Cross-domain detection via graph-induced prototype alignment. In: CVPR, pp. 12355–12364 (2020)
31. Yang, H.M., Zhang, X.Y., Yin, F., Liu, C.L.: Robust classification with convolutional prototype learning. In: CVPR, pp. 3474–3482 (2018)

32. Zhou, B., Khosla, A., Lapedriza, A., Oliva, A., Torralba, A.: Learning deep features for discriminative localization. In: CVPR, pp. 2921–2929 (2016)
33. Zhou, Y., Chen, H., Lin, H., Heng, P.-A.: Deep semi-supervised knowledge distillation for overlapping cervical cell instance segmentation. In: Martel, A.L., et al. (eds.) MICCAI 2020. LNCS, vol. 12261, pp. 521–531. Springer, Cham (2020). https://doi.org/10.1007/978-3-030-59710-8_51
34. Zhou, Y., Zhou, T., Zhou, T., Fu, H., Liu, J., Shao, L.: Contrast-attentive thoracic disease recognition with dual-weighting graph reasoning. IEEE TMI **40**, 1196–1206 (2021)

Adapting Off-the-Shelf Source Segmenter for Target Medical Image Segmentation

Xiaofeng Liu[1(✉)], Fangxu Xing[1], Chao Yang[2], Georges El Fakhri[1], and Jonghye Woo[1]

[1] Gordon Center for Medical Imaging, Department of Radiology, Massachusetts General Hospital and Harvard Medical School, Boston, MA 02114, USA
{xliu61,JWOO}@mgh.harvard.edu

[2] Facebook Artificial Intelligence, Boston, MA 02142, USA

Abstract. Unsupervised domain adaptation (UDA) aims to transfer knowledge learned from a labeled source domain to an unlabeled and unseen target domain, which is usually trained on data from both domains. Access to the source domain data at the adaptation stage, however, is often limited, due to data storage or privacy issues. To alleviate this, in this work, we target source free UDA for segmentation, and propose to adapt an "off-the-shelf" segmentation model pre-trained in the source domain to the target domain, with an adaptive batch-wise normalization statistics adaptation framework. Specifically, the domain-specific low-order batch statistics, i.e., mean and variance, are gradually adapted with an exponential momentum decay scheme, while the consistency of domain shareable high-order batch statistics, i.e., scaling and shifting parameters, is explicitly enforced by our optimization objective. The transferability of each channel is adaptively measured first from which to balance the contribution of each channel. Moreover, the proposed source free UDA framework is orthogonal to unsupervised learning methods, e.g., self-entropy minimization, which can thus be simply added on top of our framework. Extensive experiments on the BraTS 2018 database show that our source free UDA framework outperformed existing source-relaxed UDA methods for the cross-subtype UDA segmentation task and yielded comparable results for the cross-modality UDA segmentation task, compared with a supervised UDA methods with the source data.

1 Introduction

Accurate tumor segmentation is a critical step for early tumor detection and intervention, and has been significantly improved with advanced deep neural networks (DNN) [9,10,17,18,25]. A segmentation model trained in a source domain, however, usually cannot generalize well in a target domain, e.g., data acquired from a new scanner or different clinical center, in implementation. Besides, annotating data in the new target domain is costly and even infeasible [11]. To address this, unsupervised domain adaptation (UDA) was proposed to transfer knowledge from a labeled source domain to unlabeled target domains [13].

M. de Bruijne et al. (Eds.): MICCAI 2021, LNCS 12902, pp. 549–559, 2021.
https://doi.org/10.1007/978-3-030-87196-3_51

The typical UDA solutions can be classified into three categories: statistic moment matching, feature/pixel-level adversarial learning [12,14,15], and self-training [16,33]. These UDA methods assume that the source domain data are available and usually trained together with target data. The source data, however, are often inaccessible, due to data storage or privacy issues, for cross-clinical center implementation [1]. Therefore, it is of great importance to apply an "off-the-shelf" source domain model, without access to the source data. For source-free classification UDA, Liang et al. [8] proposed to enforce the diverse predictions, while the diversity of neighboring pixels is not suited for the segmentation purpose. In addition, the class prototype [13] and variational inference methods [11] are not scalable for pixel-wise classification based segmentation. More importantly, without distribution alignment, these methods relied on unreliable noisy pseudo labeling.

Recently, the source relaxed UDA [1] was presented to pre-train an additional class ratio predictor in the source domain, by assuming that the class ratio, i.e., pixel proportion in segmentation, is invariant between source and target domains. At the adaptation stage, the class ratio was used as the only transferable knowledge. However, that work [1] has two limitations. First, the class ratio can be different between the two domains, due to label shift [11,13]. For example, a disease incident rate could vary between different countries, and tumor size could vary between different subtypes and populations. Second, the pre-trained class ratio predictor used in [1] is not typical for medical image segmentation, thereby requiring an additional training step using the data in the source domain.

In this work, to address the aforementioned limitations, we propose a practical UDA framework aimed at the source-free UDA for segmentation, without an additional network trained in the source domain or the unrealistic assumption of class ratio consistency between source and target domains. More specifically, our framework hinges on the batch-wise normalization statistics, which are easy to access and compute. Batch Normalization (BN) [6] has been a default setting in the most of modern DNNs, e.g., ResNet [5] and U-Net [30], for faster and more stable training. Notably, the BN statistics of the source domain are stored in the model itself. The low-order batch statistics, e.g., mean and variance, are domain-specific, due to the discrepancy of input data. To gradually adapt the low-order batch statistics from the source domain to the target domain, we develop a momentum-based progression scheme, where the momentum follows an exponential decay w.r.t. the adaptation iteration. For the domain shareable high-order batch statistics, e.g., scaling and shifting parameters, a high-order batch statistics consistent loss is applied to explicitly enforce the discrepancy minimization. The transferability of each channel is adaptively measured first, from which to balance the contribution of each channel. Moreover, the proposed unsupervised self-entropy minimization can be simply added on top of our framework to boost the performance further.

Our contributions are summarized as follows:

- To our knowledge, this is the first source relaxed or source free UDA framework for segmentation. We do not need an additional source domain network, or the unrealistic assumption of the class ratio consistency [1]. Our method

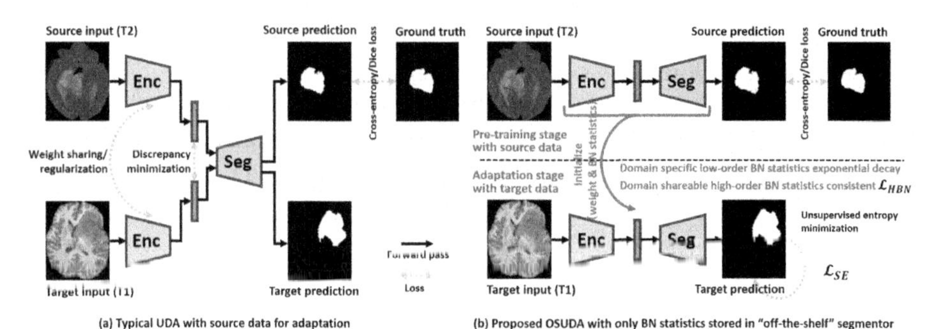

Fig. 1. Comparison of (a) conventional UDA [28] and (b) our source-relaxed OSUDA segmentation framework based on the pre-trained "off-the-shelf" model with BN. We minimize the domain discrepancy based on the adaptively computed batch-wise statistics in each channel. The model consists of a feature encoder (Enc) and a segmentor (Seg) akin to [3,32].

only relies on an "off-the-shelf" pre-trained segmentation model with BN in the source domain.
- The domain-specific and shareable batch-wise statistics are explored via the low-order statistics progression with an exponential momentum decay scheme and transferability adaptive high-order statistics consistency loss, respectively.
- Comprehensive evaluations on both cross-subtype (i.e., HGG to LGG) and cross-modality (i.e., T2 to T1/T1ce/FLAIR) UDA tasks using the BraTS 2018 database demonstrate the validity of our proposed framework and its superiority to conventional source-relaxed/source-based UDA methods.

2 Methodology

We assume that a segmentation model with BN is pre-trained with source domain data, and the batch statistics are inherently stored in the model itself. At the adaptation stage, we fine-tune the model based on the batch-wise statistics and the self-entropy (SE) of target data prediction. The overview of the different setups of conventional UDA and our "off-the-shelf (OS)" UDA is shown in Fig. 1. Below, we briefly revisit the BN in Subsect. 2.1 first and then introduce our OSUDA in Subsect. 2.2. The added unsupervised SE minimization and the overall training protocol are detailed in Subsect. 2.3.

2.1 Preliminaries on Batch Normalization

As a default setting in the most of modern DNNs, e.g., ResNet [5] and U-Net [30], Batch Normalization (BN) [6] normalizes the input feature in the l-th layer $f_l \in \mathbb{R}^{B\times H_l\times W_l\times C_l}$ within a batch in a channel-wise manner to have zero mean and unit variance. B denotes the number of images in a batch, and

H_l, W_l, and C_l are the height, width, and channels of layer l. We have samples in a batch, with index $b \in \{1, \cdots, B\}$, spatial index $n \in \{1, \cdots, H_l \times W_l\}$, and channel index $c \in \{1, \cdots, C_l\}$. BN calculates the mean of each channel $\mu_{l,c} = \frac{1}{B \times H_l \times W_l} \sum_b^B \sum_n^{H_l \times W_l} f_{l,b,n,c}$, where $f_{l,b,n,c} \in \mathbb{R}$ is the feature value. The variance $\{\sigma^2\}_{l,c} = \frac{1}{B \times H_l \times W_l} \sum_b^B \sum_n^{H_l \times W_l} (f_{l,b,n,c} - \mu_{l,c})^2$. Then, the input feature is normalized as

$$\tilde{f}_{l,b,n,c} = \gamma_{l,c}(f_{l,b,n,c} - \mu_{l,c}) / \sqrt{\{\sigma^2\}_{l,c} + \epsilon} + \beta_{l,c}, \quad (1)$$

where $\epsilon \in \mathbb{R}^+$ is a small scalar for numerical stability. $\gamma_{l,c}$ and $\beta_{l,c}$ are learnable scaling and shifting parameters, respectively.

In testing, the input is usually a single sample rather than a batch with B samples. Therefore, BN stores the exponentially weighted average of the batch statistics at the training stage and used it in testing. Specifically, the mean and variance over the training are tracked progressively, i.e.,

$$\overline{\mu}_{l,c}^k = (1 - \eta) \cdot \overline{\mu}_{l,c}^{k-1} + \eta \cdot \mu_{l,c}^k; \quad \{\overline{\sigma}^2\}_{l,c}^k = (1 - \eta) \cdot \{\overline{\sigma}^2\}_{l,c}^{k-1} + \eta \cdot \{\sigma^2\}_{l,c}^k, \quad (2)$$

where $\eta \in [0, 1]$ is a momentum parameter. After K training iterations, $\overline{\mu}_{l,c}^K$, $\{\overline{\sigma}^2\}_{l,c}^K$, $\gamma_{l,c}^K$, and $\beta_{l,c}^K$ are stored and used for testing normalization [6].

2.2 Adaptive Source-Relaxed Batch-Wise Statistics Adaptation

Early attempts of BN for UDA simply added BN in the target domain, without the interaction with the source domain [7]. Recent studies [2,19,20,26] indicated that the low-order batch statistics, i.e., mean $\mu_{l,c}$ and variance $\{\sigma^2\}_{l,c}$, are domain-specific, because of the divergence of cross-domain representation distributions. Therefore, brute-forcing the same mean and variance across domains can lead to a loss of expressiveness [29]. In contrast, after the low-order batch statistics discrepancy is partially reduced, with domain-specific mean and variance normalization, the high-order batch statistics, i.e., scaling and shifting parameters $\gamma_{l,c}$ and $\beta_{l,c}$, are shareable across domains [20,26].

However, all of the aforementioned methods [2,19,20,26,29] require the source data at the adaptation stage. To address this, in this work, we propose to mitigate the domain shift via the adaptive low-order batch statistics progression with momentum, and explicitly enforce the consistency of the high-order statistics in a source-relaxed manner.

Low-Order Statistics Progression with an Exponential Momentum Decay Scheme. In order to gradually learn the target domain-specific mean and variance, we propose an exponential low-order batch statistics decay scheme. We initialize the mean and variance in the target domain with the tracked $\overline{\mu}_{l,c}^K$ and $\{\overline{\sigma}^2\}_{l,c}^K$ in the source domain, which is similar to applying a model with BN in testing [6]. Then, we progressively update the mean and variance in the t-th adaptation iteration in the target domain as

$$\overline{\mu}_{l,c}^t = (1 - \eta^t) \cdot \mu_{l,c}^t + \eta^t \cdot \overline{\mu}_{l,c}^K; \quad \{\overline{\sigma}^2\}_{l,c}^t = (1 - \eta^t) \cdot \{\sigma^2\}_{l,c}^t + \eta^t \cdot \{\overline{\sigma}^2\}_{l,c}^K, \quad (3)$$

where $\eta^t = \eta^0 \exp(-t)$ is a target adaptation momentum parameter with an exponential decay w.r.t. the iteration t. $\mu_{l,c}^t$ and $\{\sigma^2\}_{l,c}^t$ are the mean and variance of the current target batch. Therefore, the weight of $\overline{\mu}_{l,c}^K$ and $\{\overline{\sigma}^2\}_{l,c}^K$ are smoothly decreased along with the target domain adaptation, while $\mu_{l,c}^t$ and $\{\sigma^2\}_{l,c}^t$ gradually represent the batch-wise low-order statistics of the target data.

Transferability Adaptive High-Order Statistics Consistency. For the high-order batch statistics, i.e., the learned scaling and shifting parameters, we explicitly encourage its consistency between the two domains with the following high-order batch statistics (HBS) loss:

$$\mathcal{L}_{HBS} = \sum_{l}^{L} \sum_{c}^{C_l} (1 + \alpha_{l,c}) \{ |\gamma_{l,c}^K - \gamma_{l,c}^t| + |\beta_{l,c}^K - \beta_{l,c}^t| \}, \tag{4}$$

where $\gamma_{l,c}^K$ and $\beta_{l,c}^K$ are the learned scaling and shifting parameters in the last iteration of pre-training in the source domain. $\gamma_{l,c}^t$ and $\beta_{l,c}^t$ are the learned scaling and shifting parameters in the t-th adaptation iteration. $\alpha_{l,c}$ is an adaptive parameter to balance between the channels.

We note that the domain divergence can be different among different layers and channels, and the channels with smaller divergence can be more transferable [22]. Accordingly, we would expect that the channels with higher transferability contribute more to the adaptation. In order to quantify the domain discrepancy in each channel, a possible solution is to measure the difference between batch statistics. In the source-relaxed UDA setting, we define the channel-wise source-target distance in the t-th adaptation iteration as

$$d_{l,c} = \left| \frac{\overline{\mu}_{l,c}^K}{\sqrt{\{\overline{\sigma}^2\}_{l,c}^K + \epsilon}} - \frac{\mu_{l,c}^t}{\sqrt{\{\sigma^2\}_{l,c}^t + \epsilon}} \right|. \tag{5}$$

Then, the transferability of each channel can be measured by $\alpha_{l,c} = \frac{L \times C \times (1 + d_{l,c})^{-1}}{\sum_l \sum_c (1 + d_{l,c})^{-1}}$. Therefore, the more transferable channels will be assigned with higher importance, i.e., with larger weight $(1 + \alpha_{l,c})$ in $\mathcal{L}_{l,c}$.

2.3 Self-entropy Minimization and Overall Training Protocol

The training in the unlabeled target domain can also be guided by an unsupervised learning framework. The SE minimization is a widely used objective in modern DNNs to encourage the confident prediction, i.e., the maximum softmax value can be high [1,4,8,24]. SE for pixel segmentation is calculated by the averaged entropy of the classifier's softmax prediction given by

$$\mathcal{L}_{SE} = \frac{1}{B \times H_0 \times W_0} \sum_{b}^{B} \sum_{n}^{H_0 \times W_0} \{ \delta_{b,n} \log \delta_{b,n} \}, \tag{6}$$

where H_0 and W_0 are the height and width of the input, and $\delta_{b,n}$ is the histogram distribution of the softmax output of the n-th pixel of the b-th image in a batch. Minimizing $\mathcal{L}_{SE}$ leads to the output close to a one-hot distribution.

At the source-domain pre-training stage, we follow the standard segmentation network training protocol. At the target domain adaptation stage, the overall training objective can be formulated as $\mathcal{L} = \mathcal{L}_{HBS} + \lambda \mathcal{L}_{SE}$, where λ is used to balance between the BN statistics matching and SE minimization. We note that a trivial solution of SE minimization is that all unlabeled target data could have the same one-hot encoding [4]. Thus, to stabilize the training, we linearly change the hyper-parameter λ from 10 to 0 in training.

Table 1. Comparison of HGG to LGG UDA with the four-channel input for our four-class segmentation, i.e., whole tumor, enhanced tumor, core tumor, and background. $\pm$ indicates standard deviation. SEAT [23] with the source data for UDA training is regarded as an "upper bound."

Method	Source data	Dice Score [%] ↑				Hausdorff Distance [mm] ↓			
		WholeT	EnhT	CoreT	Overall	WholeT	EnhT	CoreT	Overall
Source only	no UDA	79.29	30.09	44.11	58.44 ± 43.5	38.7	46.1	40.2	41.7 ± 0.14
CRUDA [1]	Partial[a]	79.85	31.05	43.92	58.51 ± 0.12	31.7	29.5	30.2	30.6 ± 0.15
OSUDA	**No**	**83.62**	**32.15**	**46.88**	**61.94 ± 0.11**	**27.2**	**23.4**	**26.3**	**25.6 ± 0.14**
OSUDA-AC	No	82.74	32.04	46.62	60.75 ± 0.14	27.8	25.5	27.3	26.5 ± 0.16
OSUDA-SE	No	82.45	31.95	46.59	60.78 ± 0.12	27.8	25.3	27.1	26.4 ± 0.14
SEAT [23]	Yes	84.11	32.67	47.11	62.17 ± 0.15	26.4	21.7	23.5	23.8 ± 0.16

[a] An additional class ratio predictor was required to be trained with the source data.

3 Experiments and Results

The BraTS2018 database is composed of a total of 285 subjects [21], including 210 high-grade gliomas (HGG, i.e., glioblastoma) subjects, and 75 low-grade gliomas (LGG) subjects. Each subject has T1-weighted (T1), T1-contrast enhanced (T1ce), T2-weighted (T2), and T2 Fluid Attenuated Inversion Recovery (FLAIR) Magnetic Resonance Imaging (MRI) volumes with voxel-wise labels for the enhancing tumor (EnhT), the peritumoral edema (ED), and the necrotic and non-enhancing tumor core (CoreT). Usually, we denote the sum of EnhT, ED, and CoreT as the whole tumor. In order to demonstrate the effectiveness and generality of our OSUDA, we follow two UDA evaluation protocols using the BraTS2018 database, including HGG to LGG UDA [23] and cross-modality (i.e., T2 to T1/T1ce/FLAIR) UDA [32].

For evaluation, we adopted the widely used Dice similarity coefficient and Hausdorff distance metrics as in [32]. The Dice similarity coefficient (the higher, the better) measures the overlapping part between our prediction results and the ground truth. The Hausdorff distance (the lower, the better) is defined between two sets of points in the metric space.

3.1 Cross-Subtype HGG to LGG UDA

HGG and LGG have different size and position distributions for tumor regions [23]. Following the standard protocol, we used the HGG training set (source

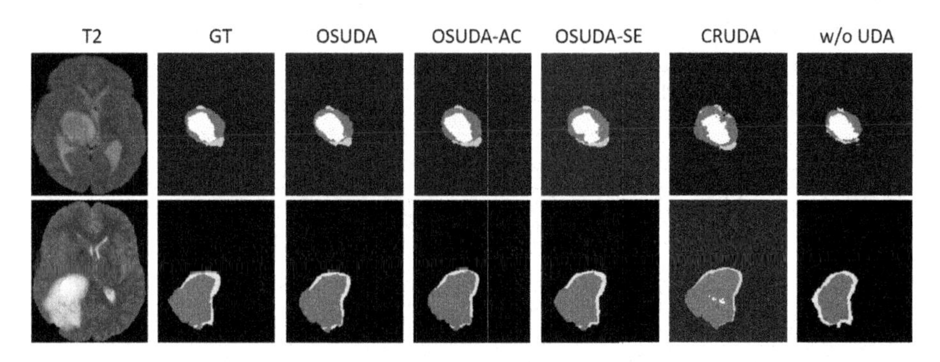

Fig. 2. The comparison with the other UDA methods, and an ablation study of adaptive channel-wise weighting and SE minimization for HGG to LGG UDA.

domain) to pre-train the segmentation model and adapted it with the LGG training set (target domain) [23]. The evaluation was implemented in the LGG testing set. We adopted the same 2D U-Net backbone in [23], sliced 3D volumes into 2D axial slices with the size of 128×128, and concatenated all four MRI modalities to get a 4-channel input.

The quantitative evaluation results are shown in Table 1. Since the pixel proportion of each class is different between HGG and LGG domains, the class ratio-based CRUDA [1] only achieved marginal improvements with its unsupervised learning objective. We note that the Dice score of the core tumor was worse than the pre-trained source-only model, which can be the case of negative transfer [27]. Our proposed OSUDA achieved the state-of-the-art performance for source-relaxed UDA segmentation, approaching the performance of SEAT [23] with the source data, which can be seen as an "upper-bound."

We used OSUDA-AC and OSUDA-SE to indicate the OSUDA without the adaptive channel-wise weighting and self-entropy minimization, respectively. The better performance of OSUDA over OSUDA-AC and OSUDA-SE demonstrates the effectiveness of adaptive channel-wise weighting and self-entropy minimization. The illustration of the segmentation results is given in Fig. 2. We can see that the predictions of our proposed OSUDA are better than the no adaptation model. In addition, CRUDA [1] had a tendency to predict a larger area for the tumor; and the tumor core is often predicted for the slices without the core.

3.2 Cross-Modality T2 to T1/T1ce/FLAIR UDA

Because of large appearance discrepancies between different MRI modalities, we further applied our framework to the cross-modality UDA task. Since clinical annotation of the whole tumor is typically performed on T2-weighted MRI, the typical cross-modality UDA setting is to use T2-weighted MRI as the labeled source domain, and T1/T1ce/FLAIR MRI as the unlabeled target domains [32]. We followed the UDA training (80% subjects) and testing (20% subjects) split

Table 2. Comparison of whole tumor segmentation for the cross-modality UDA. We used T2-weighted MRI as our source domain, and T1-weighted, FLAIR, and T1ce MRI as the unlabeled target domains.

Method	Source data	Dice Score [%] ↑				Hausdorff Distance [mm] ↓			
		T1	FLAIR	T1CE	Average	T1	FLAIR	T1CE	Average
Source only	No UDA	6.8	54.4	6.7	22.6 ± 0.17	58.7	21.5	60.2	46.8 ± 0.15
CRUDA [1]	Partial[a]	47.2	65.6	49.4	54.1 ± 0.16	22.1	17.5	24.4	21.3 ± 0.10
OSUDA	**No**	**52.7**	**67.6**	**53.2**	**57.8 ± 0.15**	**20.4**	**16.6**	**22.8**	**19.9 ± 0.08**
OSUDA-AC	No	51.6	66.5	52.0	56.7 ± 0.16	21.5	17.8	23.6	21.0 ± 0.12
OSUDA-SE	No	51.1	65.8	52.8	56.6 ± 0.14	21.6	17.3	23.3	20.7 ± 0.10
CycleGAN [31]	Yes	38.1	63.3	42.1	47.8	25.4	17.2	23.2	21.9
SIFA [3]	Yes	51.7	68	58.2	59.3	19.6	16.9	15.01	17.1
DSFN [32]	Yes	57.3	78.9	62.2	66.1	17.5	13.8	15.5	15.6

[a] An additional class ratio predictor was required to be trained with the source data.

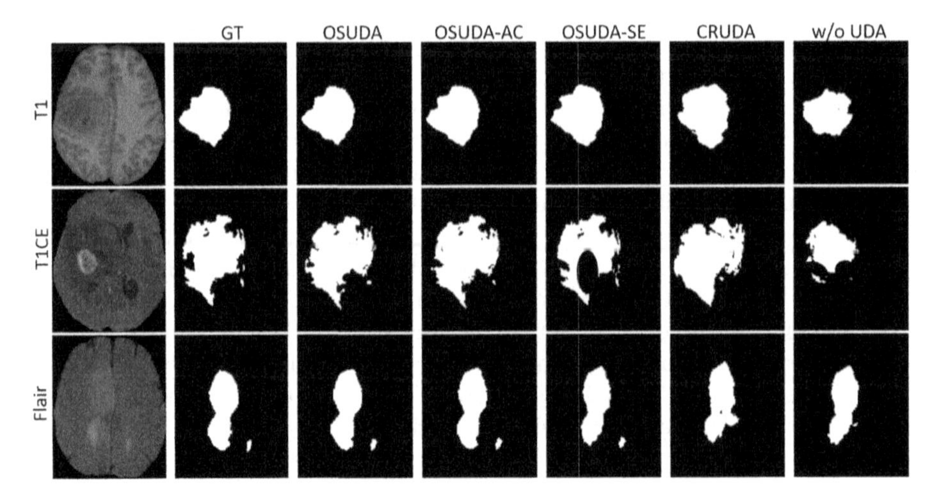

Fig. 3. Comparison with the other UDA methods and an ablation study for the cross-modality whole tumor segmentation UDA task. From top to bottom, we show a target test slice of T1, T1ce, and FLAIR MRI.

as in [32], and adopted the same single-channel input backbone. We note that the data were used in an unpaired manner [32].

The quantitative evaluation results are provided in Table 2. Our proposed OSUDA outperformed CRUDA [1] consistently. In addition, in CRUDA, the additional class ratio prediction model was required to be trained with the source data, which is prohibitive in many real-world cases. Furthermore, our OSUDA outperformed several UDA methods trained with the source data, e.g., CycleGAN [31] and SIFA [3], for the two metrics. The visual segmentation results of three target modalities are shown in Fig. 3, showing the superior performance of our framework, compared with the comparison methods.

4 Discussion and Conclusion

This work presented a practical UDA framework for the tumor segmentation task in the absence of the source domain data, only relying on the "off-the-shelf" pre-trained segmentation model with BN in the source domain. We proposed a low-order statistics progression with an exponential momentum decay scheme to gradually learn the target domain-specific mean and variance. The domain shareable high-order statistics consistency is enforced with our HBS loss, which is adaptively weighted based on the channel-wise transferability. The performance was further boosted with the unsupervised learning objective via self-entropy minimization. Our experimental results on the cross-subtype and cross-modality UDA tasks demonstrated that the proposed framework outperformed the comparison methods, and was robust to the class ratio shift.

Acknowledgement. This work is partially supported by NIH R01DC018511, R01DE027989, and P41EB022544.

References

1. Bateson, M., Kervadec, H., Dolz, J., Lombaert, H., Ben Ayed, I.: Source-relaxed domain adaptation for image segmentation. In: Martel, A.L., et al. (eds.) MICCAI 2020. LNCS, vol. 12261, pp. 490–499. Springer, Cham (2020). https://doi.org/10.1007/978-3-030-59710-8_48
2. Chang, W.G., You, T., Seo, S., Kwak, S., Han, B.: Domain-specific batch normalization for unsupervised domain adaptation. In: Proceedings of the IEEE/CVF Conference on Computer Vision and Pattern Recognition, pp. 7354–7362 (2019)
3. Chen, C., Dou, Q., Chen, H., Qin, J., Heng, P.A.: Synergistic image and feature adaptation: Towards cross-modality domain adaptation for medical image segmentation. In: Proceedings of the AAAI Conference on Artificial Intelligence, vol. 33, pp. 865–872 (2019)
4. Grandvalet, Y., Bengio, Y.: Semi-supervised learning by entropy minimization. In: NIPS (2005)
5. He, K., Zhang, X., Ren, S., Sun, J.: Deep residual learning for image recognition. In: Proceedings of the IEEE Conference on Computer Vision and Pattern Recognition (CVPR), June 2016
6. Ioffe, S., Szegedy, C.: Batch normalization: accelerating deep network training by reducing internal covariate shift. In: International Conference on Machine Learning, pp. 448–456. PMLR (2015)
7. Li, Y., Wang, N., Shi, J., Hou, X., Liu, J.: Adaptive batch normalization for practical domain adaptation. Pattern Recogn. **80**, 109–117 (2018)
8. Liang, J., Hu, D., Feng, J.: Do we really need to access the source data? Source hypothesis transfer for unsupervised domain adaptation. In: International Conference on Machine Learning, pp. 6028–6039. PMLR (2020)
9. Liu, X., et al.: Unimodal regularized neuron stick-breaking for ordinal classification. Neurocomputing **388**, 34–44 (2020)
10. Liu, X., Han, X., Qiao, Y., Ge, Y., Li, S., Lu, J.: Unimodal-uniform constrained Wasserstein training for medical diagnosis. In: Proceedings of the IEEE/CVF International Conference on Computer Vision Workshops (2019)

11. Liu, X., et al.: Domain generalization under conditional and label shifts via variational Bayesian inference, In: IJCAI (2021)
12. Liu, X., Hu, B., Liu, X., Lu, J., You, J., Kong, L.: Energy-constrained self-training for unsupervised domain adaptation. ICPR (2020)
13. Liu, X., et al.: Subtype-aware unsupervised domain adaptation for medical diagnosis. In: AAAI (2021)
14. Liu, X., Xing, F., El Fakhri, G., Woo, J.: A unified conditional disentanglement framework for multimodal brain MR image translation. In: ISBI, pp. 10–14. IEEE (2021)
15. Liu, X., Xing, F., Prince, J.L., Carass, A., Stone, M., El Fakhri, G., Woo, J.: Dual-cycle constrained bijective VAE-GAN for tagged-to-cine magnetic resonance image synthesis. In: ISBI, pp. 1448–1452. IEEE (2021)
16. Liu, X., Xing, F., Stone, M., Zhuo, J., Timothy, R., Prince, J.L., El Fakhri, G., Woo, J.: Generative self-training for cross-domain unsupervised tagged-to-cine MRI synthesis. In: MICCAI (2021)
17. Liu, X., Xing, F., Yang, C., Kuo, C.-C.J., El Fakhri, G., Woo, J.: Symmetric-constrained irregular structure inpainting for brain MRI registration with tumor pathology. In: Crimi, A., Bakas, S. (eds.) BrainLes 2020. LNCS, vol. 12658, pp. 80–91. Springer, Cham (2021). https://doi.org/10.1007/978-3-030-72084-1_8
18. Liu, X., Zou, Y., Song, Y., Yang, C., You, J., Kumar, B.V.K.V.: Ordinal regression with neuron stick-breaking for medical diagnosis. In: Leal-Taixé, L., Roth, S. (eds.) ECCV 2018. LNCS, vol. 11134, pp. 335–344. Springer, Cham (2019). https://doi.org/10.1007/978-3-030-11024-6_23
19. Mancini, M., Porzi, L., Bulo, S.R., Caputo, B., Ricci, E.: Boosting domain adaptation by discovering latent domains. In: Proceedings of the IEEE Conference on Computer Vision and Pattern Recognition, pp. 3771–3780 (2018)
20. Maria Carlucci, F., Porzi, L., Caputo, B., Ricci, E., Rota Bulo, S.: AutoDIAL: automatic domain alignment layers. In: Proceedings of the IEEE International Conference on Computer Vision, pp. 5067–5075 (2017)
21. Menze, B.H., et al.: The multimodal brain tumor image segmentation benchmark (BRATS). IEEE Trans. Med. Imaging **34**(10), 1993–2024 (2014)
22. Pan, X., Luo, P., Shi, J., Tang, X.: Two at once: enhancing learning and generalization capacities via IBN-Net. In: Ferrari, V., Hebert, M., Sminchisescu, C., Weiss, Y. (eds.) ECCV 2018. LNCS, vol. 11208, pp. 484–500. Springer, Cham (2018). https://doi.org/10.1007/978-3-030-01225-0_29
23. Shanis, Z., Gerber, S., Gao, M., Enquobahrie, A.: Intramodality domain adaptation using self ensembling and adversarial training. In: Wang, Q., et al. (eds.) DART/MIL3ID -2019. LNCS, vol. 11795, pp. 28–36. Springer, Cham (2019). https://doi.org/10.1007/978-3-030-33391-1_4
24. Wang, D., Shelhamer, E., Liu, S., Olshausen, B., Darrell, T.: Fully test-time adaptation by entropy minimization. arXiv preprint arXiv:2006.10726 (2020)
25. Wang, J., et al.: Automated interpretation of congenital heart disease from multi-view echocardiograms. Med. Image Anal. **69**, 101942 (2021)
26. Wang, X., Jin, Y., Long, M., Wang, J., Jordan, M.: Transferable normalization: towards improving transferability of deep neural networks. arXiv preprint arXiv:2019 (2019)
27. Wang, Z., Dai, Z., Póczos, B., Carbonell, J.: Characterizing and avoiding negative transfer. In: Proceedings of the IEEE/CVF Conference on Computer Vision and Pattern Recognition, pp. 11293–11302 (2019)
28. Wilson, G., Cook, D.J.: A survey of unsupervised deep domain adaptation. ACM Trans. Intell. Syst. Technol. (TIST) **11**(5), 1–46 (2020)

29. Zhang, J., Qi, L., Shi, Y., Gao, Y.: Generalizable semantic segmentation via model-agnostic learning and target-specific normalization. arXiv preprint arXiv:2003.12296 (2020)
30. Zhou, X.Y., Yang, G.Z.: Normalization in training u-net for 2-D biomedical semantic segmentation. IEEE Robot. Autom. Lett. **4**(2), 1792–1799 (2019)
31. Zhu, J.Y., Park, T., Isola, P., Efros, A.A.: Unpaired image-to-image translation using cycle-consistent adversarial networks. In: ICCV (2017)
32. Zou, D., Zhu, Q., Yan, P.: Unsupervised domain adaptation with dual scheme fusion network for medical image segmentation. In: Proceedings of the Twenty-Ninth International Joint Conference on Artificial Intelligence, IJCAI 2020, International Joint Conferences on Artificial Intelligence Organization, pp. 3291–3298 (2020)
33. Zou, Y., Yu, Z., Liu, X., Kumar, B., Wang, J.: Confidence regularized self-training. In: Proceedings of the IEEE/CVF International Conference on Computer Vision, pp. 5982–5991 (2019)

Quality-Aware Memory Network for Interactive Volumetric Image Segmentation

Tianfei Zhou[1], Liulei Li[2], Gustav Bredell[1], Jianwu Li[2(✉)], and Ender Konukoglu[1]

[1] Computer Vision Laboratory, ETH Zurich, Zurich, Switzerland
{tianfei.zhou,gustav.bredell,ender.konukoglu}@vision.ee.ethz.ch
[2] School of Computer Science and Technology, Beijing Institute of Technology, Beijing, China
{liliulei,ljw}@bit.edu.cn
https://github.com/0liliulei/Mem3D

Abstract. Despite recent progress of automatic medical image segmentation techniques, fully automatic results usually fail to meet the clinical use and typically require further refinement. In this work, we propose a *quality-aware memory network* for interactive segmentation of 3D medical images. Provided by user guidance on an arbitrary slice, an interaction network is firstly employed to obtain an initial 2D segmentation. The quality-aware memory network subsequently propagates the initial segmentation estimation bidirectionally over the entire volume. Subsequent refinement based on additional user guidance on other slices can be incorporated in the same manner. To further facilitate interactive segmentation, a quality assessment module is introduced to suggest the next slice to segment based on the current segmentation quality of each slice. The proposed network has two appealing characteristics: 1) The memory-augmented network offers the ability to quickly encode past segmentation information, which will be retrieved for the segmentation of other slices; 2) The quality assessment module enables the model to directly estimate the qualities of segmentation predictions, which allows an active learning paradigm where users preferentially label the lowest-quality slice for multi-round refinement. The proposed network leads to a robust interactive segmentation engine, which can generalize well to various types of user annotations (*e.g.*, scribbles, boxes). Experimental results on various medical datasets demonstrate the superiority of our approach in comparison with existing techniques.

Keywords: Interactive segmentation · Memory-augmented network

T. Zhou and L. Li—Contribute equally to this work.

M. de Bruijne et al. (Eds.): MICCAI 2021, LNCS 12902, pp. 560–570, 2021.
https://doi.org/10.1007/978-3-030-87196-3_52

1 Introduction

Accurate segmentation of organs/lesions from medical imaging data holds the promise of significant improvement of clinical treatment, by allowing the extraction of accurate models for visualization, quantification or simulation. Although recent deep learning based automatic segmentation engines [16,21,29,34] have achieved impressive performance, they still struggle to achieve sufficiently accurate and robust results for clinical practice, especially in the presence of poor image quality (*e.g.*, noise, low contrast) or highly variable shapes (*e.g.*, anatomical structures). Consequently, *interactive segmentation* [2,18,27,28,32,33] garners research interests of the medical image analysis community, and recently became the choice in many real-life medical applications.

In interactive segmentation, the user is factored in to play a crucial role in guiding the segmentation process and in correcting errors as they occur (often in an iteratively-refined manner). Classical approaches employ Graph Cuts [1], GeoS [4] or Random Walker [5,6] to incorporate scribbles for segmentation. Yet, these methods require a large amount of input from users to segment targets with low contrast and ambiguous boundaries. With the advent of deep learning, there has been a dramatically increasing interest in learning from user interactions. Recent methods demonstrate higher segmentation accuracy with fewer user interactions than classical approaches. Despite this, many approaches [11,22,26] only focus on 2D medical images, which are infeasible to process ubiquitous 3D data. Moreover, 2D segmentation does not allow the integration of prior knowledge regarding the 3D structure, and slice-by-slice interactive segmentation will impose extremely high annotation cost to users. To address this, many works [3,13,20,27,28] carefully design 3D networks to segment voxels at a time. While these methods enjoy superior ability of learning high-order, volumetric features, they require significantly more parameters and computations in comparison with the 2D counterparts. This necessitates compromises in the 3D network design to fit into a given memory or computation budget.

To address these issues, we take a novel perspective to explore memory-augmented neural networks [12,14,23,25] for 3D medical image segmentation. Memory networks augment neural networks with an external memory component, which allows the network to explicitly access the past experiences. They have been shown effective in few-shot learning [23], contrastive learning [7,29], and also been explored to solve reasoning problems in visual dialog [12,25]. The basic idea is to retrieve the relevant information from the external memory to answer a question at hand by using trainable memory modules. We take inspiration from these efforts to cast volumetric segmentation as a memory-based reasoning problem. Fundamental to our model is an external memory, which enables the model to online store segmented slices in the memory and later mine useful representations from the memory for segmenting other slices. In this way, our model makes full use of context within 3D data, and at the same time, avoids computationally expensive 3D operations. During segmentation, we dynamically update the memory to maintain shape or appearance variations of the target.

This facilitates easy model updating without extensive parameter optimization. Based on the memory network, we propose a novel interactive segmentation engine with three basic processes: 1) *Initialization:* an interaction network is employed to respond to user guidance on an arbitrary slice to obtain an initial 2D segmentation of a target. 2) *Propagation:* the memory network propagates the initial mask to the entire volume. 3) *Refinement:* the physician could provide extra guidance on low-quality slices for iterative refinement if the segmentation results are unsatisfactory.

Our contributions are three-fold: **First**, we propose a memory-augmented network for volumetric interactive segmentation. It is able to incorporate rich 3D contextual information, while avoiding expensive 3D operations. **Second**, we equip the memory network with a quality assessment module to assess the quality of each segmentation. This facilitates automatic selection of appropriate slices for iterative correction via human-in-the-loop. **Third**, our approach outperforms previous methods by a significant margin on two public datasets, while being able to handle various forms of interactions (*e.g.*, scribbles, bounding boxes).

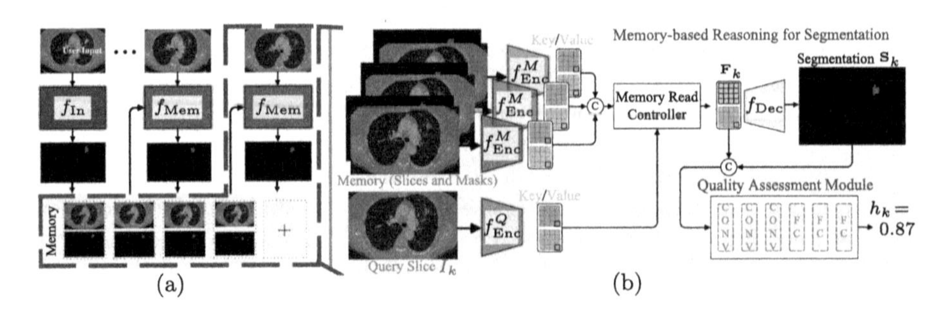

Fig. 1. Illustration of the proposed 3D interactive segmentation engine. (a) Simplified schematization of our engine that solves the task with an interaction network (f_{In}) and a quality-aware memory network (f_{Mem}). (b) Detailed network architecture of f_{Mem}. ⓒ denotes the concatenation operation. Zoom in for details.

2 Methodology

Let $V \in \mathbb{R}^{h \times w \times c}$ be a volumetric image to be segmented, which has a spatial size of $h \times w$ and c slices. Our approach aims to obtain a 3D binary mask $\mathbf{S} \in \{0,1\}^{h \times w \times c}$ for a specified target by utilizing user guidance. As shown in Fig. 1(a), the physician is asked to provide an initial input on an arbitrary slice $I_i \in \mathbb{R}^{h \times w}$, where I_i denotes the i-th slice of V. Then, an interaction network (f_{In}, Sect. 2.1) is employed to obtain a coarse 2D segmentation $\mathbf{S}_i \in [0,1]^{h \times w}$ for I_i. Subsequently, $\mathbf{S}_i$ is propagated to all other slices with a quality-aware memory network (f_{Mem}, Sect. 2.2) to obtain $\mathbf{S}$. Our approach also takes into account iterative refinement so that segmentation performance can be progressively improved with multi-round inference. To aid the refinement, the memory network has a module that estimates the segmentation performance on each slice and suggests the user to place guidance on the slice with the worst segmentation quality.

2.1 Interaction Network

The interaction network takes the user annotation at an interactive slice I_i to segment the specified target (or refine the previous result). At the tth round, its input consists of three images: the original gray-scale image I_i, the segmentation mask from the previous round $\mathbf{S}_i^{t-1}$, and a binary image $\mathbf{M}_i \in \{0,1\}^{h\times w}$ that encodes user guidance. Note that in the first round (*i.e.*, $t=0$), the segmentation mask $\mathbf{S}_i^{-1}$ is initialized as a neutral mask with 0.5 for all pixels. These inputs are concatenated along the channel dimension to form an input tensor $\mathbf{X}_i^t \in \mathbb{R}^{h\times w\times 3}$. The interaction network f_{IN} conducts the segmentation for I_i as follows:

$$\mathbf{S}_i^t = f_{\mathrm{In}}(\mathbf{X}_i^t) \in \mathbb{R}^{h\times w}. \quad (1)$$

Region-of-Interest (ROI). To further enhance performance and avoid mistakes in case of small targets or low-contrast tissues, we propose to crop the image according to the rough bounding-box estimation of user input, and apply f_{IN} only to the ROI. We extend the bounding box by 10% along sides to preserve more context. Each ROI region is resized into a fixed size for network input. After segmentation, the mask made within the ROI is inversely warped and pasted back to the original location.

2.2 Quality-Aware Memory Network

Given the initial segmentation $\mathbf{S}_i^t$, our memory network learns from the interactive slice I_i and segments the specified target in other slices. It stores previously segmented slices in an external memory, and takes advantage of the stored 3D image and corresponding segmentation to improve the segmentation of each 2D query image. The network architecture is shown in Fig. 1(b). In the following paragraphs, the superscript 't' is omitted for conciseness unless necessary.

Key-Value Embedding. Given a query slice I_k, the network mines useful information from memory $\mathcal{M}$ for segmentation. Here, each memory cell $\mathcal{M}_j \in \mathcal{M}$ consists of a slice I_{m_j} and its segmentation mask $\mathbf{S}_{m_j}$, where m_j indicates the index of the slice in the original volume. As shown in Fig. 1(b), we first encode the query I_k as well as each memory cell $\mathcal{M}_j = \{I_{m_j}, \mathbf{S}_{m_j}\}$ into pairs of *key* and *value* using dedicated encoders (*i.e.*, query f_{Enc}^Q and memory encoder f_{Enc}^M):

$$\mathbf{K}_k^Q, \mathbf{V}_k^Q = f_{\mathrm{Enc}}^Q(I_k), \quad (2)$$

$$\mathbf{K}_{m_j}^M, \mathbf{V}_{m_j}^M = f_{\mathrm{Enc}}^M(I_{m_j}, \mathbf{S}_{m_j}). \quad (3)$$

Here, $\mathbf{K}_k^Q \in \mathbb{R}^{H\times W\times C/8}$ and $\mathbf{V}_k^Q \in \mathbb{R}^{H\times W\times C/2}$ indicate key and value embedding of the query I_k, respectively, whereas $\mathbf{K}_{v_j}^M$ and $\mathbf{V}_{v_j}^M$ correspond to the key and value of the memory cell $\mathcal{M}_j$. H, W and C denote the height, width and channel dimension of the feature map from the backbone network, respectively. Note that for each memory cell, we apply Eq. (3) to obtain pairs of key and value embedding. Subsequently, all memory embedding are stacked together to build a pair of 4D key and value features (*i.e.*,$\mathbf{K}^M \in \mathbb{R}^{N\times H\times W\times C/8}$ and $\mathbf{V}^M \in \mathbb{R}^{N\times H\times W\times C/2}$), where $N = |\mathcal{M}|$ denotes memory size.

Memory Reading. The memory read controller retrieves relevant information from the memory based on the current query. Following the key-value retrieval mechanism in [12,25], we first compute the similarity between every 3D location $p \in \mathbb{R}^3$ in $\mathbf{K}^M$ with each spatial location $q \in \mathbb{R}^2$ in $\mathbf{K}_k^Q$ with dot product:

$$s_k(p,q) = \frac{\mathbf{K}^M(p)\cdot\mathbf{K}_k^Q(q)}{||\mathbf{K}^M(p)||\,||\mathbf{K}_k^Q(q)||} \in [-1,1], \tag{4}$$

where $\mathbf{K}^M(p) \in \mathbb{R}^{C/8}$ and $\mathbf{K}_k^Q(q) \in \mathbb{R}^{C/8}$ denote the features at the pth and q^{th} position of $\mathbf{K}^M$ and $\mathbf{K}_k^Q$, respectively. Next, we compute the read weight w_k by softmax normalization:

$$w_k(p,q) = \exp(s_k(p,q))/\sum\nolimits_o \exp(s_k(o,q)) \in [0,1]. \tag{5}$$

Here, $w_k(p,q)$ measures the matching probability between p and q. The memory summarization is then obtained using the weight to combine the memory value:

$$\mathbf{H}_k(q) = \sum\nolimits_p w_k(p,q)\mathbf{V}^M(p) \in \mathbb{R}^{C/2}. \tag{6}$$

Here, $\mathbf{V}^M(p) \in \mathbb{R}^{C/2}$ denotes the feature of the pth 3D position in $\mathbf{V}^M$. $\mathbf{H}_k(q)$ indicates the summarized representation of location q. For all $H \times W$ locations in $\mathbf{K}_k^Q$, we independently apply Eq. (6) and obtain the feature map $\mathbf{H}_k \in \mathbb{R}^{H\times W\times C/2}$. To achieve a more comprehensive representation, the feature map is concatenated with query value $\mathbf{V}_k^Q$ to compute a final representation $\mathbf{F}_k = \mathtt{cat}(\mathbf{H}_k, \mathbf{V}_k^Q) \in \mathbb{R}^{H\times W\times C}$.

Final Segmentation Readout. $\mathbf{F}_k$ is leveraged by a decoder network f_{Dec} to predict the final segmentation probability map for the query slice I_k:

$$\mathbf{S}_k = f_{\text{Dec}}(\mathbf{F}_k) \in [0,1]^{h\times w}. \tag{7}$$

Quality Assessment Module. While the memory network provides a compelling way to produce 3D segmentation, it does not support human-in-the-loop scenarios. To this end, we equip the memory network with a lightweight quality assessment head, which computes a quality score for each segmentation mask.

In particular, we consider *mean intersection-over-union (mIoU)* as the basic index for quality measurement. For each query I_k, we take the feature $\mathbf{F}_k$ and the corresponding segmentation $\mathbf{S}_k$ together to regress a mIoU score h_k:

$$h_k = f_{\mathrm{QA}}(\mathbf{F}_k, \mathbf{S}_k) \in [0, 1], \tag{8}$$

where $\mathbf{S}_k$ is firstly resized to a size of $H \times W$ and then concatenated with $\mathbf{F}_k$ for decoding. The slice with the lowest score is curated for next-round interaction.

2.3 Detailed Network Architecture

We follow [31] to implement the interaction network $f_{\mathrm{In}}(\cdot)$ as a coarse-to-fine segmentation network, however, other network architectures (*e.g.*, U-Net [21]) can also be used here instead. The network is trained using the cross-entropy loss. For the quality-aware memory network, we utilize ResNet-50 [8] as the backbone network for both f^{Q}_{Enc} (Eq. (2)) and f^{M}_{Enc} (Eq. (3)). The `res4` feature map of ResNet-50 is taken for computing the key and value embedding. For $f_{\mathrm{Dec}}(\cdot)$, we first apply Atrous Spatial Pyramid Pooling module after the memory read operation to enlarge the receptive field. We use three parallel dilated convolution layers with dilation rates 2, 4 and 8. Then, the learned feature is decoded with a residual refinement module proposed in [19]. The quality-aware module, $f_{\mathrm{QA}}(\cdot)$, consists of three 3×3 convolutional layers and three fully connected layers.

3 Experiment

Experimental Setup. Our experiments are conducted on two public datasets: ***MSD*** [24] includes ten subsets with different anatomy of interests, with a total of 2,633 3D volumes. In our experiments, we study the most challenging lung (64/32 for `train/val`) and colon (126/64 for `train/val`) subsets. $\boldsymbol{KiTS}_{19}$ [9] contains 300 arterial phase abdominal CT scans with annotations of kidney and tumor. We use the released 210 scans (168/42 for `train/val`) for experiments.

For comparison, we build a baseline model, named Interactive 3D nnU-Net, by adapting nnU-Net [10] into an interactive version. Specifically, we use the interaction network (Sect. 2.1) to obtain an initial segmentation, and this segment is then concatenated with the volume as the input of 3D nnU-Net. The quality-aware iterative refinement is also applied. In addition, we compare with a state-of-the-art method DeepIGeoS [28]. Several non-interactive methods are also included.

Interaction Simulation. Our approach can support various types of user interactions, which facilitates use in clinical routine. We study three common interactions: ***Scribbles*** provide sparse labels to describe the targets and rough outreach, ***Bounding Boxes*** outline the sizes and locations of targets, whereas ***Extreme Points*** [15] outline a more compact area of a target by labeling its *leftmost*, *rightmost*, *top*, *bottom* pixels. To simulate scribbles, we manually label the data in KiTS_{19} and MSD, resulting in 3,585 slices. Bounding boxes and extreme points can be easily simulated from ground-truths with relaxations. We train an independent f_{In} for each of these interaction types. In the first interaction round, we compute a rough ROI according to user input. Then we treat all the pixels out of the enlarged ROI as the background, and the pixels specified by scribbles, or in regions of bounding boxes and extreme points as foreground. We encode user guidance as a binary image $\mathbf{M}$ (Sect. 2.1) for input.

Table 1. Quantitative results (DSC %) on (left) MSD [24] and (right) KiTS_{19} [9] `val`.

method	lung cancer	colon cancer	method	kidney (organ)	kidney (tumor)
non-interactive methods:			*non-interactive methods*		
C2FNAS [30]	70.4	58.9	Mu et al. [17]	**97.4**	78.9
3D nnU-Net [10]	66.9	56.0	3D nnU-Net [10]	96.9	85.7
interactive methods:			*interactive methods*		
Interactive 3D nnU-Net [10]			Interactive 3D nnU-Net [10]		
scribbles	73.9	68.1	scribbles	94.5	86.3
bounding boxes	74.7	68.5	bounding boxes	95.3	86.8
extreme points	75.1	69.8	extreme points	95.6	87.6
DeepIGeoS [28]			DeepIGeoS [28]		
scribbles	76.6	72.3	scribbles	95.7	87.6
bounding boxes	77.2	73.0	bounding boxes	96.4	88.5
extreme points	77.5	73.2	extreme points	96.7	88.9
Ours			**Ours**		
scribbles	80.9	79.7	scribbles	96.9	88.2
bounding boxes	81.5	79.3	bounding boxes	97.0	88.4
extreme points	**82.0**	**80.4**	extreme points	97.0	**89.1**

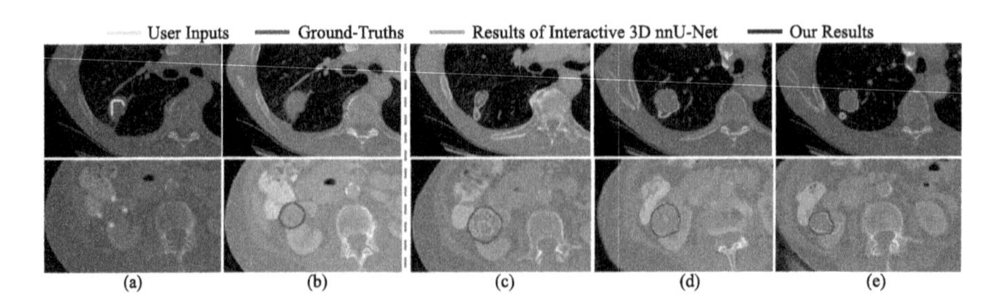

Fig. 2. Qualitative results of our approach *v.s.* Interactive 3D nnU-Net on two samples in MSD-Lung (row #1) and KITS_{19} (row #2), using scribbles and extreme points as supervision, respectively. (a) Interactive slices; (b) Results of interactive slices using the interaction network; (c)–(e) Results of other slices. Zoom in for details.

Training and Testing Details. Our engine is implemented in PyTorch. We use the same settings as [31] to train f_{In} (Sect. 2.1). The quality-aware memory network f_{Mem} (Sect. 2.2) is trained using Adam with learning rate 1e–5 and batch size 8 for 120 epochs. To make a training sample, we randomly sample 5 temporally ordered slices from a 3D image. During training, the memory is dynamically updated by adding the slice and mask at the previous step to the memory for the next slice.

During inference, simulated user hints are provided to f_{IN} for an initial segmentation of the interactive slice. Then, for each query slice, we put this interactive slice and the previous slice with corresponding segmentation mask into the memory as the most important reference information. In addition, we save a new memory item every N slices, where N is empirically set to 5. We do not add all slices and corresponding masks into memory to avoid large storage and computational costs. In this way, our memory network achieves the effect of online learning and adaption without additional training.

Quantitative and Qualitative Results. Table 1 (left) reports segmentation results of various methods on MSD `val`. For interactive methods, we report results at the 6th round which well balances accuracy and efficiency. It can be seen that our method leads to consistent performance gains over the baselines. Specifically, our approach significantly outperforms Interactive 3D nnU-Net by more than **7%** for lung cancer and **10%** for colon cancer, and outperforms DeepIGeoS [28] by more than **4%** and **7%**, respectively. Moreover, for different types of interaction, our method produces very similar performance, revealing its high robustness to user input. Table 1 (right) presents performance comparisons on KiTS_{19} `val`. The results demonstrate that, for kidney tumor segmentation, our engine generally outperforms the baseline models. The improvements are lower than seen for the MSD dataset due to the fact that the initial segmentation is already of high quality resulting in smaller dice score gains for adjustments.

Figure 2 depicts qualitative comparisons of our approach against Interactive 3D nnU-Net on representative examples from MSD and KITS_{19}. As seen, our approach produces more accurate segmentation results than the competitor.

Table 2. Ablation study of the quality assessment module in terms of DSC (%).

Variant	MSD (lung)	MSD (colon)	KiTS_{19} (tumor)
Oracle	81.4	80.4	89.1
Random	80.1	77.5	86.8
Quality assess.	81.3	79.7	88.6

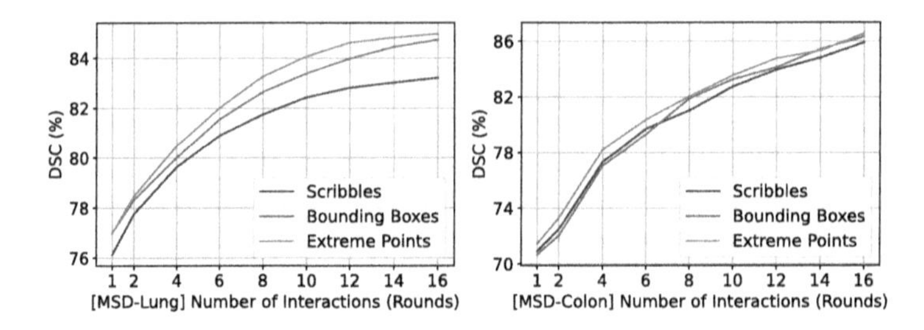

Fig. 3. The impact of number of interactions on MSD Lung (left) and Colon (right).

Efficacy of Quality Assessment Module. The quality assessment module empowers the engine to automatically select informative slices for iterative correction. To prove its efficacy, we design two baselines: 'oracle' selects the worst segmented slice by comparing the masks with corresponding ground-truths, while 'random' conducts random selection. As reported in Table 2, our method (*i.e.*, quality assessment module) significantly outperforms 'random' across three sets, and is comparable to 'oracle', proving its effectiveness.

Impact of Multi-round Refinement. Fig. 3 shows DSC scores with growing number of interactions on lung and colon subsets of MSD. We observe that multi-round refinement is crucial for achieving higher segmentation performance, and the performance becomes almost marginal at the 16th round.

Runtime Analysis. For a 3D volume with size $512 \times 512 \times 100$, our method needs 5.13 s on average for one-round segmentation on a NVIDIA 2080Ti GPU, whereas Interactive 3D nnU-Net needs 200 s. Hence our engine enables a significant increase in inference speed.

4 Conclusion

This work presents a novel interactive segmentation engine for 3D medical volumes. The key component is a memory-augmented neural network, which employs an external memory for accurate and efficient 3D segmentation. Moreover, the quality-aware module empowers the engine to automatically select informative slices for user feedback, which we believe is an important added value of the memory network. Experiments on two public datasets show that our engine outperforms other alternatives while having a much faster inference speed.

Acknowledgment. This research was supported in part by the Varian Research Grant and Beijing Natural Science Foundation (No. L191004).

References

1. Boykov, Y.Y., Jolly, M.P.: Interactive graph cuts for optimal boundary & region segmentation of objects in nd images. In: ICCV, pp. 105–112 (2001)
2. Bredell, G., Tanner, C., Konukoglu, E.: Iterative interaction training for segmentation editing networks. In: International Workshop on Machine Learning in Medical Imaging, pp. 363–370 (2018)
3. Çiçek, Ö., Abdulkadir, A., Lienkamp, S.S., Brox, T., Ronneberger, O.: 3D U-Net: learning dense volumetric segmentation from sparse annotation. In: Ourselin, S., Joskowicz, L., Sabuncu, M.R., Unal, G., Wells, W. (eds.) MICCAI 2016. LNCS, vol. 9901, pp. 424–432. Springer, Cham (2016). https://doi.org/10.1007/978-3-319-46723-8_49
4. Criminisi, A., Sharp, T., Blake, A.: GeoS: geodesic image segmentation. In: Forsyth, D., Torr, P., Zisserman, A. (eds.) ECCV 2008. LNCS, vol. 5302, pp. 99–112. Springer, Heidelberg (2008). https://doi.org/10.1007/978-3-540-88682-2_9
5. Grady, L.: Random walks for image segmentation. IEEE TPAMI **28**(11), 1768–1783 (2006)
6. Grady, L., Schiwietz, T., Aharon, S., Westermann, R.: Random walks for interactive organ segmentation in two and three dimensions: implementation and validation. In: Duncan, J.S., Gerig, G. (eds.) MICCAI 2005. LNCS, vol. 3750, pp. 773–780. Springer, Heidelberg (2005). https://doi.org/10.1007/11566489_95
7. He, K., Fan, H., Wu, Y., Xie, S., Girshick, R.: Momentum contrast for unsupervised visual representation learning. In: CVPR, pp. 9729–9738 (2020)
8. He, K., Zhang, X., Ren, S., Sun, J.: Deep residual learning for image recognition. In: CVPR, pp. 770–778 (2016)
9. Heller, N., et al.: The KiTS19 challenge data: 300 kidney tumor cases with clinical context, CT semantic segmentations, and surgical outcomes. arXiv preprint arXiv:1904.00445 (2019)
10. Isensee, F., et al.: nnU-Net: self-adapting framework for u-net-based medical image segmentation. arXiv preprint arXiv:1809.10486 (2018)
11. Kitrungrotsakul, T., Yutaro, I., Lin, L., Tong, R., Li, J., Chen, Y.W.: Interactive deep refinement network for medical image segmentation. arXiv preprint arXiv:2006.15320 (2020)
12. Kumar, A., et al.: Ask me anything: dynamic memory networks for natural language processing. In: ICML, pp. 1378–1387 (2016)
13. Liao, X., et al.: Iteratively-refined interactive 3D medical image segmentation with multi-agent reinforcement learning. In: CVPR, pp. 9394–9402 (2020)
14. Lu, X., Wang, W., Danelljan, M., Zhou, T., Shen, J., Van Gool, L.: Video object segmentation with episodic graph memory networks. In: Vedaldi, A., Bischof, H., Brox, T., Frahm, J.-M. (eds.) ECCV 2020. LNCS, vol. 12348, pp. 661–679. Springer, Cham (2020). https://doi.org/10.1007/978-3-030-58580-8_39
15. Maninis, K.K., Caelles, S., Pont-Tuset, J., Van Gool, L.: Deep extreme cut: from extreme points to object segmentation. In: CVPR, pp. 616–625 (2018)
16. Milletari, F., Navab, N., Ahmadi, S.A.: V-Net: fully convolutional neural networks for volumetric medical image segmentation. In: 3DV, pp. 565–571 (2016)
17. Mu, G., Lin, Z., Han, M., Yao, G., Gao, Y.: Segmentation of kidney tumor by multi-resolution VB-nets (2019)
18. Olabarriaga, S.D., Smeulders, A.W.: Interaction in the segmentation of medical images: a survey. MedIA **5**(2), 127–142 (2001)

19. Qin, X., Zhang, Z., Huang, C., Gao, C., Dehghan, M., Jagersand, M.: BASNet: boundary-aware salient object detection. In: CVPR, pp. 7479–7489 (2019)
20. Rajchl, M., et al.: DeepCut: object segmentation from bounding box annotations using convolutional neural networks. IEEE TMI **36**(2), 674–683 (2016)
21. Ronneberger, O., Fischer, P., Brox, T.: U-Net: convolutional networks for biomedical image segmentation. In: Navab, N., Hornegger, J., Wells, W.M., Frangi, A.F. (eds.) MICCAI 2015. LNCS, vol. 9351, pp. 234–241. Springer, Cham (2015). https://doi.org/10.1007/978-3-319-24574-4_28
22. Sakinis, T., et al.: Interactive segmentation of medical images through fully convolutional neural networks. arXiv preprint arXiv:1903.08205 (2019)
23. Santoro, A., Bartunov, S., Botvinick, M., Wierstra, D., Lillicrap, T.: Meta-learning with memory-augmented neural networks. In: ICML, pp. 1842–1850 (2016)
24. Simpson, A.L., et al.: A large annotated medical image dataset for the development and evaluation of segmentation algorithms. arXiv preprint arXiv:1902.09063 (2019)
25. Sukhbaatar, S., Szlam, A., Weston, J., Fergus, R.: End-to-end memory networks. arXiv preprint arXiv:1503.08895 (2015)
26. Sun, J., et al.: Interactive medical image segmentation via point-based interaction and sequential patch learning. arXiv preprint arXiv:1804.10481 (2018)
27. Wang, G., et al.: Interactive medical image segmentation using deep learning with image-specific fine tuning. IEEE TMI **37**(7), 1562–1573 (2018)
28. Wang, G., et al.: DeepIGeoS: a deep interactive geodesic framework for medical image segmentation. IEEE TPAMI **41**(7), 1559–1572 (2018)
29. Wang, W., Zhou, T., Yu, F., Dai, J., Konukoglu, E., Van Gool, L.: Exploring cross-image pixel contrast for semantic segmentation. arXiv preprint arXiv:2101.11939 (2021)
30. Yu, Q., et al.: C2FNAS: coarse-to-fine neural architecture search for 3D medical image segmentation. In: CVPR (2020)
31. Zhang, S., Liew, J.H., Wei, Y., Wei, S., Zhao, Y.: Interactive object segmentation with inside-outside guidance. In: CVPR, pp. 12234–12244 (2020)
32. Zhao, F., Xie, X.: An overview of interactive medical image segmentation. Ann. BMVA **2013**(7), 1–22 (2013)
33. Zhou, Y., Xie, L., Shen, W., Wang, Y., Fishman, E.K., Yuille, A.L.: A fixed-point model for pancreas segmentation in abdominal CT scans. In: Descoteaux, M., Maier-Hein, L., Franz, A., Jannin, P., Collins, D.L., Duchesne, S. (eds.) MICCAI 2017. LNCS, vol. 10433, pp. 693–701. Springer, Cham (2017). https://doi.org/10.1007/978-3-319-66182-7_79
34. Zhou, Z., Siddiquee, M.M.R., Tajbakhsh, N., Liang, J.: UNet++: a nested u-net architecture for medical image segmentation. In: Deep Learning in Medical Image Analysis and Multimodal Learning for Clinical Decision Support, pp. 3–11 (2018)

Improving Pneumonia Localization via Cross-Attention on Medical Images and Reports

Riddhish Bhalodia[1], Ali Hatamizadeh[2], Leo Tam[2], Ziyue Xu[2], Xiaosong Wang[2], Evrim Turkbey[3], and Daguang Xu[2](✉)

[1] School of Computing, University of Utah, Salt Lake City, UT, USA
riddhishb@sci.utah.edu
[2] NVIDIA Corporation, Santa Clara, USA
{ahatamizadeh,leot,ziyuex,xiaosongw,daguangx}@nvidia.com
[3] Department of Radiology and Imaging Sciences, National Institutes of Health Clinical Center, Bethesda, USA
evrim.turkbey@nih.gov

Abstract. Localization and characterization of diseases like pneumonia are primary steps in a clinical pipeline, facilitating detailed clinical diagnosis and subsequent treatment planning. Additionally, such location annotated datasets can provide a pathway for deep learning models to be used for downstream tasks. However, acquiring quality annotations is expensive on human resources and usually requires domain expertise. On the other hand, medical reports contain a plethora of information both about pnuemonia characteristics and its location. In this paper, we propose a novel weakly-supervised attention-driven deep learning model that leverages encoded information in medical reports during training to facilitate better localization. Our model also performs classification of attributes that are associated to pneumonia and extracted from medical reports for supervision. Both the classification and localization are trained in conjunction and once trained, the model can be utilized for both the localization and characterization of pneumonia using only the input image. In this paper, we explore and analyze the model using chest X-ray datasets and demonstrate qualitatively and quantitatively that the introduction of textual information improves pneumonia localization. We showcase quantitative results on two datasets, MIMIC-CXR and Chest X-ray-8, and we also showcase severity characterization on COVID-19 dataset.

Keywords: Pneumonia localization · Cross-attention · Multi-modal system

---

Electronic supplementary material The online version of this chapter (https://doi.org/10.1007/978-3-030-87196-3_53) contains supplementary material, which is available to authorized users.

M. de Bruijne et al. (Eds.): MICCAI 2021, LNCS 12902, pp. 571–581, 2021.
https://doi.org/10.1007/978-3-030-87196-3_53

1 Introduction

Pneumonia localization in chest X-rays and its subsequent characterization is of vital importance. At times pneumonia is a symptom of a disease (e.g., for COVID-19), and the severity and location of pneumonia can affect the treatment pipeline. Additionally, as manual annotations of the disease are resource-heavy, automatic localization can provide annotations or a guide to annotations for downstream deep learning-based tasks that rely on bounding-box annotations. However, to train an automatic pneumonia location detector, we need a large amount of annotated data that is not readily available and presents us with a chicken and egg problem.

Medical reports, on the other hand, are highly descriptive and provide a plethora of information. For instance, a report snippet "diffused opacity in left-lung" indicates the structure and location of opacity that can indicate pneumonia. Ideally, we can distill such information from the text and inform the localization of corresponding images without added supervision. The problem of associating textual information and leveraging it to improve/inform image domain tasks (e.g., object detection) is termed visual grounding. In computer vision literature, visual grounding methods have successfully demonstrated such a weak supervision approach of finding accurate annotation with text captions [5,21]. Such methods usually rely on a two-stage approach where the first stage is a box detector, and methods like Faster-RCNN [15] provide great box-detectors for natural images. However, the luxury of having a good initial box-detector is not available in medical imaging models.

Textual information from medical reports is hard to distill, as they have lot of specific terminologies, no clear sentence construction and many redundant/extraneous information. Therefore, in this paper, we establish a set of attributes that are informative about the location as well as characterization for pneumonia, and automatically extract them. We propose an attention-based image-text matching architecture to train a localization network, that utilizes the extracted attributes and the corresponding images. Additionally, we also have a jointly-trained attribute classification module. Attribute classification provides a more richer information as compared to a simple pneumonia classification score and the bounding box localization. The probability value for each attribute can be used for characterizing new images that might not have clinical reports associated. We showcase quantitative results on two chest X-ray datasets, MIMIC-CXR [7], and Chest X-ray-8 [18]. Additionally, we showcase qualitative results on the COVID-19 dataset [4] and demonstrate severity characterization by utilizing the predicted attributes. Although this method is extensively evaluated for pneumonia localization, it can easily be applicable to other diseases as well.

2 Related Works

Weakly supervised visual grounding methods utilize textual information from natural image captions and perform object localization. Most such methods

[5,21] are two-stage methods, where the first stage is a box-detector such as Faster-RCNN [15] trained on natural images. Other methods such as Retinanet [10] are end-to-end approaches for object localization that rely on pre-defined anchors. Weakly-supervised visual grounding is commonly achieved by enforcing similarity between image and text modalities in native space [1], or in a learned common latent space [5]. Some other works utilize attention models for vision-language alignments [11], as well as contrastive loss [6].

Class activation mapping (CAM) [22] and its variants [16] rely on a surrogate classification task to localize the regions of interest (ROI) and have been utilized in the context of disease localization [18]. More recent works incorporate image manipulation to localize the objects better [2,17]. Another class of methods utilize attention-based mechanisms to match the image and text modalities. Stacked cross-attention (SCAN) [8] is one such approach that provides a heat map of the image, based on how bounding boxes attend to the associated captions. Some recent works [19] incorporate similar ideas for image-text matching and improve upon SCAN.

Retinanet [10] and its variants have seen success in supervised pneumonia localization in Kaggle RSNA Pneumonia detection challenge [14]. Other works employ semi-supervision for disease localization and use limited amount of annotations [9]. Some recent works [13,20] use entire reports and match image and text features for disease classification and utilize CAM model for localization. The proposed method computes ROI-text similarity and weight ROIs for localization, and the attribute classification module can characterize different properties of a given X-ray image.

3 Methods

3.1 Attribute Extraction

For text features we utilize the clinical reports from the MIMIC-CXR dataset [7]. These reports are involved, complicated, and without clear sentence structures to be utilized directly. Instead, we pre-process this data to extract a dictionary of important text-attributes indicative of pneumonia location and characteristics. These attributes are keywords associated with the disease word in the clinical report. For example, the attribute set for an image with pneumonia can be *left, diffuse, basal*, etc., where each word describes the location, structure or characteristic of pneumonia. Figure 2 (green boxes and attributes) show examples of pneumonia annotation and the associated attributes. We emphasize here that the attribute-set is a set of 22 keywords that is kept constant, this is described in full detail in the supplementary material. We utilize a modified version of natural language-based label extractor provided by [18] to extract these attributes associated with each image and each disease class. A Word2Vec [12] model is trained on the entire set of medical reports. Using this trained model we extract the text-features for the limited attribute set. For a given text report T we extract M attributes, and their features are given as $\{\mathbf{m}_i\}_{i=1}^{M}$.

3.2 Box Detector and Image Features

An initial box-detector is essential for two-stage grounding methods [5]. Kaggle RSNA Pneumonia detection challenge [14] provides annotated X-ray images that can be utilized to train a box-detector. It contains ChestX-Ray-8 [18] images, with 2560 pneumonia annotations (bounding-boxes). Retinanet [10] and its variations are the best performing architectures in this challenge, therefore, we train a Retinanet with Resnet-50 backbone. This network produces regions of interest (ROIs) and pneumonia classification score. Additionally, we also get the ROI-features coming from the last convolution map of the network. For a given image I, we extract N ROIs (anchor boxes), the features are given as $\{\mathbf{r}_i\}_{i=1}^N$, the classification scores as $\{s_i\}_{i=1}^N$, and the geometric information (four box corner coordinates) as $\{\mathbf{g}_i\}_{i=1}^N$. This Retinanet box detector also acts as our *supervised baseline* for evaluating the textual information's contribution to disease localization. Once trained, we fix the weights of the box-detector.

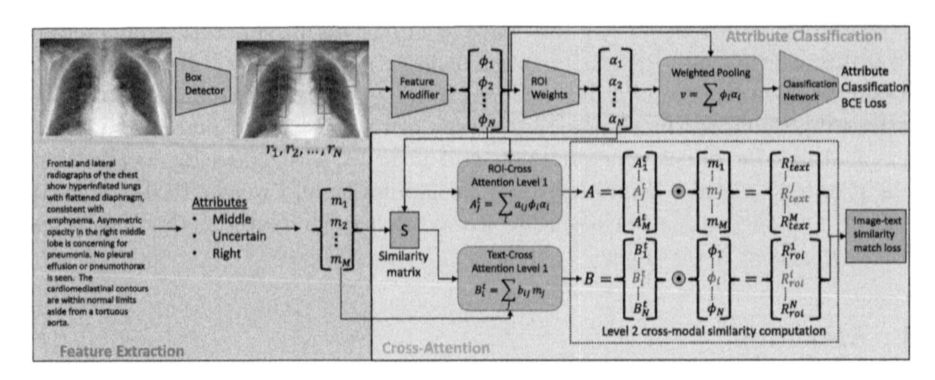

Fig. 1. Network architecture for training the attention based image-text matching for localization. (Color figure online)

3.3 Network Architecture and Attention Model

The proposed architecture is shown in Fig. 1, the architecture has three major components described as follows:

Feature Extractor. First the features are extracted for the text and ROI as described in Sects. 3.1 and 3.2, respectively. As these text and ROI features are coming from fixed architectures and not affected by training they are essentially independent and we need an agency to modify one of them to *match* to the other. For this, we modify the ROI features as, $\boldsymbol{\phi}_i = W_1\mathbf{r}_i + W_2[\texttt{LN}(W_g\mathbf{g}_i)|\texttt{LN}(W_s s_i)]$. The W's are the weights applied via fully connected layers which are to be optimized, LN denotes the layer normalization, and $[\cdots|\cdots]$ denotes concatenation. The modified ROI-features ($\boldsymbol{\phi}_i$) and the text-features $\mathbf{m}_i$ (directly from the Word2Vec model) are of the same dimension. This module is given by blue region in Fig. 1.

Attribute Classification. The method's primary goal is to provide selectivity in the ROIs, i.e., the box-detector will extract a large number of ROIs (via the anchors of Retinanet [10]) and we select the best ones. We want to discover the appropriate ROIs that can successfully classify the attribute string; for this, we have a two-stage approach (shown in green in Fig. 1). First, we compute a weight vector from the ROI-features $\boldsymbol{\phi}_i$, called α_i. These provide weights on each ROI feature, and we get an aggregate ROI feature: $\mathbf{v} = \sum_{i=1}^{N} \alpha_i \boldsymbol{\phi}_i$. Secondly, we utilize $\mathbf{v}$ to perform a multi-label attribute classification by passing $\mathbf{v}$ through a set of fully-connected layers to produce an attribute probability vector. The classification loss is given by binary cross-entropy between the target attributes and the predicted attribute probability vector, we denote it as $\mathcal{L}_{BCE}$. The ROI weights have another function; it also allows for the selection of ROIs that matches with the input attributes best in the cross-attention module described below.

Cross-Attention. The cross-attention is employed to answer the following questions: (i) How well does each weighted ROI describe the input set of text attributes? And (ii) How well does each attribute describe the set of ROIs?. A similar technique is employed in the stacked cross-attention (SCAN) [8]. First, we construct weighted contribution vectors for a given image, attribute pair as, $\mathbf{A}_j = \sum_{i=1}^{N} \alpha_i \boldsymbol{\phi}_i a_{ij}$ and $\mathbf{B}_i = \sum_{j=1}^{M} \mathbf{m}_j b_{ij}$. Where, i and j denote the index of ROI and attribute respectively. Finally, the a_{ij} and b_{ij} are given by $a_{ij} = \frac{\exp(\lambda_a s_{ij})}{\sum_j \exp(\lambda_a s_{ij})}$, $b_{ij} = \frac{\exp(\lambda_b s_{ij})}{\sum_i \exp(\lambda_b s_{ij})}$. The λs are constants, and s_{ij} represents the cosine-similarity between $\boldsymbol{\phi}_i$ and $\mathbf{m}_j$, and is normalized as given in [8]. $\mathbf{A}_j$ represents the aggregate ROI feature based on its contribution to the text attribute $\mathbf{m}_j$, and $\mathbf{B}_i$ represents the aggregate attribute feature based on its contribution to the ROI feature $\boldsymbol{\phi}_i$. Now, we match the features from two modalities, i.e., the weighted ROI feature ($\mathbf{A}_j$) should represent the text attribute ($\mathbf{m}_j$) information (and vice versa). To enforce this another level of similarity is introduced via cosine similarity, i.e., $R^j_{text} = \frac{\mathbf{A}_j^T \mathbf{m}_j}{||\mathbf{A}_j|| ||\mathbf{m}_j||}$, $R^i_{roi} = \frac{\mathbf{B}_i^T \boldsymbol{\phi}_i}{||\mathbf{B}_i|| ||\boldsymbol{\phi}_i||}$. Mean

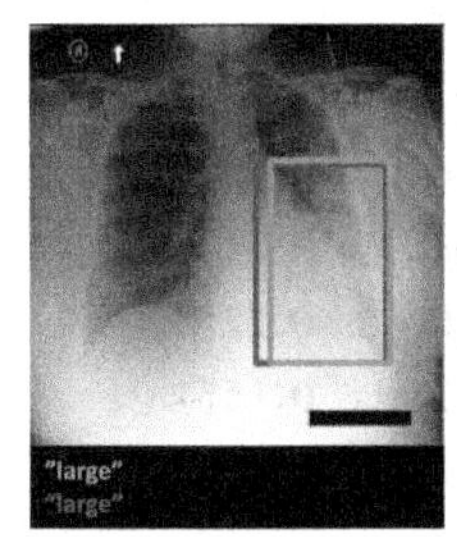

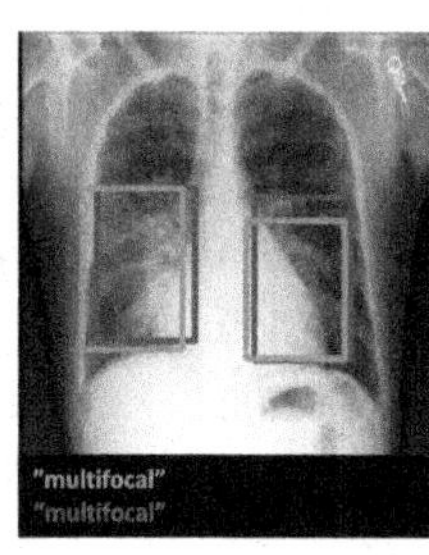

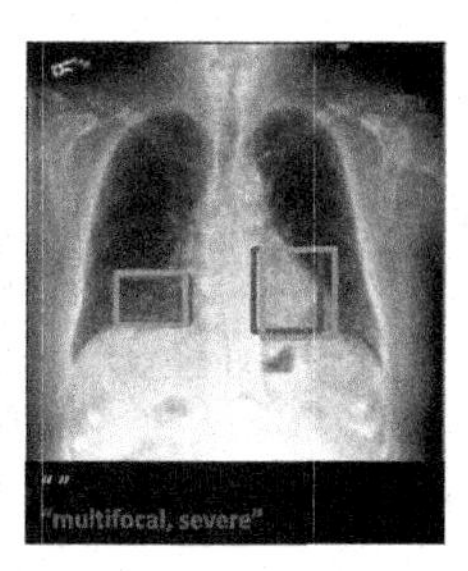

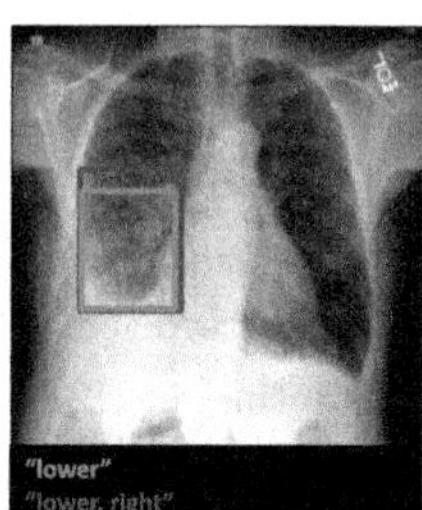

Fig. 2. Examples of localization and attribute classification from MIMIC-CXR test data. Green: expert annotated boxes and extracted attributes, Red: predicted boxes and attributes. (Color figure online)

similarity values are: $S_{roi}(I,T) = \frac{1}{N}\sum_{i=1}^{N} R^i_{roi}$ and $S_{text}(I,T) = \frac{1}{M}\sum_{j=1}^{M} R^j_{text}$. These mean similarity values reflect how well a given image I matches with the associated report T.

3.4 Loss Construction and Inference

Along with the classification loss function, we need another loss which enforces the ROI features to be matched with the input attribute features. We propose to use a contrastive formulation by defining *negative* ROIs (I_n) and attributes (T_n). We use the negative samples and the positive samples (I,T) to define similarity (S_{roi} and S_{text} from previous section) and formulate a triplet loss as follows:

$$\begin{aligned}\mathcal{L}_{trip} = &\max(\beta - S_{roi}(I,T) + S_{roi}(I_n,T), 0)\\ &+ \max(\beta - S_{text}(I,T) + S_{text}(I,T_n), 0)\end{aligned} \quad (1)$$

β is the triplet loss margin which we set as 0.8 for all experiments. *Negative ROIs* I_n are created by taking the set of lowest ranking ROIs coming from the Retinanet box detector. As Retinanet has pre-defined anchors, we are not filtering out any legitimate boxes when obtaining negative anchors/ROIs. *Negativc attributes* T_n are the negatives of individual attributes in the set, i.e., each attribute word will have its corresponding negative attribute word. These are chosen by finding the nearest word to the given attribute using the pre-trained word2vec model. Hence, the final loss is given as $\mathcal{L} = \mathcal{L}_{trip} + \mathcal{L}_{BCE}$. **Inference:** We utilize the α_is as weights for ROI selection, and following that we perform non-maximal suppression to remove redundant ROIs. As our inference only depends on α we do not require any text input during the testing.

Table 1. Pneumonia localization performance on different dataset using different methods, the Retinanet [10] refers to the supervised baseline.

Method	Dataset	IoU@0.25	IoU@0.5	IoU@0.75
CAM [22]	MIMIC-CXR	0.521	0.212	0.015
GradCAM [16]	MIMIC-CXR	**0.545**	0.178	0.029
Retinanet [10]	MIMIC-CXR	0.493	0.369	0.071
Proposed w/o classification	MIMIC-CXR	0.510	0.408	0.097
Proposed	MIMIC-CXR	0.529	**0.428**	**0.123**
Retinanet [10]	Chest X-ray-8	0.492	0.430	**0.115**
Proposed w/o classification	Chest X-ray-8	0.484	0.422	0.099
Proposed	Chest X-ray-8	**0.507**	**0.439**	**0.114**

4 Datasets and Implementation Details

The Retinanet trained on the RSNA challenge data (described in Sect. 3.2) is used as the initial box-detector in all the experiments. For training the proposed model we utilize the MIMIC-CXR dataset [7], it is a large-scale dataset that consists of 473,064 chest X-ray images with 206,754 paired radiology reports for 63,478 patients. We process each clinical report associated with a frontal X-ray image to extract the attributes (described in Sect. 3.1). As there are only limited set of clinical reports with the attributes in the fixed set of 22 keywords, we only utilize the images corresponding to pneumonia and having at least one of the attributes in this set, which results in 11,308 training samples. We train the proposed network on this subset of MIMIC images with early stopping at 185 epochs as the validation loss reaches a plateau (less than 0.1% change in 10 epochs). This trained model is used to perform quantitative and qualitative evaluations in the results section. We divide the data into 90%, 5%, 5% as a training, validation and testing split. We would also like to quantify the effect that an attribute classification module would have on the localization performance, therefore, we train another model without the classification module with just the triplet loss (Eq. 1). Additional information about the hyperparameters selection and other implementation details are in supplementary material. For evaluation on MIMIC-CXR we have a held out set of 169 images (part of test split) with pneumonia that are annotated by a board certified radiologist. We also evaluate on Chest-X-ray-8 dataset just utilizing the 120 annotations given for pneumonia as our test set. Finally, we perform a qualitative evaluation on COVID-19 X-Ray dataset [4] that contains 951 X-ray images acquired from different centers across the world and many images have associated medical reports/notes. Due to different acquisition centers, scanners and protocols across the data the intensity profiles and chest positioning has a huge variation among them.

5 Results

5.1 MIMIC and Chest X-ray-8 Dataset

We use bounding boxes annotations of MIMIC-CXR images to test pneumonia localization performance (169 annotated images as described in Sect. 4). For evaluation of localization performance, we employ intersection over union (IoU) evaluated at different thresholds of overlap. The quantitative results are provided in Table 1, where we see that introduction of the textual information improves the IoU score from the supervised baseline. Our method provides a different score for selecting ROIs compared to the supervised baseline, which is trained on a limited and controlled data that might degrade performance in cases with domain shifts. We also see that the proposed network, when trained without the attribute classification module performs worse as compared to one trained with it. Additionally, we also compare against CAM [22] and GradCAM [16] that use disease classification to perform localization using the activation heatmap.

We threshold the heatmaps to generate the bounding boxes from these methods and use same evaluation as described for proposed method. We see that the proposed method outperforms (especially at 0.5 and 0.75 thresholds indicating better localization) these other weakly-supervised methods. We also use the MIMIC-CXR trained model to evaluate the performance on ChestX-ray-8 [18] dataset, we only use data containing bounding box for pneumonia and not other diseases. The quantitative results using the supervised baseline as well as the proposed method is given in Table 1, and shows the proposed method outperforms the supervised baseline and without classification model. The attribute classification accuracy on the test set for MIMIC-CXR using the proposed method is 95.6% with an AUC of 0.84. Qualitative localization and attribute prediction is shown in Fig. 2.

5.2 COVID-19 Dataset

COVID-19 dataset of Chest X-Ray [4] is an important use-case for pneumonia characterization. We use the X-ray images from this dataset as an evaluation set on a model trained on MIMIC data. We look at specific cases and compare them with the reports; for the most part, the predicted attributes align with the information in associated reports. Two such interesting findings are shown in Fig. 3. In (a) the two scans of the same subject are shown, these are taken at day 0 and day 4. We notice that our model predicts the attribute (especially "severe") on day 4 scan, which is also suggestive from the report snippet shown below the images. In (b), despite images being very saturated, our model characterizes well and differentiates two cases as in report snippet.

Another aspect of pivotal importance is being able to characterize the severity of pneumonia. In addition to the Chest X ray images the dataset also provides a measure of severity for a subset of the overall dataset, this subset contains 94

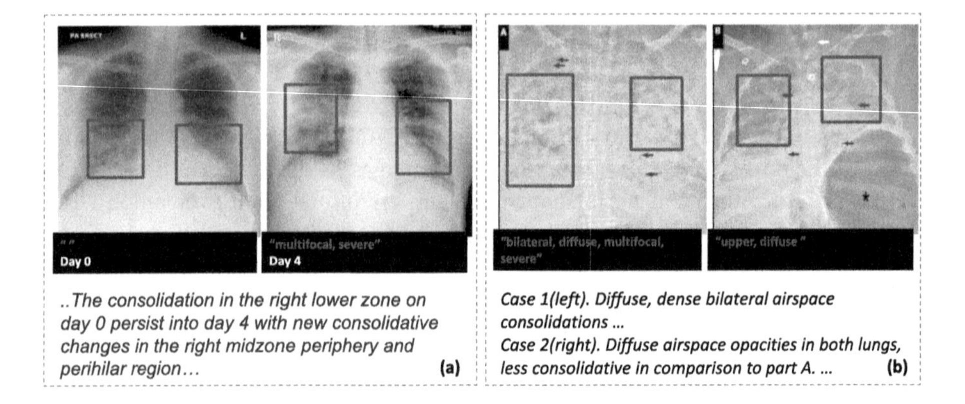

Fig. 3. Example case studies for pneumonia characterization from the COVID-19 Chest X-Ray dataset. The images, predicted attributes and localization, report snippet are shown here.

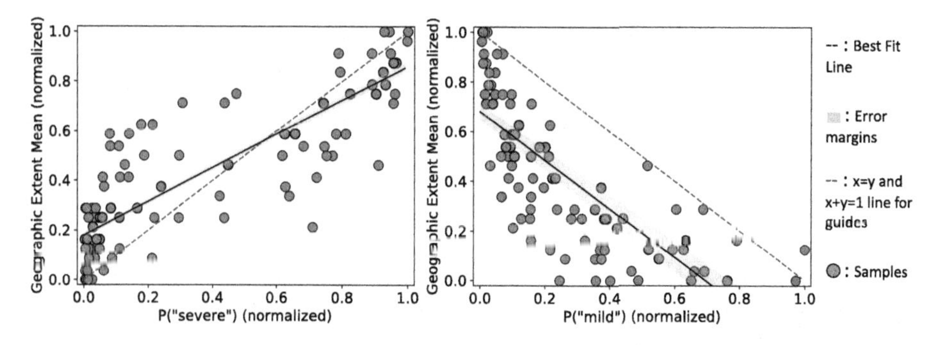

Fig. 4. Correlation between the attribute probabilities (P(*mild*) and P(*severe*)) and the ground-truth aggregate severity scores from experts.

images and their severity is quantified via *geographic extent mean* as described in [3]. It describes the extent of lung involvement by ground glass opacity or consolidation for each lung. The severity are mean of 3 expert ratings from 0–8 (8 is most severe). We hypothesize that the probability value associated with the attributes "severe" and "mild" can describe severity. We compute Pearson and Spearman correlation between the attribute probabilities and the ground truth severity scores, as well as other statistics quantified in Table 2. Statistics are evaluated using 5-fold cross validation on 94 cases. The p-values for the computation of each correlation is $< 10^{-10}$ demonstrating a near 100% confidence that these correlations are statistically significant. This is showcased in Fig. 4. A high positive correlation with P(*severe*) and a high negative correlation with P(*mild*) is in line with our hypothesis.

Table 2. Attribute probabilities compared to expert given severity. CC = correlation coefficient, R^2 = coefficient of determination, MAE = mean absolute error, MSE = mean squared error

Attribute	Pearson CC	Spearman CC	R^2	MAE	MSE
Severe	0.82 ± 0.001	0.84 ± 0.005	0.59 ± 0.09	0.14 ± 0.02	0.032 ± 0.008
Mild	-0.75 ± 0.003	-0.84 ± 0.02	0.56 ± 0.04	0.15 ± 0.01	0.035 ± 0.005

6 Conclusion

This paper introduces a novel attention-based mechanism of leveraging textual information from medical reports to inform disease localization on corresponding images. The network also comprises of a jointly trained attribute classification module. We showcase that the proposed method performs better than the supervised and other weakly supervised baselines. To showcase disease characterization, we test the model on COVID-19 dataset and qualitatively demonstrate that the attributes can characterize the images and perform localization even with extreme image variation. Furthermore, we perform severity characterization using our model that provides a statistically significant correlation with expert given severity ranking of COVID-19 X-rays.

References

1. Chen, K., Gao, J., Nevatia, R.: Knowledge aided consistency for weakly supervised phrase grounding. In: Proceedings of the IEEE Conference on Computer Vision and Pattern Recognition, pp. 4042–4050 (2018)
2. Choe, J., Shim, H.: Attention-based dropout layer for weakly supervised object localization. In: Proceedings of the IEEE/CVF Conference on Computer Vision and Pattern Recognition, pp. 2219–2228 (2019)
3. Cohen, J.P., et al.: Predicting COVID-19 pneumonia severity on chest x-ray with deep learning. Cureus **12**(7) (2020)
4. Cohen, J.P., Morrison, P., Dao, L.: COVID-19 image data collection. arXiv 2003.11597 (2020). https://github.com/ieee8023/covid-chestxray-dataset
5. Datta, S., Sikka, K., Roy, A., Ahuja, K., Parikh, D., Divakaran, A.: Align2Ground: weakly supervised phrase grounding guided by image-caption alignment. In: Proceedings of the IEEE/CVF International Conference on Computer Vision, pp. 2601–2610 (2019)
6. Gupta, T., Vahdat, A., Chechik, G., Yang, X., Kautz, J., Hoiem, D.: Contrastive learning for weakly supervised phrase grounding. arXiv preprint arXiv:2006.09920 (2020)
7. Johnson, A.E., et al.: MIMIC-CXR-JPG, a large publicly available database of labeled chest radiographs. arXiv preprint arXiv:1901.07042 (2019)
8. Lee, K.-H., Chen, X., Hua, G., Hu, H., He, X.: Stacked cross attention for image-text matching. In: Ferrari, V., Hebert, M., Sminchisescu, C., Weiss, Y. (eds.) ECCV 2018. LNCS, vol. 11208, pp. 212–228. Springer, Cham (2018). https://doi.org/10.1007/978-3-030-01225-0_13
9. Li, Z., et al.: Thoracic disease identification and localization with limited supervision. In: Proceedings of the IEEE Conference on Computer Vision and Pattern Recognition, pp. 8290–8299 (2018)
10. Lin, T.Y., Goyal, P., Girshick, R., He, K., Dollár, P.: Focal loss for dense object detection. In: Proceedings of the IEEE International Conference on Computer Vision, pp. 2980–2988 (2017)
11. Liu, X., Li, L., Wang, S., Zha, Z.J., Meng, D., Huang, Q.: Adaptive reconstruction network for weakly supervised referring expression grounding. In: Proceedings of the IEEE/CVF International Conference on Computer Vision, pp. 2611–2620 (2019)
12. Mikolov, T., Chen, K., Corrado, G., Dean, J.: Efficient estimation of word representations in vector space. arXiv preprint arXiv:1301.3781 (2013)
13. Moradi, M., Madani, A., Gur, Y., Guo, Y., Syeda-Mahmood, T.: Bimodal network architectures for automatic generation of image annotation from text. In: Frangi, A.F., Schnabel, J.A., Davatzikos, C., Alberola-López, C., Fichtinger, G. (eds.) MICCAI 2018. LNCS, vol. 11070, pp. 449–456. Springer, Cham (2018). https://doi.org/10.1007/978-3-030-00928-1_51
14. of North America, R.S.: RSNA pneumonia detection challenge (08 2018)

15. Ren, S., He, K., Girshick, R., Sun, J.: Faster R-CNN: towards real-time object detection with region proposal networks. arXiv preprint arXiv:1506.01497 (2015)
16. Selvaraju, R.R., Cogswell, M., Das, A., Vedantam, R., Parikh, D., Batra, D.: Grad-CAM: visual explanations from deep networks via gradient-based localization. In: Proceedings of the IEEE International Conference on Computer Vision, pp. 618–626 (2017)
17. Singh, K.K., Lee, Y.J.: Hide-and-seek: forcing a network to be meticulous for weakly-supervised object and action localization. In: 2017 IEEE International Conference on Computer Vision (ICCV), pp. 3544–3553. IEEE (2017)
18. Wang, X., Peng, Y., Lu, L., Lu, Z., Bagheri, M., Summers, R.M.: ChestX-ray8: hospital-scale chest x-ray database and benchmarks on weakly-supervised classification and localization of common thorax diseases. In: Proceedings of the IEEE Conference on Computer Vision and Pattern Recognition, pp. 2097–2106 (2017)
19. Wei, X., Zhang, T., Li, Y., Zhang, Y., Wu, F.: Multi-modality cross attention network for image and sentence matching. In: Proceedings of the IEEE/CVF Conference on Computer Vision and Pattern Recognition (CVPR) (2020)
20. Wu, J., et al.: Automatic bounding box annotation of chest x-ray data for localization of abnormalities. In: 2020 IEEE 17th International Symposium on Biomedical Imaging (ISBI), pp. 799–803. IEEE (2020)
21. Xiao, F., Sigal, L., Jae Lee, Y.: Weakly-supervised visual grounding of phrases with linguistic structures. In: Proceedings of the IEEE Conference on Computer Vision and Pattern Recognition, pp. 5945–5954 (2017)
22. Zhou, B., Khosla, A., Lapedriza, A., Oliva, A., Torralba, A.: Learning deep features for discriminative localization. In: Proceedings of the IEEE Conference on Computer Vision and Pattern Recognition, pp. 2921–2929 (2016)

Combining Attention-Based Multiple Instance Learning and Gaussian Processes for CT Hemorrhage Detection

Yunan Wu[1(✉)], Arne Schmidt[2], Enrique Hernández-Sánchez[2], Rafael Molina[2], and Aggelos K. Katsaggelos[1]

[1] Image and Video Processing Laboratory, Department of Electrical and Computer Engineering, Northwestern University, Evanston, IL 60208, USA
yuanwu2020@u.northwestern.edu, a-katsaggelos@northwestern.edu

[2] Depto. Ciencias de la Computacion e I.A., Universidad de Granada, 18071 Granada, Spain
{arne,rms}@decsai.ugr.es, enrique0197@correo.ugr.es,
https://ivpl.northwestern.edu/
http://decsai.ugr.es/

Abstract. Intracranial hemorrhage (ICH) is a life-threatening emergency with high rates of mortality and morbidity. Rapid and accurate detection of ICH is crucial for patients to get a timely treatment. In order to achieve the automatic diagnosis of ICH, most deep learning models rely on huge amounts of slice labels for training. Unfortunately, the manual annotation of CT slices by radiologists is time-consuming and costly. To diagnose ICH, in this work, we propose to use an attention-based multiple instance learning (Att-MIL) approach implemented through the combination of an attention-based convolutional neural network (Att-CNN) and a variational Gaussian process for multiple instance learning (VGPMIL). Only labels at scan-level are necessary for training. Our method (a) trains the model using scan labels and assigns each slice with an attention weight, which can be used to provide slice-level predictions, and (b) uses the VGPMIL model based on low-dimensional features extracted by the Att-CNN to obtain improved predictions both at slice and scan levels. To analyze the performance of the proposed approach, our model has been trained on 1150 scans from an RSNA dataset and evaluated on 490 scans from an external CQ500 dataset. Our method outperforms other methods using the same scan-level training and is able to achieve comparable or even better results than other methods relying on slice-level annotations.

Keywords: Attention-based multiple instance learning · Variational Gaussian processes · CT hemorrhage detection

This work has received funding from the European Union's Horizon 2020 research and innovation programme under the Marie Skłodowska Curie grant agreement No 860627 (CLARIFY Project) and also from the Spanish Ministry of Science and Innovation under project PID2019-105142RB-C22.

M. de Bruijne et al. (Eds.): MICCAI 2021, LNCS 12902, pp. 582–591, 2021.
https://doi.org/10.1007/978-3-030-87196-3_54

1 Introduction

Acute intracranial hemorrhage (ICH) has always been a life-threatening event that causes high mortality and morbidity rate [13]. Rapid and early detection of ICH is essential because nearly 30% of the life loss happens in the first 24 h [18]. In order to prompt the optimal treatment to patients in short time, computer-aided diagnosis (CAD) is being designed to establish a better triaging protocol.

Recently, deep learning (DL) algorithms have been proposed for the diagnosis of ICH. The most direct way is to train models on single slice to detect ICH precisely at slice-level [5,6]. For instance, Chilamkurthy et al. [6] modified ResNet-18 CNNs to predict ICH of each slice, slice-level probabilities were then combined using a random forest to provide ICH predictions at scan-level. Unfortunately, 3D spatial information is missing when each slice is trained independently. Recurrent neural networks (RNN) were introduced to link the consecutive 2D slices with feature vectors extracted from CNNs so as to enhance sequential connections among slices [3,12,14]. Although this approach achieves good performances in ICH detection, its training requires large size hand-labeled datasets at slice-level, whose generation is time-consuming and adds to the burden of radiologists. The use of scan-level annotations greatly reduces this workload. A full scan only needs one single label, which can even be automatically generated by natural language processing (NLP) methods applied to clinical radiologist reports [19]. Therefore, some studies focused on 3D CNNs to predict the existence of ICH at scan-level [9,17,19]. However, two major limitations of 3D CNNs are the highly expensive computation and the inability to localize ICH of slice-level which can serve as a instructive guidance for radiologists.

Multiple instance learning (MIL) is a weakly supervised learning method that has been recently applied to DL, especially in the domain of pathology [4]. Here, we treat ICH diagnosis as an MIL problem, where a full scan is defined as a “bag” and each slice in the scan is defined as an “instance”. A scan is classified as ICH if at least one slice in this scan has ICH and is normal if all slices are normal. Few studies use MIL method in ICH detection [15,16]. For instance, Remedios et al. [15] combine CNNs with MIL to predict ICH at scan-level, but the model was trained with a max-pooling operation so it could only select the most positive instance in a bag.

Variational Gaussian Processes for Multiple Instance Learning (VGPMIL) [7] treat the MIL problem in a probabilistic way. The model has several advantages such as robustness to overfitting (due to the non-parametric modelling provided by Gaussian processes), faster convergence and predictions for instances as well as for bags. One limitation is that the model can not be trained directly on images because of their high feature dimensionality. Therefore we use the attention-based CNN (Att-CNN) as a feature extractor and use the VGPMIL model to make the ICH predictions on slice as well as on scan levels.

To the best of our knowledge, this is the first study that combines Att-CNN for feature extraction with VGPMIL to improve hemorrhage detection using only scan level annotations. We demonstrate that (1) Att-CNN is able to predict accurate slice labels with no need for 2D slice annotations; (2) VGPMIL benefits from Att-CNN and improves the ICH predictions at both slice and scan levels; and (3) our Att-MIL approach outperforms other methods at scan-level and generalizes well to other datasets.

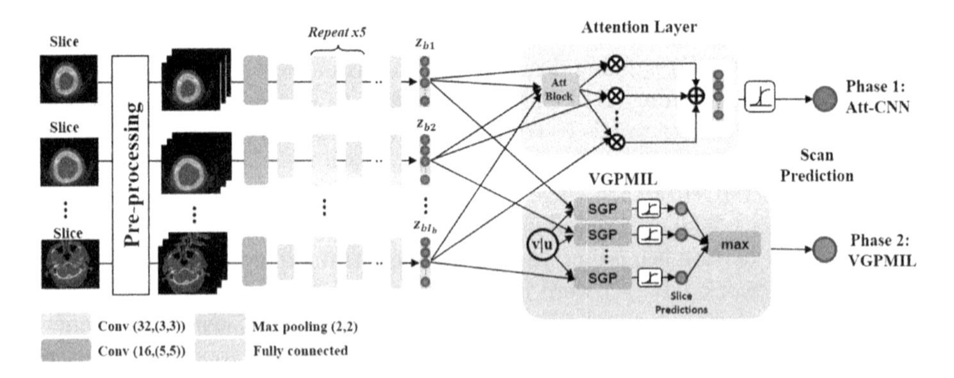

Fig. 1. The proposed architecture of Att-MIL, including the Att-CNN in Phase 1 and VGPMIL in Phase 2. Phase 1 updates parameters of Att-CNN at scan-level. Phase 2 uses features extracted from the Att-CNN to train the VGPMIL. The diagnosis are obtained using features of the trained Att-CNN and the VGPMIL for prediction.

2 Methods

2.1 Intracranial Hemorrhage Detection as a Multiple Instance Learning Problem

First we mathematically describe the detection of ICH as a problem of multiple instance learning (MIL). We treat CT slices $\{x_{b1}, x_{b2}, .., x_{bI_b}\}$ as instances and the full scan as a bag X_b. We assume that all bag labels $Y_b \in (0,1), b = 1, 2, .., B$ are available, where B describes the total number of CT scans. The true labels of slices $\{y_{b1}, y_{b2}, .., y_{bI_b}\}$ remain unobserved in the multiple instance learning setting. Notice that the number of slices I_b is bag dependent.

When a CT scan contains ICH we know that at least one slice must contain the pattern of hemorrhage while a negative scan contains only negative slices.

$$Y_b = 0 \Leftrightarrow \forall i \in \{1, ..., I_b\} : y_{bi} = 0 \quad (1)$$

$$Y_b = 1 \Leftrightarrow \exists i \in \{1, ..., I_b\} : y_{bi} = 1 \quad (2)$$

2.2 Model Description

The proposed model is defined in Fig. 1, training consists of two phases that are executed sequentially. First, the Att-CNN is trained to extract features from the slice images (Phase 1), then the VGPMIL model is trained based on these features to obtain slice and scan level predictions (Phase 2).

Phase 1. A convolutional neural network f_{cnn} serves as a feature extractor to obtain a vector of high level features z_{bi} for each instance x_{bi} in a bag b, so $z_{bi} = f_{cnn}(x_{bi}), \forall i = 1, 2, .., I_b$. The CNN model in Fig. 1 is implemented with six convolutional blocks, followed by a flatten layer and a fully connected layer. The convolutional block is used to extract discriminative features from each CT slice, including one convolutional layer and one maxpooling layer. The fully connected layer is used to decrease the size of feature vectors $z_{bi} \in R^{M \times 1}$, which are fed to the attention layer.

In order to weight each slice differently, we add an attention layer f_{att} to the CNN (i.e., a two-layered neural networks) [8], where an attention weight α_{bi} is assigned to each feature vector z_{bi}. The weights are determined by the model. Specifically, let $Z_b = \{z_{b1}, \dots, z_{bI_b}\}$ be the set of all feature vectors of bag b and $\alpha_{bi} = \alpha_1, ..., \alpha_{I_b}$ the attention weights for feature vectors z_{bi}. The weights α_{bi} add up to 1 and for each bag, there are as many coefficients as instances in the bag. The attention layer f_{att} is defined as $f_{att}(Z_b) = \sum_{i=1}^{I^b}, \alpha_{bi} z_{bi}$, where

$$\alpha_{bi} = \frac{exp\{w^\top \tanh(V z_{bi})\}}{\sum_{j=1}^{I^b} exp\{w^\top \tanh(V z_{bj})\}}. \tag{3}$$

$w \in R^{L \times 1}$ and $V \in R^{L \times M}$ are trainable parameters, M denotes the size of feature vectors and $L = 50$. The non-linearity $\tanh(\cdot)$ generates both positive and negative values in the gradient flow. After that, the weighted feature vectors pass through a classifier g_c (i.e., one fully connected layer with sigmoid function) to predict the scan labels:

$$p(Y_b|X_b) = g_c(f_{att}(Z_b)) = g_c(f_{att}(f_{cnn}(X_b))). \tag{4}$$

The feature extractor, attention layers and classifier are trained jointly using the common cross-entropy loss until convergence,

$$\mathcal{L} = \sum_{b=1}^{B} H(Y_b, p(Y_b|X_b)). \tag{5}$$

We refer to the whole process as Att-CNN. The weak labels at slice level can be obtained from the attention weights [8,16]. Specifically, if a scan is predicted as normal, all slices are normal. If a scan is predicted as ICH, the slices with min-max normalized attention weights larger than 0.5 are predicted as ICH. Notice that this is not a proper classifier but a way to obtain one from the attention weights (see [8,16] for details).

Phase 2. Once the training at Phase 1 has been completed, we no longer use the attention layer but replace it by a VGPMIL model [7] which has to be trained using z_{bi} for instance and bag predictions. It is important to note here that in the experimental section we will also analyze the use of $f_{att}(Z_b)$ as input to VGPMIL (experiment "AL-Aw" in Sect. 3.2).

A Variational Gaussian Process for Multiple Instance Learning is now used to learn to predict slice and CT scan labels (see Phase 2 in Fig. 1). Using the trained network, we calculate $Z_b = \{z_{b1}, \ldots, z_{bI_b}\}$, according to $z_{bi} = f_{cnn}(x_{bi})$. All Z_b are joined together in the matrix Z. Then, using the VGPMIL formulation we introduce a set of inducing points $U = \{u_1, \ldots, u_M\}$ and corresponding output v which are governed by the distribution $v|U \sim \mathcal{N}(v|0, K_{UU})$, where K_{UU} is the covariance matrix of the RBF kernel. Using the fully independent training conditional approximation, we utilize the conditional distribution $f|Z, U, v \sim \mathcal{N}(f|K_{ZU}K_{UU}^{-1}v, \mathcal{K}_{UU})$, where $\mathcal{K}_{UU} = \text{diag}(K_{ZZ} - K_{ZU}K_{UU}^{-1}K_{UZ_b})$. The conditional distribution of $y = \{y_{bi}, b = 1, \ldots, B, i = 1, \ldots, I_b\}$ given the underlying Gaussian process realization f is modelled using the product of independent Bernoulli distributions $\prod_{b=1}^{B}\prod_{i=1}^{Ib} \text{Ber}(y_{bi}|\sigma(f_{bi}))$, where $\sigma(\cdot)$ denotes the sigmoid function. Finally the observation model for the whole set of bag labels Y is given by

$$p(Y|y) = \prod_{b=1}^{B} p(Y_b|y_b) = \prod_{b=1}^{B} \left(\frac{H}{H+1}\right)^{G_b} \left(\frac{1}{H+1}\right)^{1-G_b}, \tag{6}$$

where $G_b = Y_b \max\{y_{bi}, i = 1, \ldots, I_b\} + (1 - Y_b)(1 - \max\{y_{bi}, i = 1, \ldots, I_b\})$ and H is a reasonably large value (see the experimental section). Notice that $G_b = [Y_b == \max\{y_{bi}, i = 1, \ldots, I_b\}]$.

To perform inference we approximate the posterior $p(y, f, v|Y, Z, U)$ by the distribution $q(v)p(f|Z, U, v)q(y)$, with $q(y) := \prod_b \prod_i q_{bi}(y_{bi})$, which is optimized by solving

$$\hat{q}(v), \hat{q}(y) = \underset{q(v),q(y)}{\text{argmin}} \; \text{KL}(q(v)p(f|Z, U, v)q(y)||p(y, f, v|Y, Z, U)). \tag{7}$$

The kernel and inducing location U parameters can be optimized by alternating between parameter and distribution optimization. The minimization of the Kullback-Leibler (KL) divergence produces a posterior distribution approximation $\hat{q}(u) = \mathcal{N}(\hat{m}, \hat{S})$, which is used to predict the instance and then the bag labels.

3 Experimental Design

3.1 Dataset and Preprocessing

A collection of 39650 slices of head CT images acquired from 1150 patients published in the 2019 Radiological Society of North America (RSNA) challenge [1] are included in this study. The number of slices ranges from 24 to 57 for each scan (512 × 512 pixels). In order to mimic the way radiologists adjust different window widths (W) and centers (C) to read CT images, we apply three windows for each CT slice in a scan with the Hounsfield Units (HU) to enhance the display of brain, blood, and soft tissue, respectively, using [W:80, C:40], [W:200, C:80] and [W:380, C:40]. The three window slices are concatenated into three channels and normalized to [0,1]. The CT scans are split into 1000 (Scan-P:411, Scan-N:589; Slice-P:4976, Slice-N: 29520) for training and validation and the rest 150 (Scan-P:72, Scan-N:78; Slice-P:806, Slice-N: 4448) for testing. Positive(P) represents the case with ICH and Negative(N) represents the normal case. In addition, the models trained on the RSNA dataset are further evaluated on 490 scans (Scan-P:205, Scan-N:285) of an external CQ500 dataset acquired from different institutions in India [6] to test the robustness of the model, each of which has 16 to 128 slices and goes through the same preprocessing steps.

3.2 Network Training

The model in Phase 1 is trained with the Adam optimizer [10] with a learning rate of 5×10^{-4}. The batch size is 16 per step. The att-CNN model is trained in two different settings: with no attention layer (nAL), where we use just the unweighted average, and with attention layer (AL). The AL setting is further divided into an experiment where we multiply the extracted features by the attention weight (AL-Aw) and another one with the raw features (AL-nAw) before feeding them into the VGPMIL model. The Att-CNN training process takes an average of about 4.5 h for 100 epochs with an early stopping operation. We report the mean and standard deviation of 5 independent runs. Both training and testing processes are performed using Tensorflow 2.0 in Python 3.7 on a single NVIDIA GeForce RTX 2070 Super GPU. The VGPMIL in Phase 2 is trained with Numpy 1.19 and runs on the 8 core AMD Ryzen 7 4800HS CPU. All experiments take less than 2 min to train the VGPMIL to converge within 200 epochs. The sparse Gaussian processes (SGP) are trained with $H = 100$, 200 inducing points (tested 50, 100, 200, 400) and a radial basis function kernel with a length scale of 1.5 (tested 1.0, 1.5, 2.0, 3.0) and a variance of 0.5 (tested 0.1, 0.5, 1.0, 2.0). The prediction time of the model takes an average of 2.5 s to predict a full scan of one patient in both phases.

4 Results and Discussions

4.1 Attention Layer vs. No Attention Layer

In order to show the capability of the attention layers to learn ICH insightful features, the results in Table 1 provide a comparison of the performances of Att-CNN and VGPMIL with and without attention layers. At scan-level, both Att-CNN and VGPMIL achieve better diagnosis performance with the attention layers. Although the recall is high for results without the attention layers, other metrics are extremely bad. Especially the accuracy score close to 0.5 means that neither the CNN nor the VGPMIL method is able to automatically detect ICH without the attention layers. Similar results are shown at slice-level predictions. Furthermore, we use t-distributed stochastic neighbor embedding (tsne) to reduce the size of the feature vectors at slice-level to two and visualize their distributions to verify our hypothesis. As shown in Fig. 2, it is evident to observe

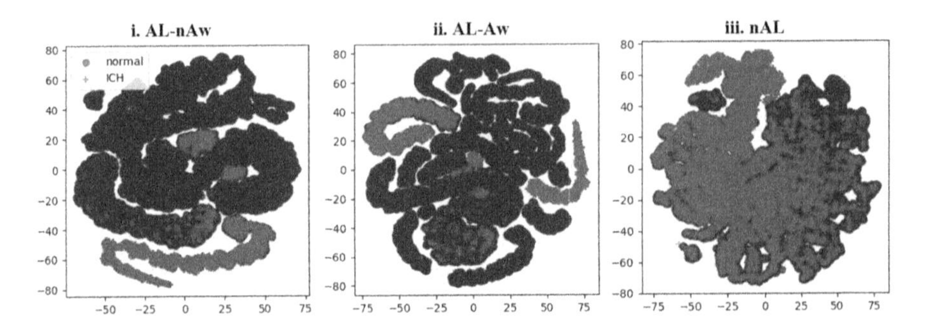

Fig. 2. The feature distribution of the RSNA instances in experiment AL-nAw (i), AL-Aw (ii) and nAL (iii). The dimensionality is reduced from 8 to 2 dimensions by t-distributed stochastic neighbor embedding (tsne). We observe that the attention layer (AL) helps the network to learn expressive features for the instances as the classes in AL-nAw and AL-Aw are better separated than that in nAL.

that feature distributions with attention layers (AL) are better separated than those with no attention layer (nAL), which demonstrates the role of attention layers in helping networks learn discriminative features at slice-level.

4.2 Attention Weights vs. No Attention Weights

As attention layers are necessary for networks to extract expressive feature vectors, our second hypothesis is that based on those expressive features, VGPMIL is able to achieve good performance even without using attention weights (AL-nAw). In Table 1, we compare the results of VGPMIL trained on feature vectors with and without attention weights. At slice-level prediction, VGPMIL without attention weights (AL-nAw) performs slightly better than with attention weights (AL-Aw). At scan-level, both results are equally good, but the overall ROC-AUC of VGPMIL without attention weights (0.964 ± 0.006) is slightly higher than with attention weights (0.951 ± 0.011). The results demonstrate that VGPMIL does not necessarily rely on attention weights to improve ICH predictions because feature separability is present at some level in both cases AL-nAw and AL-Aw, as shown in Fig. 2.

4.3 Attention-Based CNN vs. VGPMIL

We first do ablation studies with feature dimensions of 8, 32, and 128 for VGPMIL. The performance is best with 8 features while it gets worse with 32 and 128 features. Next, we do ablations with features extracted from earlier layers (i.e., convolutional layers) where the model performance drops significantly due to their higher feature dimensions. Therefore, VGPMIL chooses the optimal 8-dimension feature size extracted from the fully connected layers of CNN.

Table 1. Evaluation on the RSNA dataset at slice level and scan level. The results represent the average of 5 independent runs

Level	Slice-level				
Model	Att-CNN+VGPMIL			Att-CNN	
Metrics	AL-nAw	AL-Aw	nAL	AL	nAL
accuracy	**0.938 ± 0.003**	0.937 ± 0.006	0.797 ± 0.005	0.923 ± 0.005	0.502 ± 0.038
f1 score	**0.786 ± 0.013**	0.763 ± 0.031	0.441 ± 0.030	0.773 ± 0.008	0.353 ± 0.013
precision	**0.904 ± 0.017**	0.892 ± 0.018	0.382 ± 0.018	0.705 ± 0.023	0.221 ± 0.010
recall	0.664 ± 0.024	0.668 ± 0.053	0.529 ± 0.052	0.857 ± 0.015	**0.884 ± 0.023**
Level	Scan-level				
Model	Att-CNN+VGPMIL			Att-CN	
Metrics	AL-nAw	AL-Aw	nAL	AL	nAL
accuracy	0.780 ± 0.089	0.743 ± 0.176	0.603 ± 0.040	**0.781 ± 0.023**	0.495 ± 0.009
f1 score	**0.814 ± 0.059**	0.794 ± 0.104	0.707 ± 0.021	0.811 ± 0.017	0.656 ± 0.004
Precision	**0.705 ± 0.099**	0.705 ± 0.172	0.548 ± 0.026	0.694 ± 0.023	0.487 ± 0.004
recall	0.975 ± 0.025	0.944 ± 0.043	0.997 ± 0.006	0.975 ± 0.021	**1.000 ± 0.000**
ROC AUC	**0.964 ± 0.006**	0.951 ± 0.010	0.913 ± 0.013	0.951 ± 0.011	0.768 ± 0.047

Table 1 compares the performance of attention-based CNN and VGPMIL. At scan-level, VGPMIL shows a better AUC score (0.964 ± 0.006) than that of Att-CNN (0.951 ± 0.011). At the slice-level Att-CNN does not provide explicit predictions, but these can be derived from the attention weights as shown in [8, 16]. Therefore, we compare the performance of Att-CNN and VGPMIL at slice-level, where VGPMIL performs better, predicting slice labels at the accuracy of 0.938 ± 0.003 and the precision of 0.904 ± 0.017. This is significant because slice predictions are important for radiologists to localize ICH in a shorter time and the results show that our method is able to infer accurate slice labels even without annotating or training with any slice levels. Table 2 compares the AUC scores of our method with the state of the art training at scan-level. Although these studies use different dataset for their methods, the comparison indicates that our method outperforms them with a relatively smaller dataset.

Finally, we evaluate our method on an external testing CQ500 dataset and compare ROC scores with other methods (see Table 2). The labeling type "Scan" means the labels on scan-level in our training dataset, and the type "Slice" means the training labels on slice-level. For comparison, our method outperforms [11] predicting ICH at the same scan-level and is comparable to [6,12] training on slice labels. Notice that training on slice-level [6,12] is easier as it is fully supervised and involves more labels than scan-level methods. The results on CQ500 dataset further prove the good generalization of our method.

Table 2. Comparison of different approaches for binary ICH detection. Our results are reported as mean and standard deviation of 5 independent runs.

ICH detection at scan-level with different dataset				
Source	Dataset size	Labeling type	Method	AUC
Saab et al. [16]	4340 scans	Scan	MIL	0.91
Jnawali et al. [9]	40357 scans	Scan	3D CNNs	0.87
Titano et al. [19]	37236 scans	Scan	3D CNNs	0.88
Sato et al. [17]	126 scans	Scan	3D Autoencoder	0.87
Arbabshirani et al. [2]	45583 scans	Scan	3D CNNs	0.85
Ours (on RSNA)	1150 scans	Scan	Att-CNN + VGPMIL	0.964 ± 0.006
Evaluation on CQ500				
Source	Dataset size	Labeling type	Method	AUC
Chilamkurthy et al. [6]	490 Scans	Slice	2D CNNs	0.94
Nguyen et al. [12]		Slice	2D CNN + LSTM	0.96
Monteiro et al. [11]		Scan	CNN voxel-based Segment	0.83
Ours		Scan	Att-CNN + VGPMIL	0.906 ± 0.010

5 Conclusions

In this work, we propose an attention-based MIL method that combines attention-based CNN and VGPMIL to predict ICH at scan-level and achieves competitive AUC scores to previous works. Attention layers are important to extract meaningful features that VGPMIL relies on to improve the ICH predictions at both scan and slice levels. Importantly, our method is able to accurately predict slice labels without any slice annotations, greatly reducing the workload in future clinical research. Furthermore, the evaluations on the external dataset prove the good generalization of our method. This study paves the way for a promising weakly supervised learning method that fuses VGPMIL in CNN as a single end-to-end trainable model.

References

1. RSNA intracranial hemorrhage detection. https://kaggle.com/c/rsna-intracranial-hemorrhage-detection
2. Arbabshirani, M.R., et al.: Advanced machine learning in action: identification of intracranial hemorrhage on computed tomography scans of the head with clinical workflow integration. NPJ Digital Med. **1**(1), 1–7 (2018)
3. Burduja, M., Ionescu, R.T., Verga, N.: Accurate and efficient intracranial hemorrhage detection and subtype classification in 3D CT scans with convolutional and long short-term memory neural networks. Sensors **20**(19), 5611 (2020)

4. Campanella, G., Silva, V.W.K., Fuchs, T.J.: Terabyte-scale deep multiple instance learning for classification and localization in pathology (2018)
5. Chang, P., et al.: Hybrid 3D/2D convolutional neural network for hemorrhage evaluation on head CT. AJNR Am. J. Neuroradiol. **39**(9), 1609–1616 (2018)
6. Chilamkurthy, S., et al.: Development and validation of deep learning algorithms for detection of critical findings in head CT scans (2018)
7. Haußmann, M., Hamprecht, F., Kandemir, M.: Variational Bayesian multiple instance learning with gaussian processes. 2017 IEEE Conference on Computer Vision and Pattern Recognition (CVPR), pp. 810–819 (2017)
8. Ilse, M., Tomczak, J.M., Welling, M.: Attention-based deep multiple instance learning (2018)
9. Jnawali, K., Arbabshirani, M.R., Rao, N., Patel, A.A.: Deep 3D convolution neural network for CT brain hemorrhage classification. In: Mori, K., Petrick, N. (eds.) Medical Imaging 2018: Computer-Aided Diagnosis, p. 47. SPIE (2018)
10. Kingma, D.P., Ba, J.: Adam: a method for stochastic optimization (2017)
11. Monteiro, M., et al.: Multiclass semantic segmentation and quantification of traumatic brain injury lesions on head CT using deep learning: an algorithm development and multicentre validation study. Lancet Digital Health **2**(6), e314–e322 (2020)
12. Nguyen, N.T., Tran, D.Q., Nguyen, N.T., Nguyen, H.Q.: A CNN-LSTM architecture for detection of intracranial hemorrhage on CT scans (2020)
13. Otite, F.O., Khandelwal, P., Malik, A.M., Chaturvedi, S., Sacco, R.L., Romano, J.G.: Ten-year temporal trends in medical complications after acute intracerebral hemorrhage in the united states. Stroke **48**(3), 596–603 (2017)
14. Patel, A., Leemput, S.C.v.d., Prokop, M., Ginneken, B.V., Manniesing, R.: Image level training and prediction: intracranial hemorrhage identification in 3D non-contrast CT. IEEE Access **7**, 92355–92364 (2019)
15. Remedios, S., et al.: Extracting 2D weak labels from volume labels using multiple instance learning in CT hemorrhage detection. In: Medical Imaging 2020: Image Processing, vol. 11313, p. 113130F (2020)
16. Saab, K., et al.: Doubly weak supervision of deep learning models for head CT. In: Shen, D., et al. (eds.) MICCAI 2019. LNCS, vol. 11766, pp. 811–819. Springer, Cham (2019). https://doi.org/10.1007/978-3-030-32248-9_90
17. Sato, D., et al.: A primitive study on unsupervised anomaly detection with an autoencoder in emergency head CT volumes. In: Medical Imaging 2018: Computer-Aided Diagnosis, p. 60 (2018)
18. Sobrino, J., Shafi, S.: Timing and causes of death after injuries, vol. 26, no. 2, pp. 120–123 (2013)
19. Titano, J.J., et al.: Automated deep-neural-network surveillance of cranial images for acute neurologic events. Nat. Med. **24**(9), 1337–1341 (2018)

CPNet: Cycle Prototype Network for Weakly-Supervised 3D Renal Compartments Segmentation on CT Images

Song Wang[1], Yuting He[1], Youyong Kong[1,4], Xiaomei Zhu[3], Shaobo Zhang[5], Pengfei Shao[5], Jean-Louis Dillenseger[2,4], Jean-Louis Coatrieux[2], Shuo Li[6], and Guanyu Yang[1,4](✉)

[1] LIST, Key Laboratory of Computer Network and Information Integration (Southeast University), Ministry of Education, Nanjing, China
yang.list@seu.edu.cn
[2] Univ Rennes, Inserm, LTSI - UMR1099, 35000 Rennes, France
[3] Department of Radiology, The First Affiliated Hospital of Nanjing Medical University, Nanjing, China
[4] Centre de Recherche en Information Biomédicale Sino-Français (CRIBs), Nanjing, China
[5] Department of Urology, The First Affiliated Hospital of Nanjing Medical University, Nanjing, China
[6] Department of Medical Biophysics, University of Western Ontario, London, ON, Canada

Abstract. Renal compartment segmentation on CT images targets on extracting the 3D structure of renal compartments from abdominal CTA images and is of great significance to the diagnosis and treatment for kidney diseases. However, due to the unclear compartment boundary, thin compartment structure and large anatomy variation of 3D kidney CT images, deep-learning based renal compartment segmentation is a challenging task. We propose a novel weakly supervised learning framework, Cycle Prototype Network, for 3D renal compartment segmentation. It has three innovations: (1) A Cycle Prototype Learning (CPL) is proposed to learn consistency for generalization. It learns from pseudo labels through the forward process and learns consistency regularization through the reverse process. The two processes make the model robust to noise and label-efficient. (2) We propose a Bayes Weakly Supervised Module (BWSM) based on cross-period prior knowledge. It learns prior knowledge from cross-period unlabeled data and perform error correction automatically, thus generates accurate pseudo labels. (3) We present a Fine Decoding Feature Extractor (FDFE) for fine-grained feature extraction. It combines global morphology information and local detail information to obtain feature maps with sharp detail, so the model will achieve fine segmentation on thin structures. Our extensive experiments demonstrated our great performance. Our model achieves Dice of 79.1% and

S. Wang and Y. He—Equal contribution.

M. de Bruijne et al. (Eds.): MICCAI 2021, LNCS 12902, pp. 592–602, 2021.
https://doi.org/10.1007/978-3-030-87196-3_55

78.7% with only four labeled images, achieving a significant improvement by about 20% than typical prototype model PANet [16].

1 Introduction

3D renal compartment segmentation is the process of extracting the 3D structure of renal cortex and medulla from abdominal CTA images, which has great significance on laparoscopic partial nephrectomy [2,12,13,18]. During operation, correct segmentation of renal compartments helps doctors control the proportion of nephrectomy [14], reduce the loss of renal function. Post-operatively, it assists in monitoring the recovery of renal function [3,7], ultimately achieve the goal of reducing the cost of surgery, increasing the success rate of surgery, and providing patients with higher quality medical services.

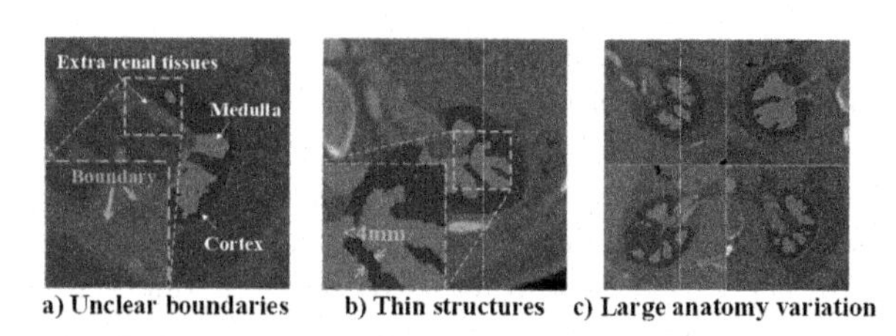

Fig. 1. Challenges of our renal compartment segmentation. (a) Unclear boundaries between renal compartments and extra-renal tissues which make model lose the ability to extract distinguishable features. (b) Thin structures of renal compartments which make feature extractor lose fine-grained features. (c) Anatomy variation between images which make model sensitive to singular structures.

Deep learning has achieved remarkable success in medical image segmentation [10,15], renal tumors segmentation [5,9] and renal artery [4] segmentation, but deep-learning based renal compartment segmentation on CT images is a challenging task owing to its particularity: **(1) The boundary of renal compartments is not clear.** As shown in Fig. 1(a), the CT values are similar between cortex, medulla and extra-renal tissues. Model will lose the ability to extract distinguishable features for compartment anatomy, therefore it is prone to over segment or under segment. **(2) The structure of renal compartments is thin.** As is shown in Fig. 1(b), the cortex extends into the kidney structure, entangles with the medulla to form several thin structures with unstable morphology. This makes feature extractors with large receptive fields easy to lose fine-grained features. The model trained with these features is not sensitive to small structures, thus be unable to segment the small part of renal compartments. **(3) The large anatomy variation and small dataset scale.** As is shown in Fig. 1(c), the renal medulla is divided into a number of random shapes. This anatomy varies between different kidneys, so fine annotation requires a lot of time for professional doctors, which limits the scale of the labeled dataset. Therefore, the labeled dataset cannot cover all compartment morphologies. Model is unable to learn generalized knowledge, will be sensitive to singular structures, and have poor segmentation capabilities for unseen morphological structures.

There is no automatic, label-efficient and high-accuracy renal compartment segmentation works on CT images being reported. Some semi-automatic works design image operation combined with manual annotation to achieve renal compartment segmentation [17], requiring a lot of labor costs. Deep-learning based renal compartment segmentation methods perform segmentation automatically, but the small scale of labeled dataset seriously limits their performance [6,8].

Therefore, we proposed an automatic and label-efficient renal compartment segmentation framework Cycle Prototype Network (CPNet), which efficiently extracts renal compartments with only a few labels. It has three innovations:

(1) We proposed a Cycle Prototype Learning framework (CPL) to learn consistency for generalization. It uses labels as guidance to extract features accurately and forms regularization through a reverse process to improve the generalization of the model. Feature maps are extracted under the guidance of the support label, the obtained feature vectors of the same compartment have smaller differences, and those of different compartments are more distinguishable. Prototype vectors that represent the features of compartments will be obtained by combining feature vectors of the same compartment. Prototypes are then used as templates to segment query images and train the network, forcing the feature vector extracted by the network to aggregate to the prototype vector. The feature vector of the unclear boundary deviates further from the cluster center, thus a higher penalty will be imposed. Therefore, the network will extract more discriminative boundary features. After that, the framework uses query prediction to reversely segment support images in the reverse process. This process uses the same feature extractor and prototype space to encourage the network to extract consistent class features on different images, forming a regularization thus improves the generalization ability of the model.
(2) We proposed a Bayes Weakly Supervised Module (BWSM) based on cross-period prior knowledge to embed prior for pseudo label generation. Different renal compartments have different reactions on contrast agents, resulting in different performances on images of different periods. We take use of this prior to use CTA and CTU images, combined with network prediction, to obtain pseudo labels through Bayes optimization. The module first obtains noisy pseudo-labels from CTA and CTU images, which contain accurate location information, but noisy morphological information. Then it includes the network prediction as likelihood, which has relatively smooth morphological information but inaccurate location information. It uses prior knowledge Bayes theory to synthesize the two, and the obtained posterior probability weakens the error components in the two and forms a more accurate pseudo-label. Embed the posterior pseudo-label into the model indirectly expands the size of the training set. A larger training set can cover more possible compartment anatomy variations, forcing the model to reduce its attention

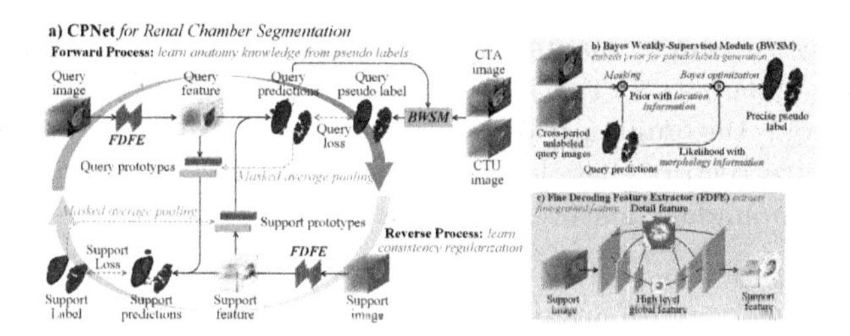

Fig. 2. The structure of our CPNet framework (a) CPL learns consistency for generalization. It forms regularization through forward and reverse processes thus enhance robustness. (b) Bayes Weakly Supervised Module. It embeds prior knowledge for pseudo label generation to enhance generalization. (c) Fine Decoding Feature Extractor. It focuses on fine-grained detail information thus extract feature maps with sharp detail.

to unstable spatial distribution features, thus improving the generalization ability of the model.

(3) We proposed a Fine Decoding Feature Extractor (FDFE) that combines location information and detail information to extract fine-grained features. In the encoder-decoder stream, the decoder restores high-resolution image with the coordinate information recorded in the encoder, thus restores the global morphological information of the high-level feature maps. The cross-layer connection directly transmits the local detail information to the decoder, adds the detail feature lost in the encoder-decoder stream. Such structure combines global and local features, has better performance for segmentation tasks of renal compartments that focus on small volumes.

2 Methodology

As shown in Fig. 2, our CPNet uses a cycle prototype learning paradigm to efficiently implement weakly supervised segmentation of renal compartments. It has three related modules: (a) The main CPL framework learns consistency with two processes. The forward process learns knowledge from pseudo labels to improve the robustness of the model. The reverse process achieves regularization and improves the generalization of the model. (b) Our BWSM extracts prior knowledge and embeds pseudo-label into learning, thus improves the robustness of the model on images with large anatomy variation. (c) Our FDFE combines the global morphological features recovered by the decoder and the local detail features passed by cross-connection transmission, thus make the resulting feature maps have sharper detail information.

2.1 CPL for Consistency Generalization

Advantages of CPL: (1) Stronger generalization ability. The reverse process in the framework forms regularization, forcing the extracted features to meet the consistency principle, making the network robust to the noise in labels. (2) Pay more attention to the boundary area. Our framework extracts class prototypes under the guidance of label, and then imposes high penalties on boundary feature vectors that deviate from the prototype vector so that more distinguishable boundary features can be extracted.

CPL Structure for Consistency Regularization: As shown in Fig. 2, our framework trains an efficient feature extraction network on weakly supervised dataset, and consists of two processes: forward process and reverse process. It first uses FDFE to extract support feature x_s and query feature x_q from the support image i_s and query image i_q. In the forward process, it uses the support label y_s to perform masked average pooling $M(\cdot)$ to obtain the support prototype $t_s = M(y_s \cdot x_s)$. Feature vectors in x_q is classified by calculating cosine similarity $CS(\cdot)$ with t_s to obtain query prediction $y_q' = CS(t_s \cdot x_q)$. Similarly, in the reverse process, it uses y_q' as query label, extracts the query prototype $t_q = M(y_q' \cdot x_q)$ and predicts the support image to obtain the support prediction $y_s' = CS(t_q \cdot x_s)$.

Forward and Reverse Learning Process: We train our model through forward and reverse process. In the forward process, our model learns from query pseudo labels, so we set query loss L_{query} to optimize the performance of our model. It is calculated between query prediction y_q' and query pseudo-label $\hat{y}_q$, and is used to measure the robustness of the model on various query images. In reverse process, our model learns consistency regularization. If the query prediction we get in forward process is accurate, reverse process will recover the correct support label with it. Therefore, the support loss $L_{support}$ calculated between support prediction y_s' and support label y_s is set to measure the generalization of the model on recovering support label. Both losses are cross-entropy loss [4], the total loss of our learning process is as follows:

$$L_{total} = \theta L_{query} + L_{support} \tag{1}$$

where θ is the query loss weight hyperparameter used to balance these losses.

2.2 BWSM for Prior Embedding

Advantages of BWSM: (1) Enlarges training dataset. It extracts prior knowledge from unlabeled data and embeds it into learning, which indirectly expands the scale of the training set and improves generalization. (2) Extracts accurate pseudo-labels. Prior pseudo-labels are optimized by network prediction, thus reduces the influence of noise and obtains more accurate pseudo-labels.

BWSM Process of Pseudo Label Generation: As shown in Fig. 3, our BWSM has a prior knowledge extraction process and a Bayes correction process. The prior knowledge extraction process uses the different appearance of compartments in CTA and CTU images to produce a prior prediction of renal compartments. It first filters the CTA and CTU images and then subtracts them to obtain the prior feature map f_q, which multiplies the kidney mask m_q generated from network prediction y'_q to obtain the prior probability for correct and wrong predictions $p_{correct}$ and p_{wrong}. The Bayes correction process combines network prediction to correct the prior pseudo-label to obtain a more accurate posterior pseudo-label. The softmax probability is used as the likelihood probability $l_{correct}$ and l_{wrong}, then it is used to modify $p_{correct}$ and p_{wrong} to obtain the posterior pseudo label $f_{correct}$ and f_{wrong}. The process of Bayes correction process is as follows:

$$\begin{cases} p_{correct} = 1/3 + \omega \\ p_{wrong} = (1 - p_{correct})/2 \\ p_{correct} = \dfrac{p_{correct} * l_{correct}}{p_{correct} * l_{correct} + p_{wrong} * l_{wrong}} \\ f_{wrong} = \dfrac{p_{wrong} * l_{wrong}}{p_{correct} * l_{correct} + p_{wrong} * l_{wrong}} \end{cases} \tag{2}$$

where ω is the prior probability difference hyperparameter to balance the influence of prior pseudo label.

2.3 FDFE for Fine-Grained Feature Extracting

Advantages of FDFE: (1) Emphasizes global morphology restoration. The decoder inherits the position information saved by the encoder thus performs spatial restoration more accurately. (2) Emphasizes the extraction of local detail features. The cross-layer connection transmits high-resolution features without downsampling, so the output of the network has sharper details. (3) Enhances segmentation on thin structures. A combination of morphology and detail information makes the output feature maps have sharp detail features. Such feature maps will make the model able to segment thin renal compartment structures.

Structure of FDFE: As shown in Fig. 2(c), our FDFE combines global information and detail information to extract fine-grained features. Global information is restored by up-pooling in the encoder-decoder stream. Detail information is retained by skip connection between convolution blocks of the same dimension. Specifically, the encoder consists of several repeated blocks. Each block contains two $3 \times 3 \times 3$ convolutional layers and a $2 \times 2 \times 2$ pooling layer, each convolutional layer is followed by a group norm layer. The decoder consists of several convolutional blocks corresponding to the decoder.

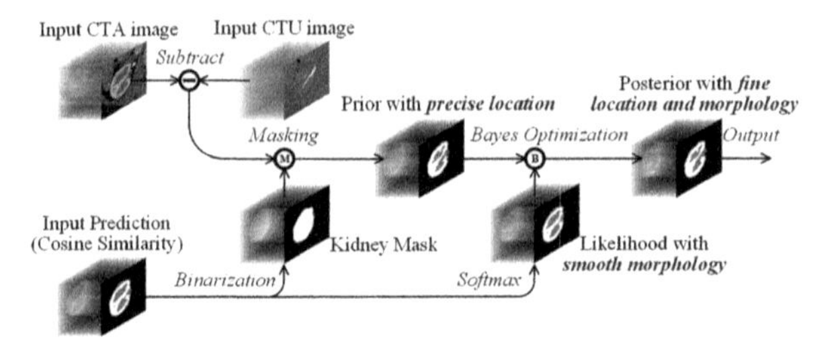

Fig. 3. BWSM for prior embedding. Our BWSM combines prior knowledge and network prediction to generate precise pseudo label.

3 Experiments and Results

Experiment Settings: Our dataset is obtained by preprocessing the abdominal enhanced CT images of patients undergoing LPN surgery. The pixel size of these CT images is between $0.59\,\text{mm}^2$ to $0.74\,\text{mm}^2$. The slice thickness is fixed at 0.75 mm, and the z-direction spacing is fixed at 0.5 mm. 60 kidney ROIs of size $160 \times 160 \times 200$ were used in the research, half of which were used as the training set and the other half as the test set. 4 images in the training set and all 30 images in the test set are fine labeled. We trained our model for 20,000 iterations. In each iteration, we randomly select two supporting images for two prototypes representing two renal compartments and one query image from the corresponding data set to form an episode for training. The support and query sets in the first 2000 iterations are all from 4 labeled training images. The source of the support set in the last 18000 iterations remains unchanged. The query set is taken from the remaining 26 unlabeled training images. During the test process, the support set is still extracted from 4 labeled training images, while 30 test images work as the query set aimed to be segmented.

We use the SGD optimizer with learning rate $lr = 0.001$, momentum of 0.9, and batchsize of 1. When using the Bayes algorithm to optimize pseudo-labels, we assign the prior probability difference hyperparameter $\omega = 0.05$. When using the fully-supervised data set to initialize the network, the query loss weight hyperparameter is set at $\theta = 1$, and when the pseudo-label is introduced for training, the θ is changed to 0.1. We use mean Dice for medulla and cortex (Dice-M and Dice-C), and Average Hausdorff Distance (AHD-M, AHD-C) [4] to parametrically measure model performance.

Comparative Analysis: Our framework has the best performance compared to other methods. As shown in Table 1, given four labeled images, our mean Dice of cortex segmentation is 78.4, and 79.1 for medulla segmentation. Some methods achieve very low performance. Segnet and Unet did not learn enough knowledge from the given four labeled images, so they cannot correctly segment renal compartments. Prototypical method PANet achieves Dice of 55.9 and 56.7. As shown in Fig. 4. Segnet and Unet judge the entire kidney structure as medulla and cannot correctly segment the cortex. PANet roughly segments renal compartments, but there is serious detail loss. Our CPNet retains detail better and achieves fine segmentation.

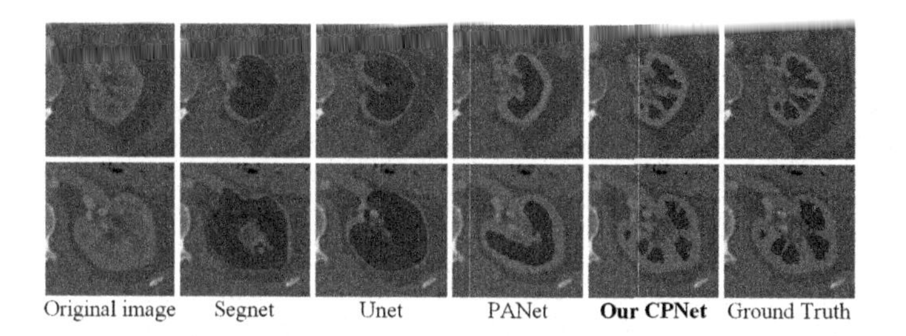

Fig. 4. The visual superiority of our CPNet. Segnet and Unet only obtain the rough boundary of the whole kidney. Prototypical method PANet can roughly segment renal compartments but have serious detail loss. Our CPNet learns detail information on the four labeled images better and achieves fine segmentation.

Table 1. Our method achieves best result on AHD and Dice. Methods with (P) are prototypical methods.

Network	Dice-M(%)	Dice-C(%)	Avg Dice(%)	AHD-M	AHD-C	Avg AHD
SegNet [1]	65.2 ± 5.6	Unable	-	6.1 ± 1.3	Unable	-
U-Net [11]	72.7 ± 8.6	2.2 ± 14.0	37.4 ± 35.4	4.1 ± 1.5	9.4 ± 2.4	6.8 ± 3.4
(P)PANet [16]	55.9 ± 9.2	56.7 ± 8.8	56.3 ± 9.0	3.7 ± 1.0	3.5 ± 1.1	3.6 ± 1.1
(P)Our CPNet-BWSM	58.1 ± 7.6	59.6 ± 6.8	58.9 ± 7.2	3.3 ± 0.9	3.1 ± 0.8	3.2 ± 0.9
(P)Our CPNet-FDFE	76.5 ± 9.0	77.0 ± 9.2	76.8 ± 9.1	2.6 ± 1.1	2.6 ± 1.1	2.6 ± 1.1
(P)Our CPNet-Total	$\mathbf{78.4 \pm 9.2}$	$\mathbf{79.1 \pm 7.9}$	$\mathbf{78.7 \pm 8.6}$	$\mathbf{1.8 \pm 0.8}$	$\mathbf{1.7 \pm 0.7}$	$\mathbf{1.7 \pm 0.8}$

Analysis of Innovations: The results show that the use of the FDFE has a significant improvement in the effect compared with the end-to-end semantic feature extractor network (VGG16), achieves an improvement on mean Dice by about 20%. Our BWSM is also important for network performance. Compared with training without this module, mean Dice improves by about 2%.

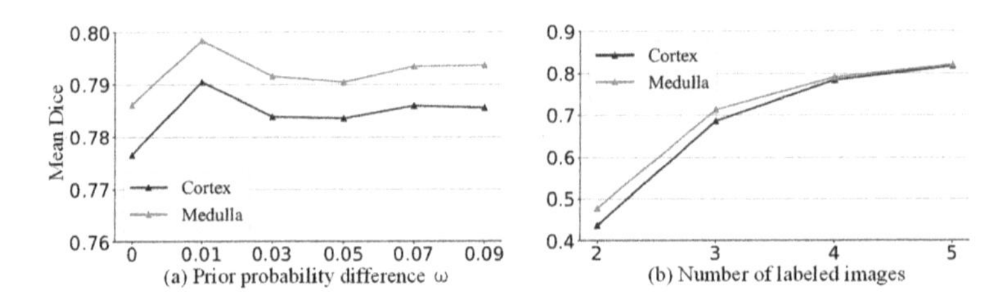

Fig. 5. Our hyperparameter analysis. (a) Model performance increases then decreases as ω increases, achieving the best performance at $\omega = 0.01$. (b) Model performance increases as the number of labeled images increases. The increase speed gradually decreases and finally stabilizes.

Hyperparameter Analysis: Prior Probability Difference ω: As shown in Fig. 5(a), Model performance increases then decreases as ω increases and achieves the best outcome at $\omega = 0.01$. $\omega = 0$ means BWSM directly takes network prediction as the pseudo label, and high ω means BWSM directly takes prior probability as the pseudo label. Both of them are biased and need correction, so we set ω at a balanced point. **Data amount analysis:** As shown in Fig. 5(b), model performance increases with the increase of the number of labels. More labels contain more information, so it is obviously better for training. In order to prove our superiority under weak supervision, we set the number of labels at a relatively small amount.

4 Conclusion

In this article, we propose a new automatic renal compartment segmentation framework Cycle Prototype Network on 3D CT images. The main Cycle Prototype Learning framework uses labels as guidance to extract features accurately and forms regularization through a reverse process to improve generalization. In addition, by embedding the Bayes Weakly Supervised Module into the framework, it can learn from unlabeled data autonomously, which improves the generalization of the framework. A unique Fine Decoding Feature Extractor is adopted to further strengthen the framework's capability to extract fine-grained detail features. The experiment results show that this method has obtained satisfactory segmentation accuracy and has potential for clinical application.

Acknowledgements. This research was supported by the National Key Research and Development Program of China (2017YFC0109202), National Natural Science Foundation under grants (31800825, 31571001, 61828101), Key Research and Development Project of Jiangsu Province (BE2018749). We thank the Big Data Center of Southeast University for providing the GPUs to support the numerical calculations in this paper.

References

1. Badrinarayanan, V., Kendall, A., Cipolla, R.: SegNet: a deep convolutional encoder-decoder architecture for image segmentation. IEEE Trans. Pattern Anal. Mach. Intell. **39**(12), 2481–2495 (2017). https://doi.org/10.1109/TPAMI.2016.2644615
2. Chen, X., Summers, R.M., Cho, M., Bagci, U., Yao, J.: An automatic method for renal cortex segmentation on CT images: evaluation on kidney donors. Acad. Radiol. **19**(5), 562–570 (2012). https://doi.org/10.1016/j.acra.2012.01.005. https://europepmc.org/articles/PMC3319195
3. Ficarra, V., et al.: Preoperative aspects and dimensions used for an anatomical (PADUA) classification of renal tumours in patients who are candidates for nephron-sparing surgery. Eur. Urol. **56**(5), 786–793 (2009). https://doi.org/10.1016/j.eururo.2009.07.040. https://www.sciencedirect.com/science/article/pii/S030228380900788X
4. He, Y., et al.: Dense biased networks with deep priori anatomy and hard region adaptation: semi-supervised learning for fine renal artery segmentation. Med. Image Anal. **63**, 101722 (2020)
5. Heller, N., et al.: The state of the art in kidney and kidney tumor segmentation in contrast-enhanced CT imaging: results of the KiTS19 challenge. Med. Image Anal. **67**, 101821 (2021)
6. Jackson, P., Hardcastle, N., Dawe, N., Kron, T., Hofman, M.S., Hicks, R.J.: Deep learning renal segmentation for fully automated radiation dose estimation in unsealed source therapy. Front. Oncol. **8**, 215 (2018). https://doi.org/10.3389/fonc.2018.00215. https://www.frontiersin.org/article/10.3389/fonc.2018.00215
7. Kutikov, A., Uzzo, R.G.: The R.E.N.A.L. nephrometry score: a comprehensive standardized system for quantitating renal tumor size, location and depth. J. Urol. **182**(3), 844–853 (2009). https://doi.org/10.1016/j.juro.2009.05.035
8. Li, J., Lo, P., Taha, A., Wu, H., Zhao, T.: Segmentation of renal structures for image-guided surgery. In: Frangi, A.F., Schnabel, J.A., Davatzikos, C., Alberola-López, C., Fichtinger, G. (eds.) MICCAI 2018. LNCS, vol. 11073, pp. 454–462. Springer, Cham (2018). https://doi.org/10.1007/978-3-030-00937-3_52
9. Lin, Z., et al.: Automated segmentation of kidney and renal mass and automated detection of renal mass in CT urography using 3D U-Net-based deep convolutional neural network. Eur. Radiol. (2021). https://doi.org/10.1007/s00330-020-07608-9
10. Litjens, G., et al.: A survey on deep learning in medical image analysis. Med. Image Anal. **42**, 60–88 (2017)
11. Ronneberger, O., Fischer, P., Brox, T.: U-Net: convolutional networks for biomedical image segmentation. In: Navab, N., Hornegger, J., Wells, W.M., Frangi, A.F. (eds.) MICCAI 2015. LNCS, vol. 9351, pp. 234–241. Springer, Cham (2015). https://doi.org/10.1007/978-3-319-24574-4_28
12. Shao, P., et al.: Laparoscopic partial nephrectomy with segmental renal artery clamping: technique and clinical outcomes. Eur. Urol. **59**(5), 849–855 (2011)
13. Shao, P., et al.: Precise segmental renal artery clamping under the guidance of dual-source computed tomography angiography during laparoscopic partial nephrectomy. Eur. Urol. **62**(6), 1001–1008 (2012)
14. Taha, A., Lo, P., Li, J., Zhao, T.: Kid-Net: convolution networks for kidney vessels segmentation from CT-volumes. In: Frangi, A.F., Schnabel, J.A., Davatzikos, C., Alberola-López, C., Fichtinger, G. (eds.) MICCAI 2018. LNCS, vol. 11073, pp. 463–471. Springer, Cham (2018). https://doi.org/10.1007/978-3-030-00937-3_53

15. Wang, G., et al.: Interactive medical image segmentation using deep learning with image-specific fine tuning. IEEE Trans. Med. Imaging **37**(7), 1562–1573 (2018). https://doi.org/10.1109/TMI.2018.2791721
16. Wang, K., Liew, J.H., Zou, Y., Zhou, D., Feng, J.: PANet: few-shot image semantic segmentation with prototype alignment. In: Proceedings of the IEEE/CVF International Conference on Computer Vision (ICCV) (2019)
17. Xiang, D., et al.: CorteXpert: a model-based method for automatic renal cortex segmentation. Med. Image Anal. **42**, 257–273 (2017). https://doi.org/10.1016/j.media.2017.06.010
18. Zhang, S., et al.: Application of a functional 3-dimensional perfusion model in laparoscopic partial nephrectomy with precise segmental renal artery clamping. Urology **125**, 98–103 (2019)

Observational Supervision for Medical Image Classification Using Gaze Data

Khaled Saab[1(✉)], Sarah M. Hooper[1], Nimit S. Sohoni[2], Jupinder Parmar[3], Brian Pogatchnik[4], Sen Wu[3], Jared A. Dunnmon[3], Hongyang R. Zhang[5], Daniel Rubin[6], and Christopher Ré[3]

[1] Department of Electrical Engineering, Stanford University, Stanford, USA
ksaab@stanford.edu
[2] Institute for Computational and Mathematical Engineering, Stanford University, Stanford, USA
[3] Department of Computer Science, Stanford University, Stanford, USA
[4] Department of Radiology, Stanford University, Stanford, USA
[5] Khoury College of Computer Sciences, Northeastern University, Boston, USA
[6] Department of Biomedical Data Science, Stanford University, Stanford, USA

Abstract. Deep learning models have demonstrated favorable performance on many medical image classification tasks. However, they rely on expensive hand-labeled datasets that are time-consuming to create. In this work, we explore a new supervision source to training deep learning models by using gaze data that is passively and cheaply collected during a clinician's workflow. We focus on three medical imaging tasks, including classifying chest X-ray scans for pneumothorax and brain MRI slices for metastasis, two of which we curated gaze data for. The gaze data consists of a sequence of fixation locations on the image from an expert trying to identify an abnormality. Hence, the gaze data contains rich information about the image that can be used as a powerful supervision source. We first identify a set of gaze features and show that they indeed contain class-discriminative information. Then, we propose two methods for incorporating gaze features into deep learning pipelines. When no task labels are available, we combine multiple gaze features to extract weak labels and use them as the sole source of supervision (Gaze-WS). When task labels are available, we propose to use the gaze features as auxiliary task labels in a multi-task learning framework (Gaze-MTL). On three medical image classification tasks, our Gaze-WS method without task labels comes within 5 AUROC points (1.7 precision points) of models trained with task labels. With task labels, our Gaze-MTL method can improve performance by 2.4 AUROC points (4 precision points) over multiple baselines.

Keywords: Medical image diagnosis · Eye tracking · Weak supervision

Electronic supplementary material The online version of this chapter (https://doi.org/10.1007/978-3-030-87196-3_56) contains supplementary material, which is available to authorized users.

M. de Bruijne et al. (Eds.): MICCAI 2021, LNCS 12902, pp. 603–614, 2021.
https://doi.org/10.1007/978-3-030-87196-3_56

1 Introduction

A growing challenge in medical imaging is the need for more qualified experts to read an increasing volume of medical images, which has led to interpretation delays and reduced quality of healthcare [25]. Deep learning models in radiology [5], dermatology [6], and other areas [7] can increase physician throughput to alleviate this challenge. However, a major bottleneck to developing such models is the need for large labeled datasets [7].

Fortunately, studies in psychology and neurophysiology have shown that human gaze data contains task-related information [38]. Specifically, past studies have shown eye movement is driven by a reward-based mechanism, which induces a hidden structure in gaze data that embeds information about the task [11]. Moreover, with recent advances in eye tracking technology and the rise of commodity augmented reality, gaze data is likely to become both ubiquitous and cheap to collect in the coming years [22,33]. Since collecting gaze data would require no additional human effort in many real-world image analysis workflows, it offers a new opportunity for cheap model supervision, which we term *observational supervision.*

Can gaze data provide supervision signals for medical imaging? While prior work from natural image classification have used gaze data for object recognition applications [15,20,35], using gaze data in the context of medical image diagnosis introduces new questions. For instance, the visual scanning process of medical image diagnosis is often more protracted than in typical object recognition tasks [17], resulting in longer gaze sequences containing richer task-related information. Therefore, it is not clear which features of the gaze data are useful for supervising image diagnosis models, and how much information can be extracted from them. To explore the above questions, we collected gaze data from radiologists as they performed two medical imaging tasks. The first is chest X-ray diagnosis for pneumothorax (i.e., a collapsed lunch), which is a life threatening event. The second is MRI diagnosis for brain metastases, which is a spreading cancer in the brain that requires quick detection for treatment. Our two novel datasets, along with a public dataset on abnormal chest X-ray detection [14], represent challenging real-world medical settings to study observational supervision.

In this work, we propose an observational supervision framework that leverages gaze data to supervise deep learning models for medical image classification. First, we use our three medical datasets to identify several critical statistics of gaze data (summarized in Table 1) such as the number of uniquely visited regions. Second, we use these gaze data statistics to weakly supervise deep learning models for medical image diagnosis (Gaze-WS)—without relying on any task labels. Finally, we propose to combine gaze data statistics and task labels using a multi-task learning framework to inject additional inductive bias from gaze data along with task labels (Gaze-MTL). See Fig. 1 for an illustration. The key intuition behind our approach is that gaze data provides discriminative information through differences in scanning patterns between normal and abnormal images. Interestingly, such signals can be explained using reward-based modeling from the neuroscience literature [29]. We theoretically show that the discriminative

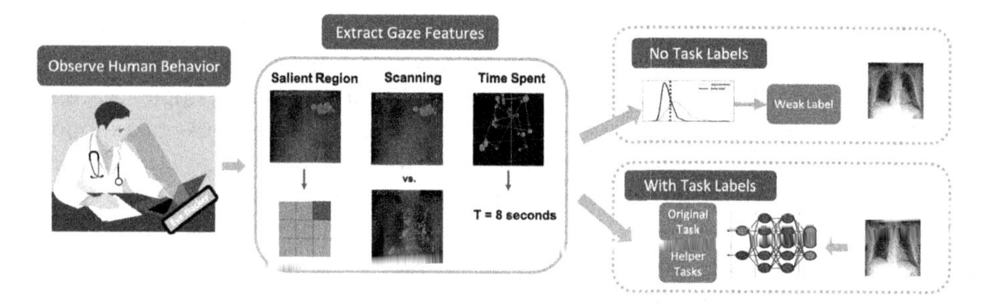

Fig. 1. Illustration of the observational supervision framework. After passively collecting gaze data from experts, we extract informative gaze data statistics which we use in two settings: we turn multiple gaze data statistics into weak labels and combine them to train a ResNet-50 CNN without task labels (top-right). We write helper tasks that predict gaze statistics in a multi-task learning framework along with task labels (bottom-right).

power of a gaze sequence scales with the size of the abnormal region and how likely the fixation occurs in the abnormal region.

We evaluate our framework on the three medical imaging tasks.[1] Using only gaze data, we find that Gaze-WS comes within 5 AUROC points (or 1.7 precision points) of models trained using task labels. We achieve these results using attention statistics of gaze data such as the number of uniquely visited regions in an image, which is smaller for abnormal images than normal images. Using gaze data and task labels, we find that Gaze-MTL improves performance by up to 2.4 AUROC points (or 4 precision points) compared to previous approaches that integrate gaze data [15,18,20,28]. We observe that measuring the "diffusivity" (see Table 1 for the definition) of an expert's attention as a statistic transfers most positively to the target task in multi-task learning.

2 Related Work

Deep learning models such as convolutional neural networks (CNNs) are capable of extracting meaningful feature representations for a wide variety of medical imaging tasks, ranging from cancer diagnosis [6] to chest radiograph abnormality detection [5]. Since clinicians do not log structured task labels in their standard workflows, these works often involve an expensive data labeling process. We instead explore a setting where we have access to gaze data collected passively from clinicians during their workflows, which we believe will become ubiquitous and cheap with recent augmented reality eyewear. Our work is a first exploration in using gaze data as the main source of supervision for medical image classification models.

[1] Our two novel datasets and code are available at https://github.com/HazyResearch/observational.

Our work is closely related to object detection research that uses gaze data as auxilliary information [15,30,31,35,37]. A common approach from these works is to turn gaze data into attention heatmaps [18,20,28]. For example, Karessli et al. demonstrates that features derived from attention heatmaps can support class-discriminative representations that improve zero-shot image classification [15]. Wang et al. [35] integrate gaze data into the optimization procedure as regularization for training a support vector machine (see also [8]). However, our work investigates gaze data collected from radiologists, which is distinct from gaze collected during natural objection detection tasks. For example, the gaze data that we have collected consists of long sequences of fixations as a result of a protracted search process. By contrast, the gaze data from object detection tasks consists of much shorter sequences [21]. This distinction brings a new challenge of how to extract meaningful features from gaze data for medical imaging. There has also been recent work investigating the integration of gaze data to improve medical diagnosis systems for lung cancer screening [1,16] and abnormality detection in chest X-rays [14]. However, these methods are hindered by their need for gaze at test time. In our work, we propose two novel approaches for using gaze data to weakly supervise deep learning pipelines using tools from both data programming [24] and multi-task learning [26,36,39].

A different approach to deal with the high cost of labeling in medical imaging is to is to extract labels from clinician reports. NLP models are trained to classify which abnormalities are described in the text and can therefore act as automated annotators when reports are present [23,34]. Such annotation tools, however, are expensive to create and are application-specific. A cheaper and more flexible alternative is data programming, where experts manually write heuristic functions that label data, which has been applied to medical reports [4,27]. We view our approach as a complement to NLP-based approaches. In scenarios where class labels are not found in medical reports [2] or when NLP-based annotators are trained for a different language or task, gaze is a viable alternative (via Gaze-WS). In scenarios where NLP-based annotators are applicable, gaze may be used alongside the labels to further improve performance (via Gaze-MTL). We believe it is a promising direction to explore combining gaze data and NLP-based labeling for medical imaging in future work.

3 Data Collection and Methods

We start by introducing the datasets and gaze data collection process. Then, we present a set of gaze features and describe a theoretical model for studying these features. Finally, we present two approaches to supervising deep learning models using gaze data.

3.1 Gaze Data Collection and Features

Since medical imaging datasets with gaze data are not readily available, we collected gaze data by collaborating with radiologists. We consider three datasets:

classifying chest X-rays for pneumothorax (CXR-P), classifying chest X-rays for a general abnormality (CXR-A), and classifying brain MRI slices for metastasis (METS) (all binary image classification). Positive and negative samples from each task can be found in Figure S.2.

For CXR-P, we use the SIIM-ACR Pneumothorax dataset [19], which contains 5,777 X-ray images (22% contain a pneumothorax). We took a stratified random sample of 1,170 images to form our train and validation sets, and collected task labels and gaze data from three board-certified radiologists. We used the remaining 4,607 images as a held-out test set. For CXR-A, Karagryis et al. [14] collected the gaze data of a radiologist reading 1,083 chest X-ray images taken from the MIMIC-CXR Database (66% abnormal) [13]. We reserved 216 random images as a held-out test set, and used the rest for train and validation sets. Importantly, this dataset also gives us the advantage of evaluating our methods on a different eye tracker. For METS, after receiving training from a board-certified radiologist, we analyzed 2,794 MRI slices from 16 cases, comprising our train and validation sets (25% contain a lesion). The held-out test set has 1,664 images.

To collect gaze data, we built custom software to interface with a screen-based Tobii Pro Nano eye tracker. This state-of-the-art eye tracker is robust to head movements, corrective lenses, and lighting conditions. At the start of each session, each radiologist went through a 9-point calibration process. While in use, the program displays a single image to the user and collects gaze coordinates. Once the user has analyzed the image, they press a key to indicate the label given to the image. The program then saves the set of gaze coordinates that overlapped with the image and displays the next image.

Gaze Data Statistics. By analyzing random samples of gaze sequences for multiple tasks, we find that the "scanning behavior," e.g., the amount of time the labeler spends scanning over the image versus fixating on a specific patch, correlates strongly with task-related information, such as the existence of an abnormal region in the image. We derive three quantitative features for scanning behavior: time on maximum patch, diffusivity, and unique visits (described in Table 1). We also consider the amount of time spent on a task, since it has been shown to be indicative of task difficulty [3]. We also considered other features such as total distance, velocity, and fixation angles, but found that they provide less signal than those in Table 1. We provide a full list and detailed descriptions of gaze features considered in Table S.4, and visualize the class-conditional distributions of our key gaze data statistics in Figure S.1.

Modeling Scanning Behavior Using a Reward-Based Search Model. We consider a labeler actively searching for salient image features that contribute to their classification. Drawing inspiration from neuroscience literature [29], we assume that the scanning behavior is directed by a latent reward map with high rewards in task-informative regions (e.g. an abnormal region in a radiograph) and low rewards in less informative regions (e.g. background). Suppose that a reward of Q_i is obtained by fixating at region i and that there are p regions

Table 1. A summary of gaze data statistics used in our framework, along with their average difference in value between classes across our three tasks (class gap).

Feature	Description	Class Gap
Time on maximum patch	Maximum time dedicated to a patch	0.09
Diffusivity	Maximum time spent on any local region, where each local region is an average of neighboring patches	0.21
Unique visits	Number of uniquely visited patches	0.12
Time spent	Total time spent looking at the image	0.10

in total. We consider the gaze sequence of a labeler as a random sequence, where the probability of visiting the i-th region is given by $\Pr(i) = \frac{\exp(Q_i)}{\sum_{i=1}^{p} \exp(Q_i)}$.

We show that the discriminative power of these gaze statistics scales with an interesting quantity that we term the "attention gap." Informally, the attention gap captures the differences of experts' scanning behaviors between normal and abnormal images. Let $Q_{no} > 0$ denote the reward of visiting a normal region and $Q_{ab} > Q_{no}$ denote the reward of an abnormal region. Let p denote the total number of regions and $s \in [0, 1]$ denote sparsity—the fraction of abnormal regions. Therefore, for a random visit, the probability that the visit lands in a particular abnormal region is equal to

$$\frac{1}{p} \cdot \frac{\exp(Q_{ab} - Q_{no})}{s \cdot \exp(Q_{ab} - Q_{no}) + (1 - s)}. \tag{1}$$

The above quantity is larger than $\frac{1}{p}$ as long as $Q_{ab} > Q_{no}$, resulting in an "attention gap" between abnormal and normal regions. Equation 1 reveals that the discriminative signal in gaze features increases as the attention gap increases and the sparsity decreases.

3.2 First Observational Supervision Method: Gaze-WS

We propose a method that combines the gaze data statistics from Table 1 to compute posterior probabilities of task labels, enabling model supervision using gaze data alone. Given training pairs consisting of an image and a gaze sequence, denoted by $\{(x_i, g_i)\}_{i=1}^{n}$, our goal is to predict the labels of test images—without being provided gaze sequences at test time.

(i) We first compute m gaze features $h_{i1} \in \mathbb{R}^{a_1}, \ldots, h_{im} \in \mathbb{R}^{a_m}$ from each gaze sequence g_i. Specifically, we use four gaze data statistics from Table 1: time on max patch, diffusivity, unique visits, and time spent. We use these features to compute labels $\{\hat{y}_i\}_{i=1}^{n}$ that approximate the true (unobserved) class labels $\{y_i\}_{i=1}^{n}$.

(ii) Using a small validation set, we fit two Gaussians to each feature—one each for the positive and negative classes—which are used to estimate the likelihoods $p(h_{im}|y)$ for each unlabeled training sample. We assume the features are conditionally independent and compute the posterior probability $\hat{y}_i = P(y_i = 1 \mid h_{i1}, \ldots, h_{im})$ using Bayes' theorem. We convert them to binary labels with a threshold selected via cross validation.

3.3 Second Observational Supervision Method: Gaze-MTL

We next consider a second setting, where we have both task labels and gaze data, and propose a multitask learning (MTL) framework known as hard parameter sharing [26]. The idea is that gaze data contains fine-grained information such as task difficulty and salient regions [10], which complements class labels. Specifically, we are given training tuples consisting of an image, a gaze sequence, and a label, denoted by $\{(x_i, g_i, y_i)\}_{i=1}^n$. Our goal is to train an image classification model that predicts the labels of test images. (Again, we do not assume access to gaze at test time.) Denote the domain of $\{x_i\}$ by $\mathcal{X}$.

(i) For each sequence g_i, we compute m gaze features $h_{i1} \in \mathbb{R}^{a_1}, \ldots, h_{im} \in \mathbb{R}^{a_m}$.
(ii) We train a feature representation model (e.g., a CNN) $f_\theta(\cdot) : \mathcal{X} \to \mathbb{R}^d$ with feature dimension d parameterized by θ, along with the target task head $A_0 \in \mathbb{R}^{k \times d}$ and *helper task heads* $A_1 \in \mathbb{R}^{a_1 \times d}, \ldots, A_m \in \mathbb{R}^{a_m \times d}$. We minimize the following loss over the training data:

$$L(\theta) = \frac{1}{n} \sum_{i=1}^{n} \left(\ell_0(A_0 f_\theta(x_i), y_i) + \sum_{j=1}^{m} \alpha_j \ell_j(A_j f_\theta(x_i), h_{ij}) \right) . \quad (2)$$

Here ℓ_0 denotes the prediction loss for the main task. $\ell_1, \ldots, \ell_m$ denote prediction losses for the m helper tasks, and $\alpha_1, \ldots, \alpha_m$ are hyperparameters that weight each helper task. In our experiments, we used the soft cross-entropy loss [24] to predict the normalized gaze features as the helper tasks.

In the above minimization problem, a shared feature representation model $f_\theta(x_i)$ is used for the target task as well as all helper tasks. This is also known as hard parameter sharing in MTL, and works as an inductive transfer mechanism from the helper tasks to the target task. Importantly, at inference time, given an image, our model predicts the label using the target task head alone. Hence, we do not require gaze data for inference.

4 Experimental Results

We validate that gaze data alone can provide useful supervision via Gaze-WS, and improves model performance using Gaze-MTL. For all models, we train a ResNet-50 CNN architecture [12] pretrained on ImageNet for 15 epochs using PyTorch v1.4.0. Common hyperparameters were chosen by random search to maximize validation accuracy. Models were trained on two Titan RTX GPU's, where a single epoch took approximately 8 s. More details are in our code. We choose the operating points that achieve the same recall scores as previously published models: 55 for CXR-P [32], 90 for CXR-A [5], and 53 for METS [9].

4.1 Using Gaze Data as the Sole Supervision Source

We validate our hypothesis that gaze features alone can be used to supervise well-performing medical image classification models. We compare the test performance of Gaze-WS to that of a supervised CNN trained with task labels. We also measure performance as the number of training samples varies (Fig. 2).

This scaling analysis shows that Gaze-WS models improve with more weakly labeled data and approach the supervised CNN performance, ultimately coming within 5 AUROC points for CXR-P and CXR-A. Moreover, we find that Gaze-WS for CXR-P has higher recall on small abnormalities (which are more difficult to detect in practice) compared to the supervised CNN, but misses more of the large abnormalities. Intuitively, a labeler may spend more time examining the smaller abnormalities, making them more easily identifiable from gaze.

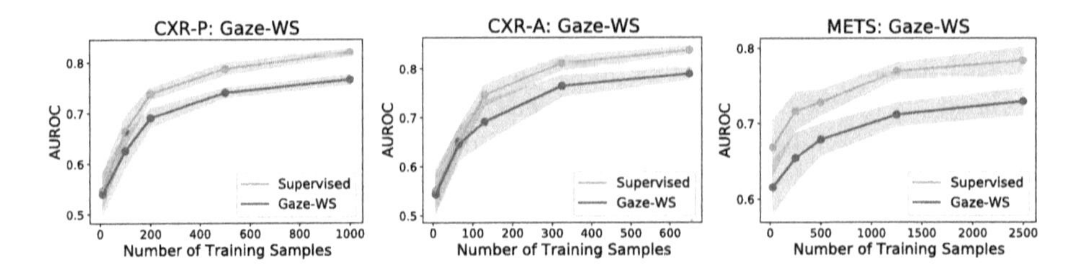

Fig. 2. Mean AUROC of Gaze-WS scales competitively to supervised learning on the same CNN. The results are averaged over 10 random seeds and the shaded region indicates 95% confidence intervals. We find that using gaze data as the sole supervision source achieves within 5 AUROC points to supervised learning on the same CNN model.

We next inspect the weak labels estimated by the gaze data for METS and CXR-P. We find that the weak labels achieve a mean AUROC of 80 and 93 on CXR-P and METS, respectively. This performance closely matches the intuition we develop in Sect. 3. Recall that we expect the separation strengths to scale with dataset sparsity and attention gap. For CXR-P and METS, we find that METS' estimated sparsity ($s = 0.03$) is lower and attention gap ($z = 6.5$) is higher than CXR-P's ($s = 0.07$, $z = 1.5$), which indicates we should expect larger separation strengths for METS.

Due to the noise in the weak labels, there is a clear tradeoff between the number of labels, the ease with which those labels are collected, and model performance. For instance, in CXR-P, Gaze-WS achieves the same performance as the supervised CNN model using about 2$\times$ as many training samples. These results suggest that Gaze-WS may be useful for passively collecting large quantities of noisy data to achieve the same performance as a model trained with fewer (but more expensive) task labels.

4.2 Integrating Gaze Data Along with Task Labels

We empirically validate our hypothesis that gaze data provides additional information beyond the task labels and can be injected into model training to improve model performance via multi-task learning. We train a CNN for each dataset using Gaze-MTL, and compare its performance to a CNN with the same architecture trained with only task labels, or trained by incorporating gaze data through the following existing methods: CAM-REG [28], HM-REG [18], Template [20], and ZeroShot [15].

Table 2. Gaze-MTL improves upon supervised learning and multiple baseline methods by up to 2.4 AUROC points on three medical imaging datasets. The results are averaged over 10 random seeds with 95% significance.

Dataset	Metric	Image-only	HM-REG	CAM-REG	Template	ZeroShot	Gaze-MTL
CXR-P	AUROC	81.5 ± 0.6	78.9 ± 1.8	78.3 ± 1.4	78.6 ± 1.4	82.1 ± 0.8	$\mathbf{83.0} \pm 0.5$
	F1-score	56.6 ± 0.7	50.2 ± 3.4	51.0 ± 3.3	53.5 ± 1.2	56.0 ± 0.8	$\mathbf{57.5} \pm 0.7$
	Precision	58.5 ± 1.4	48.7 ± 4.3	49.7 ± 4.2	52.2 ± 2.2	57.0 ± 1.7	$\mathbf{60.2} \pm 1.4$
CXR-A	AUROC	83.8 ± 0.9	83.1 ± 0.6	82.9 ± 1.0	82.3 ± 1.2	83.2 ± 0.8	$\mathbf{84.3} \pm 1.6$
	F1-score	84.1 ± 0.9	83.9 ± 0.9	82.6 ± 1.8	81.7 ± 0.7	84.2 ± 0.9	$\mathbf{84.5} \pm 0.7$
	Precision	78.9 ± 1.5	78.5 ± 1.5	76.0 ± 3.2	74.7 ± 1.3	79.1 ± 1.6	$\mathbf{79.4} \pm 1.2$
METS	AUROC	78.4 ± 1.8	77.6 ± 1.3	65.3 ± 1.8	56.0 ± 1.1	76.8 ± 1.9	$\mathbf{80.8} \pm 1.1$
	F1-score	53.4 ± 1.4	52.4 ± 1.2	49.5 ± 3.6	38.9 ± 1.4	34.7 ± 10.1	$\mathbf{55.0} \pm 1.2$
	Precision	53.5 ± 2.7	51.5 ± 2.7	48.1 ± 4.4	32.0 ± 1.1	52.7 ± 4.1	$\mathbf{57.5} \pm 2.6$

Table 2 shows that Gaze-MTL results in a performance improvement over each baseline for our three medical tasks. We also find that different gaze features result in larger performance boosts when used as auxiliary tasks for different datasets (Table S.2).

To further investigate which gaze features are most useful, we compute a task similarity score between each gaze feature and the target task by measuring the impact of transfer learning between the tasks. We find that the gaze features that have higher task similarity scores with the target tasks are the same auxiliary tasks with which Gaze-MTL achieved the largest gains (details in Table S.2).

It is common to use the class activation map (CAM), which highlights the areas of an image that are most responsible for the model's prediction, to reveal additional localization information [27]. For CXR-P, we found the CAMs of Gaze-MTL to overlap with the ground-truth abnormality regions 20% more often than the Image-only model. This suggests that models trained with gaze provide more accurate localization information.

The performance boost we see with Gaze-MTL suggests that it is a promising method for integrating gaze data into ML pipelines. Particularly in high-stakes settings where gaze data can be readily collected in conjunction with class labels, Gaze-MTL may be used to integrate additional information from the expert labeler to boost model performance, without requiring additional labeling effort.

5 Conclusions

This work introduced an observational supervision framework for medical image diagnosis tasks. We collected two eye tracking datasets from radiologists and presented two methods for incorporating gaze data into deep learning models. Our Gaze-WS results showed that using gaze data alone can achieve nearly comparable performance to fully supervised learning on CNNs. This result is rather surprising and suggests that gaze data provides promising supervision signals for medical imaging. Furthermore, our Gaze-MTL results showed that gaze data can provide additional inductive biases that are not present in human labels to improve upon the performance of models supervised with task labels alone. We hope that our novel datasets and encouraging results can inspire more interest in integrating gaze data into deep learning for medical imaging.

References

1. Aresta, G., et al.: Automatic lung nodule detection combined with gaze information improves radiologists' screening performance. IEEE J. Biomed. Health Inform. **24**(10) (2020)
2. Bosmans, J.M., Weyler, J.J., Parizel, P.M.: Structure and content of radiology reports, a quantitative and qualitative study in eight medical centers. Eur. J. Radiol. **72**(2) (2009)
3. Cole, M.J., Gwizdka, J., Liu, C., Bierig, R., Belkin, N.J., Zhang, X.: Task and user effects on reading patterns in information search. Interact. Comput. **23**(4) (2011)
4. Dunnmon, J.A., et al.: Cross-modal data programming enables rapid medical machine learning. Patterns (2020)
5. Dunnmon, J.A., Yi, D., Langlotz, C.P., Ré, C., Rubin, D.L., Lungren, M.P.: Assessment of convolutional neural networks for automated classification of chest radiographs. Radiol. **290**(2) (2019)
6. Esteva, A., et al.: Dermatologist-level classification of skin cancer with deep neural networks. Nature **542**(7639) (2017)
7. Esteva, A., et al.: A guide to deep learning in healthcare. Nat. Med. **25**(1) (2019)
8. Ge, G., Yun, K., Samaras, D., Zelinsky, G.J.: Action classification in still images using human eye movements. In: Proceedings of the IEEE Conference on Computer Vision and Pattern Recognition Workshops (2015)
9. Grøvik, E., Yi, D., Iv, M., Tong, E., Rubin, D., Zaharchuk, G.: Deep learning enables automatic detection and segmentation of brain metastases on multisequence MRI. J. Magnet. Resonance Imaging **51**(1) (2020)
10. Hayhoe, M.: Vision using routines: a functional account of vision. Visual Cognit. **7**(1–3) (2000)

11. Hayhoe, M., Ballard, D.: Eye movements in natural behavior. Trends in Cogn. Sci. **9**(4) (2005)
12. He, K., Zhang, X., Ren, S., Sun, J.: Deep residual learning for image recognition. In: Proceedings of the IEEE Conference on Computer Vision and Pattern Recognition (2016)
13. Johnson, A., Pollard, T., Mark, R., Berkowitz, S., Horng, S.: Mimic-CXR database (2019). https://doi.org/10.13026/C2JT1Q. https://physionet.org/content/mimic-cxr/1.0.0/
14. Karargyris, A., et al.: Creation and validation of a chest x-ray dataset with eye-tracking and report dictation for AI development. Sci. Data **8**(1) (2021)
15. Karessli, N., Akata, Z., Schiele, B., Bulling, A.: Gaze embeddings for zero-shot image classification. In: Proceedings of the IEEE Conference on Computer Vision and Pattern Recognition (2017)
16. Khosravan, N., Celik, H., Turkbey, B., Jones, E.C., Wood, B., Bagci, U.: A collaborative computer aided diagnosis (c-cad) system with eye-tracking, sparse attentional model, and deep learning. Med. Image Anal. **51** (2019)
17. Klein, J.S., Rosado-de-Christenson, M.L.: A Systematic Approach to Chest Radiographic Analysis. Springer (2019)
18. Lai, Q., Wang, W., Khan, S., Shen, J., Sun, H., Shao, L.: Human vs. machine attention in neural networks: a comparative study. arXiv preprint arXiv:1906.08764 (2019)
19. for Imaging Informatics in Medicine (SIIM), S.: Siim-ACR pneumothorax segmentation (2019). https://www.kaggle.com/c/siim-acr-pneumothorax-segmentation
20. Murrugarra-Llerena, N., Kovashka, A.: Learning attributes from human gaze. In: 2017 IEEE Winter Conference on Applications of Computer Vision (WACV). IEEE (2017)
21. Papadopoulos, D.P., Clarke, A.D.F., Keller, F., Ferrari, V.: Training object class detectors from eye tracking data. In: Fleet, D., Pajdla, T., Schiele, B., Tuytelaars, T. (eds.) ECCV 2014. LNCS, vol. 8693, pp. 361–376. Springer, Cham (2014). https://doi.org/10.1007/978-3-319-10602-1_24
22. Qiao, X., Ren, P., Dustdar, S., Liu, L., Ma, H., Chen, J.: Web AR: a promising future for mobile augmented reality-state of the art, challenges, and insights. Proc. IEEE **107**(4) (2019)
23. Rajpurkar, P., et al.: Chexnet: radiologist-level pneumonia detection on chest x-rays with deep learning. arXiv preprint arXiv:1711.05225 (2017)
24. Ratner, A., De Sa, C., Wu, S., Selsam, D., Ré, C.: Data programming: creating large training sets, quickly. In: Advances in Neural Information Processing Systems, vol. 29 (2016)
25. Rimmer, A.: Radiologist shortage leaves patient care at risk, warns royal college. BMJ: British Med. J. (Online) **359** (2017)
26. Ruder, S.: An overview of multi-task learning in deep neural networks. arXiv preprint arXiv:1706.05098 (2017)
27. Saab, K., et al.: Doubly weak supervision of deep learning models for head CT. In: Shen, D., et al. (eds.) MICCAI 2019. LNCS, vol. 11766, pp. 811–819. Springer, Cham (2019). https://doi.org/10.1007/978-3-030-32248-9_90
28. Saab, K., Dunnmon, J., Ratner, A., Rubin, D., Re, C.: Improving sample complexity with observational supervision. In: International Conference on Learning Representations, LLD Workshop (2019)
29. Samson, R., Frank, M., Fellous, J.M.: Computational models of reinforcement learning: the role of dopamine as a reward signal. Cogn. Neurodyn. **4**(2) (2010)

30. Selvaraju, R.R., et al.: Taking a hint: leveraging explanations to make vision and language models more grounded. In: Proceedings of the IEEE International Conference on Computer Vision (2019)
31. Stember, J., et al.: Eye tracking for deep learning segmentation using convolutional neural networks. J. Digital Imaging **32**(4) (2019)
32. Taylor, A.G., Mielke, C., Mongan, J.: Automated detection of moderate and large pneumothorax on frontal chest x-rays using deep convolutional neural networks: a retrospective study. PLoS Med. **15**(11) (2018)
33. Valliappan, N., et al.: Accelerating eye movement research via accurate and affordable smartphone eye tracking. Nat. Commun. **11**(1) (2020)
34. Wang, X., Peng, Y., Lu, L., Lu, Z., Bagheri, M., Summers, R.M.: Chestx-ray8: hospital-scale chest x-ray database and benchmarks on weakly-supervised classification and localization of common thorax diseases. In: Proceedings of the IEEE Conference on Computer Vision and Pattern Recognition (2017)
35. Wang, X., Thome, N., Cord, M.: Gaze latent support vector machine for image classification improved by weakly supervised region selection. Pattern Recogn. **72** (2017)
36. Wu, S., Zhang, H., Ré, C.: Understanding and improving information transfer in multi-task learning. In: International Conference on Learning Representations (2020). https://openreview.net/forum?id=SylzhkBtDB
37. Yu, Y., Choi, J., Kim, Y., Yoo, K., Lee, S.H., Kim, G.: Supervising neural attention models for video captioning by human gaze data. In: Proceedings of the IEEE Conference on Computer Vision and Pattern Recognition (2017)
38. Yun, K., Peng, Y., Samaras, D., Zelinsky, G.J., Berg, T.L.: Exploring the role of gaze behavior and object detection in scene understanding. Frontiers Psychol. **4** (2013)
39. Zhang, H.R., Yang, F., Wu, S., Su, W.J., Ré, C.: Sharp bias-variance tradeoffs of hard parameter sharing in high-dimensional linear regression. arXiv preprint arXiv:2010.11750 (2020)

Inter Extreme Points Geodesics for End-to-End Weakly Supervised Image Segmentation

Reuben Dorent[(✉)], Samuel Joutard, Jonathan Shapey, Aaron Kujawa, Marc Modat, Sébastien Ourselin, and Tom Vercauteren

School of Biomedical Engineering and Imaging Sciences, King's College London, London, UK
reuben.dorent@kcl.ac.uk

Abstract. We introduce *InExtremIS*, a weakly supervised 3D approach to train a deep image segmentation network using particularly weak train-time annotations: only 6 extreme clicks at the boundary of the objects of interest. Our fully-automatic method is trained end-to-end and does not require any test-time annotations. From the extreme points, 3D bounding boxes are extracted around objects of interest. Then, deep geodesics connecting extreme points are generated to increase the amount of "annotated" voxels within the bounding boxes. Finally, a weakly supervised regularised loss derived from a Conditional Random Field formulation is used to encourage prediction consistency over homogeneous regions. Extensive experiments are performed on a large open dataset for Vestibular Schwannoma segmentation. *InExtremIS* obtained competitive performance, approaching full supervision and outperforming significantly other weakly supervised techniques based on bounding boxes. Moreover, given a fixed annotation time budget, *InExtremIS* outperformed full supervision. Our code and data are available online.

1 Introduction

Convolutional Neural Networks (CNNs) have achieved state-of-the-art performance for many medical segmentation tasks when trained in a fully-supervised manner, i.e. by using pixel-wise annotations. In practice, CNNs require large annotated datasets to be able to generalise to different acquisition protocols and to cover variability in the data (e.g. size and location of a pathology). Given the time and expertise required to carefully annotate medical data, the data-labelling process is a key bottleneck for the development of automatic image medical segmentation tools. To address this issue, growing efforts have been made to exploit weak annotations (e.g. scribbles, bounding boxes, extreme points) instead of time-consuming pixel-wise annotations.

Electronic supplementary material The online version of this chapter (https://doi.org/10.1007/978-3-030-87196-3_57) contains supplementary material, which is available to authorized users.

M. de Bruijne et al. (Eds.): MICCAI 2021, LNCS 12902, pp. 615–624, 2021.
https://doi.org/10.1007/978-3-030-87196-3_57

Weak annotations have been used in different contexts. The first range of approaches uses weak annotations at test-time to refine or guide network predictions. These techniques are commonly referred to as interactive segmentation [3,11,16,19,20,23] or semi-automated tools [8,10,22]. Networks are typically trained with fully annotated images and weak annotations are only used at inference time. In contrast, weakly supervised techniques are only trained with incomplete but easy-to-obtain annotations and operate in fully-automatic fashion at inference stage [2,6,7,9,14,15,24]. Bounding boxes are the most commonly used weak annotations [6,7,14]. However, extreme points are more time-efficient while providing extra information [12]. To our knowledge, only one extreme points supervision technique has been proposed for 3D medical image segmentation [15]. This method alternates between pseudo-mask generation using Random Walker and network training to mimic these masks. This computationally expensive approach relies on generated pseudo-masks which may be inaccurate in practice.

In this work, we propose a novel weakly supervised approach to learn automatic image segmentation using extreme points as weak annotations during training, here a set of manual extreme points along each dimension of a 3D image (6 in total). The proposed approach is end-to-end and trainable via standard optimisation such as stochastic gradient descent (SGD). The contributions of this work are three-fold. Firstly, extreme points along each dimension are automatically connected to increase the amount of foreground voxels used for supervision. To prompt the voxels along the path to be within the object, we propose to use the network probability predictions to guide the (deep) geodesic generation at each training iteration. Secondly, we employ a CRF regularised loss to encourage spatial and intensity consistency. Finally, extensive experiments over a publicly available dataset for Vestibular Schwannoma segmentation demonstrate the effectiveness of our approach. With only 6 points per training image, our weakly supervised framework outperformed other weakly supervised techniques using bounding boxes and approached fully supervised performance. Our method even outperformed full supervision given a fixed annotation time budget.

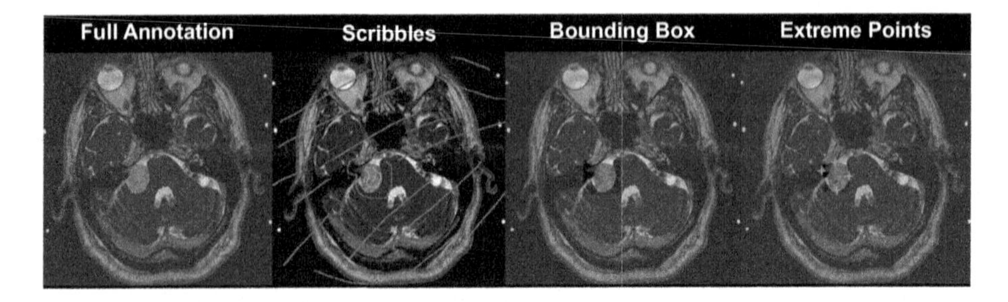

Fig. 1. Example of weak labels for our use case of Vestibular Schwannoma (VS) segmentation. For better readability, annotations are illustrated on a 2D axial slice. Magenta: Background. Green: VS. (Color figure online)

2 Related Work

Weakly Supervised Image Segmentation. Weakly supervised learning covers a large variety of techniques that aim to build predictive models using time-efficient weak annotations. Most existing weakly supervised approaches [14,15,24] adopt an iterative training strategy that alternates between pseudo-mask (proposal) generations using non-learning techniques (e.g. Random Walker, CRFs) and training a CNN to mimic these proposals. The pseudo-masks are initially generated using the weak annotations and then refined using the CNN predictions. These techniques are computationally expensive and rely on robust but not always accurate proposal generation. For this reason, other studies investigated end-to-end approaches. Direct regularisation loss derived from CRFs has been proposed [2,18], reaching almost full-supervision performance. This regularisation loss has only been employed with scribble annotations. Alternatively, constraints based on anatomical priors such as the volume or the shape have been studied for bounding box annotations [6,7]. However, these approaches have only been tested on 2D slices with a bounding box provided per slice.

Extreme Points as Weak Annotations. Different weak annotations have been proposed such as scribbles [2,9,18,19], bounding boxes [7,14] and extreme points [10,15]. Examples are shown in Fig. 1. Scribbles are user-friendly and easy to use. However, they must not only cover the foreground but also the background. Since medical images are typically mostly composed of background, most of the annotation time is dedicated to the background. In contrast, bounding boxes and extreme points are focused on the object of interest. Extreme clicks are the most time-efficient technique to generate bounding boxes and provide extra information compared to bounding boxes [12]. A recent work proposed a weakly supervised technique using extreme clicks [15]. This approach follows a computationally expensive two-step optimization strategy using Random Walker. To increase the foreground seeds, paths between the extreme points are initially computed, using only the image gradient information. In practice, this may lead to paths outside of the object and thus inaccurate initial pseudo-masks.

3 Learning from Extreme Points

3.1 Notations

Let $X : \Omega \subset \mathbb{R}^3 \rightarrow \mathbb{R}$ and $Y : \Omega \subset \mathbb{R}^3 \rightarrow \{0,1\}$ denote a training image and its corresponding ground-truth segmentation, where Ω is the spatial domain. Let f_θ be a CNN parametrised by the weights θ that predicts the foreground probability of each voxel. In a fully supervised learning setting, the parameters θ are optimized using a fully annotated training set $D = \{(X_i, Y_i)\}_{i=1}^{n}$ of size n. However, in our scenario the segmentations Y are missing. Instead, a set of 6 user clicks on the extreme points $\mathcal{E}_{\text{points}}$ is provided for each training image X. These extreme points correspond to the left-, right-, anterior-, posterior-, inferior-, and superior-most parts of the object of interest.

Accurate extreme points further provide a straightforward means of computing a tight bounding box B_{tight} around the object. While voxels outside of the bounding box are expected to be part of the background, user-click errors could result in some true foreground voxels lying outside the resulting bounding box. For this reason, a larger bounding box B_{relax} is used for the background by relaxing the tight bounding box B_{tight} by a fixed margin r. We hereafter denote $\overline{B_{\text{relax}}}$ the area outside the relaxed bounding box B_{relax}.

A straightforward training approach is to minimize the cross-entropy on the (explicitly or implicitly) annotated voxels, i.e. the 6 extreme points and $\overline{B_{\text{relax}}}$:

$$\theta^* = \arg\min_{\theta} \sum_{i=1}^{n} \left[- \sum_{k \in \overline{B_{\text{relax}}}} \log\left(1 - f_\theta\left(X_i\right)_k\right) - \sum_{k \in \mathcal{E}_{\text{points}}} \log\left(f_\theta\left(X_i\right)_k\right) \right] \tag{1}$$

This naive approach raises two issues: 1/the problem is highly imbalanced with only 6 foreground annotated voxels; and 2/no attempt is made to propagate labels to non-annotated voxels. The next two sections propose solutions to address these by using inter extreme points geodesics and a CRF regularised loss.

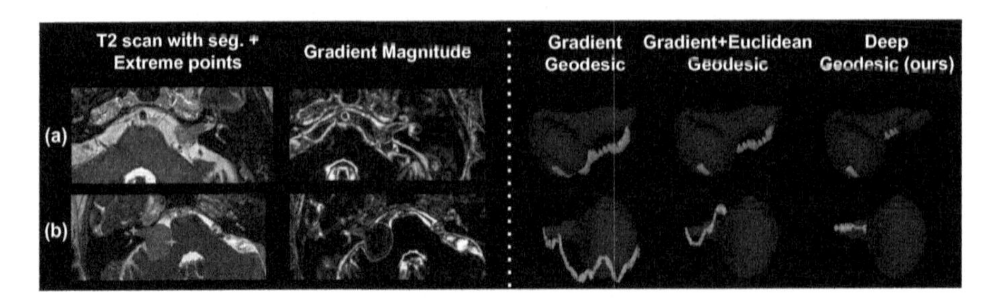

Fig. 2. Comparison of the different inter extreme points geodesics (green). Axial T2 slices with manual segmentation (blue) and image gradient magnitude are shown. (a) anterior- to posterior-most points geodesics; (b) right- to left-most points geodesics. Geodesics have been dilated for a better representation. Unlike the two other approaches, deep geodesics (ours) stay within the object. (Color figure online)

3.2 Inter Extreme Points Geodesics Generation Using CNNs Output

In this section, we propose to increase the amount of foreground voxels used for supervision by making better use of the extreme points. Let us assume that the foreground object is path-connected. Then, there is a path within the object that connects the extreme points. Voxels along this path could then be used for foreground supervision. At any given time during training, we take advantage of this connectivity hypothesis to compute inter extreme points paths which are likely to remain in the foreground class and use these for added supervision.

A popular approach to find paths connecting points within an object is based on geodesics or shortest paths with respect to a length metric. A geodesic is a length-minimising curve. In our case, the path domain is restricted to the tight bounding box B_{tight}. The length of paths intercepting edges in the image is typically penalised by accumulating the image gradient magnitudes [15,20]. However, extreme points are on object boundaries and thus likely to be on intensity edges. The shortest path could then circumvent the object without intercepting any additional edge, as illustrated in Fig. 2. Adding the Euclidean distance in the length metric helps [3] but may still be insufficient.

To help geodesics remain in the object, we propose to penalise paths passing through background voxels using the network background probabilities. The length of a discrete path Γ used to compute our deep geodesics is defined as:

$$L_{\text{deep}}(\Gamma, X) = \sum_{k=1}^{|\Gamma|-1} \underbrace{\gamma_e d\left(\Gamma_k, \Gamma_{k+1}\right)}_{\text{Euclidean}} + \underbrace{\gamma_g \| \nabla X\left(\Gamma_k\right) \|}_{\text{Image Gradient}} + \underbrace{\left(1 - f_\theta(X)_k\right)}_{\text{CNN Background Prob.}} \tag{2}$$

where $d\left(\Gamma_k, \Gamma_{k+1}\right)$ is the Euclidean distance between successive voxels, and $\| \nabla X\left(\Gamma_k\right) \|$ is a finite difference approximation of the image gradient between the points (Γ_k, Γ_{k+1}), respectively weighted by γ_e and γ_g.

We additionally propose an automated hyper-parameter policy for γ_e and γ_g by scaling each quantity between 0 and 1, i.e.:

$$1/\gamma_e = \max_{k \in B_{\text{tight}}} d(\Gamma_0, k) \text{ and } 1/\gamma_g = \max_{k \in B_{\text{tight}}} \| \nabla X\left(k\right) \|$$

Finally, the three inter extreme points geodesics $G_{\text{deep}} = (G^x, G^y, G^z)$ are computed using the Dijkstra algorithm [1] after each forward pass to update and refine our deep geodesics over training iterations.

3.3 Pseudo-Label Propagation via CRF Regularised Loss

In this section, we propose to propagate the pseudo-label information provided by the deep geodesics and the bounding box to the unlabelled voxels. Instead of using an external technique that requires a computationally expensive optimisation (e.g. CRFs, Random Walker) [14,15], we propose to use a direct unsupervised regularised loss as employed in other works [2,18]. This regularisation term encourages spatial and intensity consistency and can be seen as a relaxation of the pairwise CRF term. It is defined as:

$$R(f_\theta\left(X\right)) \triangleq \frac{1}{|\Omega|} \sum_{k,l \in \Omega} f_\theta(X)_k \exp\left(- \frac{d(k,l)^2}{2\sigma_\alpha^2} - \frac{(X_k - X_l)^2}{2\sigma_\beta^2} \right) (1 - f_\theta(X)_l)$$

where σ_α and σ_β respectively control the spatial and intensity consistency.

3.4 Final Model

To summarise, deep geodesics are computed on the fly after each forward pass during training. The voxels belonging to geodesics connecting extreme points (G_{deep}) and those outside of the relaxed bounding box ($\overline{B_{\text{relax}}}$) are used for supervision and a regularised loss ensures consistency:

$$\theta^* = \arg\min_{\theta} \sum_{i=1}^{n} \mathcal{L}(f_\theta(X_i), \overline{B_{\text{relax}}} \cup G_{\text{deep}}) + \lambda R(f_\theta(X_i)) \tag{3}$$

where $\mathcal{L}$ is a segmentation loss on the annotated voxels. We employed the sum of the Dice loss, the cross-entropy and the class-balanced focal loss.

4 Experiments

Dataset and Annotations. We conducted experiments on a large publicly available dataset[1] for Vestibular Schwannoma (VS) segmentation using high-resolution T2 (hrT2) scans as input. VS is a non-cancerous tumour located on the main nerve connecting the brain and inner ear and is typically path-connected. This collection contains a labelled dataset of MR images collected on 242 patients with VS undergoing Gamma Knife stereotactic radiosurgery (GK SRS). hrT2 scans have an in-plane resolution of approximately 0.5×0.5mm, an in-plane matrix of 384×384 or 448×448, and slice thickness of $1.0 - 1.5$ mm. The dataset was randomly split into 172 training, 20 validation and 46 test scans. Manual segmentations were performed in consensus by the treating neurosurgeon and physicist. Manual extreme clicks were done by a biomedical engineer.

Implementation Details. Our models were implemented in PyTorch using MONAI and TorchIO [13]. A 2.5D U-Net designed for VS segmentation was employed [17,21]. Mimicking training policies used in nnU-Net [4], 2.5D U-Nets were trained for 9000 iterations with a batch size of 6, i.e. 300 epochs of the full training set. Random patches of size $224 \times 224 \times 48$ were used during training. SGD with Nesterov momentum ($\mu = 0.99$) and an initial learning rate of 0.01 were used. The poly learning rate policy was employed to decrease the learning rate $(1 - \frac{\lfloor it/30 \rfloor}{300})^{0.9}$. The bounding box margin was set to $r = 4$ voxels. For the regularised loss, we used a public implementation based on [2,5]. σ_α, σ_β and λ were respectively set to 15, 0.05 and 10^{-4}. Similar results were obtained for $\sigma_\alpha \in \{5, 15\}$ and $\sigma_\beta \in \{0.5, 0.05\}$. The model with the smallest validation loss is selected for evaluation. Our code and pre-trained models are publicly available[2].

Evaluation. To assess the accuracy of each fully automated VS segmentation method, Dice Score (Dice) and 95th percentile of the Hausdorff Distance (HD95) were used. Precision was also employed to determine whether an algorithm tends to over-segment. Wilcoxon signed rank tests ($p < 0.01$) are performed.

[1] https://doi.org/10.7937/TCIA.9YTJ-5Q73.

[2] https://github.com/ReubenDo/InExtremIS/.

Table 1. Quantitative evaluation of different approaches for VS segmentation. Mean and variance are reported. Improvements in each stage of our ablation study are statistically significant $p < 0.01$ as per a Wilcoxon test.

Approach	Dice (%)	HD95 (mm)	Precision (%)
Gradient Geodesic	47.6 (15.5)	37.7 (44.9)	33.6 (15.0)
Gradient+Euclidean Geodesic	68.4 (14.9)	65.3 (41.3)	62.9 (23.1)
Deep Geodesic (ours)	78.1 (11.7)	11.9 (25.5)	88.3 (15.2)
Deep Geod. + Reg. Loss (InExtremIS)	**81.9 (8.0)**	**3.7 (7.4)**	**92.9 (5.9)**
Simulated Extreme Points (InExtremIS)	83.7 (7.7)	4.9 (14.9)	90.0 (7.5)
Fully Sup. - Equivalent Budget (2 h)	70.1 (14.2)	8.9 (13.0)	86.1 (13.1)
Fully Sup. - Unlimited Budget (26 h)	87.3 (5.5)	6.8 (19.9)	84.7 (8.2)
DeepCut [14]	52.4 (30.2)	12.9 (14.9)	52.4 (29.6)
Bounding Box Constraints [7]	56.0 (18.8)	16.4 (17.5)	49.7 (19.1)

Ablation Study. To quantify the importance of each component of our framework, we conducted an ablation study. The results are shown in Table 1. We started with geodesics computed using only the image gradient magnitude, as in [15]. Adding the Euclidean distance term significantly increased the Dice score by +20.8pp (percentage point). Adding the network probability term led to satisfying performance with a significant boost of +9.7pp in Dice score. In terms of training speed, the computation of the deep geodesics paths only took 0.36 s per training iteration. Finally, using the CRF regularised loss allowed for more spatially consistent segmentations, as shown in Fig. 3, increasing significantly the Dice by +3.8pp and reducing the HD95 by a factor 3. These results demonstrate the role of each contribution and prove the effectiveness of our approach.

Extreme Points (37 s/scan) vs. Full Annotations (477 s/scan). To compare the annotation time between weak and full supervision, a senior neurosurgeon segmented 5 random training scans using ITK-SNAP. On average it takes 477 s to fully annotate one hrT2 scan. In comparison, the full dataset was annotated with extreme points in 1 h 58 min using ITK-SNAP, i.e. 37 s/scan. Full segmentation is thus 13× more time-consuming than extreme points. Note that the reported annotation times cover the full annotation process, including the time required to open, save and adjust the image contrast (∼15 s).

Manual vs. Simulated Extreme Points. To study the robustness of our method to extreme points precision, we compared our approach using manual extreme points and extreme points obtained using the ground-truth segmentations. Since the simulated points are on the object boundaries, r was set to 0. Table 1 shows comparable results using both types of extreme points.

Segmentation Accuracy per Annotation Budget. We proposed to determine what is the best annotation approach given an annotation time budget. To do so, we trained a fully supervised approach using different time budgets. Given an annotation budget, the number of training and validation scans for full supervision was estimated by considering an annotation time of 477 s/scan. Results

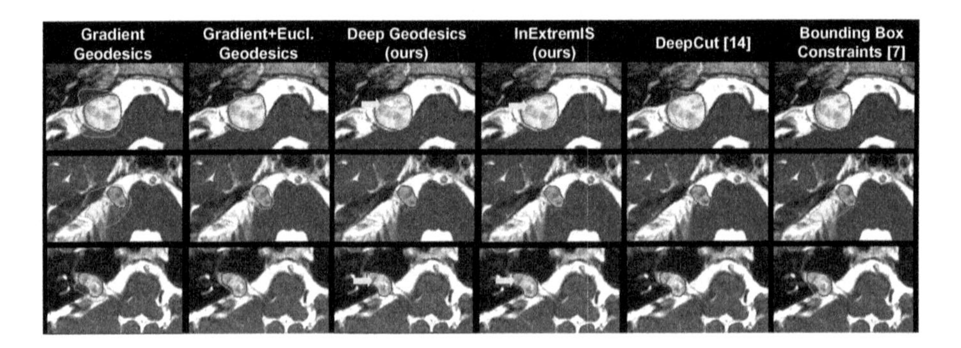

Fig. 3. Qualitative comparison of the different approaches. In green the predictions, in blue the manual segmentations. Red arrows show the effect of the regularised loss. (Color figure online)

are detailed in Appendix. First, extreme point supervision outperformed full supervision given a fixed time budget of 2 h (+11.8pp Dice). Full-supervision even required a 5× larger annotation budget than extreme point supervision to reach comparable performance. Finally, as expected, full supervision slightly outperformed weak supervision with an unlimited budget (+5.4pp Dice).

Bounding Boxes vs. Extreme Points. We compared our approach to two weakly supervised approaches based on bounding boxes: DeepCut [14] and a recently proposed 2D approach using constraints [7]. These approaches were implemented in 2D using the code from [7]. A 2D U-Net was used to allow for a fair comparison. The dilated bounding boxes were used for training. Results in Table 1 show that our approach significantly outperformed these techniques, demonstrating the benefits of our weakly supervised technique. Precision measurements and qualitative results in Fig. 3 highlight a key drawback of these approaches: They tend to over-segment when the bounding boxes are drawn in 3D and thus not tight enough for every single 2D slice. In contrast, our approach doesn't have any tightness prior and allow for large 2D box margins. Note that comparisons with semi-automated techniques requiring bounding boxes or extreme points at inference time are out-of-scope.

5 Discussion and Conclusion

In this work, we presented a novel weakly supervised approach using only 6 extreme points as training annotation. We proposed to connect the extreme points using a new formulation of geodesics that integrates the network outputs. Associated with a CRF regularised loss, our approach outperformed other weakly supervised approaches and achieved competitive performance compared to a fully supervised approach with an unlimited annotation budget. A 5 times larger budget was required to obtain comparable performance using a fully supervised approach. Our approach was tested on manual interactions and not on simulated interactions as in [7,15]. Our framework could typically be integrated in semi-automated pipelines to efficiently annotate large datasets with connected objects (e.g., VS, pancreas, liver, kidney). Future work will be conducted to identify which structures and shapes are suitable for our approach.

Although our approach has only been tested on a single class problem, the proposed technique is compatible with multi-class segmentation problems. Future work will especially investigate multi-class problems. We also plan to use our approach for weakly supervised domain adaptation problems.

Acknowledgement. This work was supported by the Engineering and Physical Sciences Research Council (EPSRC) [NS/A000049/1] and Wellcome Trust [203148/Z/16/Z]. TV is supported by a Medtronic/Royal Academy of Engineering Research Chair [RCSRF1819\7\34].

References

1. Dijkstra, E.W.: A note on two problems in connexion with graphs. Numer. Math. **1**(1), 269–271 (1959)
2. Dorent, R., et al.: Scribble-based Domain Adaptation via Co-segmentation. MICCAI (2020)
3. Gulshan, V., Rother, C., Criminisi, A., Blake, A., Zisserman, A.: Geodesic star convexity for interactive image segmentation. In: Computer Vision and Pattern Recognition (CVPR) (2010)
4. Isensee, F., Jaeger, P.F., Kohl, S.A.A., Petersen, J., Maier-Hein, K.H.: nnU-Net: a self-configuring method for deep learning-based biomedical image segmentation. Nat. Methods **18**(2), 203–211 (2021)
5. Joutard, S., Dorent, R., Isaac, A., Ourselin, S., Vercauteren, T., Modat, M.: Permutohedral attention module for efficient non-local neural networks. In: MICCAI (2019)
6. Kervadec, H., Dolz, J., Tang, M., Granger, E., Boykov, Y., Ben Ayed, I.: Constrained-CNN losses for weakly supervised segmentation. Med. Image Anal. **54**, 88–99 (2019)
7. Kervadec, H., Dolz, J., Wang, S., Granger, E., Ben Ayed, I.: Bounding boxes for weakly supervised segmentation: global constraints get close to full supervision. Med. Imaging Deep Learn. (MIDL) **121**, 365–381 (2020)
8. Khan, S., Shahin, A.H., Villafruela, J., Shen, J., Shao, L.: Extreme points derived confidence map as a cue for class-agnostic segmentation using deep neural network. In: MICCAI (2019)
9. Lin, D., Dai, J., Jia, J., He, K., Sun, J.: ScribbleSup: scribble-supervised convolutional networks for semantic segmentation. In: Computer Vision and Pattern Recognition (CVPR) (2016)
10. Maninis, K.K., Caelles, S., Pont-Tuset, J., Van Gool, L.: Deep extreme cut: from extreme points to object segmentation. In: Computer Vision and Pattern Recognition (CVPR) (2018)
11. McGrath, H., et al.: Manual segmentation versus semi-automated segmentation for quantifying vestibular schwannoma volume on MRI. Int. J. Comput. Assisted Radiol. Surgery **15**(9), 1445–1455 (2020)

12. Papadopoulos, D.P., Uijlings, J.R.R., Keller, F., Ferrari, V.: Extreme clicking for efficient object annotation. In: IEEE International Conference on Computer Vision, ICCV, pp. 4940–4949. IEEE Computer Society (2017)
13. Pérez-García, F., Sparks, R., Ourselin, S.: TorchIO: a Python library for efficient loading, preprocessing, augmentation and patch-based sampling of medical images in deep learning. Computer Methods and Programs in Biomedicine, p. 106236 (2021)
14. Rajchl, M., et al.: DeepCut: Object segmentation from bounding box annotations using convolutional neural networks. IEEE Trans. Med. Imaging **36**(2), 674–683 (2017)
15. Roth, H.R., Yang, D., Xu, Z., Wang, X., Xu, D.: Going to extremes: weakly supervised medical image segmentation. Mach. Learn. Knowl. Extract. **3**(2), 507–524 (2021)
16. Rother, C., Kolmogorov, V., Blake, A.: "GrabCut": Interactive foreground extraction using iterated graph cuts. ACM SIGGRAPH **2004**, 309–314 (2004)
17. Shapey, J., et al.: An artificial intelligence framework for automatic segmentation and volumetry of vestibular schwannomas from contrast-enhanced T1-weighted and high-resolution T2-weighted MRI. J. Neurosur. JNS **134**(1), 171–179 (2021)
18. Tang, M., Perazzi, F., Djelouah, A., Ayed, I.B., Schroers, C., Boykov, Y.: On regularized losses for weakly-supervised CNN segmentation. In: Ferrari, V., Hebert, M., Sminchisescu, C., Weiss, Y. (eds.) ECCV 2018. LNCS, vol. 11220, pp. 524–540. Springer, Cham (2018). https://doi.org/10.1007/978-3-030-01270-0_31
19. Wang, G., et al.: Interactive medical image segmentation using deep learning with image-specific fine tuning. IEEE Trans. Med. Imaging **37**(7), 1562–1573 (2018)
20. Zuluaga, M.A., et al.: DeepIGeoS: a deep interactive geodesic framework for medical image segmentation. IEEE Trans. Pattern Anal. Mach. Intell. **41**(7), 1559–1572 (2019)
21. Wang, G., et al.: Automatic segmentation of vestibular schwannoma from T2-weighted MRI by deep spatial attention with hardness-weighted loss. In: MICCAI (2019)
22. Wang, Z., Acuna, D., Ling, H., Kar, A., Fidler, S.: Object instance annotation with deep extreme level set evolution. In: Computer Vision and Pattern Recognition (CVPR) (2019)
23. Xu, N., Price, B., Cohen, S., Yang, J., Huang, T.S.: Deep interactive object selection. In: Computer Vision and Pattern Recognition (CVPR) (2016)
24. Zhang, L., Gopalakrishnan, V., Lu, L., Summers, R.M., Moss, J., Yao, J.: Self-learning to detect and segment cysts in lung CT images without manual annotation. In: International Symposium on Biomedical Imaging (ISBI) (2018)

Efficient and Generic Interactive Segmentation Framework to Correct Mispredictions During Clinical Evaluation of Medical Images

Bhavani Sambaturu[1(✉)], Ashutosh Gupta[2], C. V. Jawahar[1], and Chetan Arora[2]

[1] International Institute of Information Technology, Hyderabad, India
bhavani.sambaturu@research.iiit.ac.in
[2] Indian Institute of Technology, Delhi, India

Abstract. Semantic segmentation of medical images is an essential first step in computer-aided diagnosis systems for many applications. However, given many disparate imaging modalities and inherent variations in the patient data, it is difficult to consistently achieve high accuracy using modern deep neural networks (DNNs). This has led researchers to propose interactive image segmentation techniques where a medical expert can interactively correct the output of a DNN to the desired accuracy. However, these techniques often need separate training data with the associated human interactions, and do not generalize to various diseases, and types of medical images. In this paper, we suggest a novel conditional inference technique for DNNs which takes the intervention by a medical expert as test time constraints and performs inference conditioned upon these constraints. Our technique is generic can be used for medical images from any modality. Unlike other methods, our approach can correct multiple structures simultaneously and add structures missed at initial segmentation. We report an improvement of 13.3, 12.5, 17.8, 10.2, and 12.4 times in user annotation time than full human annotation for the nucleus, multiple cells, liver and tumor, organ, and brain segmentation respectively. We report a time saving of 2.8, 3.0, 1.9, 4.4, and 8.6 fold compared to other interactive segmentation techniques. Our method can be useful to clinicians for diagnosis and post-surgical follow-up with minimal intervention from the medical expert. The source-code and the detailed results are available here [1].

Keywords: Machine learning · Segmentation · Human-in-the-Loop

1 Introduction

Motivation: Image segmentation is a vital imaging processing technique to extract the region of interest (ROI) for medical diagnosis, modeling, and

---

Electronic supplementary material The online version of this chapter (https://doi.org/10.1007/978-3-030-87196-3_58) contains supplementary material, which is available to authorized users.

M. de Bruijne et al. (Eds.): MICCAI 2021, LNCS 12902, pp. 625–635, 2021.
https://doi.org/10.1007/978-3-030-87196-3_58

Table 1. Comparative strengths of various interactive segmentation techniques.

Capability	Description	[11]	[28]	[19]	[10]	Ours
Feedback mode	Point	✓	✓	✗	✓	✓
	Box	✗	✗	✓	✗	✓
	Scribble	✗	✗	✗	✗	✓
Training requirement	Pre-training with user interaction	✓	✓	✓	✓	✗
	Can work with any pre-trained DNN	✓	✓	✗	✗	✓
Correction modes	Correct multiple labels	✗	✗	✗	✗	✓
	Insert missing labels	✗	✗	✗	✗	✓
Generalization	Adapt: Distribution mismatch	✗	✗	✗	✗	✓
	Segment new organs than trained for	✗	✗	✗	✗	✓

intervention tasks. It is especially important for tasks such as the volumetric estimation of structures such as tumors which is important both for diagnosis and post-surgical follow-up. A major challenge in medical image segmentation is the high variability in capturing protocols and modalities like X-ray, CT, MRI, microscopy, PET, SPECT, Endoscopy and OCT. Even within a single modality, the human anatomy itself has significant variation modes leading to vast observed differences in the corresponding images. Hence, fully automated state-of-the-art methods have not been able to consistently demonstrated desired robustness and accuracy for segmentation in clinical use. This has led researchers to develop techniques for interactive segmentation which can correct the mispredictions during clinical evaluation and make-up for the shortfall.

Current Solutions: Though it is helpful to leverage user interactions to improve the quality of segmentation at test time, this often increases the burden on the user. A good interactive segmentation method should improve the segmentation of the image with the minimum number of user interactions. Various popular interactive segmentation techniques for medical imaging have been proposed in the literature [25,32,33]. The primary limitation is that it can segment only one structure at a time. This leads to a significant increase in user interactions when a large number of segments are involved. Recent DNN based techniques [11,19,28] improve this aspect by reducing user interactions. It exploits pre-learnt patterns and correlations for correcting the other unannotated errors as well. However, they require vast user interaction data for training the DNN model, which increases cost and restricts generalization to other problems.

Our Contribution: We introduce an interactive segmentation technique using a pre-trained semantic segmentation network, without any additional architectural modifications to accurately segment 2D and 3D medical images with help from a medical expert. Our formulation models user interactions as the additional test time constraints to be met by the predictions of a DNN. The Lagrangian formulation of the optimization problem is solved by the proposed alternate maximization and minimization strategy, implemented through the stochastic gradient descent. This is very similar to the standard back-propagation based training for the DNNs and can readily be implemented. The proposed technique has several advantages: (1) exhibits the capability to correct multiple structures at the same time leading to a significant reduction in the user time. (2) exploits the learnt correlations in a pre-trained deep learning semantic segmentation network so that a little feedback from the expert can correct large mispredictions. (3) requires no joint training with the user inputs to obtain a better segmentation, which is a severe limitation in other methods [11,28] (4) add missing labels while segmenting a structure if it was missed in the first iteration or wrongly labeled as some other structure. The multiple types of corrections allow us to correct major mispredictions in relatively fewer iterations. (5) handle distribution mismatches between the training and test sets. This can arise even for the same disease and image modality due to the different machine and image capturing protocols and demographies. (6) for the same image modality, using this technique one can even segment new organs using a DNN trained on some other organ type. Table 1 summarizes the comparative advantages of our approach.

2 Related Work

Conventional Techniques: Interactive segmentation is a well-explored area in computer vision and some notable techniques are based on Graph Cuts [25,29,32], Edge or Active Contours [12,30], Label propagation using Random Walk or other similar style [7,33], and region-based methods [9,26]. In these techniques, it is not possible to correct multiple labels together without the user providing the initial seeds and also not possible to insert a missing label.

DNN Based Techniques: DNN based techniques use inputs such as clicks [11,28], scribbles [18], and bounding boxes [19] provided by a user. Other notable techniques include [2,17,19,28,36]. These methods require special pre-training with user-interactions and associated images. This increases the cost of deployment and ties a solution to pre-decided specific problem and architecture.

Interactive Segmentation for Medical Images: Interactive Segmentation based methods, especially for medical image data, have been proposed in [3,10,15,34,35]. The methods either need the user inputs to be provided as an additional channel with the image [3] or need an additional network to process the user input [35]. BIFSeg [34] uses the user inputs at test time with a DNN for interactive segmentation of medical images. However, our method is significantly different in the following manner: (a) DNN - use their own custom neural networks [34]. However, our method can use pre-existing segmentation networks. This allows our method to use newer architectures which may be proposed in the future as well. (b) Optimization - use CRF-based regularization for label correction [34]. We propose a novel restricted Lagrangian-based formulation. This enables us to do a sample specific fine-tuning of the network, and allows our method to do multiple label corrections in a single iteration which is novel. (c) User Inputs - use scribbles and bounding boxes as user inputs [34]. We can correct labels irrespective of the type of user input provided.

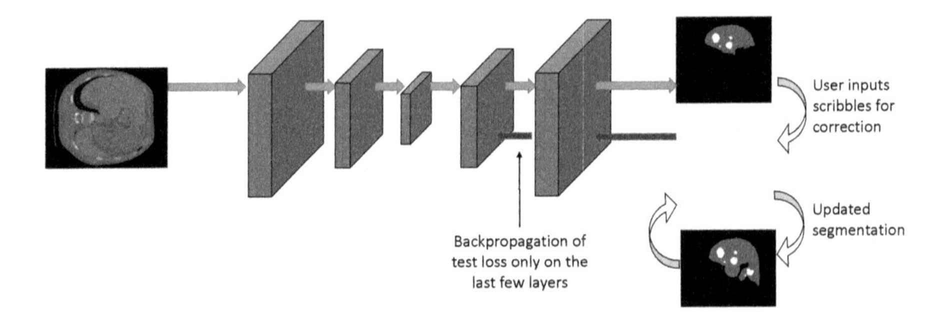

Fig. 1. The figure shows the working of our algorithm. Note that depending upon the application, our framework can use different pre-trained network architectures. Hence we do not give the detailed architecture of any particular model. The first step in our framework is to obtain an initial segmentation using the pre-trained deep learning network. The user then examines the segmentation and adds scribbles where the desired correction is required. This is then used to refine the weights of the network and the improved segmentation is obtained.

3 Proposed Framework

The goal is to design an approximate optimization algorithm that can encode the constraints arising from user-provided inputs in the form of scribbles. A simple gradient descent strategy similar in spirit to the Lagrangian relaxation proposed by [16] is optimized. The strategy allows us to use existing libraries and infrastructure built for any image modality optimizing the loss for the DNNs using the standard back-propagation procedure.

Problem Definition: A neural network with N layers is parameterized by weights W from input to output. We represent this as a function $\Psi(x, y, W) \rightarrow \mathbb{R}_+$ to measure the likelihood of a predicted output y given an input x and parameters/weights W. We also want to enforce that the output values belong to a set of scribbles $\mathbb{S}^x$ provided by the user to correct the segmentation dependent on x. Here, $\mathbb{S}^x$ encodes both the location in the image where correction is required and the desired segmentation class label.

We can express the constraint, $y \in \mathbb{S}^x$, as an equality constraint, using a function $g(y, \mathbb{S}^x) \rightarrow \mathbb{R}_+$. This function measures the compatibility between the output y and scribbles $\mathbb{S}^x$ such that $g(y, \mathbb{S}^x) = 0$ if and only if there are no errors in y with respect to $\mathbb{S}^x$. In our case, $g(y, \mathbb{S}^x)$ is the cross-entropy loss between the predicted labels y and the segmentation class label encoded in $\mathbb{S}^x$. This allows us to solve the optimization problem by minimizing the following Lagrangian:

$$\min_{\lambda} \max_{y} \Psi(x, y, W) + \lambda\, g(y, \mathbb{S}^x). \tag{1}$$

Note that the compatibility constraints in $g(y, \mathbb{S}^x)$ factorize over the pixels and one trivial solution of the optimization problem as described above is to simply change the output variables to the class labels provided by the scribbles.

Algorithm 1: Scribble aware inference for neural networks

Input : test instance x, input specific scribbles $\mathbb{S}^x$, max epochs M, pre-trained weights W, η learning rate, α regularization factor
$W_\lambda \leftarrow W$ `// reset to have instance-specific weights`
Output : Refined segmentation
while $g(y, \mathbb{S}^x) > 0$ *and iteration* $< M$ **do**
  $y \leftarrow f(x; W_\lambda)$ `// perform infererence using weights W_λ`
  $\nabla \leftarrow g(y, \mathbb{S}^x) \frac{\partial}{\partial W_{\lambda_l}} \Psi(x, y, W_{\lambda_l}) + \alpha \frac{W_l - W_{\lambda_l}}{||W_l - W_{\lambda_l}||_2}$ `// constraint loss`
  $W_{\lambda_l} \leftarrow W_{\lambda_l} - \eta \nabla$ `// update instance-specific weights with SGD`
end
return y, *the refined segmentation*

However, this does not allow us to exploit the neighborhood information inherent in the images, and the correlations learnt by a DNN due to prior training over a large dataset.

We note that the network's weights can also control the compatibility of the output configurations with the scribble input. Since the weights are typically tied across space, the weights are likely to generalize across related outputs in the neighborhood. This fixes the incompatibilities not even pointed-to by the limited scribbles given by the user. Hence, we propose to utilize the constraint violation as a part of the objective function to adjust the model parameters to search for an output satisfying the constraints efficiently.

We propose to optimize a "dual" set of model parameters W_λ over the constraint function while regularizing W_λ to stay close to the original weights W. The network is divided into a final set of layers l and an initial set of layers $N-l$. We propose to optimize only the weights corresponding to the final set of layers W_{λ_l}. The optimization function is given as:

$$\min_{W_{\lambda_l}} \Psi(x, \hat{y}, W_{\lambda_l})\ g(\hat{y}, \mathbb{S}^x) + \alpha||W_l - W_{\lambda_l}||, \tag{2}$$

where $\hat{y} = \arg\max_y \Psi(x, y, W_{\lambda_l})$. This function is reasonable by definition of the constraint loss $g(\cdot)$, though it deviates from the original optimization problem, and the global minima should correspond to the outputs satisfying the constraints. If we initialize $W_\lambda = W$, we also expect to find the high-probability optima. If there is a constraint violation in $\hat{y}$, then $g(\cdot) > 0$, and the following gradient descent procedure makes such $\hat{y}$ less likely, else $g(\cdot) = 0$ and the gradient of the energy is zero leaving $\hat{y}$ unchanged.

The proposed algorithm (see Algorithm 1) alternates between maximization to find $\hat{y}$ and minimization w.r.t. W_{λ_l} to optimize the objective. The maximization step can be achieved by employing the neural network's inference procedure to find the $\hat{y}$, whereas minimizing the objective w.r.t. W_{λ_l} can be achieved by performing stochastic gradient descent (SGD) given a fixed $\hat{y}$. We use the above-outlined procedure in an iterative manner (multiple forward, and back-propagation iterations) to align the outcome of the segmentation network with the scribble input provided by the user.

Figure 1 gives a visual description of our framework. It explains the stochastic gradient-based optimization strategy, executed in a manner similar to the standard back-propagation style of gradient descent. However, the difference is that while the back-propagation updates the weights to minimize the training loss, the proposed stochastic gradient approach biases the network output towards the constraints generated by the user provided scribbles at the test time.

Scribble Region Growing: The success of an interactive segmentation system is determined by the amount of burden on a user. This burden can be eased by allowing the user to provide fewer, shorter scribbles. However, providing shorter scribbles can potentially entail a greater number of iterations to obtain the final accurate segmentation. Hence, we propose using region growing to increase the area covered by the scribbles. We grow the region to a new neighborhood pixel, if the intensity of the new pixel differs from the current pixel by less than a threshold T.

4 Results and Discussions

Dataset and Evaluation Methodology: To validate and demonstrate our method, we have evaluated our approach on the following publicly available datasets containing images captured in different modalities: **(1) Microscopy:** 2018 Data Science Bowl (2018 DSB) [5] (nucleus), MonuSeg [14] (nucleus), and ConSeP [8] datasets (epithelial, inflammatory, spindle shaped and miscellaneous cell nuclei) **(2) CT:** LiTS [6] (liver and tumor cells) and SegThor [22] (heart, trachea, aorta, esophagus) challenges **(3) MRI:** BraTS' 15 [20] (necrosis, edema, non-enhancing tumor, enhancing tumor) and CHAOS [13] (liver, left kidney, right kidney, spleen) datasets. All the experiments were conducted in a Linux environment on a 20 GB GPU (NVIDIA 2018Tx) on a Core-i10 processor, 64 GB RAM, and the scribbles were provided using the WACOM tablet. For microscopy images, the segmented image was taken and scribbles were provided in areas where correction was required using LabelMe [31]. For CT and MRI scans, the scribbles were provided in the slices of the segmentation scan where correction was desired using 3-D Slicer [23]. For validating on each of the input modalities, and the corresponding dataset, we have taken a recent state-of-the-art approach for which the DNN model is publicly available and converted it into an interactive segmentation model. We used the same set of hyper-parameters that were used for training the pre-trained model. The details of each model, and source code to test them in our framework are available at [1]. To demonstrate the time saved over manual mode, we have segmented the images/scans using LabelMe for microscopy, and 3-D Slicer for CT/MRI, and report it as full human annotation time (F). We took the help of two trained annotators, two general practitioners and a radiologist for the annotation.

Ablation Studies: We also performed ablation studies to determine: (1) Optimum number of iterations, (2) Layer number upto which we need to update the

Table 2. User Interaction Time (**UT**) and Machine Time (**MT**) in minutes to separate structures (**F:** Full Human Annotation, **R:** Our method - Region Growing, **N:** Our Method - No Region Growing. Methods [19,21,25,28,32,33] were applied till a dice coefficient of 0.95 was reached.

Dataset	User interaction time									Machine time							
	F	**R**	**N**	[25]	[28]	[19]	[21]	[32]	[33]	**R**	**N**	[25]	[28]	[19]	[21]	[32]	[33]
2018 DSB	66	5	7	13	12	12		–	–	6	10	11	10	10		–	–
CoNSeP	30	6	8	16	18	20	–	–	–	5	7	17	20	23	–	–	–
LiTS	120	7	8	–	–	–	11	12	13	10	12	–	–	–	11	13	11
CHAOS	136	13	15	–	–	–	58	66	83	25	30	–	–	–	50	66	83
BraTS' 15	166	11	13	–	-	–	76	83	100	58	81	–	–	-	100	116	133

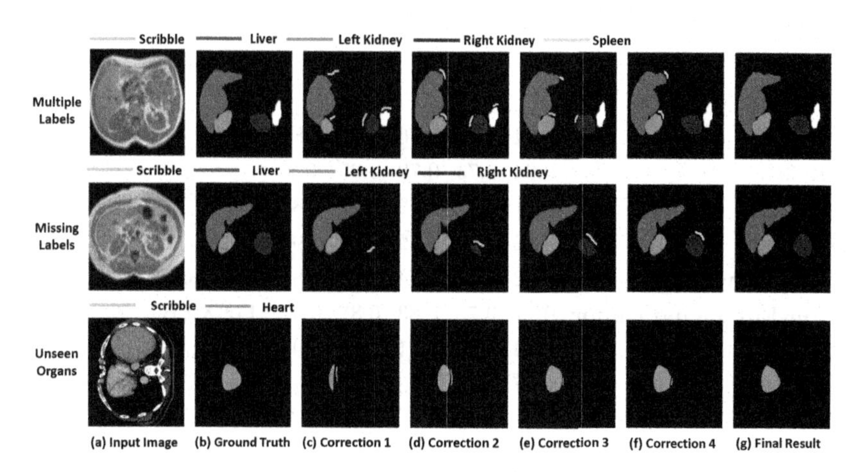

Fig. 2. (a) Correcting multiple labels (top row) (b) Inserting missing labels (middle row) (c) Interactive segmentation of organs the model was not trained for (bottom row). Incremental improvement as scribbles are added shown. No other state-of-the-art approach has these capabilities. More qualitative results are provided here [1].

weights, (3) Type of user input (point, box, scribble) and, (4) Effect of scribble length on the user interaction time. Owing to space constraints, the result of the ablation studies are provided on the project page [1]. We find scribble as the most efficient way of the user input through our ablation study, and use them in the rest of the paper.

Image Segmentation with Multiple Classes: Our first experiment is to evaluate interactive segmentation in a multi-class setting. We use two trained annotators for the experiment. We have used the validation sets of the 2018 Data Science Bowl (2018 DSB), CoNSeP, LiTS, CHAOS and the BraTS' 15 challenge datasets for the evaluation. We have used the following backbone DNNs to demonstrate our approach: [4,5,8,24,27]. The details of the networks are provided on the project webpage due to a lack of space. For the microscopy images we compare against Grabcut [25], Nuclick [10], DEXTR [19] and f-BRS

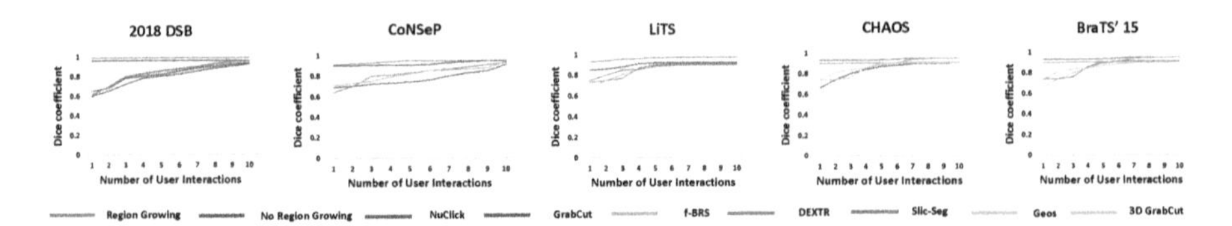

Fig. 3. Improvement in segmentation accuracy per user interaction: Our models (region and no-region growing) consistently achieve best accuracy, and in the least number of user interactions.

Table 3. Left: Dice Coefficient improvement for tissues with each interaction by medical expert. **Right:** User Interaction Time (**UT**) and Machine Time (**MT**) for distribution mismatch scenario (in mins).

Tissue Type	1	2	3	4	5
Nucleus	0.54	0.62	0.76	0.81	0.86
Healthy	0.64	0.73	0.79	0.85	0.9
Necrosis	0.61	0.65	0.72	0.81	0.85
Edema	0.72	0.75	0.82	0.89	0.92
Enhancing tumor	0.62	0.65	0.74	0.85	0.89
Non-Enhancing tumor	0.71	0.75	0.83	0.87	0.92
Liver	0.73	0.75	0.81	0.89	0.92
Tumor	0.67	0.72	0.83	0.87	0.89

Method	UT	MT
Ours	8	7
Nuclick	13	10
DEXTR	20	11
f-BRS	23	12
GrabCut	25	13

[28]. For the CT and MRI datasets, we have compared our method against 3-D GrabCut [21], Geos [32] and SlicSeg [33]. Table 2 shows that our technique gives an improvement in user annotation time of 13.3, 12.5, 17.8, 10.2 and 12.4 times compared to full human annotation time and 2.8, 3.0, 1.9, 4.4 and 8.6 times compared to other approaches for nucleus, multiple cells, liver and tumour, multiple organs, and brain segmentation respectively. We also compared the segmentation accuracy per user interaction for every method. Figure 3 shows that our method with region growing outperforms all the methods both in terms of accuracy achieved, and the number of iterations taken to achieve it.

Figure 2 shows the visual results. The top row shows the segmentation obtained by adding multiple labels in one interaction by our approach. We segment both the tumors and the entire liver by using two scribbles at the same time. One of the important capabilities of our network is to add a label missing from the initial segmentation which is shown in the middle row. Note that our method does not require any pre-training with a specific backbone for interactive segmentation. This allows us to use the backbone networks that were trained for segmenting a particular organ. This ability is especially useful in the data-scarce medical setting when the DNN model for a particular organ is unavailable. This capability is demonstrated in the bottom row of Fig. 2 where a model trained for segmenting liver on LiTS challenge [6] is used to segment the heart from SegThor challenge [22].

Distribution Mismatch: The current methods cannot handle distribution mismatches forcing pre-training on each specific dataset, requiring significant time, effort, and cost. Our method does not need any pre-training. We demonstrate the advantage on the MonuSeg dataset [14] using the model pre-trained on the 2018 Data Science Bowl [5]. Table 3 (Right) shows that our method requires much less user interaction and machine time compared to other methods.

Evaluation of Our Method by Medical Experts: Our approach was tested by medical experts: two general practitioners and a radiologist. We select five most challenging images/scans from the 2018 Data Science Bowl, LiTS, and BraTS' 15 datasets with the least dice score when segmented with the pre-trained segmentation model. The LiTS and the BraTS' 15 datasets were selected owing to their clinical relevance for the diagnosis and volumetric estimation of tumors. Table 3 (Left) gives the dice coefficient after each interaction. The improvement in user interaction and machine time are provided in the supplementary material on the project webpage.

5 Conclusion

Modern DNNs for image segmentation require a considerable amount of annotated data for training. Our approach allows using an arbitrary DNN for segmentation and converting it to an interactive segmentation. Our experiments show that we did not require any prior training with the scribbles and yet outperform the state-of-the-art approaches, saving upto 17x (from 120 to 7 mins) in correction time for a medical resource personnel.

References

1. Project page. http://cvit.iiit.ac.in/research/projects/cvit-projects/semi-automatic-medical-image-annotation
2. Acuna, D., Ling, H., Kar, A., Fidler, S.: Efficient interactive annotation of segmentation datasets with Polygon-RNN++. In: 2018 Proceedings of the IEEE Conference on Computer Vision and Pattern Recognition (CVPR). IEEE (2018)
3. Amrehn, M., et al.: UI-Net: Interactive artificial neural networks for iterative image segmentation based on a user model. In: Eurographics Workshop on Visual Computing for Biology and Medicine (2017)
4. Bellver, M., et al.: Detection-aided liver lesion segmentation using deep learning. arXiv preprint arXiv:1711.11069 (2017)
5. Caicedo, J., et al.: Nucleus segmentation across imaging experiments: the 2018 Data Science Bowl. Nat. Methods **16** (2019)
6. Christ, P.: LiTS - Liver tumor segmentation challenge (2017). https://competitions.codalab.org/competitions/17094
7. Grady, L.: Random walks for image segmentation. IEEE Trans. Pattern Anal. Mach. Intell. **28** (2006)

8. Graham, S., et al.: Hover-net: Simultaneous segmentation and classification of nuclei in multi-tissue histology images. Med Image Anal. (2019)
9. Horowitz, S., Pavlidis, T.: Picture segmentation by a tree traversal algorithm. JACM **23** (2016)
10. Jahanifar, M., Koohbanani, N.A., Rajpoot, N.: Nuclick: From clicks in the nuclei to nuclear boundaries. Med. Image Anal. **65** (2020)
11. Jang, W., Kim, C.: Interactive image segmentation via backpropagating refinement scheme. In: 2019 Proceedings of the IEEE Conference on Computer Vision and Pattern Recognition (CVPR). IEEE (2019)
12. Kass, M., Witkin, A., Terzopoulos, D.: Snakes: Active contour models. Int. J. Comput. Vis. **1** (1988)
13. Kavur, A.E., Selver, M.A., Dicle, O., Barış, M., Gezer, N.S.: CHAOS - Combined (CT-MR) healthy abdominal organ segmentation challenge data. Med. Image Anal. (2019)
14. Kumar, N., et al.: A multi-organ nucleus segmentation challenge. IEEE Trans. Med. Imaging. **39** (2020)
15. Lee, H., Jeong, W.-K.: Scribble2Label: scribble-supervised cell segmentation via self-generating pseudo-labels with consistency. In: Martel, A.L., et al. (eds.) MICCAI 2020. LNCS, vol. 12261, pp. 14–23. Springer, Cham (2020). https://doi.org/10.1007/978-3-030-59710-8_2
16. Lee, J., Mehta, S., Wick, M., Tristan, J., Carbonell, J.: Gradient-based inference for networks with output constraints. In: 2019 Proceedings of the AAAI Conference on Artificial Intelligence (AAAI), vol. 33 (2019)
17. Li, Z., Chen, Q., Koltun, V.: Interactive image segmentation with latent diversity. In: 2018 Proceedings of the IEEE Conference on Computer Vision and Pattern Recognition (CVPR). IEEE (2018)
18. Lin, D., Dai, J., Jia, J., He, K., Sun, J.: Scribblesup: Scribble-supervised convolutional networks for semantic segmentation. In: 2016 Proceedings of the IEEE conference on computer vision and pattern recognition (CVPR). IEEE (2016)
19. Maninis, K., Caelles, S., Pont-Tuset, J., Van Gool, L.: Deep extreme cut: from extreme points to object segmentation. In: 2018 Proceedings of the IEEE Conference on Computer Vision and Pattern Recognition (CVPR). IEEE (2018)
20. Menze, B., et al.: The multimodal brain tumor image segmentation benchmark. BRATS). IEEE Trans. Med. Imaging (2014)
21. Meyer, G.P., Do, M.N.: 3D GrabCut: Interactive foreground extraction for reconstructed 3d scenes. In: 2015 Proceedings of the Eurographics Workshop on 3D Object Retrieval. The Eurographics Association (2015)
22. Petitjean, C., Ruan, S., Lambert, Z., Dubray, B.: SegTHOR: Segmentation of thoracic organs at risk in CT images. In: 2019 Proceedings of the International Conference on Image Processing Theory, Tools and Applications (IPTA) (2019)
23. Pieper, S., Halle, M., Kikinis, R.: 3D Slicer. In: 2004 IEEE International Symposium on Biomedical Imaging: nano to macro (ISBI). IEEE (2004)
24. Qin, Y., et al.: Autofocus layer for semantic segmentation. In: Frangi, A.F., Schnabel, J.A., Davatzikos, C., Alberola-López, C., Fichtinger, G. (eds.) MICCAI 2018. LNCS, vol. 11072, pp. 603–611. Springer, Cham (2018). https://doi.org/10.1007/978-3-030-00931-1_69
25. Rother, C., Kolmogorov, V., Blake, A.: Grabcut interactive foreground extraction using iterated graph cuts. ACM Trans. Graph. (TOG) **23**(3) (2004)
26. Sahoo, P.K., Soltani, S., Wong, A., Chen, Y.C.: A survey of thresholding techniques. Comput. Gr. Image Process. **41** (1988)

27. Sinha, A., Dolz, J.: Multi-scale self-guided attention for medical image segmentation. IEEE J. Biomed. Health Inform. (2020)
28. Sofiiuk, K., Petrov, I., Barinova, O., Konushin, A.: f-BRS: rethinking backpropagating refinement for interactive segmentation. In: 2020 Proceedings of the IEEE Conference on Computer Vision and Pattern Recognition (CVPR). IEEE (2020)
29. Straehle, C.N., Köthe, U., Knott, G., Hamprecht, F.A.: Carving: scalable interactive segmentation of neural volume electron microscopy images. In: Fichtinger, G., Martel, A., Peters, T. (eds.) MICCAI 2011. LNCS, vol. 6891, pp. 653–660. Springer, Heidelberg (2011). https://doi.org/10.1007/978-3-642-23623-5_82
30. Top, A., Hamarneh, G., Abugharbieh, R.: Spotlight: automated confidence-based user guidance for increasing efficiency in interactive 3D image segmentation. In: Menze, B., Langs, G., Tu, Z., Criminisi, A. (eds.) MCV 2010. LNCS, vol. 6533, pp. 204–213. Springer, Heidelberg (2011). https://doi.org/10.1007/978-3-642-18421-5_20
31. Torralba, A., Russell, B., Yuen, J.: LabelMe: online image annotation and applications. Int. J. Comput. Vis. **98**(8) (2010)
32. Criminisi, A., Sharp, T., Blake, A.: GeoS: geodesic image segmentation. In: Forsyth, D., Torr, P., Zisserman, A. (eds.) ECCV 2008. LNCS, vol. 5302, pp. 99–112. Springer, Heidelberg (2008). https://doi.org/10.1007/978-3-540-88682-2_9
33. Wang, G., et al.: Slic-Seg: A minimally interactive segmentation of the placenta from sparse and motion-corrupted fetal MRI in multiple views. Med Image Anal 34 (2016)
34. Wang, G., et al.: Interactive medical image segmentation using deep learning with image-specific fine tuning. IEEE Trans. Med. Imaging. 37 (2018)
35. Wang, G., et al.: DeepiGeos: A deep interactive geodesic framework for medical image segmentation. IEEE Trans. Pattern Anal. Mach Intell 41 (2019)
36. Xu, N., Price, B., Cohen, S., Yang, J., Huang, T.: Deep interactive object selection. In: 2016 Proceedings of the IEEE Conference on Computer Vision and Pattern Recognition (CVPR). IEEE (2016)

Learning Whole-Slide Segmentation from Inexact and Incomplete Labels Using Tissue Graphs

Valentin Anklin[1,2], Pushpak Pati[1,2(✉)], Guillaume Jaume[1,3], Behzad Bozorgtabar[3], Antonio Foncubierta-Rodriguez[1], Jean-Philippe Thiran[3], Mathilde Sibony[4,5], Maria Gabrani[1], and Orcun Goksel[1,6]

[1] IBM Research-Europe, Zurich, Switzerland
[2] ETH Zurich, Zurich, Switzerland
pus@zurich.ibm.com
[3] EPFL, Lausanne, Switzerland
[4] Cochin Hospital, Paris, France
[5] University of Paris, Paris, France
[6] Uppsala University, Uppsala, Sweden

Abstract. Segmenting histology images into diagnostically relevant regions is imperative to support timely and reliable decisions by pathologists. To this end, computer-aided techniques have been proposed to delineate relevant regions in scanned histology slides. However, the techniques necessitate task-specific large datasets of annotated pixels, which is tedious, time-consuming, expensive, and infeasible to acquire for many histology tasks. Thus, weakly-supervised semantic segmentation techniques are proposed to leverage weak supervision which is cheaper and quicker to acquire. In this paper, we propose SegGini, a weakly-supervised segmentation method using graphs, that can utilize weak *multiplex* annotations, *i.e.*, *inexact* and *incomplete* annotations, to segment arbitrary and large images, *scaling* from tissue microarray (TMA) to whole slide image (WSI). Formally, SegGini constructs a tissue-graph representation for an input image, where the graph nodes depict tissue regions. Then, it performs weakly-supervised segmentation via node classification by using *inexact* image-level labels, *incomplete* scribbles, or both. We evaluated SegGini on two public prostate cancer datasets containing TMAs and WSIs. Our method achieved state-of-the-art segmentation performance on both datasets for various annotation settings while being comparable to a pathologist baseline. Code and models are available at: https://github.com/histocartography/seg-gini.

Keywords: Weakly-supervised semantic segmentation · Scalable digital pathology · Multiplex annotations · Graphs in digital pathology

V. Anklin, P. Pati and G. Jaume—Contributed equally to this work.

Electronic supplementary material The online version of this chapter (https://doi.org/10.1007/978-3-030-87196-3_59) contains supplementary material, which is available to authorized users.

M. de Bruijne et al. (Eds.): MICCAI 2021, LNCS 12902, pp. 636–646, 2021.
https://doi.org/10.1007/978-3-030-87196-3_59

1 Introduction

Automated delineation of diagnostically relevant regions in histology images is pivotal in developing automated computer-aided diagnosis systems in computational pathology. Accurate delineation assists the focus of the pathologists to improve diagnosis [33]. In particular, this attains high value in analyzing giga-pixel histology images. To this end, several supervised methods have been proposed to efficiently segment glands [8,29], tumor regions [3,7], and tissue types [5]. Though these methods achieve high-quality semantic segmentation, they demand tissue, organ and task-specific dense pixel-annotated training datasets. However, acquiring such annotations for each diagnostic scenario is laborious, time-consuming, and often not feasible. Thus, weakly supervised semantic segmentation (WSS) methods [10,43] are proposed to learn from weak supervision, such as *inexact* coarse image labels, *incomplete* supervision with partial annotations, and *inaccurate* supervision where annotations may not always be ground-truth.

WSS methods using various learning approaches, such as graphical model, multi-instance learning, self-supervised learning, are reviewed in [10]. WSS methods using various types of weak annotations are presented in [2,43]. Despite the success in delivering excellent segmentation performance, mostly with natural images, WSS methods encounter challenges in histology images [10], since histology images contain, (i) finer-grained objects (*i.e.*, large intra- and inter-class variations) [34], and (ii) often ambiguous boundaries among tissue components [39]. Nevertheless, some WSS methods were proposed for histology. Among those, the methods in [13,14,16,18,35,38] perform patch-wise image segmentation and cannot incorporate global tissue microenvironment context. While [9,28] propose to operate on larger image-tiles, they remain constrained to working with fixed and limited-size images. Thus, a WSS method operating on arbitrary and large histology images by utilizing both local and global context is needed. Further, most methods focus on binary classification tasks. Though HistoSegNet [9] manages multiple classes, it requires training images with *exact* fine-grained image-level annotations. *Exact* annotations demand pathologists to annotate images beyond standard clinical needs and norms. Thus, a WSS method should ideally be able to learn from *inexact*, coarse, image-level annotations. Additionally, to generalize to other WSS tasks in histology, methods should avoid complex, task-specific post-processing steps, as in HistoSegNet [9]. Notably, WSS methods in literature only utilize a single type of annotation. Indeed, complementary information from easily or readily available multiplex annotations can boost WSS performance.

To this end, we propose SegGini, "SEGmentation using Graphs with Inexact aNd Incomplete labels". SegGini represents a histology image using a superpixel-based tissue-graph and follows a classification approach to segment it. Our major contributions are, (i) SegGini is the first WSS method scalable to arbitrary image sizes, unlike pixel-based WSS or fully-connected graph-based WSS [26,41], (ii) to the best of our knowledge, SegGini is the first WSS method to simultaneously learn from weak multiplex supervision, *i.e.*, *inexact* image-level labels and *incomplete* scribbles. (iii) SegGini incorporates both local and global

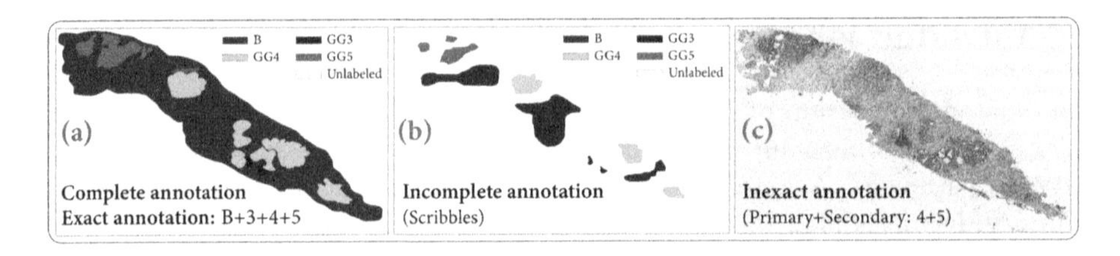

Fig. 1. Overview of various annotation types for a sample prostate cancer WSI, (a) *complete* pixel-level and *exact* image-level annotation, (b) *incomplete* scribbles of Gleason patterns, and (c) *inexact* image-level Gleason grade (P+S).

inter-tissue-region relations to build contextualized segmentation, principally in agreement with inter-pixel relation based state-of-the-art WSS method [2].

We evaluate our method on two H&E stained prostate cancer datasets [27,42] and segment Gleason patterns, *i.e.*, Benign (B), Grade3 (GG3), Grade4 (GG4) and Grade5 (GG5), by using *incomplete* scribbles of Gleason patterns and *inexact* image-level Gleason grades. Image-level grades are defined as the combination of the most common (*primary*, P) and the second most common (*secondary*, S) cancer growth patterns in the image. Figure 1 exemplifies *incomplete* and *inexact* annotations, along with *complete* pixel-level and *exact* image-level annotation.

2 Methods

This section presents the proposed SegGini methodology (Fig. 2) for scalable WSS of histology images. First, an input image is preprocessed and transformed into a tissue graph representation, where the graph nodes denote tissue superpixels. Then, a Graph Neural Network (GNN) learns contextualized features for the graph nodes. The resulting node features are processed by a *Graph-head*, a *Node-head*, or both based on the type of weak supervision. The outcomes of the heads are used to segment Gleason patterns. Additionally, a classification is performed to identify image-level Gleason grades from the segmentation map.

Preprocessing and Tissue Graph Construction. An input H&E stained image X is stain-normalized using the algorithm in [31] to reduce any appearance variability due to tissue preparation. Then, the normalized image is transformed into a Tissue-Graph (TG) (Fig. 2(a)), as proposed in [20]. Formally, we define a TG as $G := (V, E, H)$, where the nodes V encode meaningful tissue regions in form of *superpixels*, and the edges E represent inter-tissue interactions. Each node $v \in V$ is encoded by a feature vector $h(v) \in \mathbb{R}^d$. We denote the node features set, $h(v)$, $\forall v \in V$ as $H \in \mathbb{R}^{|V| \times d}$. Motivated by [6], we use superpixels as visual primitives, since rectangular patches may span multiple distinct structures.

The TG construction follows three steps: (i) superpixel construction to define V, (ii) superpixel feature extraction to define H, and (iii) graph topology construction to define E. For superpixels, we first use the unsupervised

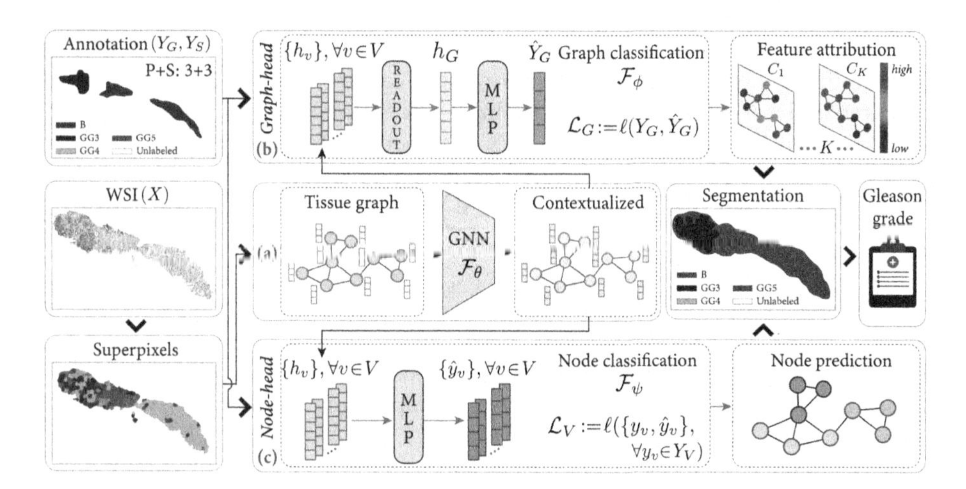

Fig. 2. Overview of the proposed SegGini methodology. Following superpixel extraction, (a) Tissue-graph construction and contextualization, (b) *Graph-head*: WSS via graph classification, (c) *Node-head*: WSS via node classification.

SLIC algorithm [1] emphasizing on space proximity. Over-segmented superpixels are produced at a lower magnification to capture homogeneity, offering a good compromise between granularity and noise smoothing. The superpixels are hierarchically merged based on channel-wise color similarity of superpixels at higher magnification, *i.e.*, channel-wise 8-bin color histograms, mean, standard-deviation, median, energy, and skewness. These then form the TG nodes. The merging reduces node complexity in the TG, thereby enabling a scaling to large images and contextualization to distant nodes, as explained in next section. To characterize the TG nodes, we extract morphological and spatial features. Patches of 224×224 are extracted from the original image and encoded into 1280-dimensional features with MobileNetV2 [23] pre-trained on ImageNet [11]. For a node $v \in V$, morphological features are computed as the mean of individual patch-level representations that belong to v. Spatial features are computed by normalizing superpixel centroids by the image size. We define the TG topology by constructing a region adjacency graph (RAG) [22] from the spatial connectivity of superpixels.

Contextualized Node Embeddings. Given a TG, we aim to learn discriminative node embeddings (Fig. 2(a)) that benefit from the nodes' context, *i.e.*, the tissue microenvironment and inter-tissue interactions. The contextualized node embeddings are further used for semantic segmentation. To this end, we use a GNN, that operates on graph-structured data [12,17,37]. In particular, we use Graph Isomorphism Network (GIN) [37] layers, a powerful and fast GNN architecture that functions as follows. For each node $v \in V$, GIN uses a *sum*-operator to *aggregate* the features of the node's neighbors $\mathcal{N}(v)$. Then, it *updates* the node features $h(v)$ by combining the *aggregated* features with the current node features $h(v)$ via a multi-layer perceptron (MLP). After T GIN layers, *i.e.*, acquiring context up to T-hops, the intermediate node features $h^{(t)}(v)$, $t = 1, \ldots, T$ are concatenated to define the contextualized node embeddings [36]. Formally, a GNN $\mathcal{F}_\theta$ with batch normalization (BN) is described for $v, u \in V$ as,

$$h^{(t+1)}(v) = \text{MLP}\Big(\text{BN}\big(h^{(t)}(v) + \sum_{u \in \mathcal{N}(v)} h^{(t)}(u)\big)\Big),\ t = \{0, .., T-1\} \quad (1)$$

$$h(v) = \text{Concat}\left(\left\{ h^{(t)}(v) \,\middle|\, t = 1, .., T \right\}\right) \quad (2)$$

Weakly Supervised Semantic Segmentation. The contextualized node embeddings $h(v)$, $\forall v \in V$ for a graph G, corresponding to an image X, are processed by SegGini to assign a class label $\in \{1, .., K\}$ to each node v, where K is the number of semantic classes. SegGini can incorporate multiplex annotations, *i.e.*, *inexact* image label Y_X and *incomplete* scribbles Y_S. Then, the weak supervisions for G are, the graph label Y_G, *i.e.*, the image label Y_X, and node labels $y_v \in Y_V$ that are extracted from Y_S by assigning the most prevalent class within each node. This is a reasonable assumption, as the tissue regions are built to be semantically homogeneous. The *Graph-head* (Fig. 2(b)) and the *Node-head* (Fig. 2(c)) are executed for using Y_G and Y_V, respectively. Noticeably, unlike [9], SegGini does not involve any post-processing, thus being a generic method that can be applied to various organs, tissue types, segmentation tasks, etc.

The *Graph-head* consists of a graph classification and a feature attribution module. First, a graph classifier $\mathcal{F}_\phi$ predicts $\hat{Y}_G$ for G. $\mathcal{F}_\phi$ includes, (i) a global average pooling *readout* operation to produce a fixed-size graph embedding h_G from the node embeddings $h(v), \forall v \in V$, and (ii) a MLP to map h_G to Y_G. As G directly encodes X, the need for patch-based processing is nullified. $\mathcal{F}_\theta$ and $\mathcal{F}_\phi$ are trained on a graph-set $\mathcal{G}$, extracted from the image-set $\mathcal{X}$, by optimizing a multi-label weighted binary cross-entropy loss $\mathcal{L}_G := l(Y_G, \hat{Y}_G)$. The class-weights are defined by $w_i = \log(N/n_i), i = 1, ..., K$, where $N = |\mathcal{X}|$, and n_i is the class example count; such that higher weight is assigned to smaller classes to mitigate class imbalance during training. Second, in an offline step, we employ a discriminative *feature attribution* technique to measure importance scores $\forall v \in V$ towards the classification of each class. Specifically, we use GraphGrad-CAM [15,21], a version of Grad-CAM [24] that can operate with GNNs. *Argmax* across class-wise node attribution maps from GraphGrad-CAM determines the node labels.

The *Node-head* simplifies image segmentation into classifying nodes $v \in V$. It inputs $h(v)$, $\forall v \in V$ to a MLP classifier $\mathcal{F}_\psi$ to predict node-labels y_v, $\forall v \in V$. $\mathcal{F}_\theta$ and $\mathcal{F}_\psi$ are trained using the multi-class weighted cross-entropy loss $\mathcal{L}_V := l(y_v, \hat{y}_v)$. The class-weights are defined by $w_i = \log(N/n_i), i = 1, ..., K$, where N is the number of annotated nodes, and n_i is the class node count. The node-wise predicted classes produce the final segmentation.

Multiplexed Supervision: For multiplex annotations, both heads are executed to perform WSS. $\mathcal{F}_\theta$, $\mathcal{F}_\phi$, and $\mathcal{F}_\psi$ are jointly trained to optimize a weighted loss

$\mathcal{L} = \lambda\mathcal{L}_G + (1 - \lambda)\mathcal{L}_V$, with which complementary information from multiplex annotations helps improve the individual classification tasks and thus improving WSS. Subsequently, we employ the classification approach in [4] to determine the Gleason grades from the generated segmentation maps.

3 Experiments

We evaluate our method on 2 prostate cancer datasets for Gleason pattern segmentation and Gleason grade classification.

UZH dataset [42] comprises of five TMAs with 886 spots, digitized at 40× resolution (0.23 μm/pixel). Spots (3100 × 3100 pixels) contain *complete* pixel-level annotations and *inexact* image-level grades. We follow a 4-fold cross-validation at TMA-level with testing on TMA-80 as in [4]. The second pathologist annotations on the test TMAs are used as a pathologist-baseline.

SICAPv2 dataset [27] contains 18 783 patches of size 512 × 512 with *complete* pixel annotations and WSI-level grades from 155 WSIs at 10× resolution. We reconstruct the original WSIs and annotation masks from the patches, containing up to 11000^2 pixels. We follow a 4-fold cross-validation at patient-level as in [27]. An independent pathologist's annotations are included as a pathologist-baseline.

We evaluate the methods for four annotation settings, *complete* ($\mathcal{C}$) and *incomplete* ($\mathcal{IC}$) pixel annotations, *inexact* image labels ($\mathcal{IE}$) as well as $\mathcal{IE} + \mathcal{IC}$. $\mathcal{IC}$ annotations with various pixel percentages are created by randomly selecting regions from $\mathcal{C}$ (see supplementary material for more details). We report per-class and average Dice scores as segmentation metrics, and weighted F1-score as a classification metric. We present means and standard-deviations on the test set for 4-fold cross-validation for all experiments.

Baselines: We compare SEGGINI with several state-of-the-art methods:

- UZH-CNN [4] and FSConv [27], for segmentation and classification using $\mathcal{C}$
- Neural Image Compression (NIC) [30], Context-Aware CNN (CACNN) [25], and CLAM [18], for weakly-supervised classification using $\mathcal{IE}$
- HistoSegNet [9], for weakly supervised segmentation using $\mathcal{IE}$.

These baselines are implemented based on code and algorithms in the corresponding publications. Baselines [18,25,30] directly classify WSI Gleason grades, and do not provide segmentation of Gleason patterns. Also, HistoSegNet [9] was trained herein with $\mathcal{IE}$, instead of *exact* image labels, since accessing the *exact* annotations would require using $\mathcal{C}$, that violates weak supervision constraints.

Implementations were conducted using PyTorch [19] and DGL [32] on an NVIDIA Tesla P100. SEGGINI model consists of 6-GIN layers, where the MLP in GIN, the *graph-head*, and the *node-head* contain 2-layers each with PReLU activation and 32-dimensional node embeddings, inspired by [40]. For graph augmentation, the nodes were augmented randomly with rotation and mirroring. A hyper-parameter search was conducted to find the optimal batch size $\in \{4, 8, 16\}$, learning rate $\in \{10^{-3}, 5\times10^{-4}, 10^{-4}\}$, dropout $\in \{.25, .5\}$, and $\lambda \in \{.25, .5, .75\}$ for each setting. The methods were trained with Adam optimizer to select the model with the best validation loss. For fair comparison, we evaluated all the baselines with similar patch-level augmentations and hyper-parameter searches.

Table 1. Results on UZH dataset as *Mean±std* using complete ($\mathcal{C}$), inexact ($\mathcal{IE}$), incomplete ($\mathcal{IC}$), and $\mathcal{IE}+\mathcal{IC}$ settings. Setting-wise best scores are in **bold**.

Annot.	Method	per-class Dice				avg. Dice	weight-F1
		Benign	Grade3	Grade4	Grade5		
$\mathcal{C}$	UZH-CNN[4]	**69.5**±**6.0**	54.7±3.9	63.6±3.2	34.6±4.6	55.6±1.8	49.2±4.3
$\mathcal{C}$	SegGini	64.2±8.0	**71.3**±**1.9**	**72.9**±**2.8**	**55.6**±**3.3**	**66.0**±**3.1**	**56.8**±**1.7**
$\mathcal{IE}$	CLAM[18]	-	-	-	-	-	45.7±4.6
$\mathcal{IE}$	NIC[30]	-	-	-	-	-	33.5±5.5
$\mathcal{IE}$	CACNN[25]	-	-	-	-	-	26.1±5.1
$\mathcal{IE}$	HistoSegNet[9]	**89.0**±**3.8**	42.4±10.9	56.8±10.4	34.8±12.9	55.7±3.2	41.6±9.3
$\mathcal{IE}$	SegGini	63.0±9.3	**69.6**±**5.6**	**67.6**±**5.4**	**55.7**±**7.0**	**64.0**±**1.8**	**52.4**±**3.2**
	Pathologist	83.33	44.53	69.29	57.28	63.60	48.98

Annot.	Method	avg. Dice			
		5% pixel	10% pixel	25% pixel	50% pixel
$\mathcal{IC}$	SegGini	58.2±3.1	62.9±2.2	63.3±2.3	**65.3**±**3.1**
$\mathcal{IC}+\mathcal{IE}$	SegGini	**63.7**±**2.9**	**65.6**±**3.0**	**63.6**±**2.0**	64.2±2.6

Results and Discussion: Table 1 and 2 present the segmentation and classification results of SegGini and the baselines, divided in groups for their use of different annotations. For the $\mathcal{C}$ setting, SegGini significantly outperforms UZH-CNN [4] on per-class and average segmentation as well as classification metrics, while reaching segmentation performance comparable with pathologists. For the $\mathcal{IE}$ setting, SegGini outperforms HistoSegNet on segmentation and classification tasks. Interestingly, SegGini also outperforms the classification-tailored baselines [18,25,30]. SegGini delivers comparable segmentation performance for *inexact* and *complete* supervision, *i.e.*, 64% and 66% average Dice, respectively. Comparing $\mathcal{IC}$ and $\mathcal{IE}+\mathcal{IC}$, we observe that $\mathcal{IE}+\mathcal{IC}$ produces better segmentation, especially in the low pixel-annotation regime. Such improvement, however, lessens with increased pixel annotations, which is likely due to the homogeneous Gleason patterns in the test set with only one or two patterns per TMA. Notably, SegGini with $\mathcal{IE}$ setting outperforms UZH-CNN with $\mathcal{C}$ setting.

On SICAPv2 dataset in $\mathcal{C}$ setting, SegGini outperforms FSConv for both segmentation and classification, and performs comparable to the pathologist-baseline for classification. SICAPv2 is highly imbalanced with a large fraction of benign regions. Thus, SegGini yields better results for benign class, while relatively poor performance for Grade5, which is rare in the dataset. For the $\mathcal{IE}$ setting, SegGini significantly outperforms HistoSegNet that trains using tile-labels, set the same as WSI-labels. This indicates HistoSegNet's inapplicability to WSIs with WSI-level supervision. For $\mathcal{IE}$, SegGini performs superior to [25,30] and comparable to [18]. Combining $\mathcal{IE}$ and $\mathcal{IC}$ for segmentation, the complementarity of annotations substantially boosts the performance. SegGini with $\mathcal{IE}+\mathcal{IC}$ consistently outperforms $\mathcal{IC}$ for various % of pixel annotations. Notably, $\mathcal{IE}+\mathcal{IC}$ outperforms $\mathcal{C}$ while using only 50% pixels. This confirms the benefit of learning from *multiplex* annotations. SegGini's inference time to process a WSI (11K × 11K pixels at 10×) is 14 ms, comparable to CLAM (11 ms). TG building takes 276.7 s including 183.5 s for superpixel detection and merging, 20.5 s for patch feature extraction and 85.7 s for RAG building. Figure 3 presents qualitative results on both datasets for various annotation settings. $\mathcal{IE}+\mathcal{IC}$ produces satisfactory segmentation while correcting any errors in $\mathcal{IE}$ by incorporating scribbles. The results indicate that SegGini provides competitive segmentation even with *inexact* supervision. Thus, we can leverage readily available slide-level Gleason grades from clinical reports, to substantially boost the segmentation, potentially together with a few *incomplete* scribbles from pathologists.

Table 2. Results on SICAPv2 as *Mean±std* using complete ($\mathcal{C}$), inexact ($\mathcal{IE}$), incomplete ($\mathcal{IC}$), and $\mathcal{IE}+\mathcal{IC}$ settings. Setting-wise best scores are in **bold**.

Annot.		per-class Dice				avg. Dice	weight-F1
	Method	Benign	Grade3	Grade4	Grade5		
$\mathcal{C}$	FSConv [27]	59.4±3.0	23.7±2.6	30.7±2.7	**9.1±2.9**	31.3±2.5	59.9±5.0
$\mathcal{C}$	SegGini	**90.0±0.1**	**39.4±3.3**	**40.2±2.7**	7.4±2.4	**44.3±2.0**	**62.0±3.6**
$\mathcal{IE}$	CLAM[18]	-	-	-	-	-	47.5±4.3
$\mathcal{IE}$	NIC[30]	-	-	-	-	-	32.4±10.0
$\mathcal{IE}$	CACNN[25]	-	-	-	-	-	21.8±4.7
$\mathcal{IE}$	HistoSegNet[9]	**78.1±1.4**	1.5±0.7	8.4±0.9	1.6±0.3	22.4±0.3	16.7±4.3
$\mathcal{IE}$	SegGini	55.9±12.0	**19.5±6.7**	**20.7±2.9**	**8.0±4.2**	**26.0±5.0**	**48.7±6.3**
	Pathologist	-	-	-	-	-	63.00

Annot.	Method	avg. Dice 10% pixel	25% pixel	50% pixel	100% pixel
$\mathcal{IC}$	SegGini	37.8±1.1	**41.9±1.0**	42.4±0.8	44.3±2.0
$\mathcal{IC}+\mathcal{IE}$	SegGini	**39.6±1.2**	41.8±0.6	**46.0±0.6**	**47.0±1.8**

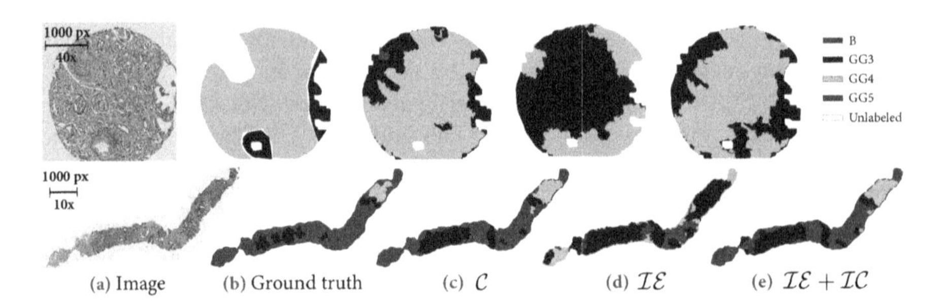

(a) Image (b) Ground truth (c) $\mathcal{C}$ (d) $\mathcal{IE}$ (e) $\mathcal{IE} + \mathcal{IC}$

Fig. 3. Example of predicted segmentation maps on UZH and SICAPv2 datasets for various annotation settings. $\mathcal{I}_\mathcal{C}$ is 10% and 25% for the datasets, respectively.

4 Conclusion

We proposed a novel WSS method, SegGini, to perform semantic segmentation of histology images by leveraging complementary information from weak multiplex supervision, *i.e.*, *inexact* image labels and *incomplete* scribbles. SegGini employs a graph-based classification that can directly operate on large histology images, thus utilizing local and global context for improved segmentation. SegGini is a generic method that can be applied to different tissues, organs, and histology tasks. We demonstrated state-of-the-art segmentation performance on two prostate cancer datasets for various annotation settings, while not compromising on classification results. Future research will focus on studying the generalizability of our method to previously unseen datasets.

References

1. Achanta, R., et al.: Slic superpixels compared to state-of-the-art superpixel methods. IEEE Trans. Pattern Anal. Mach. Intell. **34**, 2274–2282 (2012)
2. Ahn, J., et al.: Weakly supervised learning of instance segmentation with inter-pixel relations. In: IEEE CVPR, pp. 2204–2213 (2019)
3. Aresta, G., et al.: Bach: grand challenge on breast cancer histology images. Med. Image Anal. **56**, 122–139 (2019)
4. Arvaniti, E., et al.: Automated gleason grading of prostate cancer tissue microarrays via deep learning. In: Scientific Reports, vol. 8, p. 12054 (2018)
5. Bandi, P., et al.: Comparison of different methods for tissue segmentation in histopathological whole-slide images. In: IEEE ISBI, pp. 591–595 (2017)
6. Bejnordi, B., et al.: A multi-scale superpixel classification approach to the detection of regions of interest in whole slide histopathology images. In: SPIE 9420, Medical Imaging 2015: Digital Pathology, vol. 94200H (2015)
7. Bejnordi, B., et al.: Diagnostic assessment of deep learning algorithms for detection of lymph node metastases in women with breast cancer. JAMA **318**, 2199–2210 (2017)
8. Binder, T., et al.: Multi-organ gland segmentation using deep learning. In: Frontiers in Medicine (2019)

9. Chan, L., et al.: Histosegnet: semantic segmentation of histological tissue type in whole slide images. In: IEEE ICCV, pp. 10661–10670 (2019)
10. Chan, L., et al.: A comprehensive analysis of weakly-supervised semantic segmentation in different image domains. IJCV **129**, 361–384 (2021)
11. Deng, J., et al.: Imagenet: a large-scale hierarchical image database. In: IEEE CVPR, pp. 248–255 (2009)
12. Dwivedi, V., et al.: Benchmarking graph neural networks. In: arXiv (2020)
13. Ho, D., et al.: Deep multi-magnification networks for multi-class breast cancer image segmentation. In: Computerized Medical Imaging and Graphics, vol. 88, p. 101866 (2021)
14. Hou, L., et al.: Patch-based convolutional neural network for whole slide tissue image classification. In: IEEE CVPR, pp. 2424–2433 (2016)
15. Jaume, G., et al.: Quantifying explainers of graph neural networks in computational pathology. In: IEEE CVPR (2021)
16. Jia, Z., et al.: Constrained deep weak supervision for histopathology image segmentation. IEEE Trans. Med. Imaging **36**, 2376–2388 (2017)
17. Kipf, T., Welling, M.: Semi-supervised classification with graph convolutional networks. In: ICLR (2017)
18. Ming, Y., et al.: Data efficient and weakly supervised computational pathology on whole slide images. In: Nature Biomedical Engineering (2020)
19. Paszke, A., et al.: Pytorch: an imperative style, high-performance deep learning library. In: NeurIPS, pp. 8024–8035 (2019)
20. Pati, P., et al.: Hact-net: A hierarchical cell-to-tissue graph neural network for histopathological image classification. In: MICCAI, Workshop on GRaphs in biomedicAl Image anaLysis (2020)
21. Pope, P., et al.: Explainability methods for graph convolutional neural networks. In: IEEE CVPR, pp. 10764–10773 (2019)
22. Potjer, F.: Region adjacency graphs and connected morphological operators. In: Mathematical Morphology and its Applications to Image and Signal Processing. Computational Imaging and Vision, vol. 5, pp. 111–118 (1996)
23. Sandler, M., et al.: Mobilenetv2: inverted residuals and linear bottlenecks. In: IEEE CVPR, pp. 4510–4520 (2018)
24. Selvaraju, R., et al.: Grad-cam : visual explanations from deep networks. In: IEEE ICCV, pp. 618–626 (2017)
25. Shaban, M., et al.: Context-aware convolutional neural network for grading of colorectal cancer histology images. IEEE Trans. Med. Imaging **39**, 2395–2405 (2020)
26. Shi, Y., et al.: Building segmentation through a gated graph convolutional neural network with deep structured feature embedding. ISPRS J. Photogramm. Remote. Sens. **159**, 184–197 (2020)
27. Silva-Rodríguez, J., et al.: Going deeper through the Gleason scoring scale: an automatic end-to-end system for histology prostate grading and cribriform pattern detection. In: Computer Methods and Programs in Biomedicine, vol. 195 (2020)
28. Silva-Rodríguez, J., et al.: Weglenet: a weakly-supervised convolutional neural network for the semantic segmentation of Gleason grades in prostate histology images. In: Computerized Medical Imaging and Graphics, vol. 88, p. 101846 (2021)
29. Sirinukunwattana, K., et al.: Gland segmentation in colon histology images: the Glas challenge contest. Med. Image Anal. **35**, 489–502 (2017)
30. Tellez, D., et al.: Neural image compression for gigapixel histopathology image analysis. IEEE Trans. Pattern Anal. Mach. Intell. **43**, 567–578 (2021)
31. Vahadane, A., et al.: Structure-preserving color normalization and sparse stain separation for histological images. IEEE Trans. Med. Imaging **35**, 1962–1971 (2016)

32. Wang, M., et al.: Deep graph library: towards efficient and scalable deep learning on graphs. In: CoRR, vol. abs/1909.01315 (2019)
33. Wang, S., et al.: Pathology image analysis using segmentation deep learning algorithms. Am. J. Pathol. **189**, 1686–1698 (2019)
34. Xie, J., et al.: Deep learning based analysis of histopathological images of breast cancer. In: Frontiers in Genetics (2019)
35. Xu, G., et al.: Camel: a weakly supervised learning framework for histopathology image segmentation. In: IEEE ICCV, pp. 10681–10690 (2019)
36. Xu, K., et al.: Representation learning on graphs with jumping knowledge networks. ICML **80**, 5453–5462 (2018)
37. Xu, K., et al.: How powerful are graph neural networks? In: ICLR (2019)
38. Xu, Y., et al.: Weakly supervised histopathology cancer image segmentation and classification. Med. Image Anal. **18**, 591–604 (2014)
39. Xu, Y., et al.: Large scale tissue histopathology image classification, segmentation, and visualization via deep convolutional activation features. In: BMC bioinformatics, vol. 18 (2017)
40. You, J., et al.: Design space for graph neural networks. In: NeurIPS (2020)
41. Zhang, L., et al.: Dual graph convolutional network for semantic segmentation. In: BMVC (2019)
42. Zhong, Q., et al.: A curated collection of tissue microarray images and clinical outcome data of prostate cancer patients. In: Scientific Data, vol. 4 (2017)
43. Zhou, Z.: A brief introduction to weakly supervised learning. Natl. Sci. Rev. **5**, 44–53 (2017)

Label-Set Loss Functions for Partial Supervision: Application to Fetal Brain 3D MRI Parcellation

Lucas Fidon[1(✉)], Michael Aertsen[2], Doaa Emam[4,5], Nada Mufti[1,3,4], Frédéric Guffens[2], Thomas Deprest[2], Philippe Demaerel[2], Anna L. David[3,4], Andrew Melbourne[1], Sébastien Ourselin[1], Jan Deprest[2,3,4], and Tom Vercauteren[1]

[1] School of Biomedical Engineering and Imaging Sciences, King's College London, London, UK
lucas.fidon@kcl.ac.uk
[2] Department of Radiology, University Hospitals Leuven, Leuven, Belgium
[3] Institute for Women's Health, University College London, London, UK
[4] Department of Obstetrics and Gynaecology, University Hospitals Leuven, Leuven, Belgium
[5] Department of Gynecology and Obstetrics, University Hospitals Tanta, Tanta, Egypt

Abstract. Deep neural networks have increased the accuracy of automatic segmentation, however their accuracy depends on the availability of a large number of fully segmented images. Methods to train deep neural networks using images for which some, but not all, regions of interest are segmented are necessary to make better use of partially annotated datasets. In this paper, we propose the first axiomatic definition of label-set loss functions that are the loss functions that can handle partially segmented images. We prove that there is one and only one method to convert a classical loss function for fully segmented images into a proper label-set loss function. Our theory also allows us to define the leaf-Dice loss, a label-set generalisation of the Dice loss particularly suited for partial supervision with only *missing* labels. Using the leaf-Dice loss, we set a new state of the art in partially supervised learning for fetal brain 3D MRI segmentation. We achieve a deep neural network able to segment white matter, ventricles, cerebellum, extra-ventricular CSF, cortical gray matter, deep gray matter, brainstem, and corpus callosum based on fetal brain 3D MRI of anatomically normal fetuses or with open spina bifida. Our implementation of the proposed label-set loss functions is available at https://github.com/LucasFidon/label-set-loss-functions.

Electronic supplementary material The online version of this chapter (https://doi.org/10.1007/978-3-030-87196-3_60) contains supplementary material, which is available to authorized users.

M. de Bruijne et al. (Eds.): MICCAI 2021, LNCS 12902, pp. 647–657, 2021.
https://doi.org/10.1007/978-3-030-87196-3_60

1 Introduction

The parcellation of fetal brain MRI is essential for the study of fetal brain development [2]. Reliable analysis and evaluation of fetal brain structures could also support diagnosis of central nervous system pathology, patient selection for fetal surgery, evaluation and prediction of outcome, hence also parental counselling [1,4,16,22,25]. Deep learning sets the state of the art for the automatic parcellation of fetal brain MRI [9,13,18,19]. Training a deep learning model requires a large amount of accurately annotated data. However, manual parcellation of fetal brain 3D MRI requires highly skilled raters and is time-consuming.

Training a deep neural network for segmentation with partially segmented images is known as partially supervised learning [26]. Recent studies have proposed to use partially supervised learning for body segmentation in CT [5,8,23,26] and for the joined segmentation of brain tissues and lesions in MRI [6,21]. One of the main challenges in partially supervised learning is to define loss functions that can handle partially segmented images. Several previous studies have proposed to adapt existing loss functions for fully supervised learning using a somewhat ad hoc marginalization method [8,21,23]. Theoretical motivations for such marginalisation were missing. It also remains unclear whether it is the only way to build loss functions for partially supervised learning.

In this paper, we give the first theoretical framework for loss functions in partially supervised learning. We call those losses *label-set loss functions*. While in a fully supervised scenario, each voxel is assigned a single label, which we refer to as a *leaf-label* hereafter to avoid ambiguity; with partial supervision, each voxel is assigned a combined label, which we refer to as a *label-set*. As illustrated in Fig. 1, a typical example of partial supervision arises when there are missing leaf-label annotations. In which case the regions that were not segmented manually are grouped under one label-set (*unannotated*). Our theoretical contributions are threefold: 1) We introduce an axiom that label-set loss functions must satisfy to guarantee compatibility across label-sets and leaf-labels; 2) we propose a generalization of the Dice loss, leaf-Dice, that satisfies our axiom for the common case of missing leaf-labels; and 3) we demonstrate that there is one and only one way to convert a classical segmentation loss for fully supervised learning into a loss function for partially supervised learning that complies with our axiom. This theoretically justifies the marginalization method used in previous work [8,21,23].

In our experiments, we propose the first application of partial supervision to fetal brain 3D MRI segmentation. We use 244 fetal brain volumes from 3 clinical centers, to evaluate the automatic segmentation of 8 tissue types for both normal fetuses and fetuses with open spina bifida. We compare the proposed leaf-Dice to another labels-set loss [23] and to two other baselines. Our results support the superiority of labels-set losses that comply with the proposed axiom and show that the leaf-Dice loss significantly outperforms the three other methods.

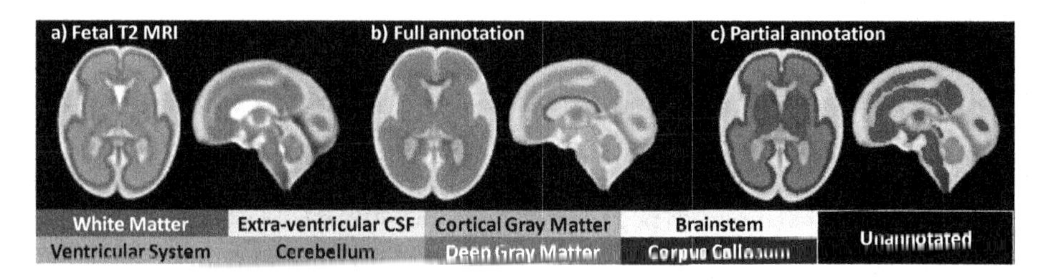

Fig. 1. Illustration of partial annotations on a control fetal brain MRI [12]. b) all the leaf-labels are annotated. c) partial annotations where cortical gray matter (CGM), deep gray matter (DGM), and brainstem (B) are not annotated. In this cases, unannotated voxels have the label-set annotation {CGM, DGM, B}.

2 Theory of Label-Set Loss Functions

In fully-supervised learning, we learn from ground-truth segmentations $\mathbf{g} \in \mathbf{L}^N$ where N is the number of voxels and $\mathbf{L}$ is the set of final labels (e.g. white matter, ventricular system). We denote elements of $\mathbf{L}$ as *leaf-labels*. In contrast, with partial supervision, $\mathbf{g} \in \left(2^{\mathbf{L}}\right)^N$, where $2^{\mathbf{L}}$ is the set of subsets of $\mathbf{L}$. In other words, each voxel annotation is a *set* of leaf-labels (a *label-set*). The label-set of an annotated white matter voxel would be simply {white matter}, and the label-set of a voxel that could be either white matter or deep gray matter would be {white matter, deep gray matter}. In both fully-supervised and partially-supervised learning, the network is trained to perform full segmentation predictions: $\mathbf{p} \in P\left(\mathbf{L}\right)^N$, where $P(\mathbf{L})$ is the space of probability vectors for leaf-labels.

2.1 Label-Set Loss Functions

A loss function $\mathcal{L}_{partial}(\cdot,\cdot)$ for partially supervised learning can be any differentiable function that compares a proposed probabilistic network output for leaf-labels, $\mathbf{p} \in P\left(\mathbf{L}\right)^N$, and a partial label-set ground-truth annotation, $\mathbf{g} \in \left(2^{\mathbf{L}}\right)^N$,

$$\mathcal{L}_{partial} : P\left(\mathbf{L}\right)^N \times \left(2^{\mathbf{L}}\right)^N \rightarrow \mathbb{R} \tag{1}$$

However, such $\mathcal{L}_{partial}$, in general, may consider all possible label-sets as independent and ignore the relationship between a label-set and its constituent leaf-labels. We claim that segmentation loss functions for partially supervised learning must be compatible with the semantic of label-set inclusion. For example, in the case of three leaf-labels $\mathbf{L} = \{l_1, l_2, l_3\}$, let voxel i be labeled with the label-set $g_i = \{l_1, l_2\}$. We know that the true leaf-label is either l_1 or l_2. Therefore the exemplar leaf-label probability vectors $\mathbf{p}_i = (0.4, 0.4, 0.2)$, $\mathbf{q}_i = (0.8, 0, 0.2)$, and $\mathbf{h}_i = (0, 0.8, 0.2)$ need to be equivalent conditionally to the ground-truth partial annotation g_i. That is, the value of the loss function should be the same whether the predicted leaf-label probability vector for voxel i is $\mathbf{p}_i$, $\mathbf{q}_i$, or $\mathbf{h}_i$.

Formally, let us define the marginalization function Φ as

$$\begin{aligned} \Phi : P(\mathbf{L})^N \times \left(2^{\mathbf{L}}\right)^N &\rightarrow P(\mathbf{L})^N \\ (\mathbf{p}, \mathbf{g}) &\mapsto (\tilde{p}_{i,c}) \end{aligned} \quad \text{s.t.} \quad \forall i, c, \begin{cases} \tilde{p}_{i,c} = \frac{1}{|g_i|} \sum_{c' \in g_i} p_{i,c'} & \text{if } c \in g_i \\ \tilde{p}_{i,c} = p_{i,c} & \text{if } c \notin g_i \end{cases}$$

In the previous example, $\Phi(\mathbf{p};\mathbf{g})_i = \Phi(\mathbf{q};\mathbf{g})_i = \Phi(\mathbf{h};\mathbf{g})_i = (0.4, 0.4, 0.2)$. We define **label-set loss functions** as the functions $\mathcal{L}_{partial}$ that satisfy the axiom

$$\forall \mathbf{g}, \forall \mathbf{p}, \mathbf{q}, \quad \Phi(\mathbf{p};\mathbf{g}) = \Phi(\mathbf{q};\mathbf{g}) \implies \mathcal{L}_{partial}(\mathbf{p}, \mathbf{g}) = \mathcal{L}_{partial}(\mathbf{q}, \mathbf{g}) \tag{2}$$

We demonstrate that a loss $\mathcal{L}$ is a label-set loss function if and only if

$$\forall (\mathbf{p}, \mathbf{g}), \quad \mathcal{L}(\mathbf{p}, \mathbf{g}) = \mathcal{L}\left(\Phi(\mathbf{p};\mathbf{g}), \mathbf{g}\right) \tag{3}$$

See the supplementary material for a proof of this equivalence.

2.2 Leaf-Dice: A Label-Set Generalization of the Dice Loss

In this section, as per previous work [5,6,8,21,23,26], we consider the particular case in which, per training example, there is only one label-set that is not a singleton and contains all the leaf-labels that were not manually segmented in this example. An illustration for fetal brain segmentation can be found in Fig. 1.

We propose a generalization of the mean class Dice Loss [10,15] for this particular case and prove that it satisfies our axiom (2).

Let $\mathbf{g} \in \left(2^{\mathbf{L}}\right)^N$ be a partial label-set segmentation such that there exists a label-set $\mathbf{L}'_{\mathbf{g}} \subsetneq \mathbf{L}$ that contains all the leaf-labels that were not manually segmented for this subject. Therefore, $\mathbf{g}$ takes its values in $\{\mathbf{L}'_{\mathbf{g}}\} \cup \left\{\{c\} \mid c \in \mathbf{L} \setminus \mathbf{L}'_{\mathbf{g}}\right\}$. We demonstrate that the leaf-Dice loss defined below is a label-set loss function

$$\forall \mathbf{p}, \quad \mathcal{L}_{Leaf-Dice}(\mathbf{p}, \mathbf{g}) = 1 - \frac{1}{|\mathbf{L}|} \sum_{c \in \mathbf{L}} \frac{2 \sum_i \mathbb{1}(g_i = \{c\})\, p_{i,c}}{\sum_i \mathbb{1}(g_i = \{c\})^\alpha + \sum_i p_{i,c}^\alpha + \epsilon} \tag{4}$$

where $\alpha \in \{1, 2\}$ (in line with the variants of soft Dice encountered in practice), and $\epsilon > 0$ is a small constant. A proof that $\mathcal{L}_{Leaf-Dice}$ satisfies (3) can be found in the supplementary material. It is worth noting that using $\mathcal{L}_{Leaf-Dice}$ is not equivalent to just masking out the unannotated voxels, i.e. the voxels i such that $g_i = \mathbf{L}'_{\mathbf{g}}$. Indeed, for all the $c \in \mathbf{L} \setminus \mathbf{L}'_{\mathbf{g}}$, the term $\sum_i p_{i,c}^\alpha$ in the denominator pushes $p_{i,c}$ towards 0 for all the voxels indices i including the indices i for which $g_i = \mathbf{L}'_{\mathbf{g}}$. As a result, when $g_i = \mathbf{L}'_{\mathbf{g}}$, only the $p_{i,c}^\alpha$ for $c \notin \mathbf{L}'_{\mathbf{g}}$ are pushed toward 0, which in return pushes the $p_{i,c}^\alpha$ for $c \in \mathbf{L}'_{\mathbf{g}}$ towards 1 since $\sum_{c \in \mathbf{L}} p_{i,c} = 1$.

2.3 Converting Classical Loss Functions to Label-Set Loss Functions

In this section, we demonstrate that there is one and only one canonical method to convert a segmentation loss function for fully supervised learning into a label-set loss function for partially supervised learning satisfying (3).

Table 1. Number of 3D MRI and number of manual segmentations available per tissue types. WM: white matter, Vent: ventricular system, Cer: cerebellum, ECSF: extra-ventricular CSF, CGM: cortical gray matter, **DGM**: deep gray matter, BS: brainstem, **CC**: corpus callosum.

Train/Test	Condition	MRI	WM	Vent	Cer	ECSF	CGM	**DGM**	BS	**CC**
Training	Atlas [12]	18	18	18	18	18	18	18	18	18
Training	Controls	116	116	116	116	54	0	0	0	18
Training	Spina Bifida	30	30	30	30	0	0	0	0	0
Testing	Controls	34	34	34	34	34	15	15	15	0
Testing	Spina Bifida	66	66	66	66	66	25	25	25	41

Formally, a label-set segmentation $\mathbf{g}$ can be seen as a soft segmentation using an injective function $\Psi : \left(2^{\mathbf{L}}\right)^N \hookrightarrow P\left(\mathbf{L}\right)^N$ that satisfies the relation

$$\forall \mathbf{g} \in \left(2^{\mathbf{L}}\right)^N, \forall i, c, \quad [\Psi(\mathbf{g})]_{i,c} > 0 \implies c \in g_i \tag{5}$$

Note however, that the function Ψ is not unique. Following the maximum entropy principle leads to choose

$$\Psi_0 : (g_i) \mapsto (\tilde{p}_{i,c}) \quad \text{s.t.} \quad \forall i, c \begin{cases} \tilde{p}_{i,c} = \frac{1}{|g_i|} & \text{if } c \in g_i \\ \tilde{p}_{i,c} = 0 & \text{otherwise} \end{cases} \tag{6}$$

We are interested in converting a loss function for fully-supervised learning $\mathcal{L}_{fully} : \ P\left(\mathbf{L}\right)^N \times P\left(\mathbf{L}\right)^N \to \mathbb{R}$ into a label-set loss function for partial supervision defined as $\mathcal{L}_{partial}(\mathbf{p}, \mathbf{g}) = \mathcal{L}_{fully}(\mathbf{p}, \Psi\left(\mathbf{g}\right))$.

Assuming that $\mathcal{L}_{fully}$ is minimal if and only if the predicted segmentation and the soft ground-truth segmentation are equal, we demonstrate in the supplementary material that $\mathcal{L}_{partial}$ is a label-set loss function if and only if

$$\forall (\mathbf{p}, \mathbf{g}) \in P\left(\mathbf{L}\right)^N \times \left(2^{\mathbf{L}}\right)^N, \quad \begin{cases} \mathcal{L}_{partial}(\mathbf{p}, \mathbf{g}) = \mathcal{L}_{fully}\left(\Phi(\mathbf{p}; \mathbf{g}), \Psi(\mathbf{g})\right) \\ \Psi(\mathbf{g}) = \Psi_0(\mathbf{g}) \end{cases} \tag{7}$$

Therefore, the only way to convert a fully-supervised loss to a loss for partially-supervised learning that complies with our axiom (2) is to use the marginalization function Φ on the predicted segmentation and Ψ_0 on the ground-truth partial segmentation. For clarity, we emphasise that the Leaf-Dice is a generalisation rather than a conversion of the Dice loss.

Related Work. When $\mathcal{L}_{fully}$ is the mean class Dice loss [10,15] and the values of $\mathbf{g}$ are a partition of $\mathbf{L}$, like in Sect. 2.2, one can prove that $\mathcal{L}_{partial}$ is the marginal Dice loss [23] (proof in supplementary material). Similarly, when $\mathcal{L}_{fully}$ is the cross entropy loss, $\mathcal{L}_{partial}$ is the marginal cross entropy loss [8,21,23]. Note that in [8,21,23], the marginalization approach is proposed as a possible method to convert the loss function. We prove that this is the only method.

3 Fetal Brain 3D MRI Data with Partial Segmentations

In this section, we describe the fetal brain 3D MRI datasets that were used.

Training Data for Fully Supervised Learning: 18 fully-annotated control fetal brain MRI from a spatio-temporal fetal brain MRI atlas [12].

Training Data for Partially Supervised Learning: 18 fully-annotated volumes from the fully-supervised dataset, combined with 146 partially annotated fetal brain MRI from a private dataset. The segmentations available for those 3D MRI are detailed in Table 1.

Multi-centric Testing Data: 100 fetal brain 3D MRI. This includes 60 volumes from University Hospital Leuven and 40 volumes from the publicly available FeTA dataset [17]. The segmentations available for those 3D MRI are detailed in Table 1. The 3D MRI of the FeTA dataset come from a different center than the training data. Automatic brain masks for the FeTA data were obtained using atlas affine registration with a normal fetal brain and a spina bifida spatio-temporal atlas [11,12].

Image Acquisition and Preprocessing for the Private Dataset. All images were part of routine clinical care and were acquired at University Hospital Leuven. In total, 74 cases with open spina bifida and 135 cases with normal brains, referred as controls, were included. Three spina bifida cases have been excluded by a pediatric radiologist because the quality of the 2D MRI resulted in corrupted 3D MRI which did not allow accurate segmentation. The gestational age at MRI ranged from 20 weeks to 35 weeks (median = 26.9 weeks, IQR = 3.4 weeks). For each study, at least three orthogonal T2-weighted HASTE series of the fetal brain were collected on a 1.5T scanner using an echo time of 133 ms, a repetition time of 1000 ms, with no slice overlap nor gap, pixel size 0.39 mm to 1.48 mm, and slice thickness 2.50 mm to 4.40 mm. A radiologist attended all the acquisitions for quality control. The fetal brain 3D MRI were obtained using `NiftyMIC` [7] a state-of-the-art super resolution and reconstruction algorithm. The volumes were all reconstructed to a resolution of 0.8 mm isotropic and registered to a standard clinical view. `NiftyMIC` also outputs brain masks that were used to define the label-sets and to mask the background [20].

Labelling Protocol. The labelling protocol is the same as in [17]. The different tissue types were segmented by a trained obstetrician and medical students under the supervision of a paediatric radiologist specialized in fetal brain anatomy, who quality controlled and corrected all manual segmentations. The voxels inside the brain mask that were not annotated by experts were assigned to the label-set containing all the tissue types that were not annotated for the 3D MRI. It is assumed that the voxels that were not annotated by experts were correctly not annotated.

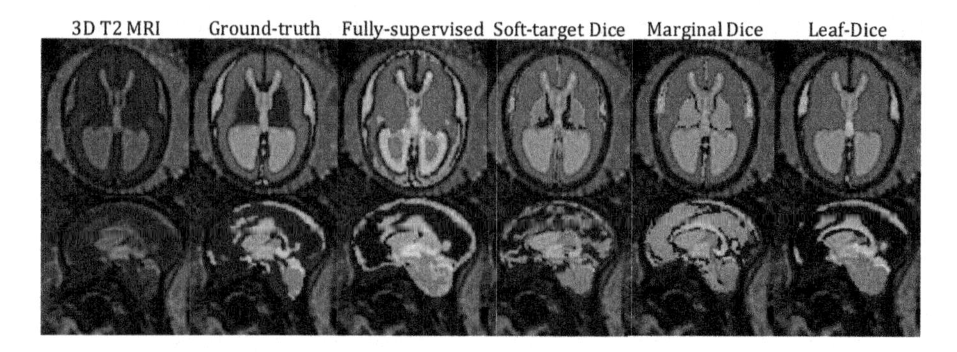

Fig. 2. Qualitative comparison on an open spina bifida case. Only the proposed Leaf-Dice loss provides satisfactory segmentations for all tissue types.

4 Experiments

In this section, we compare three partially supervised methods and one fully supervised method using the fetal brain 3D MRI dataset described in Sect. 3.

Deep Learning Pipeline. We use a 3D U-Net [3] architecture with 5 levels, leaky ReLU and instance normalization [24]. For training and inference, the entire volumes were used as input of the 3D U-Net after padding to $144 \times 160 \times 144$ voxels. For each method of Table 2, an ensemble of 10 3D U-Net is used. Using an ensemble of models makes the comparison of segmentation results across methods less sensitive to the random initialization of each network. Each 3D U-Net is trained using a random split of the training dataset into 90% for training and 10% for early stopping. During training, we used a batch size of 3, a learning rate of 0.001, and Adam [14] optimizer with default hyperpameter values. The learning rate was tuned for the fully supervised learning baseline. Random right/left flip, random scaling, gamma intensity augmentation, contrast augmentation, and additive Gaussian noise were used as data augmentation during training. Our code for the label-set loss functions and the deep learning pipeline are publicly available[1,2].

Hardware. For training we used NVIDIA Tesla V100 GPUs. Training each model took from one to two days. For inference, we used a NVIDIA GTX 1070 GPU. The inference for a 3D MRI takes between one and two minutes.

Specificities of Each Method. Baseline 1 is trained using fully supervised learning, the mean class Dice loss [10,15], referred as $\mathcal{L}_{Dice}$, and the training dataset for fully supervised learning of Sect. 3 (18 training volumes only). The

[1] https://github.com/LucasFidon/label-set-loss-functions.
[2] https://github.com/LucasFidon/fetal-brain-segmentation-partial-supervision-miccai21.

Table 2. Evaluation on the Multi-centric Testing Data (100 3D MRI). We report mean (standard deviation) for the Dice score (DSC) in percentages and the Hausdorff distance at 95% (HD95) in millimeters for the eight tissue types. Methods underlined are trained using partially supervised learning. **Loss functions in bold** satisfy the axiom of label-set loss functions. Best mean values are in bold. Mean values significantly better with $p < 0.01$ (resp. worse) than the ones achieved by the fully-supervised learning baseline are marked with a $*$ (resp. a $\dagger$).

Model	Metric	WM	Vent	Cer	ECSF	CGM	**DGM**	BS	**CC**
Baseline 1	DSC	76.4	72.1	67.3	63.9	47.3	**72.7**	56.0	51.6
		(12.5)	(18.8)	(28.6)	(31.3)	(10.9)	(8.7)	(27.7)	(10.5)
Fully-Supervised	HD95	3.3	4.8	4.9	7.3	5.3	**5.8**	9.1	5.1
		(1.2)	(3.7)	(5.2)	(7.8)	(0.9)	(2.1)	(8.2)	(3.1)
Baseline 2	DSC	89.5*	87.5*	87.2*	37.7†	31.4†	18.2†	20.2†	12.0†
		(6.5)	(9.6)	(10.3)	(34.2)	(14.8)	(20.5)	(13.1)	(10.8)
Soft-target Dice	HD95	1.7*	1.6*	2.2*	7.1	6.2†	24.2†	33.9†	29.3†
		(1.3)	(2.2)	(4.5)	(6.8)	(2.0)	(9.0)	(5.9)	(8.0)
Baseline 3 [23]	DSC	89.6*	87.7*	87.6*	66.6*	43.9	37.8†	39.4†	11.1†
		(6.7)	(10.4)	(9.5)	(27.7)	(15.1)	(11.3)	(16.8)	(12.3)
Marginal Dice	HD95	1.7*	1.6*	2.4*	6.2*	4.4*	26.7†	33.4†	28.7†
		(1.3)	(2.2)	(5.4)	(7.6)	(1.4)	(6.7)	(6.1)	(6.3)
Ours	DSC	**91.5***	**90.7***	**89.6***	**75.3***	**56.6***	71.4	**61.5***	**62.0***
		(6.7)	(8.9)	(10.1)	(24.9)	(14.3)	(8.6)	(21.7)	(10.9)
Leaf-Dice	HD95	**1.5***	**1.4***	**1.7***	**5.4***	**3.9***	7.3†	**7.9**	**2.9***
		(1.1)	(2.0)	(1.8)	(8.3)	(1.3)	(2.3)	(4.0)	(1.5)

three other methods are trained using partially supervised learning and the training dataset for partially supervised learning of Sect. 3. Baseline 2 is trained using the soft-target Dice loss function defined as $\mathcal{L}_{Dice}(\mathbf{p}, \Psi_0(\mathbf{g}))$, where Ψ_0 is defined in (6). Note that Baseline 2 does not satisfy the label-set axiom (2). Baseline 3 is trained using the marginal Dice loss [23]. Our method is trained using the proposed Leaf-Dice loss defined in (4). The loss functions of Baseline 3 and our method satisfy the axiom of label-set loss functions (2).

Statistical Analysis. We used the two-sided Wilcoxon signed-rank test. Differences in mean values are considered significant when $p < 0.01$.

Comparison of Fully Supervised and Partially Supervised Learning. The quantitative evaluation can be found in Table 2. The partially supervised learning methods of Table 2 all perform significantly better than the fully supervised learning baseline in terms of Dice score and Hausdorff distance on the tissue types for which annotations are available for all the training data of Table 1, i.e. white matter, ventricles, and cerebellum. Some tissue types segmentations are scarce in the partially supervised training dataset as can be seen in Table 1, i.e. cortical gray matter, deep gray matter, brainstem, and corpus callosum. This makes it challenging for partially supervised methods to learn to segment those tissue types. Only Leaf-Dice achieves similar or better segmentation results than the fully supervised baseline for the scarce tissue types, except in terms of Hausdorff distance for the deep gray matter. The extra-ventricular CSF is an intermediate case with almost half the data annotated. The Leaf-Dice and the Marginalized Dice significantly outperform the fully supervised baseline for all metrics for the extra-ventricular CSF, and the soft-target Dice performs significantly worse than this baseline.

Comparison of Loss Functions for Partially Supervised Learning. The Leaf-Dice performs significantly better than the soft-target Dice for all tissue types and all metrics. The Marginal Dice performs significantly better than the soft-target Dice in terms of Dice score for extra-ventricular CSF, cortical gray matter, deep gray matter, and brainstem. Since the soft-target Dice loss is the only loss that does not satisfy the proposed axiom for label-set loss functions, this suggests label-set loss functions satisfying our axiom perform better in practice.

In addition, our Leaf-Dice performs significantly better than the Marginal Dice [23] for all metrics and all tissue types. The qualitative results of Fig. 2 also suggest that Leaf-Dice performs better than the other approaches. This suggests that using a converted fully supervised loss function, as proposed in Sect. 2.3 and in previous work [8,21,23], may be outperformed by dedicated generalised label-set loss functions.

5 Conclusion

In this work, we present the first axiomatic definition of label-set loss functions for training a deep learning model with partially segmented images. We propose a generalization of the Dice loss, Leaf-Dice, that complies with our axiom for the common case of missing labels that were not manually segmented. We prove that loss functions that were proposed in the literature for partially supervised learning satisfy the proposed axiom. In addition, we prove that there is one and only one way to convert a loss function for fully segmented images into a label-set loss function for partially segmented images.

We propose the first application of partially supervised learning to fetal brain 3D MRI segmentation. Our experiments illustrate the advantage of using partially segmented images in addition to fully segmented images. The comparison of our Leaf-Dice to three baselines suggests that label-set loss functions that satisfy our axiom perform significantly better for fetal brain 3D MRI segmentation.

Acknowledgments. This project has received funding from the European Union's Horizon 2020 research and innovation program under the Marie Skłodowska-Curie grant agreement TRABIT No 765148. This work was supported by core and project funding from the Wellcome [203148/Z/16/Z; 203145Z/16/Z; WT101957], and EPSRC [NS/A000049/1; NS/A000050/1; NS/A000027/1]. TV is supported by a Medtronic / RAEng Research Chair [RCSRF1819\7\34].

References

1. Aertsen, M., et al.: Reliability of MR imaging-based posterior fossa and brain stem measurements in open spinal Dysraphism in the era of fetal surgery. Am. J. Neuroradiol. **40**(1), 191–198 (2019)
2. Benkarim, O.M., et al.: Toward the automatic quantification of in utero brain development in 3D structural MRI: a review. Hum. Brain Mapp. **38**(5), 2772–2787 (2017)
3. Çiçek, Ö., Abdulkadir, A., Lienkamp, S.S., Brox, T., Ronneberger, O.: 3D U-Net: learning dense volumetric segmentation from sparse annotation. In: Ourselin, S., Joskowicz, L., Sabuncu, M.R., Unal, G., Wells, W. (eds.) MICCAI 2016. LNCS, vol. 9901, pp. 424–432. Springer, Cham (2016). https://doi.org/10.1007/978-3-319-46723-8_49
4. Danzer, E., Joyeux, L., Flake, A.W., Deprest, J.: Fetal surgical intervention for myelomeningocele: lessons learned, outcomes, and future implications. Dev. Med. Child Neurol. **62**(4), 417–425 (2020)
5. Dmitriev, K., Kaufman, A.E.: Learning multi-class segmentations from single-class datasets. In: Proceedings of the IEEE/CVF Conference on Computer Vision and Pattern Recognition, pp. 9501–9511 (2019)
6. Dorent, R., et al.: Learning joint segmentation of tissues and brain lesions from task-specific hetero-modal domain-shifted datasets. Med. Image Anal. **67**, 101862 (2021)
7. Ebner, M., et al.: An automated framework for localization, segmentation and super-resolution reconstruction of fetal brain MRI. NeuroImage **206**, 116324 (2020)
8. Fang, X., Yan, P.: Multi-organ segmentation over partially labeled datasets with multi-scale feature abstraction. IEEE Trans. Med. Imaging **39**(11), 3619–3629 (2020)
9. Fetit, A.E., et al.: A deep learning approach to segmentation of the developing cortex in fetal brain mri with minimal manual labeling. In: Medical Imaging with Deep Learning, pp. 241–261. PMLR (2020)
10. Fidon, L., et al.: Generalised Wasserstein Dice score for imbalanced multi-class segmentation using holistic convolutional networks. In: Crimi, A., Bakas, S., Kuijf, H., Menze, B., Reyes, M. (eds.) BrainLes 2017. LNCS, vol. 10670, pp. 64–76. Springer, Cham (2018). https://doi.org/10.1007/978-3-319-75238-9_6
11. Fidon, L., et al.: A spatio-temporal atlas of the developing fetal brain with spina bifida aperta. Open Research Europe (2021)
12. Gholipour, A., et al.: A normative spatiotemporal MRI atlas of the fetal brain for automatic segmentation and analysis of early brain growth. Sci. Rep. **7**(1), 1–13 (2017)
13. Khalili, N., et al.: Automatic brain tissue segmentation in fetal MRI using convolutional neural networks. Magnet. Resonan. Imaging **64**, 77–89 (2019)
14. Kingma, D.P., Ba, J.: Adam: a method for stochastic optimization. arXiv preprint arXiv:1412.6980 (2014)
15. Milletari, F., Navab, N., Ahmadi, S.A.: V-net: fully convolutional neural networks for volumetric medical image segmentation. In: 2016 Fourth International Conference on 3D Vision (3DV), pp. 565–571. IEEE (2016)
16. Moise Jr., K.J., et al.: Current selection criteria and perioperative therapy used for fetal myelomeningocele surgery. Obstetrics Gynecol. **127**(3), 593–597 (2016)
17. Payette, K., et al.: A comparison of automatic multi-tissue segmentation methods of the human fetal brain using the FeTA dataset. arXiv preprint arXiv:2010.15526 (2020)

18. Payette, K., Kottke, R., Jakab, A.: Efficient multi-class fetal brain segmentation in high resolution MRI reconstructions with noisy labels. In: Hu, Y., et al. (eds.) ASMUS/PIPPI -2020. LNCS, vol. 12437, pp. 295–304. Springer, Cham (2020). https://doi.org/10.1007/978-3-030-60334-2_29
19. Payette, K., et al.: Longitudinal analysis of fetal MRI in patients with prenatal spina bifida repair. In: Wang, Q., et al. (eds.) PIPPI/SUSI -2019. LNCS, vol. 11798, pp. 161–170. Springer, Cham (2019). https://doi.org/10.1007/978-3-030-32875-7_18
20. Ranzini, M., Fidon, L., Ourselin, S., Modat, M., Vercauteren, T.: MONAIfbs: MONAI-based fetal brain MRI deep learning segmentation. arXiv preprint arXiv:2103.13314 (2021)
21. Roulet, N., Slezak, D.F., Ferrante, E.: Joint learning of brain lesion and anatomy segmentation from heterogeneous datasets. In: International Conference on Medical Imaging with Deep Learning, pp. 401–413. PMLR (2019)
22. Sacco, A., et al.: Fetal surgery for open spina bifida. Obstetrician Gynaecologist **21**(4), 271 (2019)
23. Shi, G., Xiao, L., Chen, Y., Zhou, S.K.: Marginal loss and exclusion loss for partially supervised multi-organ segmentation. Med. Image Anal. 101979 (2021)
24. Ulyanov, D., Vedaldi, A., Lempitsky, V.: Instance normalization: The missing ingredient for fast stylization. arXiv preprint arXiv:1607.08022 (2016)
25. Zarutskie, A., et al.: Prenatal brain imaging for predicting need for postnatal hydrocephalus treatment in fetuses that had neural tube defect repair in utero. Ultrasound Obstetrics Gynecol. **53**(3), 324–334 (2019)
26. Zhou, Y., et al.: Prior-aware neural network for partially-supervised multi-organ segmentation. In: Proceedings of the IEEE/CVF International Conference on Computer Vision, pp. 10672–10681 (2019)

Author Index

Printed in the United States
by Baker & Taylor Publisher Services